# Masonry Structures
## *Behavior and Design*

**ROBERT G. DRYSDALE**
*McMaster University, Hamilton, Ontario*

**AHMAD A. HAMID**
*Drexel University, Philadelphia*

**LAWRIE R. BAKER**
*Deakin University, Geelong, Australia*

PRENTICE HALL
Englewood Cliffs, New Jersey 07632

Library of Congress Cataloging-in-Publication Data

Drysdale, Robert G.
    Masonry structures : behavior and design / Robert G. Drysdale,
Ahmad A. Hamid, Lawrie R. Baker.
        p.    cm.
    Includes bibliographical references and index.
    ISBN 0-13-562026-0
    1. Masonry.    I. Hamid, Ahmad A.    II. Baker, Lawrie R.
III. Title.
TH1199.D78  1993
624.1'832--dc20

                                                    92-39538
                                                        CIP

**Acquisitions editor:** Doug Humphrey
**Production editor:** Bayani Mendoza de Leon
**Cover designer:** Design Source
**Copy editor:** Peter Zurita
**Prepress buyer:** Linda Behrens
**Manufacturing buyer:** Dave Dickey
**Editorial assistant:** Susan Handy

© 1994 by Prentice-Hall. Inc.
A Simon & Schuster Company
Englewood Cliffs, New Jersey 07632

Printed in the United States of America
10   9   8   7   6   5   4   3   2   1

ISBN 0-13-562026-0

Prentice-Hall International (UK) Limited, *London*
Prentice-Hall of Australia Pty. Limited, *Sydney*
Prentice-Hall Canada Inc., *Toronto*
Prentice-Hall Hispanoamericana. S.A., *Mexico*
Prentice-Hall of India Private Limited, *New Delhi*
Prentice-Hall of Japan, Inc., *Tokyo*
Simon & Schuster Asia Pte. Ltd., *Singapore*
Editora Prentice-Hall do Brasil, Ltda., *Rio de Janeiro*

Like the builders of the simple arch bridge who understood the importance of each masonry unit, we dedicate this book to our wives *Marna*, *Nevan*, and *Carol* and our families.

# Contents

Contents

Contents

Contents        

Contents                                                xiii

# Preface

To the majority of architects and engineers today, masonry falls uneasily between the two professions. Studies in architectural history give a familiarity with masonry buildings of previous years, but the potential for the modern use of masonry is often given meager treatment. Many engineering courses do not mention masonry as a structural material and those that do usually give it little attention. This situation is slowly changing, but most graduates entering the professions have a predisposition to design with steel, concrete, or timber rather than masonry even when masonry is the most appropriate material. This is probably true to many practicing designers who often had even less exposure to masonry as a modern structural material.

When used efficiently, masonry has a dual role to enclose or divide space and to form part of the load-resisting structure of the building. A masonry element is therefore of concern both to the architect, for its environmental and aesthetic qualities, and to the engineer, for its durability and strength. Research into masonry over the past 30 years or so now allows its use in building with as much confidence as the more familiar modern materials of steel and concrete.

This book attempts to give a broad understanding of masonry to designers and students, ranging from ancient beginnings to modern usage, covering planning, materials science, building science, structural design, and construction. It mostly deals with matters of international validity and is not confined by adherence to any code. Building codes and regulations vary from place to place within a country as well as internationally and are being changed with increasing rapidity. Nevertheless, such codes are an important guide and constraint and designers should become familiar with the relevant current local codes. This book gives many design examples; the majority using ACI 530/ASCE 5/TMS 402 Building Code Requirements for Masonry Structures.

# AUTHORS' GUIDE ON USE OF THE BOOK

Users of this book should find it beneficial to understand its organization and the reasons for its content and style of presentation.

**Readership.**   This book contains information common to the profession of architecture, building, and engineering with some chapters focussed principally on the interests of a particular discipline. The first four chapters dealing with history of masonry, building design, and materials are of general interest although Chapter 3 on design may be oriented slightly more to the interests of *architects*. While the chapters on masonry veneers, connectors, building science, and construction are directly related to architectural design, introductory information in the structural design chapters (i.e., infill walls) can help the architect gain an appreciation for the opportunities and limitations related to structural requirements.

*Builders, building officials,* and others involved in the building industry, will relate to the information described above for architects. Specific aspects of structural behavior and design should also be useful. A comprehensive index is provided to assist the user locate relevant information.

Most of the book should be of interest to *structural engineers* but it is anticipated that the chapters dealing with properties of materials and structural design will be of greatest interest.

**Units of Measurement.**   Dual units are used throughout this book not only because of the international nature of this book but also because Executive Order 12770 in the USA requires use of metric in the design of all new federal facilities by January 1994. US Customary units are followed by System International (SI) units in brackets. In all but the last two chapters, what is called *hard conversion* has been used, where, for instance, calculations using a 7 5/8 in. thick block are redone using the metric 190 mm block not the direct equivalent 193.7 mm thickness. The major building design examples in Chapters 16 and 17 contain only directly equivalent values, known as *soft conversion,* to avoid duplication of the extensive numerical presentations.

**Building Codes and Standards.**   Although most design books are based on a specific building code, that approach reduces the book's usefulness for design according to other codes and results in it becoming out-dated as codes change. To the extent possible, the structural design parts of the book have been based on *fundamental behavior*. To illustrate design, specific provisions of ACI 530/ASCE 5/ TMS 402, Uniform Building Code, CAN S304 (Canada), AS 3700 (Australia) and to a lesser extent BS 5628 (Britain) and NZS 4230 (New Zealand) have been presented. In the numerous design examples, use of the first two codes predominates, but other code provisions are included to illustrate alternatives particularly related to strength and limit states design. Although working stress design is emphasized, design based on strength is covered because most international codes are based on this approach and this is the direction of development of future North American codes.

**Use in Undergraduate Courses.** Practical limits on time available for classroom instruction and independent reading assignments mean that only a selected list of topics from this book are likely to be included in an undergraduate course. Such a "shopping list" approach to using the book is encouraged by organization of chapters into essentially stand alone segments.

Because this book was written to cater for the needs of a wide range of users, from the student to the practitioner to the graduate student and researcher, we have included some ideas on how it can be used for undergraduate instruction. Undergraduate students will concentrate on basic concepts whereas, later as practitioners and graduate students, they will often find the solution to their particular problem in some of the more complex sections.

The first three chapters form a basic introductory study for all readers. Time spent on formal study of this material will depend on the specific interests of the group and on the length of the course. For example, this material may be set largely as background reading for engineering students but form the major part of a detailed study for architecture students.

The following table sets out suggested courses for architecture and building students depending in the class contact hours available for the course. The table caters for courses from 6 to 36 hours of class contact time. The Basic Introduction varies from 4 hours (involving a fairly superficial treatment) to 16 hours duration, (allowing more detailed treatment and some studio applications). The Materials and Construction components each increase from 1 to 4 hours as greater knowledge is imparted for the study of other sections. Building Science is introduced in the Basic Introduction and built upon as time allows. As frequently noted in the book, engineering considerations often have implications for architectural design and construction and vice versa. Hence selected sections for structural components are listed in the table for the extended courses.

**SUGGESTED COURSES FOR ARCHITECTURAL AND BUILDING STUDENTS**

| Content selected from | Class contact for courses A to E (hrs) | | | | |
|---|---|---|---|---|---|
| | A | B | C | D | E |
| **Basic Introduction** Chapters 1, 2, 3 | 4 | 6 | 9 | 12 | 16 |
| **Materials** Chapter 4, Sec. 5.1–5.4 | 1 | 2 | 2 | 3 | 4 |
| **Construction** Chapter 15, Sec. 12.1–12.3, 12.5, 13.2, 13.3 | 1 | 2 | 2 | 3 | 4 |
| **Building Science** Chapter 14 | — | 2 | 4 | 4 | 6 |
| **Structures** Sec. 6.1, 7.1–7.3, 8.1–8.3, 9.1–9.3, 11.1–11.2, 12.4, 13.1–13.2 | — | — | 1 | 2 | 6 |
| **Total Hours** | 6 | 12 | 18 | 24 | 36 |

The following table of suggested courses for engineering students follows the same format as that above but includes a finer breakdown of specific topics. Naturally, each instructor will develop a curriculum suited to the particular need of the students. The amount of time assigned to any topic is an indicator of the depth of coverage, although, even with the longer contact hours, instructors will likely find that some sections can only be briefly introduced. We suggest a concentration on fundamental behavior and basic design and that discussions of other topics be included to broaden the perspectives of the students. The suggested contents are suited to senior undergraduate structural engineering students who have completed at least a one semester course in reinforced concrete design. If this background is not available, the time allotted to each structural design topic should be increased. The identified coverage of construction topics is augmented by related comments in the design chapters.

**SUGGESTED COURSES FOR ENGINEERING STUDENTS**

| Content topic area | Content selected from | Class contact for course (hrs) | | | | | |
|---|---|---|---|---|---|---|---|
| | | A | B | C | D | E | F |
| **Basic** | 1.1, 1.2, 1.4, 2.2, 2.5, 3.1–3.3. | 1 | 1 | 1 | | | |
| **Introduction** | All of Chaps. 1, 2, and 3 | | | | 2 | 2 | 3 |
| **Materials** | 4.1–4.4, 4.7, 5.1–5.4 | 1 | 2 | 2 | | | |
| | All of Chaps. 4 and 5 | | | | 3 | 3 | 3 |
| **Construction** | 12.1, 12.5, 13.1, 13.2, 15.1–15.3 | 1 | 1 | 1 | 1 | | |
| | All of Chaps. 12, 13, and 15 | | | | | 2 | 2 |
| **Building science** | 14.1, 14.3–14.5 | | | 1 | 1 | | |
| | All of Chap. 14 | | | | | 2 | 3 |
| **Beams** | 6.1–6.4 | 1 | 2 | 3 | 3 | 3 | |
| | All of Chap. 6 except 6.7 | | | | | | 4 |
| **Flexural walls** | 7.1–7.3 | | | 1 | 1 | 1 | 1 |
| **Loadbearing walls** | 8.4–8.6 | 2 | 2 | | | | |
| | All of Chap. 8 | | | 3 | 3 | 4 | 4 |
| **Columns and pilasters** | All of Chap. 9 | | | | | 2 | 2 |
| **Shear walls** | 10.1, 10.4, 10.6 | | 2 | 2 | | | |
| | All of Chap. 10 | | | | 3 | 3 | 4 |
| **Infill walls** | All of Chap. 11 | | | | | 1 | 1 |
| **Single-story building** | 16.1–16.4 | | 2 | | | | |
| | All of Chap. 16 | | | 4 | 6 | 6 | 6 |
| **Multistory building** | 17.1–17.4 | | | | 1 | 1 | 3 |
| | **Total hours** | 6 | 12 | 18 | 24 | 30 | 36 |

The problems sets provided at the end of each Chapter are generally graduated from fairly simple to quite difficult. Our experience is that students find the latter problems quite challenging and time consuming. Additional classroom help and working in groups can be employed to relieve this situation. Also some problems are essentially small projects, making it impractical for an individual student to undertake more than one or two such problems. A suggested approach is to have groups of students undertake different projects and report their results to the entire class.

# ACKNOWLEDGMENTS

The preparation of this book was made possible by financial support provided by the Brick Institute of America and the National Concrete Masonry Association. We are grateful for this support and the personal support and encouragement of J. Greg Borchelt, P.E., Director of Engineering and Research for BIA, and Mark B. Hogan, P.E., Vice President of Engineering for NCMA. Under their direction, review of the content greatly contributed to the quality of the book. BIA staff consisting of Greg Borchelt, Brent Gabby, Mark Nunn, Matt Scolforo, Christine Subasic, and Brian Trimble and NCMA staff consisting of Mark Hogan, Maribeth Bradfield, Kevin Callahan, and Robert Vanlaningham engaged in a comprehensive review. The critical review of the first draft and a second review of the chapters as sent to the publisher were significant factors in the improved presentation. We extend our sincere thanks to the above mentioned staff for their dedicated effort. NCMA and BIA also supplied many figures and photographs which help to make the text less formidable.

Our friend and colleague, Gary T. Suter, was initially a co-author of this book and participated in its planning. He completed the first draft of Chapter 4 before other commitments caused him to withdraw. We acknowledge his contribution and appreciate his continued assistance in providing many photographs for the book.

The manuscript was prepared by the Engineering Word Processing Center of McMaster University and the drawings were produced by James Q. Burdette, a graduate student at Drexel University. Their care and attention to detail are appreciated. Others who assisted by review of particular sections are: Ivan J. Becica from Oliver and Becica, Christas Christakis and Jon Morrison from Christakis and Kachele, A.C.C. Warnock of the Institute for Research in Construction - NRCC, Edward A. Gazzola of Morrison Hershfield Ltd.

We are indebted to many academic and masonry industry colleagues for providing photographs and other assistance. In particular we wish to thank Russell H. Brown of Clemson University for review of two chapters, and James E. Amrhein of the Masonry Institute of America and C.T. (Tom) Grimm of Austin, Texas for their unfailing assistance.

*Robert G. Drysdale*
*Ahmad A. Hamid*
*Lawrie R. Baker*

# ABOUT THE AUTHORS

**Robert G. Drysdale** is Professor of Civil Engineering at McMaster University in Hamilton, Ontario where he introduced and has taught masonry and building science courses for over twenty years. He is well known for his masonry research which has contributed to changes in building codes and standards and to product development. He has over 60 publications in the past ten years. He is Chairman of CSA Steering Committee for Masonry and is active on masonry technical committees, including chairing the Technical Advisory Committee of TMS.

Dr. Drysdale has extensive practical experience in structural design, including most types of buildings ranging from standard commercial and residential construction to heavy industrial buildings, water towers, air supported structures, and unique projects such as the roof support system for the Saddle Dome in Calgary. He has supervised and/or conducted tests for fire resistance, sound transmission, thermal resistance, rain penetration, and air and vapor barriers. He has been involved in investigations of structural failures and serviceability problems.

**Ahmad A. Hamid** is a Professor in the Department of Civil and Architectural Engineering and Director of the Structural Testing Laboratory and the Masonry Research Laboratory at Drexel University in Philadelphia, PA. He teaches masonry courses at both the undergraduate and graduate levels. His fundamental and applied masonry research includes material characterization and seismic resistance of reinforced masonry structures. He has published over 100 articles on behavior, design and evaluation of masonry materials, assemblages and structures. Dr. Hamid participates in code development as a member of the Masonry Standards Joint Committee as well as ASTM and TMS.

Dr. Hamid is involved in engineering design and consulting activities in North America and the Middle East. He is regularly engaged by government agencies and engineering firms to evaluate the structural performance and safety of concrete and masonry structures including adequacy of masonry walls in nuclear power plants to withstand seismic loads.

**Lawrie R. Baker** is Head of the School of Engineering at Deakin University in Australia. He previously held positions as Dean of the School of Architecture and as Head of the Department of Building at that University. His 28 years of teaching experience include the teaching of masonry and structures to students of architecture, building, and civil engineering, as well as to practitioners throughout Australia.

Dr. Baker has been researching and consulting in masonry for some 20 years. Both his masters and doctoral theses were in the field of masonry; he has over 50 journal articles and conference papers, over 100 confidential reports for industrial clients, and two books, *Masonry Code of Practice* and *Australian Masonry Manual*. He is Chair of the Structures Committee for the Standards Association of Australia, Masonry Code. Dr. Baker is director of the Masonry Research Centre at Deakin University in Geelong, Victoria and a Fellow of the Institution of Engineers, Australia.

# CHAPTER 1

# Ancient Masonry

The Great Pyramid of Giza, Egypt.

## 1.1 INTRODUCTION

No textbook on masonry is complete without at least a brief treatment of mankind's masonry heritage. From ancient times to the present, there are spectacular examples of construction that would be very difficult and extremely expensive to duplicate today even with our advanced design skills, modern machines, and modern materials. Review of the ancient use of masonry can help us place present-day construction in perspective and hopefully impart a measure of inspiration from the past, where examples of sophisticated structural forms from an analytical point of view go back for more than 10,000 years. The treatment in this chapter gives a broad overview of the development of masonry structures from earliest times. From a structural viewpoint, the discussion progresses from the simplest to the more complex structural forms but not necessarily chronologically, where domes of mud and ice were used from earliest times. After a brief history of masonry materials, early building elements and the development of building structure are considered. The development of masonry structures has been constrained by the available materials, construction skills, design abilities (whether intuitive or analytical), and costs. The importance of each factor has varied over time with perhaps cost being the greatest constraint on the form of masonry structure used today.

## 1.2 HISTORY OF MASONRY MATERIALS

Many materials have been used for the construction of masonry, those locally available being most convenient. Whenever civilizations developed in river plains, the alluvial deposits were used to create a brick architecture. In the Mesopotamian culture between the Tigris and Euphrates rivers, present-day mounds of crumbling soil in a treeless and stoneless desert testify to the widespread use of sun-dried mud bricks in past buildings.

Where civilizations existed in the vicinity of mountains or rocky outcrops, stone was used. Monumental stone buildings were built by the Egyptians along the rocky borders of the Nile valley.

In the Arctic regions, ice blocks are used to make igloos, and in modern cities, even glass is used to make masonry walls.

The common masonry materials used today are made from stone, clay, calcium silicate, and concrete. The background to and early uses of these materials are briefly presented in what follows. More detailed information is given in Chapter 4 and in Refs. 1.1, 1.2, and 1.3.

### 1.2.1 Stone

The first masonry was a crude stack of selected natural stones. The mortar, if any, was earth packed between them. As tools became available, stones were roughly trimmed, stacked, wedged with smaller stones, and bedded in clay.

As skills improved, stone masonry units were shaped into polygonal or squared units so that close-fitting joints were obtained. Thin lime mortar joints were used or the units were laid dry, sometimes with joints so precisely matched that a thin knife blade could not be inserted between them. Figure 1.1 shows some of these arrangements.

Sedimentary rocks (mainly sandstones and limestones) were split along their natural bedding planes using picks, crowbars, and wedges and dressed to size with chisels. Sand was used as an abrasive to grind bedding surfaces flat or, in conjunction with toothless saw blades, to cut rock to size. Toothed saw blades were used to cut soft rock.

The use of power tools and explosives has made the manufacture of stone masonry units much simpler than in the past. Most stonework in building is now a thin nonstructural veneer used solely for its aesthetic appeal. Types of stone commonly used include basalt (bluestone), granite, limestone, marble, sandstone, and slate.

### 1.2.2 Clay Units

Clay bricks have been in use for at least 10,000 years and possibly for as long as 12,000 years. Sun-dried bricks were widely used in Babylon, Egypt, Spain, South

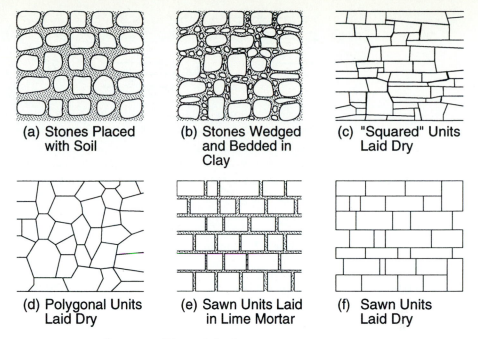

(a) Stones Placed with Soil

(b) Stones Wedged and Bedded in Clay

(c) "Squared" Units Laid Dry

(d) Polygonal Units Laid Dry

(e) Sawn Units Laid in Lime Mortar

(f) Sawn Units Laid Dry

**Figure 1.1**  Stone masonry.

America, the Indian reserves of the United States, and elsewhere. Wide usage is illustrated by the word "adobe," which is now incorporated in the English language but is a Spanish word based on the Arab word "atob," meaning sun-dried brick. The earliest bricks were made by pressing mud or clay into small lumps, sometimes cigar-shaped, and allowing them to dry in the air or the sun. These were then laid with mud mortar into walls having roughly horizontal coursing or a herringbone pattern or a combination of both, as shown in Fig. 1.2. By 3000 B.C., bricks were being made by hand in a mold with cattle dung or straw incorporated to increase strength. About this time, it was discovered that baking or *firing* brick in *clamps* greatly increased their strength and durability. An early clamp was simply a stack of dried bricks, with layers of brushwood fuel included, covered on the outside with a layer of clay to reduce heat loss. The clamp was then set on fire at several places, depending upon the direction of the wind, and allowed to burn out. These primitive firing conditions produced large variations in size and quality of bricks. Improved quality

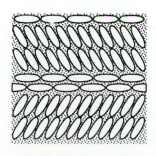

**Figure 1.2**  Cigar-shaped mud bricks laid in mud mortar.

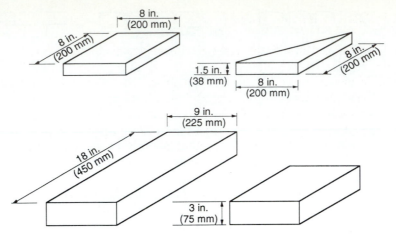

**Figure 1.3**  Roman bricks made in a mold.

was later achieved using specially constructed kilns where dried bricks were stacked on the perforated floor of an oven with fuel being introduced and burnt below in a more controlled manner.

Brickmaking in Europe, at least from Roman times until a few centuries ago, was a lengthy process. Sun-dried bricks were sometimes required to be five years old before their use in building. In making burnt bricks, it was usual to dig the clay in the autumn and leave it to weather all winter before molding it into bricks in the spring. Bricks were not set in the kiln until they were aged, preferably after two years of drying in the shade. Bricks varied greatly from broad flat bricks only 1 in. (25 mm) thick to those of modern size. Some Roman bricks are shown in Fig. 1.3.

Clay materials used for later brickmaking were mixed with water to create a homogeneous mass and then formed into brick units by pressing into a mold. Although the first brickmaking machine was patented in 1619, the turning point for the mechanical production of bricks came in 1858 with the introduction of the Hoffman kiln, which allowed the firing of bricks to be carried out in a continuous process. Today, the whole brickmaking process (mining, forming, drying, firing, cooling, and delivery) is highly mechanized and can be completed in less than a week. Typical contemporary clay units are shown in Fig. 1.4.

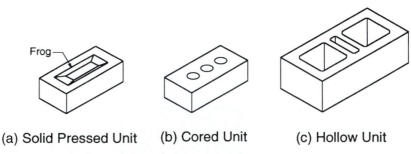

(a) Solid Pressed Unit    (b) Cored Unit    (c) Hollow Unit

**Figure 1.4**  Contemporary clay units.

### 1.2.3 Calcium Silicate Units

Calcium silicate (sand-lime) bricks were made in ancient times by molding lime mortar into brick shapes and allowing them to dry in air. Production continued into the 1880s, but air drying was slow. In 1866, accelerated hardening by the application of steam was introduced in the United States.

Although an improvement, the process was still too long. Further advances were made when steam under pressure was used to cure the units in Germany in 1894. As a result, rapid production of modern calcium silicate bricks became practical. Calcium silicate bricks have since been introduced into most countries.

Calcium silicate units are now manufactured using sand, lime, and water with sand constituting from 90 to 95% of the dry mix. The sand can be replaced by or used in combination with crushed siliceous rock or gravel and the lime can be either quicklime or hydrated lime.

### 1.2.4 Concrete Masonry Units

Concrete masonry units were first made at about the same time as the steam-cured calcium silicate units and as better quality cements were developed. The first blocks were unpopular because they were solid and, therefore, heavy to handle. Techniques for making hollow blocks in wooden molds developed about 1866. A fairly dry mixture of sand, cement, and water was placed in the mold and tamped by hand. The mold was then removed, and the masonry unit allowed to cure in the air. During the next decade, a number of shapes were patented in Britain and the United States. Methods of manufacture using simple machines gradually improved, but it was not until 1914 that power tamping replaced hand tamping. In 1924, a *stripper* machine for demolding was successfully introduced. Another major advance was made in 1939 when tamping of the concrete mixture was replaced by vibration under pressure. Further developments in materials handling have resulted in the fully automated machines of today.

Curing techniques have also progressed. Air curing with occasional water spraying has been replaced by water vapor curing, steam curing at atmospheric pressure, high-pressure steam curing (autoclaving), and burner curing systems.

Modern concrete blocks are generally manufactured by vibrating a mixture of portland cement, sand, and aggregate in a mold under pressure, curing with low-pressure or high-pressure steam, and then, in some cases, exposing them to carbon dioxide in the curing chamber to reduce subsequent shrinkage of the units. Figure 1.5(a) shows a typical standard hollow concrete block. Two of many specially shaped units used to accommodate reinforcing bars are shown in Figs. 1.5(b) and (c).

### 1.2.5 Mortars

Early mortars were basically used to fill cracks and provide uniform bedding for masonry units. Such mortars might have been clay, bitumen, or clay–straw mixtures and their weathering characteristics depended very much on local exposure conditions and the thinness of the joint.

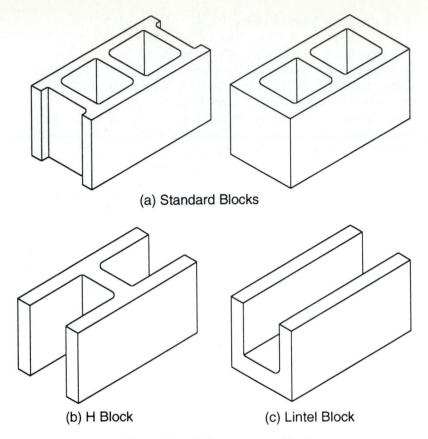

(a) Standard Blocks

(b) H Block              (c) Lintel Block

**Figure 1.5**   Hollow concrete blocks.

The forerunners of modern mortars date to the use of calcined gypsum, lime, and natural pozzolans. Following Egyptian use of calcined gypsum a few thousand years ago, the Greeks and Romans added lime and water and, with the addition of sand and crushed stone or brick, produced the earliest types of concrete. The Romans found that lime mortars did not harden under water, but that mixing ground lime with volcanic ash produced what became known as pozzolanic cement. The name "pozzolanic" is derived from the village of Pozzuoli near Vesuvius, the source of the volcanic ash. The Coliseum in Rome is an example of a Roman structure bonded with pozzolanic cement mortar that has weathered well over many centuries.

No significant developments in cements and mortars took place until the eighteenth century when John Smeaton, in the reconstruction of the Eddystone Lighthouse in England, mixed pozzolana with limestone containing a high proportion of clayey matter to produce a durable mortar that would set and harden under water. According to Neville,[1.4] Smeaton recognized the importance of clay and was the first to understand the chemical properties of hydraulic lime.

The next important development was the manufacture and patenting of portland cement by Joseph Aspdin in England in the early part of the nineteenth century. Combining portland cement with sand, lime, and water produced a much stronger mortar than previously possible and this mortar would also set and harden under water. One now speaks of hard and soft mortars, depending on the proportions of cement and lime and it is widely recognized that mortars should not have a higher compressive strength or hardness than required for a particular application. This is because softer mortars, with lower compressive strengths, accommodate small movements better. A more complete history of cements and mortars is given in Refs. 1.4 and 1.5.

## 1.3 EARLY BUILDING ELEMENTS

There are two fundamental structural problems when building, namely, how to achieve height and how to span over an opening—in a sense, how to span vertical and horizontal spaces, respectively. The former is achieved in masonry construction by using columns, towers, and walls and the latter by using lintels, beams, and arches. Some structural forms, such as vaults and domes, span vertically and horizontally at the same time. All of these structural forms are illustrated in this section by ancient examples, most of which can be found in the classic work by Sir Banister Fletcher,[1.6] first published in 1896.

### 1.3.1 Building Up

The simplest way of building is to stack masonry units one upon the other. Cro-Magnon man built piles of rock some 20,000 years ago to give concealment and protection during the hunt, and modern man still builds cairns to mark a mountain top or other places of significance. If squared stones or bricks are used, then a higher stack can be built because each unit rests squarely on the ones beneath and transfers its weight vertically downward to the ground. Simple calculations show that stone masonry units having a strength of 6000 psi (42 MPa) could be stacked to a height over 1 mile (1.6 km) before the units at the bottom would be crushed by the weight of those above. Long before this height was reached, however, a structure of practical proportions would have become unstable because of uneven bearing between the units and lack of vertical alignment. In addition, lateral forces from wind or earthquake would become a dominant factor.

**Pyramids.** The structural form of the pyramid, the logical development of the rock pile, was used extensively by early man. Examples of pyramid construction are shown in Fig. 1.6. The first crude pyramids made from about 3000 B.C. by the Egyptians were mud brick tombs, called mastabas, of rectangular plan with stepped sloping sides rising to a flat top some 30 ft (9 m) above the ground. Another early type of pyramid found in Mesopotamia was the ziggurat, an artificial mountain of mud and brick of stepped construction up to 175 ft (53 m) high with ceremonial staircases leading to a temple room at the top. Pyramid construction reached its zenith in about 2580 B.C. when the Egyptians used dressed stone laid in mortar to build a 481 ft (147 m) high structure. Although much less than the maximum theoretical height, it remained the tallest structure in the world until the twentieth century.

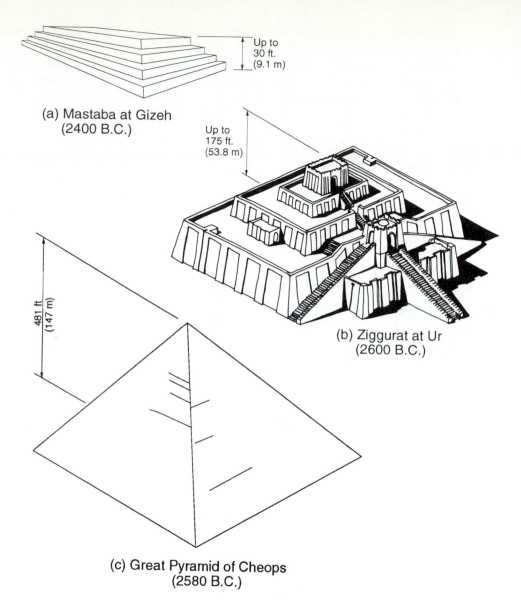

(a) Mastaba at Gizeh
(2400 B.C.)

Up to
30 ft.
(9.1 m)

Up to
175 ft.
(53.8 m)

(b) Ziggurat at Ur
(2600 B.C.)

481 ft
(147 m)

(c) Great Pyramid of Cheops
(2580 B.C.)

**Figure 1.6**   Pyramid forms of building.

Temples of pyramid form were also constructed in Central and South America from about 900 B.C. to A.D. 1400. These temples were built from mud brick or stone, often contained rooms at several levels, and reached heights of up to 187 ft (57 m).

For a given height of building, the pyramid with its wide base and tapered sides is the most stable of shapes and spreads its weight over a large area. It is, however, uneconomical in the use of materials; the Great Pyramid required about two million stone blocks averaging about $2\frac{1}{2}$ tons (2.2 tonnes) each, with some weighing as much as 15 tons (13.4 tonnes).

Further information on pyramid construction is contained in Refs. 1.7 and 1.8.

**Walls.** Walls having vertical faces contain much less material than pyramid construction even though in ancient times these walls were invariably solid and massive. Such walls were used for retaining earth, fortification of communities, and enclosing buildings. They were constructed of stone, sun-dried brick, or kiln-fired brick with construction techniques and quality ranging widely.

Early stone walls ranged from rubble masonry to ashlar masonry in which the stone was cut to shape and required only thin mortar joints or none at all. A common intermediate type of construction used outer facings finely finished by the mason with the interior later filled by laborers with uncut stone and mortar. The Romans also used this type of construction with outer faces of brick infilled with an early type of concrete. Figure 1.7 shows examples of Roman masonry walls.

An early type of masonry wall consisted of mud bricks laid in mortar made from mud or bitumen. Mortar joints ranged from only $\frac{1}{16}$ to $\frac{1}{8}$ in. (1 to 3 mm) in thickness to thicknesses of up to $1\frac{1}{2}$ in. (40 mm). Reed matting was sometimes used as reinforcing every few courses to strengthen the wall and control shrinkage. Although the exterior walls of ancient domestic buildings were stabilized by closely spaced cross walls and some larger walls were buttressed, many walls relied on their massiveness for stability.

An example of such a massive wall is found at the palace at Ctesiphon in Mesopotamia, where a mud brick wall 16.5 ft (5 m) thick at the bottom rises to a height of 113 ft (34.4 m). This wall is shown in cross-section in Fig. 1.8(a) and also in Fig. 1.18.

The stability of such a wall can be understood in terms of a thrust line. Consider the equilibrium of the wall above section *A–A*, at a distance *y* below the top of the wall, as shown in Fig. 1.8(b). The weight of the masonry above *A–A* acts through its center of gravity and exerts a downward force *W* on the masonry below. This downward force is resisted by an opposing force, reaction *R*, supplied by the masonry below. For equilibrium, the thrust *R* = *W*. By taking sections at different levels in the wall, the locations of the thrust line throughout the height of the wall can be obtained. With only the self-weight of the wall acting, the thrust line will be at or

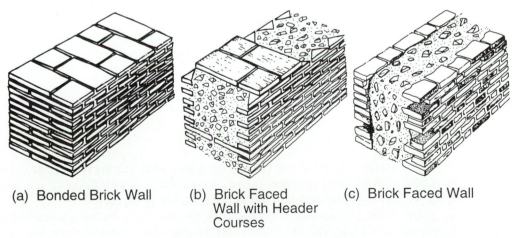

(a) Bonded Brick Wall    (b) Brick Faced Wall with Header Courses    (c) Brick Faced Wall

**Figure 1.7**  Roman masonry walls.

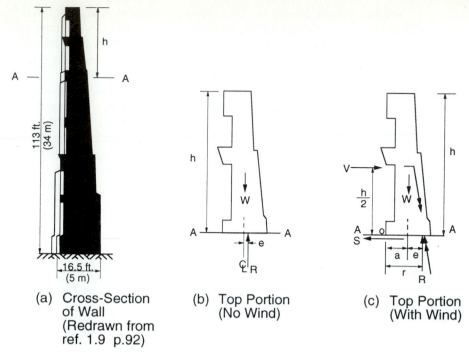

(a) Cross-Section of Wall (Redrawn from ref. 1.9 p.92)

(b) Top Portion (No Wind)

(c) Top Portion (With Wind)

**Figure 1.8** Equilibrium of mud brick wall at Palace of Ctesiphon (c. A.D. 550).

near the center line of the wall throughout its height and compressive stresses that are uniform or nearly uniform across the thickness of the wall will be obtained as shown in Fig. 1.9(a).

Suppose now that the wind blows from the left side and exerts a resultant force $V$ on the masonry above section $A–A$ in Fig. 1.8(c). To prevent the upper portion from sliding off horizontally under the action of $V$, a shear resistance $S$ must be mobilized through friction or bond in the mortar such that $S = V$.

As previously shown, for vertical equilibrium, $R = W$, but in this case, the force $R$ must be located where the resultant of $V$ and $W$ intersects the section $A–A$. This can be found graphically or by balancing moments about point 0 so that

$$Rr = Wa + Vy/2 \qquad (1.1)$$

Therefore, the vertical thrust $R$ is at a distance $r$ from the windward face, where

$$r = a + \frac{Vy}{2W} \qquad (1.2)$$

The vertical thrust at the section combines with the shear force at the section to produce an inclined resultant thrust. However, it is usually more convenient to work with the vertical and horizontal components of force.

The value of $r$ and the eccentricity of the thrust, $e$, increase as the wind force $V$ increases. Thrust lines for various increasing wind intensities are shown in Fig. 1.9.

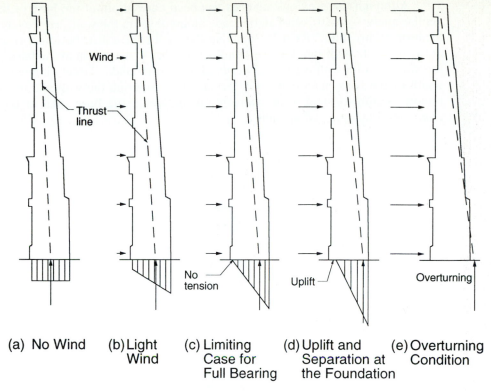

Wind →

Thrust
line

No
tension

Uplift

Overturning

(a) No Wind

(b) Light
Wind

(c) Limiting
Case for
Full Bearing

(d) Uplift and
Separation at
the Foundation

(e) Overturning
Condition

**Figure 1.9**  Stability of wall under increasing wind force.

The stresses at the base of the wall can be examined in terms of the location of the thrust line. With no wind, Fig. 1.9(a), the thrust line at the base coincides approximately with the center line at the base and, assuming a linear behavior for the foundation material, an approximate uniform distribution of stress results. With a light wind, Fig. 1.9(b), the thrust line moves toward the leeward face of the wall and hence the stresses at the base have to adjust such that the resultant of the stress distribution coincides with the thrust line. This adjustment of the stresses is accomplished by the materials deforming slightly. As wind increases in intensity, a stage is reached when the thrust line is located at one-third of the base width from the leeward edge of the base, Fig. 1.9(c). A triangular stress distribution therefore results, which extends for the full width of the base and doubles the maximum stress compared with the no-wind case. This is a significant stage as it is the limiting case that produces no uplift. Further increase in wind force produces uplift on the windward side of the wall. A triangular distribution of stress results over the contact area, with its resultant coinciding with the thrust line, as in Fig. 1.9(d). It can be seen that the increasing wind force moves the thrust line closer to the leeward face, producing ever increasing compressive stresses, which can lead to crushing of the masonry at these locations or failure of the foundation. Even if such failures do not occur, the wall will overturn when the thrust line falls outside the base, as shown in Fig. 1.9(e).

Although the previous analysis has been carried out at the base of the wall, a similar analysis of stresses can be made at any level in the wall. However, from the location of thrust line, it can be seen that the most severe distribution of stress occurs at the base. In addition, the magnitude of the thrust is greatest at the base. Because the stability of a wall is generally critical at its base and is largely influenced by the width of the base, materials could be saved by tapering the wall and incorporating hollow portions. Thus, walls were often thicker at the base than at the top and contained recesses and openings.

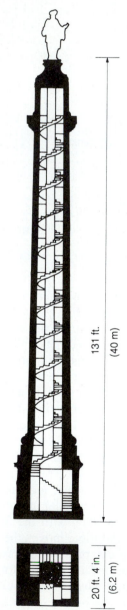

**Figure 1.10**  Trajan's Column (A.D. 113). (Redrawn from Ref. 1.9)

Ancient Masonry    Chap. 1

**Columns and Towers.** A cross-section is more efficient when the material is concentrated at the extremities. This principle was used for the construction of tower-type structures where the perimeter walls enclose a central space. An example of tower-type construction, shown in Fig. 1.10, is one of the pillars of victory, Trajan's Column, erected in Rome in A.D. 113[1.9]. The stability of this structure can be analyzed in terms of a thrust line in a similar manner to that discussed before. Strasbourg Cathedral in France was the tallest building in the world prior to the twentieth century, with its spire, completed in 1439, rising to a height of 466 ft (142 m), just slightly lower than the Great Pyramid.

Although relatively economical in the use of materials, medieval towers are still massive by modern standards, with surviving examples having walls about 6 ft (2 m) thick at the base. High compressive stresses are imposed on the foundations and many of the campanili, or bell towers, of Europe have developed a marked lean, the most noted being at Pisa.

### 1.3.2 Spanning Across

**Beams or Lintels.** Horizontal space is most easily spanned by placing a beam across an opening. This occurs naturally when a tree trunk falls across a creek or small river. Primitive man used the same technique by balancing a stone across two other stones. In this post-and-lintel form of construction, the log or the balanced stone bends to produce compressive stresses on the top face and tensile stresses on the lower face. Because the tensile strength of rock is low and it is liable to crack, the stone beam or lintel must have a relatively large cross-section and span over only a short distance.

Post-and-lintel construction is a fundamental form of masonry construction used all over the world, ranging from the ceremonial stone circle at Stonehenge to the classical Greek architecture of the Parthenon in Athens[1.10]. Stone lintels were also used to support the masonry above openings in walls, but this method was cumbersome and limited, as illustrated by the lintel over the Lion Gate at Mycenae (Fig. 1.11), which weighed between 25 and 30 tons (tonnes) but spanned little more than 10 ft (3 m).

It is instructive to consider the structural action of this large stone lintel, shown separately in Fig. 1.12(a). If we assume that the weight of the stone plus the load it carries is 30 tons (26.8 tonnes) then, because of symmetry, each of the vertical reactions is 15 tons (13.4 tonnes). The precise locations of the vertical reactions are not known as they depend upon the nature of the contact surfaces between the lintel and the abutments. If the bearing is uneven, then there will be a tendency for the most highly stressed regions to settle with time, leading to eventual uniform bearing.

By considering the forces on half of the lintel, as shown in Fig. 1.12(b), it can be seen that vertical forces balance, and no shearing force is present at midspan. Because the lintel is simply resting on the abutments, it is likely that there is no effective horizontal resistance and hence no arching action in the lintel. The external couple produced by the weight, therefore, must be resisted by an equal and opposite internal couple:

$$\text{Internal couple} = Td = Cd \qquad (1.3)$$

**Figure 1.11**  Lion Gate (c. 1250 B.C.).

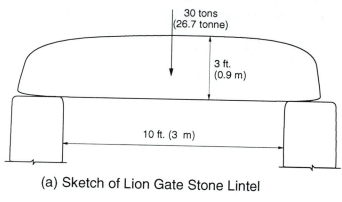

(a) Sketch of Lion Gate Stone Lintel

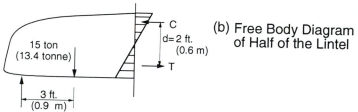

**Figure 1.12**  Equilibrium of stone lintel.

and

$$T = C \text{ for horizontal equilibrium}$$

where $T$ = the resultant tensile force in the lintel
$C$ = the resultant compressive force in the lintel
$d$ = the lever arm between T and C

Because there is no resultant horizontal force, the concept of thrust line, used previously, is not applicable here. If one adopts the usual assumption of linear variation of stress over the cross-section, then the resultant forces $T$ and $C$ act at the centroids of triangular distributions. In this case, the lever arm distance $d$ is two-thirds the depth of the lintel at midspan, that is, 2 ft (0.6 m).

Assuming a reaction located 6 ft (1.8 m) from midspan and balancing the internal couple at midspan with the opposing external couple produced by the weight gives

$$T \times 2 = 15 \times 3.0$$

Therefore, $T$ = 22.5 tons (20 tonnes) and $C$ = 22.5 tons (20 tonnes).

The top portion of the stone lintel can easily resist the compressive force, but, because stone is relatively weak in tension, the 22.5 ton (20 tonne) tensile force in the bottom portion is the critical factor. It is this tensile force that has led to cracking and collapse of many stone lintels.

**Primitive Arch.** A greater span than is possible in bending can be achieved using two inclined stone slabs resting against each other to form a primitive arch. An example of this is at the temple of Apollo, Delos (c. 426 B.C.), where a distance of 20 ft (6 m) is spanned. In this case, solid rock abutments provide horizontal restraints and the arrangement can be analyzed using the concept of the thrust line.

For the inclined lintel shown in Fig. 1.13(a), let $W$ denote the weight of each stone and $L$ the distance spanned. To prevent the stones spreading, horizontal thrusts

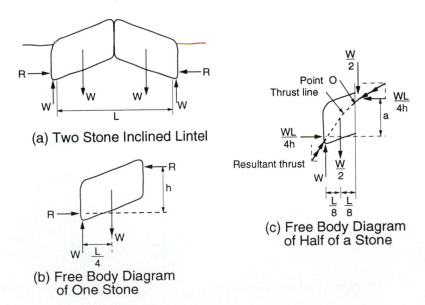

(a) Two Stone Inclined Lintel

(b) Free Body Diagram of One Stone

(c) Free Body Diagram of Half of a Stone

**Figure 1.13**   Equilibrium of inclined lintel (primitive arch).

$R$ are applied by the abutments at locations determined by the nature of contact between the surfaces. Due to the symmetry of the arrangement, the upward reaction force at each abutment is $W$. By considering the horizontal equilibrium of one of the stones, as in Fig. 1.13(b), it can be seen that a horizontal thrust of $R$ is required at the crown. Its location is determined by the contact between the two stones. For rotational equilibrium, the clockwise couple of $WL/4$ must be balanced by the anti-clockwise couple $Rh$, leading to the following expression for the magnitude of the horizontal thrust:

$$R = \frac{WL}{4h} \tag{1.4}$$

where $h$ = vertical distance between the horizontal forces as determined by the inclination of the stone and the vertical contact between the end surfaces.

The position of the thrust line at midlength of one of the stones can be found by considering the equilibrium of one half of the stone, as indicated in Fig. 1.13(c). Taking moments about 0 gives

$$\frac{WL}{4h} a + \frac{W}{2}\left(\frac{L}{8}\right) = W\frac{L}{4}$$
$$a = \tfrac{3}{4} h \tag{1.5}$$

By considering other sections, it can be shown that the complete thrust line is parabolic. Two of many possible arrangements of thrust line are shown in Figs. 1.14(a) and (b).

The thrust line shown in Fig. 1.14(a) would apply if the abutting faces at the abutments and crown were bearing evenly against each other. If, however, abutting

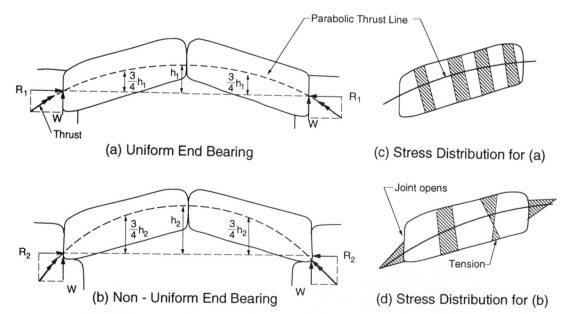

**Figure 1.14**   Thrust lines and stresses in inclined lintels (primitive arches).

Ancient Masonry    Chap. 1

faces were aligned with each other while the stone slabs were temporarily supported during construction, then, upon removal of the supports, deformation of the stone abutments under load would lead to a thrust line more like that shown in Fig. 1.14(b). It follows from Eq. 1.4 that if $h_2$ is $1\frac{1}{2}$ times $h_1$, then $R_2$ is only $\frac{2}{3}$ of the thrust $R_1$.

The stresses in the stones can be examined in terms of the thrust line. Where the thrust line coincides with the centroid of the cross-section of the arch, the stresses are uniform over the section. Where the thrust line is eccentric to the centroid of an arch section, the stone has to deform slightly until the resultant of the stress distribution coincides with the thrust line.

Where the thrust line is at one-third the depth of a member with a rectangular section, a triangular distribution of stress results, and where the thrust line lies outside the middle third of the member, tensile stresses result on the opposite face of the stone. The rock can resist a small amount of tension, but the joints cannot. Therefore, if the thrust line lies outside the middle third of the thickness at the abutments or the crown, the joints will open up and the thrust line will coincide with the third point of the contact zone. Stress distributions corresponding to the thrust lines in Figs. 1.14(a) and (b) are shown in Figs. 1.14(c) and (d), respectively. This reasoning leads to the middle third rule, which requires the thrust line to fall in the middle third of a rectangular section throughout the member length so as not to produce tensile stresses or opening of joints.

**Corbelled Arches.**　　Stone lintels are cumbersome and it may be advantageous to span some openings using masonry units. This was achieved in many parts of the ancient world by corbelling. In corbelling, subsequent courses of stones or bricks on each side of an opening project beyond the previous course until the inclined sides meet in the middle. Until they are joined, corbelled sections are precariously balanced in position but are usually temporarily supported. The weight of additional courses is resisted by arching action of the corbelled masonry. An example is the Mycenaean fortress at Tiryns, as shown in Fig. 1.15. The method dates back to about 2900 B.C., but, because of its inherent limitations, it never developed to span more

**Figure 1.15**　Corbelled arch in wall, Tiryns, Greece (c. 600 B.C.).

than about 10 ft (3 m), and in this case, large stones were needed. The Lion Gate (Fig. 1.11) had corbelling over the stone lintel to relieve the load on the lintel.

**True Arches.** A significant structural advance was made with the introduction of the first true arches in about 1400 B.C. These were built of wedge-shaped stones or bricks, called voussoirs, arranged to form a semicircle. Although the earliest known example at Ur in Mesopotamia spanned only 32 in. (0.8 m), this form of construction had the potential to span large distances. The reason for this can be seen by examining the thrust line of the primitive arch shown in Fig. 1.14. If the stone was hewn away so that only a narrow thickness, centered around the thrust line, remained, then the resulting arch would be entirely in compression and uniformly stressed over its thickness. Masonry under uniform compressive stress is very strong and hence this efficient structural shape has the potential to span large distances.

Simple calculations show that stone masonry having a compressive strength of 6000 psi (42 MPa) could arch more than ½ mile (0.8 km) before the stone would crush under its own weight. However, an arch of practical proportions would buckle under its own weight at a much shorter span. (The longest arch bridge built, using hollow concrete box sections, is in Gladesville, Australia, with a clear span of 1000 ft (305 m).)

The shape of the thrust line depends upon the loads applied to the arch and can be determined mathematically. However, the arching action is more readily visualized by analogy. A cable when suspended between two points takes up a catenary shape under its own weight and is in pure tension, as shown in Fig. 1.16(a). If the cable is then imagined to be made rigid, inverted, and supported at the same two points, its weight would act in the reverse direction and the cable would be in pure compression rather than tension. An inverted catenary is, therefore, the correct shape for an arch that must resist only its own weight. If the arch resists other applied loads, then the required shape can be obtained by applying equivalent weights to the

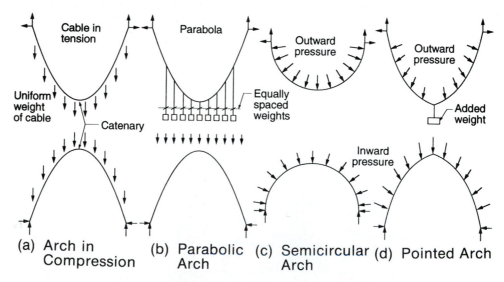

**Figure 1.16** Cable-arch analogy.

Ancient Masonry   Chap. 1

cable model. The shape taken by the cable will depend upon the positions of the added weights and the relative magnitudes of these weights compared with that of the cable. In many practical cases, the weight of the arch (cable) predominates, but it is instructive to note that if very large additional forces are applied so that the weight of the arch (cable) becomes insignificant, three common geometric shapes can be obtained. The first is a parabola, obtained by applying equal weights at closely spaced equal horizontal intervals across the cable span. The second is a circular arc obtained by applying closely and evenly spaced radial loads to the cable. And finally a *Gothic-arch* shape is obtained if in addition to evenly spaced radial loads a concentrated load is applied to the midspan of the cable. These cases are shown in Figs. 1.16(b), (c), and (d), respectively.

Although the geometries of the catenary and parabola are similar as are the loadings required to produce them, the semicircular shape and its associated radial loading are quite different. The Gothic shape is closer to the catenary and parabola than is the semicircle.

Most practical loadings on arches in buildings are vertical uniform loads and hence the thrust line in most arches is roughly parabolic. The semicircle is, therefore, not the best shape to use in building. However, semicircular arches were commonly used up to and throughout the Roman period (300 B.C.–A.D. 365) probably due to the ease of setting out the shape rather than any notion of the thrust line configuration[1.9].

The semicircular arches that survived evidently did so either because they were thick enough for the roughly parabolic thrust line to be contained within the arch or because the arch was surrounded by walls that allowed the thrust line to safely pass outside the arch, as shown in Figs. 1.17(a) and (b). Despite these limitations, semicircular arches were used to span large distances, as illustrated by a bridge near Aosta, in northwest Italy, which spanned 117 ft (35.7 m).

An advance on the semicircular shape did not occur until around the fourth century A.D. The outstanding, though massive, example of arch construction of approximately parabolic shape is at Ctesiphon in Iraq (c. A.D. 550), where a mud brick arch remains today. It is 24 ft (7.3 m) thick at the base, rises to 120 ft (36.7 m), and spans 83 ft (25.3 m). Figure 1.18 shows this ancient site.

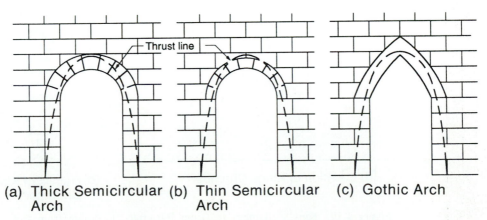

(a) Thick Semicircular Arch   (b) Thin Semicircular Arch   (c) Gothic Arch

**Figure 1.17**   Thrust lines in common arches.

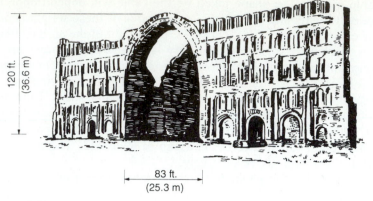

120 ft.
(36.6 m)

83 ft.
(25.3 m)

**Figure 1.18** Mud brick remains of Palace at Ctesiphon, Iraq (c. A.D. 550). (Redrawn from Ref. 1.9)

Pointed arches introduced in Syria during the sixth century[1.11] were later developed into a fine art by the Gothic master builders[1.12] during the twelfth to sixteenth centuries. Although the Gothic arch is not parabolic in shape, it is easily set out using circular arcs and, as shown in Fig. 1.17(c), can accommodate a parabolic thrust line far better than semicircular arches.

Though more efficient in shape and more finely constructed, the Gothic arches did not exceed the spans of earlier arches. The largest Gothic arch, completed in 1598, spans 73 ft (22.3 m) over the nave of the Gerona Cathedral in Spain. Most Gothic arches have considerably shorter spans.[1.9]

### 1.3.3 Enclosing Space

The structural elements so far considered do not by themselves enclose a significant space, if any. For example, in the Great Pyramid, the king's chamber occupies only about 1/4000 of the volume, as illustrated in Fig. 1.19. This space in fact is created

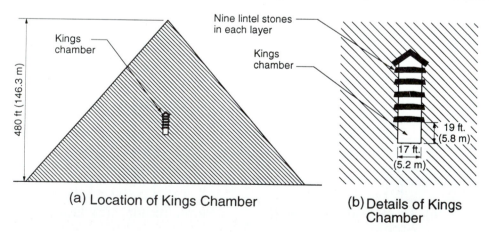

480 ft (146.3 m)

Kings chamber

Nine lintel stones in each layer

Kings chamber

19 ft. (5.8 m)

17 ft. (5.2 m)

(a) Location of Kings Chamber

(b) Details of Kings Chamber

**Figure 1.19** King's chamber in Great Pyramid. (Redrawn from Ref. 1.9)

by repeating the post-and-lintel construction considered earlier. In each layer, nine huge stone lintels spanning 17 ft (5.2 m) are placed side by side to form the roof of a chamber 34 ft (10.4 m) long. In a way, this structural form can be considered as a lateral translation of the post-and-beam construction.

Structural elements that span both vertically and horizontally and hence enclose space can be obtained by the lateral translation of an arch shape to form a barrel vault or by rotation of an arch about a central axis to form a dome.

**Barrel Vaults.** Early elements were constructed by the corbelling technique. The upper courses of two parallel walls were progressively projected beyond the course beneath until two walls met at the middle. As with corbelled arches, corbelled vaults are severely limited in the distance they can span without temporary supports during construction.

Vaults of true arch construction were made by laying stones or bricks in horizontal courses with their beds radiating from the center, as illustrated in Fig. 1.20(a). Temporary support using timber framework, called centering, was required until the vault, built simultaneously from each abutment, was completed at the crown.

The Romans employed this method often, using a combination of brick and concrete. An impressive example completed in A.D. 313 is the semicircular vault at the Basilica of Constantine, which spans 83 ft (25 m).

Another method of building vaults was to lay the masonry units in a series of rings or arches side by side so that the vault grew longitudinally from one end to the other. This latter technique allowed mud brick vaults to be constructed without timber centering. Construction began by laying units on a slight lean against an end wall until a complete inclined arch was laid. An additional layer of thin mud bricks could then be added, the adhesion of the clay mortar being sufficient to hold the units in position on the sloping plane until the arch layer was completed. Hence, the completed vault was literally a series of arches constructed side by side, as shown in Fig. 1.20(b). An early example of this construction is the brick vaulted drains in the Palace of Sargon in Persia. These vaults, built about 720 B.C., were slightly pointed. A grander example existed at the Palace of Ctesiphon where only an end arch now remains (Fig. 1.18).

(a) Construction of Vaults Requiring Temporary Support

(b) Construction of Vaults Using End Walls and Previous Construction for Stability During Construction

**Figure 1.20**   True arch construction.

**Domes.**    Another structural element that encloses space is the dome. This can be thought of as the shape formed by the rotation of an arch about its vertical axis.

Once again the earliest domes were formed by the corbelling technique, which leads to a fairly pointed dome. In this case, each course of masonry forms a horizontal ring in which, to some extent, each masonry unit is prevented from overbalancing by adjacent units forming a compression ring. For this action to be effective, the mortar joints or friction must provide sufficient resistance to prevent each horizontal ring of units from enlarging. Where a corbelled dome is underground, this resistance is improved by the lateral earth pressure on the ring, thus allowing domes of considerable span to be built. An example of this latter type is the Tomb of Agamemnon, built in 1325 B.C. It is 48 ft (14.5 m) in diameter and 44 ft (13.5 m) high inside, as shown in Fig. 1.21.

True domes consisted of horizontal courses of masonry units with more or less radial joints. The masonry units were tapered to allow them to be laid either dry or with only thin joints or, alternatively, thick mortar joints were used to accommodate the use of nontapered units. True domes had to be temporarily supported during construction, particularly near the crown.

The stability of true vaults and domes can be assessed in a similar fashion to arches. In fact, the behavior of a barrel vault is examined by considering a unit length of it as an arch. The thrust line of the arch becomes a thrust surface in the vault.

Stability of domes can be thought of in terms of a thrust surface. The simplest case to imagine is the semicircular dome or hemisphere. A free-floating soap bubble resists the internal pressure imparted during its formation by taking up a spherical shape and developing surface tension forces. The soap film cannot resist compressive or shear forces and therefore contains only tensile stresses that, because of symmetry, are the same throughout the bubble in all directions. By analogy, a table tennis ball submerged in deep water is subjected to uniform external pressure and is, therefore,

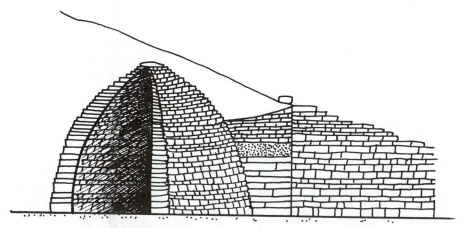

**Figure 1.21**    Tomb of Agamemnon (c. 1325 B.C.). (Redrawn from Ref. 1.9)

Ancient Masonry    Chap. 1

in pure compression. Hence, a spherical thrust surface results from uniform external radial pressure. Although the compressive stresses under such a loading are uniform in all directions, it is convenient and sufficient for a complete mathematical treatment to consider only two components—one in the direction of a horizontal ring, the other at right angles to this direction. If the sphere is split into hemispheres, then the compressive stresses in the horizontal ring direction will still be internally balanced, but the compressive stresses in the vertical direction will have to be balanced by external supports. Hence, half a table tennis ball (negligible weight) resting on a smooth table and subject to uniform radial external pressure is in pure compression. It follows that the hemisphere is not the thrust surface for types of loading other than radial compression. In practice, a uniform thin hemispherical dome subjected to its own weight will not have uniform stresses throughout and will develop tension in the horizontal rings of the lower 60% of the dome height.[1.2] This is illustrated in Fig. 1.22.

There are several ways of overcoming the problem of tension stresses. The usual way in early times was to use a thick dome in relation to its span so that the true thrust surface was contained within the thickness of the hemispherical dome. Another method was to thicken the dome over its lower portions so that, although the inside surface remained spherical, the midthickness corresponded more closely with the true thrust surface. These solutions were used by the Romans for the construction of the Pantheon in A.D. 123.[1.9] The brick and concrete dome in this building spanned 143 ft (43.6 m) and rested on walls 23 ft (7 m) thick. Its span has not been exceeded by any masonry dome since. Figure 1.23 shows a cross-section of this dome.

Another possible way to counteract the ring tension in the lower portion of hemispherical domes is to provide reinforcement. Although the Romans were known to have used metal clamps of iron or bronze in masonry arches to fix stones to each other, the method was not apparently used by them to stabilize hemispherical domes. Ring reinforcement was first used in a Gothic-shaped dome on the Florence Cathedral completed in 1462.[1.13]

Tensile stresses can be avoided if only the upper 40% of a hemisphere is used as a dome. In this case, horizontal inward acting forces must be supplied by buttresses at the base of the dome in addition to vertical support. The shallow dome is characteristic of Byzantine architecture and an excellent sample of this is the dome built over the church of St. Sophia in Constantinople (Istanbul) in A.D. 537.[1.14]

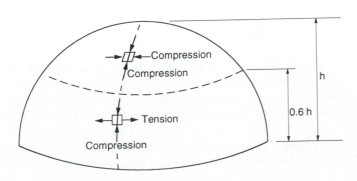

**Figure 1.22** Stress in a hemispherical dome under its own weight.

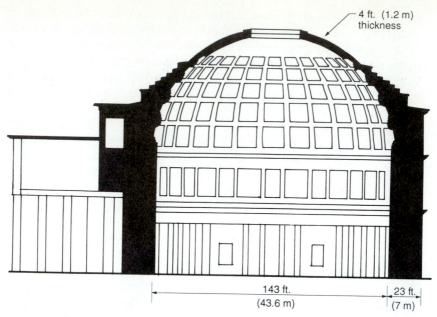

**Figure 1.23** The Pantheon in Rome (A.D. 123). (Redrawn from Ref. 1.9)

The ultimate solution from an engineering point of view is to make the shape of the dome follow the thrust surface. This sophisticated solution to the problems posed by the hemispherical domes used exclusively by the Romans was in fact that used by the earliest builders of domes.[1.15] Beehive-shaped mud huts were used as long ago as 9000 B.C. and numerous mud brick houses of parabolic shaped domes were built in Cyprus around 5650 B.C. Figure 1.24(a) gives an idea of the scale of these buildings. A later example is the stone dome laid without mortar, shown in Fig. 1.24(b), on the Church at Ezra, Syria, built in 515 B.C. and still used today.

The thrust surface is approximated by the Gothic shape in dome form. This was commonly used in Europe from the fifteenth to nineteenth centuries partly because of the ease of laying it out. An example of this form of construction, shown in Fig. 1.24(c), is the dome of St. Peter's Cathedral in Rome, completed in 1590. Nonetheless, no less than ten iron chains have been inserted around the perimeter of the dome at different times to prevent it from spreading.[1.2]

## 1.4 DEVELOPMENT OF BUILDING STRUCTURE

Most, if not all, ancient masonry buildings were loadbearing in the sense that the masonry resisted all of the imposed load in addition to its self-weight. Nonloadbearing masonry in buildings is a fairly recent concept in which masonry is used for its aesthetic and durable qualities, but not for supporting the floors and the roof of a building.

From earliest times, people built shelters to satisfy their physical needs, whereas places of worship, burial, government, and commerce were constructed for cultural

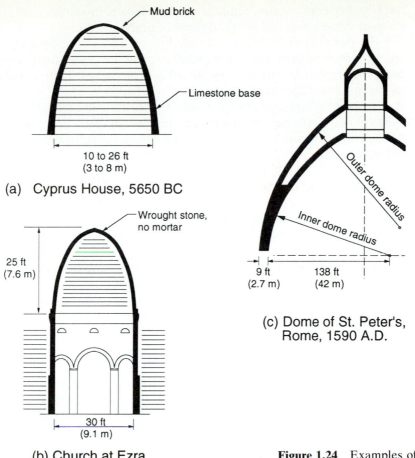

(a)  Cyprus House, 5650 BC

Mud brick

Limestone base

10 to 26 ft
(3 to 8 m)

Wrought stone,
no mortar

25 ft
(7.6 m)

30 ft
(9.1 m)

(b) Church at Ezra,
Syria, 515 B.C.

Outer dome radius

Inner dome radius

9 ft        138 ft
(2.7 m)    (42 m)

(c) Dome of St. Peter's,
Rome, 1590 A.D.

**Figure 1.24**  Examples of domes approximating thrust surfaces.

needs. Apart from simply utilizing caves, primitive people also built screens and huts of branches, reeds, skins, earth, and stone. The earliest buildings seem to have been round in plan with a transition to rectangular planning taking place between 9000 and 7000 B.C.

As their building skills developed, they eventually came to use only timber and masonry for the supporting structure of all substantial buildings. The difficulty of spanning horizontally with masonry constrained early builders to nearly always use timber for this function. Hence, masonry walls had timber lintels over openings. Intermediate floors were timber and roofs were timber-framed and clad with turf, thatching, clay-plastered reed matting, and the like.

### 1.4.1 Posts and Lintels

Many buildings were made entirely of timber using the basic elements of columns or posts to span vertically and beams or lintels to span horizontally. Because timber structures burn and decay, they have a short life and, needless to say, none of the early timber construction remains today. More permanent cultural buildings were made of stone posts and lintels instead of timber.

In these stone buildings, closely spaced columns supported a web of stone lintels, which in turn supported thinner lintels or stone slabs to form a flat roof deck. Alternatively, the construction might be topped with a timber roof system supporting burnt clay tiles.

This substitution of stone for timber elements is illustrated by masonry columns in Egypt and Crete carved to represent bundles of reeds or to imitate earlier timber styles, and by the *posts and lintels* at Stonehenge, which were connected by mortice and tenon joints.

### 1.4.2 Vaults and Domes

The development of the arch form replaced cumbersome stone and vulnerable timber lintels in walls with stone or brick masonry spanning wider openings. Similarly, barrel vaults and domes permitted the construction of large-span fire-resistant roofs although timber was still often used to support an outer covering of tiles. Great ingenuity was often shown in combining barrel vaults to create pleasing buildings. The culmination of this development was the cross vault, or groin vault, so named after the lines created by its intersecting surfaces. Examples of vaults are shown in Fig. 1.25.

Hemispherical domes were set upon cylindrical walls or more interestingly over polygonal or even square supporting walls. In these latter cases, the supporting walls were made thick enough to contain the base of the dome or thinner walls were used with lintels or arches spanning across the corners to support the dome. An elegant solution, to support a dome over a square plan area, utilized portions of another hemispherical surface called a pendentive dome. In this solution, the top dome had a diameter equal to the side of the square area below. The pendentive dome was a hemisphere encompassing the square below, but truncated at the base of the top dome and at the vertical sides of the square, as indicated in Fig. 1.26. This arrangement concentrated the weight of the dome on the four corner points, but allowed the square floor space to be further extended by the addition of barrel vaults of half hemispheres.

Intricate and daring arrangements of domes and vaults were attempted. The outstanding example is the Church of St. Sophia in Constantinople (A.D. 537) where the pendentive dome is extended at opposite ends by two half domes, each of which in turn are extended by two or three lesser half domes placed around their bases.[1.2]

All of these building forms generated from barrel vaults and domes, though elegant in geometry, were massive in construction. They were essentially plate structures, with roofs in single or double curvature, transmitting loads to walls below. The large thicknesses were not determined by the crushing strengths of the materials, but by the need to accommodate thrusts within the plates.

### 1.4.3 Gothic

Three developments took place to convert this heavy construction to the lightness and openness of the later Gothic buildings. First, arch ribs were incorporated in roof or ceiling structures, thus allowing the thickness of the masonry spanning between

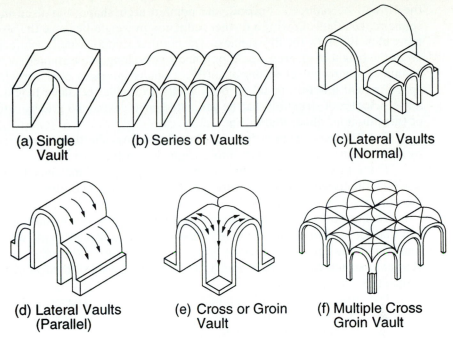

(a) Single Vault

(b) Series of Vaults

(c) Lateral Vaults (Normal)

(d) Lateral Vaults (Parallel)

(e) Cross or Groin Vault

(f) Multiple Cross Groin Vault

**Figure 1.25**   Examples of combined barrel vaults.

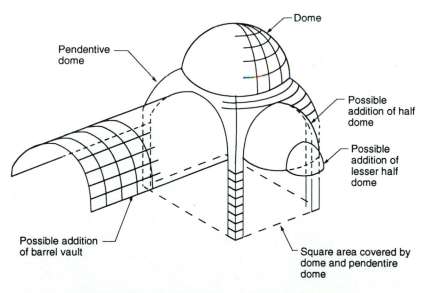

Dome

Pendentive dome

Possible addition of half dome

Possible addition of lesser half dome

Possible addition of barrel vault

Square area covered by dome and pendentire dome

**Figure 1.26**   Combinations of domes and vaults.

the ribs to be reduced. Second, the pointed arch shape was used instead of the semicircular, thus allowing a further reduction in weight because the structure could more closely follow the thrust lines. The pointed arch provided architectural flexibility as the height of an arch, vault, or dome was no longer determined by its span. For example, where barrel vaults of different spans intersect over a rectangular floor space, a neat junction results if both vaults are made the same height. Interesting and efficient roof structures can be generated on a rectangular grid by using variations of this junction, illustrated in Fig. 1.27.

Finally, heavy supporting walls running across the thrust lines were replaced by flying buttresses and towers more aligned with the thrusts.

These three developments in combination produced masonry-framed buildings in which the framing elements were the arch rib, the flying buttress, the column, and the buttress wall or tower.

This masonry framework was infilled with thin masonry vaulting in the ceiling and a latticework of arches, masonry panels, and stained glass in the walls. A timber-framed roof was generally used to protect the ceiling from the weather. An example of such a structure, shown in Fig. 1.28, is the Beauvais Cathedral in France, commenced in A.D. 1220.

The structural load path can be explained entirely in terms of thrust lines. The weight of the masonry ceiling was distributed to the arch ribs, which directed the thrust down through walls and columns and laterally, via flying buttresses, to wall or tower buttresses. These wall or tower buttresses in turn diverted the thrust to the ground, often helped by the stabilizing weight of spires or statues at their peaks.[1.2]

Early forms of masonry building structures using arches, vaults, and domes are rarely used today partly because of the high labor costs in constructing such intricate

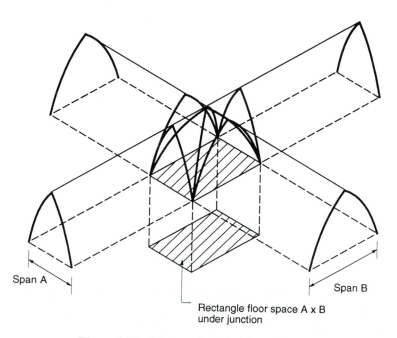

Span A

Span B

Rectangle floor space A x B
under junction

**Figure 1.27**  Intersection of pointed vaults.

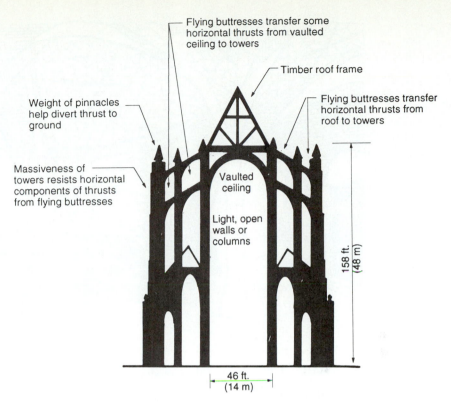

Figure 1.28, caption labels:

Flying buttresses transfer some horizontal thrusts from vaulted ceiling to towers

Timber roof frame

Weight of pinnacles help divert thrust to ground

Flying buttresses transfer horizontal thrusts from roof to towers

Massiveness of towers resists horizontal components of thrusts from flying buttresses

Vaulted ceiling

Light, open walls or columns

158 ft. (48 m)

46 ft. (14 m)

**Figure 1.28**  Beauvais Cathedral, France (A.D. 1220–1573). (Redrawn from Ref. 1.9)

shapes and because of modern styles. One of the few attempts to integrate these forms into contemporary styles was taken by Gaudi, a Spanish architect, at the end of the nineteenth century.[1.16,1.17] He built free-form structures that followed thrust surfaces and lines. An example is the Colonia Güell near Barcelona, Spain, shown in Fig. 1.29. He obtained the shapes experimentally by hanging weights from draped cloth and wire in a three-dimensional version of the cable analogy described earlier. This concept has not been further developed and many of the efficient forms of early masonry construction have yet to be successfully adapted to contemporary materials and conditions.

### 1.4.4 Single-Story Loadbearing Buildings

Contemporary single-story masonry buildings have developed from simple ancient domestic buildings such as those built by the Romans in the first four centuries A.D. The smallest type of dwelling at the time consisted of a single open-fronted room formed by masonry walls that supported a timber loft and roof covered with fired clay tiles. Although the masonry walls in such buildings were loadbearing in that they supported the weight of the roof, the magnitudes of the resulting compressive stresses in the walls were practically negligible. The critical structural function of these walls was to resist lateral forces from wind or earthquake. This was achieved

**Figure 1.29**  Crypt of the Colonia Güell, Spain (1914).

by using thick walls that were buttressed by end walls to improve their stability under lateral load.

### 1.4.5 Multistory Loadbearing Buildings

A multistoried version of the small domestic dwelling was also built by the Romans during the first century A.D.[1.6] These apartments blocks were up to five or more stories high. An apartment *block* or *insula* was generally of rectangular plan subdivided by brick-faced concrete walls about 3 ft (1 m) thick into rooms, passageways, and stairways, as illustrated in Fig. 1.30. The cellular nature of this plan ensured adequate stability against the lateral forces of wind and earthquake. Because the construction is similar to that used today, these ancient buildings have a surprisingly modern appearance. Indeed, for low-rise domestic buildings, the structural principles have hardly changed.

**Traditional.**  Where multistory commercial buildings were required with open floor space, internal walls were replaced by timber columns to support the floors. In this case, the cellular nature of the wall system was lost and, because the internal timber construction was relatively flexible, the external masonry walls alone resisted lateral loads. This form of construction, a perimeter of thick masonry walls that received little or no lateral support from an interior frame of timber or iron, continued until the early years of the twentieth century. Figure 1.31(a) illustrates this principle.

In a long building, the end walls have negligible effect on the lateral support of the side walls at midlength of the building and hence the lateral stability of this portion

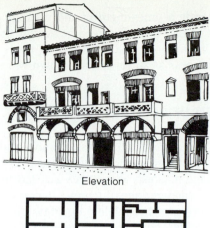

Elevation

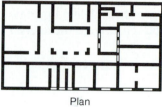

Plan

**Figure 1.30** Early multistory load-bearing construction in Ostia. (Redrawn from Ref. 1.6)

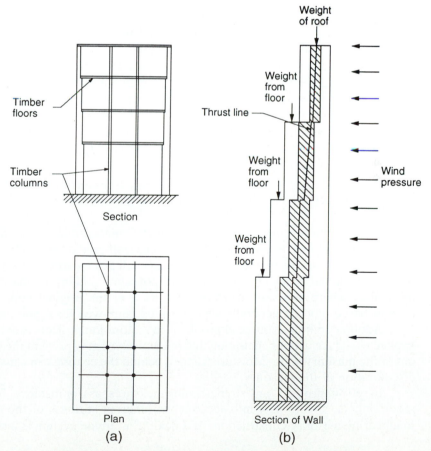

Timber floors

Timber columns

Section

Plan

(a)

Weight of roof

Weight from floor

Thrust line

Weight from floor

Weight from floor

Wind pressure

Section of Wall

(b)

**Figure 1.31** Traditional multistory loadbearing buildings.

can be assessed by considering a unit length of the wall. A possible rationale for design of such buildings, if in fact any was used, was that the thrust line resulting from the weight of the wall, the floor loads, and wind loads should lie within the middle third of the wall thickness at all levels, as shown in Fig. 1.31(b).

## 1.5 RESTORATION AND RETROFIT OF HERITAGE STRUCTURES

Because the restoration of historic structures is a major study in itself, only brief introductory comments are provided here. Particularly over the past few decades, interest in preserving heritage masonry structures has led to the need to develop appropriate architectural and engineering strategies. In carrying out any such strategies, it is of paramount importance that qualified professionals be engaged and that all parties involved in a project be sensitive to the special status and requirements of a heritage structure.

In the restoration or retrofit of a heritage structure, the following key issues are frequently encountered:

- What is the makeup and condition of foundations, walls, floors, roofs, chimneys, and other building envelope elements such as windows?
- What are the causes of any problems exhibited by the building?
- What is the end use of the structure and what code regulations pertaining to structural and fire safety must be satisfied?
- Especially in adaptive reuse projects, what levels of interior temperature and humidity are encountered? New requirements may necessitate the installation of thermal insulation and air/vapor barriers that in themselves can alter the behavior and performance of the building. The solution devised for one problem should not trigger new problems.
- What are sources of excessive moisture and how can such moisture be directed away from masonry walls and foundations?
- What aspects of vegetation, especially large trees and climbing vines, have had or will have an effect on the condition of the masonry?

In restoration projects, the most frequently encountered remedial measure is the repointing of mortar joints that have either weathered excessively or cracked. It is important to ensure that repointing mortar matches existing mortar as closely as possible to avoid local overstressing and potential spalling of masonry units. In particular, where soft lime mortars were used in the original construction, hard cement-based mortars generally should not be employed for repointing.

Another common source of problems in restoration projects is exterior coatings applied to reduce water penetration that in themselves may prevent the natural drying out of the masonry wall. Moisture trapped behind the coating can cause spalling and discoloration of the masonry.

In retrofit projects, any strengthening measures, as, for instance, for earthquake requirements, should take into account the inherent stiffness of the often massive loadbearing masonry construction. If a flexible alternate system is installed, it must be recognized that for this flexible system to resist significant load, the masonry must undergo large deformations that will usually result in substantial cracking or other damage to the masonry. If this is not acceptable, either the alternate system must be correspondingly stiffened or the existing masonry must be strengthened to incorporate a specified degree of ductility.

## 1.6 CLOSURE

Much can be learned from study of ancient masonry structures, and modern designers and builders must be in awe of the accomplishments of their predecessors. It may be in large part because of such ancient successes that the tradition of masonry is so strong and that many aspects of masonry design and construction used today have developed from standard construction practice rather than from rationalized design requirements. However, many of our contemporary uses of masonry, which are often mundane in comparison to ancient structures, have resulted in design conditions requiring a more complete understanding of masonry behavior.

## 1.7 REFERENCES

1.1   J. Bower, *History of Building,* Crosby, Lockwood Staples, London, 1973.

1.2   H. J. Cowan, *The Master Builders,* Wiley, New York, 1977.

1.3   K. Ward-Harvey, *Fundamental Building Materials,* Sakoga, Sydney, 1984.

1.4   A. M. Neville, *Properties of Concrete,* Pitman Publishing, 1972.

1.5   H. J. McKee, *Introduction to Early American Masonry,* National Trust for Historic Preservation and Columbia University, New York, 1973.

1.6   J. Musgrove (Ed.), *A History of Architecture,* 19th ed., Butterworths, London, 1987.

1.7   S. Lloyd and H. W. Miller, *Ancient Architecture,* Rizzoli, New York, 1986.

1.8   D. Heyden and P. Gendrop, *Pre-Columbian Architecture of Mesoamerica,* Rizzoli, New York, 1988.

1.9   J.B. Ward-Perkins, *Roman Architecture,* Faber & Faber, London, 1988.

1.10  R. Martin, *Greek Architecture,* Faber & Faber, London, 1988.

1.11  J. D. Hoag, *Islamic Architecture,* Rizzoli, New York, 1987.

1.12  L. Grodecki, *Gothic Architecture,* Rizzoli, New York, 1985.

1.13  P. Murray, *Renaissance Architecture,* Rizzoli, New York, 1985.

1.14  C. Mango, *Byzantine Architecture,* Rizzoli, New York, 1985.

1.15  E. Guidoni, *Primitive Architecture,* Rizzoli, New York, 1987.

1.16  C. Martinelli, *Gaudi—His Life, His Theories, His Work,* The MIT Press, Cambridge, MA, 1975.

1.17  M. Tafuri and F. Dalco, *Modern Architecture I,* Faber & Faber, London, 1986.

## 1.8 PROBLEMS

**1.1 (a)** Using clay bricks or wood blocks, construct a corbelled arch 2 ft (0.6 m) high where subsequent courses are "corbelled out" one-third of the thickness of the unit. Describe and comment on the construction method and the strength and stability of the structure after completion.

**(b)** Demonstrate or explain how improved types of arches could be built. What aspects produce the most significant improvements? Explain.

**1.2** Using string, paper, or other flexible material, construct two examples of true arches as discussed in Sec. 1.3.2. (Fast-setting epoxy or other glues, paints, or other coatings can be used to stiffen the catenary shape.) Test and comment on the stability and load-carrying capacity of these arches.

**1.3** Choose one of the buildings mentioned in this chapter and by consulting Ref. 1.7 and others, write a brief essay covering historical context, materials, structural elements, and building form.

**1.4** Many of the structural forms considered in this chapter exist in present-day civic buildings, churches, heritage buildings, etc. Photograph or sketch an example of each different form in your locality, noting the building name, location, date of construction, and giving a brief description of the element and its structural action.

# CHAPTER 2

# Contemporary Masonry

Loadbearing brick building.

## 2.1 INTRODUCTION

As discussed in Chap. 1, many traditional masonry buildings were designed utilizing the weight of the floors and the massive walls to offset tensile stresses due to moments from eccentricity of vertical loads and from lateral loads. It has been recognized that the achievement of lateral stability by gravity places a practical economic limit on the size of loadbearing masonry structures. This has led designers and builders to seek ways in which these massive bearing walls could be decreased in thickness while maintaining their structural stability.

The significant improvements in masonry materials and advances in manufacturing have contributed to the growth of masonry construction as a competitive and cost-efficient contemporary building system. High-strength units are now available with a variety of shapes, colors, and textures. Moisture, sound, and thermal characteristics have been improved and ready-mix mortars and grout are available for better quality control and speed of construction. Many admixtures for coloring and workability are available and have been used in masonry construction. Details of present-day masonry materials and their properties are given in Chaps. 4 and 5.

This chapter presents a broad view of contemporary masonry buildings, covering masonry elements, types of masonry construction, masonry systems, and design.

## 2.2 MASONRY ELEMENTS

The common masonry elements used today differ greatly in detail from those used in ancient masonry buildings. Various types of walls, columns, pilasters, beams, and lintels are considered in this section.

### 2.2.1 Walls

The function of a wall, the load it carries, and its location in the building are some of the factors to be considered in choosing the wall type. Walls are normally constructed in *running* or *half bond*, that is, with each head joint (vertical mortar joint) positioned half-way over the unit below. This produces the familiar symmetrical pattern over the face of the wall and normally permits a corner to be made without cutting units as most masonry units have a length twice their thickness (see Fig. 2.1(a)). Hollow

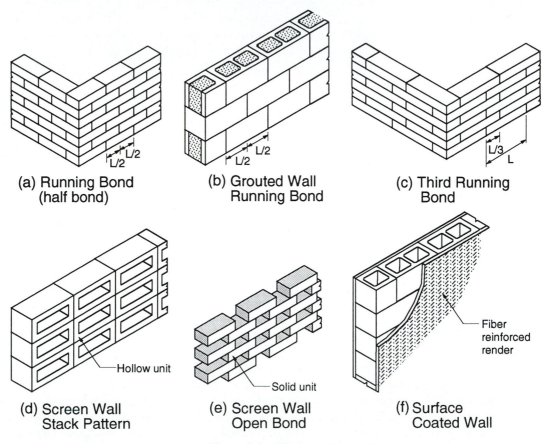

(a) Running Bond
(half bond)

(b) Grouted Wall
Running Bond

(c) Third Running
Bond

(d) Screen Wall
Stack Pattern

Hollow unit

(e) Screen Wall
Open Bond

Solid unit

(f) Surface
Coated Wall

Fiber reinforced render

**Figure 2.1** Single-wythe walls.

masonry units are made such that the cores line up in a wall if running bond is used, thus facilitating placement of grout and reinforcement (Fig. 2.1(b)). Other patterns or bonds are also used. A one-third overlap is sometimes used in running bond, particularly where the thickness of the masonry unit is one-third of its length as corners can then be made without cutting (Fig. 2.1(c)). For decorative purposes, a stack pattern (Fig. 2.1(d)), in which there is no overlap of units, or even an open bonded pattern (Fig. 2.1(e)) can be used. Unreinforced masonry walls built in these latter patterns have very low flexural strengths for bending about a vertical axis. Discussions of the common types of masonry walls are presented in what follows.

**Single-Wythe (Leaf) Walls.**    Single-wythe walls have a thickness equal to the width of the masonry unit and can be constructed of solid, cored, or hollow units with or without reinforcement.

Another type of single-wythe wall introduced to reduce laying costs is made by dry stacking units and then rendering both faces of the wall with a thin layer of fiber-reinforced cement plaster, as shown in Fig. 2.1(f), to achieve stability. However, the efficiency of masonry is such that this construction method is not commonly used.

In general, single-wythe walls are used in both loadbearing and nonloadbearing applications. When used on the exterior of a building, special precautions are often necessary to prevent moisture penetration and limit thermal transmission. They also have limited resistance to out-of-plane lateral loads, when unreinforced.

**Solid and Composite Walls.**    In the past, solid walls were often made by constructing two wythes of stonework or brickwork and infilling the intervening space with rubble or concrete. Modern solid walls have two (or more) closely spaced wythes joined together by header units or metal ties and the vertical collar joint or cavity between them is solidly filled with mortar or grout. The greater thickness of solid walls relative to single-wythe walls gives them greater strength and resistance to moisture penetration.

Sufficient headers or ties must be used to make the walls monolithic. Wythes are usually spaced $\frac{3}{8}$–$\frac{3}{4}$ in. (10–20 mm) apart and the collar joint between them filled with mortar as construction progresses. In some cases, a cavity 2–3 in. (50–75 mm) wide is constructed and later filled with grout to form a grouted wall. Where metal ties (including joint reinforcement) are used, both wythes are generally laid in running bond.

Where headers are used, a variety of bond patterns are available. In the *common bond*, each wythe, laid in running bond, is joined at each sixth course with a row of headers. From a structural point of view, this is preferable to the *Flemish bond*, which results in planes of vertical weakness as units are placed with little overlap. Figures 2.2(a) to (d) illustrate these types of construction.

It is usual for solid walls to be constructed with the same masonry units throughout to form a homogeneous masonry element. However, for economic or architectural reasons, different units can be used in each wythe of a solid wall to form a composite wall. For example, clay brick can be used on the exterior face and concrete block on the inside face of the wall. For such a case, the different physical properties of the units must be taken into account in design. In particular, the long-term differential movements that tend to take place due to shrinkage of the concrete units and

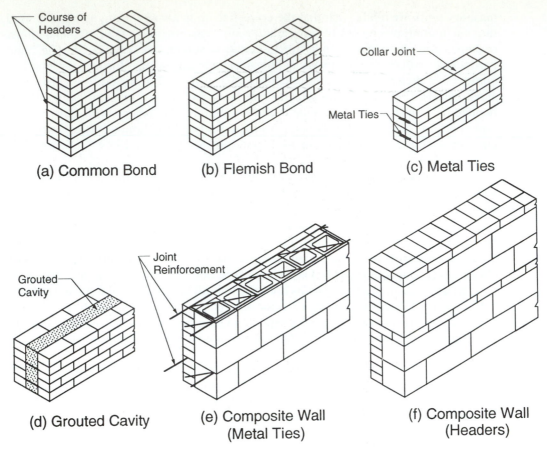

(a) Common Bond    (b) Flemish Bond    (c) Metal Ties

(d) Grouted Cavity    (e) Composite Wall (Metal Ties)    (f) Composite Wall (Headers)

**Figure 2.2**    Methods of joining wythes to form solid walls.

moisture expansion of the clay units can lead to bond failure at the collar joint. Where metal ties are used (Fig. 2.2(e)), integrity of the two wythes can be maintained even if debonding occurs. Where headers are used (see Fig. 2.2(f)), differential movement may shear the headers or lead to debonding underneath the headers, depending upon the direction of the relative movement. In either case, the integrity of the composite action of the wall would be destroyed. The potential for such disruption should be investigated if composite construction using markedly different materials for the two wythes is proposed.

Single-wythe and solid walls can be reinforced. Reinforcement can be placed in the bed and collar joints, the grouted cavity, the grouted pockets of walls made with solid units, and in the grouted cores of hollow units. Figure 2.3 shows examples of reinforced walls. Vertical reinforcement is not easily laterally restrained by lateral ties and therefore it is not normally counted on to increase resistance to axial compressive loads. Where reinforcement is spaced at large intervals, the wall is considered to consist of reinforced strips of masonry with unreinforced masonry spanning between them. This system is referred to as *partially* or *lightly* reinforced masonry. A common arrangement is for a hollow block wall, supported along its top and bottom

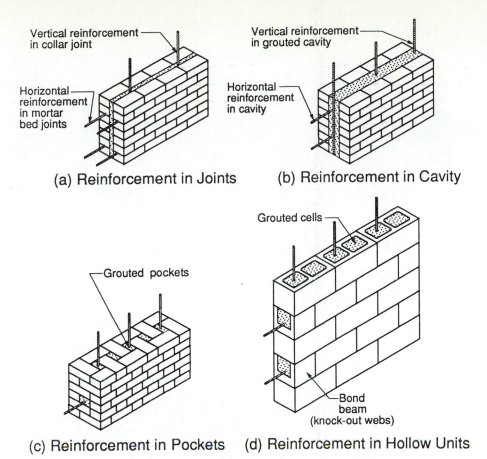

(a) Reinforcement in Joints   (b) Reinforcement in Cavity

(c) Reinforcement in Pockets   (d) Reinforcement in Hollow Units

**Figure 2.3**   Reinforcement locations in masonry walls.

edges, to have vertical reinforcement grouted into some of its vertically aligned cells. This reinforcement acts with a narrow strip of masonry on each side of it to form a vertical reinforced beam. The unreinforced masonry spans horizontally between these reinforced strips to resist lateral loads. Other possibilities are to have horizontally reinforced strips with unreinforced masonry spanning vertically between them, or a combination of both vertically and horizontally spanning reinforced strips with the unreinforced masonry between them resisting lateral loads by two-way bending action. These cases are illustrated in Fig. 2.4.

**Cavity Walls.**   Many early cavity walls consisted of two wythes separated by a cavity, but connected together with header units. If the header units were closely spaced, the wall acted as a hollow wall with composite action. The contemporary version of this is the diaphragm wall. The contemporary cavity wall was introduced in Britain during the early nineteenth century and commonly consisted of two 4½ in. (110 mm) wythes of masonry spaced 2 in. (50 mm) apart and connected by iron ties. Similar walls are used today (Fig. 2.5), except that the wall ties are often made of galvanized steel and increasingly of stainless steel to resist corrosion.

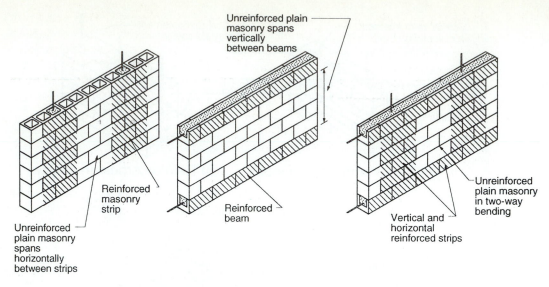

**Figure 2.4** Vertical and horizontal reinforcing strips in masonry walls.

In cavity wall construction, the two wythes can be of similar or dissimilar materials, a common example of the latter case being an external wythe of clay brick and an internal wythe of concrete block. If properly constructed, the cavity ensures complete waterproofness as any water penetrating the outer wythe runs down the inner face of the outer wythe and is collected and diverted by flashing and weep holes at the bottom of the cavity to the exterior of the building (see Chap. 12). The air space in the cavity improves the wall's thermal performance, which is often further enhanced by inserting insulation in the cavity.

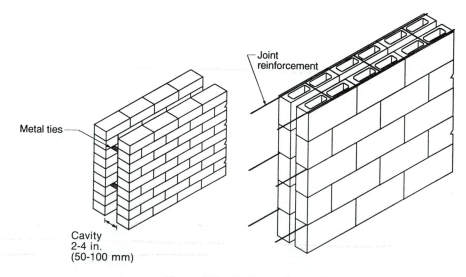

**Figure 2.5** Cavity walls.

In most applications, the inner wythe supports the weight of floors and roofs and the outer wythe is nonloadbearing. It is difficult to make both wythes loadbearing because differential vertical movements between the two wythes and deflection of the supported floor slab will fully or partially relieve one wythe of load.

Lateral loads are resisted by bending of both wythes as they are forced by the wire ties to deflect laterally by nearly the same amount. Thus, for the case of axially stiff wall ties, the individual wythes share the total lateral load roughly in proportion to their flexural stiffnesses. In some cases, the inner wythe may be designed to resist lateral forces by itself, in which case the outer wythe can be regarded as a nonstructural veneer.

**Veneer Walls.** A masonry veneer wall consists of a nonstructural cladding anchored to a structural backup wall and used for its aesthetic and durable qualities. The backup may be a masonry or concrete wall or a framed wall of wood or metal studs. If the veneer is thin, it can be either adhered to the backup wall or anchored to it with metal ties, but in either case, it is not relied on to resist lateral or axial loads. Where an adhered thin veneer is used, great care is necessary in selecting the adhesive and in allowing for differential movements between it and the backing. Design of thin veneers is a specialized area of construction not usually associated with unit masonry and is not dealt with further in this text.

Where a masonry veneer is used, it is anchored to the backup with metal ties usually with a minimum air space of 1–2 in. (25–50 mm) to prevent water penetration. Depending upon environmental conditions, the veneer wall system can also incorporate insulation and a water vapor barrier. Typical veneer walls are shown in Fig. 2.6.

The brick veneer wall was developed in Britain in the 1780s as cladding to wood-framed buildings and it is now also used in multistory construction with steel stud frames as well as masonry backing. The design and performance of such systems are discussed in detail in Chap. 12.

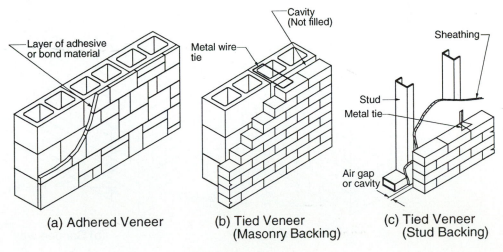

(a) Adhered Veneer

(b) Tied Veneer
(Masonry Backing)

(c) Tied Veneer
(Stud Backing)

**Figure 2.6**   Veneer walls.

**Diaphragm Walls.** A diaphragm wall (also referred to as a utility wall) can be thought of as a form of hollow wall or as a cavity wall in which the two wythes are tied together by masonry webs, either tied or bonded to them so as to achieve monolithic structural action in compression and vertical bending.

Example diaphragm walls are shown in Fig. 2.7. Tied webs have an advantage over bonded webs in that cutting of units is minimized, the two wythes can be constructed relatively independently, and the running bond appearance can be maintained. It is important, however, that tying or bonding of the webs to the wythes develops sufficient vertical shear resistance at the interface to maintain integrity of the full cross-section during vertical out-of-plane bending.

Composite action of the two wythes, as opposed to independent action in cavity walls, greatly enhances flexural resistance and allows diaphragm walls to span larger vertical distances and resist greater lateral loads. The bending resistance increases approximately in proportion to the spacing of the two wythes. All masonry material throughout the wall should be similar so as to avoid potential problems with differential movements.

### 2.2.2 Columns and Pilasters

Masonry columns are isolated vertical structural members that transmit concentrated dead and live loads to the foundation. Although terminology is by no means uniform, the term "pier" is sometimes used to describe a masonry column that is really a short length of wall. A column is usually of rectangular cross-section with length no more than three times its overall thickness. Other shapes can also be used. Ideally, the reactions from other structural members should be applied concentrically to columns so as to minimize buckling and eccentricity effects. Increased load-carrying capacity can be achieved by increasing the unit compressive strength, increasing the mortar

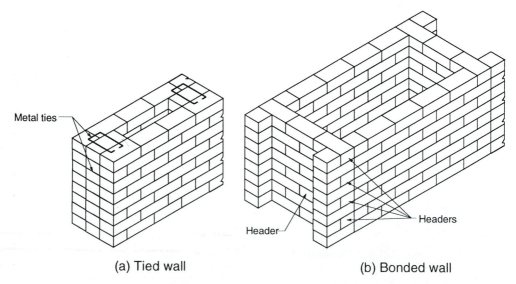

Metal ties

Header

Headers

(a) Tied wall    (b) Bonded wall

**Figure 2.7**  Diaphragm walls.

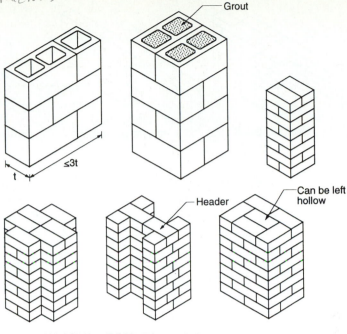

**Figure 2.8** Columns.

strength, or by filling the cells of hollow units with grout. Some possible configurations of columns are shown in Fig. 2.8.

A pilaster is a thickened wall section built integral with the wall and is sometimes described as an engaged column. The pilaster can project out from the wall on one or both sides and can be connected to the wall by headers (bonded pilaster) or by metal ties (unbonded pilaster). Examples of pilasters are illustrated in Fig. 2.9. Pilasters can perform four functions. First, a pilaster can act as a column to support beams, trusses, and other structural members that apply a concentrated load to the wall. The significant eccentricity in this case is in the direction transverse to the line of the wall. Second, a pilaster can help support a slab or other uniform load by increasing the wall's resistance to buckling. Third, pilasters can also act with the wall, as a beam between top and bottom supports, to resist lateral loads on the wall. Finally, pilasters can be used as small buttresses to increase the stability of a free-standing wall.

Columns and pilasters can be reinforced. Reinforcement increases their resistance to axial load provided that lateral ties are placed at close centers to confine the longitudinal reinforcement and prevent local buckling of the reinforcement. Such a reinforcing cage also confines the central core of grout or concrete and increases the ductility of the member. If the load is applied centrally, the increase in axial load capacity is not large, particularly for low percentages of steel. Where substantial bending occurs, the reinforcement markedly increases the column's resistance to flexure.

The main reinforcement can be placed in the cores of hollow units or within grouted pockets formed by the solid or cored masonry units, as shown in Fig.

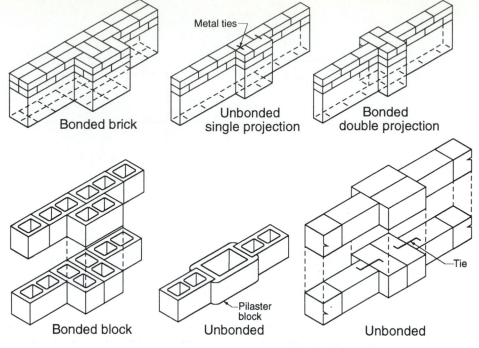

**Figure 2.9** Pilasters.

2.10. Lateral ties are preferably placed within the grouted core, but can be placed within bed joints of the surrounding masonry. Where a pilaster is used to give lateral support to a wall rather than resist compressive loads, lateral ties may not be necessary as the member acts as a vertical beam rather than as a column. Design of columns and pilasters is the focus of Chap. 9.

### 2.2.3 Beams and Lintels

In masonry construction, there are roof beams, floor beams, bond beams, and grade beams. Bond beams are usually located at the roof and floor levels and serve the dual functions of tying the building around the perimeter and acting as a chord member in floor or roof diaphragm action to transmit lateral loads to the shear walls. A lintel is a horizontal beam spanning over a door or a window opening and supports loads from the wall above. Nonmasonry lintels are sometimes used.

All masonry beams must be reinforced because the limited tensile bond strength of masonry should not be relied upon to resist flexural tensile stresses from sustained gravity loads. For lightly loaded small spans, such as lintels over doorways, longitudinal reinforcement is often placed in bed joints. Reinforcement can also be placed in a grouted pocket or cavity formed by solid masonry units or in the grouted cores of hollow units, or in special shaped lintel units, as shown in Fig. 2.11.

Where shear stresses in a beam are low, stirrups may not be required, but where shear becomes excessive, stirrups must be incorporated to resist diagonal tension. The contribution of compression reinforcement should not be considered in

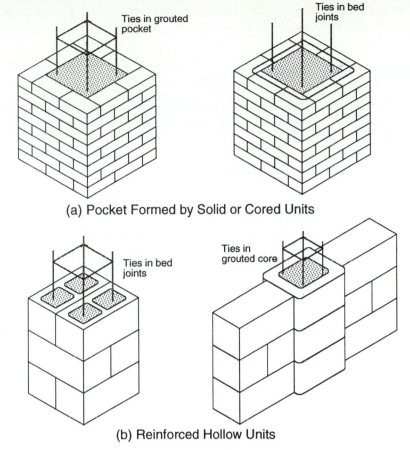

**(a) Pocket Formed by Solid or Cored Units**

**(b) Reinforced Hollow Units**

**Figure 2.10**   Column reinforcement.

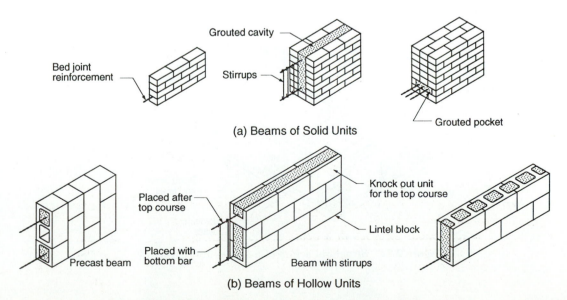

**(a) Beams of Solid Units**

**(b) Beams of Hollow Units**

**Figure 2.11**   Reinforced masonry beams and lintels.

design unless it is properly confined by stirrups or ties. Design of beams and lintels is covered in Chap. 6.

## 2.3 MASONRY BUILDING SYSTEMS

The masonry elements considered before can be combined into masonry building systems that can be conveniently classified as loadbearing single or multistory buildings and hybrid buildings.

### 2.3.1 Single-Story Loadbearing Buildings

Single-story buildings make up the majority of loadbearing masonry construction. Typical examples are warehouses, industrial buildings, gymnasiums, banks, and commercial stores. Single family detached housing also employs single-story loadbearing walls. In these applications, loadbearing masonry walls are the exterior envelope walls. Figure 2.12 illustrates the main features of a one-story masonry building.

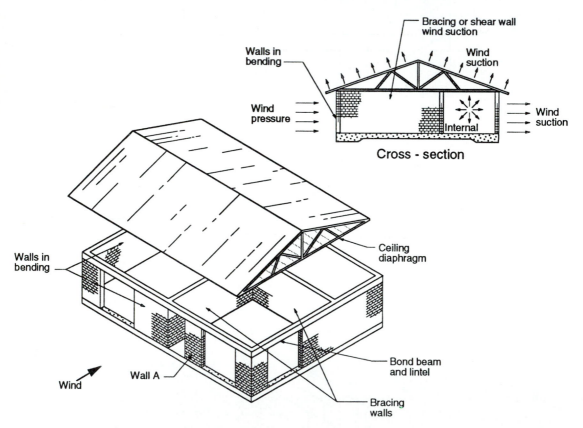

**Figure 2.12**   Structural action of a single-story structure.

Thin walls are used for economic reasons and, since unreinforced free-standing thin walls have negligible stability, they must be laterally supported in some way. Stability is achieved by using end walls, intermediate cross walls, supports along the top edge of the wall, or a combination of these. A lateral support along the top edge of the wall is usually provided by the roof or ceiling system. In domestic construction, a plaster ceiling is sufficiently strong in its own plane to give support to the outside wall by spanning between cross walls. In commercial or industrial buildings, a braced ceiling or flat roof provides a strong enough diaphragm to span between end walls of the building. The structural action of these buildings when resisting lateral wind loads is illustrated in Fig. 2.12.

Wind pressures (or suctions) acting on the exterior walls are transferred by them to supports provided by the floor, roof, end walls, and cross walls. The portion of load transferred to the roof level is in turn transferred by the roof or ceiling system, acting as a stiff diaphragm in its own plane, to the cross walls and end walls. The components of load distributed to the end walls and cross walls are then transmitted through these walls, by shearing action to the foundation.

Wall A (Fig. 2.12) is likely to be a critical element in resisting wind pressure because it is only supported along the top and bottom. With lightweight roof construction and reduced weight of thin walls, out-of-plane vertical bending in tall walls may produce tensile stresses that require the wall to be reinforced. Axial compressive loads and horizontal shear are usually small in single-story buildings and can easily be resisted even though thin walls are used.

### 2.3.2 Multistory Loadbearing Buildings

Many conventional low-rise and high-rise masonry buildings were designed and built utilizing self-weight to counteract tensile stresses from lateral loads. The tallest was the Monadnock building erected in Chicago from 1889 to 1891. This 16-story building had an internal pin-jointed iron frame (unbraced) and exterior walls of solid loadbearing masonry. It was noted for the simplicity of its architectural elevational treatment. However, the walls were nearly 6 ft (1.8 m) thick at the foundation, occupied valuable floor space, imposed a heavy load on the foundations, and by 1940 had settled 20 in. (500 mm) into the soft clay. This was the last high-rise loadbearing masonry building constructed in Chicago for many decades. Steel-framed buildings, which had been introduced a few years earlier, and later concrete frames took over as the structural element of multistory buildings. The masonry cladding, now supported by the frame acted as a weather barrier and provided the aesthetic effect. Frames were designed to be structurally sufficient by themselves.

It was not generally recognized until the 1930s that infill brickwork within the steel frame acted as shear panels resisting the lateral distortion of the frame in its own plane. Twenty years later, it was realized that masonry cross walls in a multistory building would act as shear walls whether or not a steel frame was present. Thus, in the 1960s, many multistory loadbearing buildings were constructed in several countries using masonry shear walls instead of concrete or steel frames to achieve stability. In this construction, masonry walls support a concrete floor, which

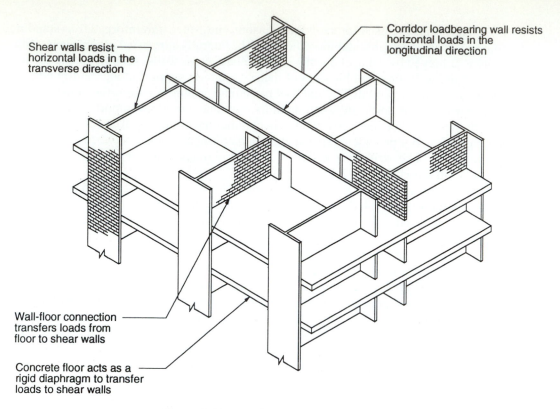

Shear walls resist horizontal loads in the transverse direction

Corridor loadbearing wall resists horizontal loads in the longitudinal direction

Wall-floor connection transfers loads from floor to shear walls

Concrete floor acts as a rigid diaphragm to transfer loads to shear walls

**Figure 2.13**   Multistory shear wall system.

in turn supports the next story of masonry walls placed directly above those below. The concrete floor slabs act as rigid diaphragms to distribute the lateral load to the shear walls, which in turn transmit them to the foundation. A schematic of a multistory shear wall system and the functions of key elements are shown in Fig. 2.13. The integration of walls and floors to act as a system that resists lateral loads relies upon the adequacy of the connections.

The overturning effect of wind on a traditional, loadbearing, multistory masonry building is resisted by the walls facing the wind. Contemporary loadbearing masonry construction resists overturning by walls placed parallel to the wind load direction. In the former case, the thickness of the wall is the critical factor, whereas in the latter case, the length of the wall is the critical factor, as illustrated by the simplified comparison in Figs. 2.14(a) and (b). For the latter case, large window areas are possible in external walls, thus overcoming one of the disadvantages of traditional construction.

Stability must be provided against wind or earthquake from all directions. This is usually attained by using a system of internal shear walls in both longitudinal and transverse directions as illustrated in Fig. 2.13. Because of their increased structural efficiency, modern loadbearing buildings have thinner walls than traditional buildings. For example, the 21-story twin Liberty Park East Towers in Pittsburgh (Fig. 2.15) have walls 15 in. (380 mm) thick.

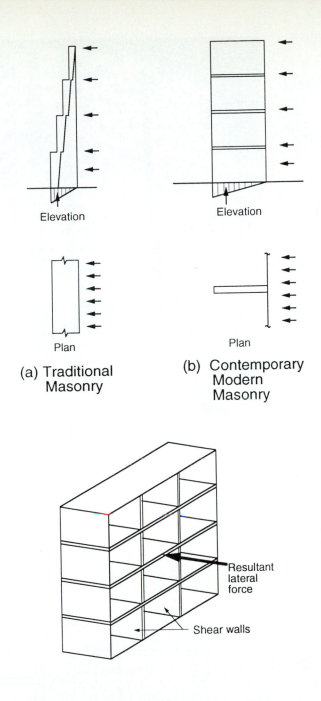

Elevation

Elevation

Plan

Plan

(a) Traditional
Masonry

(b) Contemporary
Modern
Masonry

Resultant
lateral
force

Shear walls

(c) Perspective
Showing
Window Areas

**Figure 2.14** Lateral load resistance of masonry walls.

**Figure 2.15** Liberty Park East Towers, Pittsburgh. (*Courtesy of Brick Institute of America.*)

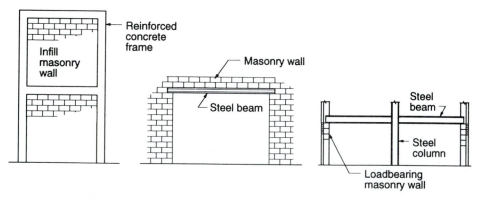

**Figure 2.16** Hybrid building systems with masonry walls.

### 2.3.3 Hybrid Buildings

Masonry can be used with other materials to form a hybrid composite building system. Loadbearing masonry shear walls have been utilized in steel framing systems as service cores and stairways as well as to carry lateral shear loads. Another example is infill frame buildings where masonry infill walls provide stiffness to control building drift. Great care is necessary in detailing and construction to allow for the long-term differential movements of the masonry and the framework that can lead to overstressing and failure. Masonry infill elements should be properly designed to carry loads, otherwise they can crack, thereby greatly reducing their stiffness, and result in increased deformations and stresses in the framing system. Examples of hybrid building systems with masonry walls are shown in Fig. 2.16.

## 2.4 TYPES OF MASONRY CONSTRUCTION

Each of the masonry building systems described in the previous section have, until recent times, been constructed of unreinforced masonry. As wall thicknesses have been reduced and distances between supports increased, reinforced and prestressed masonry have been introduced.

### 2.4.1 Unreinforced Masonry

Unreinforced masonry has been commonly used in low and medium-rise buildings in areas of low seismic activity. Plain masonry elements are the simplest to construct as they contain no reinforcement other than the possible inclusion of light joint reinforcement to control shrinkage cracking. Therefore, they rely on the strength of the masonry alone to resist loads. Because masonry is strong in compression but weak in tension, unreinforced masonry has great resistance to compressive loads, but only limited resistance to loads causing tensile stresses. Therefore, tensile stresses in unreinforced masonry must be designed to values below the tensile strength or the section is assumed to crack. The actions of unreinforced masonry under axial compressive loads and under lateral loads are considered in turn in the following paragraphs.

Consider, first, the building elevation shown in Fig. 2.17(a). Using the concepts introduced in Chap. 1, one could attempt to trace the thrust lines down through the walls of the building, assess their stability, and compute compressive stresses. However, there are difficulties in such an analysis.

Let us suppose that for a particular wall, the location of the thrust line and the magnitude of the thrusts are known at both the top and bottom of the wall, as shown in Fig. 2.17(b). If no lateral loads are applied and the weight of the wall is small compared with the total thrust, then the thrust line will be approximately straight. From this thrust line and the original geometry of the wall, one can determine where tension (and potential cracking) would occur and also determine stress distributions at various locations in the wall as shown on the left part of Fig. 2.17(b). The thrust line is at the centroid of the stress distribution at each location.

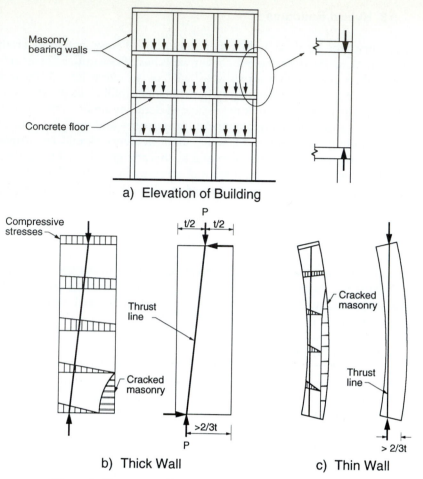

a) Elevation of Building

b) Thick Wall

c) Thin Wall

**Figure 2.17**   Thrust lines in unreinforced masonry walls.

Consider now an unreinforced wall such as shown in Fig. 2.18(a). This vertical strip of wall is laterally supported only at the roof or ceiling line and at the floor level. The weight of the roof, $P$, and the weight of the wall itself, $W$, are resisted by a vertical reaction, $R = P + W$, at the base of the wall, assumed to act centrally in this illustration.

From symmetry, the total lateral force of $ph$ acting on the wall is resisted equally by lateral supports at top and bottom. By considering the equilibrium of the top half of the wall, the location of the thrust line at this section can be calculated.

Taking moments about the section centroid (point c) in Fig. 2.18(b),

$$R \cdot e = \frac{ph}{2} \cdot \frac{h}{2} - \frac{ph}{2} \cdot \frac{h}{4} \tag{2.1}$$

and

$$e = \frac{ph^2}{8R} = \frac{ph^2}{8} \left( \frac{1}{P + W/2} \right) \tag{2.2}$$

By considering other sections, the complete thrust line can be located.

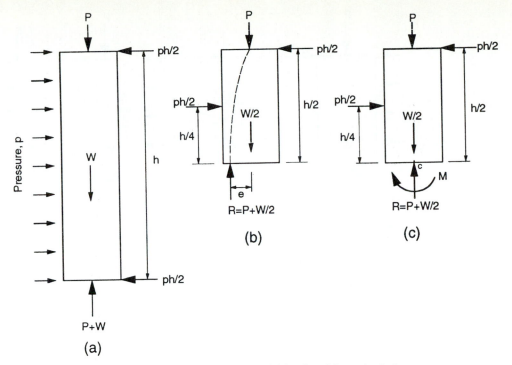

**Figure 2.18** Masonry wall under axial load and lateral wind pressure.

It is likely that with the thick walls of the past, the thrust line was contained within the wall, such as in Fig. 2.19(a), with compressive stress resulting through out. However, from Eq. 2.2, it can be seen that the eccentricity of the thrust line is dependent upon the relative values of lateral and axial forces.

This analysis is approximately correct for the massive walls used in the past because the deformations due to the thrust are small in comparison with the initial geometry. The deformations of modern slender walls, however, can significantly change the geometry of the wall in relation to the thrust line, as shown to an exaggerated scale in Fig. 2.17(c). Representative stress distributions and extents of cracking are also shown. The actual analysis is complicated because the actions of the applied thrusts change as the wall deforms and the resisting capacity of the masonry is reduced as cracking progresses (Fig. 2.17(c)). A further complication is that the location of the thrust line at the top and bottom of the wall can be influenced by the deformed shape of the wall.

The concept of the thrust line is still valid in today's bearing-wall design. However, deformations associated with slender walls have led to practical design techniques that, because of wall deflections, do not directly relate to the thrust line within the wall, but do rely on an assessment of the location of the thrust line at the top and bottom of the wall.

With today's lightweight roof construction and the reduced weight of thin walls, there is an increased likelihood that the value of the eccentricity calculated from Eq. 2.2 results in the thrust line falling outside the thickness of the wall as in Fig. 2.19(b) leading to the need for reinforcement.

Sec. 2.4    Types of Masonry Construction

53

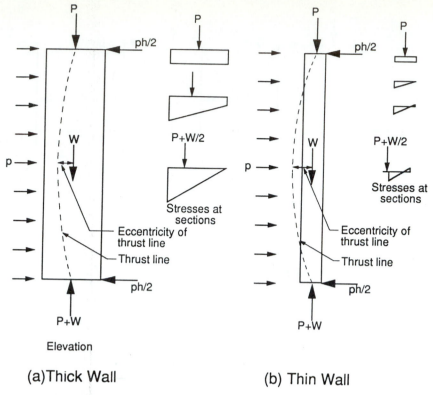

**Figure 2.19** Thrust lines in thick and thin walls.

(a)Thick Wall        (b) Thin Wall

### 2.4.2 Reinforced Masonry

Although all ancient masonry was essentially unreinforced masonry, metal ties were sometimes used to anchor one masonry unit to another. An early instance of the systematic use of wrought iron bars embedded in stone masonry as armature was in the Church of St. Genevieve in Paris built around 1770. One of the first known uses of iron reinforcement in brick masonry was in two shafts associated with the Blackwall Tunnel under the Thames River in 1825. However, reinforced masonry, in a modern sense, was not commonly used in buildings until about a century later, when it was mainly used in the seismically active areas of India, Japan, and the United States. Reinforced masonry is widely used today in many countries.

During the 1933 Long Beach, California, earthquake, many unreinforced masonry buildings collapsed, indicating the need for reinforcement for adequate performance. Since then, the use of unreinforced masonry has been prohibited in the Pacific Coast region of the United States and present codes in the United States require that all masonry structures in Seismic Zones 3 and 4 be reinforced. For Seismic Zone 2, some minimum of reinforcement is specified.

Reinforcement is mainly incorporated in masonry to resist tensile and shear stresses and to provide adequate ductility. The general basis of design is illustrated by the loaded wall shown in Fig. 2.20(a).

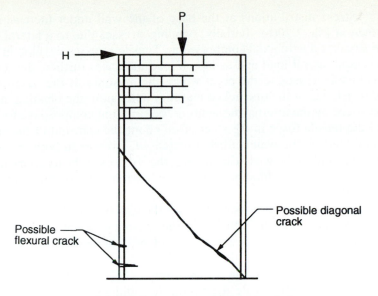

**(a) Wall under combined axial and lateral loads.**

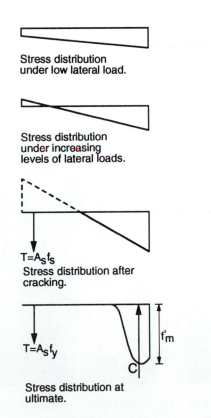

Stress distribution under low lateral load.

Stress distribution under increasing levels of lateral loads.

$T=A_s f_s$
Stress distribution after cracking.

$T=A_s f_y$

$f'_m$

C

Stress distribution at ultimate.

**(b) Stress distributions.**

**Figure 2.20**  Stress distributions in a reinforced masonry shear wall.

Stress distributions at the base of the wall under increasing lateral load are shown in Fig. 2.20(b). Initially, bending stresses due to a lateral load are less than the stresses due to axial compression. Tensile stresses develop in the masonry with increased lateral load and as lateral load increases further, the tensile stress in the masonry increases to the point where the masonry at the base cracks. Neglecting the resistance of the uncracked masonry in tension, the bending moment at the base is resisted by the internal moments of the resultant compressive force in the masonry and the tensile force in the steel taken about the centroid of the section. When the lateral load on the wall is further increased, stresses in both the masonry and steel increase, and as the wall nears failure, the stresses in the masonry take on a nonlinear distribution, up to a limiting strain. The tension reinforcement may or may not yield before failure.

There is another possible way for the wall to fail. The high shear force near the base produces diagonal tension in the masonry, which may cause it to crack, as shown in Fig. 2.20(a). A small amount of diagonal tension is usually allowed in masonry walls, but when this value is exceeded, horizontal steel must be placed at these locations to resist the tensile force.

Reinforcement made it possible to build the 28-story loadbearing Excalibur Hotel in Las Vegas, Nevada (Seismic Zone 2) in 1989 (see Fig. 2.21). The walls at the bottom floors were constructed with 12 in. (300 mm) concrete masonry units. The compressive strength of the masonry was 4000 psi (27.5 MPa). Vertical reinforcing bars, placed in block cores and grouted, provided the strength and ductility to adequately resist the gravity and earthquake loads.

### 2.4.3 Prestressed Masonry

Placing a weighty statue on top of a column to increase its resistance to lateral thrust from an arch roof can be regarded as an early form of prestressing. Using high-

**Figure 2.21**  The Excalibur Hotel, Las Vegas. (*Courtesy of James E. Amrhein.*)

strength steel rods to prestress masonry is, however, a recent innovation not yet widely used. In this form of construction, steel rods are inserted at appropriate locations in an unreinforced masonry element and then tightened down against end plates so as to compress the element. In almost all masonry applications, the steel is centrally located in the element so that the induced compressive stresses are uniform over its cross-section. The internal tensioned steel, rather than an external statute, prestresses the element before additional loads are applied. The advantage of prestressing is that any subsequent tensile stresses that tend to develop are suppressed by the precompression, as shown in Fig. 2.22.

Prestressed masonry elements are typically designed to be free of tension under service loads. This contrasts with reinforced elements, which must crack for the reinforcement to become effective. If minor cracks do occur under load, the prestressed steel closes them again when the load is removed. Other advantages are that prestressing rods can be inserted before or after construction of the masonry and they do not need to be grouted, provided that they and the anchorage are protected against corrosion. Many elements that are reinforced could be prestressed instead. Figure 2.23 shows examples of prestressed masonry elements.

The general principles used for post-tensioned prestressed concrete are applicable to prestressed masonry, but differences in the two forms of construction should not be overlooked. In particular, influence of construction methods, losses, end anchorages, and shear in prestressed masonry require special attention.

The initial prestress induced in the masonry is modified by long-term deformations. Creep of the masonry under the action of the prestressing force can usually be estimated as being in the order of 2–3 times its initial elastic deformation but differences between clay and concrete masonry can be very large. This shortening of the masonry and, consequently, of the prestressing tendon leads to a loss of

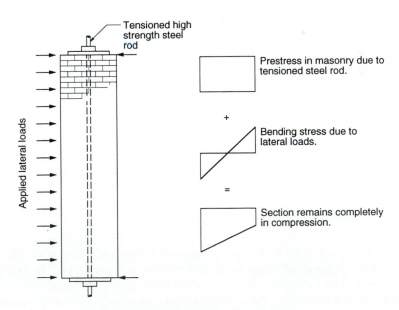

**Figure 2.22**  Principles of prestressing.

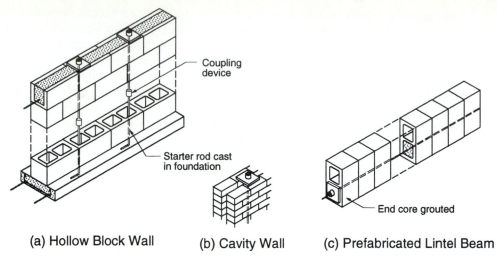

(a) Hollow Block Wall    (b) Cavity Wall    (c) Prefabricated Lintel Beam

**Figure 2.23**  Examples of prestressed masonry elements.

prestress. Shrinkage of concrete or calcium silicate masonry will lead to a further loss of prestressing force, and, conversely, any long-term expansion of clay masonry will lead to a corresponding gain in prestress. In assessing loss of prestress, it is recommended that changes in strain in the prestressing tendon be calculated for the specific case rather than using a global percentage loss, which can be grossly in error. The types and forms of calculation performed for post-tensioned prestressed concrete[2.1,2.2] can serve as a model for prestressed masonry.

When loaded to failure, prestressed masonry acts similarly to reinforced masonry. The member is assumed to be cracked in flexure and compressive stresses in the masonry become nonlinear, as was shown in Fig. 2.20(b).

Further consideration is not given to this topic in this book. The reader is referred to Refs. 2.3 to 2.7 for detailed analysis and design information.

## 2.5 STRUCTURAL DESIGN

Design of contemporary masonry buildings is based on rational analysis methods derived from research and expressed in building codes and masonry standards. The aim of codes has remained essentially the same for thousands of years. Stated in modern terms, it is to proportion and arrange walls and other elements to safely resist applied forces.

### 2.5.1 Analysis and Design Methods

The theory of structures as we know it today only began to develop about two centuries ago and yet many impressive buildings were erected before this time. Early masonry structures were built by skilled craftsmen who relied on experience accumulated by trial and error. Most of these early structures were massive and overdesigned by modern standards.

Conservativeness in early buildings of substance was encouraged by the lack of rational theory, the availability of cheap labor, plentiful materials, and, in some cases, the threat of death to the builder if the structure fell down and killed someone.[2.8]

On the other hand, Gothic craftsmen in later centuries, still without any knowledge of statics, exploited the potential of the pointed arch and its derivatives to an extent that, in many cases, cannot be justified by present-day standards of structural safety. However, many buildings did collapse as builders attempted lighter structures, pressed by limited resources as a result of constant warfare and encouraged by an aesthetic sense and the religious thought at the time that ascribed structural success or failure to Divine action. At that time, structural design rules were those of geometry and proportion based on experience and perhaps an intuitive feel for thrust lines. It is interesting to note that, for massive masonry structures subjected mainly to self-weight induced axial forces, the concept of design by proportion is basically valid.

Until quite recently, all masonry design was done empirically without structural theory. Except for relatively simple 1 to 3 story structures, empirical design is now largely replaced by rational design based on working stress, ultimate strength, or limit states methods.

The elastic theory of structures was introduced in the midnineteenth century. This gave rise to the working stress design method, first applied to wrought iron in railway bridges. In this method, the tested ultimate strength of the steel was divided by a safety factor (usually about 4) to arrive at a permissible stress under working or *service* loads. Stresses in the structure under working loads, computed by elastic theory, were not to exceed this permissible value. This highly successful method of design was widely adopted for other materials including masonry. One of the problems with this method was that the permissible stress was based on the ultimate strength of a small test specimen which did not necessarily reflect the ultimate strength of the structure. Safety factors of 10 to 20 may have to be applied to the material strength when designing structural elements such as columns. These high factors are obtained by applying an additional safety factor (say, 4) to the tested strength of a column. In this way, the working stress method of design has been adapted to take into account such things as the slenderness of a column or wall.

In the 1920s, attention was given to developing structural theories that could predict the load at which a member or a structure would collapse, and in the 1950s, the ultimate strength methods of design were introduced. In this concept, a safety factor is placed on the load that causes the member or structure to collapse rather than limiting stresses under working conditions. Although ultimate strength design methods and working stress design methods can be matched to give identical answers in some situations, in general, they do not. This is largely because the interactions between combinations of loads (axial load and bending) and combinations of material (masonry and steel) are not constant because of geometric and material nonlinearities. Also, in terms of actual collapse, many members and structures still have a significant reserve of strength beyond the limits of local section capacity. Such nonlinear behavior cannot be readily incorporated into a working stress approach. In these cases, ultimate strength theories and limit states design provide a much more reliable and consistent measure of safety.

The previous points are illustrated in the following examples of a Gothic arch carrying a vertical load at its apex and a modern masonry wall panel subjected to wind loading.

**Gothic Arch.** Consider the pointed arch, consisting of two segments of a circle, shown in Fig. 2.24(a). The weight of the arch is uniform along the length of the two arcs and hence the resulting thrust line is very close to an inverted catenary. The arch in this example was made thick enough for a catenary to be just contained within its middle third. This arch is unable to sustain a very heavy load at the crown because with increasing load, the thrust lines become approximately straight and introduce bending into the curved arcs.

To illustrate the behavior, the arch is examined when a small load is applied at the crown and increased until collapse. First, before any load is added, the structure can be considered as a continuous member and analyzed by elastic theory because it is expected that, under its own weight, the material is entirely in compression. Analysis should confirm that the thrust line is in the middle third of the thickness, as shown in Fig. 2.24(a). In this example, the elastic thrust line falls outside the middle third of the arch near the base (see Fig. 2.24(b)) and therefore a strict application of the working stress method (which does not permit tension in this case) would require the arch to be thickened. However, the arch shown will open slightly at joints *A* under its own weight as tension tends to develop. Suppose now that a small load *P* is applied and is gradually increased until the arch collapses.

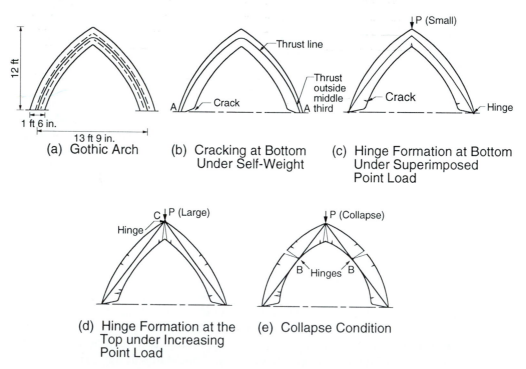

**Figure 2.24** Gothic arch behavior under increasing point load.

As $P$ increases, the thrust line gradually straightens out and reaches the edge of the arch thickness at $A$, as shown in Fig. 2.24(c). There is now virtually point contact at these locations as the joints open. Additional cracking can occur near the supports and under load $P$. One can expect that a further increase in $P$ will drive the thrust line outside the thickness of the arch at the supports (points $A$) and that it will collapse. However, these points of contact simply act as rotation points and the arch now behaves as a two-hinge arch. The thrust line further straightens as $P$ is increased until it reaches the edge of the arch at the crown, as in Fig. 2.24(d). This joint now completely opens with point $C$ acting as a hinge in a three-hinge arch. More load straightens the thrust lines further and causes bending in the arch until the thrust lines touch the inside edges of the arch at points $B$, creating other hinges as shown in Fig. 2.24(e). The hinges now form a mechanism in the arch that collapses, as shown in Fig. 2.25(a). Figure 2.25(b) shows approximately how the crown of the arch deflects under increasing load. Calculations show that a load of $P$ equal to about

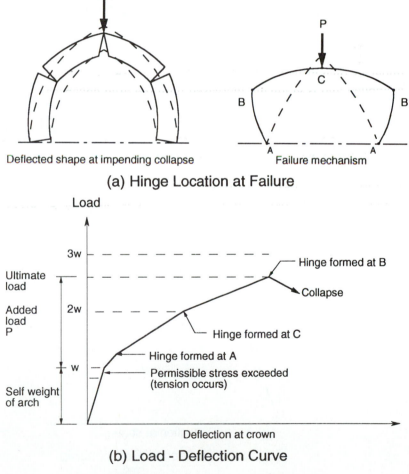

Deflected shape at impending collapse

Failure mechanism

**(a) Hinge Location at Failure**

**(b) Load - Deflection Curve**

**Figure 2.25** Gothic arch collapse.

1.5 times the weight of the arch can be added before it collapses. This collapse capacity should be judged in the context of working stress design, which indicates that the arch has an unsatisfactory factor of safety even under its own weight. Because builders of the Gothic arches had no concept of stresses, they were concerned only with the collapse condition—found by trial and error.

**Wall Panel.**    A modern example of structural behavior is a single wythe masonry wall supported laterally around all its edges and subjected to uniform wind pressure. As increasing wind pressure is applied, the wall deflects laterally, bending both vertically and horizontally between the edge supports. This bending, superimposed on the self-weight effects of the wall, produces tensile stresses on the leeward face of the wall. The wall eventually cracks along a bed joint near midheight, as

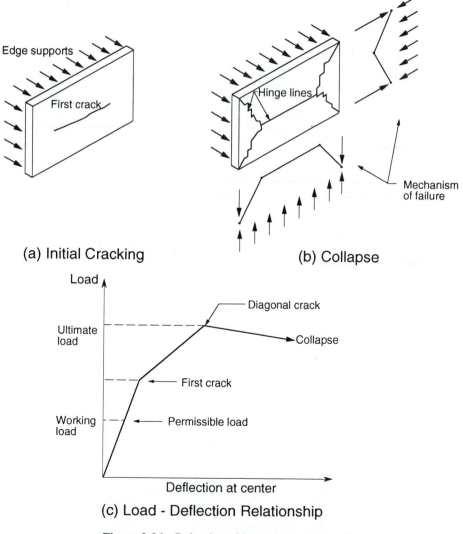

**Figure 2.26**   Behavior of laterally loaded wall.

shown in Fig. 2.26(a), because the flexural strength of masonry bending vertically (normal to bed joints) is less than when bending horizontally (parallel to bed joints). Additional wind load is now resisted by the uncracked portions of masonry spanning in the stronger horizontal direction assisted to some extent by the top and bottom supports. Eventually, the wall cracks again along the diagonals, as shown in Fig. 2.26(b), where the tensile stresses from this combined bending reaches a critical value. The crack pattern now forms a mechanism leading to collapse of the wall. Figure 2.26(c) shows approximately how the center of the wall deflects laterally as wind pressure increases.

In working stress design, a safety factor is applied to the material strengths, resulting in

$$\text{Working load} = \frac{\text{load at first crack}}{\text{working stress safety factor}}$$

whereas in capacity or collapse design,

$$\text{Working load} = \frac{\text{collapse load}}{\text{combined overload/understrength safety factor}}$$

Once again, these two approaches cannot be simply related. The ultimate strength approach based on section capacity is intermediate between the working stress and collapse approaches and gives a truer indication of safety than the working stress method.

**Limit States Design.** Ultimate strength methods of design were introduced in the 1960s, but it soon became evident that they were inadequate by themselves. Although a structure may have adequate resistance against collapse, it may perform inadequately under working loads. For example, deflections or cracking may be excessive. A complete design, therefore, should check both the safety of a structure and its performance under working loads.

The term "limit states design" was adopted in the 1960s to describe this broader design philosophy. Here a limit state is defined as a particular condition at which a structure becomes unfit for its intended purpose. Limit states include those for fire, fatigue, durability, water penetration, and progressive collapse, but the main ones for the structural design of buildings are as follows:

*Strength limit state:* Corresponds to collapse or loss of structural integrity.

*Stability limit state:* Corresponds to loss of static equilibrium of a structure (or part), considered as a rigid body, such as overturning.

*Serviceability limit state:* Corresponds to the minimal acceptable in-service conditions relating to such concerns as deflection, cracking, vibration, and local deformation.

The first two states are concerned with safety and are sometimes referred to as the ultimate limit states. Limit states design is a convenient and comprehensive way of ensuring that the various limit states do not occur during the designed life of the building.

## 2.5.2 Research, Codes, and Standards

Efforts have been made to develop rational masonry design as an effective means to achieve competitive and cost-efficient loadbearing masonry structures. Although structural theory was available in the midnineteenth century, rational design rules were not incorporated into codes until a century later and after considerable research into masonry.

Early research concentrated on the compressive strength of masonry and this was hampered by the inadequate capacities of testing machines. However, toward the end of the nineteenth century, such equipment became available in some universities and government agencies in Britain, Canada, Sweden, and the United States. Even though many aspects of masonry construction, including reinforced masonry, were investigated, the results of this work were slow to be adopted. Although some modifications to existing codes were made, the depression of the 1930s and World War II stalled developments.

By the time the post–war building boom started, steel and reinforced concrete were firmly established as the normal structural materials. Loadbearing masonry was not prohibited, but masonry building codes were very much the same as they had been at the turn of the century. The very thick walls that were required prevented masonry from being economically viable for most construction. Masonry began to be thought of as being mainly useful for cladding, minor buildings, and architectural decoration.

One of the first breaks with tradition occurred when British Standard CP 111 was revised in 1948. Based on an evaluation of research, a number of new features was introduced to masonry design. Arbitrary minimum thicknesses were replaced by the concept of *slenderness*, defined as the ratio of effective wall height to effective thickness. Effective thickness was defined as the actual thickness of the wall except for cavity walls, where $\frac{2}{3}$ of the sum of the thickness of two wythes was to be taken. Slenderness was limited to 18. Allowable compressive stresses had to be reduced for ratios greater than 2, but a 25% increase was permitted for peak values under eccentric loading. Designers were not allowed to rely on masonry tensile strength but rules were given for reinforced masonry. A great deal of conservatism was intentionally provided as it was thought that more testing was needed regarding eccentric and lateral loads and the effects of deflections on the strengths of members.

In North America, the rule-of-thumb type of design continued to apply, reaching a peak in 1953 with the publication of American Standards Association Building Code Requirements for Masonry (ASA A41.1-1953). This was an excellent document, and, in many ways, one of the best and most comprehensive treatises on masonry design ever written, certainly up to that time. It forms the general basis for conventional design or empirical design, as it is often known today. One paragraph in the Commentary to the code is interesting because it shows the direction of committee thinking even at that time:

> Although the committee is aware of the desirability of more rational methods of design than the arbitrary limitations on the ratios of wall thickness to distance between lateral supports, it was considered that more basic information is needed before they can be formulated. Quantitative data are needed on the chief factors which affect lateral stability

and the resistance of walls to lateral loads. To fill this need, many more tests will have to be made, some of which will provide additional information on the effects of conditions at the boundaries of walls.

Actually, by the mid-1950s, it was obvious to many that masonry could help satisfy the world's demands for housing if its technology could be upgraded. A major study in this direction took place in Switzerland under the direction of Haller.[2.9] Again, it was the availability of a 2,000,000 lb. (8.90 MN) testing machine at Dubendorf that made it possible. After some 10 years' of research in collaboration with consultants and contractors, he supervised (about 1953) the construction of several apartment buildings up to 18 stories in height having walls only 6 in. (152 mm) thick in some cases.

After extensive testing, he reached a number of conclusions regarding the strength of masonry. Bricks should be uncracked in manufacture, not too highly perforated, of uniform height, not too warped, and have low absorption. Mortar should consist of high-strength cement-lime mortars because pure cement mortars were too sensitive to brick absorption effects. Bricks should be laid square, by trained operators, with all joints of minimum practical thickness ($\frac{1}{2}$ in. (13 mm)) and filled completely with mortar. The designer should give full details of piers, corners, intersections and openings, should not permit the cutting of chases, and should provide careful inspection for major buildings.

Because of all these factors as well as the necessary assumptions in analysis, Haller advocated load factors of 4 and 5, the lower factor to be used in case of full inspection and testing of materials. He also advocated that design be based on prior tests of short walls as the best integrator of all the uncertainties in materials and construction. After an initial period, when each individual building was undertaken as a research project, the knowledge and experience gained was eventually codified in Swiss standards.[2.10]

During the 1960s, several countries produced masonry codes. The Standards Association of Australia published separate codes for concrete blockwork (AS CA32-1963) and brickwork (AS CA47-1969). The blockwork code was based on information from North America and the brickwork code was based on the British code and the work of Haller. Both were written in terms of working stresses.

The bare essentials of a unified analytical approach to masonry design, modelled on the British code, were introduced into the 1965 National Building Code of Canada. These elementary rules for the design of plain and reinforced masonry by analysis of forces, moments, slenderness, and stresses have become known as "engineered masonry." This is as opposed to traditional arbitrary rules regarding minimum thickness and support spacing known as empirical design.

In the United States, as a result of research data produced in the 1960s, developments in unreinforced masonry design was evident in the Brick Institute of America (BIA) and the National Concrete Masonry Association (NCMA) publications on design of masonry structures. They were adopted by International Conference of Building Officials (ICBO) and Building Official and Code Administration (BOCA) building codes. After the Long Beach earthquake of 1933, reinforced masonry began its continuous growth in the Uniform Building Code published by ICBO. In 1978, the ACI 531 Committee published "Building Code Requirements for Concrete Masonry," which includes both unreinforced and reinforced masonry design based on the working stress concept. In 1988, a new standard and specification was published jointly by the ACI 530 Committee and the ASCE-5 Committee. This standard covers both concrete and clay masonry and was intended to replace the ANSI A41.1, ANSI A41.2, BIA-69, NCMA TR-75B, and ACI 531 masonry codes. BOCA adopted the new standard in June 1989. A new version of this standard was published by the Masonry Standards Joint Committee[2.11]. This standard, which represented the state-of-the-art in design practice in North America in 1985, has unique features, which include the elimination of arbitrary slenderness limits, minimum steel ratios in seismic areas, coefficients for thermal expansion, moisture expansion, shrinkage and creep, and design guidelines for multi-wythe composite and noncomposite action.

All codes in the United States up to this point were based on working stress concepts. However, there has been a continuing trend toward developing a limit states masonry code. The 1991 edition of the Uniform Building Code (UBC-91)[2.12] in the United States contains a strength design method for slender reinforced masonry walls, which takes into account the effect of additional (secondary) moment from axial load due to wall deflection. It also contains a strength design method for reinforced masonry shear walls.

In the 1960s, the USSR adopted a limit states approach to design[2.13] and in 1984 in Germany, design provisions were introduced based on tests.[2.14] In 1978, the British Standards Institution published a new masonry code (BS 5628: Part I–1978),[2.7] for unreinforced masonry, expressed in limit states format. Part 2 covering reinforced and prestressed masonry was published in 1985. The Standards Association of Australia adopted a limit states design code in 1988 (AS 3700-1988)[2.15] and the current New Zealand standard for masonry[2.16] uses the strength design approach with special emphasis on seismic design requirements.

The continuing research effort will result in further data that will facilitate the development of limit states design methodology.


### 2.5.3 Sources of Information

To produce efficient but safe, high-quality buildings, masonry designers and contractors need to be well-informed. It is essential to be familiar with the requirements of local building codes, but a knowledge of international standards and model codes is also useful. This is particularly so because of the current trend of publishing an explanatory commentary with standards and codes.

As previously mentioned, there has been a great deal of research into masonry over the past 20 years and this is still very active. The results of this work, including many practical aspects, are in conference proceedings, seminar notes, and various journals. In addition, this work is eventually published in technical notes and publications by various masonry associations and organizations. These associations exist in all countries and the local association can be of great help to the designer or builder. They can usually provide an answer to specific problems or give reference to appropriate authorities. Particular reference to all of these sources of information are given in Appendix A.

## 2.6 CLOSURE

Whereas Chap. 1 gave a brief historical account of masonry from a structural point of view, Chap. 2 introduced the reader to contemporary masonry elements and to design and construction methodologies. It should be apparent that present-day masonry structures and methods of design are usually quite different from what has gone before. However, the so-called modern multistory loadbearing masonry building is simply an adaption of the apartment buildings of ancient Rome and it is surprising that this form took so long to develop. There are probably other ancient forms that could be adapted to present-day needs and conditions. With the advantages of engineered analysis and design methods, materials science, and mechanized construction methods, there is the opportunity to exploit the full potential of masonry materials in buildings. Further detailed information to assist in the planning, design and construction of masonry buildings is given in subsequent chapters.

## 2.7 REFERENCES

2.1   A. Nilson, *Design of Prestressed Concrete,* John Wiley, New York, 1978.

2.2   J. Libby, *Modern Prestressed Concrete,* Van Nostrand Reinhold, New York, 1977.

2.3   W. G. Curtin, G. Shaw, J. K. Beck, and W. A. Bray, *Structural Masonry Designers Manual,* Granada, London, 1982.

2.4   "Post-Tensioned Masonry Structures," VSL Report, Series 2, VSL International, Berne, Switzerland, 1990.

2.5   M. E. Phipps, and T. I. Montague, *The Design of Prestressed Concrete Blockwork Diaphragm Walls,* Aggregate Concrete Block Association, Leicester, England.

2.6   A. E. Shultz, and M. J. Scolforo, "An Overview of Prestressed Masonry," The Masonry Society Journal, Vol. 10, No. 1, Boulder, CO, Aug. 1991, pp. 6–21.

2.7   British Standards Institution, "Code of Practice for Use of Masonry. BS5628, 1978, Part I: Structural Use of Unreinforced Masonry," (confirmed April 1985). BS5628:1985, Part 2: Structural Use of Reinforced and Prestressed Masonry, London.

2.8   J. Amrhein, *Reinforced Masonry Engineering Handbook,* Fifth Edition, Masonry Institute of America, Los Angeles, 1992.

2.9   P. Haller, "Load Capacity of Brick Masonry," in *Designing, Engineering and Construction with Masonry Products,* F. Johnson, Ed., Gulf Publishing, Houston, TX, 1969, pp. 129–149.

2.10  *"Design of Masonry Walls,"* Swiss Standard SIA Recommendations V 177/2, Swiss Society of Engineers and Architects, Zurick 1989.

2.11  Masonry Standards Joint Committee, *Building Code Requirements for Masonry Structures, ACI 530/ASCE 5/TMS 402,* American Concrete Institute, Detroit; American Society of Civil Engineers; New York, The Masonry Society, Boulder, 1992.

2.12  *Uniform Building Code*, Chapter 24: Masonry, International Conference of Building Officials, Whitter, CA, 1991.

2.13  Canada Institute for Scientific and Technical Information, "Building Standards and Regulations (U.S.S.R.), Part II, Section V, and Chapter 2, Plain and Reinforced Masonry Structures Design Standards," D. E. Allen, trans., Ottawa, 1976.

2.14 *Masonry; masonry designed on the basis of suitability tests; Design and Construction,* DN 1053: Part 2, Institute for Building Technology, Berlin, 1984.

2.15 Standards Association of Australia, "Masonry in Buildings," SAA Masonry Code (AS 3700), North Sydney, N.S.W., 1988.

2.16 Standards Association of New Zealand, "Code Practice for the Design of Masonry Structures," NZS 4230: Part 1: 1991, Wellington, N.Z.

## 2.8 PROBLEMS

**2.1** Discuss and illustrate examples where masonry pilasters can be effectively used in the design of masonry walls. From a review of pilaster configurations (on site if possible), discuss how the introduction of pilasters into walls would affect the construction process.

**2.2** Devise a simple experiment or perform simple bending calculations to demonstrate the relative strengths and stiffnesses of a cavity wall and solid wall under out-of-plane loading. Each is to be composed of two wythes of equal thickness.

**2.3** Today's masonry buildings resist lateral loads using structural systems that differ from ancient masonry. Discuss the differences and illustrate or show examples to help clarify your answer.

**2.4** Identify the masonry codes and standards applicable in your locality and state whether they are based on the working stress design, strength design, or the limit states design concept.

**2.5** Discuss and compare how safety requirements are satisfied using working stress and ultimate strength design methods.

**2.6** This chapter contains descriptions of many types of masonry walls. Identify and photograph or sketch four different types in your locality (preferably under construction), giving a short description of each and paying particular attention to construction details.

**2.7** Compile a list of the nearest manufacturers or materials associations and contractors or masonry associations that can provide information on masonry.

# CHAPTER 3

# Building Design

Brick veneer with polychromatic pattern.

## 3.1 INTRODUCTION

Masonry in buildings normally acts as structural elements to support or resist loads and as architectural elements to divide or enclose space. In both cases, it can provide the finished surface or be covered with a render or veneer. This dual role usually requires close collaboration between the architect, the engineer, and the builder in the planning, design, and construction phases of the building. Therefore, those involved in the design of masonry buildings must have a good general knowledge of the structural, architectural, and constructional requirements. In the following chapters, detailed information on these matters is provided, but an overview is given here so that they can be seen in the context of overall planning and design.

The design process is sequential and iterative where decisions are made to choose the best alternative. The first phase involves definition of the client's needs, which include functional, aesthetic, and budgeting requirements. A conceptual design is then developed and possible building layouts are developed to satisfy the defined needs. A final choice is made based on prioritized needs within the available budget. During this latter phase, the structural system is selected and a preliminary design is prepared to estimate cost. The final design of the structure includes integration of the mechanical and electrical systems. Construction drawings and specifications are prepared during this stage.

In this chapter, structural, environmental, and aesthetic requirements and their integration are first considered, followed by related planning and economic implications. Consideration is limited to present-day design. A history of masonry construction and design from ancient Egyptian to the nineteenth century is briefly presented in Chap. 1.

## 3.2 STRUCTURAL REQUIREMENTS

### 3.2.1 Design Criteria

To be structurally sound, a building and the individual elements must satisfy criteria for strength, stability, and serviceability, and in seismic areas, they must also satisfy criteria for ductility and energy-absorption capabilities. Each of these criteria must be checked individually for each element and for the building as a whole. These design criteria are illustrated in Fig. 3.1 using an idealized load-deflection curve for a vertically spanning reinforced masonry wall under out-of-plane loads. The serviceability cracking or deflection limits, the ultimate strength, and the energy absorption potential of this wall depend only on the properties of the wall. The energy-absorbing properties of the wall are not necessarily limited to its ultimate strength as redundancy within the building or load dissipation may allow additional advantage of energy absorption up to the collapse state of the element. Ductility, expressed by the deformation after yielding, is a measure of this desirable quality. The same type of deformation relationship applies to the overall building except that the useful energy absorbing properties cannot extend beyond the ultimate strength of the building.

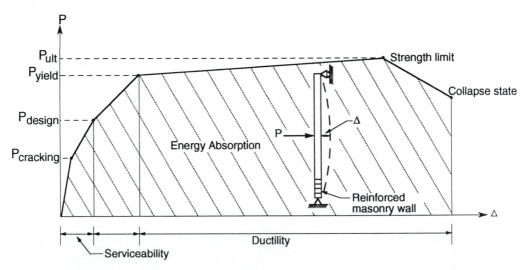

**Figure 3.1** Structural design criteria illustrated by an idealized load-deflection curve.

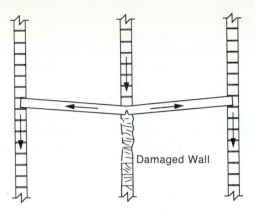

**Figure 3.2** Alternate load path.

The overall building must be strong enough to transfer all loads through the structure to the ground without collapsing or losing structural integrity by rupture of the material at critical sections, by transformation of the whole or parts into mechanisms, or by instability. This strength criterion is valid for all loads that will normally be applied to the building during its lifetime, but the structure cannot be economically designed to resist gross overloads or severe accident without sustaining some damage. There is a danger that a localized incident, such as an explosion or a collision, will initiate a chain-reaction effect such that the failure of one element progressively induces the failure of other elements until the whole building, or a substantial part of it, totally collapses. Masonry buildings, therefore, should be planned so that, if for some reason a loadbearing element such as a wall fails, the damage is localized around that area and the building does not progressively collapse. A building having this quality is said to be robust and normally has alternate paths or mechanisms to transfer load to the foundation. Figure 3.2 illustrates this idea where a floor system is shown to act as a catenary to transfer part of the load away from the damaged wall.

A building must also be stable so that the building as a whole or any of its parts do not lose static equilibrium as a rigid body. This stability criterion requires, for example, that neither the building as a whole nor any of its component shear walls will overturn when subjected to lateral wind forces. (See Fig. 3.3.)

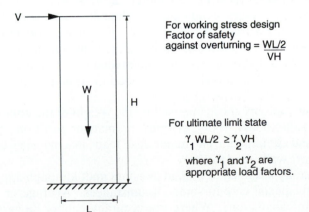

For working stress design
Factor of safety
against overturning = $\dfrac{WL/2}{VH}$

For ultimate limit state

$$\gamma_1 WL/2 \geq \gamma_2 VH$$

where $\gamma_1$ and $\gamma_2$ are appropriate load factors.

**Figure 3.3** Stability criterion (overturning).

In high-risk seismic zones, structures should be ductile and capable of dissipating energy through inelastic actions. The ductility capacity of masonry shear walls affects the magnitude of inertia-induced forces during earthquakes. Economic design in seismic areas necessitates adequate ductility in the masonry shear walls and the connecting elements.

Serviceability criteria require that certain minimal acceptable in-service conditions be achieved. Serviceability is not usually thought of as being related to safety, but in fact it is often associated with strength, such as the influence of crack width on corrosion of reinforcement. Serviceability is also concerned with deformations that affect the efficient use or appearance of structural or nonstructural elements, excessive vibrations that produce discomfort or affect the satisfactory operation of equipment, and local damage or cracking that affects the appearance or durability of the structure.

There are a number of uncertainties in satisfying the previous criteria for strength, stability, ductility, and serviceability. First, there is the uncertainty associated with determining the maximum loads and load combinations that the building is likely to experience in its lifetime. Also, there is uncertainty when assessing the variable strengths of materials to be used in construction. The range of possible masonry strengths is quite large, but it is usually a simple matter to ensure that the as-built strength is greater than required. An exception to this is the tensile strength of unreinforced masonry, which should be carefully evaluated. Then there are the uncertainties associated with the theories that relate material properties and loads to strength and ductility. Finally, there is the difficulty of determining the appropriate degree of safety or serviceability that should normally be provided or the extent of damage acceptable in extreme circumstances.

Building codes in all countries give minimum requirements regarding most of these matters. Even where these requirements are given quantitatively, however, the designer usually has to exercise judgment in application. Fortunately, the inherent strength of masonry and the large wall areas required for other purposes usually ensure that strengths are greater than required.

### 3.2.2 Loads

The various loads and effects that must be considered in the structural design of a building include dead loads, live loads, wind loads, earthquake loads, snow and rain loads, accidental loads, earth pressure, liquid pressure, movements resulting from foundation settlements, temperature changes, shrinkage, expansion, elastic shortening, and creep. These loads and actions and their combined effects are considered in what follows. References to design codes are included in Appendix A.

**Dead Loads.**    Dead loads are those imposed by the weight of the components of a building, including walls, floors, roofs, ceilings, permanent partitions, permanently fixed machinery, and equipment. The major dead loads in a masonry building are the weights of the roof, walls, and floors.

Dead loads are generally assumed to remain constant and, although apparently simply calculated, require special care in some circumstances. Average or typical values are usually used in calculations. Where the accuracy cannot be assured, a conservative estimation of dead load should be used. That is, a high estimate should be made where the dead load increases an action or reduces stability. Where the dead load reduces the effects of other loads (i.e., overturning), care should be taken not to overestimate its value and only that portion of dead load that cannot be removed from the structure should be taken into account.

Manufacturing and some other buildings can have future increases in dead loads due to new installations or occupancy changes. For such buildings, it may be wise to include some allowance for additional load in the original design, otherwise the future utility of the building may be too limited. Typical values of self-weight for different masonry walls are given in Appendix B and Ref. 3.1 contains design dead loads for different construction materials.

**Live Loads.**    Live loads arise from the use of the building, including the weight of people and movable objects and materials, impact, and inertia loads. They usually consist of two components: a sustained component that remains relatively constant for a particular occupancy, and a variable component that arises from the clustering of people or objects.

Design live loads normally correspond to peak values likely to occur over a 50-year period. By using this type of definition, characteristic loads have been developed in building codes for certain well-defined occupancies or uses. Where impact is a design consideration, such as from moving vehicles in parking buildings or cranes and lifting equipment in industrial buildings, this is usually accounted for by increasing the static loads by some defined ratio. The major live loads apply vertical gravity forces to the structure which are transferred with the dead loads by the walls to the foundations. Live loads, however, are movable and the following effects should be taken into account.

Live loads should be arranged to maximize the design effect under consideration. For example, fully loading all floor spans may be less critical in the design of a wall than loading alternate spans because of the larger eccentricity of load produced in the latter case. Other arrangements of live load may be important. Because the full live load is unlikely to occur over a large floor area, a reduction factor, based on floor area, is often applied by codes. Also, since it is unlikely that all floors of a multistory building will have the full live load occur simultaneously, a reduction factor is often applied to the cumulative effect when designing the loadbearing walls in the lower stories.

For example, standards[3.1-3.2] allow reductions in live loads for structural members having an influence area of 400 ft$^2$ (37 m$^2$) or more. The reduced live load $L$ can be calculated from the unreduced value $L_0$ by the equation

$$L = L_0 \left( 0.25 + \frac{15}{\sqrt{A}} \right) \tag{3.1}$$

where $A$ is the influence area in square feet and is equal to four times the tributary area for a column and two times the tributary area for a beam or a wall. In this case, the value of $L$ should not be less than $0.5L_0$ for members supporting one floor nor less than $0.4L_0$ for multistory construction. Other standards may use different formulas. [In Eq. 3.1, the number 15 changes to 4.57 when $A$ is in m$^2$.]

Some live loads, such as vehicles and traveling cranes, produce horizontal inertia forces, and these must be taken into account when assessing the lateral stability of the building.

**Wind Loads.** Wind loads on a building depend upon the design wind speed, which is determined statistically. For example, in some cases it is derived from peak (3 second gust) wind speeds that are likely to occur over a 50-year period with a 5% chance of being exceeded. The design wind speed depends upon many factors including geographic location, the roughness of the terrain, the height of the building above the ground, shielding by other buildings of equal or greater height, wind direction, and the presence of escarpments, ridges, hills, or valleys. Designers should anticipate changes to the terrain and surrounding structures, including the permanence of obstructions such as a wooded area in cyclonic conditions. Most building codes contain procedures for establishing design wind velocities for most common conditions, exceptions being during tornadoes or downbursts that occur in thunderstorms. As an example of a wind load provision, ASCE 7[3.1] specifies that the velocity pressure $q_z$ at height $z$ be calculated from the formula

$$q_z = 0.00256K_z (IV)^2 \qquad (3.2)$$

where $V$ = basic wind speed in mph, depending on the geographic location
$K_z$ = velocity exposure coefficient usually given in building codes in terms of height and exposure conditions
$I$ = importance factor

Design pressures on a building are determined from the design wind speed and pressure coefficients with some allowance being made for the importance of the building or building element. The pressure coefficients applicable to a particular building depend upon the shape of the building, the slope and shape of the roof, and the disposition of openings in the building. Designers need to exercise discretion as to which windows and doors can be open during high winds and whether or not windows and doors can be broken by flying debris or simply fail under the associated high pressures. The inside of the building can be subjected to either internal pressure or suction and various external surfaces can also be subjected to either pressure or suction. Local external suctions along small areas near windward corners and roof edges can be $1\frac{1}{2}$ to 2 times the general values. Figure 3.4 shows forces due to wind pressure and suction on buildings.

All portions of the structure must be designed to resist these pressures and suctions, and the structure as a whole must have stability against the summation of all these effects (except for the increased local external suctions). A static analysis for wind load, as opposed to a dynamic analysis, is appropriate for most masonry buildings.

A common critical factor for buildings having a lightweight roof is the uplift force that tends to lift the roof from its supports. It is necessary in this case to provide proper anchorage to the lower parts of the wall or preferably to the floor below. This topic is covered in more detail in Chap. 16.

The wind acting on external masonry walls must be resisted by bending action of the walls. It is preferable that these panels span vertically between floors as well

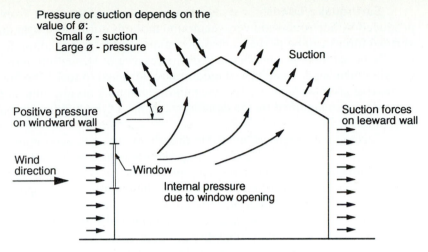

**Figure 3.4** Wind forces on building.

as horizontally between cross walls or columns so that they have enhanced bending in two way flexure (see Fig. 3.5). Unless window and door frames are designed to span between floors, the wind pressure on them is distributed to the adjacent masonry wall panels. Eventually, all wind loads are transferred to roof and floor levels and to cross walls. Roofs (braced) and floors act as diaphragms in their own plane and distribute loads to the cross walls.

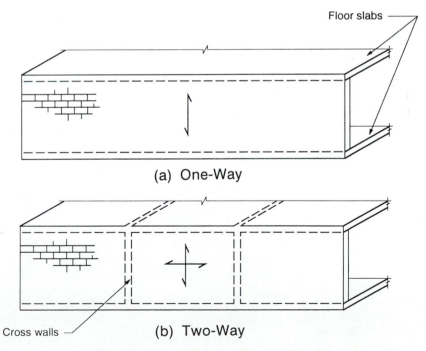

**Figure 3.5** Lateral load resistance of walls.

**Earthquake Loads.**    Seismic forces on a building result from inertia loads produced by horizontal and vertical ground motions during an earthquake. Although vertical forces can be of the same order of magnitude as horizontal forces, they can usually be safely neglected because of the reserve of strength in most vertical load-carrying members. Buildings, therefore, are designed to resist the horizontal inertia forces that arise when the mass of the building components are subjected to horizontal ground motion modified by the dynamic response of the building and its foundations (Fig. 3.6(a)).

Seismic risk maps are available that divide an area geographically into zones ranging from Zone 0, where no earthquake damage is expected to Zone 4, where major damage has occurred in close proximity to major fault systems. A more comprehensive and quantitative geographic description for design purposes is one giving contours of peak horizontal ground accelerations and contours of peak horizontal ground velocities. These values are obtained statistically from records, for example, on the basis of a 10% probability of being exceeded in a 50-year period.

The type of soil or ground conditions at the building site can have a major influence on the shaking of the structure. As motions propagate from the bedrock to the surface, less rigid materials such as loose sands and soft clays can greatly amplify this motion or tend toward a soil–structure resonance.

Building parameters that affect seismic forces include the fundamental period of vibration, which is mainly dependent on the type of structural system: frame or shear wall. The length, width, and height of the building, the materials and structural system used, and the energy-absorption capacity as affected by damping and ductility also influence the response.

The magnitude of seismic forces are dependent upon the mass of the building as well as live loads and other loads present during the earthquake. Normally, only a percentage of the live load is included and this only for specific types of buildings such as those used for storage.

The importance of the building is also taken into account with forces being increased for post-disaster buildings such as hospitals and for special buildings such as schools.

A dynamic analysis is necessary to more accurately predict the response of a building to earthquake loading, but, particularly for uniform buildings, design is

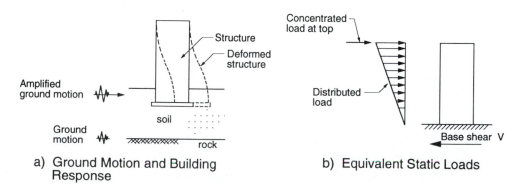

a) **Ground Motion and Building Response**

b) **Equivalent Static Loads**

**Figure 3.6**   Earthquake effect on buildings.

usually carried out using static equivalent loads to check for lateral stability along each major axis. The general procedure in this case is to calculate the total horizontal seismic force at the base of the building (base shear) as determined by the appropriate building code. This force is then distributed in a prescribed manner over the height of the building, approximately corresponding to a triangular distribution with maximum intensity at the top and zero at the ground, as shown in Fig. 3.6(b) with the possibility of an additional concentrated load at the top of the building.

As an example for seismic loading, ASCE 7-88[3.1] specifies the total lateral seismic force using the base shear formula:

$$V = ZIKCSW \qquad (3.3)$$

where Z = seismic zone coefficient (depending on seismic Zones 0, 1, 2, 3, or 4)
$\quad$ I = occupancy importance factor
$\quad$ K = horizontal force factor (depending on the structural system)
$\quad$ C = response coefficient (depending on the fundamental period of the building)
$\quad$ S = soil factor
$\quad$ W = total dead load

The base shears in the two orthogonal directions are calculated independently.

Lateral seismic forces on structural and nonstructural elements can be calculated using the formula

$$F_p = ZIC_pW_p \qquad (3.4)$$

where $F_p$ = lateral seismic force
$\quad C_p$ = horizontal force factor for elements
$\quad W_p$ = weight of the element

These lateral forces must be resisted in the same manner as the lateral wind forces described earlier. Wind and seismic forces are not considered to act simultaneously and, although seismic forces will control in many seismic areas, wind forces are likely to be critical in the less severe seismic areas. It is not uncommon for the design at different elevations of a building to be controlled by different lateral loads. Shear and moment envelopes from wind and seismic analysis are illustrated in Fig. 3.7 to show this situation.

**Snow and Rain Loads.** The roof of a building should be designed to withstand the greatest weight of snow or rain that is likely to accumulate. To determine the design snow load, a statistical approach is used to first estimate the depth of snow likely to accumulate on the ground in a particular location for a given time period. This depth is then converted to a ground snow load by taking into account the density of snow and making an allowance for absorbed rainwater. The snow load on the roof is found by modifying the ground snow load by various factors that take into account that wind will blow snow from the roof in exposed locations, that snow will slide off roofs depending upon slope, friction, and obstructions, and that the snow will accumulate in protected areas of the roof depending upon roof configurations, obstructions, and wind direction.

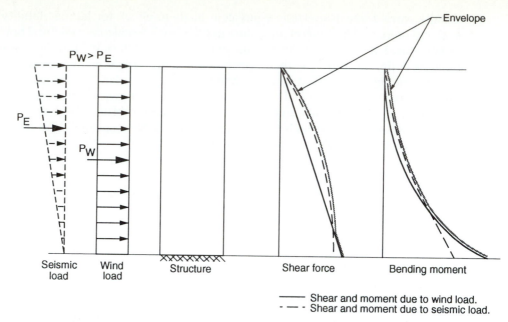

**Figure 3.7**  Shear and moment envelopes for wind and seismic loads.

The amount of rain that can accumulate on a roof depends on the intensity and duration of rainfall, the slope of the roof, and the effectiveness of roof drains. Large loads can occur when flat roofs permit ponding or when box gutters with inadequate drainage or overflows are used. A dangerous situation can develop where increased deflection of the roof under the weight of rain water leads to greater deflection and ponding of water. Severe cases may results in failure of the roof system. When only light members are necessary for strength requirements, deflection of primary and secondary roof members should be checked. In addition, unless the flat roof is actually cambered or sloped to facilitate drainage, drains should be located at natural low points at midspan rather than near supports.

**Accidental Loads.**   Accidental or abnormal loads are those considered to have such a low probability of occurrence as not to warrant inclusion in design. Such loads include violent changes in air pressure from high explosives or service system explosion; accidental impact from vehicles, aircraft, or a crane; faulty practice such as gross construction errors, unauthorized structural alteration, or lack of maintenance; fire; flood; tornado; severe subsidence or erosion of foundations; and hail and snow in areas not normally exposed to them.

No structure can be expected to resist all abnormal loads without damage, but the damage incurred should not be disproportionate to the original cause. Damage should not spread to locations in the structure remote from the original cause, that is, progressive collapse should not occur.

The appropriate degree of confinement of consequential damage depends on the structure and its occupancy. In many cases, it will be sufficient if the damage is confined to spans immediately adjacent to and floors immediately above and below the incident.  Although consequences of progressive collapse may be more severe in high-rise buildings, the principle of damage control should be applied to all buildings.  This can often be achieved by good design and detailing at no extra construction cost.

Three ways of dealing with accidental loading are: reduce the probability of accidental loading, generally employ good design practices, and carry out quantitative design.  The probability of accidental loading can be reduced, for example, by preventing vehicle impact by the use of landscaping pylons, by providing adequate venting to areas where explosive materials are stored or used, and by adequate building maintenance.

Good design includes providing lateral support to loadbearing elements in the form of intersecting walls, buttresses, and returns and providing alternative load paths utilizing continuous construction and adequate tying of other building components together.  For masonry in particular, significant advantages can be gained by this approach.  Many examples exist, such as shown in Fig. 3.8, where masonry has been able to bridge over large damaged areas by arching action.  In important cases, specific checks should be made to ensure that the structure can remain stable after damage.  In such cases, ultimate strength methods of design with reduced load factors are appropriate.

Accidental loading, by definition, cannot be adequately dealt with by codes or regulations.  Designers, therefore, have a particular responsibility in all of their work to consider the possibility of abnormal load conditions and to take reasonable steps to minimize the consequences.

**Figure 3.8**  Bridging over damaged area in a brick masonry wall. (*Courtesy of Clayford T. Grimm.*)

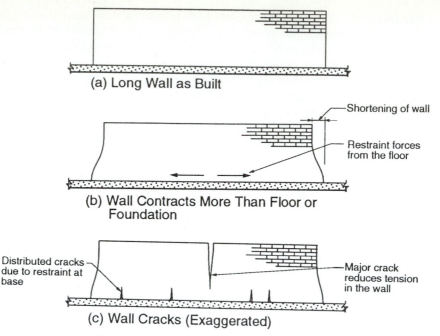

**(a) Long Wall as Built**

Shortening of wall

Restraint forces from the floor

**(b) Wall Contracts More Than Floor or Foundation**

Distributed cracks due to restraint at base

Major crack reduces tension in the wall

**(c) Wall Cracks (Exaggerated)**

**Figure 3.9**  Effect of restraint of movement.

**Other Loads and Effects.**    Various other loads and actions such as earth pressure, groundwater pressure, and liquid pressure may need to be considered in designing a structure. Also, consideration should be given to loads that can occur during construction. Construction loads can be critical because of incomplete restraint of masonry elements and lower strength of recently laid masonry. Progressive collapse has a higher probability of occurring during construction and temporary bracing of walls may be necessary (see Chap. 15).

Differential movements of foundations can result from frost heave, seasonal moisture change in expansive soils, and uneven settlement. These movements plus temperature change, shrinkage, expansion, and creep are effects that can induce large forces in a structure. Figure 3.9 illustrates forces and cracking generated due to restraint of movement from shrinkage of a masonry wall.

Masonry walls with openings are particularly susceptible to these movements, which often cause cracks to propagate from the corners of the openings. As discussed in Chap. 15, an effective approach to accommodate movement without damage to walls with openings is to separate the wall sections at the window and door openings over the full story or building height. Advice on spacing and detailing of movement joints is also given in this chapter.

Of particular concern is the case where clay masonry is built within a surrounding concrete frame. The effects of expanding clay masonry and shrinking concrete are compounded, and, as shown in Fig. 3.10, the infill can develop high compressive forces in both vertical and horizontal directions while the concrete frame is correspondingly stressed in tension. Figure 3.11, showing cracking through the full depth of a beam, is a classic example of just such a situation.

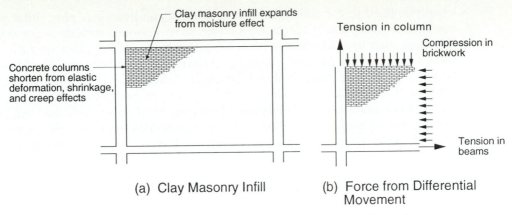

(a) Clay Masonry Infill

(b) Force from Differential Movement

**Figure 3.10** Effect of differential movement on infill panels.

Precautions should also be taken to ensure that movements of nonmasonry elements do not damage adjacent masonry. Such movements include the thermal expansion and contraction of roof trusses and roof slabs, shrinkage and creep of concrete slabs or frames, and the creep or long-term deflections and rotations of connecting or supporting members.

**Combinations of Loads and Other Effects.** Several of the previous loads and other effects can occur simultaneously and the combined effects on the structure must be considered. However, because it is unlikely that the maximum value of each type of load will occur at the same time, increased permissible stresses or reduced load factors are appropriate for some combinations. For example, a factor of 0.75 is

**Figure 3.11** Beam cracking in reinforced concrete frame due to differential movement.

usually applied to the combined effects of dead and live loads plus either wind or earthquake load. Under combined loading, most masonry structures are subjected simultaneously to axial loads, in-plane lateral loads, and out-of-plane lateral loads. In some cases, such as combined bending and axial load, the combination of loads may be less critical than the individual loads, and, therefore, it is necessary to establish whether the loads are dependent or independent of each other. An example of a critical load combination is bending moment from wind with no live load and only the minimum dead load present. Just such a situation does occur for tension-controlled flexural capacity where axial compression is effectively a beneficial prestress.

## 3.3 ENVIRONMENTAL REQUIREMENTS

A building can be structurally sound but fail to provide a satisfactory internal environment. In the extreme, a building may be abandoned and demolished because it is functionally unacceptable. Fortunately, problems with the internal environments created by some buildings do not usually have such drastic consequences. Nevertheless, the thermal and acoustic characteristics of work or living spaces and the adequacy of light and access have important influences on the physical and mental well being of the occupants. In addition, health and safety are dependent upon the ability of the building to exclude moisture and to resist fire.

In this section, the general environmental requirements in relation to temperature, sound, moisture, and fire for masonry buildings are briefly considered. These topics are covered in greater depth in Chap. 14.

### 3.3.1 Temperature Control

Thermal considerations for buildings include the comfort of users and the energy requirements of heating and air conditioning equipment. Such considerations can influence building orientation and configuration, color and texture of surfaces, window size, orientation and position, and structural system used. Heat gain or loss in a building is dependent upon many factors, including transmission through roofs and floors and air flow through openings, but emphasis is given here to the role of masonry walls.

One of the obvious functions for masonry walls forming part of the building envelope is to provide some degree of thermal insulation between the interior and exterior environments. Heat is gained or lost through exterior walls by conductance. Masonry materials have fairly high conductivities or, in other words, the thermal resistances of masonry walls are not high. However, the thermal resistance can be increased by several means. Resistance increases as density of the material decreases; hence, a wall's resistance is increased by using masonry units made of less dense or aerated materials. A rough surface on a unit traps a thicker air film at the surface and increases resistance. Resistance can also be increased by increasing the length of the transmission path by using thicker walls or by using hollow or cored units. Overall resistance of a wall system can be improved by introducing an air space. Incorporating insulation in the cavity or in cores of units or on the inside face of the wall further increases overall thermal resistance. Insulating the cavity may require the cavity size to be increased with the consequent need to use stiffer wall ties or design each wythe independently to resist lateral loads.

In areas where the outside daily temperature remains fairly constant, for example, in some tropical areas or some very cold areas, the performance of the wall is very largely determined by its insulation value. For these cases, a static analysis of thermal performance of the walls is valid.

On the other hand, where there is even a mild change between day and night temperatures, the *thermal inertia* effect of the wall can dominate. (A wall with high thermal inertia will be slow to heat up and slow to cool down.) For these cases, a dynamic thermal analysis is required as a static analysis will seriously underestimate the thermal performance of a masonry wall. Thermal inertia increases with increasing density, mass, and specific heat, but decreases with increasing conductance. The high density and mass of masonry walls means that they have high thermal inertia. It is this property of the external masonry walls that greatly improves their thermal performance by reducing the variation of the inside temperature over a 24 hour period. Heat is stored as temperature rises and is released as temperature drops, thus reducing heating and cooling loads. The thermal inertia of internal masonry walls, not part of the building envelope, increases this effect, thus further reducing inside temperature fluctuations.

It is this property of masonry walls, sometimes referred to as *thermal mass*, that can be utilized in both passive and active solar-heating design of rooms or buildings.

### 3.3.2 Sound Control

It is desirable to limit both the amount of sound within a room that is reflected by walls and also the amount that passes through walls to adjacent areas, as shown diagrammatically in Fig. 3.12.

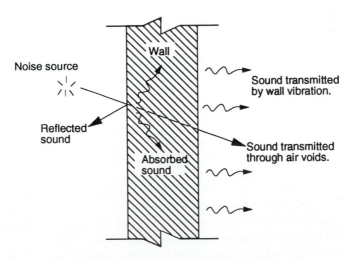

**Figure 3.12** Sound reflection and transmission.

Reduction of sound within a room is determined by the sound-absorption properties of the wall surface. The greater the porosity of the material and the rougher the surface texture, the more sound will be absorbed by the surface. Hence, masonry units with an open structure and a split or rough face or special units with surface slots can be used to control reflected noise within a room. Hard smooth surfaces obtained by plastering and painting can give an undesirably high level of reflected sound. Sound levels within a room can be greatly modified by changing acoustical properties of the ceiling, floor coverings, and furnishings.

Sound may be transmitted through a wall by forced vibrations or by porosity mechanisms. Impact sound is passed through a wall by forced vibrations alone. Cushioning the impact will reduce the impact at the source. The degree of sound insulation against forced vibrations offered by a wall depends upon its inertia or mass, its stiffness, and its internal damping properties. An intervening air space may also isolate forced vibrations. Airborne sound passes through a wall by both forced vibration and porosity mechanisms. Hence structural walls often have masonry units selected on the basis of porosity, density, and surface texture as well as strength.

Noise control can be greatly enhanced by proper building layout, as is discussed in Chap. 14.

### 3.3.3 Moisture Control

The building envelope, including the masonry walls, is required to act as a barrier to prevent the ingress of unwanted moisture into the building. Sources of moisture from outside the building include rain and snow, water vapor contained in the air, and groundwater. Poor workmanship in construction of the building envelope or subsequent cracking of the masonry in service can result in the penetration of rain through the walls to cause dampness on the inside surface, particularly under high winds. The addition of an outer wythe of masonry to the building envelope to form a drained cavity can be used to help resist wind-driven rain, as discussed in Chaps. 12 and 14. Groundwater can soak through basement walls or be drawn into walls above ground by capillary action. Water vapor in the air can infiltrate through the walls, condense in cavities, and then soak through to the inside surface. Dampness in masonry is often accompanied by *efflorescence*—the deposition of white salts on the surface of the masonry as the moisture evaporates.

There is a long history of problems caused by water vapor condensation in masonry walls due to air leakage and vapor transmission. However, the phenomenon is well understood and practices are available that, if properly followed, result in a watertight and durable building. Where dampness or deterioration occurs in walls, it is commonly due to inadequate design, poor detailing, poor workmanship, inappropriate use of materials, lack of maintenance, or a combination of these factors. Flashings, cavities, vapor barriers, air barriers, waterproofing membranes, and sealants are commonly used to improve the moisture resistance of a wall. However, these can influence its structural behavior where, for example, flashing at the base of a wall changes the lateral support conditions of the wall.

More detailed information on moisture control is given in Chap. 14.

### 3.3.4 Fire Control

Fire is one of the major causes of loss of life and loss of property, and all buildings should be designed and assessed with regard to fire safety. Although it is desirable to minimize fire damage costs, the main concern is for the safety of occupants.

In a balanced approach to the problem of fire, there are four basic ways of reducing fire risk to occupants and fire fighters. First, materials can be selected that reduce the risk of fire. This is accomplished by reducing the amount of combustible material present and excluding materials that generate excessive smoke. Masonry, being noncombustible, is excellent in this regard. Second, an early warning system for both occupants and fire fighters can be provided by smoke and combustion detectors. Then, once the fire has started, the occupants must be able to find a safe haven, either in fire-resistant compartments within the building or by escape to the outside. Tenable conditions, practically smoke-free, must be maintained in these fire compartments and in the escape routes for a significant period of time during a fire emergency. Masonry is commonly used for this purpose, even where the structural system is not masonry. Sprinkler systems, fire fighting facilities, access for fire fighters, and other means to assist in extinguishing the fire should be provided. This balanced concept provides a backup to detection and fire fighting systems.

Building regulations at national or local levels usually prescribe the minimum fire ratings to be used in a building, depending upon building type, wall function, etc. Tabulated values of fire ratings are also often given for common wall types (see Chap. 14).

In North America, the standard test used to determine fire ratings, although common, is fairly crude. Like all standard tests, the standard fire test is not intended to accurately represent the performance of masonry members in actual fires, although to a limited extent test conditions are typical for a member in service. Basically, it provides a means of comparing performance of different components and systems. The fire rating, like the compressive strength test for masonry, gives the value of an important parameter that should be modified by many factors in practical application.

A fully rational approach to design for fire would involve three phases. First is an estimation of the fire load to which a member could be exposed based on the amount of combustible material available, the likely combustion rate, and similar factors. Second is an evaluation of the required performance of the masonry member. For example, isolated piers and walls containing large unprotected openings cannot contain the spread of fire, and in these cases, the walls need only to have structural adequacy. Masonry providing fire protection to a steel column must satisfy both structural adequacy and insulation criteria.

Finally, an assessment of the fire resistance of the wall (insulation, structural adequacy, or both, as required) must take into account the thermal properties of the materials, the form and dimensions of the member, the support and restraint conditions to the member, the applied forces, and any applied finishes.

Although such an approach is not yet fully available, developments in this direction are occurring. For example, the Australian masonry code AS3700-1988 separates requirements for insulation and structural adequacy. It gives maximum slenderness ratios (taking into account support and restraint conditions of the member) that ensure structural adequacy for given fire resistance periods.

## 3.4 AESTHETICS

For a building to be fully functional, both the building as a whole and the habitable spaces within must satisfy human aesthetic needs. Visual comfort can be just as important as physical comfort.

Internally, a pleasing and adequately lit working or living environment can be created by assembling masonry units of various textures, colors, sizes, and patterns into masonry elements that form open or closed spaces. The rectangular format of the masonry unit is most simply and logically extended to form rectangular wall elements that are either simple or compound. These elements in turn are easily and logically extended to form rectilinear spaces, either simple or complex, that are pleasing, functional, and stable. Furthermore, despite its rectangular shape, the small size of the masonry unit and the hand-built site-placed character of masonry enables curvilinear elements to be readily built. This feature accommodates a wide range of interesting building shapes. When viewed from nearby, the masonry units establish a subtle reference pattern that imparts a comfortable human scale to the enclosed space. This pattern can be accentuated by varying the color of both the units and mortar, by recessing joints, or by using wall lighting to form a decorative background for plants, furnishings, etc.

The external appearance of a building usually consists of a combination of a limited number of geometric forms organized in a pleasing visual relationship with a sensitivity to scale, proportion, symmetry, gradation, and repetition. In masonry buildings, the external appearance can be intimately linked to the internal division of space and structural action, or completely mask it. For example, in framed construction, the nonloadbearing nature of masonry infill panels can be expressed visually by exposing and emphasizing the framing elements or the masonry can completely cover the frame so as to make it externally indistinguishable from a loadbearing building. Similarly, polychromatic masonry (using masonry units of different color) can define patterns on external walls that may or may not relate to the internal structure. These points are illustrated in Fig. 3.13.

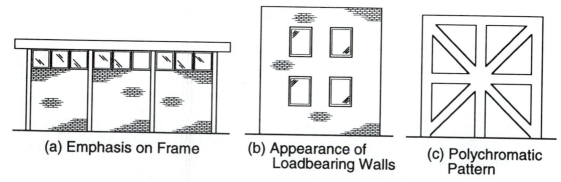

(a) Emphasis on Frame    (b) Appearance of Loadbearing Walls    (c) Polychromatic Pattern

**Figure 3.13**  Elevation treatment of framed buildings.

The two essential structural elements in masonry buildings are the loadbearing masonry walls and the floor system. The appearance of the building is largely determined by the relative emphasis given to these elements. Expression of the loadbearing walls gives a vertical emphasis to the building, whereas expression of the horizontal loadbearing elements, the floors, gives a horizontal emphasis to the building. Intermediate treatments are possible where both horizontal and vertical lines have a similar emphasis.

From afar, the individual identity of the masonry units is lost to the larger building elements and the treatment of windows has an important role in conveying the desired visual impression. As shown in Fig. 3.14(a), the continuous vertical lines of windows can minimize the impression of horizontal floors and accentuate the vertical loadbearing elements. An expanse of glass between exposed floors minimizes the impression of vertical load-carrying members (see Fig. 3.14(b)). Punched windows in the masonry wall allow varying weight to be given to vertical and horizontal elements and can provide a neutral impression, as shown in Fig. 3.14(c). Windows at the top of the panels in framed construction, as shown in Fig. 3.13(a), add to the feeling that the panels are nonloadbearing.

Large windows set at the face of the masonry give the impression of lightness of construction, windows set back from the face create shadows that can accentuate their presence, and small windows recessed into the masonry give the impression of massive construction. These points are illustrated in Fig. 3.15.

Many decisions made on the basis of aesthetics have other consequences that must be evaluated in reaching decisions. For example, window treatment in an end wall will determine whether or not it behaves structurally as a coupled shear wall. The raking of joints to emphasize the modular nature of masonry will reduce the strength of the masonry and can increase susceptibility to moisture penetration and corrosion.

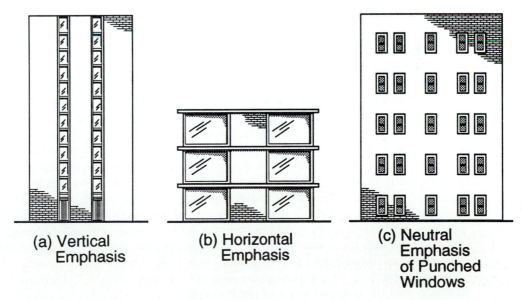

(a) Vertical Emphasis

(b) Horizontal Emphasis

(c) Neutral Emphasis of Punched Windows

**Figure 3.14** Influence of window treatment on structural emphasis.

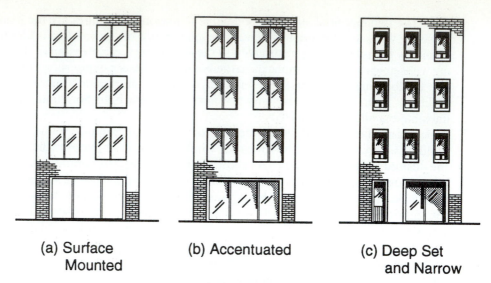

| (a) Surface Mounted | (b) Accentuated | (c) Deep Set and Narrow |

**Figure 3.15** Window settings.

## 3.5 INTEGRATION OF REQUIREMENTS

Integration of all building requirements is needed to achieve good quality and economical buildings. Inevitably, the integration of structural, environmental, aesthetic, and other requirements eventually occurs in construction of the building. This integration will be poor if the planning of the separate requirements has been carried out in isolation. Poor integration is evidenced by such things as electrical conduits and water pipes mounted on wall surfaces; exposed heating and air conditioning ducts passing through unplanned holes in the walls at structurally sensitive locations; unsightly coverstrips, sealants, leakage, or stains at connections that have been poorly designed or improvised on-site; an unsatisfactory living or working environment resulting from poor acoustics, poor lighting, or an uncomfortable climate.

Integration of functions will be best achieved if there is close collaboration between the various professionals involved in the project. Good integration of requirements is evidenced by placing electrical and water services in wall cavities, the cells of hollow units, or casting within floor slabs; by placing air conditioning ducts within ceiling spaces or service shafts or exposing them with architectural effect with properly designed penetrations through walls and floors; by neatly designing connections that fit in with the masonry coursing, are waterproof, and do not lead to staining of the masonry; by a pleasing external appearance; by a well-lit, quiet, pleasant, and comfortable working or living internal environment. Although good integration of functions may add to the initial design cost, the solving of poor integration problems on-site during construction can lead to large costs associated with delays, demolition, reconstruction, construction cost "extras," increased maintenance, and the reduced value of the building.

Traditionally, the task of integrating the various requirements for masonry buildings has been the province of the architect. In loadbearing construction, particularly high-rise construction, the structural engineer must be involved from the early planning stages. Because the building contractor is not usually engaged until planning is complete, his important contribution only becomes available at a late stage. Increasingly, it is being realized for all buildings that the input of those familiar with construction management and construction methods is essential at the planning and design stages for optimum integration of functions. It is at these early stages of the project that decisions are made that are likely to have a major influence on construction methods and project costs.

The terms "buildability" and "constructibility" are now being used to refer to the optimum integration of construction knowledge and experience into the planning and design phases of a project. However, a well-thought-out rational design can result in an expensive and poor-quality building if its construction is poorly managed, whereas, an inadequate design can be partially remedied by competent construction management. Ideally, achieving constructibility or buildability is a continuing process that brings together the expertise of the various designers, builders, contractors, and trades people from the inception of the project to its completion. Its concern is with cost, time, and quality objectives.

Such elements of constructibility that the masonry contractor or tradesman can advise on are: construction-driven planning and design; preferred masonry systems; optimum work sequences; industrial relations implications; safety of construction; practicality of technical specifications and documentation; construction details and mockups; site training of masons for special requirements; detailed drawing of service penetrations; modular layouts and dimensional coordination; standardization and repetition of construction details; construction under adverse weather conditions; achievable tolerances; accessibility during construction; procurement of materials; and quality control.

It is clear that the three professionals involved in a building project (architect, engineer, and builder) judge building quality differently. The architect views quality in terms of aesthetics and the degree to which the building fulfills the client's functional requirements. Quality is determined by periodic visual inspections of the finished product by the architect and (for aesthetic judgment) comparison with sample masonry panels constructed for that purpose at the beginning of the project. The engineer views quality in terms of the adequacy of the structure and building services. In this case, quality is determined by inspection and quantitative testing of materials and components during construction. The builder assesses quality in terms of constructibility or buildability and the functional performance of the building as a whole and in particular the building envelope. To ensure that the quality requirements of the architect and engineer are likely to be met during construction, the builder generally provides a supervisor to ensure that the correct materials, proportions, procedures, and workmanship are being employed on a day-to-day basis.

These different approaches should be integrated in a quality-assurance scheme. Quality assurance is a management tool, defined as those planned and systematic actions necessary to assure all concerned that the finished building will perform its intended function. A quality-assurance scheme involves planning those actions necessary to be taken in relation to quality, carrying out those actions, documenting the actions, and resolving any noncomplying conditions. It incorporates quality control, which is a production tool to measure and control the physical properties of materials and the construction methods in quantitative terms.

In summary, the involvement of architect, engineer, and builder is ideally required at an early stage of the project if the various building requirements are to be successfully integrated in the planning and design of the building. These professionals should also be integrated into a quality-assurance scheme to ensure that the constructed building achieves the design requirements.

## 3.6 PLANNING THE BUILDING

The need for the integration of structural, environmental, aesthetic, and constructional requirements of a building from project inception to completion was generally considered before. Here, the specific implications for the planning phase of masonry buildings are discussed in some detail. Consideration is first given to the form, elevation, and plan of the building, followed by wall configurations and layouts, floors and roofs, connections of elements, and foundations.

### 3.6.1 Building Form

The external form of a building is largely determined by the type of internal space required such as office, commercial, industrial, or residential and can be single or multistory of either framed or loadbearing wall construction.

Planning of a building usually involves development of a building module that is the least common denominator that fits into optimum room sizes and other elements such as standard windows, ceiling, and lighting modules. For masonry construction, the building module adopted should take into account the dimensions of the masonry unit or units to be used. The basic module for masonry is 4 in. (100 mm) and conformance with multiples of this module should include the actual height of the units plus the thickness of bed joints (usually ⅜ in. (10 mm)). Heights of door and window openings and the clear story height between floors should be compatible with this modular characteristic. Where double wythe walls of different units are used, the course heights of both the wythes should preferably correspond at the opening heights. Use of adjustable ties between wythes has reduced the importance of having course heights correspond at more frequent intervals to allow rigid wall ties or headers to be installed. Adjustable ties must be sufficiently stiff to connect the wythes.

The modular length of a unit is its actual length plus the thickness of a head joint. Lengths of walls and widths of openings in the wall should correspond to a multiple of the half-modular length for a half-running bond or a third-modular length for a third-running bond, and so on. The usual ⅜ in. (10 mm) mortar joint can be varied slightly to give a building module that fits the building dimensions better and minimizes the need to cut units. Masons often adjust the joint thickness on site to suit the planned dimensions, and this is normally quite acceptable if uniformly carried out. However, making joints too thick or too thin can lead to inferior masonry, and in such cases, it is better to cut the units.

In framed construction, the spacing of the framing members should be based on the modular dimensions of the masonry cladding or infill walls. Modular planning results in more economical construction and a more pleasing appearance than when units have to be specially cut. Nevertheless, units can be readily cut for special purposes where necessary. Apart from these considerations, the form of a framed building is largely determined by the capabilities of the steel or concrete framing members.

In loadbearing masonry buildings, whether they be single or multistory, the walls serve as structural elements to support or resist loads, as architectural elements to divide or enclose space, or as a finish material. Within economic constraints, both the geometric configuration and structural form are determined by a complex interplay of requirements for functional space, environmental comfort, structural strength, and architectural expression. A comprehensive treatment is not intended here, however, some factors influencing the elevation, plan, and wall layout of loadbearing masonry buildings are considered.

### 3.6.2 Elevation

Elevations are the most obvious views of a building and in many cases form the first and lasting impression. One of the strongest influences on elevation is the height of the building. In general, the most economical building is one that has as few stories as possible to meet the owner's floor-area requirements. This, of course, has to be offset by the cost of the land, and, in general, the higher the cost of the land, the higher the building must be for economic viability. Masonry buildings are conveniently classified according to height as single-story, low-rise, medium-rise, or high-rise.

The single-story building is usually domestic, industrial, commercial, or public building with easy access at ground level. The greater part of the building envelope in such buildings is likely to be the roof area, which should correspondingly receive appropriate structural and environmental considerations. The low-rise building is conveniently described as having from two to four stories, this being the commonly accepted limiting height that can be adequately serviced by stairways. To be suitable for economical loadbearing masonry construction, these buildings (together with the medium- and high-rise cases) should have identical floor plans of loadbearing walls at each story.

Medium-rise masonry buildings have from five to about ten stories. The lower limit is determined by the need to have elevators for vertical transportation and the upper value is the approximate limit that can be built with the masonry sizes determined by environmental requirements (i.e., fire and sound) rather than structural requirements.

Above this height, wall sizes will probably be determined by structural requirements. As building height increases, the area taken up by the service core, including elevators, becomes a more significant factor in building design, structural action becomes more critical, and structural costs increase.

With increasing height, buildings become more sensitive to the performance of individual elements. This general decrease in structural robustness with increasing height is shown diagramatically in Fig. 3.16. This figure also illustrates that, particularly in earthquake areas, decreasing robustness results if discontinuities are introduced in the elevation and especially if these discontinuities concentrate mass at the upper levels of the building, a critical concern in earthquake areas. Where significant

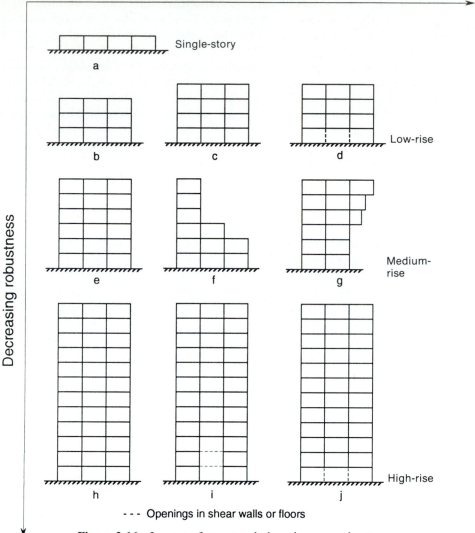

**Figure 3.16** Impact of structural elevations on robustness.

changes in elevation occur, for example, in Fig. 3.16(f), a vertical isolation joint can be conveniently inserted to form two separate buildings. In this case, sufficient separation between the two must be allowed to prevent the two buildings from hitting each other under high wind or earthquake loads, and flexible connections (including those for mechanical and electrical services) should be designed to accommodate differential structural movements. Elimination of floor panels or shear walls at lower stories, such as shown in Figs. 3.16(i) and (j), have a dramatic impact on the robustness of a building.

Windows in the external walls affect the structural action of the wall to varying degrees. Where small punched windows are used, as in Fig. 3.17(c), the whole wall

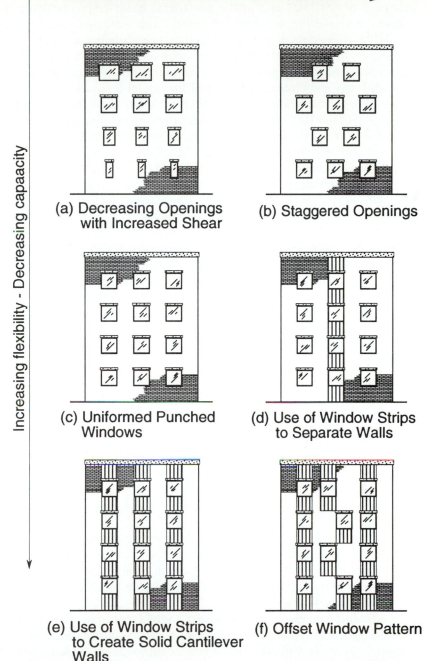

**Figure 3.17** Placement of windows and its effect on wall flexibility.

may act monolithically in resisting lateral racking loads. Critical areas exist at the corners of windows and in the horizontal beam elements beneath or above windows. Because the accumulated shear force increases from top to bottom of the building, these effects can be reduced by using smaller windows in the lower stories as shown in Fig. 3.17(a). Staggering openings, as shown in Fig. 3.17(b), as opposed to aligned openings produces better coupling of the vertical sections, which improves the stiffness and strength of the wall for lateral resistance.

Using structurally isolated infill panels above and below window opening, as in Figs. 3.17(d) and (e), effectively divides the wall into uncoupled shear walls. Although there is a progressive decrease in lateral strength from Fig. 3.17(a) through to Fig. 3.17(e) (for unreinforced or equally reinforced sections), stress concentrations at windows are eliminated in Fig. 3.17(e), and the increased flexibility has the desirable effect of eliminating the stiff beam elements, which are vulnerable to brittle shear failure. Such cantilever systems have performed well without excessive cracking in resisting lateral forces, including earthquakes. The discontinuities introduced by offsetting windows in Fig. 3.17(f) seriously weakens the shear wall and such walls commonly suffer severe earthquake damage at changes in section.

For economic reasons, the clear height from floor to ceiling in a multistory building is usually set at or near the minimum consistent with the use. Vertical spacing of the horizontal building elements (floors and the roof) is then determined by this requirement for clear space, the depths of the floor or roof elements, and the need (if any) to incorporate mechanical and electrical services in ceiling spaces.

### 3.6.3 Plan

The overall shape of the building plan is determined by many factors including the shape, size, and orientation of the site, relation to neighboring buildings, internal communication routes, natural lighting requirements, and the functions for which the building is designed. One important parameter is the ratio of perimeter wall length to floor area. This ratio gives an indication of the expense of the perimeter walls in relation to the usable area enclosed. Because most heat is lost or gained through the external walls of multistory buildings, this ratio also gives an indication of relative insulation required and operating costs for the heating and air conditioning of various plan shapes.

Figure 3.18 was developed to show the relative efficiency of various building shapes in this regard. The percentages have been obtained by comparing the length of the perimeter walls required for the particular shape with the circumference of a circle having the same floor area. It can be seen that the square is the most efficient of the rectilinear shapes.

Building shape in plan is also an important indicator of resistance to lateral loads. Symmetric plans and shapes of roughly equal dimensions along each of the major axes are less likely to have significant torsional effects. The spacing between walls in the two perpendicular directions is a function of the floor type and the permissible span-to-depth ratio for deflection control. The total length of walls along a major axis is directly related to the lateral stiffness of the building. Higher lateral loads require a larger wall-length ratio (total wall length divided by floor area) to control lateral drift and maximize shear strength. Total lateral force in a given direc-

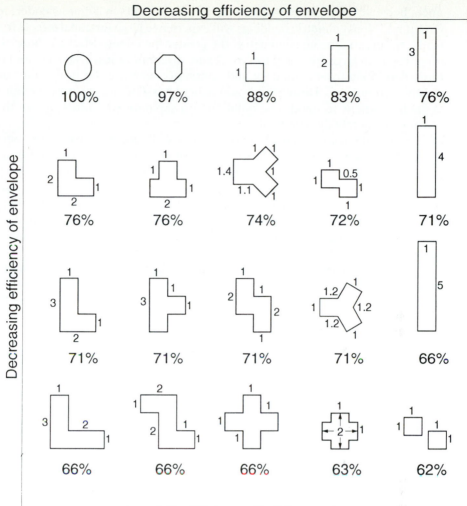

**Figure 3.18** Efficiency of building envelopes.

tion divided by the plan area of all effective walls in that direction provides an approximate average shear stress. If this is greater than the maximum permissible shear stress assigned by the building code, it is an indication that too few shear walls have been provided to resist lateral load in that direction. Walls oriented perpendicular to the direction of the lateral force contribute negligible resistance unless they form flanges with parallel walls. Lateral stability must be checked in the directions of both the major axes.

Symmetric plans provide good resistance to lateral loads because the locations of resultant loads are near the centroid of the building and therefore produce little, if any, torsional moment. Particularly in earthquake areas, as a rectangular plan becomes elongated, nonuniform lateral loading will produce some torsional effects because the resultant of the external loads does not coincide with the center of resistance of the building. Furthermore such a building has reduced torsional resistance, because the reduced width of the building results in a more slender, less stable structure. Nonsymmetric buildings will experience large torsional effects from lateral loads; the greater the nonsymmetry, the greater the effect. Marked changes of shape often result in large torsional forces. Thus, discontinuities, setbacks, and other features detracting from symmetry and continuity increase the probability of damage during earthquakes. Designers should, where possible, incorporate movement joints at such locations to divide the building into separate structures that in themselves are approximately symmetrical.

For buildings with equal areas of shear walls located around the perimeter of the building, increased irregularity of the shape results in decreased resistance to torsion, as indicated in Fig. 3.19. Therefore, not only do elongated and nonsymmetric buildings experience larger torsional loads, but they make less effective use of shear walls to resist torsion. A shear wall is most effective in resisting torsion when its distance from the center of rigidity is greatest. The nonsymmetric buildings in Fig. 3.19 have large shear wall areas relatively close to the center of resistance (point of rotation) and are thus less effective.

Similar remarks apply to the location of stairwells or service shafts in the building. In general, the greater the degree of nonsymmetry, the greater the torsional effects. The corresponding locations of openings or irregularities in the floor slabs

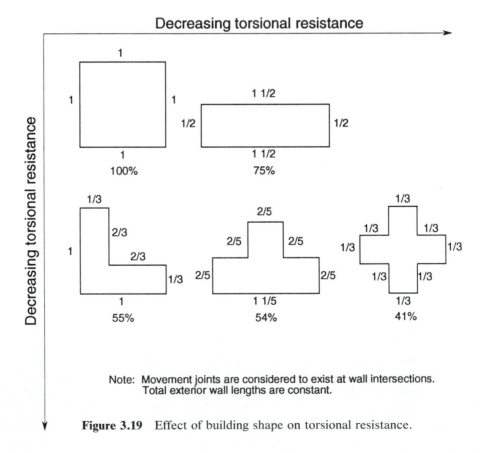

Note: Movement joints are considered to exist at wall intersections.
Total exterior wall lengths are constant.

**Figure 3.19**   Effect of building shape on torsional resistance.

Decreasing effectiveness of floor diaphragm

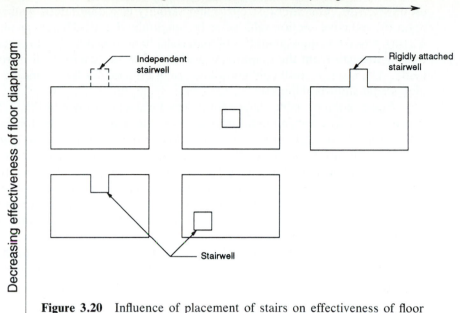

**Figure 3.20** Influence of placement of stairs on effectiveness of floor diaphragms.

also affect the effectiveness of the floor as a diaphragm. The decreasing effectiveness of floor diaphragms with increased eccentricity of openings is illustrated in Fig. 3.20.

### 3.6.4 Wall Configuration and Layout

Bearing walls form the vertical boundaries of spaces within the building and their arrangement is a complex problem involving structural, functional, and environmental considerations requiring close collaboration between the architect and the engineers for the solution. Constraints to wall layout include the need to comply with local building regulations regarding minimum room sizes, window areas, means of egress, and building services.

Gravity loads are distributed from the roof and the floors to the walls, which direct them vertically to the foundation. In multistory construction, cross walls or shear walls also distribute the lateral load from story to story down to the foundation. Loadbearing and shear walls, therefore, should be placed in good vertical alignment so as to minimize any eccentricity of load on them. Unless other means are used, cross walls should be arranged to give stability in all directions.

The possibility of future changes in wall layouts to accommodate different use of the building should be considered. If all internal walls are loadbearing, maximum structural efficiency but minimum layout flexibility are the likely results. Greatest flexibility of space is achieved by using the minimal number of loadbearing walls consistent with structural sufficiency and using nonloadbearing partitions for further subdivision.

In most applications, the configuration of a wall in plan has a marked influence on its strength. The lateral strength and stability of a continuous wall is dependent upon the effective section relative to its longitudinal axis. Hence, as shown in Fig. 3.21(a), based on equal quantities of materials, there is generally a progressive reduction in strength from the diaphragm wall down to the cavity wall in resisting face loads. The stronger wall configurations are often used for free-standing walls or for high single-story buildings.

Some of the possible shapes for isolated walls are shown in Fig. 3.21(b). These shapes are easily arranged to form rooms in single or multistory buildings where they also act as shear walls in resisting lateral overturning forces. Configurations with material concentrated at the extremities are generally most efficient from a structural point of view because these result in a higher section modulus and consequently lower bending stresses. For equal cross-sectional areas and equal length in the direction of applied lateral load, the sections in Fig. 3.21(b) are arranged in order of decreasing bending strength. However, other factors are important in selection of shear walls such as torsional resistance, flexural strength about the lateral axis of bending, and the fact that the thickness of a wall is usually limited to the width of a masonry unit or a multiple of it. Taking these factors into account, the strengths of shear walls range from a maximum for the hollow rectangular section, which has "flanges" effective in resisting forces about both axes, to a minimum for the straight wall, which does not have the benefit of flanges about its major axis and has practically no bending resistance about its minor axis. A limitation is that material in a flange can become ineffective if it is too remote from a connecting web.

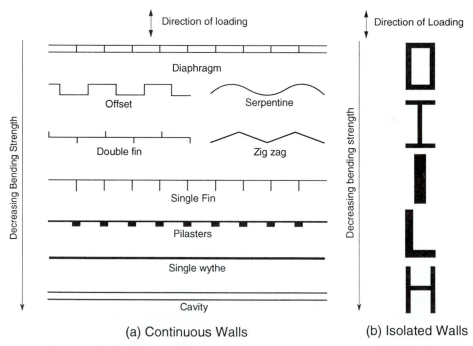

**Figure 3.21**  Effect of wall configuration on bending strength. (Equal quantities of material.)

From a structural point of view, walls must be properly arranged to act as compressive members carrying vertical loads from the floors to the ground, act as shear walls in resisting lateral forces on the whole building, and act locally as flexural members in resisting face loads. Flanged walls and walls with returns have superior lateral stiffness, resistance, and ductility compared to planar walls and therefore are recommended for high-demand situations. Symmetry of cross-section provides similar wall response under reversal of lateral forces.

Shear walls should be designed to resist the transverse and longitudinal lateral forces and torsional forces that act on the whole building. Most floor systems do not have sufficient moment connection with the shear wall system to contribute to the lateral load resistance of the building except to ensure that the lateral load is distributed to the shear walls. Hence, the uncoupled shear walls can be designed relatively independently of the floor system used except for the connection to it.

Some simplified loadbearing wall arrangements are shown by heavy lines in Fig. 3.22 with nonloadbearing walls or partitions shown by light lines. The nonloadbearing veneers on external walls (if any) are not shown because they do not contribute significantly to the robustness of a building. The cellular arrangement shown in Fig. 3.22(a) is the most robust and stable as it provides adequate shear walls in both directions and the floor slabs, which have two-way action, provide alternate load paths should one of the wall elements be removed or damaged. Reduced two-way action of the floor slabs occurs in Fig. 3.22(b) and in the simple cross-wall arrangement shown in Figs. 3.22(c) and (d). The floor slabs shown in Fig. 3.22(c) span in one direction in the central portion of the building; however, the end bays have some two-way action because of returns on the end walls. Some two-way action is developed at the central section with intersecting walls, shown in Fig. 3.22(d).

The previous buildings are more robust than that shown in Fig. 3.22(e), where the end walls have no returns and, therefore, only span vertically between floors in resisting lateral loads. The arrangement of walls shown in Fig. 3.22(f) is unstable as there are no longitudinal shear walls to resist any lateral forces acting in that direction. The longitudinal partition walls not built with masonry usually have negligible resistance. Masonry partitions could be designed to resist lateral loads.

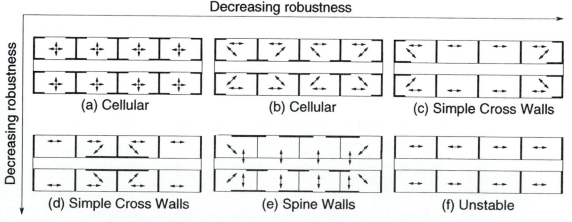

**Figure 3.22** Effect of wall arrangement on robustness.

It can be seen from these wall arrangements that there is a tendency for the provision of increased flexibility of wall layout (more partitions) to result in less robustness of the building. Careful design, therefore, is necessary to achieve an appropriate balance.

The importance of the plan shape of a building in determining its resistance to torsional forces was considered earlier in this section. It was pointed out that a symmetrical building plan (preferably square) has best resistance. However, internal wall layout is also an important factor in determining torsional resistance. For example, a square building plan has the resultant lateral force located approximately at the center of the square, but if shear walls are placed only around three boundaries, the center of rigidity of these walls will be located near the edge of building, resulting in a large torsional moment. Generally, a symmetrical wall layout with concentration of walls at the extremities of the building gives the best arrangement for resisting torsional forces.

Figure 3.23 shows various arrangements of loadbearing walls that have inherently different resistances to torsion, assuming that intersecting walls are connected to form flanged shapes. In this illustration, the same total length of longitudinal wall is distributed in a different way for each layout. Greatest torsional resistance is obtained by concentrating the longitudinal walls at the corners of the building, as in Fig. 3.23(a). The center of rigidity is at the center of the plan (from symmetry) and the longitudinal walls, being placed as distant as possible from this center, produce the greatest torsional resistance. Although the position of the center of rigidity of the symmetrical arrangement in Fig. 3.23(b) remains at the center of the plan, the longitudinal walls are not entirely placed at the extremities thus resulting in a reduced torsional resistance. Because of lack of symmetry about one axis in Fig. 3.23(c), the center of rigidity will move slightly off that axis and lateral forces, which act approximately through the center of the plan area, will have an increased torsional effect due to this offset of the center of rigidity. Also the distances from the center of rigidity of the flanged sections created with longitudinal walls have been reduced, thus reducing the torsional resistance.

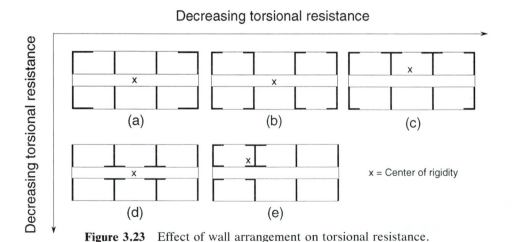

**Figure 3.23**   Effect of wall arrangement on torsional resistance.

Although the arrangement of walls in Fig. 3.23(d) is symmetric, the longitudinal walls have been moved close to the center of rigidity and the sections produced have a greatly reduced influence on the torsional resistance of the total arrangement.

A very poor arrangement of longitudinal walls is shown in Fig. 3.23(e). Here they are clustered toward one corner, displacing the center of rigidity a large distance from the center of the plan, and greatly increasing the torsional effects of the lateral loads. In addition, the longitudinal walls are at a short distance from the center of rigidity and therefore contribute less to the overall torsional resistance.

Although the principles of lateral load resistance have been discussed in relation to multistory loadbearing buildings, they are also applicable to smaller single-story buildings. Lateral forces are lower in these buildings, but there may also be fewer shear walls to resist these loads. In these cases, care must be taken to see that adequate shear resistance and torsional resistance are achieved. Figure 3.24 shows some arrangements of walls that are unable to resist torsional forces. In each case, the building would twist about the center of rigidity at point 0, assuming no interaction between intersecting walls.

Openings for windows and doors affect lateral stiffness of walls and their ability to resist lateral forces and to control lateral drift. The effect is dependent upon size and locations of openings but is not as critical for low-rise buildings as in high-rise buildings, where greater strength and stiffness are required. In this case, care should be exercised in selection of opening sizes and locations. A detailed discussion is presented in Chap. 10 on design of masonry shear walls. In addition, because the requirements for openings are often greatest at the ground floor of multistory buildings, there is a natural tendency to have a *soft story* at this level. For a time, designers attempted to use this more flexible level as a means for reducing the transfer of seismic motion to the upper parts of the building. However, in severe earthquakes, these *soft stories* have sustained extensive damage because of the very high shear demand on the piers at this level. In several cases, this resulted in collapse of the building. Care should be taken to avoid introducing a soft story.

Environmental factors also influence wall layout. The need for adequate natural light, the use of passive solar heating, fire-protection requirements, and noise control are examples of constraints on the placement of walls. Information on these topics is presented in Chap. 14.

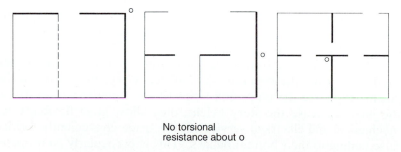

No torsional
resistance about o

**Figure 3.24** Unacceptable wall arrangements.

### 3.6.5 Floors and Roofs

The major structural elements of a loadbearing masonry building are the loadbearing walls, the floor and roof systems, and the foundations. An important structural function of both the floor and roof systems is to distribute lateral forces applied to the building by wind or earthquake to the loadbearing walls. Apart from in-plane diaphragm action, the floors do not take part in resisting lateral loads and, therefore, the floor system can be optimized for resisting gravity loads alone. Floor systems can be categorized by structural response as either two-way or one-way systems. Two-way systems have greater robustness as alternate load paths are possible if a support is removed or damaged. Also, floor systems can be categorized by method of construction. They may be completely cast-in-place concrete, a composite construction of precast slabs with an in-situ topping, or simply precast concrete. In another type of floor system, steel joists or concrete beams span in one direction with a cast-in-place slab spanning across them with either independent or composite action. A timber floor system can also be used. Examples of these floor systems are illustrated in Fig. 3.25 in roughly decreasing order of robustness.

Cast-in-place slabs are generally the thinnest system and will provide the minimum story height for a given clear headroom. They require extensive formwork and shoring during construction. Although they have the advantage that a ceiling finish, if any, can be directly applied to the underside of the slab, services must be placed within the slab, exposed under the slab, or other special arrangements must be made. Cast-in-place slabs, because of their continuity, provide good diaphragm action.

Precast slabs eliminate the need for formwork and can be manufactured with good surface finishes requiring no further treatment. Close tolerances are usually required in construction of the building and careful attention must be given to tying the units together to achieve effective diaphragm action. Services such as electrical and communication lines can be placed in the topping (if any), in the cores of hollow slabs, or under the slab.

When a structural topping is to be applied to precast units, they can be easily shored if necessary. The topping can also be used to produce a level surface and to help the precast units act together and achieve effective diaphragm action. Precast Tees and Double Tees can be used for long spans or heavy loads. Where mesh reinforced topping is not added, weld plates for connecting the individual elements along their sides and reinforcing rods grouted in keyways between the slabs may be necessary to create sufficient strength and stiffness in the diaphragm. Precast planks set dry on walls without positive connections are not as robust as other floor systems because of the lack of redundancy.

Where joists or beams are used, either reusable or permanent forms can be used to support the concrete slab. If necessary, the joists can be shored during construction and the props and forms (if any) removed soon afterwards. Although the joists increase the story height, they allow great flexibility for installation of mechanical and electrical services in the space conveniently formed by attaching a false ceiling to their bottom flanges. This is particularly so if open-web steel joints are used, allowing services to pass through the beams as well as between them. Dia-

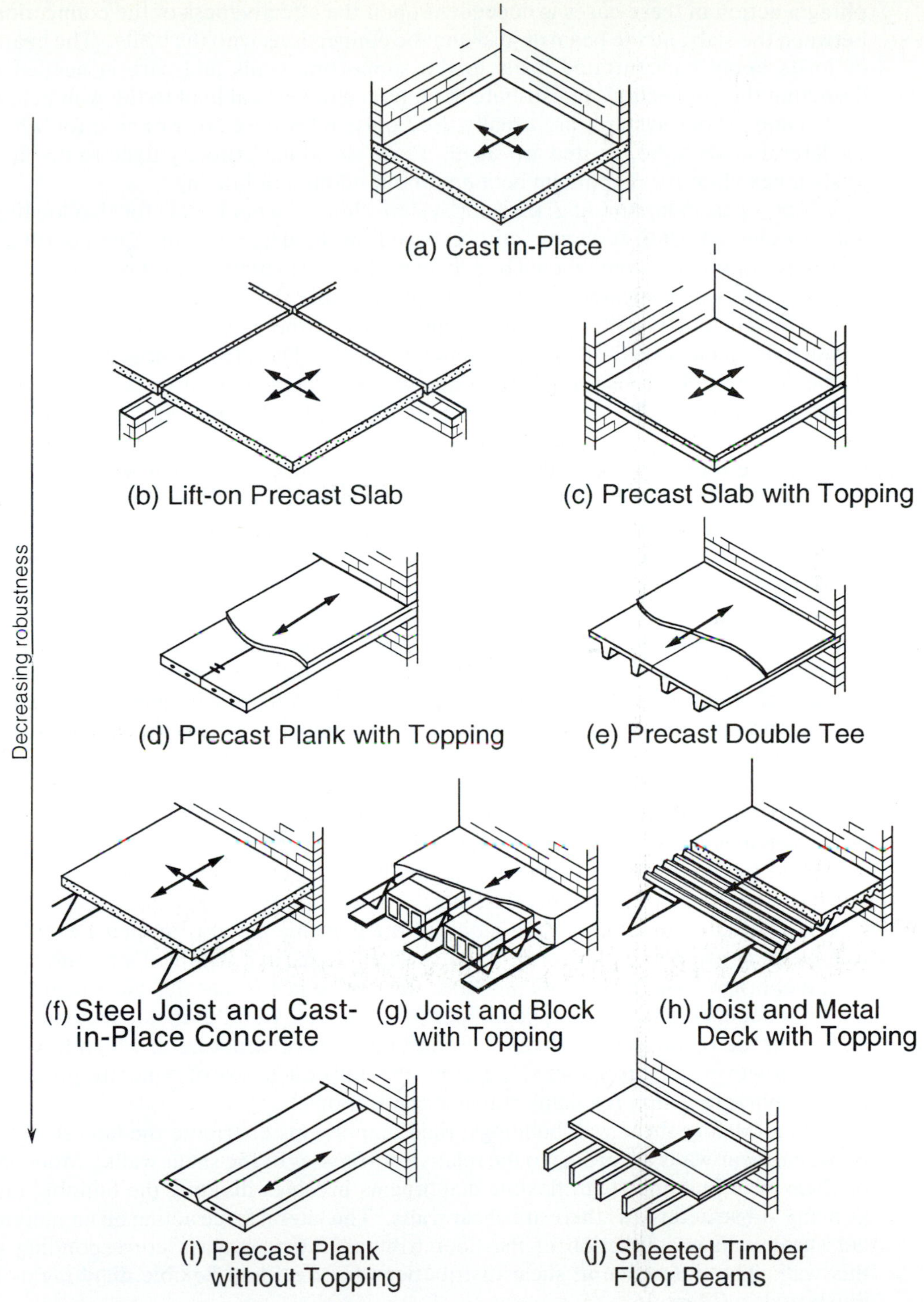

(a) Cast in-Place

(b) Lift-on Precast Slab

(c) Precast Slab with Topping

(d) Precast Plank with Topping

(e) Precast Double Tee

(f) Steel Joist and Cast-in-Place Concrete

(g) Joist and Block with Topping

(h) Joist and Metal Deck with Topping

(i) Precast Plank without Topping

(j) Sheeted Timber Floor Beams

Decreasing robustness

**Figure 3.25** Floor systems.

phragm action in these cases is dependent upon the effectiveness of the connections between the slab and the beam or joist and the connections with the walls. The beams or joists apply concentrated loads to the supporting walls and care is needed in designing the connection to distribute the lateral and vertical load to the wall below.

Timber floor systems are usually used only in low-rise construction for which the lateral loads to be resisted are small. Floor sheathing properly fixed to the floor joists gives effective diaphragm action without additional bracing.

For a particular building, the floor system should be applicable for that location and be selected using economic criteria based on the desired span. The possibility of using combined systems should be considered—for example, using precast balcony and stair units in association with cast-in-place slabs.

The flat roof system of a masonry building distributes loads to the shear walls in much the same way as do the floors at lower levels. Therefore, stiffness is required in the plane of the roof for effective diaphragm action. A roof deck of the same construction as the floor system is commonly used. This provides continuity of construction and permits additional stories or other construction to be added later. These roofs behave in the same way as the floors except that waterproofing, drainage, and thermal movements must be considered.

Where an alternate roof system is used, it usually consists of trusses or joists with a lightweight roofing. If the roof is sheathed with corrugated steel decking or substantial plywood (or equivalent) properly fixed to the structural members, then the diaphragm action developed may be adequate to transfer lateral loads to the shear walls. However, as is discussed in Chap. 16, the relative stiffness of the roof diaghragm affects this distribution and should be considered in the design. Where no such sheathing is used or where doubt exists as to its effectiveness, then cross bracing in the plane of the roof should be incorporated to stiffen the roof and help distribute lateral load to the walls.

A roof system may also have to provide lateral support to the tops of external masonry walls and both the diaphragm bracing and the connections to the walls must be designed accordingly. In addition, a lightweight roofing system will probably be subjected to net uplift forces and consequently must be adequately tied down.

The response of multistory buildings to lateral forces, particularly earthquake forces, is often a function of the rigidity of the floor diaphragms in their own plane relative to the rigidity of the shear walls.[3,4] Diaphragms are classified structurally as rigid diaphragms and flexible diaphragms. It is to be noted that this classification is not only dependent upon the type of floor (concrete vs. wood or metal decking), but also upon the rigidity of the shear walls and the type of structure in which they are used. However, as a rough guide, a $2\frac{1}{2}$ in. (63 mm) thickness of concrete produces rigid diaphragm action for many framing applications.

In modeling shear wall buildings, rigid diaphragms distribute the lateral forces between shear walls according to the relative stiffnesses of the shear walls. Modeling of shear wall buildings with flexible diaphragms involves dividing the building into as many substructures as there are shear walls. The lateral force acting on an individual shear wall is a function of the floor tributary area or mass corresponding to this wall. The difference in shear distributions for rigid and flexible diaphragms is illustrated in Chap. 16.

### 3.6.6 Connections

The proper connections of wall to roof, wall to floor, and wall to wall are crucial in achieving satisfactory performance of a building. They must be structurally adequate, protected against adverse environments, and satisfy aesthetic requirements.

From a structural point of view, the connections transfer loads or permit movement between the joining elements and determine to a very large extent the validity of assumptions made in design. It is, therefore, of utmost importance that the connection details used in construction provide for the structural actions and movements upon which the design is based and vice versa. In loadbearing construction, the roof and floors are generally assumed to act as diaphragms in distributing lateral loads to the walls. Hence, the connections of the wall to the roof and floors must be strong and stiff enough to ensure this action. Alternatively, if an element is not designed to take part in this action, the connections must be sufficiently flexible to minimize load transfer in this plane. Figure 3.26 shows lateral load transfer from diaphragm to shear walls via floor-to-wall connections. On the other hand, a roof system is usually assumed to be simply supported on the walls, and, if this is so, then the roof-to-wall connections must be such that the roof system does not introduce unexpected bending moments into the wall. The type of floor-to-wall connection not only determines the degree of fixity appropriate in analysis, but also the appropriate values of eccentricities of load on the wall to be used in the design. There is a great difference in load eccentricity when a floor system rests centrally on a wall compared with one that is attached to the face of a wall.

Complete load transfer is often assumed in design at intersecting walls and this can be achieved in several ways as discussed in Chap. 10. On the other hand, provisions to allow some relative movement between two masonry walls may be desirable. In this case, there should be no bonding at the junction, but rather metal ties are installed to provide the required restraint and yet permit the desired move-

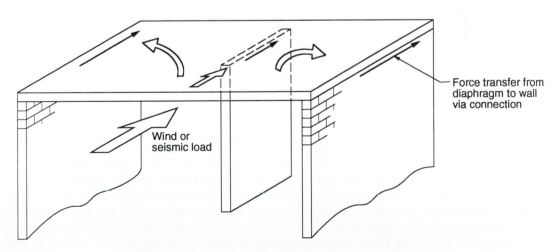

**Figure 3.26**   Load transfer from diaphragms to shear walls.

ment. This can be achieved by utilizing the axial stiffness of the tie to provide restraint and the out-of-plane flexibility of the tie to allow movement. Alternately, special ties designed to allow only limited movement can be installed. Where a masonry wall is connected to a steel or concrete frame, special connections are often used to provide lateral out-of-plane support but allow in-plane movement between the masonry and the frame. This topic is discussed in more detail in Chap. 11.

### 3.6.7 Movement Joints and Joints Between Adjoining Elements

Joints between elements are often weak links where moisture, sound, heat, wind, and fire can penetrate and they must be protected against these environmental factors. Cracking of joint filler or opening of gaps at joints between adjoining elements can occur due to differential movements.

Protection against the ingress of moisture is an important design consideration for masonry buildings. The construction of joints is often complicated by the need to install flashings and drainage that can modify the structural characteristics of any connection at the joint. Flashings usually destroy the tensile bond of mortar at the joint and reduce its shear resistance. Such effects must be assessed in the design phase. Similarly, the provision for differential movement at movement joints will modify structural behavior. Where a concrete floor slab rests on a clay masonry wall, a slip joint is necessary to allow the shrinking concrete slab to slide over the expanding clay masonry. Yet the slab must provide lateral support to the wall. These conflicting requirements are usually met by providing a smooth interface between the slab and the wall that has sufficient friction to resist the relatively small out-of-plane lateral load forces, but allows differential movements between the two elements. In cases where out-of-plane forces can be large or where positive connection is deemed necessary, metal connectors with in-plane flexibility or other out-of-plane restraints can be added. When designing joints, account should be taken of the differential movements that occur between masonry, concrete, steel and timber, and, indeed, between two different masonry materials.

Finally, joints between elements are important aesthetic features either because they are exposed and form key elements of architectural expression or because they are required to have minimal visual impact or be hidden from view. Joints should fit within the normal dimensions of masonry. (See Chap. 15 for movement joint information.)

### 3.6.8 Foundations

Loads are transmitted to the foundations of masonry buildings by the loadbearing walls and hence continuous rather than isolated footings are usually appropriate.

Where soil conditions permit, simple strip footings are the most economical. The design width of the footing should not only satisfy allowable stresses in the foundation material, but should also be varied to ensure that neither excessive settlement nor differential settlement will occur. Seasonal movement from shrinkage or swelling of some clays or from frost heave may require the footing to be located at sufficient depth (normally specified in building codes) to avoid these variations.

Where the bearing capacity of the soil is insufficient for strip footings, the area of the footing can be increased to a mat (raft) covering the whole building area, thereby generally reducing soil stresses to an acceptable level. However, in order to spread the building load over the whole plan area, the mat must be quite strong and stiff. Such construction is expensive, particularly for masonry buildings, where the independent cantilever nature of the superstructure normally contributes little to the combined building/mat stiffness. Special treatment of the basement wall layout can, in some cases, provide an economical solution in which the basement itself acts as a deep raft.

Where strip footings are inadequate, another possible solution is to support the strip footings by piles or by piers. In these cases, the wall and footing can be connected to become a composite beam spanning between the piles or piers with no support assumed from the soil beneath it. In fact, where swelling and shrinkage of clays or where frost heave occurs, special precautions are usually necessary to ensure that the soil does not lift the beam off the piles or piers. This can be achieved by pouring the beam on a compressible forming material, or, in the case of frost heave, utilizing the heat leaking from the building to prevent freezing under the beam.

Detailed design of footings is not considered here. However, footing design can be simplified by making the masonry less susceptible to cracking from differential movements. This can be achieved by including movement joints to permit a limited amount of movement between different parts of the building and by designing the footings to have comparable settlement.

## 3.7 ECONOMIC ASPECTS

The economic performance of any building involves costs associated with land, foundations, superstructure, cladding, construction, surface finishes, fire proofing, heat and sound insulation, services, maintenance, insurance, taxes, and design. These are conveniently divided into initial construction costs and long-term maintenance and operation costs. Where the building layout suits a masonry structure, experience has shown that other structural alternatives are usually more expensive.[3.5,3.6]

Foundations for masonry buildings are usually simple because loads are evenly distributed along walls rather than concentrated at columns, as is the case for framed buildings. The superstructure consists of a roof and floors that distribute both vertical and lateral loads to masonry shear walls. Moment connections between walls and floors are not generally required and hence there is minimal interaction between wall design and floor design. Therefore, the most economical floor system can be chosen without significantly affecting wall performance. Continuity of floors over supports is possible and desirable to reduce midspan bending moments and to improve robustness. Similarly, optimization of the wall system, for example, whether or not to couple shear walls, can be independent of the floor system. Internal walls that subdivide space frequently can be arranged efficiently for structural purposes without compromising the function or the aesthetics of the building. The small masonry unit allows a great variety and complexity of wall layouts without significantly increasing costs.

Masonry cladding can be incorporated into the structural system of some buildings to help resist lateral loads and, in the case of a single-wythe wall, to act as a vertical loadbearing element.

The construction of masonry buildings, though labor-intensive, can be quite economical. Masonry units are readily assembled on site without the need for expensive equipment and generally require no formwork or temporary construction facilities other than scaffolding. For masonry buildings, there are fewer materials and processes on site to handle and schedule and fewer trades to engage and coordinate. Masonry gains strength rapidly and a minimum of time is needed before proceeding to the next stage of construction. Unlike some framed buildings, other trades can follow story by story as the construction progresses. Laying of units is most economical when the units have a relatively large face area, do not require cutting, and can be laid from one side of the wall.

Masonry units come in a wide range of textures, colors, and shapes and can be used with various bonding patterns and joint treatments to give a variety of surface finishes. The use of colored mortar and raked or recessed joints slightly increases the cost of the masonry. Masonry is most economical when no additional surface finish is used, but, where needed, a masonry wall can be painted, rendered, or covered with a veneer. Walls incorporating more than one type of masonry unit are more expensive.

The contribution that masonry makes to the fire resistance, thermal and sound insulation, and finish of a building is substantial even in buildings that are framed in other materials. In loadbearing masonry buildings, minimum wall-thickness requirements specified in building regulations for these environmental factors usually meet structural requirements for heights of up to ten stories. The multifunctional use of walls for environmental, structural, aesthetic finish, and subdivision purposes makes the greatest contribution to the economic competitiveness of loadbearing masonry construction. These positive attributes should be optimized in the planning and detailed design stages for local conditions.

Properly constructed masonry is very durable, as is attested by the many historic buildings still standing today. It has very low maintenance requirements and the small size of masonry units facilitates repair work.

## 3.8 CLOSURE

This chapter dealt with the appreciation of the main factors affecting the design of masonry buildings. An understanding of the separate and interrelated requirements is important, particularly at the planning stage of building design. As indicated, many different (and sometimes conflicting) considerations and requirements must be included to arrive at a design that combines well the functional, structural, environmental, and aesthetic aspects in an integrated manner aimed at producing an initially economical building and one that is durable and requires little maintenance.

The overview provided in this chapter is supplemented by information in subsequent chapters containing more detailed treatment of the design and construction of masonry building elements and, in the cases of Chaps. 16 and 17, buildings as a whole. The importance of initial planning cannot be overemphasized. Decisions made at this stage will have significant impact on subsequent stages of design and often are key factors affecting the cost and performance of the building.

## 3.9 REFERENCES

3.1 *Minimum Design Loads for Buildings and Other Structures,* American Society of Civil Engineers, ASCE 7-88, ASCE, New York, 1988.

3.2 Masonry Standards Joint Committee, *Building Code Requirements for Concrete Masonry Structures, ACI 530-92/ASCE 5-92/TMS 402-92,* American Concrete Institute and American Society of Civil Engineers, New York, 1992.

3.3 W. C. Panarese, S. Kosmatka, and F. A. Randall, *Concrete Masonry Handbook for Architects, Engineers, Builders,* Portland Cement Association, Skokie, Illinois, 1991.

3.4 R. E. Englekirk, G. C. Hart, and Concrete Masonry Association of California and Nevada, ''Earthquake Design of Concrete Masonry Buildings,'' Vol. 1, Prentice Hall, Englewood Cliffs, 1982.

3.5 National Concrete Masonry Association, ''Concrete Masonry Warehouse Walls—How to Buy a Warehouse,'' NCMA-TEK Notes No. 32, NCMA, Herndon, Virginia, 1971.

3.6 National Concrete Masonry Association, ''Concrete Masonry Walls in Multi-Family Housing—An Investment Decision,'' NCMA-TEK Notes No. 29, NCMA, Herndon, Virginia, 1971.

## 3.10 PROBLEMS

**3.1** Discuss the different phases of the design process.

**3.2** What are the main design criteria for buildings?

**3.3** What design factors affect the wind pressure and earthquake forces on buildings?

**3.4** How do lateral wind or earthquake forces generate torsional moments on a building? Discuss features of design that can be used to minimize torsional effects.

**3.5** Propose what you consider to be the best wall arrangements for the apartment buildings with footprints as shown in Fig. P3.5. The maximum floor span is 20 ft (6 m) and the required wall-length-to-floor-area ratio is 0.3 in./ft$^2$ (80 mm/m$^2$) in each of the two orthogonal directions. Discuss your reasons for suggesting that these arrangements are the best solution to the combination of requirements.

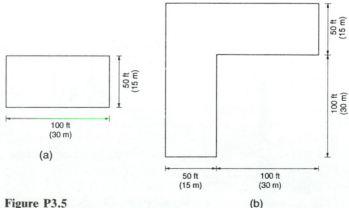

**Figure P3.5**

**3.6** For a 100 ft (30 m) high, 10-story loadbearing masonry building having the floor area shown in Fig. P3.5(a), determine the factor of safety against overturning due to an average lateral wind pressure of 25 lb/ft² (1.2 kN/m²). The average dead weight (including wall weights) is 100 lb/ft² (4.8 kN/m²) and the live load is 40 lb/ft² (1.9 kN/m²). For Seismic Zone 2, and using your local building code, compare this with overturning due to earthquake forces.

**3.7** Draw the envelope for shear and moment diagrams for the building shown in Fig. P3.7 due to N-S lateral loads. The average uniform wind pressure is 20 lb/ft² (1.0 kN/m²) and the base shear from earthquake loads is 500 kips (2240 kN).

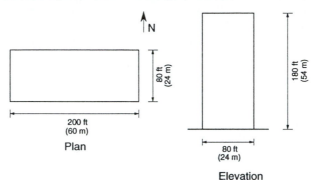

**Figure P3.7**

**3.8** Photograph or sketch a loadbearing masonry building (preferably multistory) in your locality and critically evaluate the design in terms of wall layout and aesthetics as discussed in this chapter.

**3.9** The preliminary design for a prefabricated, prestressed masonry wall element for building a group of high-quality low-cost small homes is shown in Fig. P3.9. They are C-shaped in

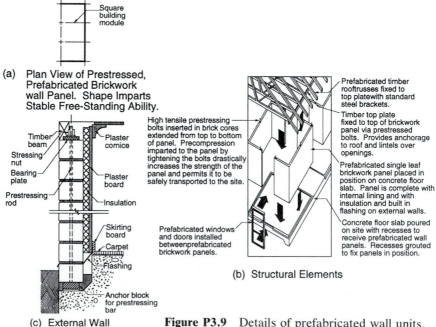

**Figure P3.9** Details of prefabricated wall units.

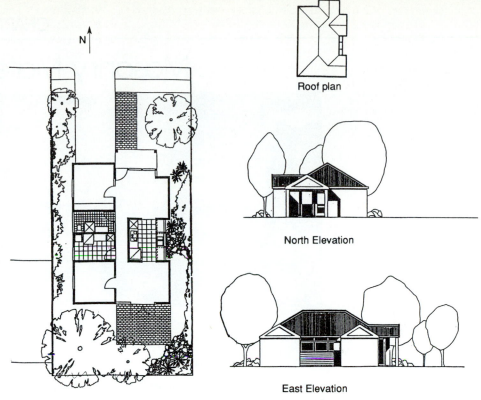

Roof plan

North Elevation

East Elevation

(d) House using prefabricated elements

**Figure P3.9** (*continued*)

plan, Fig. P3.9(a), and will be located and then grouted into preformed recesses in a concrete floor slab cast on the ground. A timber top plate ties the wall units together and provides attachment for prefabricated timber trusses to form the roof (Fig. P3.9(b)). External walls have insulation, plaster-board lining, and flashing incorporated with the prefabricated unit to give thermal resistance and prevent penetration of rain (Fig. P3.9(c)). One layout of the wall units to form a house is shown in Fig. P3.9(d).

**(a)** Check the masonry units available in your locality to see if they are suitable for this construction (aligning cores) and determine suitable dimensions of the prefabricated unit taking into account the modular dimensions of the masonry unit.

**(b)** Suggest ways in which services (water, electrical, telephone, sewage) can be integrated with the building shown in Fig. P3.9(d).

**(c)** Determine at least one alternative floor plan to that shown in Fig. P3.9(d) using prefabricated wall elements.

# CHAPTER 4

# Masonry Materials

Use of masonry materials in construction of a brick veneer wall with a concrete block backup wall.

## 4.1 INTRODUCTION

Masonry construction employs masonry units, mortar, grout, reinforcement, and a number of specialized materials such as flashing, dampproofing, and coatings. In this chapter, the characteristics of the individual materials are discussed with special emphasis on properties that influence strength and long-term performance. The interactions of combinations of masonry materials in masonry assemblages are presented in Chap. 5. Standards exist that govern the manufacture and use of masonry materials. References are made to relevant ASTM standards, and, in some cases, where substantial differences exist, comparisons to standards of other countries are made.

Masonry units can be made of clay or shale, concrete, calcium silicate (or sand lime), stone, and glass, but the most commonly available units are bricks and blocks made from clay or concrete. The first step in becoming knowledgeable about masonry is to understand the basic properties of the units. The need for determining the common material properties is briefly discussed in the next section. Subsequent sections provide background information for specific types of material.

## 4.2 COMMON PROPERTIES OF MASONRY UNITS

### 4.2.1 Description of Masonry Units

To aid in later discussions, it is convenient at this point to define some common terms used to describe masonry units. As shown in Fig. 4.1(a), most masonry units are rectangular-shaped and are laid with the long dimension (*length*) horizontal in the plane of the wall. The exposed vertical surface is the *face* of the unit and the *ends* are the vertical faces perpendicular to the plane of the wall. The length and height of the unit are usually standard for each type. For example, a concrete block unit has a standard actual length of $15\frac{3}{8}$ in. (390 mm) and a standard actual height of $7\frac{5}{8}$ in. (190 mm). The variable is the unit *width*, commonly referred to as size or thickness and is expressed as a *nominal dimension* (actual or *specified dimension* plus a joint thickness of typically $\frac{3}{8}$ in. (10 mm)). For example, a 6 in. (150 mm) concrete unit has an actual width of $5\frac{5}{8}$ in. (140 mm). The top and bottom horizontal surfaces that provide the bed for the mortar are known as *bedding areas*. Units laid with their length oriented along the wall are known as *stretchers*.

Traditionally, the properties of all masonry units were based on gross area, regardless of the degree to which voids had been incorporated in the units. However, increased sophistication in the structural design methods used to account for slender-

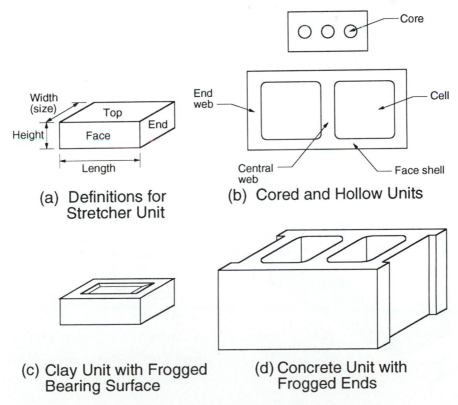

(a) **Definitions for Stretcher Unit**

(b) **Cored and Hollow Units**

(c) **Clay Unit with Frogged Bearing Surface**

(d) **Concrete Unit with Frogged Ends**

**Figure 4.1**  Shapes and terminology for masonry units.

ness and bending has resulted in use of the net area of hollow units as the area effective in resisting various forces. Because masonry is typically designed to resist vertical axial load, the *net area* is usually defined as the area of a horizontal section through the unit. This is sometimes expressed as the percent solid (100 x net area ÷ gross area) of the unit.

Voids, such as shown in Fig. 4.1(b), that extend through the full height of the unit are referred to as *cells* or *cores* whereas voids that extend only part way are called *frogs*. Cores are normally defined as having areas less than 1.5 in.² (960 mm²). As shown in Fig. 4.1(c), frogs in the bed joint can be shallow troughs or a series of conical shapes. In other cases, indentations over the full height of the ends of units are also known as frogs, and the unit is said to have *frogged ends*, such as shown in Fig. 4.1(d).

For a unit such as shown in Fig. 4.1(d), the parts on opposite faces separated by the cells are known as the *face shells* and the solid parts crossing between the face shells are called *webs*. In this illustration, there are two end webs and a center web. Extruded units tend to have uniform cross-sections over the height of the unit, whereas units formed in molds tend to have tapered cells or cores to aid in the demolding process. For hollow units, the webs and face shells can be flared out near the top to provide a larger surface for laying mortar and to create a better handhold for lifting the unit.

As illustrated in Fig. 4.2, small cores or cells near midthickness of units have very little influence on moment of inertia $I$ or section modulus $S$ for bending about the $x$ axis (out-of-plane bending). Therefore, units that are more than 75% solid are generally referred to as *solid units* with structural calculations and properties based on the dimensions of the gross section. As can be seen in Fig. 4.2, the reduction in moment of inertia is relatively small and is much less than the reduction in area. With unit strength based on gross area, this simplification is conservative but not highly wasteful for calculation of bending moment capacity. However, the "solid" terminology can sometimes create confusion where the intent is to refer to a unit without

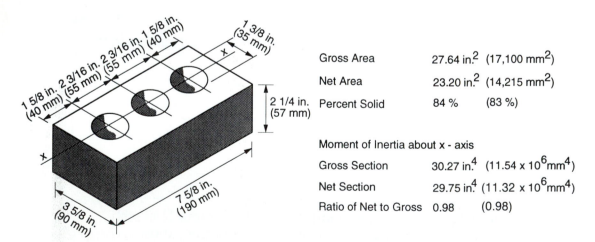

| | |
|---|---|
| Gross Area | 27.64 in.² (17,100 mm²) |
| Net Area | 23.20 in.² (14,215 mm²) |
| Percent Solid | 84 % (83 %) |

Moment of Inertia about x - axis

| | |
|---|---|
| Gross Section | 30.27 in.⁴ (11.54 x 10⁶mm⁴) |
| Net Section | 29.75 in.⁴ (11.32 x 10⁶mm⁴) |
| Ratio of Net to Gross | 0.98 (0.98) |

**Figure 4.2**  Effect of percent solid on section properties.

cores (i.e., 100% solid). Units less than 75% solid are called *hollow units* and calculations are based on the net area.

It is interesting to note that units can be classified as solid based on a 75% solid bedding area, but have a vertical cross-section much less than 75% solid either in the plane of the wall or through the thickness. This feature has some impact on the anisotropic behavior of masonry.

### 4.2.2 Properties of Masonry Units

**Compressive Strength.** Compressive strength of masonry units has long been used as a measure of quality and as a means for predicting other properties such as the compressive strength of masonry assemblages. Compression testing of individual units normally requires that the loaded faces be capped to reduce the effects of roughness and lack of plane surfaces on these faces. *Hard capping* using either a thin layer of molten sulfur compound or a gypsum plaster compound is normally required (ASTM C67 and C140).[4.1,4.2] Alternately, *soft capping* using fiberboard, plywood, or other relatively soft materials has been used. Soft capping has the advantage of decreasing the test preparation time and it is argued that it reduces the confining effects of the end platens of the test machine and therefore produces a result more representative of the actual properties of the unit.[4.3] Hard capping of concrete block generally produces higher compression strength results, as shown in Fig. 4.3, but no fixed relationships are yet available because of the influences of variable properties of the capping materials and the different effects depending upon the thickness of the capping material and failure stress. Soft capping that has become completely compressed at loads well below the failure load for the unit can provide a very "hard capping" condition at that stage.

The masonry unit's height to width ratio can lead to large differences in the apparent compressive strength of the material. This is mainly due to the relative influence of the confining effects of the loading platens of the test machine. Under a

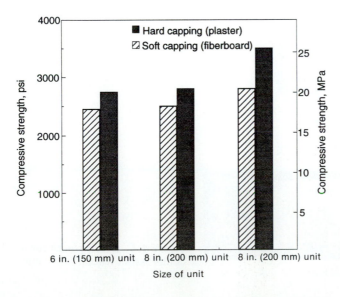

**Figure 4.3** Effect of capping material on unit compressive strength (from Ref. 4.4).

uniaxial compression load, materials tend to expand in the transverse directions (Poisson's effect). Where this expansion is restrained, transverse compressive confining stresses are built up, resulting in a triaxial compression stress state. As has been known for some time in concrete technology,[4.5] axial compressive strength increases and axial compressive strain decreases as a result of such confinement. As long as this is recognized, there is no harm in using such results as a quality control test and as standard measures of compressive strength. However, for research and other more accurate modeling, more representative measures of the compressive properties are desirable. Use of soft capping, teflon sheets, and greased platens have to various degrees reduced the amount of friction restraint at the loading platens. Another approach is to use brush platens such as shown in Fig. 4.4, where the steel cantilevers forming the fibers of the brush have very little resistance to lateral displacement and therefore do not significantly confine the unit as axial compression causes it to expand laterally.

Tests of small prisms or cores taken from masonry units normally provide different properties than tests of the units. The differences can generally be attributed to the reduced effects of platen restraint, elimination of unit geometric effects, and general size effects. Compressive strengths of units loaded normal to the ends generally differ from those loaded normal to the bedding area. In many cases, this is attributed to the different geometric and platen restraint effects. In other cases, the material may have different inherent properties as is the case for units of sedimentary stone.

**Modulus of Elasticity and Stress–Strain Relationship.**   Although modulus of elasticity values and stress–strain relationships for masonry assemblages are more useful for design calculations, these properties are often obtained for masonry units in order to provide more complete information on their characteristics. Such informa-

(a) Brush platen

(b) Failure of brick

**Figure 4.4**   Brush platen for axial compression testing of masonry units.

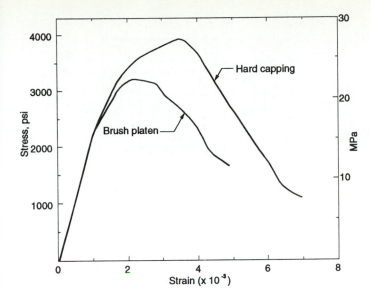

**Figure 4.5** Typical stress–strain curves for hollow concrete blocks (from Ref. 4.3).

tion tends to be mainly used in research, but other applications, such as for advanced analytical work, also exist. Because there is no standardized method for determining the modulus of elasticity for units, attention should be paid to the applicable stress range. Typically, a secant modulus of elasticity described by the slope of a line from zero stress to approximately 33% of the material strength is used.

Despite the fact that masonry materials tend to be thought of as brittle, nonlinear stress–strain curves, including a descending or strain softening branch, such as shown in Fig. 4.5, are usually found for concrete block units. Stress-strain curves with a sudden failure near the peak stress are generally the result of the uncontrolled release of energy stored in the specimen and test machine rather than being a characteristic of the material.

**Tensile Strength.** Knowledge of the tensile strength of masonry units can be important for the development of a proper understanding of failure mechanisms. As is discussed in more detail in Sec. 5.2.3, for masonry assemblages loaded under uniaxial compression, cracking parallel to the line of action of the load implies some influence of the tensile strength of the masonry units. In other situations, horizontal bending of a wall about a vertical axis can result in a flexural crack passing through the masonry units in alternate courses. For this case and the case where restraint to longitudinal shortening causes horizontal in-plane tension, the tensile strength of the units may control the capacity of unreinforced elements.

Direct tension tests can be performed on whole units or on coupons cut from the units. However, such tests are difficult to perform and produce quite variable results because of alignment complications in the test apparatus and stress concentrations associated with gripping the test specimen. Alternately, modulus of rupture (M.R.) tests[4.1] for the flexural strength based on linear elastic theory can be performed, as shown in Fig. 4.6(b), where an in-plane stress gradient is applied. As is the case for other materials, the modulus of rupture representing the flexural tensile strength

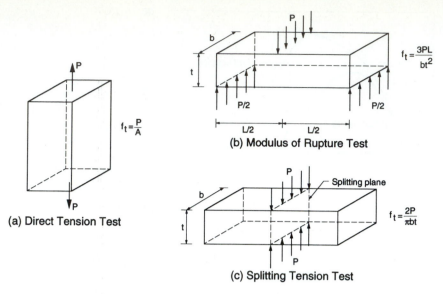

Figure 4.6  Tension test methods.

is found to be considerably higher than the direct tension result.[4.5] The reason for this difference is that the strain gradient results in a smaller fraction of the section being highly stressed.[4.6]

The foregoing two test methods take considerable time to perform properly and produce results with characteristically high variability. Therefore, splitting tension tests[4.6–4.8] across the thickness of the unit are often performed to obtain a comparative measure of tensile strength. The nearly constant tension developed over the central part of the unit height between the line loads is closer to a direct tension condition and the calculated average strengths tend to be close to but slightly higher than the direct tension results but with much less variability. Because no specimen preparation is involved other than placing the line loads on opposite faces, such tests are quite easy to perform.

**Weight, Density, Volume, and Area.**   The weight of masonry units can be determined simply by weighing the unit after it has been dried by a standardized procedure[4.1,4.2] to remove most of the free water that may be present. If the volume is well defined, such as for a rectangular, 100% solid unit, then the density of the material is simply the weight divided by the volume or, in metric units, the mass divided by the volume.

When cells, cores and frogs incorporate curved configurations that vary over the height of the unit, it is very difficult to obtain sufficiently accurate measurements of dimensions to accurately calculate volume. Therefore, volume can be experimentally determined. A reasonably accurate yet simple approach is to fill around the unit with lead shot in a container of known volume. The difference between the known volume and the volume of the lead shot is the volume of the unit. Alternately, immersion in water is theoretically a very accurate method in which a unit is immersed in room temperature water for 24 hours. The suspended immersed weight, F, is determined. The unit is then removed from the water and allowed to drain for 1 minute on a coarse wire mesh and visible surface water is removed with a damp cloth. The weight taken immediately at this stage is the wet weight E. Subsequently, the unit is dried in a ventilated oven at 212 to 239°F (100 to 115°C) for not less than 24 hours and until two successive weighings at intervals of 2 hours show an incremental loss not greater than 0.2% of the last previously determined weight. The weight at this point is the dry weight C. Then the density D of the material is

$$D = \frac{C}{E - F} \times D_{\text{water}} \qquad (4.1)$$

where $D_{\text{water}} = 62.4$ lb/ft$^3$ (1000 kg/m$^3$) and the net volume $A = C/D$ (ft$^3$ or m$^3$).

This test is relatively easy to perform, but can be quite inaccurate if water drains rapidly from the unit immediately after it is removed from the immersed condition. In such cases, the wet weight E is underestimated, resulting in an overestimation of the density.

For units with uniform voids, the net area of any section parallel to the bearing surface is constant over the height of the unit. When practical, the net area can be calculated from measured dimensions or by dividing the net volume $A$ by the unit height $H$. This calculation also gives the average net area for units with variable cross-section. As a result, it has become standard practice to describe the area of units by this average area based on volume. However, in some engineering calculations, it is necessary to know the minimum areas of the unit so that the effective mortared area, which in some cases may be only the face shells, can be used.

**Moisture Content and Absorption Properties.** The moisture content of a unit at any particular time can be expressed in absolute terms as simply the sample weight $W$ minus the weight after drying $C$ (lb or kg), or expressed per unit volume as (W-C) divided by the volume of the unit. Existing standards also place limits on the absorption of units. For different types of material, the purposes of these limits can be to minimize the potential for freeze–thaw damage, excessive volumetric change, or excessive permeability to water penetration. From the previously described tests involving immersing units in water for 24 hours, this property can be expressed either as

$$\text{Absorption (lb/ft}^3 \text{ or kg/m}^3) = \frac{E - C}{E - F} \times D_{\text{water}} \qquad (4.2)$$

or

$$\text{Absorption (\%)} = \frac{E - C}{C} \times 100 \qquad (4.3)$$

Moisture content is also expressed as a percent of the absorption as follows:

$$\text{Moisture content (\%)} = \frac{W - C}{E - C} \times 100 \qquad (4.4)$$

In addition to absorption, the rate of absorption can have an important effect on the interaction between freshly laid mortar and the masonry units. A standardized measure of this characteristic is the *initial rate of absorption* (IRA)[4.1] which is determined by measuring the weight of water absorbed by a unit in 1 minute when the unit is set into a $\frac{1}{8}$ in. (3 mm) depth of water. Because the standard clay brick had a cross-sectional area of about 30 in.$^2$, the units of measurement were traditionally established as grams/min./30 in.$^2$, but in the metric system, this was changed to kg/min./m$^2$.

**Volumetric Changes.** Dimensional change corresponding to temperature change is described by the *thermal coefficient of expansion* $\alpha_t$ which for most building materials is reasonably constant over the normal range of temperatures. It can be determined by measuring the length change corresponding to a change in temperature. It is important to ensure that the entire specimen has reached a uniform temperature. Also temperature effects on the mechanical or electrical instrumentation should be avoided.

By using demountable mechanical gages or large high-precision micrometers, length changes on masonry units can be used to document expansion or shrinkage properties. Expansion or shrinkage is usually expressed as strain by dividing the length change by the gage length. Similarly, creep strain under sustained constant load can be measured and is usually expressed as a multiple of the corresponding elastic strain. For example, a specimen loaded to an initial elastic strain of $100 \times 10^{-6}$ would undergo a further $200 \times 10^{-6}$ strain under sustained loading if the *creep coefficient* is 2.0. Various empirical formulas exist to describe the rate of creep over time. The relationship is generally fairly linear with respect to the logarithm of time.

In addition to accounting for the volumetric deformation of masonry units in design, the modifying effects of other materials such as mortar and grout should also be considered.

**Efflorescence.** Salts and other soluble materials often exist in the raw materials used to make masonry units, mortar, and grout. These salts may be carried to the surface of the masonry by water migrating through the masonry and then deposited there as solids as the water evaporates. This is known as efflorescence. When this only occurs in the initial stages of drying out of the wall, rain or mechanical spray may be successful in removing the resulting stains. However, persistent and extensive efflorescence results in staining that substantially detracts from the appearance of the masonry.

An alternate problem with migration of soluble salts toward the exposed surface of the masonry is that rapid drying can result in crystallization of salts within the body of the unit. The internal stresses resulting from this crystallization, can cause mechanical damage to the unit, leading to spalling and cracking in extreme cases.

The performance history for a particular type of unit is likely the best test for possible efflorescence problems. However, the source of the salts can be determined by standardized laboratory testing.[4.1] One method is to brush adhering dirt from units and dry them. One of a pair of units is partially immersed in distilled water to a depth of approximately 1 in. (25 mm) for 7 days in a drying room and the other is stored in a drying room without contact with water. After drying the pair of specimens in an oven for 24 hours, the surfaces are compared. Any noted difference indicates existence of efflorescence.

**Durability.** Adequate performance of a particular type of unit under similar conditions of use is the best indicator of satisfactory durability. However, some tests such as the freeze-thaw test[4.1] can be performed to give reasonable confidence that durability problems are unlikely to occur. However, even products that have excellent durability histories can fail under extreme conditions. For example, saturated units that pass 50 or 100 freeze–thaw cycles may fail after 1000 cycles. In many such cases, assurance of adequate durability is more a matter of proper design, detailing and construction (see Chaps. 12, 14, and 15) than the choice of unit.

## 4.3 CLAY MASONRY UNITS

Clay brick is the most extensively used type of masonry unit throughout the world. Their wide use is based on economic advantages that are closely tied to the fact that clays and shales are widely available both in developed and developing countries. The historic record of durability and maintenance-free aesthetically pleasing appearance is also important in this continuing widespread use. Fired clay products derive their strength from a ceramic fusion process that takes place at high firing temperatures. Clay brick, tile, and larger hollow units are employed in a wide variety of loadbearing and nonloadbearing applications.

### 4.3.1 Manufacture

**Raw Materials.** Clay masonry units actually can be made from surface clays that are upthrusts of older deposits, recent sedimentary formations, or shales formed from clays under high pressures. A third material, known as fire clay because of its refractory qualities, is mined at deeper levels and tends to have more uniform chemical and physical properties and fewer impurities than surface clays or shales. All three raw materials have similar compositions of silica and alumina compounds with different types and amounts of metallic oxides and other impurities. Metallic oxides, although technically impurities, act as fluxes to promote fusion at lower temperatures as well as affect the final color.

Because the raw material is variable, it is often necessary to minimize this variability by mixing materials from different locations in the pit and blending materials from different clay sources. Raw materials are usually put through a primary crusher to remove large lumps prior to storage. During preparation of the clay or shale for use, additional crushing is used to break up lumps and stones, after which the material is ground using huge grinding wheels.

**Manufacturing Process.** The basic manufacturing procedure has not changed much over the past two centuries. Changes have mainly involved better control over raw materials, better control of firing, and more complete mechanization and automation of manufacturing plants.

The three basic processes for forming clay units differ in the moisture content of the clay material after it has been converted into a homogeneous plastic mass, usually by the addition of water in a pug mill where blades on revolving shafts provide thorough mixing.

In developed countries, the *stiff-mud process*, involving extrusion of brick and tile units, represents the dominant method of manufacture. The water content of the mix is typically about 12 to 15% of the dry weight of the material, but only enough water is added to provide plasticity. In a fully mechanized extrusion process, an auger or screw, turning in a drum, pushes the plastic clay from the inlet end of the machine through a shredder plate to form small sections of uniform size. These sections fall through a vacuum chamber where the entrapped air is removed from the clay. A second auger in the bottom of the chamber carries the clay into a cylinder that tapers at the exit end to a rectangular die.

The reduction in cross-section of the cylinder results in increased pressure as the clay moves toward the die, and individual sections of plastic clay are pressed together to form a dense mass. If the brick are to have holes (cores or cells) in them, as most extruded brick do, an arrangement of hardened metal cores, called bridges, are suspended in the clay column. The extruded column of clay or *slug*, as it is sometimes called, then emerges from the die, as illustrated in Fig. 4.7. The smooth texture produced by the die can be altered by attachments that cut, rake, brush, or otherwise roughen the top and sides of the slug to form a variety of textures, as shown in Fig. 4.8. Coloring agents and other surface treatments such as sand can be sprayed on at this stage. A series of wires slices the slugs into individual units, as shown in Fig. 4.9. Next the brick are conveyed to an automatic setting machine for placement into kiln cars.

The *soft-mud* process is used to produce pressed brick and is suited to situations where the moisture content of the clay is high (usually in the order of 20 to 30% water

**Figure 4.7**   Column of clay or slug after extrusion from the die. (*Courtesy of Brick Institute of America.*)

**Figure 4.8** Brick textures (*Courtesy of Brick Institute of America*).

by weight). The units are formed in molds "lubricated" with fine sand or water to prevent the clay from sticking to the molds. Hand molding of a soft-mud mix is the oldest method of producing bricks, and in some developing countries is still in use, as shown in Fig. 4.10. In this case, wooden molds are used to form brick that are immediately demolded and placed in the sun to dry. In certain regions of developed countries, soft-mud machines are the most cost-effective manufacturing method to produce a molded brick.

**Figure 4.9** Extruded slugs sliced into units by a wire cutter. (*Courtesy of Brick Institute of America*.)

Sec. 4.3    Clay Masonry Units

**Figure 4.10** Hand-mold brickmaking in India (*Courtesy of Gary T. Suter*).

At the other extreme of the manufacturing processes is the *dry-press* process, which was developed for clays having relatively poor plastic qualities. With water content generally less than 10% by weight, the very stiff material is pressed into steel molds under pressures ranging from 500 to 1500 psi (3.45 to 10.35 MPa). This process is not often used to manufacture clay units today. However, some brick manufacturers do produce a dry-press brick on demand.

Before firing, the brick must be carefully dried to remove excess water. Drying normally takes from 24 to 48 hours at temperatures ranging from 100 to 400°F (38 to 151°C). In modern plants, firing is typically done in a tunnel kiln (see Fig. 4.11(a)) where the process may take 60 to 80 hours and the clay units to be fired are exposed to peak temperatures ranging from about 1700 to 2400°F (930 to 1320°C). The firing time in the kiln and the peak temperatures vary according to the properties of the plastic clay. At the peak temperature, a ceramic fusion process, called *vitrification*, takes place. Proper vitrification ensures a strong and durable brick. If lower or higher peak temperatures, than those required for proper vitrification, are reached, the brick are underfired or overfired. Such brick, though still fairly strong, may exhibit serious durability problems after some time in exposed exterior conditions.

The tunnel kiln provides a continuous firing process whereas older methods of firing, such as the beehive kiln shown in Fig. 4.11(b), involved periodic methods. Differences in temperature over the height of the beehive kiln and variations through the stacked units on the cart result in significant differences in color and mechanical properties of the units. For extruded clay brick, the presence of cores reduces cracking during drying and firing and allows more uniform firing of the clay body.

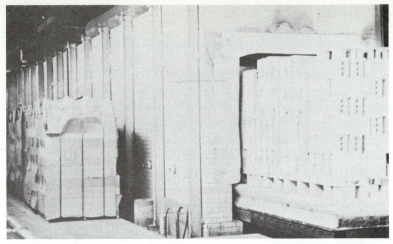

(a) Modern tunnel kiln

(b) Bee-hive kiln

**Figure 4.11**  Methods of firing clay brick. (*Courtesy of Brick Institute of America*).

When the newly fired brick emerge from the cooling phase at a temperature of about 194 to 212°F (90 to 100°C), they are bone dry. Exposure to the humidity of the environment allows the dry brick to absorb some moisture. Just as a dry sponge expands as it is wetted, the brick expand slightly with time. This so-called *moisture expansion* of clay masonry products is an important deformation characteristic that a designer must take into account. It will be treated in more detail in Sec. 4.3.10 and in Chaps. 12 and 15.

Raw materials and the manufacturing process affect the density of clay brick. The density of solid clay brick ranges between 80 to 140 lb/ft$^3$ (1300 to 2200 kg/m$^3$) with 120 lb/ft$^3$ (1950 kg/m$^3$) normally used for structural design purposes.

### 4.3.2 Grades

To ensure durability of brick for various applications, material standards typically categorize building brick into a number of grades. The three grades of building brick, SW, MW, and NW, are covered by ASTM C62.[4.9] The two grades, SW and MW, are specified in ASTM C652[4.10] for hollow brick. SW brick exhibits a high degree of resistance to frost action even when the unit is permeated with water. MW brick is used where a moderate degree of weathering resistance is required and where the unit is unlikely to be exposed to frost action when permeated with water. Typically, the use of NW brick is restricted to interior applications where exposure to weather is not an issue. For the United States, the brick standards[4.9, 4.10] contain a weathering index map and define a weathering index, which for any locality is the product of the average annual number of freezing-cycle days and the average winter rainfall in inches. The effect of weathering on brick durability and brick grades is related to the weathering index.

The grades of brick are established based on compressive strength, absorption characteristics, and saturation coefficient. Freeze–thaw testing is an option for added assurance of durability or as an alternative where the brick does not meet the minimum requirements. This topic is covered in more detail in Sec. 4.3.7.

Definitions of brick durability in other countries may be affected by factors other than frost action. For instance, Australian practice defines five exposure classifications that the masonry units and the mortar, both separately and in their interaction with each other, must be able to meet. The exposure classifications in the arid, moderate, and tropical climatic zones of Australia emphasize exposure to industrial atmospheres or salt water coastal spray zones.

### 4.3.3 Sizes and Shapes

Brick are available in many sizes based generally on a 4 x 4 in. (100 x 100 mm) module that is part of the internationally accepted planning grid of 24 x 24 in. (600 x 600 mm). A few examples of brick types are shown in Fig. 4.12. As indicated, with allowance for manufacturing tolerances, specified dimensions represent the actual dimensions which differ from the nominal dimensions by the thickness of a standard $\frac{3}{8}$ in. (10 mm) mortar joint. The dimensions are referred to as *modular dimensions* if the units, including the mortar joints, fit into the planning grid, as illustrated in Fig. 4.13. For instance, for the standard modular brick, three courses usually result in an 8 in. (200 mm) height, and because concrete block units comply with the same planning grid, it is easy to align courses when constructing a multiwythe wall. The modular planning grid facilitates interfacing with window and door openings and many other construction requirements.

*Hollow* brick, as defined by ASTM C652,[4.10] are units whose net cross-sectional area in every plane parallel to the bearing surface is between 40 and 75% of its gross cross-sectional area measured in the same plane. The effect of the reduction in bearing area is significant enough that the properties of the net cross-sectional area are used in stress calculations. As indicated in Fig. 4.12, coring patterns vary widely.

The development of larger-sized units containing a greater percentage of voids has led to increased economy in masonry construction over the past few decades.

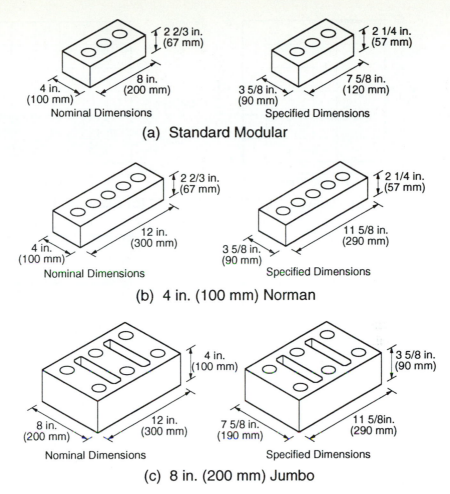

**Figure 4.12** Nominal and specified dimensions for three typical clay bricks.

Larger face dimensions require fewer units to achieve a given wall area, yet the units are not heavier because of the increased volume of cells. The presence of two or three large cells enables easy placement of reinforcing steel and grouting. Figure 4.14(a) shows a typical hollow unit. Clay tile units are manufactured from the same materials and by the same manufacturing processes as clay brick, but differ from hollow brick units in that thinner webs and face shells and lower compressive strengths are allowed. Figure 4.14(b) shows an example cell configuration for load-bearing structural clay tile. The standard specifications for loadbearing tiles are covered by ASTM C34.[4.11] Specifications for nonloadbearing tile and ceramic glazed units are covered, respectively, in ASTM C56[4.12] and ASTM C126.[4.13]

Fired clay units vary in size as a combined effect of initial shrinkage due to drying and additional shrinkage during the firing process with the total varying between roughly 5 to 15%. Therefore, dimensional tolerances and allowances for warping of the fired clay units are specified in standards according to different classifica-

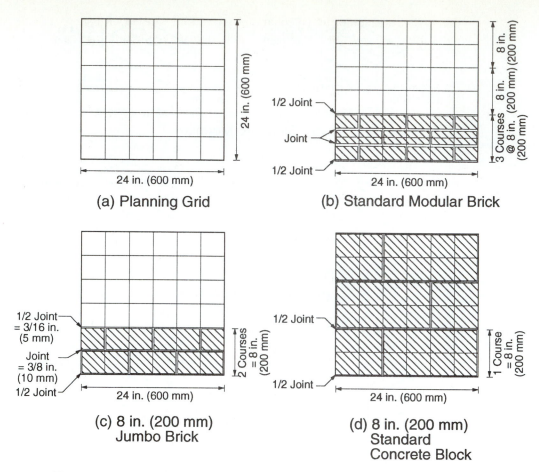

**Figure 4.13** Coursing examples of masonry units to fit into a modular planning grid.

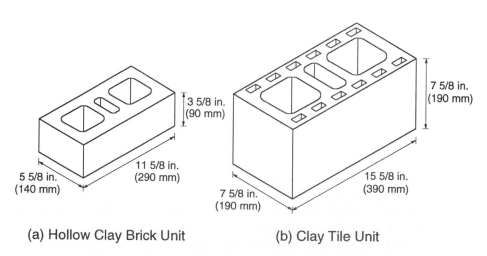

**Figure 4.14** Examples of hollow clay brick and tile units.

tions in terms of consistency and mechanical perfection required. Ranges of amounts of chippage and variations in texture and color are also included in the specifications for these classifications.

### 4.3.4 Compressive Strength

Modern clay units often have compressive strengths much higher than required to satisfy required product specifications and generally exceed by large margins the requirements for member design strengths. However, because performance records generally show that durability is linked to unit compressive strength, the latter property is very important from a serviceability as well as from an overall strength point of view.

For solid and hollow brick, the ASTM standards require the minimum compressive strengths given in Table 4.1 for different grades. The strength requirements for structural clay loadbearing wall tile are presented in Table 4.2. Five half-brick specimens must be tested flatwise to give a reasonable measure of the average strength and also to provide an indication of variability. Figure 4.15 shows a half brick capped ready for testing and the compressive failure after testing. The typical failure mode involves lateral splitting of the brick into a number of "columns" which in turn crush.

Tests[4.14] of cores drilled from clay units in the three orthogonal directions resulted in different compressive strengths. This indicates material anisotropy, which may be attributed to nonuniform firing. As discussed in Sec. 4.2.2, platen restraint, influence of capping materials, and the aspect ratio of the unit can all affect the measured compressive strength. What is important to realize is that this "standard" measure of strength is useful for ensuring conformance with properties that have been previously established on this basis.

The measured compressive strength of clay brick can vary from 3000 to 21,000 psi (20 to 145 MPa), depending on such factors as the constituents of the plastic clay, firing conditions, coring pattern, and the size and shape of units. In North America,

**Table 4.1** Physical Requirements for Solid and Hollow Brick (from Refs. 4.9 and 4.10)

| Brick grade | Minimum Compressive strength *, psi (MPa) | | Maximum water absorption by 5-h boiling (%) | | Maximum saturation coefficient (C/B) | |
|---|---|---|---|---|---|---|
| | Average of 5 bricks | Individual | Average of 5 bricks | Individual | Average of 5 bricks | Individual |
| SW | 3000 (20.7) | 2500 (17.2) | 17.0 | 20.0 | 0.78 | 0.80 |
| MW | 2500 (17.2) | 2200 (15.2) | 22.0 | 25.0 | 0.88 | 0.90 |
| NW** | 1500 (10.3) | 1250 ( 8.6) | No limit | No limit | No limit | No limit |

\*   Based on gross area for solid units and net area for hollow and face units.
\*\*  Not applicable to hollow or facing brick.

**Table 4.2** Physical Requirements for Clay Load-Bearing Wall Tile (from Ref. 4.11)

| Grade | Minimum compressive strength,* psi (MPa) | | | | Maximum water adsorption by 1-h boiling | |
|---|---|---|---|---|---|---|
| | End construction tile | | Side construction tile | | | |
| | Average of 5 specimens | Individual | Average of 5 specimens | Individual | Average of 5 specimens | Individual |
| LBX | 1400 (9.6) | 1000 (6.8) | 700 (4.8) | 500 (3.4) | 16 | 19 |
| LB | 1000 (6.8) | 700 (4.8) | 700 (4.8) | 500 (3.4) | 25 | 28 |

\* Based on gross area.

most units are in the 6000-10,000 psi (40-70 MPa) range. Tests by Hilsdorf[4.15] indicate increased moduli of elasticity with increased brick strengths, which, as is discussed in Chap. 5, can affect the failure mechanism for clay masonry assemblages under compressive load.

### 4.3.5 Tensile Strength

Extensive testing over many years has shown that tensile strength, as measured by the modulus of rupture (See ASTM C67[4.1]) or splitting tension (ASTM C1006[4.8]), increases with brick compressive strength, but is generally a smaller fraction of the compressive strength at higher strengths. According to Sahlin's review of test data[4.16], the ratio of modulus of rupture to compressive strength varies between about 0.10 and 0.32. Figure 4.16 contains data for tensile strength to compressive strength

**Figure 4.15** Clay brick compression test specimen and failure mode. (*Courtesy of Gary T. Suter.*)

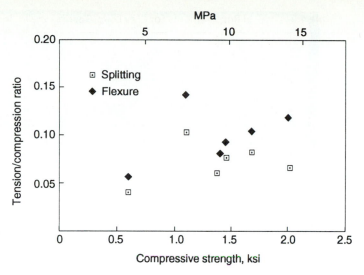

**Figure 4.16** Tensile strength of clay masonry units.

ratio for tension determined by either flexure (modulus of rupture) or splitting tension. As can be seen, the modulus of rupture values are 20 to 50% higher than the values obtained from splitting tension tests.

### 4.3.6 Absorption Properties

**Initial Rate of Absorption.** The absorption of moisture by capillary action in the unit produces a *suction* effect that draws water from mortar and grout. As described in Sec. 4.2.2, this characteristic is defined by the initial rate of absorption (IRA).[4.1] Too high or too low an IRA can be detrimental to achieving a good bond due to failure to draw mortar into sufficiently intimate contact on the one extreme versus formation of a weak dry layer of mortar at the surface of the brick at the other extreme. Figure 4.17 illustrates suction taking place in clay masonry construction. IRA values generally range between 5 and 40 g/min./30 in.$^2$ (0.25 to 2.05 kg/min./m$^2$).

If the IRA value is low, say, less than about 5 g/min./30 in.$^2$ (0.25 kg/min./m$^2$), one speaks of low-absorption or low-suction units. Such units do not absorb much water from the mortar and may tend to float on the mortar, particularly in cold weather or if the brick is damp. This can produce a poor initial and final bond, not only affecting the masonry's flexural strength, but also its water tightness and hence its durability. Construction with low-suction units typically requires adjustment in the mortar to achieve a lower water retentivity (see Sec. 4.7.3), which generally results in a drier, harsher mortar consistency.

If the IRA value is high, say, in excess of 30 g/min./30 in.$^2$ (1.5 kg/min./m$^2$), the unit is considered highly absorptive. This may result in poor bond if a dry thin layer of mortar next to the unit occurs when too much moisture is drawn from the mortar. Also, the mortar may stiffen so rapidly that there is difficulty with proper setting of the units. To overcome this problem in practice, units with IRA values

**Figure 4.17**  Clay unit suction.

exceeding about 30-40 g/min./30 in.$^2$ (1.5-2.05 kg/min./m$^2$) typically are wetted prior to laying. This will decrease their absorption of water during construction, but has the handicap that compatibility with mortar can be quite variable because of variable degrees of wetness. Therefore, wetted units should be surface dry when laid. Wet bricks can also be hard on the mason's hands.

Masonry units that are hot because of exposure to solar radiation will absorb water much more rapidly than is indicated by the IRA test. In hot weather, keeping such units in the shade or wetting may be necessary.

The British Code[4.17] relates tensile bond strengths to specific absorption characteristics, but for wide ranges of brick and mortar, unique relationships between IRA and mortar bond do not seem to exist.[4.18] However, tests have indicated that IRA values between 5 and 30 g/min./30 in.$^2$ (0.25 to 1.5 kg/min./m$^2$) generally produce good bond strength with compatible mortar. Fortunately, many modern brick and tile units exhibit IRA values in this range.

**Water Absorption and *C/B* Ratio.**  Two *submersion* tests, namely, the 5 and 24 hour tests are of interest. The 24 hour test involves determining the absorption of a unit when submerged in water at room temperature, referred to as cold water absorption, *C*. In the 5 h water absorption test, the same specimen, now partially saturated from the previous cold water submersion test, is immersed in boiling water for 5 h and its boiling water (or *B*) absorption is determined. The first test measures the relatively freely absorbed moisture and the second provides an indication of the additional pore space available for water in a more extreme moisture environment involving a higher temperature and some pressure. The ratio of the two values (*C* ÷ *B*) is known as the *saturation coefficient*, or *C/B* ratio. The *C/B* ratio is a measure of the easily filled to total fillable pore space and is an indication of the relative volume of open pore space present in the unit after free absorption has taken place. The presence of an adequate volume of open pore space is important during freeze–thaw action to accommodate the volume change in water as it freezes. Otherwise, the unit can be damaged by the buildup of internal pressures and eventually disintegrate. Lower $C/B$ ratios, therefore, are specified for greater durability. In addition, to help ensure satisfactory durability, there are limits on the total pore space, as indicated by the 5 h boiling absorption. ASTM requirements for water absorption and saturation coefficient are given in Tables 4.1 and 4.2.

### 4.3.7 Durability

The ASTM strength and absorption requirements for brick (ASTM C62[4.9]) and tile (ASTM C34[4.11]) are reproduced in Tables 4.1 and 4.2. The strength requirements are discussed in Sec. 4.3.4. Of particular interest are the saturation coefficient or $C/B$ ratio limits. The most durable brick, grade SW, may not exceed 0.78 based on the average of five tests. For MW brick, the $C/B$ requirement is 0.88. The saturation coefficient limit, in fact, can be waived if some other condition is met such as units exhibiting high compressive strength, say, in excess of 8000 psi (56 MPa)[4.19]; relatively low 24 h cold water absorption, say, less than 8%; or satisfactory performance in a standard freeze–thaw test involving perhaps 50 cycles.

Figure 4.18 illustrates a case of spalling brick due to freeze–thaw action in a severe weathering region. Although the brick was correctly specified as grade SW and met the physical requirements listed in Table 4.1, it is obviously not sufficiently durable. Such a case is relatively rare in practice, but it does sometimes occur. What are the reasons for the lack of performance? Brick durability experts today generally agree that the ASTM requirements and especially the saturation coefficient are semi-empirical and as such cannot be the sole measures of clay brick durability for the variety of brick encountered. Research indicates that besides the total porosity of a brick, the pore size distribution may be important in determining the potential frost resistance of brick.[4.20] Studies cited in the reference show good frost resistance when a large percentage of the pores exceeds 3 $\mu$m in size. It appears, therefore, that the saturation-coefficient approach, which relies solely on porosity, is not completely

**Figure 4.18** Spalling failure of clay brick from freeze–thaw action.

satisfactory and that future standards may also have to include tests for pore size. A second potential reason for lack of performance may lie in the fact that brick can be exposed to hundreds of freeze–thaw cycles over a number of years, whereas typical laboratory tests terminate after only 50 freeze–thaw cycles. Where brick will be exposed to particularly severe weathering conditions, it is also good practice to choose a brick that has performed well under similar environmental conditions. Detailing to minimize the frequency of saturation is also an effective approach to improved durability.

### 4.3.8 Thermal Movement

The *coefficients of thermal expansion* for brick generally range between $2.5 \times 10^{-6}$ and $4.0 \times 10^{-6}$ in./in./°F (0.0045 and 0.0072 mm/m/°C). Lower values pertain to units produced from fire clays, intermediate values to units manufactured from shales, and higher values to units produced from so-called surface clays. Summarizing research from a variety of sources, Grimm[4.21] reports coefficient of expansion values varying between $1.7 \times 10^{-6}$ and $6.7 \times 10^{-6}$ in./in./°F (3.1 and 12.4 $\mu$m/m/°C). He also found that brick exhibit up to a 22% increase for thermal expansion in the vertical direction as compared to the horizontal direction.

### 4.3.9 Moisture Expansion

Immediately after firing, clay products begin to absorb moisture from the environment. This moisture absorption causes complex chemical reactions within the vitrified clay itself, which lead to a slow irreversible *moisture expansion*. Typical plots of moisture expansion versus time for clay and shale bricks are shown in Fig. 4.19. Although Ritchie[4.22] determined expansions of from about 0.016 to 0.028% after approximately 500 days for Canadian bricks, testing and field evidence indicate that expansion continues very slowly even after many years. For use in practice, it is reasonable to assume that a linear relationship exists between expansion and the logarithm of time.[4.23] This means that about 50% of the 5 year growth occurs in the first 6 months and it would take 500 years for the 5 year growth to double.

In Australia, clay brick expansion has been classified[4.23] as

| | |
|---|---|
| low: | up to 0.06% |
| medium: | up to 0.12% |
| high: | up to 0.18% |

Clay brick in North America typically fall more into the low expansion category whereas, Australian brick usually exhibit low to medium expansions. Raw materials, firing temperature, and firing time affect the amount of moisture expansion. As is discussed in Chaps. 12 and 15, long or high walls are designed with movement joints so that stresses due to moisture and thermal expansions cannot build up to cause distress.

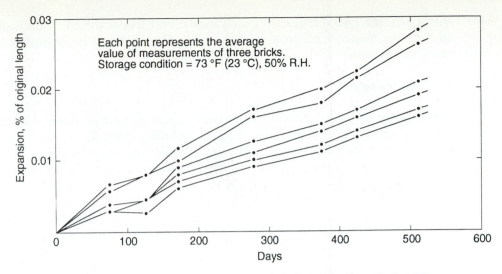

**Figure 4.19**  Exapansion of clay and shale bricks (from Ref. 4.22).

### 4.3.10 Creep

Although creep in clay brick has been reported in the literature, its magnitude at stress levels normally encountered in practice is so small that it can usually be disregarded. Clay brick masonry can exhibit significant creep deformations under high stress levels largely due to creep of the mortar in the joints. Such creep will be dealt with in Chap. 5.

### 4.3.11 Freezing Expansion

The very limited test results obtained to date indicate that when saturated clay brick freezes, expansion may take place.[4.24] Palmer[4.25] reported a mean value of freezing expansion for stiff-mud bricks of 0.000067 in./in. (0.07 mm/m) with a standard deviation of 0.000027 in./in. (0.03 mm/m).

## 4.4 CONCRETE MASONRY UNITS

The most common concrete masonry products are block and brick. Just as concrete has evolved as a modern construction material over the past 100 years, the history of concrete masonry products largely dates from this century; concrete brick, in fact, have been extensively used in North America since perhaps only the 1960s. Because the binder in concrete products is portland cement, units derive their strength from the cement hydration process and much of concrete technology is applicable. In

North America, concrete block is widely used in both loadbearing and nonloadbearing applications, whereas concrete brick is commonly employed in nonloadbearing veneers and as pavers.

### 4.4.1 Manufacture

Concrete masonry products are basically manufactured from portland cement, aggregate, and water, but other blended cements including, blast-furnace cement, and fly ash, and other inert fillers, are sometimes added. Additives such as air entraining agents, pozzolanic materials, other workability agents, and coloring pigments also can be included.

Modern mass production is carried out in highly automated plants. In the manufacturing process, a very dry zero-slump concrete is delivered to a block machine. The material is pressed and/or vibrated into steel molds having the shape of the desired unit. Molding and compacting three or more units at one time, automatic machines can produce 1,000 or more 8 x 8 x 16 in. (200 x 200 x 400 mm) blocks per hour. Figure 4.20 shows the type of modern automatic block machine used in many parts of the world.

In the block machine, high productivity requires that the sides and internal parts of the mold be removed immediately after the concrete is consolidated. Therefore, the consolidated mix must have sufficient strength and stiffness to withstand this

**Figure 4.20** Modern concrete block machine. (*Courtesy of National Concrete Masonry Association.*)

**Figure 4.21** Semi-mechanized block machine in a developing country. (*Courtesy of S. Prasetyo.*)

stage and transportation to the curing stage without slumping, crumbling, or cracking. Especially in the more rural environments of many developing countries, such a mass production facility is not economically viable; instead, manual compacting of concrete in steel molds or use of partly mechanized plant facilities can produce concrete products that meet local demands. Figure 4.21 shows a power-driven single block making machine in Indonesia where the concrete mix is manually fed into the hopper and molded units are carried by hand to the curing and drying area.

The hydration or hardening process of concrete blocks in modern plants is achieved either with accelerated curing: by steam curing at atmospheric pressure for periods up to 18 hours (low pressure curing) or by autoclaving. Autoclaving involves high pressure steam curing for periods ranging from 4 to 12 hours in cylindrical pressure vessels as shown in Fig. 4.22. Whereas low pressure steam-cured blocks require a drying period to reach specified moisture contents, autoclaved blocks typically do not. Mix designs vary with large differences in fines, density of aggregate, and amount of cement. Some blocks have a very porous open structure and others display a smooth, dense finish.

Automatic stacking of units on pallets and, when required, wrapping with plastic or other protective membranes are also part of the operation in a modern plant.

Concrete brick can be made using the extrusion process in much the same manner as clay brick except that cores are not usually included. Surface texture, wire

**Figure 4.22** Concrete blocks stacked in an autoclave for high-pressure steam curing. (*Courtesy of Gary T. Suter.*)

cutting, and stacking the units for curing follow the same general approach. Pressed concrete bricks often have a frogged bearing surface.

Concrete masonry units have a cement-grey color. However, a variety of unit colors can be produced during the manufacturing process through the addition of the proper color pigment. Different surface textures can be produced by using different types of aggregates and also by sandblasting. Split-face blocks, as shown in Fig. 4.23, can be obtained by splitting apart pairs of blocks at the middle.

### 4.4.2 Grades, Types, and Density

Classification of concrete units is covered by ASTM Standards C90,[4.26] C129[4.27] and C145[4.28] for block and C55[4.29] for brick. The two *grades* for both block and brick are N and S. Grade N is used where higher strength and resistance to moisture penetration and to severe frost action are required, whereas grade S can be used where moderate strength and moderate resistance to frost action and moisture penetration are required. Grades are not assigned to nonloadbearing concrete units. Table 4.3 contains the listing of these grades for the different types of units.

Two *types* of brick and block exist for each grade. Type I is a moisture controlled unit, whereas type II is not a moisture controlled unit. Table 4.4 lists moisture

**Figure 4.23** Split-face block.

**Table 4.3**  Strength and Absorption Requirements for Concrete Masonry Units

| Type of unit | ASTM designation | Grade of unit | Minimum compressive strength, psi (MPa), on average gross area | | Maximum water absorption, lb/ft³ (kg/m³), based on oven-dry unit weight | | | |
|---|---|---|---|---|---|---|---|---|
| | | | Average of 3 units | Individual unit | Lightweight concrete | | Medium-weight concrete, 105 to 125 (1680 to 2000) | Normal weight concrete, 125 (2000) or more |
| | | | | | Less than 85 (1360) | Less than 105 (1680) | | |
| Concrete brick | C55 | N | 3500* (24.1)* | 3000* (20.7)* | 15 (240) | 15 (240) | 13 (208) | 10 (160) |
| | | S | 2500* (17.3)* | 2000* (13.8)* | 18 (288) | 18 (288) | 15 (240) | 13 (208) |
| Solid loadbearing block | C145 | N | 1800 (12.4) | 1500 (10.3) | – | 18 (288) | 15 (240) | 13 (208) |
| | | S | 1200 (8.3) | 1000 (6.9) | 20 (320) | – | – | – |
| Hollow loadbearing block | C90 | N | 1000 (6.9) | 800 (5.5) | – | 18 (288) | 15 (240) | 13 (208) |
| | | S | 700 (4.8) | 600 (4.1) | 20 (320) | – | – | – |
| Hollow non-loadbearing block | C129 | – | 600 (4.1) | 500 (3.5) | – | – | – | – |

* Concrete brick tested flatwise.

content limits. Moisture content control is required because concrete masonry products shrink and excessive shrinkage can lead to cracking, particularly in mortar joints. The requirements of Table 4.4 are set forth on the basis that, although a greater moisture content will cause increased shrinkage, more humid environments will reduce overall shrinkage effects because less moisture will have to be expelled by the unit to reach moisture equilibrium.

The *density* of concrete block can be controlled by the use of either lightweight aggregates or foaming agents. Accordingly, the three densities or weight classifications used in ASTM standards are:

normal weight: $> 125$ lb/ft$^3$ ($> 2000$ kg/m$^3$)

medium weight: 105 to 125 lb/ft$^3$ (1680 to 2000 kg/m$^3$)

lightweight: $< 105$ lb/ft$^3$ ($< 1680$ kg/m$^3$)

$< 85$ lb/ft$^3$ ($< 1360$ kg/m$^3$)

Concrete masonry products are defined as *solid* or *hollow*, depending on whether they contain more or less than 75% net solid horizontal cross-sectional area. Choice of hollow or solid, loadbearing or nonloadbearing, grade N or S, type I or II, weight classification, and moisture content all are important design decisions. In practice, hollow block are most frequently used because of their reduced weight, ease of handling, ease of reinforcing, and overall economy. The percent solid typically is in the range from 50 to 60% and hence the void areas represent 50 to 40%.

### 4.4.3 Sizes and Shapes

Concrete masonry products come in a large variety of sizes and shapes that fit into the same modular planning grid discussed for clay masonry products in connection with Fig. 4.13. The most common concrete block has a nominal size of 8 x 8 x 16 in. (200 x 200 x 400 mm); for convenience, this size is referred to as the *standard* block. Block is available in nominal widths of 4 to 12 in. (100 to 300 mm) in 2 in. (50 mm) increments. Although the typical block height is 8 in. (200 mm), 4 in. (100 mm) high units can be obtained by special order or by cutting units. (Recall that specified block dimensions are $\frac{3}{8}$ in. (10 mm) less than the nominal values to allow for a standard mortar joint thickness.) Two- and three-cell units are both common. As shown in Fig. 4.24, the cells of hollow units are tapered and some molds also introduce flared webs and face shells. This facilitates stripping the molds and aids gripping the block during laying. The increased top area also is beneficial for mortar bedding. For loadbearing units, ASTM C90[4.26] requires the minimum face shell and web thicknesses given in Table 4.5.

Some typical concrete masonry units are illustrated in Fig. 4.25: (a) to (d) are used in wall construction; (e) and (f) in flexural members, such as bond beams, lintels, and the bottom course of other beams; (g to i) for reinforced masonry, where the open end facilitates placing around vertical steel and the larger cells facilitate grouting; and (j) and (k) in pilasters and columns. Many other shapes and sizes can be produced in a variety of textures and colors. Designers should consult their local manufactur-

| Linear shrinkage, % | Moisture content,* max., % of total, absorption (average of 3 units) | | |
|---|---|---|---|
| | Humid** | Intermediate† | Arid‡ |
| 0.03 or less | 45 | 40 | 35 |
| From 0.03 to 0.045 | 40 | 35 | 30 |
| 0.045 to 0.065, max. | 35 | 30 | 25 |

*Humidity conditions at job site or point of use
** Average annual relative humidity above 75%.
† Average annual relative humidity 50 to 75%.
‡ Average annual relative humidity less than 50%.

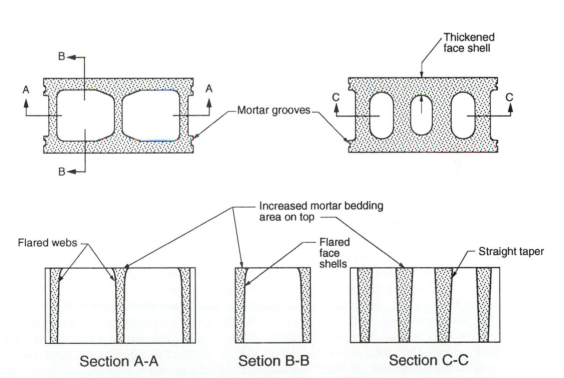

**Figure 4.24** Variations in core pattern and core shape for a standard block. [Note the presence of taper and possible flares.]

**Table 4.5** Minimum face shell and web thicknesses for load-bearing block (from Ref. 4.26)

| Nominal width of units, in. (mm) | Min. face shell thickness, in. (mm)* | Web Thickness | |
|---|---|---|---|
| | | Webs,* min, in. (mm) | Min. equivalent web thickness, in./linear ft** (mm/linear m) |
| 3 (76.2) and 4 (102) | $\frac{3}{4}$ (19) | $\frac{3}{4}$ (19) | 1-$\frac{5}{8}$ (136) |
| 6 (152) | 1 (25) | 1 (25) | 2-$\frac{1}{4}$ (188) |
| 8 (203) | 1-$\frac{1}{4}$ (32) | 1 (25) | 2-$\frac{1}{4}$ (188) |
| 10 (254) | 1-$\frac{3}{8}$ (35) <br> 1-$\frac{1}{4}$ (32)† | 1-$\frac{3}{8}$ (29) | 2-$\frac{1}{2}$ (209) |
| 12 (305) | 1-$\frac{1}{2}$ (38) <br> 1-$\frac{1}{4}$ (32)† | 1-$\frac{3}{8}$ (29) | 2-$\frac{1}{2}$ (209) |

\* Average of measurements on three units taken at the thinnest point.
\*\* Sum of the measured thickness of all webs in the unit, multiplied by 12, and divided by the length of the unit.
† This face shell thickness is applicable where allowable design load is reduced in proportion to the reduction in thickness from the basic face-shell thickness shown.

ers' product catalogs for the range of units available. Figure 4.26 shows an application of split-face block masonry that can be chosen by the designer to create a rough-texture ribbed effect.

### 4.4.4 Compressive Strength

The compressive strength of a concrete masonry *unit* is important from two points of view: first, the higher the strength, the better the durability under severe weathering conditions, and, second, unit strength tests together with mortar strength tests can serve as the basis for satisfying the required *masonry* compressive strength.

Tests of standard hard capped units loaded over the net area result in a conical shear–compression failure, as illustrated in Fig. 4.27, and is similar to the failure of concrete cylinders. As discussed in Sec. 4.2.2, this failure mode is largely due to the effects of end platen restraint. Although ASTM C140[4.2] calls for capping and hence loading over the net area, other standards may require capping and loading of face shells only. The logic is that face-shell mortar bedding is typical of hollow block construction and thus the latter approach may give a more realistic indication of the effect of unit strength on wall strength. At this time, direct conversion of results from face-shell capping tests to those from net-area capping tests is not possible. It

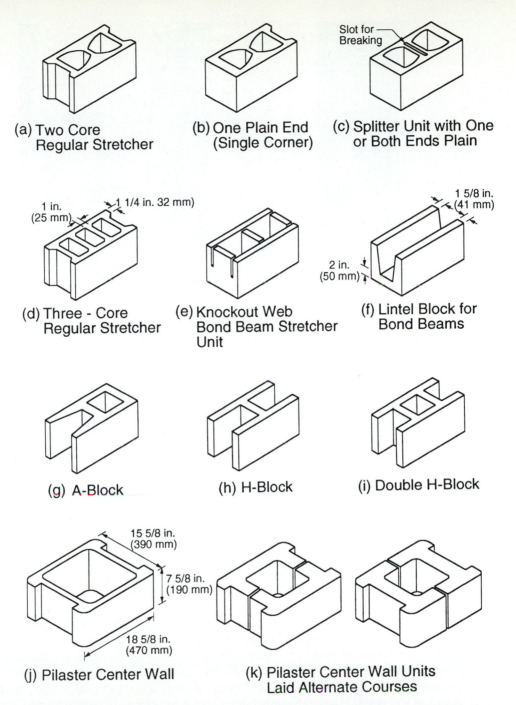

(a) Two Core Regular Stretcher

(b) One Plain End (Single Corner)

(c) Splitter Unit with One or Both Ends Plain

Slot for Breaking

(d) Three - Core Regular Stretcher

(e) Knockout Web Bond Beam Stretcher Unit

(f) Lintel Block for Bond Beams

1 in. (25 mm)

1 1/4 in. 32 mm)

1 5/8 in. (41 mm)

2 in. (50 mm)

(g) A-Block

(h) H-Block

(i) Double H-Block

(j) Pilaster Center Wall

15 5/8 in. (390 mm)

7 5/8 in. (190 mm)

18 5/8 in. (470 mm)

(k) Pilaster Center Wall Units Laid Alternate Courses

Note that (a) to (i) are shown in the 8 × 8 × 16 in. (200 × 200 × 400 mm) standard block size but other standard widths are available. Half block heights and other heights can be obtained.

**Figure 4.25** Typical concrete masonry units.

**Figure 4.26**  Application of split-face block. (*Courtesy of Gary T. Suter.*)

has been suggested[4.30] that for hollow units, *coupon tests* (small prisms cut from the face shell) provide more realistic data on compression properties.

In reviewing the compressive strength requirements of column 4 in Table 4.3, the net area versus gross area and unit shape must be kept in mind. Although ASTM C140[4.2] bases strength on gross area, engineering calculations and design codes are based on net or effective area. Basically, the concrete strengths for the first three types of units are the same. Strengths for concrete brick and solid block differ because of the two units' widely differing shapes. For the squat brick, the platen restraint will result in a significantly higher measured strength. By comparing requirements for hollow and solid block, the strengths of hollow blocks based on gross area are different but are very similar when based on net area. For example, for a 1500 psi (10.3 MPa) strength based on gross area, a hollow block with 56% solid or net area would require an average strength of 1500/0.56 = 2679 psi (10.3/0.56 = 18.4

**Figure 4.27**   Compressive failure mode of a hollow concrete block.

MPa) based on net area. It should be noted again that *North American industry practice commonly quotes masonry unit strengths based on gross area, whereas design codes use strengths based on net area.* To avoid errors, the designer must always check which strength basis is being used. Hollow blocks are used much more extensively than solid blocks and can be manufactured in strengths, ranging from 1500 to about 4000 psi (10 to 30 MPa) based on net area, to suit both low-rise and high-rise construction.

Typical stress–strain curves for hollow concrete blocks under axial compression are shown in Fig. 4.5. The shapes of the curves are similar to those of concrete. Non-linearity starts at roughly 35–50% of the strength. The shape of the curve, particularly at higher stress levels, is sensitive to the type of platen restraint.

Except in the higher strength range, lightweight blocks can be economically produced to the same strength as regular weight blocks. The modulus of elasticity of a concrete block can range from approximately 500 to 1000 times the unit compressive strength. Poisson's ratio can be taken as about 0.2.

### 4.4.5 Tensile Strength

For concrete brick, the modulus of rupture test[4.1] discussed in Sec. 4.2.2 is used to give an indication of tensile strength. Flexural tensile strengths typically range between 10 and 20% of the unit compressive strengths.

There is no widely accepted test method for determining the tensile strength of concrete block. Many direct and indirect tensile tests are employed, which range from epoxy gluing steel plates with gripping rods to the block end faces and pulling the block apart in a tensile testing machine, to epoxy gluing three blocks together lengthwise and then performing a modulus-of-rupture test.[4.6] Alternately, splitting tension tests across the face shells can be used. Such tests indicate that the block tensile strength is of the order of 10% of the unit compressive strength. Figure 4.28 contains a plot of the ratio of tensile strength to compressive strength versus compressive strength. As can be seen, the ratio of splitting tensile strength to compressive strength ranges from about 0.10 to 0.16.

### 4.4.6 Absorption

Absorption of concrete masonry units is an important property because it is related to shrinkage and, in some degree, to durability. ASTM C140[4.2] limits for absorption involving 24 hour immersion in water at room temperature are summarized in Table 4.3. No rate of absorption test like the IRA test for clay units has been standardized. Nevertheless, the absorption characteristics are important to achieve good bond between unit and mortar. For instance, if the unit is too absorptive for the mortar being used, the mortar will stiffen resulting in poor bond. More about the appropriate matching of unit absorption and mortar properties is presented in Sec. 4.7.

Because the absorption of units is an indication of the volume of pore space, permeability may have some relationship to this quantity, but as yet no direct correlation has been established. Lightweight units typically are more absorptive and in general more permeable than normal weight units.

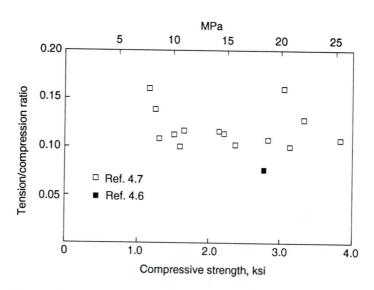

**Figure 4.28** Splitting tensile strength of concrete masonry units.

### 4.4.7 Durability

Where durability under freezing and thawing cycles is required, type N units are used because it is believed that lower absorption units with higher strengths display better durability under severe conditions. For retaining walls and external walls of buildings subject to de-icing salts, the types of spalling and cracking deterioration problems encountered in concrete construction may result. The use of high strength dense block possibly combined with special surface treatments can help overcome such problems. Overall, properly used concrete masonry units have a very good durability record.

### 4.4.8 Thermal Movement

The coefficients of thermal expansion of concrete masonry units depend to some degree on density and on the type of aggregate used. Normal weight units have an expansion coefficient of about $5 \times 10^{-6}/°F$ (0.009 mm/m/°C) and the value for lightweight units is about $4 \times 10^{-6}/°F$ (0.0072 mm/m/°C). The thermal expansion values for clay brick and normal weight concrete block are significantly different, which means that thermally induced stresses will be developed if the two materials are bonded together.

### 4.4.9 Shrinkage

In common with other cement products, concrete masonry units shrink with time. Control of shrinkage is important because in long walls, it can cause cracking, and in high walls, the overall shortening can affect the performance of other structural or nonstructural elements.

Two types of shrinkage referred to as *drying shrinkage* and *carbonation shrinkage* occur. The former relates to an overall shortening of the unit as the cement hydration process takes place and the initially moist unit comes into equilibrium with a lower humidity environment over the long term. It should be noted that upon wetting, the unit will expand to nearly its original size, and repeated drying and wetting of units, therefore, can cause reversible shortening and expansion. However, under normal service conditions, the net effect of drying shrinkage is an overall shortening of the unit. Carbonation shrinkage takes place due to a reaction of the hydration product of portland cement with carbon dioxide present in the air. This type of shrinkage is not reversible and takes place slowly over many years.

The ultimate shrinkage strain is dependent mainly on the curing conditions, cement content, aggregate type, and relative humidity of the environment. Much of the total shrinkage will occur in the factory during steam curing, autoclaving, or specialized precarbonation treatments. Typical ultimate shrinkage values are listed in Table 4.6. The important points to recognize from the tabular information and related publications are

- autoclaving can significantly reduce the amount of subsequent shrinkage
- denser aggregates produce less shrinkage than lightweight aggregates

**Table 4.6** Typical Shrinkage of Concrete
Masonry Products

| Product | Aggregate | Curing | Shrinkage, % |
|---------|-----------|--------|--------------|
| Block | Dense gravel | Low pressure steam | 0.02 – 0.05 |
|  | Dense gravel | autoclave | 0.01 – 0.04 |
|  | Lightweight | Low pressure steam | 0.04 – 0.08 |
|  | Lightweight | autoclave | 0.02 – 0.06 |
| Brick | Dense | Low pressure steam | 0.02 – 0.05 |

- the amount of shrinkage is highly variable. In critical design situations, involving, for instance, long walls, designers may want to request up-to-date shrinkage information from manufacturers in their region.

As a means of limiting shrinkage, limits on moisture content of units delivered to the job site have been established (see Table 4.4). Note that *moisture content* is the quantity of free moisture present in the unit and *absorption* measures the unit's total capacity to absorb moisture. The important point to recognize from the tabular information and previous discussion is that a concrete unit must be carefully specified to meet certain job requirements including humidity conditions.

Shrinkage is related to absorption in more than one way. Because denser units typically absorb less water, shrinkage is reduced. Units wetted prior to construction will expand and subsequently shrink a greater amount in the wall. Concrete masonry products must never be wetted prior to construction except possibly in an extremely hot and arid environment, but even then, wetting should be limited to a light misting of the units.

### 4.4.10 Creep

In common with other portland cement products, concrete masonry units continue to shorten with time under load, that is, they creep. Because concrete brick are typically not used in loadbearing applications, they are subjected to such low compressive loads that creep is not significant. The following discussion, therefore, deals with creep of concrete block.

In spite of the fact that concrete block is extensively used in loadbearing applications up to 20 and even 30 stories in height, relatively little is known about creep of concrete block masonry. Nevertheless, this much is known about creep of concrete block[4.31,4.32]:

- it is approximately of the same order of magnitude as creep in ordinary concrete made from similar aggregates
- autoclaved blocks creep less than low-pressure steam-cured block
- lightweight aggregate blocks creep more than dense aggregate block
- creep increases approximately proportional to the applied stress level

- after about 1 year at a particular stress level, most of the creep will have occurred
- about 20% of the creep occurs in mortar joints, implying that creep strains in the mortar are four to five times those occurring in the block itself.

Although no quantitative information about creep magnitudes of block alone can be cited, creep of concrete masonry is significant and will be discussed in Chap. 5.


## 4.5 CALCIUM SILICATE UNITS

### 4.5.1 Manufacture

Calcium silicate units, sometimes referred to as *sand lime units*, are manufactured by mixing sand and hydrated lime, pressing the product, and then autoclaving it to produce a tightly grained unit. The binder being lime, sand lime units are in a special category by themselves and should not be confused with concrete units. They are extensively used in many European countries, in Australia, and at an increasing rate also in North America. Colors can be introduced using various pigments and splitting or other mechanical treatment can be used to create different textures.

### 4.5.2 Grades and Durability

ASTM C73[4.33] specifies two grades of brick: SW and MW. As an aid to remembering the intended use, SW can be thought of as appropriate for severe weathering, whereas MW would be satisfactory for more moderate weathering conditions. These grades are identical to the first two grades discussed for clay brick. As an example application, SW brick should be used for foundation courses and parapets in the northeastern quarter of the United States or along the eastern coast of Canada when freezing temperatures in the presence of moisture can be anticipated.

### 4.5.3 Sizes and Shapes

Although calcium silicate units come in a variety of shapes and sizes, depending on country and regional preferences, the most common width and height dimensions center around those of typical clay or concrete bricks. Nominal lengths may vary between 8 and 12 in. (200 and 300 mm). Units in North America are typically 100% solid.

### 4.5.4 Compressive and Tensile Strengths

Compressive strengths commonly range between 2500 and 5000 psi (17 and 35 MPa), but can exceed even 10,000 psi (70 MPa). As for other brick, a measure of the tensile strength is obtained by a modulus of rupture test[4.1] (see Sec. 4.2.2). Typical results fall in the range between 300 and 700 psi (2 and 5 MPa). ASTM C73[4.33] requirements for compressive and modulus of rupture strengths are listed in Table 4.7.

**Table 4.7** Strength Requirements for Calcium Silicate Brick
(from Ref. 4.33)

| Brick grade | Minimum compressive strength (brick flatwise), psi (MPa), average gross area | | Minimum modulus of rupture (brick flatwise), psi (MPa), average gross area | |
|---|---|---|---|---|
| | Average of 5 bricks | Individual | Average of 5 bricks | Individual |
| SW | 4500 (31) | 3500 (24) | 600 (4) | 400 (3) |
| MW | 2500 (17) | 2000 (14) | 450 (3) | 300 (2) |

Both compressive and tensile strengths of calcium silicate bricks depend not only on the quantity of binder (or lime), but also on the pressure of the press and the autoclaving conditions.

### 4.5.5 Absorption

IRA values for calcium silicate brick can be misleading. Although the initial rate, as measured by the absorption within the 1 minute standard test, typically falls within reasonable limits of, say, 10 to 20 g/min./30 in.$^2$ (0.5 to 1.0 kg/min./m$^2$), the nature of the material is such that considerable absorption continues for longer periods than for clay or concrete bricks. To ensure good bond between mortar and unit in such a case, specially formulated mortar mixes may be required. As is discussed in Sec. 4.7, mortars with high water retentivity should be used to offset the unusually high suction rate of calcium silicate units. Overall absorption generally is in the range of 7 to 12%.

### 4.5.6 Thermal Movement, Shrinkage, and Creep

The coefficients of thermal expansion are in the order of 6 x 10$^{-6}$ to 12 x 10$^{-6}$ in./in./°F (0.011 to 0.022 mm/m/°C) for calcium silicate units. In common with concrete masonry products, calcium silicate brick shrink with time with a shrinkage value in the order of 0.03% being representative.[4.34]

Although calcium silicate brick deform with time under load,[4.35] they are rarely subjected to high enough stress levels that creep becomes significant.

## 4.6 BUILDING STONE

Stone has been used in masonry construction for thousands of years. Earlier massive loadbearing applications have given way to extensive use of stone in nonloadbearing veneer applications. Although stone today is also widely used as a flooring material, the presentation here will focus on building stone for exterior wall applications.

### 4.6.1 Groups

Rock formations used for the production of building stone can be of sedimentary, igneous, or metamorphic origin: *Sedimentary* means the rock has been formed by the action of water depositing minerals that have come from the decay of igneous rocks; *igneous* rocks are produced by solidification of volcanic matter; and *metamorphic* indicates the previously sedimentary or igneous rock has been recrystallized under the action of heat, pressure, or other chemical/physical agents. According to ASTM C119,[4.36] based on various origins, *groups* of building stone can be classified as follows:

- Sedimentary origin: limestone group, sandstone group
- Igneous origin: granite group
- Metamorphic origin: marble group, slate group, greenstone group

Each group may encompass a number of commercially recognized types of building stone. For exterior wall applications, limestone, sandstone, granite, and marble are likely the most commonly used building stone materials; hence, these will be dealt with in more detail in the following sections.

### 4.6.2 Sizes and Shapes

Building stone is available in a virtually limitless range of sizes and shapes, varying from brick-size units to story-height panels. In a quarry, large blocks of stone are typically split from the native rock, as illustrated in Fig. 4.29 for limestone. Metamorphic and especially sedimentary rock often contain cleavage planes or weaker bedding planes, which greatly aid in the economic splitting and removal of large blocks of stone. The blocks typically are transported to production facilities where saws, grinders, and other specialized machinery are used to arrive at the various sizes and shapes of building stone employed in construction. Required textures and finishes can be achieved by a variety of polishing or other machine tools.

### 4.6.3 Physical Requirements

As with other masonry units, building stone must meet certain physical requirements such as absorption, density, compressive strength, modulus of rupture, and hardness. For limestone, sandstone, granite, and marble, Table 4.8 lists ASTM requirements with the exception of hardness. Hardness is of vital importance where stone is to be used on horizontal wearing surfaces and hence should be carefully assessed for such applications as floors and stairs.

Regarding absorption, the tabular information shows very large variations, not only between various groups of stone, but also within a group such as limestone. The absorptive property of a limestone or sandstone seems to be dependent on the particular density of the stone. As will be further discussed in Sec. 4.7, this has implications on the mortar properties required to achieve compatibility between stone and mortar.

a) Saw cutting of sedimentary layer

b) Prying blocks of stone loose

**Figure 4.29** Limestone quarry operation. (*Courtesy of Gillis Quarries Ltd.*)

Compressive and tensile strengths generally are more than adequate to meet modern construction requirements. Strength, however, can be very much influenced by the stone's natural bedding plane or *rift*. For example, applying the load parallel to the rift, as indicated in Fig. 4.30(b), generally results in a lower compressive strength due to ease of vertical splitting. Also because rock formations may contain many weak cleavage planes, quality control must ensure that planes of gross weakness are not present, otherwise strengths in compression, tension, and shear may be only fractions of the expected values.

It is not possible to give a meaningful range of values for the modulus of elasticity for various types of building stone. The designer should contact local quarries or

**Table 4.8** Physical Requirements of Building Stone

| Group of stone | Max. absorption by weight, % | Min. density, lb/ft³ (kg/m³) | Min. compressive strength, psi (MPa) | | Min. modulus of rupture, psi (MPa) | | ASTM standard |
|---|---|---|---|---|---|---|---|
| Limestone | 12 | 110 (1760) | 1800 | (12) | 400 | (2.9) | C568 |
| | 7.5 | 135 (2160) | 4000 | (28) | 500 | (3.4) | |
| | 3 | 160 (2560) | 8000 | (55) | 1000 | (6.9) | |
| Sandstone | 20 | 140 (2240) | 2000 | (14) | 300 | (2.1) | C616 |
| | 3 | 150 (2400) | 10000 | (69) | 1000 | (6.9) | |
| | 1 | 160 (2560) | 20000 | (138) | 2000 | (13.9) | |
| Granite | 0.4 | 160 (2560) | 19000 | (131) | 1500 | (10.3) | C615 |
| Marble | 0.75 | 144 (2305) to 175 (2800) | 7500 | (52) | 1000 | (6.9) | C503 |

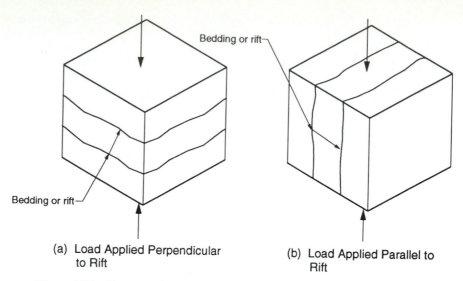

(a) Load Applied Perpendicular to Rift

(b) Load Applied Parallel to Rift

**Figure 4.30** Compressive test of building stone indicating the importance of rift direction.

industry representatives to obtain specific information. As an industry example, for Indiana limestone having a minimum compressive strength of 4000 psi (28 MPa), the modulus is reported to vary between $3.3 \times 10^6$ and $5.4 \times 10^6$ psi ($23 \times 10^3$ to $37 \times 10^3$ MPa).[4.37]

### 4.6.4 Durability

Stone is widely recognized as a durable building material that can withstand the effects of the environment for many decades and even hundreds of years. What is, of course, vital is that the right building stone is chosen for a particular application and that design, construction, and periodic maintenance are carried out properly. Although strength, hardness, absorption characteristics, and density all have an effect on durability, it is often poor design and construction details as well as inadequate maintenance that are the major contributors to any building-stone durability problem.

### 4.6.5 Thermal Movement

The coefficients of thermal expansion for building stones vary significantly not only between groups, but also within a group. Typical values are as follows:

Limestone: $4.4 \times 10^{-6}/°F$ (0.0079 mm/m/°C)
Sandstone: $6.0 \times 10^{6}/°F$ (0.011 mm/m/°C)
Granite: $4.6 \times 10^{-6}/°F$ (0.0083 mm/m/°C)
Marble: $5.0 \times 10^{-6}/°F$ (0.009 mm/m/°C)

Because of large variations in the coefficient that must be expected within each group, the designer should obtain detailed information from building stone product catalogues. To illustrate the variability, the coeffecient of thermal expansion for Indiana Limestone is reported[4.37] to vary between $2.4 \times 10^{-6}$ and $3.0 \times 10^{-6}/°F$ (0.0043 and 0.0055 mm/m/°C), values, which in themselves vary considerably from the typical limestone value quoted before.

## 4.7 MORTAR

### 4.7.1 Functions of Mortar

The main function of mortar is to bond individual masonry units into a composite assemblage that will withstand the imposed conditions of loads and weather. Mortar also serves to bond joint reinforcement and metal ties so that they can act integrally with masonry. The achievement of strength, durability, and weathertightness, therefore, is the key requisite of hardened mortar. Mortar in its plastic state also facilitates ease of construction and allows for tolerances of units and dimensions.

### 4.7.2 Mortar Types

Following centuries of use of lime mortar, combinations of portland cement, lime, sand, and water mixed in specified proportions were developed to be suitable for particular applications. Basically, the cement added strength, the lime contributed to workability, and the sand provided an inexpensive filler.

Mortars can be identified by the relative volumes of the materials. For instance, a 1:1:6 portland cement-lime mortar has equal parts of cement and lime with sand volume six times the portland cement volume or three times the volume of the total cementitious material. Rather than referring to proportions, it is convenient to have classifications; but classifications such as types 1, 2, and 3 or A, B, and C imply an order of merit or preference. To avoid this perception so that the mortar most suited to the application would be specified, classifications M, S, N, O, and K were adopted in North America. As can be seen, these correspond with every second letter in the words MaSoN wOrK.

The ASTM C270[4.38] listing of the volumetric proportions for the first four mortar types have been reproduced in the upper part of Table 4.9. In the lower part of the table are corresponding proportions where the portland cement and/or the lime has been replaced by various types of masonry cement. Although a standard (ASTM C91)[4.39] does exist for masonry cements, the performance requirements allow for fairly large variations in composition by different manufacturers.

The ASTM standard identifies mortar type by either *proportion* specification (Table 4.9) or *property* specification (Table 4.10).

Table 4.11 provides a guide for the use of masonry mortars for various building elements. For engineered masonry, types M, S, and N are the key mortars. Because type S falls between the high-strength poorer-workability type M mortar and the lower-strength good-workability type N mortar, type S provides adequate flexibility and is commonly specified for engineered masonry structures.

Besides the types of mortar discussed, there are a variety of specialty mortars on the market to address specific applications. These include refractory mortars for

**Table 4.9** Proportion Specification Requirement for Mortar
(from Ref. 4.38)

| Mortar | Type | Portland cement or blended cement | Masonry cement M | Masonry cement S | Masonry cement N | Hydrated lime or lime putty | Aggregate ratio (Measured in damp, loose conditions) |
|---|---|---|---|---|---|---|---|
| Cement-lime | M | 1 | – | – | – | Over $\frac{1}{4}$ to $\frac{1}{2}$ | Not less than $2\frac{1}{4}$ and not more than 3 times the sum of the separate volumes of cementitious materials |
| | S | 1 | – | – | – | Over $\frac{1}{2}$ to $1\frac{1}{4}$ | |
| | N | 1 | – | – | – | Over $1\frac{1}{4}$ to $2\frac{1}{2}$ | |
| | O | 1 | – | – | – | – | |
| Masonry cement | M | 1 | – | – | 1 | – | |
| | M | – | 1 | – | – | – | |
| | S | $\frac{1}{2}$ | – | – | 1 | – | |
| | S | – | – | 1 | – | – | |
| | N | – | – | – | 1 | – | |
| | O | – | – | – | 1 | – | |

**Table 4.10** Property Specification Requirement for Mortar*
(from Ref. 4.38)

| Mortar | Type | Min. average compressive strength at 28 days, psi (MPa) | Min. water retention, % | Max. air content, % | Aggregate ratio (measured in damp, loose conditions) |
|---|---|---|---|---|---|
| Cement-lime | M | 2500 (17.2) | 75 | 12 | Not less than $2\frac{1}{4}$ and not more than 3 times the sum of the separate volumes of cementitious materials. |
| | S | 1800 (12.4) | 75 | 12 | |
| | N | 750 (5.2) | 75 | 14** | |
| | O | 350 (2.4) | 75 | 14** | |
| Masonry cement | M | 2500 (17.2) | 75 | ...† | |
| | S | 1800 (12.4) | 75 | ...† | |
| | N | 750 (5.2) | 75 | ...† | |
| | O | 350 (2.4) | 75 | ...† | |

\* Laboratory-prepared mortar only.

\*\* When structural reinforcement is incorporated in cement-lime mortar, the maximum air content shall be 12%.

† When structural reinforcement is incorporated in masonry cement mortar, the maximum air content shall be 18%.

**Table 4.11** Guide for the Selection of Mortar Type
(from Ref. 4.38)*

| Location | Building segment | Recommended mortar type |
|---|---|---|
| Exterior, above grade | Loadbearing wall<br>Non-loadbearing wall<br>Parapet wall | N<br>O**<br>N |
| Exterior, at or below grade | Foundation wall, retaining wall, manholes, sewers, pavements, walks, and patios | S† |
| Interior | Loadbearing wall<br>Non-loadbearing partitions | N<br>O |

* This table does not provide for many specialized mortar uses, such as chimney, reinforced masonry, and acid-resistant mortars.
** Type O mortar is recommended for use where the masonry is unlikely to be frozen when saturated or unlikely to be subjected to high winds or other significant lateral loads. Type N or S mortar should be used in other cases.
† Masonry exposed to weather in a nominally horizontal surface is extremely vulnerable to weathering. Mortar for such masonry should be selected with due caution.

masonry fireboxes, chimneys, and flue liners, and chemically resistant mortars for a variety of industrial environments.

What is important in the choice of material proportions for a particular mortar is to obtain just the right proportions for good workability and strength. To be precise, there are a number of mortar properties both in the plastic state and in the hardened state that contribute to good workability and strength. These properties are discussed in the following sections.

### 4.7.3 Plastic Mortar Properties

Relevant properties of the plastic mortar include workability, water retentivity, and rate of hardening. There is no simple definition of *workability* because this property depends on the suction rate of units, mortar retentivity, setting time and site environmental conditions. While no on-site workability test is available, the experience of the mason in judging workability is invaluable. Good workability means the mortar will adhere to the trowel yet slide off easily; will spread readily; will adhere to vertical surfaces; and will squeeze out of joints so that it can be struck off cleanly. This permits ready laying of the units. The mortar must support the weight of additional courses without shifting. Because all these operations must be performed under varying environmental site conditions using units of differing suction rates, the experienced mason is the best judge of good workability.

A measure of workability in the laboratory is obtained by the flow test. In this test, the *flow* of mortar is measured by the increase in diameter of a cone of mortar after 25 drops on a standard flow table, as illustrated in Fig. 4.31. For laboratory mortar, a flow of 100 to 115% may be required for the development of good bond strength, depending on the absorption of the units. Values up to 130 are usually required for site-produced mortars to satisfy the mason's requirements. Flow measurements provide a reasonable indication of the physical characteristics of fresh mortar in terms of spreading and consistency for supporting successive courses of masonry units.

*Water retentivity* is the property of the mortar that prevents rapid loss of mixing water to absorptive masonry units and to the air. It is measured in the laboratory by repeating the flow test on the mortar after water has been removed by subjecting the mortar to a standard vacuum pressure of 2 in. (51 mm) of mercury (Hg) for one minute. Water retentivity is the flow after suction expressed as a percentage of the flow measured before suction. Good water retention is important for three reasons: first, water is prevented from bleeding out of the mortar; second, the mortar is prevented from stiffening too much before the masonry units are laid, and; third, sufficient water is retained to ensure proper hydration of the cement. A water retentivity of 90% or more is preferable for most masonry units except that concrete masonry units, having lower suction rates, tend to require mortar having a water retentivity of about 80%. No common on-site test exists for water retentivity.

Mortar on the board stiffens with time as a result of evaporation of water and the initial set of cementitious materials. Adding water to the mortar is known as *retempering*. Retempering is widely practiced to replace water lost by evaporation. Although the resulting water–cement ratio may increase slightly, the consequent loss of compressive strength of mortar is small and typically acceptable for the

a) Cone of mortar

b) Measurement of flow of mortar

**Figure 4.31**  Flow table setup.

sake of workability and improved bond strength. Some retempering of the mortar to restore workability in the early stages of initial set is generally permissible provided that the amounts of water are carefully controlled. Depending on environmental conditions, any mortar that is not used within between $1\frac{1}{2}$ and $2\frac{1}{2}$ hours should be discarded. The maximum time from mixing to placing of mortars should be reduced at higher temperatures because of the increased rate of set.

An appropriate *rate of hardening* of plastic mortar is important so that masonry construction can proceed economically. If the rate is too slow, mortar can extrude out of joints as laying proceeds. If the rate of hardening is too fast, mortar cannot be strung out along the laying bed for any reasonable length. The proper rate of hardening ensures good bond and easy tooling of joints. The rate of hardening depends on site environmental conditions, a subject that is explained further in Chap. 15.

### 4.7.4 Properties of Hardened Mortar

Important properties of the hardened mortar that affect masonry construction are bond, compressive strength, volume changes and durability. Of these properties bond is perhaps the most critical factor because it influences both the long term strength and the serviceability of the finished masonry. Discussion of these properties follows.

*Bond.*    Mortar should have sufficient bond for water tightness and to resist tensile stress due to external loads and internal forces arising from shrinkage and temperature changes. Bond between mortar and masonry units is achieved by mechanical interlocking and possibly to some degree by chemical adhesion, but the exact nature of the bond is still largely unknown. The multitude of parameters affecting mortar bond may be categorized under mortar type, water-cement ratio, properties of masonry units, workmanship, and curing conditions. However, the overall *compatibility* of a mortar mix with the unit is most critical. Good compatibility means that the mortar's workability and water retentivity are matched to the unit's initial rate of absorption and surface roughness. This compatibility should ensure that good initial bond is achieved within a day or less after laying. As the plastic mortar makes contact with the absorbent unit, the cementitious water slurry at the interface is drawn into intimate contact with the unit. A rapid initial bond results, which, normally, in turn leads to a satisfactory final bond. If the units are disturbed or shifted at this early stage, the intimate contact is broken and bond is impaired.

Although bond is known to be an important property, few job specifications call for bond tests and the existence of adequate bond, as reflected by design stresses in design standards, depends on experience. Alternately, for engineered masonry requiring significant tensile strengths, the Australian building code contains provisions for flexural tensile strength tests as a quality control measure. It should be emphasized that mortar bond is not just a function of the mortar but can be very much influenced by the properties of the units and other factors. For this reason, more detailed discussion of this topic is left to Sec. 5.4.4 in the next chapter.

One quick field test to check on initial bond and compatibility of mortar with brick size units involves laying up two units flatwise with one mortar joint between the two. If the assembly stays together when being lifted by the top unit about 1 or 2 minutes after laying, then good compatibility is indicated. Laboratory and field tests of final bond are discussed in Chap. 5. Such tests have shown that bond is improved by using the maximum amount of water in the mortar mix that still produces a workable consistency. This means the mortar flow should be as high as possible while still achieving good workability, good water retentivity, and an appropriate rate of hardening.

Under special circumstances, additives can be included in mortar to improve bond. For example, as discussed in Sec. 4.5.5, calcium silicate units tend to have a continuing high suction beyond the 1 minute IRA test and thus the addition of some special water-retaining agent may be required to produce high bond. The use of a type of methyl cellulose additive has indicated promising results by preventing moisture from being prematurely sucked out of the mortar and by moderately increasing the workability.[4.23]

Mortars made with masonry cement products typically have air content in the 12–18% range. Entrained air can be used to improve the workability of the mortar, and, like concrete, should result in improved durability. However, recent tests[4.40] have shown that the use of masonry cement with high air content can result in reduced bond strength. Dry curing conditions are also known to have a more marked negative effect on bond of masonry cement mortars. Inadequately defined bond requirements in North American codes have been in some part responsible for controversy regarding suitability of some of the newer mortars. Some building codes[4.41,4.42] have responded by adopting lower design stresses for flexural tensile bond for masonry cement mortars than are specified for more traditional portland cement-lime mortars. Curing conditions matter because cement-based mortars require moisture for complete hydration and gain of bond strength. On the other hand, under very wet conditions, lime-based mortars cannot absorb the necessary carbon dioxide from the air for hardening. Freshly laid masonry, therefore, must be protected to some degree in order to achieve proper bond under varying conditions of temperature, humidity, and wind. These construction conditions are more fully discussed in Chap. 15.

**Compressive Strength.** Mortar compressive strength is important because it has an influence on masonry compressive strength and because it is typically used as a measure of quality control. According to ASTM C109[4.43], the standard mortar specimen is a 2 in. (50 mm) cube cast in a nonabsorbent mold. A minimum of three specimens is typically required and these are cured under specified conditions before testing at ages of 7 or 28 days. The typical failure involves a pyramidal shape, as illustrated in Fig. 4.32.

Table 4.10 from ASTM C270[4.38] indicates minimum cube compressive strengths corresponding to the different types of mortar. The strength requirements are to be met under laboratory conditions. Long experience has shown that when the volume proportions of Table 4.9 are used, average strengths in excess of those specified in Table 4.10 are usually achieved. However, it should be clearly understood that mortar mixed in accordance with the proportion specification (Table 4.9) does not have to satisfy these property requirements. Design specifications must choose either the proportion or the property specification—not both.

**Figure 4.32** Compressive failure of mortar cubes.

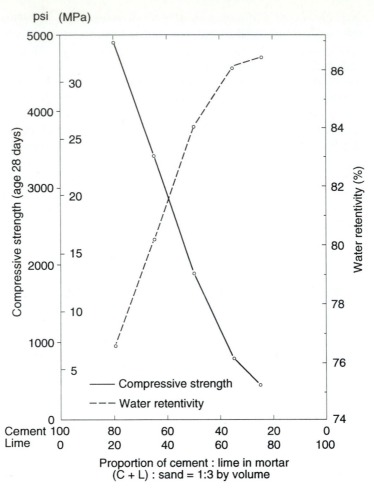

**Figure 4.33** Relation of mortar composition, compressive strength, and water retentivity (from Ref. 4.44).

Although mortar strength tests are especially important from a quality control point of view, it should be kept in mind that mortar compressive strength is frequently less important than bond.

Compressive strength is greatly influenced by the amount of cement present in the mix. As shown in Fig. 4.33, strength decreases rapidly as the cement-to-lime proportion is lowered; alternatively, water retentivity and, therefore, workability increase significantly with increasing amounts of lime.[4.44] Clearly, cement is present to achieve strength and lime contributes workability and water retentivity. One speaks of *harder* mortars when cement predominates and of *softer* mortars when lime predominates. In practice, strength and workability requirements are typically satisfied by selecting mortar proportions near the middle of the curves shown in Fig. 4.33. Workability, in fact, is generally more important than compressive strength, and the general rule is not to choose a mortar with a higher compressive strength than is required. This is so for several key reasons: first, high strengths exceeding, say, 3000 psi (21 MPa) are not required even in high-rise loadbearing masonry structures; second, there is no proportionate increase in masonry compressive strength with increasing mortar strength. (As is discussed in Sec. 5.2.4, a 100% increase in mortar strength can lead to an increased wall strength of as little as 10%). Better workability will produce better bond and contribute to more durable masonry construction. Finally, the mortar in the wall is generally stronger than the cube strength because water absorbed by the units reduces the water-cement ratio in the mortar.

**Volume Changes.** Mortar undergoes volume changes such as elastic and creep shortening under compressive load, shrinkage, and thermal movement. Because these volume changes are generally considered (and measured) as part of the overall volume changes of masonry, they are dealt with in Chap. 5 and in other chapters as discussion is warranted. However, brief points about shrinkage and extensibility of softer mortars are appropriate.

Mortar shrinkage is generally thought to be similar to concrete (0.01 to 0.08% according to Ref. 4.16) and, therefore, is a significant potential contributor to shrinkage cracking in masonry walls. However, available data may not be valid because mortar shrinkage research has typically employed specimens cast in nonabsorbent molds. Such specimens are not representative of the mortar in narrow joints where the absorbent masonry units produce significantly lower water-cement ratios and corresponding less shrinkage.

Softer mortars have more ability to extend before cracking than harder mortars. This means high-lime mortars have more flexibility and are more forgiving when subjected to small movements. It has also been reported[4.45] that, after cracking, leached-out material from high-lime mortars can lead to what is referred to as *autogeneous* healing of a fine crack. Because lime generally was the binder in mortar before the advent of cement, many older masonry structures have survived well because of the flexible nature and autogeneous healing property of softer lime mortars.

**Durability.** To be durable, mortar must be able to withstand the climatic conditions without premature deterioration. The main cause of mortar deterioration is frost action. Frost action can be particularly destructive where moisture saturated masonry undergoes many freeze–thaw cycles. Preventative measures include detailing to shed moisture, use of weather-resistant tooled joints (see Chap. 12), and separation of masonry from rain-soaked soil. Also air entraining agents added to the mortar mix will produce a very large number of tiny voids that can accommodate the increased volume of water as it freezes. Although it is generally believed that mortars having high compressive strengths are more durable, this assumption may be questionable where severe weathering conditions including freeze–thaw cycles occur. Limited field performance records indicate that medium strength, more workable mortars containing about 16% air exhibit better durability under severe environmental conditions.[4.46] However, mortar bond may be lower with high air content mortars.

### 4.7.5 Mortar Aggregates

Natural sand and manufactured sand obtained by crushing stone are generally used as mortar aggregate. Sand is the largest component of mortar and its properties

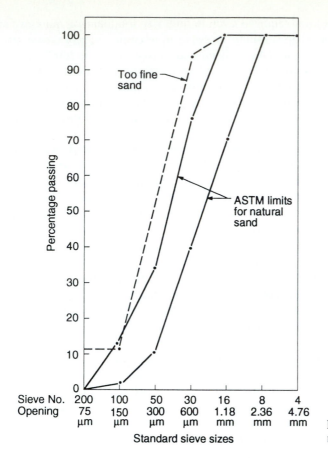

**Figure 4.34** Grading envelope for natural mortar sand.

(sizes and grading of particles) have a significant influence on the properties of mortar in both the plastic and hardened states. Coarse sands generally produce harsher mortar mixes and therefore masons may prefer using a finer sand for improved workability.

Sand gradation is usually described by the percentage of the mass or weight of grains that pass each sieve of a standard group of sieves. A range of proportions of each size is permitted by a specification, and when this range is plotted graphically, it produces a *grading envelope*, such as shown in Fig. 4.34 for natural sand.[4.47] When sand does not comply with such specified grading envelopes, laboratory tests should be conducted to show that it will produce masonry of adequate strength, especially bond strength. In practice, the use of finer sands requires additional cement, yet the amount of cement to be added must be carefully limited to control mortar shrinkage. Blending of sands may be necessary to improve overall grading.

### 4.7.6 Admixtures and Colors

**Admixtures.** The many admixtures on the market range from plasticizers and workability agents to accelerators and retarders. The inclusion of such additives frequently has an effect on strength. Hence, their use should be carefully considered in the light of other mix adjustments that may produce the desired effect without affecting strength. In general, they should only be used if specified by the designer and then strictly in accordance with the manufacturer's instructions. Overdosing or overmixing can cause large reductions in both strength and durability.

A specific additive is briefly discussed here. *Calcium chloride* has been frequently used as an accelerator under cold weather conditions. Its use in mortar is controversial because it can contribute to efflorescence staining, and accelerate corrosion of embedded steel. To ensure that no corrosion of joint reinforcement, anchors, ties, and other steel can be attributed to the presence of chlorides, it is good practice to forbid altogether the use of calcium chloride in mortar.[4.48] If metals are not present, a maximum amount of calcium chloride of 2% by weight of portland cement or 1% by weight of masonry cement is generally regarded as an upper limit. Where it is to be used, it must be added in a soluble form.

**Colors.** The appearance of masonry can be significantly altered by the use of colored mortars. The use of white cement and light colored sand can produce a pleasing contrast to darker colored units or blend with lighter colored units. The color of mortar generally is changed by means of pigments. Mineral oxides are especially suitable for pigmenting mortars. They are inert and chemically stable and are therefore unlikely to cause problems such as discoloration and efflorescence. Because they are finely ground and therefore add to the already large quantity of fines in a mix, the amount should be limited to no more than about 10% of the weight of the cement. Oxide pigments mix with the cement in mortar to form a pigmented slurry that coats the sand grains and does not fade.

Oxide pigments are sold in dry powder form and as a liquid suspension. Whichever form is used, it is essential that pigments are uniformly dispersed within the mix and that only the minimum amount of pigment is employed to achieve the desired effect. The use of too much pigment in a mix can reduce both strength and durability. To achieve a very dark mortar color, carbon black is typically used. Experience has shown that the amount should not exceed 2% by weight of portland cement, otherwise durability can be unduly affected.

## 4.8 GROUT

### 4.8 1 Workability Requirements

Grout for masonry construction is a form of high slump concrete consisting of cementitious materials, aggregate, and water. It is poured or pumped into place. Grout is used to fill cells in hollow masonry units to increase capacity and hold reinforcing steel in place. Grout can also be used to fill cavities in multiwythe masonry to produce grouted reinforced wall or beam sections.

In contrast to normal concrete, grout must always be a high-slump (8 to 10 in., or 200 to 250 mm), high water–cement ratio mix. Such a fluid mix ensures good *flowability* and complete filling of voids. This is necessary because the grout spaces typically are small and further confined by mortar fins protruding out of joints, and because water from the grout is rapidly taken up by the absorptive masonry units, thus lowering the water–cement ratio. Figure 4.35 shows the fluid grout after a slump test and Fig. 4.36 illustrates absorption of grout water by the masonry units. Note

**Figure 4.35** Slump test of fluid grout.

**Figure 4.36** Water from the fluid grout is absorbed into the block and results in surface wetness. (*Courtesy of Gary T. Suter.*)

that absorption of the cementitious water slurry is desirable to help achieve good bond between grout and units. The slump of a particular grout must be adjusted to the sizes of the spaces to be filled, the absorption characteristics of units, and to the environmental conditions of temperature and humidity. For instance, a lower slump of 8 in. (200 mm) is appropriate for relatively low-absorption units and reasonably open spaces whereas highly absorptive units or small spaces typically require a 10 in. (250 mm) slump mix.

Other important grout details relate to the types of grout, admixtures, compressive strength, and methods of placement. The first three aspects are dealt with in the following sections and the influence of methods of placement is discussed in Chap. 15.

### 4.8.2 Types

In most masonry construction, grout spaces are small and the size of aggregate should be chosen accordingly. Where grout spaces are 2 in. (50 mm) or less in their minimum dimensions, sand aggregate with a maximum aggregate size of about $\frac{1}{4}$ in. (6 mm) should be used. For grout spaces over 2 in. (50 mm), coarse aggregate with a maximum aggregate size of about $\frac{3}{8}$ in. (10 mm) should be employed. [For very large grout spaces, it may be appropriate to use a $\frac{3}{4}$ in. (19 mm) maximum aggregate size producing a grout that is more closely associated with concrete.] Based on grout space requirements, specifications generally outline two *types* of grout, *fine* and *coarse*. Table 4.12 contains the requirements according to ASTM C476.[4.49]

In North American practice today, lime is seldom added to grout. Grout for projects, requiring large volumes, is generally delivered by truck from a ready-mix concrete plant and pumped into place. It is important that the fluid mix does not exhibit segregation whereby aggregate would tend to settle out.

As a substitution for grout in minor applications, some building codes permit the use of mortar types M and S (or equivalent) to which additional water is added

**Table 4.12**  Grout Types by Volume Proportions
(from Ref. 4.49)

| Type | Parts by volume of portland cement or blended cement | Parts by volume of hydrated lime or lime putty | Aggregate, measured in a damp, loose condition | |
|---|---|---|---|---|
| | | | Fine | Coarse |
| Fine grout | 1 | 0–1/10 | $2\frac{1}{4}$–3 times the sum of the volumes of the cementitious materials | |
| Coarse grout | 1 | 0–1/10 | $2\frac{1}{4}$–3 times the sum of the volumes of the cementitious materials | 1–2 times the sum of the volumes of the cementitious materials |

to achieve required workability. Because of difficulty in controlling the mix proportions, it is recommended that this grout substitute only be allowed where relatively minor grouting is required on projects having a high level of site supervision.

### 4.8.3 Admixtures

Many grout admixtures are on the market that may be useful on projects employing highly absorptive units. Such admixtures can help control the rapid loss of water from the grout and potential shrinkage cracking both within the grout and at the grout-unit interface. Admixtures typically contain a slight expansive agent to offset shrinkage.

Calcium chloride should not be used as an accelerator under cold weather conditions because of its potential corrosive effect on any embedded steel.

### 4.8.4 Compressive Strength

Compressive strength tests of grout are important from a quality control point of view. Also grout strength has an influence on development length for reinforcing bars and on the strength of grouted masonry. The latter point is discussed in Chap. 5.

The mix proportions, listed in Table 4.12, produce grouts with cylinder (nonabsorbent) compressive strengths ranging from 1000 to 2500 psi (6.9 to 17.3 MPa)[4.48] depending on the amount of water used. However, the actual in-place grout strength will exceed these values because of the reduction in the water–cement ratio due to water absorbed by the units. ASTM C1019[4.50] specifies block (or brick) molding (Fig. 4.37) for sampling of grout to represent the effect of water-absorbed in actual walls. The use of absorbent molds provides a more realistic measure of the strength

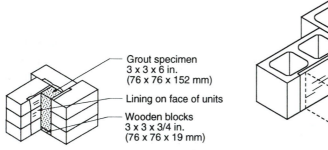

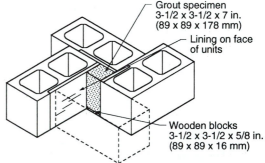

Note: Front masonry unit stack not shown to allow view of specimen.

(a) Grout Mold (Units 6 in. (152 mm) or Less in Height, 2-1/4 in. (57 mm) High Brick Shown)

Note: Front masonry unit not shown to allow view of specimen.

(b) Grout Mold (Units Greater than 6 in. (152 mm) in Height, 8 in. (203 mm) High Blocks Shown)

**Figure 4.37** Sampling of grout (from ASTM C1019).

**Figure 4.38** Core-drilled grout specimen.

of the grout. Paper towel is used to line the mold to facilitate demolding of these grout prisms.

Cores (2 in. (50 mm) in diameter) drilled from grout placed in the cells of masonry units (Fig. 4.38) have been used in research to provide a more accurate measure of grout strength properties.

The ACI 530/ASCE 5/TMS 402 masonry standard[4.41] requires that strength of grout, determined in accordance with the provisions of ASTM C 1019[4.50] should be at least equal to the specified compressive strength of the masonry but not less than 2000 psi (13.8 MPa).

## 4.9 REINFORCEMENT

As in concrete structures, reinforcement is used in masonry construction to resist tensile and shear stresses induced by various load conditions, to tie wythes of masonry together, to increase axial load-carrying capacity and to provide ductility under seismic and abnormal loadings. Additionally, reinforcement is employed to control cracking due to shrinkage, temperature effects, and applied loads.

Steel employed as reinforcement in masonry structures includes reinforcing bars, joint reinforcement, connectors, and prestressing steels.

### 4.9.1 Reinforcing Bars

The same steel reinforcing bars as used in concrete construction are employed to reinforce masonry walls, pilasters, bond beams, and lintel beams. Figure 4.39 illustrates two applications. Lists of available deformed steel bars are provided in

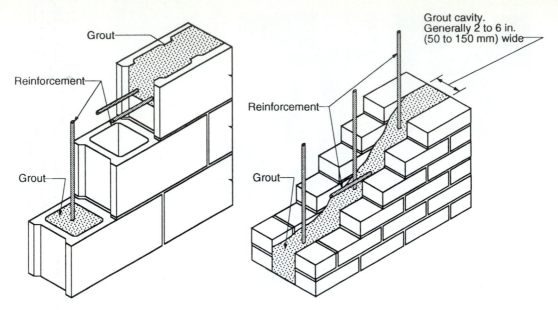

**Figure 4.39**  Use of reinforcing bars in grouted construction.

Appendix B. Because of grout space limitations, the maximum bar size is usually limited to No. 10 (30 M). Where bars must be spliced, use of smaller bars is usually necessary.

In the U.S.A., customary sizes and grades 40, 50, and 60 are used. The grades refer to minimum specified yield strengths of 40, 50, and 60 ksi (276, 345, and 414 MPa), respectively. The allowable tensile design stress for grades 40 and 50 is 20,000 psi (138 MPa), and for grade 60, it is 24,000 psi (165 MPa).[4.42] The allowable axial compressive stress is 40% of the yield strength of the steel with a maximum value of 24,000 psi (165 MPa).[4.41,4.42]

In the S.I. sizes, grades 300 and 400 are typically used where the grades refer to the minimum specified yield strengths of 300 and 400 MPa (43.5 and 58 ksi), respectively; allowable tensile design stresses are 140 and 165 MPa (20.3 and 24 ksi), respectively. The allowable axial compressive stress is again limited to 40% of the yield strength with a maximum value of 165 MPa (24 ksi).[4.51]

The modulus of elasticity of steel is taken as $29 \times 10^6$ psi (200,000 MPa).

### 4.9.2 Joint Reinforcement

Prefabricated wire joint reinforcement is extensively used in otherwise unreinforced masonry for crack control and in reinforced masonry to satisfy reinforcing requirements in the horizontal direction. It is also useful in tying an outer wythe to an inner backup, as illustrated in Fig. 4.40. Also shown in the figure are the two main types of joint reinforcement configurations used in practice, the *ladder* and *truss* types. Joint reinforcement is manufactured from wire, the most common wire size being No. 9

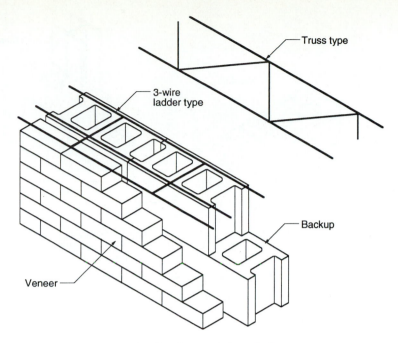

**Figure 4.40** Continuous joint reinforcement.

(0.144 in. or 3.66 mm diameter), but other sizes are available to meet various cross-sectional area demands. Table B.3 in Appendix B lists common wire gage combinations for longitudinal side rods and cross rods.

Manufacturers' product catalogues should be consulted for the variety of joint reinforcement configurations available.

Wire yield strengths frequently exceed 70,000 psi (483 MPa). Typically, the allowable stress is taken as 50% of the minimum yield strength, but no higher than 30,000 psi (207 MPa).[4.41,4.42]

### 4.9.3 Connectors

Various types of connectors are used in masonry construction. Connector is a general term for *ties*, *anchors*, and *fasteners*.[4.52] A *tie* is a device for connecting two or more wythes, or for connecting a masonry veneer to its structural backup. *Anchor*, on the other hand, refers to a device used to connect masonry walls at their intersections, or to attach walls to their supports or to other structural members or systems. *Fasteners* are devices used for securing equipment, fixtures, or parts of connectors to buildings. It is clear from the definitions that the joint reinforcement discussed in the previous section can serve as a connector and specifically a tie when used as shown in Fig. 4.40. Figures 4.41 and 4.42 illustrate different tie and anchor applications. Because connectors are very important elements of masonry construction and a large variety are encountered in practice, Chap. 13 deals specifically with their design and use in construction.

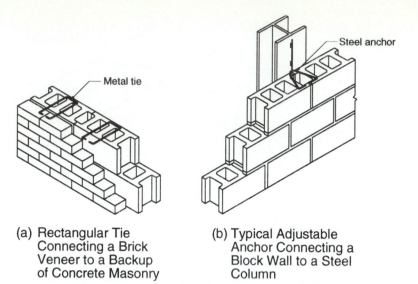

(a) Rectangular Tie
Connecting a Brick
Veneer to a Backup
of Concrete Masonry

(b) Typical Adjustable
Anchor Connecting a
Block Wall to a Steel
Column

**Figure 4.41**  Connector example showing tie and anchor applications.

**Figure 4.42**  Strap anchor connecting two intersecting walls.

### 4.9.4 Prestressing Steels

Masonry, being a compression-strong, tension-weak material, can benefit from prestressing in much the same way as concrete. The prestressing steels used are the same as in prestressed concrete. Although the technology for prestressing masonry has been largely established by research, and recent data has contributed to an assessment of strength and of prestress loss effects, applications have been few.[4.53] Because the widest use of prestressed masonry appears to have been made in Britain, the interested practitioner may wish to consult Curtin's book based on British experience.[4.54]

### 4.9.5 Corrosion Protection

All types of steel reinforcement used in masonry construction must be protected against corrosion. Corrosion occurs in the presence of oxygen and moisture either by direct oxidation or by galvanic action. Where steel reinforcement is fully embedded in dense, contaminant-free grout, it is probably as resistant to corrosion as steel in concrete. This situation is typically met in grouted reinforced masonry construction where reinforcing steel is placed in well-consolidated grout with adequate cover. Such reinforcing bars usually do not require protective coatings such as epoxy or galvanizing. Connectors crossing cavities and joint reinforcement, on the other hand, can be exposed to high moisture levels for extended periods of time and therefore require special corrosion protection. The provision of adequate corrosion resistance of connectors is an important topic, because a long service life of masonry elements such as veneer tied back by connectors must be assured. Evidence of an increasing number of connector failures in service indicates that premature corrosion takes place all too frequently. Corrosion resistance of connectors is dealt with in Chap. 13.

## 4.10 ASSOCIATED MATERIALS

A host of additional materials are closely associated with masonry construction. In a broad sense, the function of these materials is to help achieve movement control as well as moisture and thermal control. Besides fulfilling their particular functions, they must be durable, noncorroding, nonstaining, and easy to install. Moisture control is important and is achieved by incorporating a variety of materials. Examples are *dampproofing, parging, flashing, weep holes, air barriers, vapor barriers*, and *coatings*. Although detailed discussions of associated materials are presented in various chapters such as Chap. 15 dealing with movement control and Chap. 14 dealing with moisture control, a few introductory remarks are useful in this section.

### 4.10.1 Movement Joint Filler Material

Movement control is achieved by the provision of joints that do not offer significant resistance to expansion or contraction. Materials used in these joints must allow such movements and can be designed to improve resistance to water penetra-

**Figure 4.43** Use of premolded fillers in movement joints. (*Courtesy of Gary T. Suter*)

tion. Premolded elastomeric and rubber *fillers* for instance can be used, as shown in Fig. 4.43. Other materials for expansion and control joints are dealt with in Chap. 15.

### 4.10.2 Dampproofing

Dampproofing is achieved by applying a continuous coating or membrane of impervious or near impervious material on the exterior of a wall to resist moisture penetration. Below grade, existence of a high water table or poor soil drainage conditions may require that the quality of this membrane be improved to provide waterproofing. Use of materials that are easily sealed at joints, can span cracks, are not readily damaged during construction, and have long term durability is advisable.

### 4.10.3 Parging

Parging, a mortar coat, is typically applied in two layers. A base scratch coat and a finishing coat applied below ground and extending to slightly above ground level, as shown in Fig. 4.44, can serve as the base for dampproofing material such as bituminous coatings. By using types M or S mortar, each layer should be a minimum thickness of $\frac{1}{4}$ in. (6 mm). For better adhesion of the finishing coat, the scratch coat is roughened when partially set, allowed to harden for 24 hours, and then moistened just prior to applying the second coat. The finishing coat should be moist cured for at least 2 days prior to backfilling. Above grade parging reduces the air leakage through masonry and can form the base for some types of finishes.

### 4.10.4 Flashings and Weep holes

Flashings and weep holes represent measures to control moisture in walls above ground level. *Flashing* is a sheet of impervious material built into a structure to

**Figure 4.44** Parging being applied to a block masonry foundation wall. (*Courtesy of National Concrete Masonry Association.*)

prevent moisture penetration or to direct moisture to the outside. Flashing details are presented in Chaps. 12, and 15. Flashing materials range from inexpensive plastic sheets to more expensive sheet metals including lead and copper. For top quality work, copper having a minimum thickness of 0.014 in. (0.36 mm) is a preferred material. For many standard applications, layered products such as a thin layer of copper laminated to felt or kraft paper are used. *Weep holes* are openings placed in mortar joints immediately above flashing levels to permit escape of moisture through the exterior wythe. Weep holes are frequently formed simply by omitting the mortar in vertical joints at spacings of about 24 in. (600 mm). Alternatively, proprietary materials can be used to form the weep holes.

### 4.10.5 Air Barriers and Vapor Barriers

Air barriers and vapor barriers are essential wall components to control the flow of moist air through a building enclosure and to prevent condensation. There is a large variety of materials on the market that act as air barriers and vapor barriers. Constructability, durability, and the interaction of these materials with other functions of the wall are key factors relating to their selection. Their functions and requirements are dealt with in Chap. 14.

### 4.10.6 Coatings

Coatings on masonry surfaces can help control moisture movement and can alter appearance. Masonry coatings can be classified either as *solvent-based* coatings or as *water-based* coatings. The former contain or are soluble in organic solvents whereas the latter dissolve in or are dispersed in water. Examples of solvent-based coatings are oil-based paints, bituminous coatings, and epoxy coatings; examples of water-based coatings are latex paints, parging, and metallic-oxide waterproofing compounds.

In applying coatings to masonry, several masonry material properties are of key importance. These are *porosity*, *surface roughness*, *alkalinity*, and the presence of *soluble salts.* If no attention is paid to these properties, not only can premature coating failure take place, but the masonry itself can deteriorate.

*Porosity* can be defined as any space that is greater than normal atomic dimensions so that foreign molecules such as water can penetrate.[4.55] According to this definition, all common masonry units as well as mortar are porous, although the porosity varies greatly from very dense natural stones to relatively porous lightweight concrete blocks. Although all these materials are porous to some degree to water, coatings generally contain much larger molecules than water, with the result that many denser masonry materials are nonporous to coatings.[4.56]

*Surface roughness* influences the performance of masonry coatings. Because the chemical constituents of the masonry substrate and coatings differ markedly, little chemical bonding takes place and much of the adhesion between them is due to mechanical keying.[4.54] Thus, polished stone and glazed brick represent surfaces too smooth for good adhesion.

*Alkalinity* of masonry has a significant effect on the performance of coatings. Whereas masonry units, other than concrete units, are normally neutral, mortar is highly alkaline. Concrete units are always alkaline but less so when autoclaved. Due to carbonation of a relatively thin surface layer of cementitious material in the mortar and concrete block, the surface typically becomes less alkaline with time. This reduced alkalinity, however, can be offset again by fresh alkali deposits drawn to the surface by moisture evaporation. The main coating problem caused by alkalinity in the masonry has been a chemical reaction called *saponification* between oil-based coatings and alkali. As a result of saponification, a formerly dry coating film becomes soft and tacky and could, in the extreme, revert to a liquid.[4.56] It is therefore recommended that nonsaponifiable coatings be used; unless special surface preparation measures are carried out.

The presence of *soluble salts* in the masonry units or in the mortar can also affect coating performance. Depending on the properties and composition of units and mortar, as is evident from *efflorescence*, both can contain soluble salts in various concentrations. If masonry exhibiting repeated efflorescence staining is covered with a coating, perhaps to control moisture infiltration, coating failure can ensue. Efflorescence causes coating failure by mechanically destroying the film. The force exerted during crystallization of the salts behind the coating is strong enough to overcome the bond to the substrate. This implies that a source of moisture still remains despite the coating and that the coating is tending to trap the moisture in the wall. Where efflorescence problems are encountered, it is important to remedy the source of moisture and not rely on a coating to perform a function for which it is not designed.

So-called "breathing-type," heavily applied coatings have been used extensively because they can hide minor imperfections and their water vapor permeability, although normally relatively low, does allow some minor amounts of moisture in the masonry to be vented. They can be either solvent-based or water-based and typically contain a large amount of pigment mixed with a low amount of binder. Because adhesion is mostly mechanical, good surface roughness is required for good bond. *Silicones* are used as clear water repellents for masonry. Because silicones are synthetic materials made up of the elements silicon and oxygen in combination with organic groups, the word "silicone" does not refer to a single chemical product, but rather to a family of materials that can have very different properties depending on the chemical composition.

Whenever coatings are considered for masonry surfaces, it should be clearly understood that coating failure will take place when large amounts of moisture are able to enter the masonry through defects, poor detailing, or moisture-laden exfiltrating air. As this moisture builds up behind the coating, adhesive failure of the coating occurs. Even breathing-type coatings, including silicones, are not designed to handle the passage of large quantities of moisture without distress. Under freeze–thaw conditions in particular, highly saturated masonry can fail by delamination of the surface including the coating. Figure 4.45 illustrates such as case. In applying coatings to masonry, it is also important to closely follow manufacturers' instructions, otherwise coating performance can be seriously impaired.

Thermal control is achieved by the provision of *insulation* in masonry wall assemblies. Insulation materials include mineral fiber batts, rigid glass-fiber board, expanded mica loose fill, expanded polystyrene beadboard, extruded polystyrene board, urethane and isocyanurate board, phenolic foam board, and foamed-in-place urethane. Although detailed coverage of the composition and characteristics of these materials is beyond the scope of this book, such information can be readily obtained from manufacturers' and from construction handbooks. Typical applications in wall assemblies are dealt with in Chap. 14.

## 4.11 CLOSURE

This chapter provides an overview of the many materials commonly used in masonry construction. It is seen that each group of materials such as masonry units or mortar

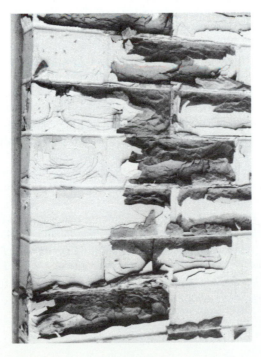

**Figure 4.45** Masonry delamination and coating failure due to freeze–thaw action.

itself includes various materials that can have widely differing physical properties. For masonry constructions to provide the required strength and serviceability, it is important to select the appropriate materials from each materials group with care and understanding. Compatibility of materials is of great importance to the long-term satisfactory performance of masonry.

Strength and deformation characteristics of masonry units, mortar and grout are emphasized in the chapter. When these materials are combined into masonry assemblages, their strength and deformation behaviors are very much influenced by the interaction of the various materials. Chapter 5 deals with behavior and strength of masonry assemblages.

## 4.12 REFERENCES

4.1   American Society for Testing and Materials, "Standard Test Method of Sampling and Testing Brick and Structural Clay Tile," ASTM C67-89a, ASTM, Philadelphia, 1990.

4.2   American Society for Testing and Materials, "Standard Methods of Sampling and Testing Concrete Masonry Units," ASTM C140-75 (reapproved 1980), ASTM, Philadelphia, 1980.

4.3   P. Guo, and R. G. Drysdale, "Stress–Strain Relationship for Hollow Concrete Block in Compression," *in Proceedings of the Fifth Canadian Masonry Symposium*, Vancouver, 1989, pp. 599–608.

4.4   G. N. Chahine, *Behavior Characteristics of Face Shell Mortared Block Masonry Under Axial Compression*, M.Eng. thesis, McMaster University, Hamilton, Ontario, Canada, 1989.

4.5   A. M. Neville, *Properties of Concrete*, Pitman Publishing, New York, 1972.

4.6   A. Hamid and R. Drysdale, "Effect of Strain Gradient on Tensile Strength of Concrete Block," ATSM STP 778—Masonry: Materials, Properties and Performance, J. G. Borchelt, Ed., ASTM, Philadelphia, 1982.

4.7   J.G. Borchelt and R. Brown, "An Indirect Tensile Test for Masonry Units," *ASTM Journal of Testing and Evaluation*, Vol. 6, 1978, pp. 134–143.

4.8   American Society for Testing and Materials, "Standard Test Method for Splitting Tensile Strength of Masonry Units," ASTM C1006-84, ASTM, Philadelphia, 1984.

4.9   American Society for Testing and Materials, "Standard Specifications for Building Brick (Solid Masonry Units Made from Clay or Shale)," ASTM C62-89a, ASTM, Philadelphia, 1989.

4.10  American Society for Testing and Materials, "Standard Specification for Hollow Brick (Hollow Masonry Units Made from Clay or Shale)," ASTM C652-89a, ASTM, Philadelphia, 1989.

4.11  American Society for Testing and Materials, "Standard Specifications for Structural Clay Load-Bearing Wall Tile," ASTM C34-84, ASTM, Philadelphia, 1984.

4.12  American Society for Testing and Materials, "Standard Specification for Structural Clay Non-Load-Bearing Tile," ASTM C56-71 (reapproved 1986), ASTM, Philadelphia, 1986.

4.13  American Society for Testing and Materials, "Standard Specifications for Ceramic Glazed Structural Clay Facing Tile, Facing Brick, and Solid Masonry Units," ASTM C126-86, ASTM, Philadelphia, 1986.

4.14 P. Rad, "Inherent Compressive and Tensile Strengths of Structural Brick," *Proceedings of the North American Masonry Conference*, Boulder, Colorado, August 1976, pp. 9/1–9/8.

4.15 H. K. Hilsdorf, "Untersuchungen über die Grundlagen der Mauerwerksfestigkeit," Bericht No. 40, Materialprüfungsamt für das Bauwesen der Technischen Hochschule, Munich, 1965.

4.16 S. Sahlin, *Structural Masonry*, Prentice-Hall, Englewood Cliffs, NJ, 1971.

4.17 British Standard Institution, "Code of Practice for Use of Masonry, Part I, Structural Use of Unreinforced Masonry," BS 5628, BSI, London, 1978 (confirmed April 1985).

4.18 A. Yorkdale, "Initial Rate of Absorption and Mortar Bond," ASTM STP 778—Masonry: Materials, Properties and Performance, J. G. Borchelt, Ed., ASTM, Philadelphia, 1982.

4.19 Canadian Standards Association, "Burned Clay Brick," A82.1, CSA, Rexdale, Ontario, 1984.

4.20 R. W. Crooks, C. L. Kilgour, and D. N. Winslow, "Pore Structure and Durability of Bricks," *in Proceedings of the Fourth Canadian Masonry* Symposium, Fredericton, Canada, June 1986, pp. 314–323.

4.21 C. T. Grimm, "Thermal Strain in Brick Masonry," *in Proceedings of the Second North American Masonry Conference*, College Park, Maryland, 1982, pp. 34-1 to 34-21.

4.22 T. Ritchie, "Moisture Expansion of Clay Bricks and Brickwork," Division of Building Research Publication No. 103, NRCC, Ottawa, Ontario 1975.

4.23 L. R. Baker, (Ed.), S. J. Lawrence, and A. W. Page, "Australian Masonry Manual", The Joint Committee of the New South Wales Public Works Department and the Assoc. of Consulting Engineers of New South Wales, Sydney, 1991.

4.24 J. I. Davidson, "Linear Expansion due to Freezing and Other Properties of Bricks," *in Proceedings of the Second Canadian Masonry Symposium*, Ottawa, Ontario June 1980, pp. 13–24.

4.25 J. A. Palmer, "Volume Changes in Brick Masonry Materials," *Journal of the American Ceramics Society*, Vol. 14, 1931, 559.

4.26 American Society for Testing and Materials, "Standard Specifications for Hollow Load-Bearing Concrete Masonry Units," ASTM C90-85, ASTM, Philadelphia, 1985.

4.27 American Society for Testing and Materials, "Standard Specification for Non-Load-Bearing Concrete Masonry Units," ASTM C129-85, ASTM, Philadelphia, 1985.

4.28 American Society for Testing and Materials, "Standard Specification for Solid Load-Bearing Concrete Masonry Units," ASTM C145-85, ASTM, Philadelphia, 1985.

4.29 American Society for Testing and Materials, "Standard Specification for Concrete Building Brick," ASTM C 55-85, ASTM, Philadelphia, 1985.

4.30 O. Senbu, and A. Baba, "Mechanical Properties of Masonry Units," in *Proceedings of the First Joint Technical Coordinating Committee on Masonry Research*, Tokyo, August 1985, pp. 3.1/1–3.1/19.

4.31 E. L. Jessop, "Moisture, Thermal, Elastic and Creep Properties of Masonry: A State-of -the-Art Report," *in Proceedings of the Second Canadian Masonry Symposium*, Ottawa, Ontario June 1980, pp. 505–520.

4.32 G. A. Fenton, *Differential Movements and Stresses Arising in Masonry Veneers of Highrise Structures*, M.Eng. thesis , Carleton University, Ottawa, Ontario 1984.

4.33 American Society for Testing and Materials, "Calcium Silicate Face Brick (Sand-Lime Brick)," ASTM C73-85, ASTM, Philadelphia, 1985.

4.34 P. Schubert, "Zum Schwindverhalten von Mauerwerk," in *Proceedings of the Sixth International Brick Masonry Conference*, Paper No. 5, Rome, 1982, pp. 57–64.

4.35 J. J. Brooks, "Time-Dependent Behaviour of Calcium Silicate and Fletton Clay Brickwork Walls," in *Proceedings of the British Masonry Society*, No. 1, November 1986, pp. 17–19.

4.36 American Society for Testing and Materials, "Definitions of Terms Relating to Dimensioned Stones" ASTM C119-87, ASTM, Philadelphia, 1990.

4.37 Indiana Limestone Institute of America, *Handbook*, ILI, Bedford, IN.

4.38 American Society for Testing and Materials, "Standard Specification for Mortar for Unit Masonry," ASTM C 270-89, ASTM, Philadelphia, 1989.

4.39 American Society for Testing and Materials, "Standard Specification for Masonry Cement," ASTM C91-91, ASTM, Philadelphia, 1991.

4.40 S. Ghosh, "Flexural Bond Strength of Masonry: An Experimental Review," in *Proceedings of the Fifth North American Masonry Conference*, University of Illinois, Urbana, Illinois, 1990, pp. 701–711.

4.41 "Building Code Requirements for Masonry Structures," ACI 530/ASCE 5/TMS 402, Masonry Standards Joint Committee, American Concrete Institute and American Society of Civil Engineers, Detroit, New York, 1992.

4.42 International Conference of Building Officials, *Uniform Building Code*, Chapter 24, ICBO Whittier, CA, 1991.

4.43 American Society for Testing and Materials, "Test Method for Compressive Strength of Hydraulic Cement Mortar," ASTM C109, ASTM, Philadelphia, 1988.

4.44 J. I. Davison, "Masonry Mortar", Canadian Building Digest 163, National Research Council of Canada, Ottawa, Ontario 1974.

4.45 C. T. Grimm, "Durability of Brick Masonry: A Review of the Literature", Masonry: Research, Application and Problems, ASTM STP 871, Grogan/Conway (Eds.), ASTM, Philadelphia, 1985.

4.46 A. Fontaine, "Durability of Masonry Mortar in a Nordic Climate," *APT*, Vol. 17, No. 2, 1985, 49–66.

4.47 American Society for Testing and Materials, "Aggregate for Masonry Mortar," ASTM C144-84, ASTM, Philadelphia, 1984.

4.48 W. C. Panarese, S. Kosmatka, and F. A. Randall, *Concrete Masonry Handbook for Architects, Engineers, Builders*, Portland Cement Association, Skokie, Illinois, 1991.

4.49 American Society for Testing and Materials, "Standard Specification for Grout for Masonry," ASTM C476-83, ASTM, Philadelphia, 1983.

4.50 American Society for Testing and Materials, "Standard Method of Sampling and Testing Grout," ASTM C1019, ASTM, Philadelphia, 1984.

4.51 Canadian Standards Association, "Masonry Design for Buildings," CAN3-S304-M84, CSA, Rexdale, Ontario, 1984.

4.52 Canadian Standards Association, "Connectors for Masonry," CAN3-A370-M84, CSA, Rexdale, Ontario, 1984.

4.53 "Reinforced and Prestressed Masonry," in *Proceedings of the International Symposium on Reinforced and Prestressed Masonry*, Edinburgh, 1984.

4.54 W. G. Curtin, G. Shaw, J. K. Beck, and W. A. Bray, *Structural Masonry Designers' Manual*, Granada Publishing, London, 1982.

4.55 P. J. Sereda, "The Structure of Porous Building Materials," CBD 127, National Research Council of Canada, Ottawa, Ontario 1970.

4.56 H.E. Ashton, "Coatings for Masonry Surfaces," CBD 131, National Research Council of Canada, Ottawa, Ontario 1970.

## 4.13 PROBLEMS

**4.1**  What are the physical properties that are commonly used to classify clay brick? State why these properties are important.

**4.2**  Why are actual unit dimensions smaller than nominal dimensions?

**4.3**  Estimate the tensile strength of 4000 psi (27.5 MPa) brick units. Discuss the possible importance of this tensile strength.

**4.4**  Discuss the factors affecting compressive strength of brick.

**4.5**  Why are clay brick units cored? How much coring is allowed for solid and for hollow clay brick units?

**4.6**  What are the minimum face shell and web thicknesses specified for 8 in. (200 mm) hollow concrete blocks?

**4.7**  Discuss how test methods can affect the measured tensile and compressive strengths of masonry units?

**4.8**  Define IRA and explain its significance as a material property. If possible, perform an IRA test and report the result.

**4.9**  Describe the $C/B$ ratio and its significance as a material property.

**4.10**  Specify a brick unit to be used in veneer construction in the northeast coast of North America and a brick unit to be used in southern California. Discuss the reasons for differences in these specifications.

**4.11**  Can brick or block units be considered to be isotropic materials? Explain.

**4.12**  What are the desirable properties of mortar and grout?

**4.13**  What are the different types of mortar and how are they classified?

**4.14**  **(a)**  Arrange a tour of a local clay-brick manufacturing plant and prepare a four- to six-page report outlining the source of raw material, manufacturing process, available products, and costs.

**(b)**  Arrange a tour of a concrete block manufacturing plant and prepare a two- to four-page report on materials used, manufacturing process, available products, and costs.

**4.15**  In groups of two to four students, mix two small batches of type S mortar, one made with portland cement and lime and the other with type S masonry cement.

**(a)**  Adjust the water content to provide a flow of 115% and make three cubes for each batch. Add additional water to reach a flow of 130% and make additional sets of three cubes for each batch. Perform compression tests at seven days of age.

**(b)**  Using available clay or concrete bricks, investigate the initial bond occurring immediately after laying.

**(c)**  Determine the air content and, if equipment is available, the water retentivity for these last batches of mortar.

**(d)**  Write a short report to present and discuss your observations.

**4.16**  Prepare small batches of fine and coarse grout with a slump of approximately 10 in. (250 mm). Prepare and test standard cylinders formed in nonabsorbent molds and grout prisms formed in block molds. Report and discuss your observations.

# CHAPTER 5

# Masonry Assemblages

Compression test of a concrete block prism.

## 5.1 INTRODUCTION

A masonry assemblage is an element composed of some or all of the constituent masonry materials: units, mortar, grout, and reinforcement. Knowledge of the interactions between these materials and of other factors affecting the physical and mechanical properties of the composite is needed to understand the fundamental behavior of masonry. With this background, including awareness of the limitations of current knowledge, it is hoped that professionals will be able to use judgement in the application of building codes. No simple set of rules can be sufficiently comprehensive to anticipate all situations.

As was discussed in Chap. 4, determination of the physical and mechanical properties of individual masonry materials is important for maintaining manufacturing standards and for quality control on the job site. In addition, provisions that ensure that the properties of the individual materials fall within established limits are necessary if the characteristic behaviors of masonry assemblages are to be predicted based on the properties of the individual materials. For example, to be reasonably assured that an expected flexural strength will be achieved, the properties of the unit and mortar (and construction conditions) must reasonably match those originally used to establish the specified bond strength. In terms of saving time and money during both the design and construction stages, it is desirable to ensure that specified properties of masonry assemblages are satisfied using simple economic tests of the individual materials.

As with concrete, it is usual to describe the structural quality of masonry in terms of a specified compressive strength $f'_m$. As a result, attempts have been made to relate other properties and design stresses to this property. However, as is illustrated in this chapter, masonry assemblages exhibit very complex interactions. Accurate prediction of their behaviors using properties of the constituent materials depends on using the appropriate properties and on correct measurements of these properties. At this stage of their development, the varying degrees of conservativeness and other differences among masonry design codes are at least partially due to the interactions between the individual materials being too complex to be accurately modeled using a few basic material properties. The large ranges of material properties and unit geometries allowed within building standards and variations in test methods also contribute to apparent differences in behavior. In addition, the potentially significant effects of workmanship, age, and curing are difficult to quantify.

To illustrate various behaviors or interactions in this chapter, attempts have been made to use representative data. However, the reader should be aware that most data are specific, not generic, and they are used here only to illustrate general trends. Where appropriate, standard test methods are reviewed.

## 5.2 AXIAL COMPRESSION

### 5.2.1 Introduction

With the modern use of high strength materials and thinner elements, compressive strength is often of prime importance in loadbearing structures. Compression tests of masonry prisms are used as the basis for assigning design stress and, in some cases, as a quality control measure. Its importance has made prism compressive behavior a major research focus and potential correlations with other strength characteristics have been investigated. Test machine capacity and specimen height limits as well as other practical considerations have led to use of prisms as the main type of compression test specimen rather than full scale specimens.

### 5.2.2 Standard Prism Tests

ASTM E447[5.1] describes test equipment, test procedures, and reporting for prism tests. As shown in Fig. 5.1, a standard prism is usually one masonry unit long, one unit thick, and can be built to various heights (usually between 1.5 to 5 times the thickness) using different numbers of units and joints. Although it is preferable that the bond pattern and mortar bedding represent the actual construction, for convenience, the stack pattern shown is commonly used. Depending on the purpose of the test, prisms can be constructed as shown for either Method A or B. Method

(a) Brick Masonry

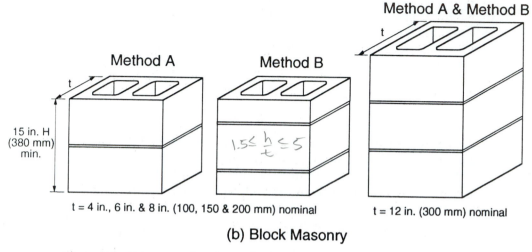

Method A & Method B

Method A          Method B

15 in. H
(380 mm)
min.

$1.5 \leq \dfrac{h}{t} \leq 5$

t = 4 in., 6 in. & 8 in. (100, 150 & 200 mm) nominal      t = 12 in. (300 mm) nominal

(b) Block Masonry

**Figure 5.1**  Prisms used to determine axial compressive strength of masonry.

A incorporates full mortar bedding in $\frac{3}{8}$ in. (10 mm) flush joints.  It is used to provide comparative data in the laboratory.  Alternatively, Method B incorporates at least two mortar joints, tooled, and with mortared areas representing actual construction.  Such prisms are intended to represent the compressive strength of masonry as built in the structure.

A key feature of the test provisions is the load transfer mechanism at the top and bottom of the prism.  This includes provisions for capping to provide flat bearing surfaces and for thickness of loading plates to distribute the load uniformly to the specimen.  The bearing surfaces of the prism can be capped with a thin layer of mortar, hard capping compounds such as hydrostone (a fast-setting gypsum-based material), or melted sulfur capping material.  As shown in Fig. 5.2, most commercial test machines have a test bed of sufficient area and stiffness to provide uniform load transfer.  However, at the top, the bearing surface of the spherical seat is usually 8 to 10 in. (200–250 mm) in diameter, which may fully cover the bearing area of a brick prism, but not for larger units such as concrete blocks.  In this case, a sufficiently stiff end plate must be used to ensure planar displacement at the top of the prism.  ASTM specifies use of a steel plate with a thickness $t_b$ of not less than half the distance $a$ from the edge of the spherical seat to the corner of the prism.  For a 9 in. (229 mm) spherical head, a plate thickness of just over 2 in. (51 mm) is required for a $7\frac{5}{8} \times 15\frac{5}{8}$ in. (190 × 390 mm) block prism.

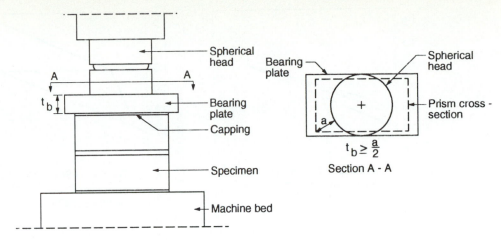

**Figure 5.2** Bearing plate requirements for the prism test (from Ref. 5.1).

The flatness of the base of the test machine and bearing plate must also conform to tight tolerances.

Important features of the prism test are as follows:

1. A *spherical seat* is necessary to apply uniform load to prisms when the ends are not perfectly parallel. It also allows different displacements to occur on opposite faces of the prism. Therefore, accidental eccentricities in loading or local flaws (weaker or less stiff zones) will result in different strains and allow failure to initiate in those areas.

2. Practical limits on the *bearing plate thickness* mean that it cannot be infinitely stiff. The specified plate thickness is recognized as being sufficient to reduce the effects of bending to an acceptably minor factor. However, test data, particularly those that have been gathered some time ago, should be scrutinized quite carefully for potentially significant effects of inadequate bearing plate thickness.

    As an illustration, test data for hard capped, hollow, nominal 8 in. (200 mm) concrete block prisms showed a 9% increase in strength when a 3 in. (76 mm) steel bearing plate was used instead of a 2 in. (51 mm) plate. As an alternative, 5 to 8 in. (127 to 203 mm) thick aluminum bearing plates, having the advantage of low weight, have been used for prism testing.[5.2,5.3]

3. A majority of the machines used for commercial testing of prisms were designed for testing concrete cylinders and have only 16 to 18 in. (400 to 450 mm) clear headroom. Therefore, two-block high and five- to seven-brick high prisms have most commonly been tested. Because it is known that the height (and number of mortar joints) can have significant influences on measured strength, correction factors for height-to-thickness ratios are normally applied to the test data. Figure 5.3 shows correction factors for concrete block and clay brick prisms from several national standards. In most cases, these can be traced back to some preliminary research by Krefelt[5.4] in 1938. The correction factors are reviewed later in Sec. 5.2.4.

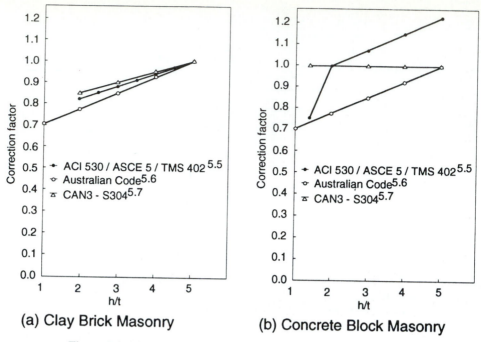

**Figure 5.3**  Correction factors for prism compressive strength.

### 5.2.3 General Failure Mechanisms

Compression tests of prisms with low height-to-thickness ratios (i.e., less than about 2:1) tend to produce the conical type shear–compression failure shown in Fig. 5.4(a). This is similar to conical failures observed for concrete cylinder tests and is similarly related to the effects of end confinement at the bearing plates. Alternately, prisms with sufficient height to minimize the end effects exhibit vertical cracking through the masonry units, as shown in Fig. 5.4(b). The same type of failure occurs in shorter prisms if end confinement is eliminated such as by using brush platens, as is discussed in Sec. 5.2.4. This pattern of cracking is consistent with failures observed in tests of full-scale walls.[5.8,5.9]

Detailed failure hypotheses have been proposed[5.10–5.16] to explain the failure mechanism and to quantitatively estimate the compressive strengths of masonry prisms. A brief description of the generally accepted explanations covering the case where the uniaxial compressive strength of the mortar is less than the strength of the masonry units are presented here. Situations where the units are significantly weaker than the mortar could result in quite different behavior, but this is an area that has not been evaluated in detail.

**Prisms of Hollow or Solid Units with Full Mortar Bedding.**  Compressive strengths of masonry prisms are higher than compressive strengths of mortar cubes and consistently lower than measured strengths of masonry units. This phenomenon has been explained by considering the dissimilarities between the two materials.[5.10,5.11] Under compression, the weaker mortar tends to expand laterally at a

a) Typical shear failure

**Figure 5.4** Failure modes of masonry prisms.

Hollow block

Grouted block

Clay brick

b) Typical splitting failure

greater rate than the masonry unit. As shown in Fig. 5.5(a), where the vertical strains in the units and mortar are different for the same stress level, the lateral strains will be proportionally different. As is illustrated in Fig. 5.5(b) for a prism made of strong units and weak mortar, the masonry units restrict the expansion of the mortar, thereby confining the mortar and creating a triaxial state of compression that enables the

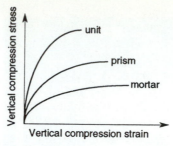

(a) Stress-Strain Relationship for Materials and Prism

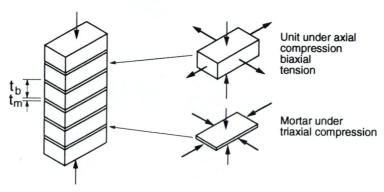

(b) States of Stress of Units and Mortar

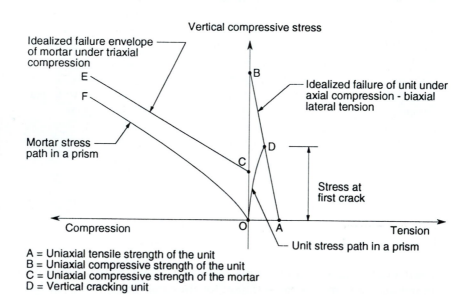

A = Uniaxial tensile strength of the unit
B = Uniaxial compressive strength of the unit
C = Uniaxial compressive strength of the mortar
D = Vertical cracking unit

(c) Failure Mechanism

**Figure 5.5** Behavior of solid prism under axial compression.

mortar to resist axial compressive stresses much higher than its uniaxial strength. The failure envelope for mortar under triaxial compression is represented by line $CE$ in Fig. 5.5(c), where point $C$ represents the uniaxial compressive strength.

In conjunction with confinement of the mortar, to maintain equilibrium, lateral tensile stresses are introduced in the units on both sides of the bed joint. The development of lateral biaxial tension with increasing axial compression is represented by line $OD$ in Fig. 5.5(c). When the combination of vertical axial compression and horizontal biaxial tension reaches the compression–biaxial tension failure envelope at point $D$ on the idealized failure envelope $AB$ of the unit, vertical cracking through the units, as shown in Fig. 5.4(b), occurs. In this representation, the mortar does not reach failure.

Opinion and mathematical models differ as to the magnitude and relative effect of the lateral tension in the unit. It has been suggested[5.16] that flaws in the units can initiate cracking at stresses well within the failure envelope. However, regardless of the details of the failure hypothesis, development of vertical cracks is a recognized phenomenon. The development of some cracks at loads well below the recorded prism capacity indicates that the separate vertical columns created by these cracks have an inherent strength and some critical combination of cracks must occur to produce failure.

**Prisms of Hollow Units Filled with Grout.** Data on grouted concrete block[5.17] and clay brick[5.12] prisms have shown that superposition of the capacities of the hollow prism (fully mortared bed joints) and the columns of grout formed in the cells can result in overestimated strengths for grouted prisms. Incomplete grout compaction, plastic and drying shrinkage of the grout, incompatibility between the stress–strain properties of the materials, and geometric factors are suggested causes.

Initial tension due to restrained drying shrinkage or the effects of gaps due to incomplete compaction or plastic shrinkage can result in the grout resisting a smaller share of the compressive load. Alternately, models incorporating the incompatibility of the stress–strain properties of the grout with the block (including different Poisson's ratios resulting in different lateral expansion for the same axial compression) identify this as a cause of additional lateral forces on the block units.[5.12,5.18] These larger lateral forces correspond to measured larger lateral strains in the units at an early stage of loading[5.17] (compared to ungrouted prisms). However, this latter result could also be associated with geometric effects such as the tapered or flared shape of the face shells and webs (see Fig. 4.24) that could result in the grout acting as a kind of wedge.

**Prisms of Hollow Units with Face Shell Mortar Bedding.** Prisms constructed with only the face shells mortared exhibit vertical cracking through the webs, as shown in Fig. 5.6. Studies of stress conditions indicate that the load transferred to the hollow units along the face shells produces nonuniform vertical compressive stress over the height and width of the unit.[5.19] The effect is to produce transverse principal tension stresses near the top and bottom of the web, as indicated in Fig. 5.7. Because the pattern of lateral stress at the middle of the web, over half of the height of the web is similar to flexural stresses in deep beams, this is sometimes referred to as analogous to deep-beam behavior.[5.20]

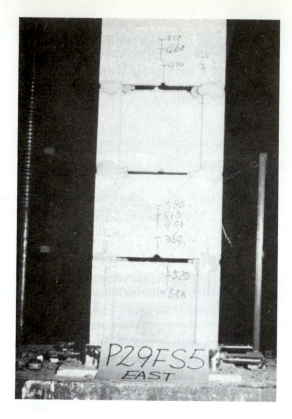

**Figure 5.6** Splitting cracks at the webs of face shell bedded block prism.

### 5.2.4 Factors Affecting Prism Strength

For solid, hollow, and grout-filled hollow masonry, there are many geometric and strength parameters that affect the relationship between prism strength and unit strength. Brief discussions and example test data follow.

**Unit Geometry and Mortar Bedded Area.** Because vertical cracks in prisms are related to development of lateral tensile stresses in the masonry units, it follows that the magnitudes of these stresses are affected by unit geometry. For example, clay bricks having vertical coring (perforations) will produce lower prism strengths than solid bricks of equal strength. This difference can be explained by considering the relationship between the mortar bedded area subject to compressive vertical stress and the minimum vertical section resisting lateral tension. As shown in Fig. 5.8, a 15% reduction in mortar bedded area can result in over a 45% reduction at the minimum vertical section through the cores. On a simplistic basis, ignoring stress concentrations and other factors, the reduction to the net stress-producing area along the bed joint of 85% compared to net stress-resisting area of 55% through the unit would suggest that coring would reduce prism strength to $55 \div 85 = 0.65$ of the value for solid units. This type of reduction is observed in the test results shown in Fig. 5.9. For the $\frac{3}{8}$ in. (10 mm) mortar joint, the perforated and solid brick prisms have

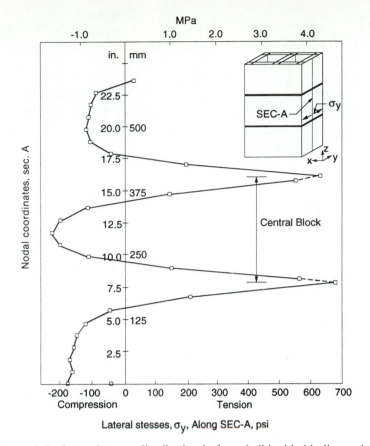

**Figure 5.7** Lateral stress distribution in face shell bedded hollow prism under axial compression (from Ref. 5.19).

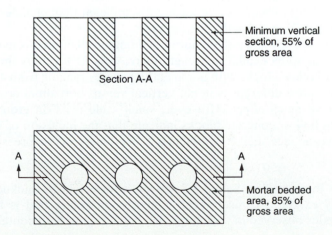

**Figure 5.8** Effect of coring on minimum sections of brick.

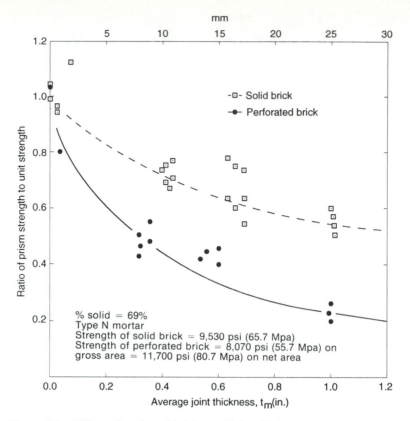

**Figure 5.9**  Effect of coring of bricks and joint thickness on compressive strength (from Ref. 5.10).

relative strengths of 0.49 and 0.74, respectively, resulting in a ratio of 0.66. Based on coring that produces a 69% solid brick with a 47% solid vertical section along the length, the ratio 0.47 ÷ 0.69 = 0.68 is very close to the 0.66 value observed.

Because of the increased strength of the material due to better firing of the clay in cored units, the net effect of the coring is usually decreased. In addition, the improved durability due to better firing may be far more significant than changes in the already quite high compressive strength of these units.

For concrete block prisms with face shell mortar bedding, tests of prisms made of hollow blocks having four webs exhibit a strength increase is 13% versus those having three webs.[5.21] For hollow blocks with nominal thicknesses ranging from 4 to 12 in. (100 to 300 mm), the small differences in net vertical versus horizontal cross-sectional areas did not have much effect. However, small scale tests on grouted prisms made with two-cell hollow concrete blocks indicate that the inclusion of the flare at the top of the webs and face shells resulted in a 27% increase in compressive strength.[5.22]

**Prism Height.**   As mentioned in Sec. 5.2.2, prisms with height-to-thickness ratios around two usually exhibit a conical shear–compression failure that is not consistent with failure modes observed in structural elements. Figure 5.10 contains

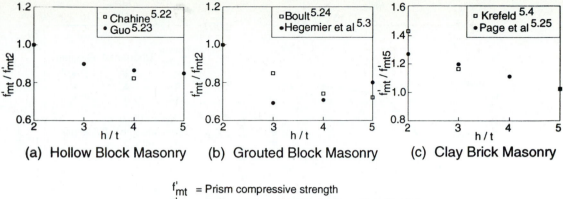

f'mt = Prism compressive strength
f'mt2 = Compressive strength of prism with h / t = 2.0
f'mt5 = Compressive strength of prism with h / t = 5.0

**Figure 5.10**  Effect of h/t ratio on prism strength.

data from tests of hollow concrete block and solid clay brick prisms constructed with different height-to-thickness ratios ($h/t$). Decreasing strength with increasing height is associated with having a sufficient portion of the prism away from the confining effects of the end platens to allow the characteristic vertical splitting to occur. In addition, two-unit high hollow block prisms incorporating only one mortar joint, cannot simulate the full interaction between the units and mortar representative of full scale wall behavior. Some research[5.24] shows that having an adequate number of bed joints in the prism may be as important as the height-to-thickness ratio.

The correction factors used in building codes, see Fig. 5.3, convert results from various prism heights to a height-to-thickness ratio of 5, except that a ratio of 2 is used for concrete masonry in North America. For working stress design purposes, compressive strengths defined in this way do not create a problem because tests of walls and columns indicate that design provisions using these values provided more than adequate safety. Use of two-block-high prisms is a practical limit for quality control using field constructed prisms both in terms of transportation and availability of commercial test machines. For quality control, this is a satisfactory and meaningful standard test. On the other hand, tests for research, product development, or other special situations should use prisms with a sufficient height-to-thickness ratio and number of bed joints to permit the correct failure mode to occur.

**Strength of Mortar.**    As discussed in Sec. 5.2.3, mortar properties affect the stresses in axially compressed prisms. The prism test data shown in Fig. 5.11(a) indicate the correlation of lower strength mortars with lower prism compressive strengths. This effect is particularly apparent where high strength units are used. Although increasing the mortar strength does increase prism strength, the increases are less pronounced at higher mortar strengths. Therefore, although some minimum strength of mortar is required for strength and to ensure adequate durability, other considerations such as better workability of the fresh mortar and a more deformable mortar to accommodate differential movements have led to the common advice of not using a higher strength mortar than is required for the job. Test results[5.17] show

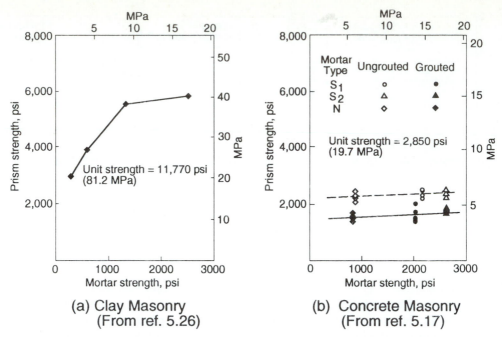

**Figure 5.11** Effect of mortar strength on prism compressive strength.

that grouting reduces the influence of mortar properties on prism strength (Fig. 5.11(b)) because of the continuity provided by the grouted cores.

**Unit Strength.** As shown in Fig. 5.12, compressive strengths of masonry prisms are related to the compressive strengths of the units. However, because the prism failure mechanism is much different from the compressive failure of the units, it is not surprising that other factors such as tensile strength and geometry can result in significant scatter of results as is shown. For the data in Fig. 5.12(b), the geometries were identical for hollow concrete block supplied from 29 different plants. By looking at a group of block in the same strength range, it was observed[5.21] that the scatter of the prism strength for face-shell mortar bedding could be related to differences in tensile strengths of the blocks. Because prisms fail when the units are sufficiently weakened by cracking, it is to be expected that both the compressive and tensile strengths of the units have significant effects.

Of necessity, generic design tables for prism compressive strength must be conservatively based on the lowest results for the units and mortar allowed within the specifications for those materials. Therefore, prism tests using specific units almost always result in a significantly higher compressive strength than the listed values in building codes.

**End Platen Restraint.** The effect of prism height discussed earlier in this section is in part an effect of end platen restraint. In Fig. 5.13, the photographs of concrete block and the corresponding concrete block prisms tested between solid steel plates and between brush platens provide a further indication of this effect. The

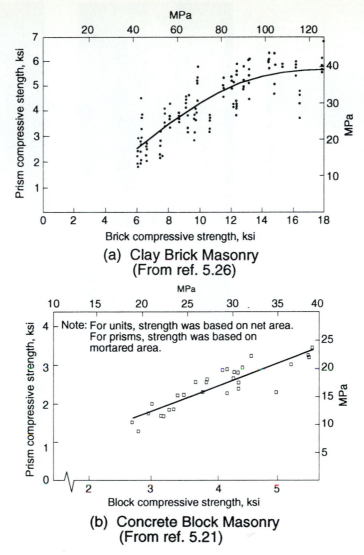

**Figure 5.12** Effect of unit strength on compressive strength of masonry prisms.

failure modes for tests between solid steel plates are also shown where the conical failures for the unit and short prism are not representative of failures in walls. Alternately, the very low lateral restraint provided by the brush platens results in vertical cracking failure of units, short prisms, and tall prisms. The failure loads of units and short prisms are significantly lower than when solid steel plates are used to transfer the load.

**Bond Pattern.** It is desirable that prisms be tested with bond patterns and mortar bedding representing field construction. This usually means that running bond should be used. Again, with prisms only two units high, it is impossible to represent

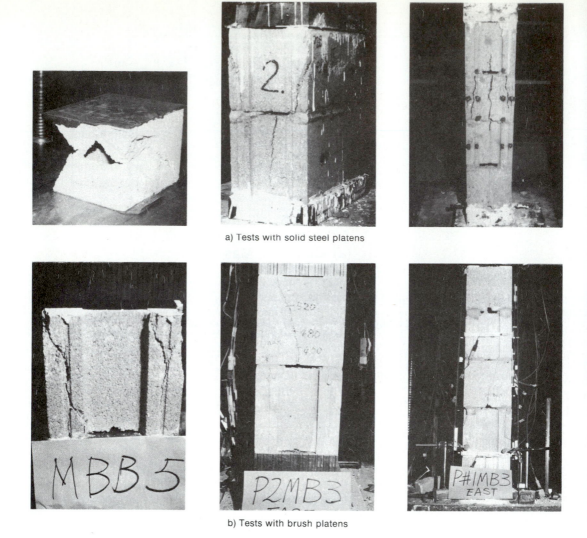

a) Tests with solid steel platens

b) Tests with brush platens

**Figure 5.13**  Effect of end platen restraint on failure mode of masonry prisms.

this condition. Tests of prisms at least four courses high are necessary to allow uncontrolled vertical cracking to develop. Compression tests of hollow clay brick prisms[5.27] and grouted concrete block prisms[5.3] showed slightly lower strengths for running bond compared to prisms built in the stack pattern. The reduction was 15% for the hollow clay brick prisms and 4% to 13% for the grouted concrete block prisms.

**Thickness of Mortar Joints.**    Practical and aesthetic considerations have resulted in the thickness of mortar joints being standardized as $\frac{3}{8}$ in. (10 mm).    However, workmanship and construction tolerances of other components do sometimes lead to noticeable variations in joint thickness. As can be seen in Fig. 5.9, the influence of thicker joints on the compressive strength of clay brick prisms is quite

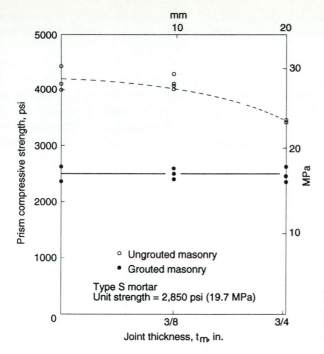

**Figure 5.14** Effect of joint thickness on compressive strength of hollow and grouted concrete masonry (from Ref. 5.17).

significant. As shown in Fig. 5.14, it is less pronounced for concrete masonry, which has properties more similar to mortar and, therefore, is less affected by differences between the materials. For fully grouted masonry, the effects of mortar joint thickness are much reduced due to the continuity of the grout.[5.12]

Another way of looking at the effect of mortar joint thickness is to express it in terms of the ratio of *unit height-to-joint thickness*.[5.28] In Fig. 5.15, from tests on clay brick prisms, it is apparent that lower strengths correspond to smaller ratios of brick height to mortar thickness. Therefore, as would be expected from the discussion of

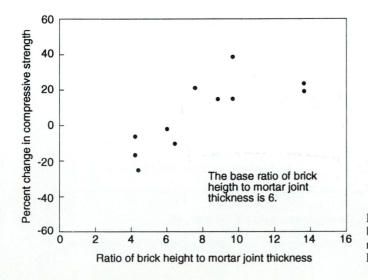

**Figure 5.15** Effect of ratio of brick height to mortar joint thickness on compressive strength (from Ref. 5.28).

failure mechanisms in Sec. 5.2.3, it is not the absolute unit height or mortar joint thickness but the unit height-to-mortar joint thickness ratio that affects the assemblage compressive strength.[5.11,5.14,5.19]

**Mortar Bedded Area.**  The failure mechanism for prisms made with hollow units is quite different, depending on whether full mortar bedding or face shell bedding is used.[5.21]  For full mortar bedding, the increase in prism failure load over face shell bedding is less than proportional to the increase in mortared area.  Therefore, based on the effective mortared area, the strengths for face shell mortared prisms are higher than for fully mortared prisms.  This result confirms the need to test prisms with mortar bedding and bond patterns conforming to the actual construction.

**Grout Strength.**  As mentioned in Sec. 5.2.3, grouted masonry has failure loads lower than those predicted using superposition of the capacity of the grouted area and the capacity of the mortared block area.  Therefore, strengths, based on gross area for grouted prisms, have been found to be considerably lower than strengths based on net area of similar hollow prisms (see Fig. 5.16).  Other researchers have found, particularly for two block high prisms, that prism strengths exceeding the strength of the unit can be achieved by using suitably strong grout.

The net result of this is that some codes such as the Canadian[5.7] and Australian[5.6] codes contain provisions that result in reduced values of compressive strength for grouted blockwork.  In ACI 530/ASCE 5/TMS 402[5.5] no distinction is made between grouted and ungrouted masonry, but a grout with 2000 psi (13.8 MPa) minimum

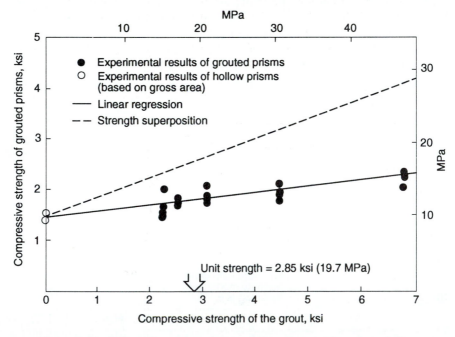

**Figure 5.16**  Effect of grout strength on prism compressive strength (from Ref. 5.17).

strength and not less than the specified compressive strength of the masonry is required.

A factor not accounted for is that fully grouted specimens have a lower scatter of results than hollow or solid prisms, which to some extent offsets any tendency for lower average strengths. It is important that designers not overreact to this aspect of behavior and attempt to place high-strength low-slump concrete as the grout. It is far more important to use a very fluid grout (slump around 10 in. (250 mm)), which is easily compacted to ensure that all voids are filled. As mentioned before, the increased reliability is a significant benefit for grouted masonry construction.

### 5.2.5 Stress–Strain Relationships

Because masonry is not homogeneous in any direction, local deformations may differ over the height, length, and width of an assemblage. Therefore, strain must be measured over a sufficient length to give representative average values. A gage length of 8 in. (200 mm) allows inclusion of one block height and one bed joint. Alternatively, three $2\frac{1}{4}$ in. (57 cm) high bricks and three mortar joints can be accommodated. For other sized units, gage lengths should be multiples of unit height plus joint thickness.

Figure 5.17 contains plots of typical stress–strain data for concrete block and clay brick prisms. The nonlinear behavior beyond about 50% of the peak stress is evident, particularly for concrete masonry. The results from most test machines indicate a sudden brittle failure shortly after reaching the ultimate strain. However, where displacement control is used, there is evidence of a descending branch to the stress–strain curves.[5.29]

Because of the nonlinear shape of the stress–strain curve, the modulus of elasticity can be defined as the chord modulus for a line joining the curve at 5% of

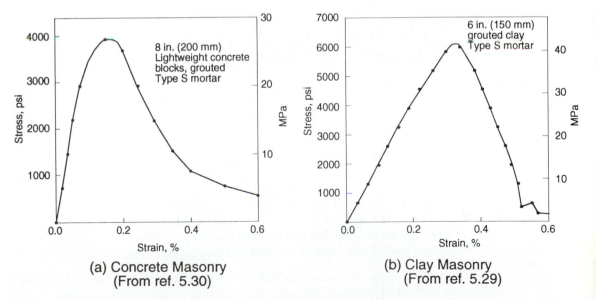

(a) Concrete Masonry
(From ref. 5.30)

(b) Clay Masonry
(From ref. 5.29)

**Figure 5.17** Stress–strain curves of masonry under axial compression.

the strength to 33% of the strength.[5.5] This region usually lies well within the reasonably linear part of the curve. The lower part of the curve is ignored because it often represents a relatively flat region associated with closing up the interfaces between the mortar and the units. In real structures, these interfaces would close up due to self-weight. [The definition of modulus of elasticity $E_m$ was adopted in the early 1980s and not used universally. The literature often refers to the tangent modulus, secant modulus, or chord modulus for different stress levels.]

Laterally confined masonry has increased strength and a more stable postpeak behavior with less degradation.[5.30,5.31]

Traditionally, the modulus of elasticity for masonry $E_m$ is calculated by the equation $E_m = kf'_m$, where $k = 1000$, and $f'_m$ = specified compressive strength. However, recent data indicate that this is a substantial overestimation for most masonry. For concrete blockwork, a k value in the order of 750 has been used as a better estimate.[5.32,5.34] Values for brickwork in the order of 500 to 600 are more typical for North American brick. The results in Fig. 5.18 illustrate the significant scatter observed, part of which can be attributed to variations in test methods including prism configuration, loading setup, instrumentation, and method of calculation. It is worth mentioning that, in the absence of known values of $E_m$, design calculations that may be sensitive to the value of $E_m$ should be checked with a range of representative values to evaluate and account for the potential for overstress.

Because the stress conditions near failure are not the same as at lower loads, no strong correlation between compressive strength and the modulus of elasticity should be expected. As an alternative, ACI 530/ASCE 5/TMS 402[5.5] relates the modulus of elasticity to the unit compressive strength and the type of mortar.

Based on strain compatibility, equilibrium, and linear elastic behavior, the modulus of elasticity $E_m$ can be calculated by equating the assemblage deformation $\Delta_m$ (center line to center line of units) to the sum of deformation in a unit $\Delta_b$ and in a joint $\Delta_j$.[5.18,5.35]

$$\Delta_m = \Delta_b + \Delta_j \tag{5.1}$$

For a uniform compressive stress $\sigma$, Eq. 5.1 can be rewritten as

$$\frac{\sigma}{E_m}(t_b + t_j) = \frac{\sigma}{E_b}t_b + \frac{\sigma}{E_j}t_j$$

From which

$$E_m = \frac{1}{\dfrac{\delta}{E_b} + \dfrac{1 - \delta}{E_j}} \tag{5.2}$$

where $\delta = t_b/(t_b + t_j)$

$t_b, t_j$ = height of unit and mortar joint thickness, respectively
$E_b, E_j$ = modulus of elasticity of unit and mortar joint, respectively

To be more accurate, calculations of the vertical deformations $\Delta_b$ and $\Delta_j$ should reflect the multiaxial states of stress in the units and mortar joints (see Sec. 5.2.3).

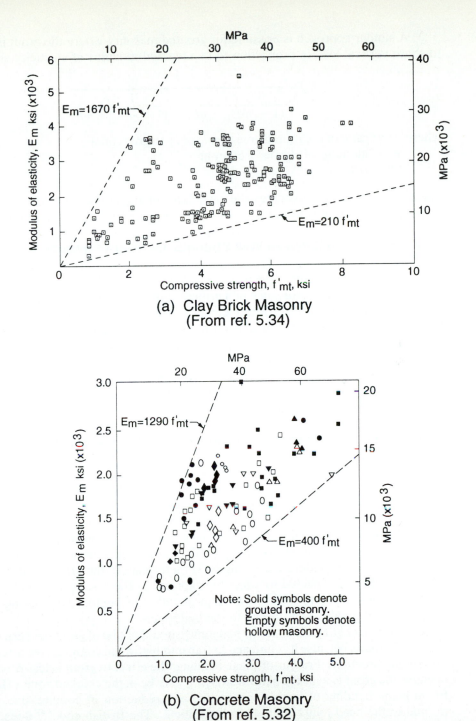

**(a) Clay Brick Masonry**
(From ref. 5.34)

**(b) Concrete Masonry**
(From ref. 5.32)

Note: $f'_{mt}$ is prism compressive strength.

**Figure 5.18** Modulus of elasticity of masonry.

A similar approach is possible for grouted masonry where the grout is allowed to share vertical load in proportion to its area and modulus of elasticity so that

$$E_m = \cfrac{1}{\cfrac{\delta}{(1 - \eta)E_g + \eta E_b} + \cfrac{1 - \delta}{(1 - \eta)E_g + \eta E_j}} \qquad (5.3)$$

where  $\eta$ = ratio of ungrouted area of masonry to gross area

$E_g$ = modulus of elasticity of grout

$\delta = t_b/(t_b + t_m)$

The above expression indicates the sensitivity of $E_m$ to the properties and area of grout. [Note that substituting $\eta = 1$ in Eq. 5.3 gives Eq. 5.2 for solid units.]

### 5.2.6 Relationship Between Wall Strength and Prism Strength

For most materials, larger specimens result in decreased strength because of the greater probability of occurrence of a critical combination of minor flaws. Full scale wall tests normally result in lower compressive strengths than prisms because of factors such as slenderness or uneven distribution of load along the length of the wall. However, in many cases when short masonry walls are loaded uniformly, tests indicate compressive strengths higher than the corresponding unconfined (no platen restraint) prism strength.[5.36–5.39] A suggested explanation is that the effect of a significant flaw is less important when it forms only a small part of the wall section versus a large part of a prism section.

### 5.2.7 Compressive Strength for Loading Parallel to the Bed Joint

Because most loadbearing masonry elements are designed to carry vertical loads in compression, compressive strength has been associated with compression normal to the bed joints. Tests of masonry prisms with compression parallel to the bed joints[5.39–5.42] resulted in the failure modes shown in Fig. 5.19. Compared to loading normal to the bed joints for hollow units and grout-filled hollow units, lower strengths and different failure modes are likely, mainly due to geometric differences. If the units themselves have different strength characteristics in the orthogonal directions, this would also be expected to affect the capacity of the assemblages for loading in these directions. Figure 5.20 shows typical ratios of compressive strengths from prisms loaded parallel and normal to the bed joints.

Lower strength parallel to the bed joints should not be of great concern for most design situations because it is unusual to transmit high axial compressive forces in the horizontal direction. For out-of-plane bending, the effect of grout is less pronounced because the grout is near the neutral axis or it may be in the cracked zone. However, for in-plane bending of beams and lintels, any reduction in compressive strength normal to the head joints could be more critical. The British code[5.43] does, in fact, specify different design stresses for compression normal and parallel to the bed joints. This has not been done in North American codes because the allowable compressive stresses are relatively low. Further, as is discussed in Chap. 6, tests on masonry beams have not shown this to be very significant—perhaps because of the

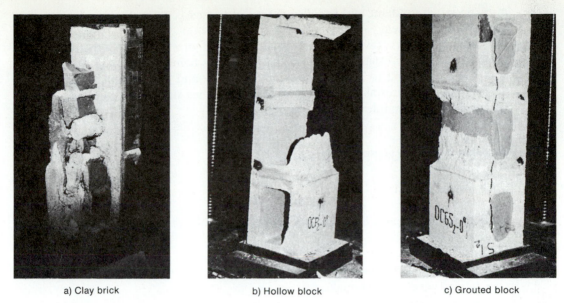

a) Clay brick       b) Hollow block       c) Grouted block

**Figure 5.19** Failure modes of masonry prisms tested under compression parallel to bed joints.

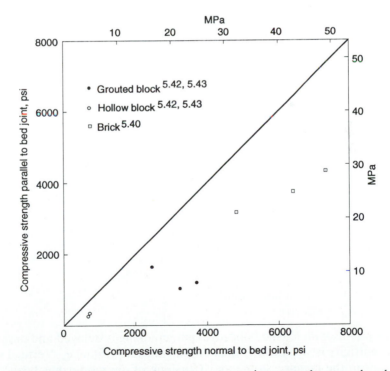

**Figure 5.20** Relationships between compressive strengths normal and parallel to bed joint.

in-plane strain gradient and because design of beams is normally based on tension controlled capacities.

For solid units, fewer mortar joints (i.e., head joints), result in the modulus of elasticity for compression parallel to the bed joints being higher than the value for compression normal to the bed joints. For hollow and grout-filled units, different geometries and material properties can result in either similar[5.32] or opposite results indicating the no simple relationship exists.

## 5.3 COMBINED AXIAL COMPRESSION AND FLEXURE

### 5.3.1 Introduction

Walls designed to carry axial load are usually also subject to in-plane or out-of-plane bending resulting from

- eccentrically applied load
- lateral wind or earthquake forces
- continuity with other framing elements
- simply accidental eccentricity of load or variation in the properties of the wall.

The combination of axial load and flexure means that sections are subject to strain gradients along their lengths or through their thicknesses. The effect of the strain gradient on behavior and on design criteria is discussed in the following sections.

### 5.3.2 Prism Tests

Tests on standard prisms loaded eccentrically with equal eccentricities and pinned end conditions have been used to assess the effect of strain gradient on compression capacities and stress–strain relationships. The factors discussed in Sec. 5.2.2 are also applicable here.

Figure 5.21 contains photos of failures of clay brick and concrete block prisms loaded under eccentric compression. Vertical splitting similar to that for concentric compression occurs for cases when the eccentricity is small enough to keep the entire section under compression. For eccentricities greater than the kern distance $e_k$, cracking occurs at the mortar joints as load is applied for the case of zero tensile bond strength.

$$e_k = S/A_n \tag{5.4}$$

where   $S$ = the section modulus
$A_n$ = effective area

For prisms with tensile bond strength, cracking is delayed and may only occur if the eccentricity is significantly greater than the kern distance. Vertical splitting may also occur in the compression zone. Crushing of the mortar has been observed at large eccentricities (approaching $t/2$).[5.45,5.46]

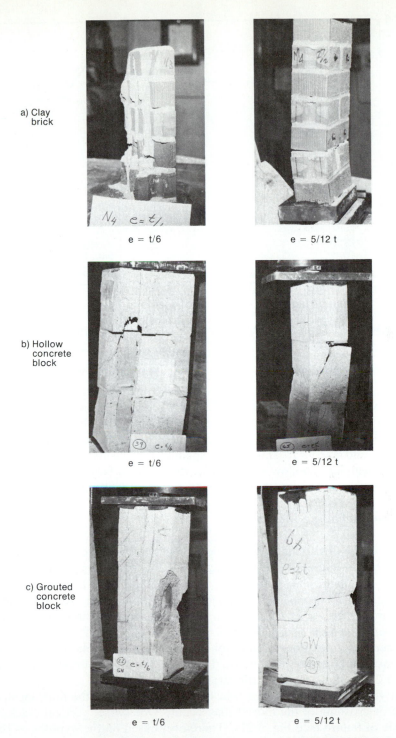

a) Clay brick

e = t/6        e = 5/12 t

b) Hollow concrete block

e = t/6        e = 5/12 t

c) Grouted concrete block

e = t/6        e = 5/12 t

**Figure 5.21** Failure modes of masonry prisms under eccentric loading.

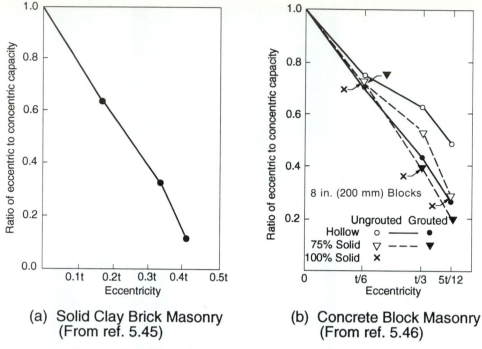

**(a) Solid Clay Brick Masonry**
**(From ref. 5.45)**

**(b) Concrete Block Masonry**
**(From ref. 5.46)**

**Figure 5.22**  Effect of eccentricity on compression capacity.

As expected, test results in Fig. 5.22 show decreasing capacities associated with increasing eccentricities. As can be seen in Fig. 5.22(b), the reduction in capacity for hollow concrete masonry at higher eccentricities is not as pronounced as for grouted or solid masonry. This can be explained by considering that the kern distance is much larger for a hollow section and, therefore, a larger fraction of the hollow section remains under compression at these high eccentricities.

### 5.3.3 General Failure Mechanisms

Analyses of the effects of strain gradient have been based on assumptions of linear elastic behavior and zero tensile strength. On this basis, for prisms made with solid units, the calculated maximum stress at the extreme fiber at failure is significantly higher than the compressive strength for concentric loading. As shown in Fig. 5.23, this apparent increase in strength can be quite substantial and has been referred to as the *strain gradient effect*. Strengths ranging from 1.3 to 1.5 times the strength for zero eccentricity have been found for prisms made with solid units and from 1.5 to 2.0 for grouted concrete blockwork. However, for hollow masonry, as can be seen in Fig. 5.23(b), there is very little strain gradient effect.

At very high eccentricities near half the thickness of the unit, local spalling of the unit or crushing of the mortar may occur. Otherwise, vertical cracking typifies the mode of failure (see Fig. 5.21). Therefore, based on the hypothesis that triaxial confinement of the mortar produces lateral tension in the unit (see Sec. 5.2.3), the strength gain has been attributed to a smaller part of the mortar joint being under high

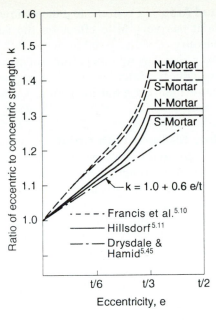

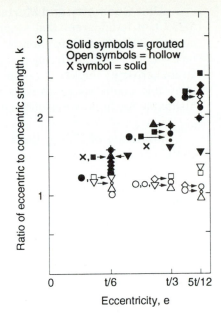

(a) Solid Clay Brick Masonry

(b) Hollow and Grouted Concrete Masonry (From ref. 5.46)

**Figure 5.23**   Ratio of eccentric to concentric compression strength of masonry.

stresses. This in turn results in lower lateral tension stresses in masonry units.[5.11] The relatively uniform stress across a face shell may explain the minor influence of this strain gradient factor on hollow masonry.

Within the limitations of the assumptions, the foregoing hypothesis provides a reasonably rational explanation for the observed behavior. However consideration of nonlinear behavior and tensile strength can also be used to partially account for this effect.

For working stress design based on linear elastic analysis, the strain gradient effect has been included in codes by increasing the theoretical eccentricity coefficient by some amount such as the 30% applied in the SCPI (now Brick Institute of America) recommendations.[5.47] Alternately, higher allowable stresses have been commonly specified for allowable compression due to flexure compared to the allowable compression due to axial load. With the introduction of strength design methods, this approach is being replaced with nonlinear empirical relationships to predict equivalent magnitudes and locations of the compression force resultant.

### 5.3.4 Factors Affecting the Influence of Strain Gradient

In addition to the magnitude of the eccentricity, the material and geometric factors discussed in what follows can influence the strain gradient effect on masonry compressive strength.

**Unit Geometry and Mortar Bedded Area.**    From the test data shown in Fig. 5.23(b), it is apparent that strain gradient has very little effect on the calculated compressive strength for hollow unit masonry. As discussed in Sec. 5.3.3, this can be explained by considering that a thin face shell has relatively uniform compressive strain across its thickness except at very high eccentricities. For this reason, higher allowable compressive stresses for flexure than for axial load do not seem to be justifiable for hollow masonry construction.

**Influence of Filling Cells of Hollow Masonry with Grout.**    Figure 5.23(b) contains test data for eccentric loading of grouted hollow block prisms. There is a considerable increase in apparent compression strength with increased eccentricity. However, the apparent increase is partly due to the strength of the grouted prism under concentric load being about 30% lower than the corresponding strength for the hollow prism. With increasing eccentricity, the grout in the middle of the prism is less highly strained and carries a lower share of the load. This in turn means that shrinkage or incomplete compaction, lateral force from the grout acting on the units, or wedging action of the grout within the units (see Sec. 5.2.3) would all have less influence. Hence, a major part of the apparent strength gain due to the strain gradient effect is attributable to the reduction in the influence of the grout on capacity.

In addition to the nonlinear effects of the stress–strain relationship, the grout column has a much higher tensile strength than is usually found for bond between the mortar and units. Therefore, the smaller effect of nonuniformity of the grout with the block, the nonlinear stress–strain relationships for the grout and block, and the increased area of material in tension all combine to contribute to the apparent strength increase.

**Material Strengths.**    Within normal ranges of strengths and types of units, grout, and mortar, they do not have any significant effect on the capacity ratios for solid or grouted masonry. However, at high eccentricities, the local spalling of the units or crushing of the mortar would be expected to be influenced to a greater extent by the strengths of the materials, particularly the mortar.

### 5.3.5 Compression Stress–Strain Relationship

**Shape and Modeling of the Relationships.**    Concentrically loaded uniaxial compression tests have traditionally been used to define the stress–strain relationships for materials loaded under strain gradients. The implicit assumption is that there is no effect of strain gradient on the shape of the stress-strain curve. Because the real stress at any fiber across the section is not known, the only rational check on whether this approach is satisfactory is to check the magnitude and location of the resultant force for various strain gradients.

For concrete, the rectangular Whitney stress block illustrated in Fig. 5.24(b) has been used to simplify calculations of the compression force resultant at ultimate strength. Because this is a well established and convenient format, similar developments have been undertaken for masonry.[5.48,5.49] As shown in Fig. 5.25, for grouted masonry, a reasonably consistent location of the result force, defined by $k_2$, has been found. The compression stress block, defined by $k_1 k_3 f'_m$, in the actual stress diagram

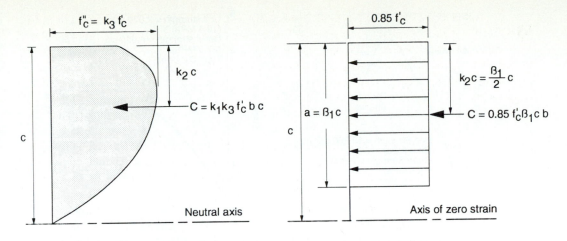

(a) Actual Stress Distribution and Stress Parameters

(b) Whitney Stress Block

**Figure 5.24** Flexural strength design of reinforced concrete members.

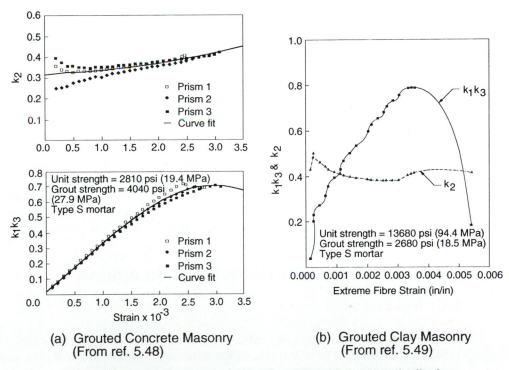

(a) Grouted Concrete Masonry (From ref. 5.48)

(b) Grouted Clay Masonry (From ref. 5.49)

**Figure 5.25** Stress parameters for the compressive stress distribution under eccentric loading.

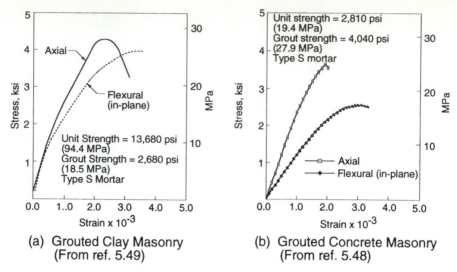

**(a) Grouted Clay Masonry**
**(From ref. 5.49)**

**(b) Grouted Concrete Masonry**
**(From ref. 5.48)**

**Figure 5.26** Stress–strain curves for 8-in. grouted masonry under concentric and eccentric compression.

(Fig. 5.24(a)), can be seen as being quite sensitive to the extreme fiber strain. For this reason, new strength design provisions for masonry codes[5.33,5.50] have tended to adopt the rectangular stress block.

Linear stress–strain relationships have also been proposed for ultimate limit state design. In either case, the empirical representations incorporate various degrees of approximation. Most building codes permit the designer to use any realistic model of the actual stress–strain relationship.

**Maximum Compressive Strain.** As is the case with concrete, the maximum extreme fiber compressive strain under eccentric load is higher than under concentric load. Typical test results are shown in Fig. 5.26. For tests of grouted hollow prisms,[5.46,5.49] the strain at maximum stress ranges from 0.0025 to 0.004 for concrete and clay masonry prisms.

Tests under strain controlled conditions indicate that, like concrete, much higher strains can occur in the postpeak load range prior to dramatic load reduction or stiffness degradation. This can be important for consideration of ductility and energy dissipation. From a practical viewpoint, a maximum masonry compressive strain of 0.003 as specified in the Uniform Building Code[5.33] is justifiable.

## 5.4 FLEXURAL TENSILE STRENGTH FOR OUT-OF-PLANE BENDING

### 5.4.1 Introduction

The flexural strength of masonry subject to out-of-plane bending relates to the resistance of walls subject to lateral loads from wind, earthquake, or earth pressures, and to eccentric load or direct bending due to gravity loading. Depending on the support conditions and wall geometry, lateral loads can result in bending about the vertical axis, horizontal axis, or both. Flexural tensile strength is usually referred to in terms of direction of the tension, that is, either normal to the bed joints, $f'_{tn}$, or parallel to the bed joints, $f'_{tp}$.

The measurement of flexural strengths, factors affecting the strengths, and illustrative values are presented in the next parts of this section.

### 5.4.2 Flexural Test Methods

**Tension Normal to the Bed Joints.** The bond between mortar and the bed joint face of masonry units has been of interest both for strength and water permeance. The basic test methods developed to determine flexural tensile strengths, illustrated in Fig. 5.27, are as follows:

**1.** *ASTM E518*.[5.51] This method uses beam tests of stack-bonded prisms with prism heights of at least 18 in. (450 mm) and either third-point loading or uniform loading using an air bag, as shown in Fig. 5.27(a). Because of the characteristic high scatter of tensile bond strengths, the minimum five tests leads to fairly high coefficients of variation and results in difficulty with statistically quantifying the confidence with which comparisons can be made.

The beam test provides a somewhat biased result because it fails at the joint with the critical combination of high bending moment and low joint strength. Several other joints in the specimen usually have higher strengths.

**2.** *ASTM C1072*.[5.52] The bond wrench test, shown in Fig. 5.27(b), was developed to provide a larger sample where each joint of a prism is tested. Testing all the joints also eliminates any bias in the test results. The ASTM bond wrench employs a relatively short lever arm and is intended to be used in a test machine. However, research regarding the influence of the length of the lever arm, which affects the ratio between axial compression and bending and thus the strain gradient, shows that there is no discernible effect. Therefore, lever arms of 4 ft (1.2 m) or longer have been used both for laboratory as well as field tests.

**3.** *ASTM E72*.[5.53] Quarter-point or uniform load tests of wall specimens, as shown in Fig. 5.27(c), have been used to provide data that may not be so directly affected by low bond strength at a particular mortar–unit interface. The difference is that the failure plane is several units long.

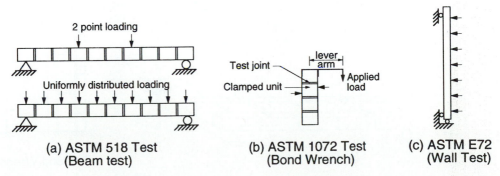

**Figure 5.27** Standard tests for flexural tension normal to bed joints.

The base support conditions for this test are very critical to obtaining meaningful results. If friction or other means of horizontal restraint exist at the bottom roller, which supports the weight of the wall, the wall is statistically indeterminate and a restraining bending moment will be developed at the lower horizontal support. The self-weight of the wall should also be accounted for when calculating the flexural tensile strength.

*Comparison of Results.* Tests on clay brick prisms[5.54,5.55] indicate that mean flexural bond strengths determined using the bond wrench (ASTM C1072) are higher than strengths of similar prisms determined by the beam tests (ASTM E518). Average flexural strengths and the corresponding coefficients of variation determined from tests of walls with well defined support conditions[5.55,5.56] are lower than values from bond wrench tests of similar prisms.

**Tension Parallel to the Bed Joints.** The foregoing test methods were developed to determine tensile bond strength normal to the bed joints. However, with some modifications, researchers have adapted them to tests for tension parallel to the bed joints. For the prism beam tests, the length parallel to the bed joints must usually be at least four units, to accommodate possible failure in a stepped or toothed pattern along head and bed mortar joints, as shown in Fig. 5.28(a). An even number of courses should be used to have an equal distribution of head joints and units through any section taken normal to the bed joints. As shown in the bottom drawing of Fig. 5.28(a), use of half-height units in the first and last courses is recommended, for symmetry and to prevent twisting.

For bond wrench testing, symmetry of the joints is very important to ensure even stress distribution. Specimens such as shown in Fig. 5.28(b) can be used as long as the distance from the bond wrench to the holding frame is sufficient to permit all potential failure modes. When failure occurs along a combination of head and bed joints, it is difficult to obtain more than one result per specimen. This reduces the advantage of using the bond wrench test method for tension parallel to the bed joints.

Walls tested in accordance with ASTM E72 with tension parallel to the bed joint provide results that are again similar to full-sized walls. However, to ensure simple support conditions, testing of the walls as horizontal beams may be preferable.

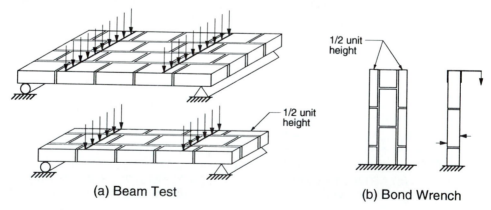

(a) Beam Test        (b) Bond Wrench

**Figure 5.28** Flexural tension tests for tension parallel to bed joints.

### 5.4.3 Failure Mechanisms

As illustrated in Fig. 5.29, there are four basic modes of flexural tensile failure. These are briefly discussed in what follows.

**1.** *Cracking (Debonding) Along the Bed Joint.* For flexural tension normal to the bed joints, failure consists of debonding of the mortar from the unit along the bed joint. The relative influences of chemical bonding, adhesion, and mechanical interlock are not well understood and no reliable model has yet been developed to predict bond based on the properties of the materials.

A typical flexural tension failure for solid brick masonry is shown in Fig. 5.29(a) and for hollow and grout-filled hollow block masonry in Fig. 5.29(b). In the latter case, the continuity provided by the continuous grout column becomes the dominant factor in flexural cracking. Cracking occurs through the minimum grout section, as shown in the lower specimen in Fig. 5.29(b).

**2.** *Cracking Through Head Joints and Masonry Units in Alternate Courses.* For flexural tension parallel to the bed joints, cracks normally initiate in the head joints and can pass directly through the units in alternate courses, as illustrated in Fig. 5.29(c). For specimens made of hollow or solid units, failure through the units is associated with units that are relatively weak compared to the shear strength of the unit-mortar interface for the toothed failure described in the next section. For fully grouted hollow masonry, the continuous columns of grout tend to force the crack to propagate in a straight line through the units in line with the head joint in alternate courses, as shown in Fig. 5.29(d). In this case, the grout itself does not contribute significantly to the strength except by forcing the failure through the unit in order that the crack does not have to cross the grout column.

**3.** *Cracking in a Toothed Pattern Along a Combination of Head and Bed Joints.* For masonry with relatively strong units and weak mortar joints, flexural failure for tension parallel to the bed joints can occur through a combination of tensile and shear debonding in the head and bed joints, as illustrated in Fig. 5.29(e).

**4.** *Diagonal Cracking—Either Stepped Along Combinations of Head and Bed Joints or Along the Shortest Path Through Units.* A stepped cracking pattern along head and bed mortar joints such as illustrated in Fig. 5.29(f) is a variation of that in Fig. 5.29(e). In cases where the masonry units are relatively weak or where continuous columns of grout significantly strengthen the bed joints, the diagonal crack can pass through the units along the shortest path. In practice, these failure patterns are usually associated with biaxial bending.

### 5.4.4 Factors Affecting the Tensile Bond Between Masonry Units and Mortar

The flexural tensile strength normal to the bed joints for solid and hollow masonry depends on the tensile bond at the unit–mortar interface. Test data[5.54–5.58] using the bond wrench test apparatus shows a large range of strengths from 30 to 250 psi (0.2 to 1.75 MPa). The factors affecting bond between mortar and units are briefly discussed in what follows. However, it should be understood that the relative effect

a) Clay brick: Cracking along bed joint

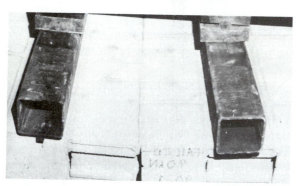

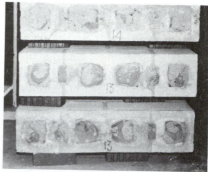

b) Concrete block: Cracking along bed joints

c) Clay brick: Cracking through head joint

d) Grouted concrete block: cracking through head joints

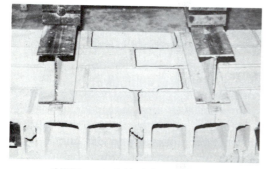

e) Hollow concrete block: Toothed cracking

f) Hollow concrete block: Stepped cracking

**Figure 5.29** Typical failure modes under flexure.

of any one factor can be very much affected by other factors. It should also be noted that the achievement of bond strengths higher than those needed in design may not provide any significant enhancement of performance, and satisfying other desirable properties such as freeze–thaw durability and adequate workability for good workmanship is also important.

**Mortar Type and Constituent Materials.** For most masonry units, a higher tensile bond corresponds to mortar with larger proportions of portland cement as shown in Fig. 5.30. In addition to the relative amounts of portland cement and lime associated with the different mortar types, it is clearly evident that the type of unit and the type of lime have significant effects.

The comparison of bond strengths between portland cement–lime mortars (PCL) and masonry cement mortars (MC) has been a very controversial issue. Although it is tempting to simply avoid this issue, some background comment should be provided. Some researchers[5.57,5.58] have found lower bond strength for masonry cement mortars, and masonry codes in the United States[5.5,5.33] specify lower allowable flexural tensile stresses than for comparable PCL mortars. Data in Fig. 5.31[5.57] is indicative of the types of differences observed. However, a major problem encountered when interpreting data and in attempts to compare data from various sources is also illustrated in this figure. The bond tests using type A clay brick obviously differ markedly from similar tests done with type B bricks. Different mortar mixing methods, wetting of brick B, and dry laboratory curing for brick A, all contributed to the differences between comparable tests and the somewhat different trends for the types of mortar used.

Because the nature of the bond is not completely understood, it is not possible at this time to offer a quantitative explanation. However, it is generally acknowledged

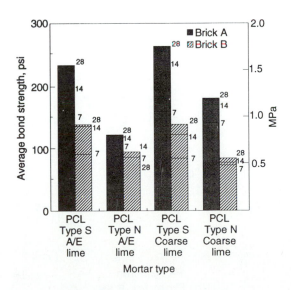

**Figure 5.30** Effect of mortar type on bond strength (from Ref. 5.57).

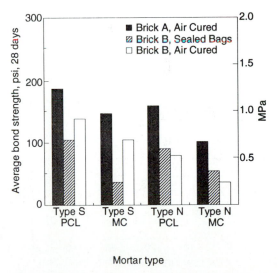

**Figure 5.31** Comparison of bond strengths using portland cement-lime and masonry cement mortars (from Ref. 5.57).

that higher contents of the entrained air result in a lower bond strength. The increased workability due to the entrained air reduces the water content. Thus, it is especially critical that MC mortars not be used for bricks that dry the mortar out rapidly and that moist curing be provided to ensure that sufficient moisture is present for hydration of the cementing components of the mortar.

**Initial Rate of Absorption, Moisture Content, and Temperature of the Masonry Unit.** During the initial contact between mortar and a unit, the unit absorbs some of the water from the mortar and in the process creates the "intimate" contact that is associated with a good bond between the two materials. As a result, there is evidence to suggest that masonry units that do not draw the mortar into intimate contact through initial absorption of water will not develop a good bond. Conversely, units with a very high suction rate may completely dry out the layer of mortar next to the unit, again resulting in poor bond. Using the initial rate of absorption (IRA) (see Sec. 4.3.6) as the measure of absorption characteristics of clay masonry, it seems justifiable to limit the maximum initial rate of absorption, but it is more difficult to establish some minimum value.[5.59] As shown in Fig. 5.32, the scatter of data are very high. Perhaps because it has not been possible to eliminate the effects of other parameters, no optimum level of IRA has been established. Although the IRA test is only a standard test for clay units, values for concrete and calcium silicate units have produced similarly scattered results.

The moisture contents of masonry units affect the bond[5.60] by altering the absorption characteristics, and it is for this reason that some bricks are wetted prior to laying. Temperature also affects the absorption properties of the units. Chapter 15 contains a more detailed discussion regarding construction in extreme weather conditions.

**Mortar Strength, Flow, and Water Retentivity.** Because higher mortar strength normally correspond to higher portland cement content, there may be some correlation between compressive strength of mortar cubes and mortar bond. How-

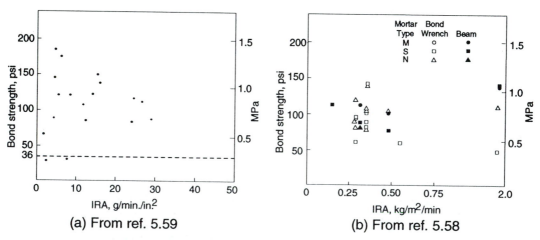

(a) From ref. 5.59          (b) From ref. 5.58

**Figure 5.32**  Effect of IRA on bond strength of clay brick masonry.

ever, other factors such as workability can have as significant an effect on bond. As shown in Fig. 5.33(a), higher flows associated with higher water contents result in better bond for clay bricks having high suction or absorption rates. Optimum flows might be lower for units with lower absorption. As shown in Fig. 5.33(b), some compromise between compressive strength and bond may be desirable where increased water content reduces compressive strength.

Water retentivity is a measure of the relative rate at which mortar allows water to be absorbed from it. Thus, it is logical to expect that, depending on the initial flow and the absorption characteristics of the units, an optimum water retentivity exists. In ASTM C270, the recommended minimum limit for water retentivity is 75%.

**Workmanship and Surface Condition of Units.** Tests indicate that contamination of the surface of the unit (e.g., layers of dust, caked mud, loose material) tends to reduce the bond between the mortar and the unit.[5.63] Water penetration inhibitors sprayed on the units or contained within the material can dramatically affect bond strength. In cases where water penetration inhibitors produce very low absorption, the bond is generally reduced. However, for untreated units that are very absorptive, such treatments have proven to be very beneficial to bond. Attempts to improve tensile bond by roughening the surface of the units have not been very successful. However, there are benefits for laying the unit. A brick with a frogged bed is less likely to move after initial placement compared to a brick with a smooth face.

The incomplete filling of joints, rocking, or otherwise shifting the units after the initial absorption of water from the mortar has occurred and the use of mortar that has begun to set, are examples of workmanship related factors that can reduce the tensile bond between the units and the mortar. The time between spreading the mortar and placing the unit, the thickness of the spread mortar relative to the final joint thickness, and the pressure applied when laying a unit are other factors that can cause variations in bond.

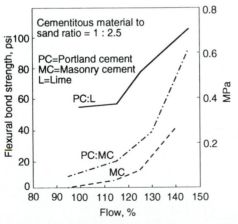

(a) Effect of Mortar Flow (From ref. 5.61)

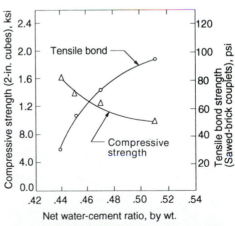

(b) Effect of Water-Cement Ratio (From ref. 5.62)

**Figure 5.33** Effect of mortar flow and water-cement ratio on bond strength.

**Curing.** Conditions during construction and the subsequent curing can have dramatic effects on the tensile bond. Because most laboratory work has been done on specimens made and stored in dry laboratory conditions, in some cases, the full potential for bond strength has not been realized. An example[5.64] from concrete block research showed that walls stored for 4 months outside under rain, snow, and large variations in temperature had an average tensile bond strength of 126 psi (0.87 MPa) compared to 54 psi (0.37 MPa) for similar specimens stored in the laboratory.

The effect of age is another interesting parameter affecting bond. Tests indicate that the bond at 28 days may not be significantly different from the 7 day bond for dry curing conditions. Data shown in Fig. 5.34 suggests that there is no clear relationship between bond strength and age in the 28 day time period. However, for clay brick prisms tested over a 5 year period, the gradual additional curing even in a dry laboratory environment resulted in significant improvements to the bond strength.[5.64] A mortar bond increase of 50% over a 6 year period is reported for clay masonry.

The data in Fig. 5.31 clearly illustrates the influence of curing. The specimens cured at 100% relative humidity by placing the prisms in sealed bags were significantly stronger than those allowed to dry out. Poor initial curing conditions such as freezing or high temperatures are especially critical and, although later curing will improve the bond, the poor initial curing can have a permanent effect.

### 5.4.5 Factors Affecting the Flexural Tensile Strength of Grout Filled Hollow Masonry

Grouting the cells in hollow masonry significantly increases the flexural tensile strength normal to the bed joints because the tensile strength of the grout is typically much higher than the bond between the mortar and the units. As is shown in Fig. 5.35, the benefit of the grout decreases as the volume of grout decreases.[5.65] However, for hollow blocks with 52% solid, a more than three fold increase in strength normal to the bed joints was found. Considering that the section modulus based on the minimum face shell area is only 71% of that for the solid wall, the moment capacity was increased by over 400%. For flexural tensile strength parallel to the bed joints, the more than two fold increase obtained was mainly associated with changing the

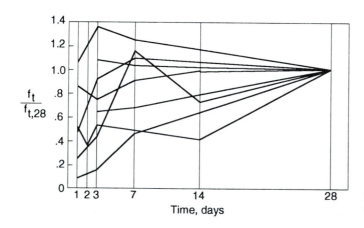

**Figure 5.34** Effect of age on bond strength (from Ref. 5.61).

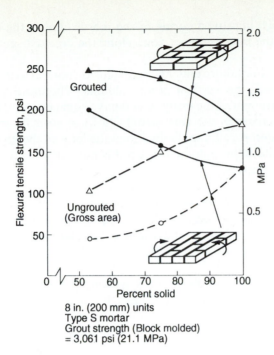

**Figure 5.35** Flexural tensile strength of hollow and grouted concrete masonry (from Ref. 5.65).

8 in. (200 mm) units
Type S mortar
Grout strength (Block molded)
= 3,061 psi (21.1 MPa)

failure mode from cracking along combinations of head and bed joints to cracking through head joints and units in alternate courses. The factors affecting the influence of grouting on the flexural tensile strength are briefly discussed in what follows.

**Relative Volume of Grout.** From the tests using solid, 75% solid, and hollow blocks, it can be seen in Fig. 5.35 that, for tension normal to the bed joints, the flexural tensile bond strength increases with higher percent solid for the ungrouted case. The stress calculations are based on the effective mortared areas. This increase in bond may be due to larger mortar areas retaining more moisture for better curing. Also, the weakening of the bond due to shrinkage near the boundaries of the mortar joints is less significant for larger areas. For the grouted case, the much higher tensile strength of the grout has a pronounced beneficial effect even though the grout is not near the extreme fiber during bending. Even for the 75% solid block, in which case the volume of grout is much smaller and is located further from the extreme fiber than for hollow block, grouting resulted in more than a 150% increase in moment capacity.

For tension parallel to the bed joints, the ungrouted hollow walls failed by cracking in a toothed pattern along combinations of head and bed joints, as shown in Fig. 5.29(e). For the 75% solid and solid blocks, failure was normally along a straight line through head joints and blocks in alternate courses, which resulted in substantial increases in tensile strength and moment capacity over hollow masonry. Filling the cells of hollow block with grout forced the failure to pass through blocks in alternate courses, which produced over a 150% increase in tensile strength and over a 200% increase in moment capacity. For the walls made of 75% solid units, the increase in strength for the smaller volume of grout is related to the improved tensile strength along the head joint through improved curing of the mortar due to the presence of water in the grout and to the larger contact zone along the cracked section.

**Strength of Grout.** For failure under flexural tension normal to the bed joints, the crack must pass through the vertical grout columns. Therefore, it would be expected that the tensile strength of the grout would have some effect. However, as shown in Fig. 5.36, the increase is not very significant for a normal range of grout strength. For flexural tension parallel to the bed joint, as long as the grout is strong enough to force flexural failure through the blocks, it is not expected that the actual strength would have much impact.

**General Discussion.** Even for reinforced masonry, where the steel is designed to resist the tensile forces, the modulus of rupture of grouted or partially grouted walls is important for deflection and stiffness calculations. Tests[5.66] on full scale block walls have shown modulus of rupture values for tension normal to the bed joint ranging from 130 psi (0.90 MPa) for cells grouted at 24 in. (600 mm) on center to 320 psi (2.20 MPa) for solidly grouted masonry. The Uniform Building Code[5.33] specifies the modulus of rupture as 2.5 times the square root of the specified compressive strength of the grouted hollow unit masonry. Most other codes do not make allowances for the grout and treat fully grouted masonry simply as solid masonry.

### 5.4.6 Orthogonal Strength Ratio

The ratio of flexural tensile strength parallel to bed joints versus normal to the bed joints, referred to as the orthogonal strength ratio R, is affected by many factors. It

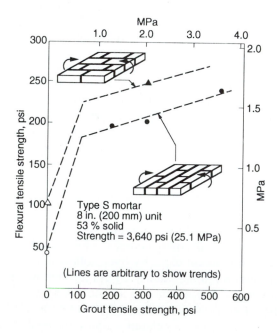

**Figure 5.36** Influence of grout strength on flexural strength of masonry (from Ref. 5.65).

is a measure of the degree of anisotropy of the material and is set at 2.0 in most masonry codes. However, except for stack pattern construction, this ratio ranges from 1.5 to 8 for clay masonry.[5.67,5.68] The reason for this range is that, for tension parallel to the bed joint, cracking must either follow the much longer path around units or alternatively pass through some units. Some factors affecting the orthogonal strength ratio are briefly discussed in what follows.

**Unit Strength.** Where tension parallel to the bed joints produces cracks through the units in alternate courses, the tensile strength of the unit should logically have some effect. However, the orthogonal strength ratio also depends on the relative magnitude of mortar bond versus unit strength. Figure 5.37 shows the variation of orthogonal strength ratio with unit strength for different flexural bond strengths.[5.67]

**Filling the Cells of Hollow Units with Grout.** As is shown in Fig. 5.38, the continuity provided by the grout reduces the degree of anisotropy and produces a more uniform strength for bending in any direction. Grouting more dramatically increases the tensile strength normal to the bed joints, which results in a decrease in the orthogonal strength ratio.

**Percent Solid of the Unit.** In general, when failure passes through the units in alternate courses under tension parallel to the bed joints, the orthogonal strength ratio decreases with the increase in percent solid of the unit.

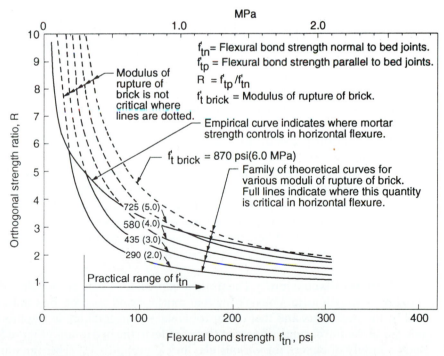

**Figure 5.37** Orthogonal strength ratio of brick masonry (from Ref. 5.67).

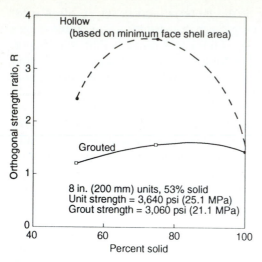

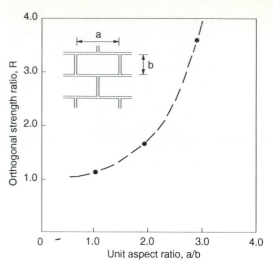

**Figure 5.38** Orthogonal strength ratio of hollow and grouted concrete masonry (from Ref. 5.65).

**Figure 5.39** Effect of aspect ratio of units on orthogonal strength ratio of brick masonry (from Ref. 5.69).

**Aspect Ratio of Units.** Test results[5.69] show that for solid brick masonry, the ratio of flexural strengths parallel versus normal to the bed joints can be highly sensitive to the aspect ratio (length versus height) of the unit. In running bond, the aspect ratio directly affects the length of the failure path around units. For tension parallel to the bed joints, higher aspect ratios correspond to higher strengths due to the increased length of the failure crack. For failure along a straight line through head joints and units, the aspect ratio of the units has no effect. Figure 5.39 illustrates the influence of aspect ratio on the orthogonal strength ratio for cases when a toothed failure occurs for tension parallel to the bed joints.

**Vertical Compression Stress.** Test results[5.70] for walls subject to combined axial load and flexure indicate that the increased bending capacity for tension normal to the bed joints is directly related to the compressive stresses caused by the axial load. For tension parallel to the bed joints, it has also been found that the compression in the bed joints increases the torsional shear resistance to toothed cracking around the units. The effect of compression normal to the bed joints is discussed in Sec. 5.4.7 and shown in Fig. 5.40. Where the crack passes through the unit, this precompression would not be expected to increase the strength.

**General Discussion.** Theoretical expressions can be developed for the orthogonal strength ratio, considering that mortar bond controls flexural capacity normal to the bed joints and that either torsional shear cracking in the mortar joints or unit rupture control capacity for tension parallel to the bed joints. Figure 5.37 displays such a family of curves for various clay brick strengths.[5.67] The modulus of rupture of the masonry unit affects the orthogonal strength ratio particularly for higher values of flexural bond strength (normal to bed joints). An empirical curve is shown in Fig.

Masonry Assemblages    Chap. 5

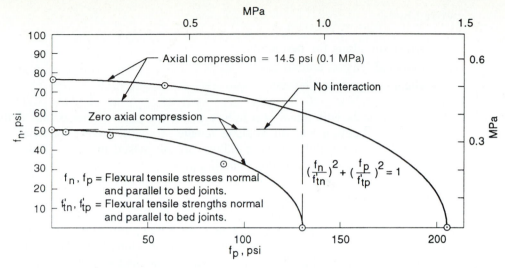

**Figure 5.40** Biaxial bending strength of brick masonry (from Ref. 5.70).

5.37 to estimate the orthogonal strength ratio R where mortar bond strength controls the horizontal flexural strength.

### 5.4.7 Biaxial Strength

Out-of-plane loading of masonry walls often produces bending in both the vertical and horizontal directions. Where the horizontal span is not more than 2 to 3 times the vertical span, biaxial bending can result in increased load carrying capacity compared to simple vertical bending. As illustrated in Fig. 5.40, the usual design assumption is to ignore interaction between the tensile stresses under biaxial bending. However, the test data presented[5.70] indicate that parabolic relationships exist to describe failure under this biaxial bending condition. The effect of axial compression is to increase horizontal and vertical bending capacities by approximately the same percentage for vertical toothed cracking.

When data are lacking, a linear interaction diagram would produce a simple and conservative approximation of this interaction. However, where tensile *stress* parallel to the bed joints does not exceed the *strength* normal to the bed joints, it seems reasonable to ignore this interaction. For special conditions such as stack pattern, grouted walls, and failure through units, more complex failure criteria may be required to accurately predict biaxial bending capacities.

## 5.5 SHEAR STRENGTH ALONG MORTAR BED JOINTS

### 5.5.1 Introduction

Masonry shear walls are intended to resist shear forces due to in-plane lateral loads plus to the effects of axial load and bending. Depending on the form of construction and the combined effects of axial load and bending, the shear failure mode is characterized by

- shear slip along the bed joints,
- diagonal tension cracking, or
- shear compression failure.

This section deals with shear slip failure along the plane of weakness defined by the bed joint. Section 5.6, on in-plane tensile strength, covers diagonal cracking strength. Other potential modes of failure are discussed in Sec. 5.7 dealing with strengths under combined loading conditions.

### 5.5.2 Test Methods

Tests to measure the shear strength along mortar bed joints have not been standardized and, as a result, many variations have been developed. Figure 5.41 illustrates some types of test that have been used. For specimens (a) to (c), the intent was to minimize bending stresses on the mortar joint and to produce uniform shear deformation along the joint. In addition to the shear force produced by the axial compression load, it is possible to add compression normal to the bed joints using hydraulic jacks and spring assemblies.[5.71,5.72]

Construction of specimens with mortar joints at various angles to the loading axis, such as shown in Fig. 5.41(d), produces different combinations of compressive force normal to the bed joints and shear force parallel to the bed joints.

### 5.5.3 Failure Modes

Figure 5.42 shows the failure modes of off-axis compression test prisms for clay brick and concrete block masonry. In Fig. 5.42(a), the shear stress parallel to the bed joints is greater than the compression force normal to the bed joints and a pure slip failure occurs. For the prisms with bed joints oriented closer to perpendicular to the applied load, as shown in Fig. 5.42(b), higher compression forces normal to the bed joints

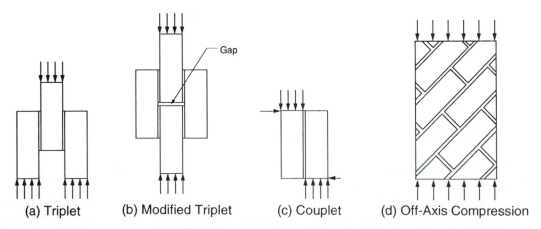

(a) Triplet    (b) Modified Triplet    (c) Couplet    (d) Off-Axis Compression

**Figure 5.41**   Test specimens for bed joint shear.

Clay brick                    Concrete block

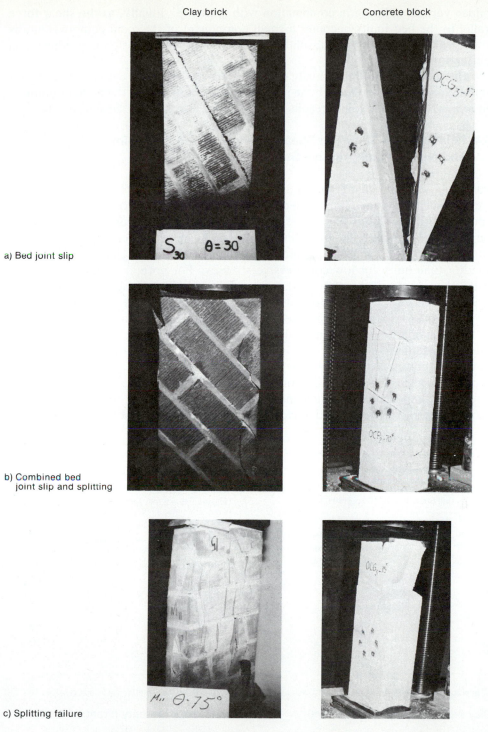

a) Bed joint slip

b) Combined bed
   joint slip and splitting

c) Splitting failure

**Figure 5.42**  Failure modes for masonry under combined shear and axial loads along the bed joints.

produce vertical splitting in combination with joint slip. Finally, as the shear force parallel to the bed joints becomes a small component of the axial compression, as shown in Fig. 5.42(c), a typical splitting failure due to compression is produced, as described in Sec. 5.2.3.

The joint slip failure normally occurs along the interface between the mortar and the unit rather than through the mortar even in cases where the higher compression normal to the bed joints results in substantial increases in shear capacity.

### 5.5.4 Relationships Between Shear Strength Along Bed Joints and Normal Compressive Stress

Figure 5.43 contains data from off-axis compression tests[5.39,5.40] for clay brick and concrete block masonry. Increased compression force normal to the bed joints corresponds with increased shear strengths. This increased shear capacity can be thought of as being similar to an increased frictional resistance due to compression. The slope for low axial compression equates to the coefficient of friction and is greater than unity for low compressive stresses. For normal compressive stress greater than about 30 to 40% of the prism compressive strength, the change to decreasing shear strength corresponds to the change to a compression splitting failure mode. This behavior is also observed for grout filled hollow masonry.[5.39]

For shear failures, experimental investigations[5.71–5.74] have shown that the shear corresponding to slip along one or more bed joints, is strongly related to the combined shear and compressive stresses. The relationship most commonly adopted to model this phenomenon is a Coulomb friction relationship. This assumes that the joint shear strength is composed of an initial shear bond strength between the mortar and the masonry unit plus a shear friction capacity, which is considered to be proportional

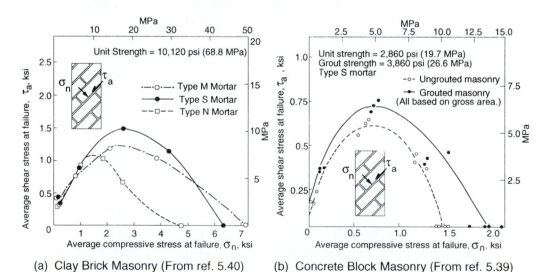

(a) Clay Brick Masonry (From ref. 5.40)   (b) Concrete Block Masonry (From ref. 5.39)

**Figure 5.43**  Relationships between shear stress at failure along the bed joint and normal compressive stress.

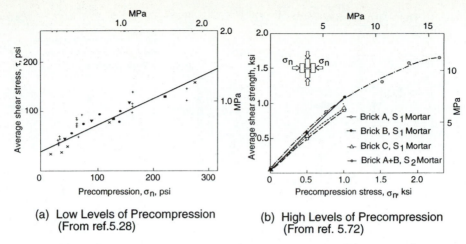

**(a) Low Levels of Precompression**
(From ref.5.28)

**(b) High Levels of Precompression**
(From ref. 5.72)

**Figure 5.44** Bed-joint shear strength versus precompression stress for clay masonry.

to the compressive stress applied normal to the bed joints. This relationship is expressed as

$$\tau = \tau_0 + \mu\sigma_n \qquad (5.5)$$

where
$\tau$ = joint shear strength
$\mu$ = coefficient of friction
$\tau_0$ = shear bond strength for $\sigma_n = 0$
$\sigma_n$ = compressive stress normal to the bed joints

The test results in Fig. 5.44 indicate the validity of this concept at low levels of compression. However, this formulation does not apply to failure modes other than slip along the mortar joints.[5.71]

### 5.5.5 Factors Affecting the Shear Strength Along Mortar Bed Joints

It is anticipated that many of the factors affecting the flexural tensile bond between mortar and masonry units (see Sec. 5.4.4) also affect the shear slip strength along mortar bed joints. However, not as much research has been done on this topic. Test results[5.72] show that shear bond strength for solid masonry is affected by the surface condition and initial rate of absorption of the units. Values ranging from 35 to 100 psi (0.24 to 0.69 MPa) are reported but with high coefficients of variation similar to those for flexural tensile bond. No strong correlation is evident between mortar or prism compressive strength and shear bond strength.

As shown in Fig. 5.45, filling the cells of hollow masonry with grout has been found[5.71,5.75] to significantly increase the shear strength along the bed joints. The magnitude of this increased strength is influenced by the tensile strength of the grout and the percent solid of the units. For example, as shown in Fig. 5.45(b), at a net-to-gross area ratio of about 0.6, an increase in average shear strength (based on

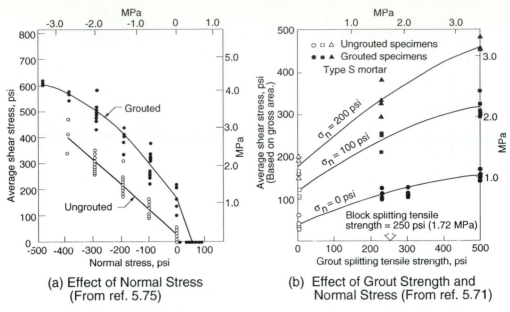

**Figure 5.45** Bed-joint shear strength for hollow and grouted concrete masonry.

shear resisting area) ranging from 50 to 100% can be achieved by grouting the cores. Therefore, this is a very effective means of improving the shear capacity along the bed joints. The increased shear resisting area also increases the shear capacity.

Test results typically indicate average coefficients of friction ranging from 0.6 to 1.0 depending on material properties and surface roughness.[5.18,5.28,5.72] Currently, masonry codes specify allowable shear bond stresses in terms of mortar type and an added component due to friction ranging from 0.2 to 0.45 times the normal compressive stress due to dead loads.

### 5.5.6 Load-Displacement Response

Following initial slip along the bed joints, the friction component of shear resistance remains, although usually at a lower value than calculated from capacities prior to slip. This information is important for modeling the shear force-displacement behavior of unreinforced solid and hollow masonry walls under reversed cyclic loading. An example of such behavior is shown in Fig. 5.46 for displacement across a mortar bed joint in brick masonry.[5.76] This figure shows a very stiff shear load-displacement response up to the peak load followed by a decrease in the shear resistance in the post-peak region corresponding to shear slip. A steady-state value of the residual shear resistance is reached that is not significantly affected by the number of cycles of loading. For reinforced masonry, the residual friction following slip is useful for calculating the shear friction associated with the clamping (dowel) action of the reinforcement resulting from the slip along the bed joints.

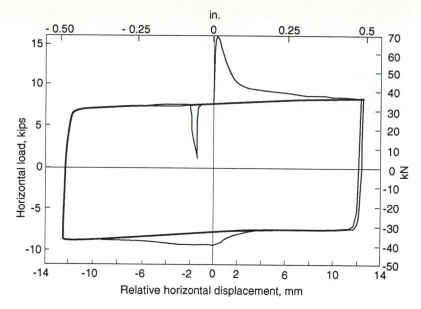

**Figure 5.46** Load-displacement response of a bed joint under cyclic loading (from Ref. 5.76).

### 5.5.7 Shear Strengths of Collar Joints and Head Joints

Composite action in multiwythe masonry walls depends on adequate shear connection between the adjacent wythes. However, limited data[5.77,5.78] on the shear bond strength of collar joints indicates a range of 5 to 100 psi (0.03 to 0.69 MPa) of the shear bond strength. The shear bond strength depends on the type and condition of the interface, effectiveness of consolidation of the mortar, and type of loading. As a result of the high coefficient of variability (over 25%), a lower-bound allowable shear value of only 5 psi is adopted in ACI 530/ASCE 5/TMS 402[5.5] for mortar-filled collar joints.

It has been shown that slushed collar joints with type S mortar have better shear bond strengths than type N mortar. Test results[5.76] show that joint thickness, unit absorption, and joint reinforcement have negligible effects on shear bond strength. It is also reported[5.77] that grouted collar joints are significantly better consolidated than mortar-filled collar joints. As a result, a higher allowable bond stress of 10 psi (0.07 MPa) is specified in ACI 530/ASCE 5/TMS 402[5.5] for grouted composite masonry. A specified minimum arrangement of ties across the collar joint is necessary for application of these design values.

Shear strength along head joints has not normally been a factor in design because of the benefit of the overlapping of units in running bond. However, this is *not* the case for stack pattern or for a continuous vertical mortar joint at the intersection between wall elements. Here, it is customary to assume that cracking due to differential shrinkage and other movements will result in effectively a zero initial bond between the units and the mortar head joint. In such cases, shear transfer along these joints is based entirely on shear friction due to clamping forces.

## 5.6 IN-PLANE TENSILE STRENGTH

### 5.6.1 Introduction

The combination of relatively low tensile strength and brittle behavior results in masonry being susceptible to tensile cracking. In fact, the cause of most masonry failures is tensile cracking. In loadbearing masonry buildings, shear walls carry vertical loads and resist the lateral in-plane loads due to wind or earthquakes. This combined loading creates principal tension stresses in the wall leading to tensile cracking when the tensile strength of the masonry is exceeded.

In addition to the potential for developing horizontal or vertical cracks corresponding to tension normal or parallel to the bed joints, various forms of diagonal cracking can occur.[5.79] Therefore, it is important that this type of failure can be predicted for various combinations of principal stresses, orientation of principal stress with respect to the mortar joints, and various combinations of material properties. Although in-plane tension normal to bed or head joints can result from in-plane flexure or from axial restraint to shrinkage and thermal movements, the main emphasis of this section relates to principal tension resulting from combined shear and axial loads.

### 5.6.2 Test Methods

ASTM describes two test methods to determine the capacity of masonry under conditions that can produce diagonal cracking. The diagonal compression test shown in Fig. 5.47(a) is based on subjecting a 4 ft (1.2 m) square section of wall to diagonal compression through steel shoes (loading plates) on two diagonally opposite corners of the specimen, as described in ASTM E519.[5.80] The failure mode for this test is usually through formation of diagonal cracks parallel to the line of action of the compression force. The diagonal tensile stress, $f_d$ is calculated from the equation

$$f_d = \frac{0.707P}{A} \tag{5.6}$$

where $P$ is the applied load, and $A$ is the average gross or net area of the cross section along the bed and head joints. Axial load normal to the bed joints can also be applied.

A difficulty with the diagonal tension test is that the stress field tends to force the cracks to follow the line of action of the compression load. This may not be the path of least resistance for other boundary conditions. In addition, the loading shoes on opposite ends of the diagonal can transfer compression load through a fairly large compression strut formed after the appearance of the diagonal cracks. Sometimes this compression strut can carry higher loads than those required to produce diagonal cracking.

The difficulty of relating the strengths and behaviors from diagonal tests to diagonal cracking in walls has resulted in an alternate test, the ASTM E72[5.53] racking

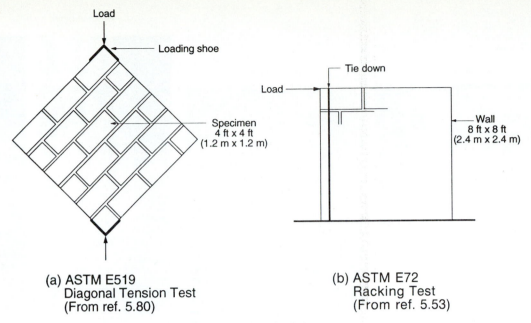

**(a) ASTM E519**
**Diagonal Tension Test**
**(From ref. 5.80)**

**(b) ASTM E72**
**Racking Test**
**(From ref. 5.53)**

**Figure 5.47** Standard shear tests.

test method. As shown in Fig. 5.47(b), an 8 ft (2.4 m) square wall panel is subjected to a horizontal force along the top. An axial load can be included. Vertical restraint by a tie down is required to prevent overturning of the specimen. Results obtained from this test are only relevant for the particular loading conditions and wall geometry used in the test. However, they can be used to confirm shear capacity provisions in codes, at least for cases similar to the test conditions.

Splitting tension test of disks and square masonry assemblages have been used to study the parameters that affect in-plane tensile strength.[5.81-5.83] As was the case for the diagonal compression test, the stress field tends to force the crack to follow the line of action of the splitting load rather than the line of least resistance. However, despite this reservation, the splitting tension test has been very useful for developing an understanding of the factors affecting in-plane tensile strength of masonry.

### 5.6.3 Failure Modes

It is expected that both the orientation of the principal tension stress and the relative magnitudes of the principal stresses affect the in-plane tensile strength of masonry. However, any comprehensive investigation of in-plane tensile cracking requires a great number of large test specimens to include all of the effects. Therefore, a compromise between direct representation and practicality has led to use of the splitting tension test to provide reasonable approximations of strength and failure modes.[5.83,5.84]

The nonisotropic nature of masonry, normally associated with the weak mortar joints, results in splitting tensile strengths and failure modes that are sensitive to the orientation of the principal stresses with respect to the bed joints. Figure 5.48 shows

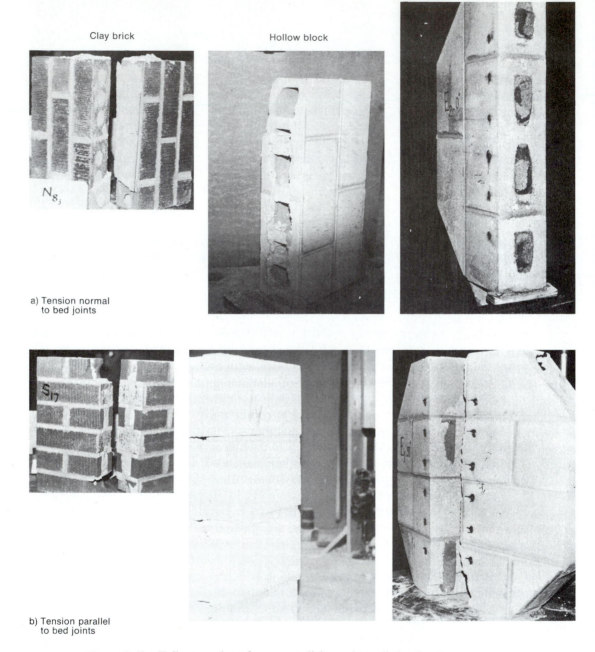

**Figure 5.48** Failure modes of masonry disks under splitting loads.

Clay brick          Hollow block          Grouted block

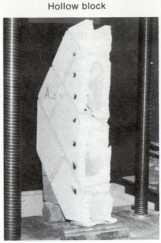

c) Diagonal tension

**Figure 5.48**   (continued)

failures for clay brick and concrete block disks tested under line loads across the thicknesses of the specimens.

For line loads parallel to the bed joints, the splitting tension normal to the bed joints typically results in debonding between the units and the mortar along the bed joint, as shown in Fig. 5.48(a). In the case of hollow masonry with grout filled cells, the columns of grout crossing the failure zone must also fail in tension.

For line loads normal to the bed joints, the resulting splitting tension parallel to the bed joints leads to the failure plane following a straight line through a combination of head joints and masonry units in alternate courses, as shown in Fig. 5.48(b). For hollow masonry filled with grout, this failure plane runs parallel to and does not cross the continuous columns of grout formed in the cells. Therefore, large increases in strength are not expected.

For line loads oriented at 45° from the bed joints, the splitting tension stress results in diagonal tension at 90° to the line load. Typical failures are shown in Fig. 5.48(c). In this case, the tension crack may have a rougher surface as the weakest path along combinations of head and bed joints and through units is followed to some extent. Grout-filled cells tend to reinforce the mortar joints at those locations and force the crack through the units.

### 5.6.4 Factors Affecting In-Plane Tensile Strength

**Orientation of the Principal Tension Stress.**   The anisotropic nature of masonry is illustrated in Fig. 5.49. For ungrouted masonry, the tensile strength parallel to the bed joints ($\theta = 0°$) is roughly double the tensile strength for tension normal to the bed joints ($\theta = 90°$). The degree of anisotropy is dependent on the properties of the constituent materials and particularly the difference between mortar bond and tensile strength of the unit. However, when the cells of hollow masonry are filled

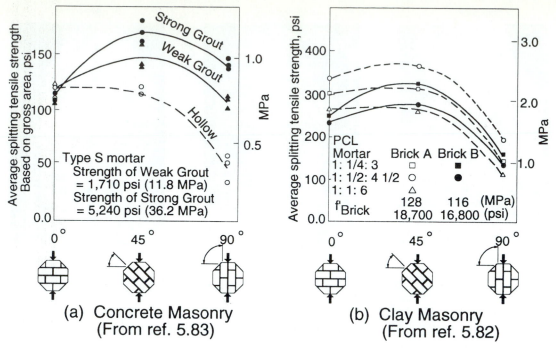

**Figure 5.49** In-plane splitting tensile strength of masonry disks.

with grout, the strengthening of the bed joints due to the continuity of the grout results in more uniform strengths, as shown in Fig. 5.49(a).

The increase in capacity due to grouting is significant for tension normal to the bed joints ($\theta = 90°$), in which case, the failure plane must pass through the grout. For tension parallel to the bed joints ($\theta = 0°$), no increase in capacity occurs because the failure plane does not cross the grout columns; see Fig. 5.48(b). In fact, although the failure load remains essentially the same, strengths recalculated using the gross area for grouted specimens and the face shell area for ungrouted specimens results in a significantly lower ultimate strength for the grouted condition.

For diagonal cracking, the strength (and failure mode) is much the same as for tension parallel to the bed joints. However, the presence of grout filled cells, which must be crossed by the diagonal crack, results in an increased strength for this case. As has been shown, the strength is sensitive to the orientation of the principal tension stress.[5.84]

**Mortar Type.** The splitting tensile strength is affected by mortar type because of the direct influence of tensile bond between the mortar and the units along the bed or head joints. For clay brick specimens, the results shown in Fig. 5.49(b) indicate that higher strengths are found for both type S and type M mortars than for type N mortar.

**Grout Strength.** For tension normal to the bed joints ($\theta = 90°$) and for diagonal tension ($\theta = 45°$), the failure plane crosses the columns of grout and, therefore, the observed increase in splitting tensile strength with increase in grout strength, shown in Fig. 5.49(a), is expected. For tension parallel to the bed joints ($\theta = 0°$), the only effect of grout strength would be on the resistance to debonding of the block from the grout column along the failure plane, but this does not appear to be significant.

**Strength of Masonry Units.** When the failure plane passes through the units, as can be the case for diagonal cracking and for tension parallel to the bed joints, the in-plane tensile strength should be affected by the tensile strength of the unit.[5.84]

## 5.7 COMBINED LOADING AND BIAXIAL STRENGTH

### 5.7.1 Introduction

Loadbearing masonry shear walls subject to combined in-plane vertical and lateral loads are in a state of biaxial stress. As illustrated in Sec. 5.6, a composite masonry assemblage exhibits distinct directional properties related to the planes along the mortar joints. Therefore, failure criteria cannot be defined solely in terms of principal stresses.[5.18,5.83–5.85] Directional variation in properties must be considered to predict the capacity of masonry under biaxial stress.

### 5.7.2 Test Methods

Tests of masonry beams and shear walls can provide a basis on which to evaluate design provisions for combined loading conditions. However, because of the nonuniform biaxial stress conditions in these specimens, it is difficult to develop fundamental relationships for failure. Therefore, researchers[5.85,5.86] have used assemblages, such as shown in Fig. 5.50, to create uniform stress conditions over larger areas. Different combinations of principal stress can be developed by altering the ratios and signs of the perpendicular stresses $\sigma_1$ and $\sigma_2$. Various orientations of these principal stresses with respect to the mortar joints can be introduced by building or cutting the specimens so that the mortar joints are at various angles to the direction of the applied loads.

### 5.7.3 Failure Modes

Tests of solid brick masonry panels[5.85] under various combinations of biaxial stresses produced the following failure modes:

Mode 1: Debonding between mortar and units along the bed and/or head joints in either a straight line or a stepped pattern.

Mode 2: Fracture of the brick units and debonding between the units and mortar at the head joints.

Mode 3: A combination of the foregoing two modes.

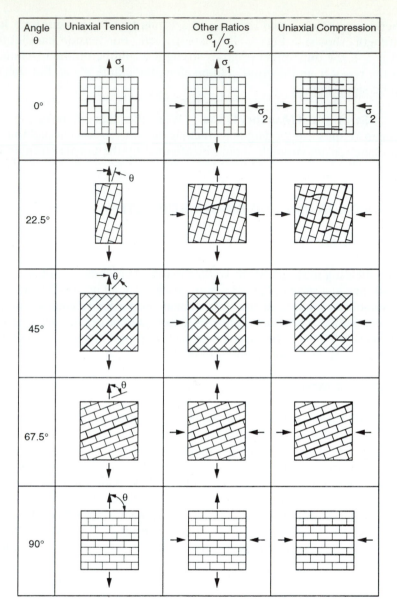

**Figure 5.50** Failure modes of brick masonry under biaxial stresses (from Ref. 5.85).

The foregoing failure modes are shown in Fig. 5.50 for brick masonry. Filling the cells of hollow masonry with grout provides continuity across the relatively weak bed joint planes, resulting in higher strength and the change of some mode 1 failures to mode 3. Off-axis compression tests, such as discussed in Sec. 5.5.2 for shear investigations, can also be used to investigate a limited range of biaxial stresses. As previously discussed, a slip type of debonding along bed joints or along a combination of head and bed joints gradually changes to a compression type of failure as shown in the upper two drawings in the right-hand column of Fig. 5.50.

### 5.7.4 Factors Affecting Failure Loads Under Biaxial Compression–Tension Stresses

A compression–tension state of stress occurs in walls as a result of combined vertical compression load and horizontal shear. Because of the anisotropic nature of masonry, determination of principal stresses is not sufficient to predict failure. The failure surface for brick masonry in Fig. 5.51, proposed by Page,[5.85] indicates that there is strong interaction between principal stresses and sensitivity to orientation of these stresses with respect to the bed joints. The shape of the surface is not unique, and all of the physical and material properties that affect the compressive, tensile, and shear strengths can result in changes to this shape.

Grouting of hollow masonry greatly reduces the influence of the weaker bed joint plane on the failure mode and on the strength under biaxial stress.[5.86] This results in behavior that is closer to being isotropic, particularly for cases where the properties of the grout and masonry units are reasonably similar.

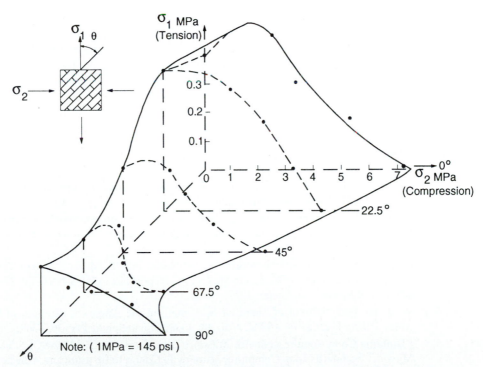

**Figure 5.51** Failure surface of brick masonry under biaxial stress (from Ref. 5.85).

## 5.8 CLOSURE

Information on behavior of masonry assemblages is given to help in developing a better understanding of the basic properties of masonry assemblages as a composite material.

The reader is reminded that the data included and the hypotheses and theories presented were selected to provide a framework for discussing the various properties of masonry assemblages. As such, they are illustrative and do not necessarily represent average properties. Whereas typical behaviors can be shown, the large scatter of test results and differences in interpretation of these results point to the need for continuing research on behavior of masonry assemblages.

The complexities associated with combining several different materials, each with its own ranges of properties, into assemblages that exhibit anisotropic behavior make it difficult to develop simple models that have a wide range of application. Therefore, it is understandable that building codes cannot directly account for all the parameters that affect behavior. To be a practical and useful tool, a building code must incorporate simplifications that take into account the normal ranges of properties of the materials and the effects of these on the interactions under various combinations of load.

Notwithstanding the need for simplified design provisions, it is important that the designer have a more in-depth knowledge of the behavior of masonry assemblages so that proper judgment can be exercised. For instance, in this chapter, it has been shown that stronger grout generally produces increased strengths for a masonry assemblage, but that this increase falls far short of being proportional to the increase in grout strength. Therefore, a designer would know that attempting to place a strong concrete in a masonry wall instead of a fluid grout would be a mistake because complete filling of the cells is far more important than a sizable increase in grout strength.

## 5.9 REFERENCES

5.1 American Society of Testing and Materials, "Compressive Strength of Masonry Assemblages," ASTM E447-84, Philadelphia, PA, 1984.

5.2 A. H. P. Maurenbrecher, "Effect of Test Procedure on Compressive Strength of Masonry Prisms," in *Proceedings of the Second Canadian Masonry Symposium*, Ottawa, Canada, 1980, pp. 119–132.

5.3 G. A. Hegemier, G. Krishomoorty, R. O. Nunn, and T. Morty, "Prism Tests for the Compressive Strength of Concrete Masonry," Report No. AMES-NSF-TR-77-1, University of California, San Diego, 1977.

5.4 W. J. Krefeld, "Effect of Shape of Specimen on the Apparent Strength of Brick Masonry," *Proceedings of ASTM*, Vol. 38, 1938, Philadelphia, PA, 363–369.

5.5 "Building Code Requirements for Masonry Structures," ACI 530/ASCE5/TMS 402-92, Masonry Standards Joint Committee, American Concrete Institute and American Society of Civil Engineers, Detroit, New York, 1992.

5.6 "Masonry in Buildings," Australian Standard 3700-1988, Standards Association in Australia, SAA, North Sydney, N.S.W., 1988.

5.7 "Masonry Design for Buildings," Canadian Standards Association, CSA Standard CAN3-S304-M84, CSA, Rexdale, Ontario, 1984.

5.8 J. Marr, "Capacity of Unreinforced Concrete Block Walls," M. Eng. thesis, McMaster University, Hamilton, Canada, 1992.

5.9 Structural Clay Products Research Foundation, "Compressive and Transverse Strength Tests of Eight-Inch Brick Walls," Research Report No. 10, SCPRF, Geneva, IL, 1966.

5.10 A. J. Francis, C. B. Horman, and L. E. Jerrems, "The Effect of Joint Thickness and Other Factors on the Compressive Strength of Brickwork," in *Proceedings of the Second International Brick Masonry Conference*, Stoke-on-Trent, England, 1970, pp. 31–37.

5.11 H. K. Hilsdorf, "Investigation into the Failure Mechanism of Brick Masonry under Axial Compression," in *Designing, Engineering and Construction with Masonry Products*, F. B. Johnson, Ed., Gulf Publishing, Houston, TX, 1969, pp. 34–41.

5.12 R. Brown and A. Whitlock, "Compressive Strength of Grouted Hollow Brick Prisms, Masonry: Materials, Properties and Performance," J. G. Borchelt, Ed., ASTM STP 728, ASTM, Philadelphia, PA, 1982.

5.13 R. Atkinson, J. Noland, and D. Abrams, "A Deformation Failure Theory for Stack-Bond Brick Masonry Prisms in Compression," in *Proceedings of the Seventh International Brick Masonry Conference*, Melbourne, Australia, 1985, pp. 577–592.

5.14 A. A. Hamid and R. G. Drysdale, "Suggested Failure Criteria for Grouted Concrete Masonry Under Axial Compression," *ACI Journal*, Vol. 76, No. 10, 1979, pp. 1047–1062.

5.15 C. L. Khoo and A. W. Hendry, "A Failure Criterion for Brickwork Under Axial Compression," in *Proceedings of the Third International Brick Masonry Conference*, Essen, West Germany, 1973, pp. 141–145.

5.16 N. G. Shrive, "A Fundamental Approach to the Fracture of Masonry," in *Proceedings of the Third Canadian Masonry Symposium*, Edmonton, Alberta, 1983, pp. 4-1–4-16.

5.17 R. G. Drysdale and A. A. Hamid, "Behaviour of Concrete Block Masonry under Axial Compression," *ACI Journal*, Vol. 76, No. 6, 1979, pp. 707–721.

5.18 A. A. Hamid, "Behaviour Characteristics of Concrete Masonry," Ph.D. thesis, McMaster University, Hamilton, Ontario, 1978.

5.19 A. Hamid and A. Chukwunenye, "The Compression Behaviour of Concrete Masonry Prisms," *Journal of the Structural Division, Proceedings of ASCE*, Vol. 112, No. 3, March 1986, pp. 605–614.

5.20 N. B. Shrive, "The Failure Mechanism of Face-Shell Bedded (Ungrouted and Unreinforced) Masonry," *International Journal of Masonry Construction*, Vol. 2, No. 3, 1982, pp. 115–128.

5.21 G. N. Chahine, "Behaviour Characteristic of Face Shell Mortared Block Masonry Under Axial Compression," M. Eng. thesis, McMaster University, Hamilton, Ontario, 1989.

5.22 A. Hamid and B. Abboud, "Effect of Block Geometry on the Compressive Strength of Concrete Block Masonry," in *Proceedings of the Fourth Canadian Masonry Symposium*, University of New Brunswick, Fredericton, 1986, pp. 290–298.

5.23 P. Guo, "Investigation and Modelling of the Mechanical Properties of Masonry," Ph.D. thesis, McMaster University, Hamilton, Ontario, 1990.

5.24 B. F. Boult, "Concrete Masonry Prism Testing," *ACI Journal*, Vol. 76, No. 4, April 1979, pp. 513–536.

5.25 A. Page and D. Brooks, "Load Bearing Masonry-A Review," in *Proceedings of the Seventh International Brick Masonry Conference*, Melbourne, Australia, 1985, pp. 81–100.

5.26 C. B. Monk, "A Historical Survey and Analysis of the Compressive Strength of Brick Masonry," Report No. 12, Structural Clay Products Research Foundation, Geneva, IL, 1967.

5.27 R. Brown and G. J. Borchelt, "Compression Tests of Hollow Brick Units and Prisms," Masonry: Components to Assemblages, ASTM STP 1063, J. Matthys, Ed., ASTM, Philadelphia, PA, 1990.

5.28 A. W. Hendry, *Structural Brickwork*, Macmillan, London, 1981.

5.29 R. H. Atkinson and R. G. Kingsley, "Comparison of the Behavior of Clay and Concrete Masonry Prisms in Compression," Technical Report, Atkinson-Noland & Assoc., Boulder, CO, 1975.

5.30 M. J. N. Priestley, "Seismic Design of Concrete Masonry Shear Walls," *Journal of the American Concrete Institute*, Vol. 83, January-February 1986, pp. 58–68.

5.31 G. Hart, J. Noland, G. Kingsley, R. Englekirk, and N. Sajjad, "The Use of Confinement Steel to Increase the Ductility in Concrete Masonry Shear Walls," *Proceedings of the Masonry Society Journal*, Vol. 7, No. 2, July-December 1988, pp. T19–T42.

5.32 A. A. Hamid, G. Ziab, and O. El Nawany, "Modulus of Elasticity of Concrete Block Masonry," in *Proceedings of the Fourth North American Masonry Conference*, Los Angeles, CA, 1987, pp. 7/1–7/13.

5.33 International Conference of Building Officials, "Uniform Building Code," Chapter 24, 1991 ed., Whittier, CA.

5.34 C. T. Grimm, "Elastic Modulus of Clay Brick Masonry," paper presented at the Reinforced and Prestressed Masonry Symposium, University of Edinburgh, Scotland, 1984.

5.35 S. Sahlin, *Structural Masonry*, Prentice-Hall, Englewood Cliffs, NJ, 1971.

5.36 J. J. Roberts, "The Effect of Different Test Procedures Upon the Indicated Strength of Concrete Blocks in Compression," *Magazine of Concrete Research*, Vol. 25, No. 83, June 1973, pp. 87–98.

5.37 J. B. Read and S. W. Clements, "The Strength of Concrete Block Walls Phrase II: Under Axial Loading," Technical Report, Cement and Concrete Association, London, 1972.

5.38 M. Hatzinikolas, J. Longworth, and J. Warwaruk, "Concrete Masonry Walls," Structural Engineering Report No. 70, University of Alberta, Edmonton, Canada, 1978.

5.39 A. A. Hamid and R. G. Drysdale, "Concrete Masonry Under Shear and Compression Along Mortar Joints," *ACI Journal*, Vol. 77, No. 5, September 1981, pp. 51–64.

5.40 A. A. Hamid and R. G. Drysdale, "Behavior of Brick Masonry Under Combined Shear and Compression Loading," in *Proceedings of the Second Canadian Masonry Symposium*, Ottawa, 1980, pp. 51–64.

5.41 A. A. Hamid, C. Chia, and H. Harris, "Joint Reinforced Block Masonry Walls Under Out-of-Plane Lateral Loading," Technical Report STL 01/88, Department of Civil Engineering, Drexel University, Philadelphia, PA, 1988.

5.42 R. Lee, J. Longworth, and J. Warwaruk, "Concrete Masonry Prism Response Due to Loads Parallel to Bed Joints," in *Proceedings of the Third North American Masonry Conference*, Arlington, TX, 1985, pp. 26/1–26/14.

5.43 British Standards Institution, "Code of Practice for Use of Masonry, Part I. Structural Use of Unreinforced Masonry," BS5628: Part 1: 1978 (confirmed April 1985).

5.44 A. H. P. Maurenbrecher, "Compressive Strength of Eccentrically Loaded Masonry Prisms," in *Proceedings of the Third Canadian Masonry Symposium*, Edmonton, Canada, June 1983, pp. 10.1–10.13.

5.45 R. G. Drysdale and A. A. Hamid, "Effect of Eccentricity on the Compressive Strength of Brickwork," *Journal of the British Ceramic Society*, No. 30, September 1982, pp. 140–149.

5.46 R. G. Drysdale and A. A. Hamid, "Capacity of Concrete Block Masonry Prisms Under Eccentric Compressive Loading," *ACI Journal*, Vol. 80, No. 11, March-April, 1983, pp. 102–108.

5.47 Structural Clay Products Institute, "Recommended Practice for Engineered Brick Masonry," SCPI, McLean, VA, 1969.

5.48 G. Assis, A. A. Hamid, and H. G. Harris, "Material Models for Grouted Block Masonry," Report No. 1.2(a)-2, U.S.–Japan Coordinated Program on Masonry Building Research, Drexel University, Philadelphia, PA, 1989.

5.49 R. H. Brown and J. M. Young, "Compressive Stress Distribution of Grouted Hollow Clay Masonry Under Strain Gradient," Report No. 1.2(b)-1, U.S.–Japan Coordinated Program for Masonry Building Research, Department of Civil Engineering, Clemson University, Clemson, SC, 1988.

5.50 Standards Association of New Zealand, "Code of Practice for the Design of Masonry Structures," NZS 4230: Part 1, Wellington, 1990.

5.51 American Society for Testing and Materials, "Test Method for Flexural Bond Strength of Masonry," ASTM E518-80, ASTM, Philadelphia, PA, 1987.

5.52 American Society for Testing and Materials, "Measurement of Masonry Flexural Bond Strength," ASTM C1072-82, ASTM, Philadelphia, PA, 1986.

5.53 American Society of Testing and Materials, "Standard Methods for Conducting Strength Tests on Panels for Building Construction," ASTM E72, ASTM, Philadelphia, PA, 1989.

5.54 A. Sarker and R. Brown, "Flexural Strength of Brick Masonry Using the Bond Wrench," Research Report No. 20, Brick Institute of America, Reston, VA, 1987.

5.55 L. Baker and G. Franklin "Variability Aspects of the Flexural Strength of Brickwork," in *Proceedings of the Fourth International Brick Conference*, Brugge, Belgium, 1976, pp. 2.b.4–2.b.4-11.

5.56 R. Drysdale and A. Essawy, "Out-of-Plane Bending of Concrete Block Walls," in *Proceedings of ASCE, Structural Journal*, Vol. 114, No. 1, January 1988, pp. 121–133.

5.57 S. Ghosh, "Flexural Bond Strength of Masonry—An Experimental Review," in *Proceedings of the Fifth North American Masonry Conference*, University of Illinois at Urbana-Champaign, 1989, pp. 701–712.

5.58 E. A. Gazzola, D. Bagnariol, J. Toneff, and R. G. Drysdale, "Influence of Mortar Materials on the Flexural Tensile Bond Strength of Block and Brick Masonry," ASTM STP 871-Masonry: Research, Application and Problems, ASTM, Philadelphia, PA, 1985.

5.59 A. Yorkdale, "Initial Rate of Absorption and Mortar Bond," ASTM STP 778-Masonry: Materials Properties and Performance, ASTM, Philadelphia, PA, 1982.

5.60 R. G. Drysdale and E. A. Gazzola, "The Flexural Tensile Bond Strength of Concrete Brickwork," in *Proceedings of the Fourth North American Masonry Conference*, Los Angeles, 1987, pp. 53/1–53/14.

5.61 L. R. Baker, "Some Factors Affecting the Bond Strength of Masonry," in *Proceedings of the Fifth International Brick Masonry Conference*, Washington, DC, 1979, pp. 84–89.

5.62 A. Isberner, "Properties of Masonry Cement Mortars," *Designing, Engineering and Constructing with Masonry Products*, F. Johnson, Ed., Gulf Publishing, Houston, TX, 1969, pp. 42–50.

5.63 A. A. Hamid, "Bond Characteristics of Sand-Molded Brick Masonry," in *Proceedings of the Masonry Society Journal*, Vol. 4, No. 1, January-June 1985, pp. T18–T22.

5.64 R. G. Drysdale, "Influence of Age and Curing Conditions on Flexural Bond Between Units and Mortar," Masonry Report, McMaster University, Hamilton, Ontario, 1990.

5.65 A. A. Hamid, and R. G. Drysdale, "Flexural Tensile Strength of Block Masonry," *Journal of the Structural Division, Proceedings of ASCE*, Vol. 114, No. 1, January 1988, pp. 50–66.

5.66 A. A. Hamid, B. E. Abbond, M. Farah, M. Hatem, and H. Harris, "Out-of-Plane Response of Block Masonry Walls Under Static Loads," TCCMAR Report No. 3.2(a)-1, U.S.–Japan Coordinated Program on Masonry Building Research, 1989.

5.67 L. R. Baker, "The Lateral Strength of Brickwork–An Overview," in *Proceedings of the Sixth International Symposium on Loadbearing Brickwork*, London, 1977, pp. 169–187.

5.68 S. Lawrence, "Flexural Strength of Brickwork Normal to and Parallel to the Bed Joints," *Journal of the Australian Ceramic Society*, Vol. 11, No. 1, May 1975, pp. 5–6.

5.69 A. A. Hamid, "Effect of Aspect Ratio of the Unit on Flexural Tensile Strength of Brick Masonry," *Journal of the Masonry Society*, Vol. 1, No. 1, January-June 1981, pp. T11–T16.

5.70 L. R. Baker, "A Failure Criterion for Brickwork in Biaxial Bending," in *Proceedings of the Fifth International Brick Masonry Conference*, Washington, DC, 1979, pp. 71–78.

5.71 A. A. Hamid, R. G. Drysdale, and A. C. Heidebrecht, "Shear Strength of Concrete Masonry Joints," *Journal of the Structural Division, Proceedings of ASCE*, Vol. 105, ST7, July 1979, pp. 1227–1240.

5.72 R. G. Drysdale, R. Vanderkyle, and A. Hamid, "Shear Strength of Brick Masonry Joints," in *Proceedings of the Fifth International Brick Masonry Conference*, Washington, DC, 1979, pp. 106–113.

5.73 R. Jolly, "Shear Strength: A Predictive Technique for Masonry Walls," Ph.D. thesis, Brigham Young University, Provo, UT, 1975.

5.74 F. Yokel and G. Fattal, "Failure Hypothesis for Masonry Shear Walls," *Journal of the Structural Division, Proceedings of ASCE*, ST3, Vol. 102, March 1976, pp. 515–532.

5.75 G. Hegemier, S. Arya, G. Krishnamoorthy, W. Nachbar, and R. Furgeson, "On the Behavior of Joints in Concrete Masonry," in *Proceedings of the North American Masonry Conference*, Boulder, CO, 1978, pp. 4/1–4/21.

5.76 R. H. Atkinson, B. P. Amaddi, S. Saeb, and S. Sture, "Response of Masonry Bed Joints in Direct Shear," *Journal of the Structural Division, Proceedings of ASCE*, Vol. 115, September 1989, pp. 2276–2296.

5.77 J. McCarthy, R. Brown, and T. Cousins, "An Experimental Study of the Shear Strength of Collar Joint in Grouted and Slushed Composite Masonry Walls," in *Proceedings of the Third North American Masonry Conference*, Arlington, TX, 1985, pp. 39.1–39.16.

5.78 R. Williams and L. Geschwindner, "Shear Stress Across Collar Joints in Composite Masonry Walls," in *Proceedings of the Second North American Masonry Conference*, College Park, MD, 1982, pp. 8.1–8.17.

5.79 C. T. Grimm, "Strength and Related Properties of Brick Masonry," *Journal of the Structural Division, Proceedings of ASCE*, Vol. 101, ST11, November 1975, pp. 2385–2403.

5.80 American Society for Testing and Materials, "Standard Test Method for Diagonal Tension (Shear) in Masonry Assemblages," ASTM E519-81 (reapproved 1988), ASTM, Philadelphia, PA, 1988.

5.81 F. B. Johnson and J. N. Thompson, "Development of Diametral Testing Procedures to Provide a Measure of Strength Characteristics of Masonry Assemblages," in *Designing,*

*Engineering and Construction with Masonry Products*, F. B. Johnson, Ed., Gulf Publishing, Houston, TX, 1969, pp. 51–57.

5.82 R. G. Drysdale and A. A. Hamid, "Tensile Strength of Brick Masonry," in *Proceedings of the International Journal of Masonry Construction*, Vol. 2, No. 4, 1982, pp. 172–177.

5.83 R. G. Drysdale, A. A. Hamid, and A. C. Heidebrecht, "Tensile Strength of Concrete Masonry," *Journal of the Structural Division, Proceedings of ASCE*, Vol. 105, ST7, July 1979, pp. 1261–1276.

5.84 R. G. Drysdale and A. A. Hamid, "Tension Failure Criteria for Plain Concrete Masonry," *Journal of the Structural Division, Proceedings of ASCE*, Vol. 110, No. 2, February 1984, pp. 228–244.

5.85 A. W. Page, "An Experimental Investigation of the Biaxial Strength of Brick Masonry," in *Proceedings of the Sixth International Brick Masonry Conference*, Rome, 1982, pp. 3–15.

5.86 G. A. Hegemier, R. O. Nunn, and S. K. Arya, "Behaviour of Concrete Masonry Under Biaxial Stress," in *Proceedings of the North American Masonry Conference*, Boulder, CO, 1978, pp. 1/1–1/28.

## 5.10 PROBLEMS

**5.1** In your own words, describe the failure mechanism for clay brick masonry under axial compression.

**5.2** Using the elastic formula for determination of the modulus of elasticity of masonry, $E_m$, calculate the following:

(a) The modulus of elasticity for a prism made of 4 in. (100 mm) (nominal) clay brick having a compressive strength of 5000 psi (35 MPa) and type S mortar. Compare this with the design value from your local building code.

(b) Plot the relationship between the elastic modulus of masonry and the elastic modulus of mortar for different unit height-to-joint thickness ratios. Comment on the plot.

**5.3** Find the following for grouted masonry made with a normal-weight 8 in. (200 mm) (nominal) hollow concrete block having a compressive strength of 3000 psi (20 MPa), type S mortar, and 4000 psi (27.5 MPa) grout:

(a) What strength would be expected using the principle of superposition of strengths? Explain. Compare with code values.

(b) What would you expect the value of the modulus of elasticity to be? Explain.

(c) Why have some researchers found the strengths calculated using superposition to overestimate compressive strength?

**5.4** (a) Determine the orthogonal strength ratio for clay brick masonry having the following properties: Type S mortar; flexural tensile bond strength normal to the bed joints = 50 psi (0.35 MPa); and flexural tensile strength of the brick = 1000 psi (6.9 MPa).

(b) What is the effect of an axial compressive stress of 100 psi (0.69 MPa) on the orthogonal strength ratio?

**5.5** "Grouting the cells of hollow concrete block masonry reduces the degree of anisotropy." Explain this statement considering compression, tension, and shear strengths. You may use drawings to help explain your answer.

**5.6** Determine the thickness required according to ASTM E447 for the steel bearing plate to test a 12 in. (300 mm) (nominal) concrete block prism. The diameter of the spherical head is 9 in. (230 mm).

**5.7** Using the experimental data presented in this chapter, estimate the following:

    **(a)** The percent increase in joint shear capacity due to grouting of hollow 8 in. (200 mm) concrete block masonry. Adhesion bond, $\tau = 50$ psi (0.35 MPa), and grout compressive strength = 3000 psi (20 MPa).

    **(b)** The percent increase in joint shear strength of clay brick masonry due to the application of a 200 psi (1.38 MPa) axial compressive stress. Adhesion bond = 60 psi (0.41 MPa).

    **(c)** The percent increase in moment carrying capacity due to grouting of hollow concrete masonry spanning in the vertical direction. Grout strength = 2000 psi (13.8 MPa), unit is 55% solid, and mortar bond strength = 60 psi (0.41 MPa) based on the net area.

    **(d)** The orthogonal in-plane tensile strength ratio for hollow block masonry and for grout filled block masonry. The 6 in. (150 mm) (nominal) concrete block units are 60% solid and have a compressive strength of 3000 psi (20 MPa). The grout has a compressive strength of 4000 psi (27.5 MPa).

**5.8** Conduct a literature search on failure theories for clay brick masonry under axial compression. Compare the strength predicted using these theories for prisms made of materials having the following properties.

Clay brick units: Compressive strength = 8000 psi (55 MPa), tensile strength = 600 psi (4 MPa), and height = $2\frac{1}{4}$ in. (57 mm). Mortar: Compressive strength = 2000 psi (14 MPa) and joint thickness = $\frac{3}{8}$ in. (10 mm).

# CHAPTER 6

# Reinforced Beams and Lintels

Construction of reinforced masonry beams over window areas (*Courtesy of National Concrete Masonry Association*).

## 6.1 INTRODUCTION

The modulus of rupture of masonry (flexural tensile strength) is a small fraction of its compressive strength. This means that unreinforced masonry beams fail on the tension side long before the compression strength can be fully utilized. Therefore, masonry beams and lintels are usually reinforced in much the same way and for the same reasons as reinforced concrete members: to increase the flexural strength. Reinforcement also helps control cracking and deflection. The behavior and design of reinforced masonry beams and lintels are the main themes of this chapter. However, many of the principles covered are equally applicable to reinforced masonry flexural walls (Chap. 7) and they are extended in Chaps. 8, 9, and 10 to cover combined axial load and bending.

Reinforced masonry beams and lintels are horizontal members used to span openings in masonry walls, as shown in Fig. 6.1(a). They can be constructed with single-wythe brickwork, double-wythe brickwork with a grouted cavity (Fig. 6.1(b)), special lintel concrete masonry units, bond beam units (Fig. 6.1(c)), or by using

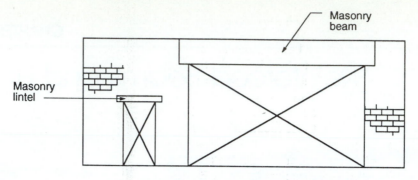

(a) Masonry Beams and Lintels

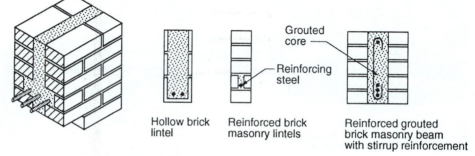

(b) Brick Masonry Beams and Lintels

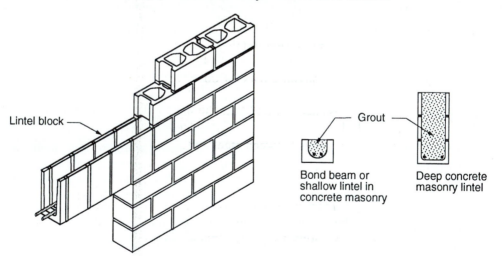

(c) Concrete Masonry Beams and Lintels

**Figure 6.1** Typical masonry beams and lintels.

units with depressed, knock-out, or slotted webs to accommodate placement of the longitudinal reinforcement.

Elements known as bond beams are courses of masonry units within a wall that are horizontally reinforced with longitudinal bars and grouted. Reinforced bond beams spaced over the height of the wall can be used to resist horizontal out-of-plane bending and in-plane tension or shear. Also, they are typically located at the top of the foundation and at floor and roof levels. Unless they form part of a lintel or beam, their main functions at these locations are to distribute vertical loads from floors and to tie elements of the building together.

Design of flexural members must take into account both strength and serviceability. Flexure, shear, and anchorage are strength considerations, while deflection and crack control are serviceability concerns.

## 6.2 FLEXURAL BEHAVIOR AND DESIGN

### 6.2.1 Fundamental Assumptions

For shallow beams (span-to-depth ratio more than 5 for simply supported beams), the fundamental assumptions on which the elastic analysis of reinforced masonry is based are:

- Internal forces at any section of a member are in equilibrium with the effects of external loads.
- Plane sections before bending remain plane after bending. That is, the strains in the masonry and the reinforcement are assumed to be directly proportional to their distance from the neutral axis.
- After cracking, the contribution of tension in the masonry is ignored and tensile forces are resisted entirely by steel reinforcement.
- Linearly elastic behavior exists for both steel and masonry within the working stress range. This implies that the neutral axis passes through the centroid of the cracked cross-section and that the magnitude of bending stress increases directly with the distance from the neutral axis.
- Complete bond exists between steel and grout.

These assumptions are also used in the working stress analysis of reinforced concrete beams. Test results indicate that within the working stress range of loading, short-term behavior of reinforced masonry is predicted quite closely. At higher loads, nonlinear material behavior, localized effects of cracking, and bond slip of the reinforcement cause deviations from the "plane sections remaining plane" assumption. However, with the addition of nonlinear stress–strain relationships, reasonably accurate predictions of the ultimate flexural capacity of a beam section can be achieved.

### 6.2.2 Behavior Under Load

When the load on a reinforced masonry beam is increased, several different stages of behavior can be clearly distinguished:

**Stresses Elastic, Section Uncracked.** For extreme fiber flexural tensile stresses less than the modulus of rupture of the masonry, no tension cracks develop and the stress and strain distributions are linear as shown in Fig. 6.2(a). Section properties can be calculated using the transformed section as shown. In this transformed section, the actual area of the reinforcement is replaced with an equivalent masonry area equal to $nA_s$ located at the level of the steel, where n is the modular ratio.

$$n = \frac{E_s}{E_m}$$

(6.1)

where  $E_s$ = modulus of elasticity of steel
$E_m$ = modulus of elasticity of masonry

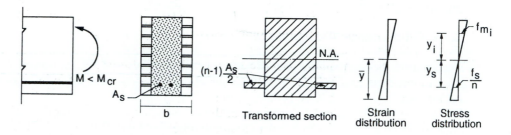

**(a) Stresses Elastic, Section Uncracked**

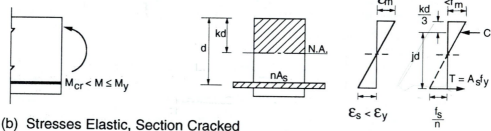

**(b)  Stresses Elastic, Section Cracked**

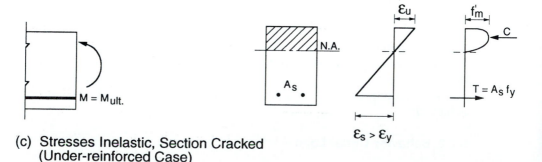

**(c)  Stresses Inelastic, Section Cracked
(Under-reinforced Case)**

**Figure 6.2**   Analysis of reinforced masonry beams.

The moment–curvature response is shown as curve A in Fig. 6.3. The bending stress in the masonry $f_{m_i}$ at any given point $i$ in the cross section is determined by

$$f_{m_i} = My_i/I_{tr} \tag{6.2}$$

where M = applied moment
$y_i$ = distance from the neutral axis to the point in question
$I_{tr}$ = moment of inertia of the transformed section

The stress in the steel $f_s$ is calculated by the equation

$$f_s = n(My_s/I_{tr}) \tag{6.3}$$

where $y_s$ is the distance from the neutral axis to the centroid of the reinforcing steel.

**Stresses Elastic, Section Cracked.** The masonry cracks when the extreme fiber tension stress exceeds the modulus of rupture. If the masonry compressive stress is less than about half the compressive strength and the steel does not reach the yield stress during cracking, both materials continue to behave elastically; see Fig. 6.2(b). This is normally the case under service load for which the working stress design method is applied. Curve B in Fig. 6.3 represents the elastic behavior for the cracked section, which now has reduced section properties.

A transformed section consisting of the masonry on the compression side of the neutral axis, and $n$ times the steel area on the other, can be used to calculate stresses

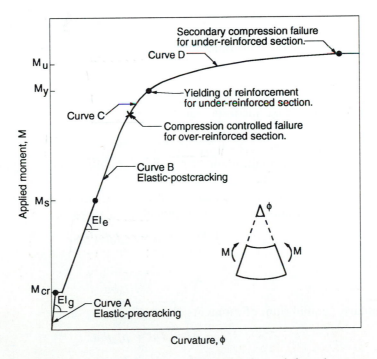

**Figure 6.3** Moment-curvature relationship for reinforced masonry beams.

and strains. Equilibrium and compatibility of strains between the masonry and the steel are considered to determine the location of the neutral axis (N.A.). As shown in Fig. 6.2(b), the compressive force resultant is calculated as

$$C = f_m kbd/2 \tag{6.4}$$

and the tension force is

$$T = A_s f_s = \rho bd f_s \tag{6.5}$$

where the steel ratio $\rho$ is defined as

$$\rho = A_s/bd \tag{6.6}$$

For equilibrium, $C = T$, and from Eqs. 6.4 and 6.5,

$$\rho f_s bd = f_m kbd/2$$

$$k = 2\rho f_s/f_m \tag{6.7}$$

Substituting for $f_s$ and $f_m$ in Eq. 6.7 in terms of strain ($\varepsilon_m = f_m/E_m$ and $\varepsilon_s = f_s/E_s$), we obtain

$$k = 2\rho \frac{\varepsilon_s E_s}{\varepsilon_m E_m} \tag{6.8}$$

By similar triangles in Fig. 6.2(b),

$$\frac{\varepsilon_s}{\varepsilon_m} = \frac{d - kd}{kd} = \frac{1 - k}{k}$$

Hence, using the modular ratio, $n = E_s/E_m$ from Eq. 6.1, Eq. 6.8 can be rewritten as

$$k = \frac{2n\rho(1 - k)}{k} \tag{6.9}$$

Solving the resulting quadratic equation for $k$:

$$k = \sqrt{2n\rho + (n\rho)^2} - n\rho \tag{6.10}$$

Thus, for a known reinforcement ratio and modular ratio, the depth of the neutral axis $kd$ may be determined by Eq. 6.10.

For a given applied moment $M$, the stresses in the masonry and the steel can be calculated. Equilibrium of moments about $T$ gives

$$M = Cjd = f_m kjbd^2/2 \tag{6.11}$$

and

$$f_m = 2M/kjbd^2 \tag{6.12}$$

Similarly, equilibrium of moments about $C$ gives:

$$M = Tjd = \rho f_s jbd^2 \tag{6.13}$$

$$f_s = M/\rho jbd^2 \tag{6.14}$$

where the moment arm between $C$ and $T$ is

$\rho_{min}$ Grade 60 $\geq .0015$

$$jd = d - kd/3 \qquad (6.15)$$

or in dimensionless form

$$j = 1 - k/3$$

Equation 6.11 can be used to calculate the allowable moment based on allowable compressive stress in masonry $F_b$ (typically equal to about 0.33 of the masonry compressive strength $f_m'$). Similarly, the maximum allowable moment may be calculated by Eq. 6.13 based on allowable steel stress, $F_s$ (typically 20 ksi (140 MPa) for grade 40 (grade 300) and 50 steel and 24 ksi (165 MPa) for grade 60 (grade 400) steel). The lesser of the two moments is the allowable design moment $M_s$, as shown in Fig. 6.3.

In working stress design, the balanced condition is a state where both the steel and masonry simultaneously reach the specified allowable stresses. For this condition:

$f_m = F_b =$ allowable masonry compressive stress in flexure

$f_s = F_s =$ allowable steel stress

From similar triangles, see Fig. 6.2(b),

$$\frac{k_b d}{d} = \frac{F_b}{F_b + F_s/n} \qquad (6.16)$$

$F_b = .33 f_m'$ (Grade 60) 24 ksi

$F_s = .4 f_y$

where $k_b$ is the $k$ value at balanced conditions. Thus,

$$k_b = \frac{F_b}{F_b + F_s/n} \qquad (6.17)$$

To obtain the balanced steel ratio $\rho_b$, equating $C$ and $T$ gives;

$$\rho_b b d F_s = F_b k_b d b/2 \qquad (6.18)$$

$C_b = T_b$

By substituting for $k_b$ and solving,

$$\rho_b = \frac{n F_b}{2 F_s (n + F_s/F_b)} \qquad (6.19)$$

This ratio, $\rho_b$, is used to indicate whether tension or compression controls the design of reinforced masonry beams. If the actual steel ratio $\rho$ is less than $\rho_b$, the steel stress will reach its allowable value, $F_s$, before the masonry reaches its allowable limit, $F_b$, and the design moment will be governed by the steel stress. On the other hand, if $\rho$ is greater than $\rho_b$, the masonry will reach its allowable compressive stress $F_b$ first and the allowable moment will be governed by this value.

**Stresses Inelastic, Section Cracked.** At or near the ultimate load, stresses are in the inelastic range and they are no longer proportional to strain; see Fig. 6.2(c). Two possible modes of flexural failure exist for reinforced masonry

$$bd^2 = \frac{m}{\rho j F_s}$$

beams. Depending on the amount of steel present in the section, yielding of steel reinforcement may or may not occur before compression failure of the masonry.

Curve C in Fig. 6.3 represents the beginning of nonlinear behavior as the stress in the masonry reaches the nonlinear region of the stress–strain curve illustrated in Fig. 5.26. If the section is *over-reinforced,* compression failure of the masonry will occur. However, for *under-reinforced* beams, yielding of the reinforcement defines the change in behavior where the gradual increase in moment defined by curve D (Fig. 6.3) is associated with the gradual increase in moment arm. Even for this tension-controlled failure, compression failure is the final mode of failure as the compression zone becomes too small to provide the required compression force. Because masonry tends to be lightly reinforced, the usual failure mode is steel yielding followed by crushing of the masonry. Yielding of the tension steel defines the maximum tensile force, and from equilibrium of tension and compression, the maximum compression force resultant is also defined.

After yielding of the reinforcement, increases in the externally applied moment result in a shift in the neutral axis toward the compression face. The increased moment is resisted by the increased moment arm due to movement of the compression force resultant. The decreased depth of the compression zone requires an increase in the maximum compressive stress to maintain equilibrium. Finally, as the crushing strain $\varepsilon_u$ is reached at the extreme compression fiber of the masonry, spalling of the masonry units or crushing of the grout, or both occurs; see Fig. 6.4(a). This type of failure is called ''secondary compression failure'' and is preceded by large deformations (see Fig. 6.4(b)) and opening of cracks. This under-reinforced failure mode is desirable because of the warning and gradual nature of its occurrence.[6.1]

When a masonry beam is heavily reinforced, the tension reinforcement does not reach its yield stress before the masonry reaches its ultimate strain and compression failure of the masonry takes place. This type of failure occurs almost without warning as failure takes place at very low deflection. As a result, it is not a desirable mode of failure.

In a balanced situation, the onset of steel yielding and the onset of masonry crushing both take place at the same time.

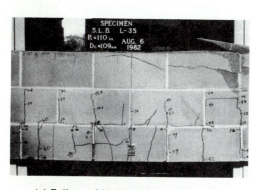

(a) Failure of Under-reinforced Beam
(*Courtesy of V.V. Neis*)

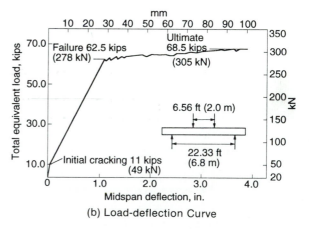

(b) Load-deflection Curve

**Figure 6.4** Flexural behavior of concrete block masonry beams (from Ref. 6.1).

At the ultimate moment capacity for an under-reinforced section such as shown in Fig. 6.5(a), the tensile force is given as

$$T = A_s f_y$$ (6.20)

As discussed in Sec. 5.3.5, an elaborate relationship would be required to precisely define the compression stress–strain behavior of the masonry under a compressive strain gradient. However, the real objective is to calculate the magnitude and location of the resultant compressive force C. Thus, mathematically simpler representations that achieve this objective are quite acceptable.

Various shapes of the stress block, such as the triangular shape shown in Fig. 6.5(c), can be used. In North America and many other parts of the world, a rectangular stress block (often referred to as the Whitney stress block), such as shown in Fig. 6.5(b), is familiar to structural designers and it is natural that its application to masonry be tested using the same parameters as for concrete, i.e., $\gamma_1 = 0.85$ and

$$\beta_1 = 1.05 - 0.05 f'_m \le 0.85 \qquad \text{(for } f'_m \text{ in ksi)} \tag{6.21}$$

or

$$\beta_1 = 1.05 - 0.0073 f'_m \le 0.85 \qquad \text{(for } f'_m \text{ in MPa)}$$

[Note: For the investigated range of masonry strengths, this essentially results in a constant value of $\beta_1 = 0.85$.]

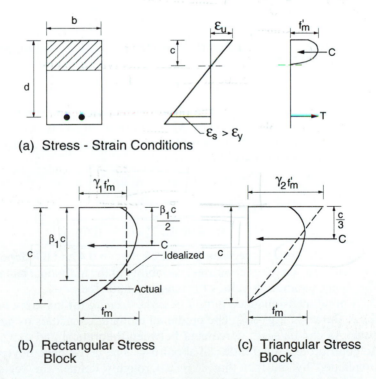

(a) Stress - Strain Conditions

(b) Rectangular Stress Block

(c) Triangular Stress Block

**Figure 6.5** Ultimate moment for under-reinforced masonry beams.

The compression resultant force is

$$C = \gamma_1 f'_m \beta_1 cb \qquad (6.22)$$

For equilibrium with $T = C$, Eqs. 6.20 and 6.22 give

$$c = \frac{A_s f_y}{\gamma_1 f'_m \beta_1 b} \qquad (6.23)$$

Thus, the nominal or theoretical ultimate moment capacity $M_n$ can be calculated as

$$M_n = T\left(d - \frac{\beta_1 c}{2}\right)$$

$$= A_s f_y \left(d - \frac{A_s f_y}{2\gamma_1 f'_m b}\right) \qquad (6.24)$$

If we now consider the steel ratio $\rho = A_s/bd$ and define a steel index $\omega$ as

*mechanical steel ratio:*

$$\omega = \rho \frac{f_y}{f'_m} \qquad (6.25)$$

Eq. 6.24 becomes

$$M_n = bd^2 f'_m \omega \left(1 - \frac{\omega}{2\gamma_1}\right) \qquad (6.26)$$

[Note that for $\gamma_1 = 0.85$ the equation is identical to that used for reinforced concrete except that $f'_m$ is substituted for $f'_c$.]

$$M_n = bd^2 f'_m \omega(1 - 0.59\omega) \qquad (6.27)$$

In the United States, variability of flexural strength in concrete beams is allowed for in the ultimate design moment $M_u$ by introducing a capacity reduction factor $\Phi$ for flexure. Thus, from safety considerations, the ultimate design moment due to the factored loads must satisfy the relation

$$M_u \leq \Phi M_n$$

or substituting from Eq. 6.27

*Assume best conditions and use 80%*

$$M_u \leq \Phi bd^2 f'_m \omega(1 - 0.59\omega) \qquad (6.28)$$

The UBC code[6.3] specifies $\Phi$ values between 0.4 and 0.8 depending on the type of construction. In other countries, variability of the individual materials is accounted for by using separate $\Phi$ values for masonry and steel.

A nondimensionalized plot of flexural strength for masonry beams versus steel index is shown in Fig. 6.6. The predicted moment capacities using Eq. 6.27 are also shown. As can be seen, agreement between theoretical and experimental values is quite good up to a steel index of about 0.4. The divergence beyond this value may be explained by the fact that $\omega = 0.4$ roughly defines the region beyond which compression controlled failure is reached. Any errors in evaluation of $f'_m$ or in repre-

*$M_u$ = factored moment* → $1.4D + 1.7L$

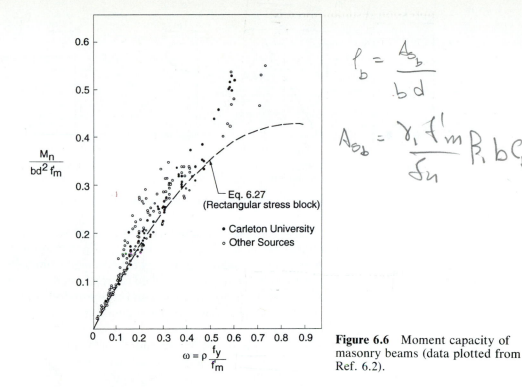

Figure 6.6  Moment capacity of masonry beams (data plotted from Ref. 6.2).

Handwritten annotations on figure:

$$\ell_b = \frac{A_{sb}}{bd}$$

$$A_{sb} = \frac{\gamma_1 f'_m}{f_y} \beta_1 b c_b$$

senting the compression force resultant have proportional effects on $C$ and $c$ in the moment capacity equation rewritten for compression controlled failure as

$$M_n = C\left(d - \frac{\beta_1 c}{2}\right) \tag{6.29}$$

In this case, it is suggested that the underestimation of strength relates more to the underestimation of the effective $f'_m$ value. Values of $f'_m$ determined by axial load tests of two block high grouted prisms do not accurately model the different contribution of the grout for the beam configurations (see Sec. 5.3). For lower amounts of reinforcement associated with tension controlled capacity, the flexural capacity is less sensitive to calculations related to the compression resultant force. It is suggested that the conservatism indicated for larger amounts of reinforcement is justifiable because of the desire to avoid compression controlled failure. The foregoing approach has been adopted by the UBC[6.3] for design of slender walls and is proposed by TMS[6.4] for limit states design.

The balanced reinforcement ratio $\rho_b$ describing the condition for which yielding of steel occurs at the same load as the masonry reaches it ultimate strain, is shown in Fig. 6.7. From equilibrium (i.e., $T = C$), $\rho_b$ can be expressed as

$$\rho_b = \beta_1 \gamma_1 (f'_m/f_y)\left(\frac{\varepsilon_u}{\varepsilon_u + \varepsilon_y}\right) \tag{6.30}$$

Handwritten annotation: $\rho_{max} = .5 \, \rho_b$

Handwritten annotations at bottom: $\varepsilon_u = .003$  $\varepsilon_u = .00207$

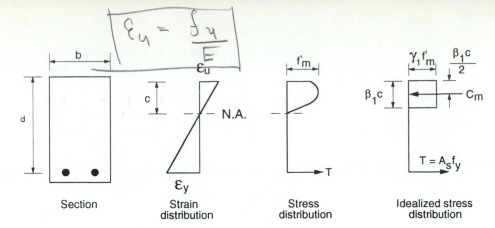

Figure 6.7   Balanced condition at ultimate moment capacity.

Data on the ultimate strain $\varepsilon_u$ for masonry under a strain gradient was discussed in Sec. 5.3.5. A representative group of data[6.2] is reproduced in Fig. 6.8 to illustrate the distribution from beam tests. Suggested design values have ranged from 0.002 to 0.0035 with the UBC opting for the 0.003 value also specified for reinforced concrete.[6.5] This value influences the differentiation between tension and compression controlled failure. Lower values of $\varepsilon_u$ result in lower $\rho_b$ values and in reduced moment capacities for compression-controlled failures.

Limiting the percent of steel avoids brittle failure and ensures some ductility.[6.6] Allen[6.7] recommended that $\rho_{max}$ be $0.5\rho_b$ for adequate ductility under seismic loading. Based on reinforced concrete masonry beam tests, Khalaf et al.[6.8] noticed reduced ductility in flexure compared to reinforced concrete beams and suggested a maximum steel ratio of $0.6\rho_b$. The Australian code[6.9] limits the steel index to $\omega \leq 0.38$.

The Canadian masonry code[6.10] specifies a minimum steel ratio of $0.55/f_y$ for $f_y$ in MPa. (This is equivalent to $80/f_y$ for $f_y$ in psi.) This lower limit is to safeguard against abrupt failure upon cracking of the masonry beam. In the Australian code, a minimum reinforcement ratio $\rho_{min} \geq 0.0015$ is specified.

Li and Neis[6.11] tested reinforced concrete block beams under cyclic reversed loading and compared their behavior with similar reinforced concrete beams. Beam

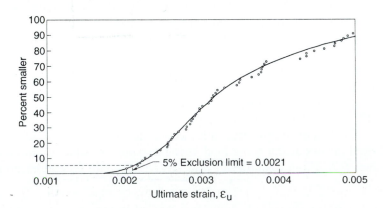

Figure 6.8   Distribution of ultimate strain (from Ref. 6.2).

details are shown in Fig. 6.9(a) and an example of load-deflection hysteresis loops is shown in Fig. 6.9(b). It was concluded that reinforced concrete masonry beams can have a high resistance to reversed cyclic loading. It was also observed that debonding between the grout and the units is the single most important factor in determining the failure of masonry beams subjected to inelastic load reversals.

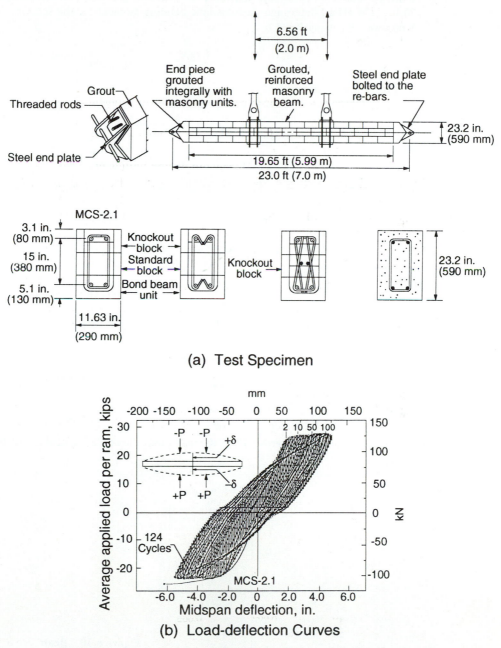

**(a) Test Specimen**

**(b) Load-deflection Curves**

**Figure 6.9** Cyclic test of reinforced masonry beams (from Ref. 6.11).

## 6.2.3 Flexural Design Examples

**Example 6.1**

A three course high, $7\frac{5}{8}$-in. (190 mm) wide grouted concrete block beam (Fig. 6.10) is reinforced with one No. 8 (25 M) bar at an effective depth $d = 20$ in. (500 mm). If the modular ratio $n = 15$ and the allowable stresses are $F_b = 850$ psi (6.0 MPa) and $F_s = 20$ ksi (140 MPa), determine the allowable bending moment for the section.

**Solution.** By using Eq. 6.6,

$$\rho = \frac{A_s}{bd} = \frac{0.79}{7.625 \times 20} = 0.0052 \quad \left(\frac{500}{190 \times 500} = 0.0053\right)$$

and from Eqs. 6.10 and 6.15

$$k = \sqrt{2n\rho + (n\rho)^2} - n\rho$$

$$= 0.324 \quad (0.327)$$

and

$$j = 1 - k/3 = 0.892 \quad (0.891)$$

The resisting moment, based on allowable compressive stress in the masonry, is

$$M = Cjd = \frac{1}{2}F_b kjbd^2 \quad \text{(from Eq. 6.11)}$$

$$= \frac{850(0.324)(0.892)(7.625)(20)^2}{2(1000)(12)} = 31.1 \text{ ft-kip.}$$

$$\left(M = \frac{6.0(0.327)(0.891)(190)(500)^2}{2 \times 10^6} = 41.5 \text{ kN} \cdot \text{m}\right)$$

From Eq. 6.13, the resisting moment, based on allowable stress in the steel, is

$$M = Tjd = \rho F_s jbd^2$$

$$= \frac{0.0052(20000)(0.892)(7.625)(20)^2}{1000(12)} = 23.6 \text{ ft-kip}$$

$$\left(M = \frac{0.0053(140)(0.891)(190)(500)^2}{10^6} = 31.4 \text{ kN} \cdot \text{m}\right)$$

The smaller moment governs. Thus, the maximum allowable moment is 23.6 ft-kip (31.4 kN · m).

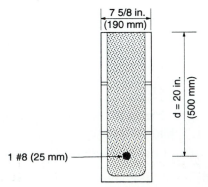

**Figure 6.10** Beam cross section for Example 6.1.

**Example 6.2**

For the masonry cross-section and allowable stresses of Example 6.1, determine the balanced amount of reinforcing steel.

**Solution.** Because the stresses in the two materials are known at balanced conditions, we can draw the stress distribution and from it determine the location of the neutral axis.

By similar triangles (Fig. 6.11),

$$k = \frac{kd}{d} = \frac{850}{850 + 1333} = 0.389$$

$$\left( k = \frac{6}{6 + 9.33} = 0.391 \right)$$

The neutral axis is

$$kd = 0.389(20) = 7.78 \text{ in. (196 mm) from the top}$$

$$j = 1 - k/3 = 0.87 \qquad (0.87)$$

*Stress Distribution.* From Eq. 6.4,

$$C = \frac{1}{2}f_m kbd = \frac{1}{2}(850)(0.389)(7.625)(20)(10)^{-3} = 25.21 \text{ kips}$$

$$\left( C = \frac{1}{2}(6)(0.391)(190)(500)(10)^{-3} = 111.4 \text{ kN} \right)$$

But

$$C = T = A_s f_s$$

$$A_s = \frac{25,210}{20,000} = 1.26 \text{ in.}^2 \qquad \left( \frac{111400}{140} = 796 \text{ mm}^2 \right)$$

Alternatively, from Eq. 6.19,

$$\rho_b = \frac{nF_b}{2F_s(n + F_s/F_b)} = \frac{15(850)}{2(20000)(15 + 20000/850)} = 0.0083$$

$$\left( \rho_b = \frac{15(6)}{2(140)(15 + 140/6)} = 0.0084 \right)$$

$$A_s = \rho_b bd = 0.0083(7.625)(20) = 1.26 \text{ in.}^2 \qquad (0.0084(190)(500) = 798 \text{ mm}^2)$$

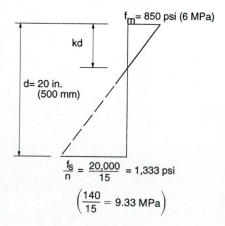

**Figure 6.11** Stress distribution for Example 6.2.

## Example 6.3

The grouted clay brick masonry beam shown in Fig. 6.12 is to carry a 10 kip (45 kN) midspan load on a 12 ft (3.6 m) simple span. Assuming material strengths of $f'_m$ = 3000 psi (20 MPa) and $F_s$ = 20 ksi (140 MPa) and a depth $d$ = 28 in. (710 mm), determine the required amount of tension reinforcement.

**Solution.** The weight of the beam, $w_0$ = 120 × 2(3⅝) × 31.63/144 for the brick and 145(2¾(31.6))/144 for the grout = 273 lb/ft. (2000(9.8)(0.18)(0.80) + 2400(9.8)(0.07)(0.80) = 4140 N/m)

The moment at midspan is

$$M = \frac{Pl}{4} + \frac{w_0 l^2}{8}$$

$$= \frac{10 \times 12}{4} + \frac{0.273 \times (12)^2}{8} = 30 + 4.9 = 34.9 \text{ ft-kips}$$

$$\left( \frac{45(3.6)}{4} + \frac{4.14(3.6)^2}{8} = 47.2 \text{ kN} \cdot \text{m} \right)$$

To arrive at an approximate amount of steel, assume $j$ = 0.90 and check later. [Note that $j$ does not vary a great deal in reinforced masonry beams.] Assuming that tensile stress in the steel controls, from Eq. 6.13,

$$M = Tjd = A_s f_s jd$$

$$A_s = \frac{M}{f_s jd} = \frac{34.9(12)(1000)}{20000(0.90)(28)} = 0.83 \text{ in.}^2$$

$$\left( \frac{47.2(10)^6}{140(0.90)(710)} = 528 \text{ mm}^2 \right)$$

Try two No. 6 (2-20 M) bars, giving $A_s$ = 0.88 in.² (600 mm²), and check the section stresses.

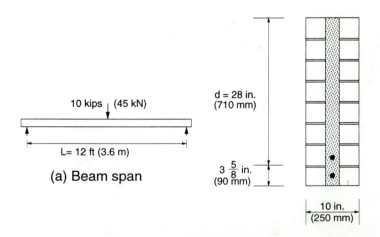

(a) Beam span

(b) Beam Cross-section

**Figure 6.12** Loading and cross-section for Example 6.3.

Using Eqs. 6.6, 6.1, 6.10, and 6.15, respectively

$$\rho = \frac{A_s}{bd} = \frac{0.88}{10 \times 28} = 0.00314 \qquad \left(\frac{600}{250 \times 710} = 0.00338\right)$$

$$n = \frac{E_s}{E_m} = \frac{29,000,000}{3000 \times 750} = 12.9 \qquad \left(\frac{200,000}{20 \times 750} = 13.3\right) \text{ (where } E_m = 750 f'_m)$$

$$750(f'm)$$

$$k = \sqrt{2n\rho + (n\rho)^2} - n\rho = 0.247 \qquad (0.258)$$

$$j = 1 - k/3 = 0.918 \qquad (0.914) \qquad \text{(fairly close to the 0.90 value assumed)}$$

However, from Eq. 6.11

$$M = \frac{1}{2} f_m kjbd^2$$

$$f_m = \frac{2(34.9)(12)(1000)}{0.247(0.918)(10)(28)^2} = 471 \text{ psi} \quad < \quad \frac{1}{3} f'_m$$

$$(f_m = 2(47.2)(10)^6/0.258(0.914)(250)(710) = 3.18 \text{ MPa})$$

Because $f_m$ is less than the allowable flexural compressive stress ($F_b = \frac{1}{3} f'_m = 1000$ psi) (6.67 MPa), the design is satisfactory.
From Eq. 6.14

$$f_s = \frac{M}{\rho jbd^2} = \frac{34.9 \times 12 \times 1000}{0.00314 \times 0.918 \times 10 \times (28)^2} = 18,530 \text{ psi} < 20 \text{ ksi}$$

$$\left(f_s = \frac{47.2(10)^6}{0.00338(0.914)(250)(710)^2} = 121.2 \text{ MPa} < 140 \text{ MPa}\right)$$

To consider minimum reinforcement,

$$\rho_{\min} = \frac{80}{f_y} = \frac{80}{40,000} = 0.0020$$

$$\left(\frac{0.55}{f_y} = \frac{0.55}{300} = 0.0018\right)$$

Because $\rho > \rho_{\min}$, the use of two No. 6 (2-20M) bars is adequate. ∎


## Example 6.4

Determine the factor of safety against collapse for the beam of Example 6.3. Use $\Phi = 0.8$.
**Solution.** By using the UBC,[6.3] the ultimate flexural strength from Eq. 6.24 is $M_u = \Phi T(d - \beta_1 c/2)$, with

$$\beta_1 c = \frac{A_s f_y}{0.85 f'_m b} = \frac{0.88(40,000)}{0.85(3000)(10)} = 1.38 \text{ in.} \qquad \left(\beta_1 c = \frac{600(300)}{0.85(20)(250)} = 42.4 \text{ mm}\right)$$

and $T = A_s f_y$

$$\therefore M_u = (0.8)(0.88)(40)(28 - 1.38/2) = 769 \text{ in.-kip} = 64.0 \text{ ft-kip}$$

$$\left( M_u = 0.8(600)(300) \left( 710 - \frac{42.4}{2} \right) (10)^{-6} = 99.2 \text{ kN} \cdot \text{m} \right)$$

$$\therefore \text{Factor of safety} = \frac{\text{ultimate strength}}{\text{applied moment}}$$

$$= \frac{64.0}{34.9} = 1.84 \qquad (99.2 \div 47.2 = 2.10 \text{ for metric values})$$

Check whether $\rho = 0.00314 \ (0.00338)$ is less than, say, $0.5\rho_b$. From Eq. 6.30

$$\rho_b = (0.85)(0.85) \left( \frac{3}{40} \right) \left( \frac{0.003}{0.003 + \dfrac{40}{29,000}} \right) = 0.037$$

$$\left( \rho_b = 0.85(0.85) \left( \frac{20}{300} \right) \left( \frac{0.003}{0.003 + 300/200,000} \right) = 0.032 \right)$$

Because $\rho < 0.5\rho_b$, the beam is under-reinforced and will exhibit adequate ductile behavior. ∎

## 6.3 SHEAR BEHAVIOR AND DESIGN

### 6.3.1 Development of Design Requirements

Reinforced masonry beams must be designed for shear as well as for bending. Maximum shear forces generally occur near the supports. Shear failure, sometimes referred to as diagonal tension failure, is a brittle mode of failure with very little deformation (see Fig. 6.13) and therefore should be avoided. Under increasing load, several different stages of behavior can be defined.

**Beam Uncracked.** At low bending moments, when the extreme fiber tension stress due to flexure is less than the modulus of rupture of the masonry, it can be

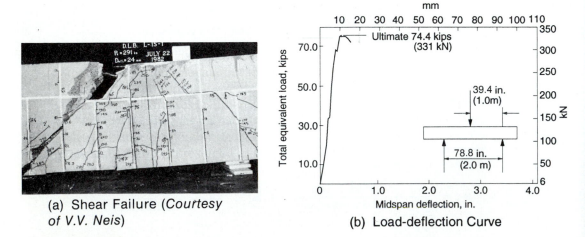

(a) Shear Failure (*Courtesy of V.V. Neis*)

(b) Load-deflection Curve

**Figure 6.13**  Shear behavior of concrete block masonry beams (from Ref. 6.1).

assumed that the beam behaves as a homogeneous elastic beam. The shear stresses $f_v$ can be calculated as

$$f_v = \frac{VQ}{Ib} \tag{6.31}$$

where $V$ = shear force

$Q$ = first moment about the neutral axis of that portion of the cross-section lying between the plane under consideration and the extreme fiber

$I$ = moment of inertia

$b$ = width of cross section at the plane under consideration

The maximum shear stress occurs at the neutral axis. As shown in Fig. 6.14(a), a parabolic stress distribution results for rectangular sections.

**Beam Cracked, No Web Reinforcement.** The shear stress calculated by Eq. 6.31 results in principal tension stress at a 45° angle at the neutral axis. When this diagonal tension stress exceeds the diagonal tensile strength of the masonry, shear cracks develop, originating at the neutral axis. The resulting diagonal crack either

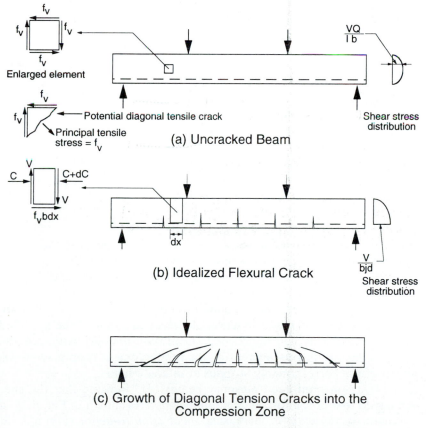

**Figure 6.14** Analysis of shear of masonry beams.

passes through both units and mortar joints or is a stepwise crack along a combination of head and bed joints.

At the extreme tension fiber, the principal tension stress is parallel to the beam and is due to flexural tension. Therefore, as shown in Fig. 6.14(b), flexural cracks will initially be oriented perpendicular to the beam. However, in the shear zone, the shear stress away from the top or bottom of the beam causes inclined principal tensile stress. Thus, the initially vertical flexural crack becomes inclined roughly perpendicular to the principal tension to form a diagonal crack, as illustrated in Fig. 6.14(c) and as shown in Fig. 6.13(a). If the diagonal growth of these cracks is not restricted, they can penetrate up into the compression zone, resulting in a shear-compression failure, as shown in Fig. 6.13(a).

Rigorous analyses, taking into account nonlinear behavior, the actual crack height, dowel action of the reinforcement, and friction forces, are beyond the scope of practical design tools. Therefore, a simplified approach was developed to calculate shear stresses.

By referring to the beam element to the left of Fig. 6.14(b), the equation for calculating shear stress can be derived from

$$\sum F_x = 0$$

$$f_v(b\ dx) = (C + dC) - C = dC$$

From $M = Cjd$ (Eq. 6.11),

$$dC = \frac{dM}{jd}$$

$$f_v bjd = \frac{dM}{dx} = V$$

$$f_v = \frac{V}{bjd} \tag{6.32}$$

This is the current approach in ACI 530/ASCE 5/TMS 402[6.12] and the UBC.[6.3]

An alternative to this is simply to describe the average shear as

$$f_v = \frac{V}{bd} \tag{6.33}$$

This formulation is used in several codes.[6.9,6.10,6.13] Because the $j$ term in Eq. 6.32 is relatively insensitive to the material properties or steel ratio, Eqs. 6.32 and 6.33 have a nearly constant ratio. In fact, common practice in the use of Eq. 6.32 is to equate $j = \frac{7}{8}$ as a constant.

In working stress design, building codes[6.3,6.10,6.12] specify allowable shear stress for flexural members as a function of the square root of the specified compressive strength decreases with increasing *shear span-to-depth ratio* a/d. Similar to reinforced concrete beams, increasing the reinforcement ratio leads to increased shear down of the corresponding stresses at ultimate load.

Tests on reinforced masonry beams[6.6,6.14] indicate that the shear behavior is similar to that of reinforced concrete beams. As shown in Fig. 6.15, masonry shear strength decreases with increasing *shear span-to-depth ratio* a/d. Similar to reinforced concrete beams, increasing the reinforcement ratio leads to increased shear

Reinforced Beams and Lintels    Chap. 6

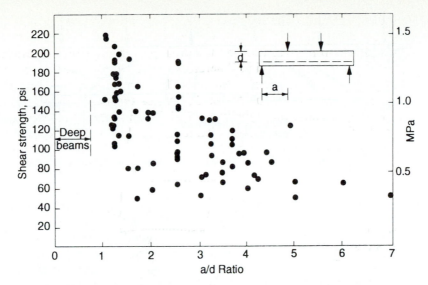

**Figure 6.15**  Shear strength of masonry versus a/d ratio (from Ref. 6.14).

strengths of reinforced masonry beams. Test results[6.6] also show that for beams normally encountered in practice (a/d ≥ 3), shear strength is not strongly correlated to the masonry compressive strength. For this reason, some codes specify constant shear stress values[6.13] or relate the shear design stress to the tension design stress.[6.9]

Despite extensive research and debate regarding shear design in reinforced concrete, most concrete codes still contain similar empirical design provisions that basically draw a constant lower bound under a large number of test results, such as shown in Fig. 6.15 for masonry.

The tendency for higher strengths corresponding to short spans (low a/d ratio) can be related to arch action. If a tied arch analogy[6.5] is used for cracked beams without web reinforcement (see Fig. 6.16), it can be readily appreciated that stronger arches should correspond to lower a/d ratios (i.e., higher a/d ratios indicate shallower, weaker arches). A stronger, more rigid tie provided by increasing the amount of tension reinforcement, expressed as the steel ratio $\rho$, should also increase the capacity of the tied arch. Although current practice, ignoring these factors, may be somewhat conservative for some beams, this is justifiable because of our strong desire to avoid shear failure. The formation of a tied arch is a very late stage of capacity development and is considered to be a particularly dangerous mode of failure because of lack of warning.

Based on a significant number of tests, Suter et al.[6.6] proposed a shear strength design equation for masonry beams without web reinforcement in terms of a/d ratio and steel ratio $\rho$. As shown in Fig. 6.17, lower a/d ratios and/or higher $\rho$ values correspond with higher masonry shear strength.

**Beam Cracked, With Web Reinforcement Included.**  Web reinforcement, normally in the form of stirrups, is used to cross diagonal cracks and control their propagation before they can lead to a shear–compression failure. As shown in Fig.

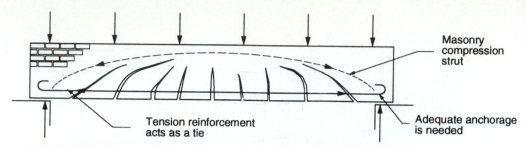

**Figure 6.16** Tied arch analogy for reinforced masonry beams with no web reinforcement.

6.18, tests indicate that web reinforcement prevents diagonal tension failure and allows the full flexural capacity to be developed.[6.15] Although inclined stirrups and bent-up bars can be used, as is done in reinforced concrete, usual practice involves vertical stirrups for ease of construction. Hence, the discussion presented is restricted to vertical stirrups (i.e., perpendicular to the axis of the flexural number). Stirrups are normally spaced at a distance $s$, as indicated in Fig. 6.19.

The tensile force in a stirrup is calculated by

$$T_s = A_v f_s \tag{6.34}$$

where $A_v$ = steel area of a stirrup

$f_s$ = steel stress in a stirrup

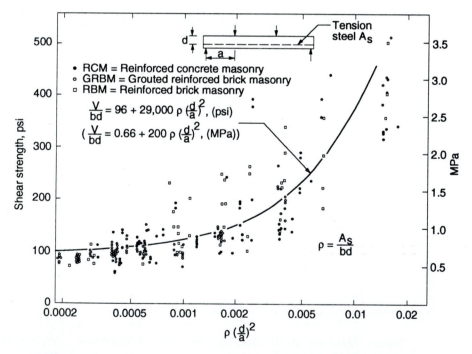

**Figure 6.17** Shear strength of masonry beams without web reinforcement (from Ref. 6.6).

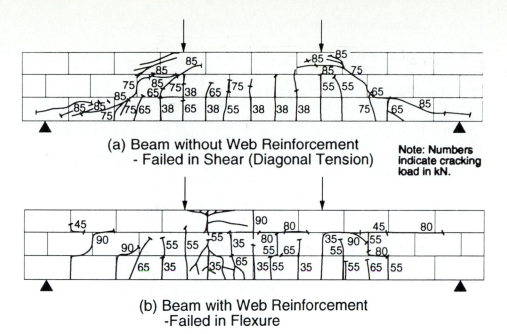

(a) Beam without Web Reinforcement
- Failed in Shear (Diagonal Tension)

Note: Numbers indicate cracking load in kN.

(b) Beam with Web Reinforcement
-Failed in Flexure

**Figure 6.18**  Effect of web reinforcement on failure mode of masonry beams (from Ref. 6.15).

For an idealized diagonal crack at 45°, Fig. 6.19(b), the vertical tensile force provided by all of the stirrups crossing the crack over the depth $d$ is

$$\sum T_s = \frac{d}{s} A_v f_s \qquad (6.35)$$

In North American codes,[6.3,6.10,6.12] when stirrups are required, they are required to resist the entire shear.

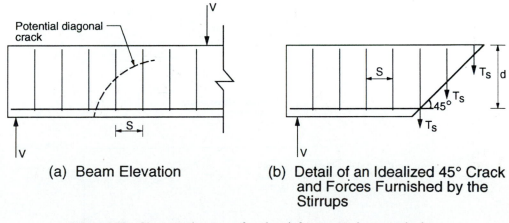

(a)  Beam Elevation

(b)  Detail of an Idealized 45° Crack and Forces Furnished by the Stirrups

**Figure 6.19**  Shear resistance of web reinforcement in a cracked reinforced masonry beam.

Summing the vertical forces in Fig. 6.19(b) gives

$\frac{f_V}{\phantom{x}}$ = lesser of $3\sqrt{8m}$ or 150 psi

$$V = \frac{d}{s}A_v f_s \tag{6.36}$$

from which the required area of each stirrup is determined by

$$A_v = \frac{Vs}{f_s d} \tag{6.37}$$

Alternately, for a particular area of stirrup $A_v$, the required stirrup spacing can be calculated from

$$s = \frac{A_v f_s d}{V} \tag{6.38}$$

The maximum spacing of vertical stirrups is limited to $d/2$ to ensure that each diagonal crack is crossed by a stirrup before it can propagate too high into the beam. Stirrups should be provided for a distance d beyond the point where the masonry can resist the entire shear (as shown in Fig. 6.20). To avoid the possibility of shear-compressive failure, the amount of web reinforcement that can be used is limited by maximum limits on the calculated shear stress in the masonry.

if $f_V$ is > $f_{V max}$, increase the section

$$S_{max} = \frac{d}{2}$$

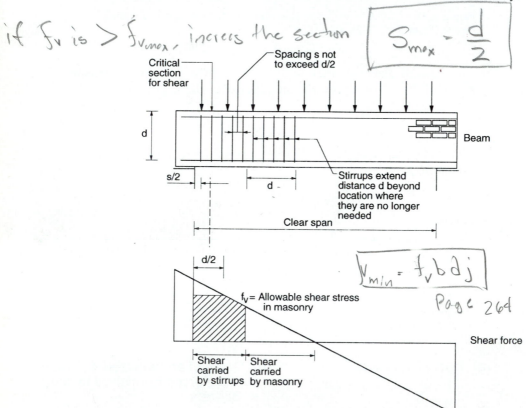

$$V_{min} = f_V b d j$$

Page 264

**Figure 6.20** Shear design of masonry beam (working stress design).

Code provisions differ regarding location of the critical section for shear with some[6.12] specifying a distance of d/2, as shown in Fig. 6.20, while others[6.9,6.10] recommend d from the face of the support. For certain support conditions such as simply supported beams, it may be more appropriate to define the support for both shear and bending as the center line of the bearing area.

Provisions for termination or splicing of tension reinforcement in shear zones, similar to those for reinforced concrete, are also specified in most masonry codes.

Cyclic tests conducted by Li and Neis[6.11] on reinforced masonry beams with web reinforcement showed that the shear strengths of these beams were lower than similar reinforced concrete beams. They attributed this to a number of possible effects such as poor bond between grout and units and the inherent weakness of the mortar joint planes that have very little mechanical interlock to increase shear resistance. Such results support the practice of neglecting the contribution of the masonry for cases where shear reinforcement is required.

### 6.3.2 Shear Design Examples

#### Example 6.5

For the beam described in Example 6.3, check to see if web reinforcement is required.
**Solution.** For working stress design, the maximum shear force $V$ at a distance $d/2$ from the face of the support is determined as

$$V = \frac{10}{2} + 0.273\left(\frac{12}{2} - \frac{14}{12}\right) = 6.31 \text{ kips}$$

$$\left(V = \frac{45}{2} + 4.14\left(\frac{3.6}{2} - \frac{0.71}{2}\right) = 28.5 \text{ kN}\right)$$

The shear stress is found using Eq. 6.32

$$f_v = \frac{V}{bjd} = \frac{6.31(10)^3}{10(0.875)28} = 25.8 \text{ psi}$$

$$\left(f_v = \frac{28.5(10)^3}{250(0.875)(710)} = 0.18 \text{ MPa}\right)$$

By using ACI 530/ASCE 5/TMS 402 values,[6.12] the allowable masonry shear stress $F_v$ is the lesser of $\sqrt{f'_m}$, or 50 psi (35 MPa). Because $\sqrt{3000} = 54.8$ psi $> 50$ psi, therefore, $F_v = 50$ psi (0.35 MPa). Because $F_v > f_v$, no shear reinforcement is required. ∎

#### Example 6.6

For the beam analyzed in Example 6.5, assume that the shear force is increased by 150% and determine the vertical stirrup requirements.
**Solution.** The shear stress, $f_v$ is now

$$f_v = 25.8 \times 2.5 = 64.5 \text{ psi} > 50 \text{ psi allowable } F_v$$

$$(f_v = 0.18 \times 2.5 = 0.45 \text{ MPa} > 0.35 \text{ MPa})$$

Hence, shear reinforcement is required to carry all the shear. The maximum allowable shear stress, $F_{v_{max}}$ must be checked.

By using the ACI 530/ASCE 5/TMS 402 values,[6.12]

$$F_{v_{\max}} = 3.0 \sqrt{f'_m} = 3.0 \sqrt{3000} = 164 \text{ psi} \qquad (1.13 \text{ MPa})$$

$$F_{v_{\max}} = 150 \text{ psi} \qquad (1.03 \text{ MPa}) \qquad \text{which controls.}$$

Because $F_{v_{\max}} > f_v$, the beam section is adequate if sufficient stirrups are included.
Try a No. 4 (10 M) single leg stirrup with $A_v = 0.20 \text{ in.}^2$ (100 mm²).
From Eq. 6.38

$$s = \frac{A_v f_s d}{V} = \frac{0.20(20{,}000)28}{2.5(6.31)(10)^3} = 7.1 \text{ in.} < \frac{d}{2} = 14 \text{ in.} \qquad \text{O.K.}$$

$$\left( s = \frac{100(140)(710)}{2.5(28.5)(10)^3} = 140 \text{ mm} < 355 \text{ mm} \qquad \text{O.K.} \right)$$

Hence, use No. 4 (10 M) single leg stirrups at 7 in. (140 mm) spacing. Stirrup spacing could be increased up to d/2 at the point where the shear force is approximately half of the maximum value. ∎

## 6.4 BOND AND ANCHORAGE OF REINFORCEMENT

To achieve full interaction between the reinforcing steel and the masonry, adequate bond must be developed. This is achieved by providing an adequate development length (or anchorage) to develop the required stress in the tension steel. In working stress design, the flexural bond stress $u$ must not exceed the specified allowable value.

### 6.4.1 Bond Stress

To determine the flexural bond stress $u$, consider an element removed from a beam, shown in Fig. 6.21. For the moment $M = Tjd$, the change in bar force, $\Delta T$, is caused by a change in moment, $\Delta M$, from one face of the element to the other face. Hence,

$$\Delta T = \frac{\Delta M}{jd} \tag{6.39}$$

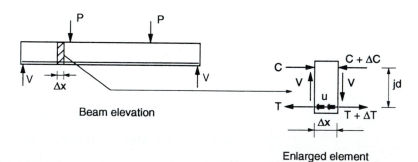

**Figure 6.21** Development of flexural bond stress in reinforced masonry beams.

and for equilibrium,

$$\Delta T = \frac{V \, \Delta x}{jd} \tag{6.40}$$

The bond between steel and grout resists the change in bar force $\Delta T$, such that

$$\Delta T = u \sum_o \Delta x$$

where $u$ is the bond stress, and $\sum o$ is the bar perimeter. By equating the two relations for $\Delta T$,

$$u \sum_o \Delta x = \frac{V \, \Delta x}{jd}$$

or

$$\boxed{u = \frac{V}{\sum o \, jd}} \qquad \text{bons stress} \; (<(160 \text{ psi})) \tag{6.41}$$

Equation 6.41 is given explicitly in some codes[6.10] as the expression for flexural bond stress. Allowable bond stress values such as 160 psi (1.1 MPa) can be specified[6.10] for deformed bars. Alternately, adequate bond can be implicitly ensured by satisfying anchorage requirements for the reinforcement.

### 6.4.2 Development length

Building codes require proper anchorage of the reinforcement. Except for use of mechanical anchors, proper anchorage means that the development length $l_d$ provided in a member must be sufficient to develop the design stress in the steel without exceeding the design value for bond stress. If the development length provided is not sufficient, hooks may be added or a greater number of smaller diameter bars can be substituted.

The equation for development length is derived from the equilibrium of the free body diagram of a length of reinforcing bar (see Fig. 6.22),

$$\pi d_b u l_d = A_s f_s = \frac{\pi d_b^2}{4} f_s$$

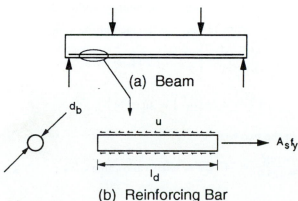

(a) Beam

(b) Reinforcing Bar

**Figure 6.22** Development length.

so that

$$\boxed{l_d = \frac{f_s d_b}{4u}} \Rightarrow l_d = \frac{f_s d_b}{640} \qquad (6.42)$$

$$\boxed{\frac{l}{d} = .0015\, d_b f_s}$$

where $f_s$ = calculated stress in the steel at some location
 $u$ = code allowable bond stress
 $d_b$ = bar diameter
 $l_d$ = development length

For working stress design, ACI 530/ASCE 5/TMS 402[6.12] recommends a development length $l_d$ equal to 0.0015 $d_b f_s$ (in.), which can be derived from Eq. 6.42 using an allowable bond stress $u$ of 160 psi.

The basic development length for strength and limit states design as proposed by Suter et al.,[6.6,6.16] with $A_b$ = area of the bar, is:

$$l_d = k_d A_b f_y / \sqrt{f_m'} \qquad (6.43)$$

where $k_d$ = 0.040 or 0.019 for units of inches and pounds or N and mm, respectively, which is similar to that for reinforced concrete.[6.5] Some reduction in bond for multiple bars in a layer or for bundled bars is also appropriate.

The length of lap splice is normally expressed in terms of the development length $l_d$ for reinforced concrete. ACI 530/ASCE 5/TMS 402[6.12] specifies a lap splice in tension or compression equal to 0.002 $d_b f_s$ (in.), which amounts to 1.33$l_d$. A minimum splice length of 12 in. (300 mm) is required.[6.12]

In general, where specific requirements for reinforcement are not provided, reference should be made to corresponding concrete design provisions.

### 6.4.3 Bond and Development Length Example

**Example 6.7**

For the beam in Example 6.3, check the adequacy of bond stress and development length. Consider the allowable bond stress to be 160 psi (1.1 MPa).
**Solution.** For a No. 6 (20 M) bar, $\Sigma o$ = 2.36 in. (61.4 mm). From Example 6.3,

$$jd = 0.918 \times 28 \text{ in.} = 25.7 \text{ in}$$

$$(0.914 \times 710 = 649 \text{ mm})$$

From Example 6.5, but for the maximum shear at the face of the support,

$$V = \frac{10}{2} + 0.273 \left(\frac{12}{2}\right) = 6.64 \text{ kips}$$

$$\left(\frac{45}{2} + 4.14 \left(\frac{3.6}{2}\right) = 30.0 \text{ kN}\right)$$

The flexural bond stress in the two bars is

$$u = \frac{V}{\Sigma_o jd} = \frac{6.64(10)^3}{2(2.36)(25.7)} = 54.7 \text{ psi} < 160 \text{ psi allowable stress} \qquad \text{O.K.}$$

$$\left(\frac{30.0(10)^3}{2(61.4)(649)} = 0.38 \text{ MPa} < 1.1 \text{ MPa}\right)$$

Where $f_s = 18{,}530$ psi (121.2 MPa) from Example 6.3, the development length required from Eq. 6.42 is

$$l_d = \frac{f_s d_b}{4\,u} = \frac{18530(0.75)}{4(160)} = 21.7 \text{ in.}$$

$$\left(\frac{121.2(10)}{4(1.10)} = 551 \text{ mm}\right)$$

The development length provided is

$$l_d = \frac{\text{span}}{2} = \frac{12 \times 12}{2} = 72 \text{ in. (1800 mm)} > \text{required } l_d \qquad \text{O.K.}$$

## 6.5 SERVICEABILITY REQUIREMENTS

Serviceability requirements for reinforced flexural members are deflection, crack control, and corrosion protection. Corrosion protection is specified in building codes by providing adequate cover or, in some cases, corrosion resistant reinforcement. Flexural cracks normally originate at the mortar joints. For example, beams constructed of 16 in. (400 mm) long concrete block, typically crack at 16 in. (400 mm) spacings. Crack spacing in reinforced concrete is more in the order of 4 to 8 in. (100 to 200 mm). Consequently, crack widths in masonry beams can be significantly wider than for corresponding reinforced concrete members. A corrosive exposure condition is a situation that a designer may wish to take into account by limiting steel stress. Such precautions may become more important for strength design where steel stresses under service loads are generally higher than those for working stress design.

ACI 530/ASCE 5/TMS 402[6.12] requires that deflection of beams and lintels due to dead plus live load be limited to the span divided by 600, or 0.3 in. (8 mm), when providing vertical support to unreinforced masonry. This empirical requirement is intended to limit excessive deflections which may result in damage to the supported masonry.

In the absence of other evidence or code recommendations, deflections of reinforced masonry beams have been calculated using methods developed for reinforced concrete members. By adopting the ACI 318[6.17] method for reinforced concrete, deflections of reinforced masonry beams were calculated by Lee et al.[6.15] using the following equation for the effective moment of inertia $I_e$:

$$I_e = \left(\frac{M_{cr}}{M_a}\right)^3 I_g + \left[1 - \left(\frac{M_{cr}}{M_a}\right)^3\right] I_{cr} \le I_g \tag{6.44}$$

where   $I_g, I_{cr}$ = gross and cracked moments of inertia
          $M_{cr}, M_a$ = cracking and applied moments

They compared the predicted deflections using this equation with measured deflections. A reasonable agreement was achieved using the modulus of elasticity of the masonry parallel to the bed joints (stress in the compression zone is parallel to bed joints).

Alternately, Horton and Tadros[6.18] compared different methods for calculating deflection of reinforced masonry members and proposed the following expression which provided a good fit with available data:

$$\Delta = \Delta_{cr}(1 - \alpha) \tag{6.45}$$

where $\Delta$ = maximum deflection of masonry member

$\Delta_{cr}$ = maximum deflection using moment of inertia of the cracked section.

$$\alpha = \left(\frac{M_{cr}}{M_a}\right)^2 \left(2 - \frac{M_{cr}}{M_a}\right)\left(1 - \frac{I_{cr}}{I_g}\right) \tag{6.46}$$

$$\text{for } M_{cr} \le M_a \le M_y$$

$$\alpha = \left(1 - \frac{I_{cr}}{I_g}\right) \qquad \text{for } M_{cr} > M_a \tag{6.47}$$

$M_y$ = yield moment, and the other terms are the same as for Eq. 6.44.

## 6.6 LOAD DISTRIBUTION ON LINTEL BEAMS

### 6.6.1 Behavior

Lintel beams span over openings in masonry walls such as shown in Fig. 6.23(a). The function of the lintel is to transfer vertical loads from above to the end supports.

There are two types of vertical loads carried by lintels:

1. Vertical distributed loads from self-weight of the wall above plus service loads from the floors and roof.
2. Concentrated loads from the floor beams, roof joists, and other beams framing into the wall.

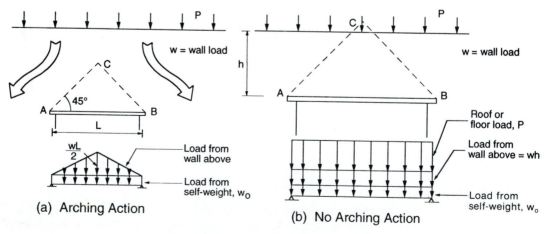

**Figure 6.23**  Load distribution on lintels.

Because of arching action (refer to Sec. 1.3.2), lintels may not carry the full load above an opening. As shown in Fig. 6.23(a), the part of masonry contained in triangle ABC is usually assumed to be carried by the lintel. Recommended practice[6.19] is to use a triangle formed by 45° to 60° angles to determine the tributary wall area for design: 45° angles are shown in Fig. 6.23. Any loads above point C (apex of the triangle) and the masonry on either side can be neglected. For arching action to be utilized, there must be enough masonry on each side of the opening to resist lateral thrust from the arching action. Arching action also requires an adequate depth of masonry above point C to carry the horizontal compressive forces from the arching thrusts. If floor or roof loads are applied below point C, Fig. 6.23(b), no arching is considered and the lintel should be designed for the full load above it.

Concentrated loads from roof and floor beams and trusses are considered to be transferred to the lintel assuming a 60° angle[6.19] triangular dispersion of the force, as shown in Fig. 6.24. The part of the load $w_p$ distributed over the length DB, is considered in the analysis and design of the lintel in addition to loads from the self-weight and the tributary area weight of wall above the opening.

An alternative to the above simplified approach is to use the analytical method given in Chap. 11.

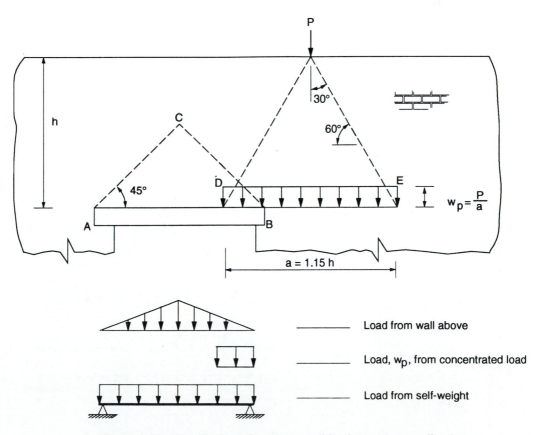

**Figure 6.24**   Distribution of concentrated loads on masonry lintels.

## 6.6.2 Lintel Beam Loading Example

**Example 6.8**

Determine the load distribution on the lintel beam shown in Fig. 6.25 due to wall weight and a concentrated load of 10 kips (45 kN) from roof trusses spaced at 8 ft (2.4 m). Consider the lintel as simply supported. Assume 100 lb/ft² (4.8 kN/m²) for the wall weight.

**Solution.** Loads from the self-weight of the lintel (uniformly distributed load) is

$$w_0 = 100 \, (8/12) = 67 \text{ lb/ft} \quad \left( 4.8 \left( \frac{200}{1000} \right) = 0.96 \text{ kN/m} \right)$$

Load from the wall above has a triangular distribution with a maximum intensity of

$$w_1 = 100 \, (3) = 300 \text{ lb/ft} \quad (4.8(0.9) = 4.32 \text{ kN/m})$$

The uniform load at the base of triangle from the concentrated load is distributed over a length $= 2 \, (\tan 30°) \, (4 \text{ ft}) = 4.62 \text{ ft}$ $(2 \tan 30°(1.2 \text{ m}) = 1.386 \text{ m})$.

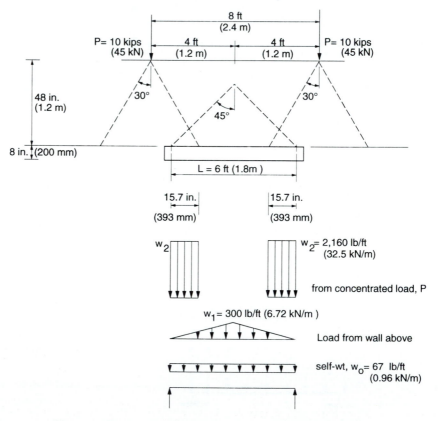

**Figure 6.25** Load distribution on lintels for Example 6.7.

Therefore,

$$w_2 = 10 \times 1000/4.62 = 2160 \text{ lb/ft } (45/1.386 = 32.5 \text{ kN/m})$$

The loading on the lintel for design is shown in Fig. 6.25. ■

## 6.7 PRESTRESSED MASONRY BEAMS

Masonry beams have been limited to relatively short spans due to cracking and deflection constraints. The use of the post-tensioned prestressing technique, used in concrete, allows increased span lengths. Also, prestressing reduces the occurrence of shrinkage cracks in mortar, which have been linked to water infiltration.

The potential for prestressed masonry in North America is of current interest.[6.20] Prestressed masonry was introduced in England for brick masonry construction during the 1980s.[6.21,6.22] Research at the University of Edinburgh[6.23] compared the behavior of prestressed brickwork and concrete beams. The load-deflection curves for the two types of beams are shown in Fig. 6.26. The similarity in behavior of the brickwork and concrete beams indicates that the concepts of prestressed concrete can be applied to prestressed masonry. The load-deflection curves show two distinct phases related to responses before and after cracking. The first phase is

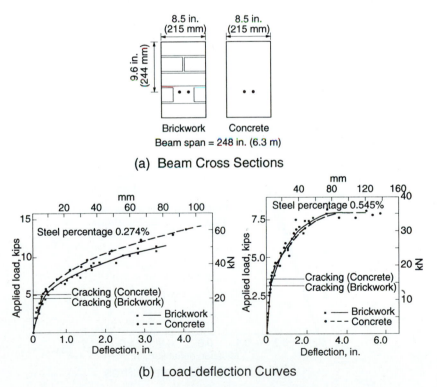

(a) Beam Cross Sections

(b) Load-deflection Curves

**Figure 6.26** Behavior of prestressed brickwork and concrete beams (from Ref. 6.23).

characterized by linear elastic response, whereas the second phase is characterized by nonlinear inelastic behavior.

Beam details are shown in Fig. 6.27(a) for prestressed concrete masonry beams tested by Ng and Cerny.[6.24] Load-deflection curves are presented in Fig. 6.27(b). It was concluded that prestressing of concrete masonry using high strength tendons is feasible and that deflection and strength can be reasonably predicted using the ACI 318[6.17] equations for post-tensioned concrete beams. If specific code requirements for design of prestressed masonry beams are lacking, it is suggested that prestressed concrete practice be followed but with substitution of masonry properties and design stresses.

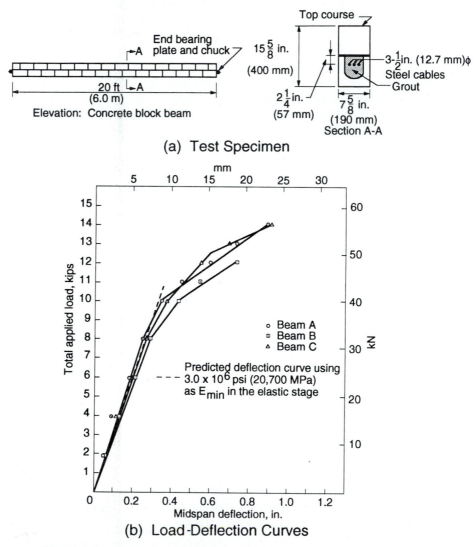

**Figure 6.27** Behavior of prestressed block masonry beams (from Ref. 6.24).

## 6.8 CLOSURE

This chapter on reinforced masonry beams and lintels has included behavior and design for flexure, shear, bond, and development length. The concepts introduced and many of the design provisions are also applicable to the design of walls (Chaps. 7, 8, and 10) and columns and pilasters (Chap. 9).

The behavior and design of deep beams are not covered and there is very little information on this topic. To some extent this is covered for the design of walls supported on beams included in Chap. 11.

For those familiar with the design of reinforced concrete beams, the similarities for reinforced masonry are apparent. Behavior under service loads and at ultimate load has been discussed to assist designers adapt to strength or limit states design methods.

Construction details, including placement of reinforcement, are discussed in Chap. 15.

## 6.9 REFERENCES

6.1   V. V. Neis and R. J. Loeffler, "Results of Ultimate Flexural and Shear Tests on Reinforced Masonry Beams," in *Proceedings of the Third Canadian Masonry Symposium*, University of Alberta, Edmonton, 1983, pp. 13/1–13/16.

6.2   G. T. Suter and G. A. Fenton, "Flexural Capacity of Reinforced Masonry Members," *ACI Journal*, Vol. 83, No. 1, January–February 1986, 127–136.

6.3   International Conference of Building Officials, "Uniform Building Code, Chapter 24: Masonry," UBC, Whittier, CA, 1991.

6.4   The Masonry Society, "Limit States Design of Masonry," draft, TMS, Boulder, CO, 1990.

6.5   J. MacGregor, "Reinforced Concrete—Mechanics and Design," Prentice-Hall, Englewood Cliffs, NJ, 1988.

6.6   G. T. Suter, H. Keller, and G. A. Fenton, "Summary of a Decade of Reinforced Masonry Research at Carleton University," Department of Civil Engineering, Carleton University, Ottawa, 1984.

6.7   D. E. Allen, "Probabilistic Study of Reinforced Concrete in Bending," Technical Paper No. 311 (NRC 11/39), Division of Building Research, National Research Council of Canada, Ottawa, 1970.

6.8   F. Khalaf, J. Glanville, and M. El Shahawi, "A Study of Flexure in Reinforced Masonry Beams," *Concrete International Journal*, Vol. 5, No. 7, July 1983, 46–55.

6.9   Standards Association of Australia, "SAA Masonry Code," Australian Standard 3700, SAA, North Sydney, 1988.

6.10  Canadian Standards Association, "Masonry Design and Construction for Building," CSA Standard CAN-S304-1984, Rexdale, Ontario, 1987.

6.11  D. Li and V. V. Neis, "Performance of Reinforced Masonry Beams Subjected to Reversed Cyclic Loadings," in *Proceedings of the Fourth Canadian Masonry Symposium*, Vol. 1, Fredricton, N.B., June 1986, pp. 351–365.

6.12  "Building Code Requirements for Masonry Structures," ACI 530/ASCE 5/TMS 402, The Masonry Joint Committee, American Concrete Institute and American Society of Civil Engineers, Detroit, New York, 1992.

6.13 British Standards Institution, "Code of Practice for Use of Masonry, BS 5628: Part 2: Structural Use of Reinforced and Prestressed Masonry," BSI, London, 1985.

6.14 G. T. Suter and H. Keller, "Shear Strength of Reinforced Masonry Beams and Canadian Code Implications," in *Proceedings of the First Canadian Masonry Symposium*, Calgary, Alberta, 1976, pp. 149–160.

6.15 R. Lee, J. Longworth, and J. Warwaruk, "Behavior of Restrained Masonry Beams," in *Proceedings of the Third Canadian Masonry Symposium*, Edmonton, Alberta, 1983, pp. 37/1–37/16.

6.16 G. T. Suter and H. Keller, "Bond and Development Length in Reinforced Concrete Masonry Beams," International Journal of Masonry Construction, Vol. 2, No. 4, 1982, 139–145.

6.17 American Concrete Institute, "Building Code Requirements for Reinforced Concrete," ACI 318-89, ACI, Detroit, MI, 1989.

6.18 R. Horton and M. Tadros, "Deflection of Reinforced Masonry Members," ACI Structural Journal, Vol. 87, No. 4, July-August 1990, pp. 453–463.

6.19 Brick Institute of America, "Reinforced Brick and Tile Lintels," BIA Technical Note No. 17H, BIA, Reston, VA, 1964.

6.20 A. Schultz and M. Scolforo, "An Overview of Prestressed Masonry," The Masonry Society Journal, Vol. 10, No. 1, August 1991, pp. 6–21.

6.21 A. W. Hendry, *Structural Brickwork*, MacMillan, London, 1981.

6.22 W. G. Curtin, G. Shaw, J. K. Beck, and W. A. Bray, *Structural Masonry Designers' Manual*, Granada Publishing, 1982.

6.23 B. Sinha and R. Pedreschi, "Can Prestressed Brickwork Beams Be Used as an Alternative to Prestressed Concrete Beams?" in *Proceedings of the 9th International Brick/Block Masonry Conference*, Berlin, 1991, 442–449.

6.24 L. Y. Ng and L. Cerny, "Post-Tensioned Concrete Masonry Beams," in *Proceedings of the Third North American Masonry Conference*, Arlington, TX, 1985, pp. 77-1–77-12.

## 6.10 PROBLEMS

**6.1** A reinforced masonry beam is composed of masonry with a permissible compressive stress of 1000 psi (6.9 MPa) and reinforcing steel having a permissible tensile stress of 20,000 psi (140 MPa). The modular ratio of steel to masonry is n = 15 and the steel is located at a depth of 25 in. (635 mm). For balanced design, sketch a diagram of the bending stress distribution of the beam showing:
   (a) distance to the neutral axis
   (b) position of the resulting masonry compression force
   (c) level arm distance between the compression and tension forces
   Check the answers obtained graphically using equations from the text.

**6.2** For the reinforced concrete masonry beam shown in Fig. P6.2, perform the following analyses:
   (a) Determine the superimposed load w required to cause cracking if modulus of rupture $f_t = 200$ psi (1.4 MPa). Calculate the midspan deflection at this load.
   (b) Using an allowable compressive stress in the masonry of $0.33f'_m$ and an allowable tensile stress in the steel of 24,000 psi (165 MPa), determine the moment capacity using working stress design. Is the capacity tension or compression controlled? What

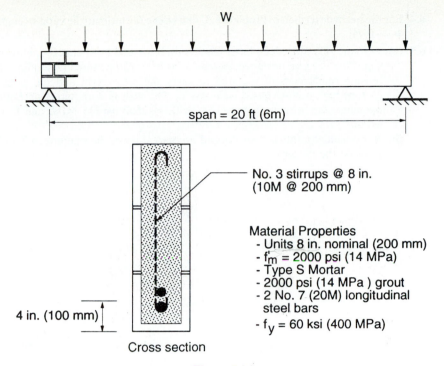

**Figure P6.2**

maximum superimposed uniformly distributed load w can the beam carry for working stress design?

(c) Calculate the midspan deflection at this working stress design load.

(d) Are the stirrups provided sufficient to resist the shear corresponding to the working stress design load?

(e) At what location along the length of the beam would stirrups not be required?

(f) Check the flexural bond stress and determine the required development length using u = 160 psi (1.10 MPa).

(g) Determine the ultimate moment capacity $M_u$ for this beam using ($\Phi = 0.75$) and calculate the uniformly distributed ultimate load (factored load) that can be carried.

**6.3** The 12 in. (300 mm) wide, 24 in. (600 mm) high double-wythe brick beam with the cross section shown in Fig. P6.3 is to be designed to carry its own weight plus a superimposed load of 2 kips/ft (30 kN/m) over a 16 ft (5 m) span. Determine the amounts of longitudinal steel and stirrups required. ($f'_m = 4000$ psi (27.5 MPa), allowable $f_s = 20,000$ psi (140 MPa), and allowable $F_b = 0.33 f'_m$.)

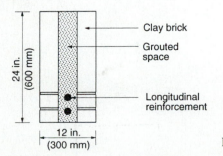

**Figure P6.3**

**6.4** For the beam described in Problem 6.3, determine the ultimate flexural strength considering $\Phi = 0.75$.

**6.5 (a)** A 16 in. (400 mm) deep lintel made with 8 in. (200 mm) concrete blocks as shown in Fig. P6.5 is to be designed to span a 10 ft (3 m) opening. Determine the amount of tension reinforcement required to carry the self-weight and the weight of the 8 ft (2.4 m) high hollow block wall above. The wall is continuous on both sides of the opening. Are stirrups required? Use $f'_m = 2000$ psi (15 MPa) and $f_y = 60$ ksi (400 MPa).

**(b)** If movement joints are introduced on both sides of the opening, what effect will this have on the design?

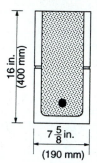

16 in.
(400 mm)

$7\frac{5}{8}$ in.

(190 mm)    **Figure P6.5**

**6.6** For advanced students familiar with prestressed concrete, design the beam in Problem 6.3 as a prestressed masonry beam using $\frac{1}{2}$ in. (13 mm) diameter seven wire prestressing strand ($A_s = 0.153$ in.$^2$ (100 mm$^2$) per strand) with an ultimate strength of 270 ksi (1860 MPa).

# CHAPTER 7

# Flexural Walls

Construction of a flexural wall
(*Courtesy of National Concrete
Masonry Association*).

## 7.1 INTRODUCTION

All masonry walls may be subjected to lateral loads normal to the faces of the walls causing them to bend out-of-plane. The loads can be permanent, such as earth pressure against a retaining wall or basement, or they can be transient resulting from wind or earthquake. Walls must also be sufficiently robust not to be damaged by incidental forces from people and equipment and they should resist accidental forces without suffering disproportionate amounts of damage or failure. These latter loads are commonly considered as equivalent uniform lateral loads and are analyzed in the same manner as wind and seismic forces.

In the ancient examples of masonry, where the masonry units often were not bonded together, walls were generally sufficiently thick that flexural stresses from the lateral loads were much lower than axial compressive stresses from self-weight and other gravity loads. In some cases, this natural "prestressing" was augmented by the addition of towers or other decorative masses to the tops of walls. In addition, openings in walls tended to be fairly small and cross-walls, buttresses, or other thickened wall sections allowed walls to span both horizontally and vertically.

It is only in the latter half of this century that the tensile strength of masonry has become relied upon to provide the flexural resistance to out-of-plane lateral loads. Gradually, rules of thumb were developed to describe accepted practice regarding height-to-thickness or length-to-thickness limits. In general, these empirical provisions have worked well except where very high wind or seismic loads have been experienced or where the normal redundancy of masonry construction was eliminated from the designs. Engineered masonry often relies on reinforcement.

The usage of masonry as exterior walls has changed to include walls that are only required to carry very light vertical axial force (often just the self-weight) plus lateral wind pressures or inertia forces due to earthquakes. This has meant that, unless simple rules incorporate large margins of safety to cover a wide range of applications, rational analysis is required to produce efficient design.

This chapter looks at the behavior and design of flexural walls, which are defined as walls subject to out-of-plane bending where self-weight and small superimposed axial loads do not have any significant secondary structural effects (buckling or additional bending moments). The only effects of the axial loads are delay of cracking due to initial precompression and some increase in compressive stress that could have a minor effect for cracked reinforced walls. When restrained boundaries allow arching to be developed, large axial compression forces will exist. Analysis methods to account for the benefits of arching are presented.

## 7.2 LOAD RESISTING MECHANISMS FOR MASONRY WALLS

Unreinforced masonry walls depend on the tensile strength of masonry and/or in-plane compressive forces to resist lateral loads. Masonry walls in low-rise buildings and the cladding of framed buildings are examples of this type of wall. An in-plane compressive force can have an important influence where even the self-weight of a single-story wall can significantly increase flexural strength.

Until recently, the material properties and flexural behavior of masonry were not sufficiently understood to enable rational design rules to be applied. In many cases, application of design methods, using measured flexural tensile strengths of masonry, did not justify common forms of construction that have performed satisfactorily. Some walls that crack are still able to resist lateral loading. This capacity is attributed to the existence of mechanisms other than simple bending. Internationally, the design provisions for flexural walls in various codes[7.1–7.6] differ markedly in the specified tensile design stresses and in the analytical methods presented.

The basic types of behavior and forms of analysis are briefly introduced in what follows.

**Flexural Tensile Strength.**   Walls that are simply supported and span in only the vertical direction must resist lateral load by bending action. If tensile strength is neglected, the bending capacity is directly related to the axial compression, and, as shown in Fig. 7.1(a), the moment capacity is

$$M = f_a S$$

(7.1)

maximum moment

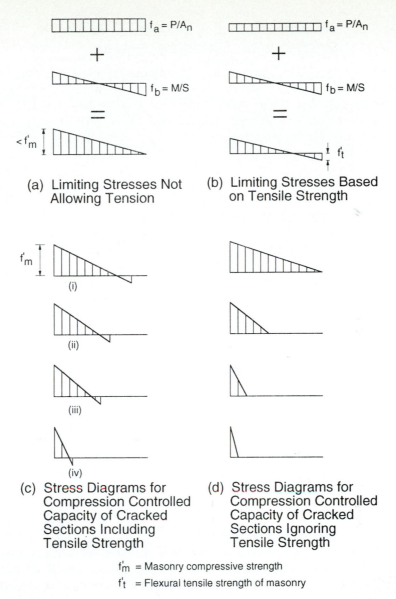

**Figure 7.1** Flexural behavior of unreinforced sections (linear elastic material).

The figure shows:

$f_a = P/A_n$    +    $f_b = M/S$    =    $< f'_m$

(a) Limiting Stresses Not Allowing Tension

$f_a = P/A_n$    +    $f_b = M/S$    =    $f_t$

(b) Limiting Stresses Based on Tensile Strength

$f'_m$    (i), (ii), (iii), (iv)

(c) Stress Diagrams for Compression Controlled Capacity of Cracked Sections Including Tensile Strength

(d) Stress Diagrams for Compression Controlled Capacity of Cracked Sections Ignoring Tensile Strength

$f'_m$ = Masonry compressive strength
$f'_t$ = Flexural tensile strength of masonry

where   $f_a$ = the axial compressive stress ($P/A_n$)
      $A_n$ = effective mortar bedded area
      $S$ = section modulus for out-of-plane bending

As can be seen in Fig. 7.2, line *A* represents the relationship between flexural capacity and axial compression ignoring tensile strength. For a specified level of axial load (point *a*), application of lateral load increases the moment until the neutral axis is at the tension face. At this point (point *b*), the relationship between moment

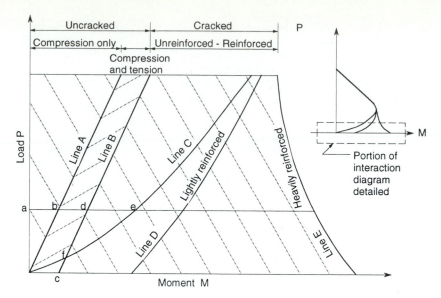

**Figure 7.2** Interaction of axial load and moment for low levels of axial load.

and load, described by eccentricity

$$e = M/P \tag{7.2}$$

is defined as the kern point,

$$e_k = S/A_n \tag{7.3}$$

Inclusion of the flexural tensile strength of masonry in the calculation of moment capacity results in the stress distribution shown in Fig. 7.1(b), where the moment capacity is

$$M = (f_a + f'_t)S \tag{7.4}$$

where $f'_t$ is the flexural tensile strength.

For cases with no axial compression, the moment capacity is indicated by point $c$ in Fig. 7.2. With the addition of axial compression, line $B$ represents the moment capacity (considering tensile strength) and, for a specified axial load $a$, the moment capacity is defined by point $d$.

**Flexural Compressive Strength.** When the tensile stress reaches the limiting tensile strength, this is usually defined as the *flexural tensile capacity*. However, if cracking is allowed, additional lateral load can be applied where the moment resistance is provided by the resultant internal compression force C shifting toward the compression face at a distance e from the centroid so that the couple Ce balances the bending moment M. [For unreinforced sections, C = P.]

The capacity is controlled by compressive strength in this case and is called the *flexural compressive capacity*. For different magnitudes of axial load, the amount of moment that can be resisted depends on the compressive strength. This is illustrated in Fig. 7.1(c), where from cases (i) to (iv), decreasing axial load P corresponds to the increasing eccentricity up to near the limiting eccentricity of $t/2$. However, the moment capacity decreases for lower axial loads. Where cracking is allowed the small remaining tension zone is usually ignored because it does not contribute much to the flexural compressive capacity. Also, because of stress reversals or unbonded masonry, tensile strength may not exist. The behavior, ignoring tensile strength, is as shown in Fig. 7.1(d).

The moment capacity (ignoring tensile strength) is illustrated as line $C$ in Fig. 7.2, and for the axial load $a$, the moment capacity is at point $e$. If tensile strength is included, the interaction relationship starts at point $c$ and, as axial load increases, it follows line $B$ to point $f$ after which the moment capacity follows line $C$.

**Flexural Capacity for Reinforced Sections.** To complete the flexural capacity options shown in Fig. 7.2, the interaction relationship for lightly reinforced sections is shown as line $D$ where addition of axial compression increases moment capacity by delaying the tension controlled failure. Alternately, line $E$ represents the case for heavily reinforced sections. In this case, the compression controlled capacity is reduced by the addition of axial load. These relationships are discussed in more detail in Sec. 7.3.

**Lateral Load Resistance through Arching.** When a wall is built between supports that restrain outward movement of any part of the wall in the plane of the wall, axial compressive forces are induced as the wall bends. These in-plane compression forces can delay cracking and, following cracking, can produce an arching action that in many cases has several times the capacity of the wall in pure flexure. The existence of this load-resisting mechanism has been responsible for walls designed by empirical rules being able to resist loads much higher than predicted by flexural calculations.[7.7,7.8]

## 7.3 FLEXURAL BEHAVIOR OF UNREINFORCED WALLS

### 7.3.1 Background

Unreinforced flexural masonry walls designed to resist lateral loads can be constructed from uncored (solid), cored, or hollow units, can be laid in full or face shell mortar bedding, and can have single or multiple wythes. In addition, the support conditions can be simple or complex and walls can be supported on two, three, or four sides. The spans for the walls in low-rise construction are usually limited by their ability to resist lateral loads because they commonly rely on the relatively small tensile strength of the assemblage. An accurate assessment of the lateral strength and safety is, therefore, essential if the maximum possible spans are to be utilized. Relatively small effects, such as self-weight, can become important in determining these maximum sizes and a good understanding of wall behavior is necessary for design.

As was indicated in Chap. 5, flexural tensile bond between mortar and individual units is quite variable. The strength can be reasonably approximated as normally distributed. The variability can be described by the standard deviation, which is commonly in the order of 20–30% of the mean strength for laboratory prepared test specimens. As was mentioned in Chap. 5, beam tests of prisms result in lower mean strengths than when each joint is tested because the critical combination of moment and tensile strength controls the capacity. For walls with many courses and many masonry units in each course, it is obvious that one weak joint will not control the capacity. Even with removal of several units from a region of the wall (an opening), the wall still has considerable capacity. Tests[7.9] comparing the flexural strengths of walls to results for individual joints between units indicate that the mean strengths from wall tests are slightly lower but that the variability is also reduced. Current practice is to define characteristic flexural tensile strengths in terms of the mean strength and standard deviation for the particular test method used. Australian practice[7.6] is to convert beam test results to equivalent joint strength data.

In the remainder of this section, the simplest cases of one-way bending of single-wythe walls are first discussed and serve as a basis for considering two-way bending. This is then extended to the consideration of multiple-wythe walls. The formulation of analysis and design procedures follows this section.

### 7.3.2 Vertical Flexure (Single-Wythe)

Walls laterally supported along their top and bottom edges resist lateral forces such as wind by spanning vertically between these supports. Flexural stresses normal to the bed joints eventually result in tensile failure when a crack occurs along a course of bed joints. The failure mechanism can be quite complex and depends upon the type of supports provided at the top and bottom edges and the magnitude of axial forces from self-weight and any superimposed loads.

**Effect of Self-Weight**   Consider the case of a masonry wall of height h and thickness t built on a concrete foundation slab and supported laterally along the top edges, as shown in Fig. 7.3(a). The weight of the wall produces a uniform compressive stress, $f_a$ = (unit weight of wall)(h/t), at the base of the wall resulting in an upward reaction, $f_a t$, per unit length of wall. It is instructive to study the wall's behavior when subjected to increasing wind pressure. Initially, any tensile stresses that tend to develop from bending of the wall will be suppressed by the compressive stresses due to the weight of the wall.

As wind pressure increases further, a stage is reached where tension is about to develop at the base of the wall on the windward side. This limiting condition is shown in Fig. 7.3(b) with a triangular distribution of stress at the base. It is unlikely that there is any tensile strength at this junction. A dampproof membrane or flashing is often placed under the wall during construction. Therefore, the wall is not bonded to the concrete slab. Even if the masonry is initially bonded directly to the concrete slab, it is likely that differential movements due to temperature and shrinkage/expansion will cause debonding at a later stage. A crack, therefore, opens at the base and the upward reaction moves closer to the leeward edge of the wall, as shown in Fig. 7.3(c).

Normally, this reaction will be very close to the edge because of elastic bending of the wall.[7.10] The weight of the wall has a stabilizing moment about this point of

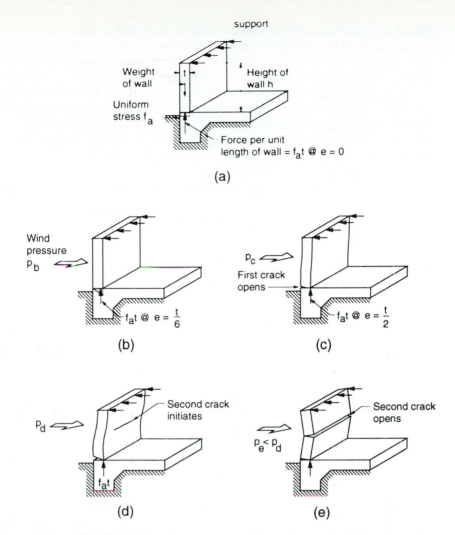

**Figure 7.3** Unreinforced masonry wall (spanning vertically) subjected to increasing wind pressure.

contact but a further increase in wind pressure is resisted by the wall acting as a simply supported member between the slab and the top supports. The location of the maximum tensile stress (initially near the top of the wall) moves down the wall towards mid-height as wind further increases, and finally a crack initiates at a bed joint, as shown in Fig. 7.3(d).

The usual method of designing the wall to resist wind pressure p is based on the maximum vertical tensile stress at mid-height of the wall, assuming it to be simply supported. For a unit length of wall, at mid-height the lateral pressure that can be resisted is

$$p \leq S \left( f'_{tn} + \frac{f_a}{2} \right) \frac{8}{h^2} \qquad (7.5)$$

where $f'_{tn}$ = flexural tensile strength normal to the bed joints

    $S$ = section modulus based on net area

Because the flexural strength of mortar joints is quite variable, it is possible that the crack will initiate where the tensile stress is slightly less than the maximum if relatively weak joints occur there. Research[7.11,7.12] has shown that the crack initiates along two or three units in the course, and then immediately propagates for the full length of the wall, as shown in Fig. 7.3(e). If the wind pressure $p_d$ is maintained, the wall will collapse as a mechanism; however, a reduced wind pressure $p_e$ can still be resisted by the stabilizing effect of the weight of the wall. Lateral deflection of the wall decreases the stabilizing effect.

The lateral loading stages described before can be evaluated as follows:

*Stage (b)*. The wall acts as a propped cantilever with bending moments determined from elastic theory at the base:

$$M_b = p_b h^2 / 8$$

The resulting tensile stress at the base is

$$f_t = \frac{p_b h^2}{8S} - f_a \qquad \qquad f_t = 0 \qquad (7.6)$$

*find the $p_b$*

where $f_a$ = wall weight ÷ effective mortared area $A_n$.

For zero tensile strength at the base, the limiting wind pressure for cracking is

$$p_b = \left(\frac{8S}{h^2}\right) f_a$$

*no crack up to this point*

*Stage (c)*. When $f_t$ reaches the flexural tensile strength normal to bed joints, $f'_{tn}$, a crack develops at the base, which then acts as a pin support.

*Stage (d)*. For simplicity, it is assumed that the upward reaction at the base is at the leeward face of the wall and that the second crack occurs at mid-height. The structural action is considered in two parts, as shown in Fig. 7.4(a), namely, the self-weight effect and the wind effect.

The resulting maximum tensile stress $f_t$ at mid-height is given by

$$f_t = \frac{p h^2}{8S} - f_a\left(\frac{A_n}{S}\frac{t}{4}\right) - \frac{f_a}{2} \qquad (7.7)$$

where self-weight at mid-height, $\frac{1}{2}(f_a A_n)$, produces a counteracting moment approximated with a lever arm of $t/2$. When $f_t$ calculated from Eq. 7.7 reaches the flexural tensile strength normal to bed joints, $f'_{tn}$, cracks occur at load $p_d$. A crack will not necessarily form at mid-height, but will occur at the critical combination of load and capacity.

*Stage (e)*. A mechanism with a crack located at a variable distance below the top support is shown in Fig. 7.5. If it is assumed that the lateral displacement of O is negligible, then considering the stability of the top section of the wall about O

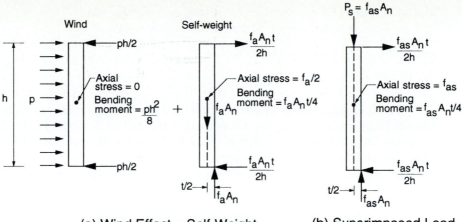

**Figure 7.4** Laterally loaded uncracked wall.

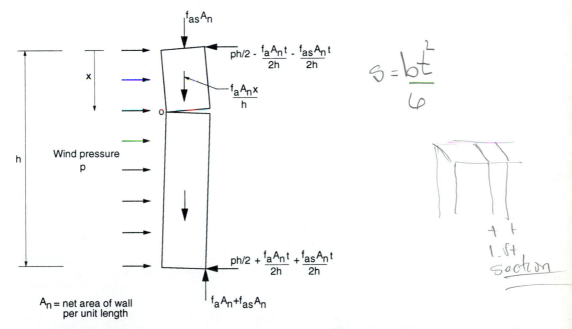

$A_n$ = net area of wall per unit length

$f_a$ = compressive stress due to wall self-weight at the base

**Figure 7.5** Failure mechanism of cracked wall under lateral wind load.

(without surcharge, $P_s = f_{as} A_n$) gives

$$\left[ p_e \left( \frac{h}{2} \right) - \frac{f_a A_n t}{2h} \right] x = p_e \frac{x^2}{2} + \frac{f_a A_n t x}{2h}$$

from which

$$p_e = \frac{2f_a A_n t}{(h - x)h} \tag{7.8}$$

and if the wall cracks at mid-height, $x = h/2$

$$p_e = \frac{4f_a A_n t}{h^2} \tag{7.9}$$

The foregoing values of lateral wind load p have been used to plot an idealized load-deflection relationship for a solid wall subjected to increasing lateral load. The response for a wall with an unbonded base is shown by the solid line in Fig. 7.6. The behavior can be thought of in terms of the thrust line. (Refer to Sec. 1.3.2 for background information on thrust lines.) At point *a* on the graph, no lateral load is

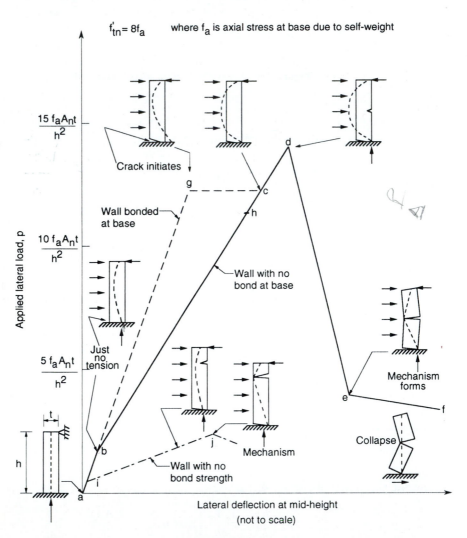

**Figure 7.6** Behavior of vertically spanning wall with simple top support (no superimposed compressive load).

keorn point @ $\left(\frac{t}{3}\right)$     Flexural Walls     Chap. 7

applied and the thrust line coincides with the centerline of the wall. At point $b$, the thrust line is at the kern point of the section at the base. At point $c$, the joint at the base has opened and the thrust there is located at the edge of the wall. The parabolic shape of the thrust line extends outside the thickness of the wall at mid-height, but the tensile strength of the masonry is sufficient to resist the tensile stresses due to wind. At point $d$, the thrust line has moved further outside the wall and the wall cracks. For the wall to remain stable the thrust line must lie within the wall at the crack location, and this can only be achieved by a reduction in applied pressure, as at point $e$. At point $f$, the deflection of the wall has reduced the lever arm of the stabilizing weight of the wall and the wind load that can be sustained is further reduced. When the lateral deflection is equal to half the wall thickness, the thrust line can no longer be contained within the wall and, even with no wind pressure, the wall collapses. The capacity using the usual design method (Eq. 7.1) is shown at point $h$.

In the foregoing behavior, it was assumed that there was no effective bond strength at the base. There are circumstances where the wall will have bond strength at its lower support. (An example is where the wall is continuous past an intermediate beam that gives lateral support.) In this case, the wall will behave the same as the unbonded wall up to point $b$. As further load is applied, tension develops at the base and the wall behaves as a propped cantilever until the tensile bond at the base is reached. Point $g$ of the dashed line in Fig. 7.6 represents this stage. The load at this point is found from Eq. 7.4. This is the ultimate load that would be calculated by normal design methods. Before cracking, the thrust line at the base falls outside the wall, and hence when a crack occurs, to maintain stability, the thrust line moves back to the edge of the base and at the same time produces greater tension near mid-height, as represented by point $c$. From this point onwards, the wall behaves the same as the wall that was unbonded at the base.

For comparative purposes, a third graph, shown as a dashed-dotted line, is included in Fig. 7.6. This is for a wall with no bond strength as could be expected with dry stacked construction. In this case, the first joint below the top support opens to form a mechanism. Point $j$ on the curve, when the mechanism forms, is found by substituting x = 0 in Eq. 7.8 to give

$$p_e = \frac{2f_a A_n t}{h^2} \tag{7.10}$$

From a consideration of the behavior illustrated in Fig. 7.6, it can be seen that the usual methods of assessing ultimate capacity of vertically spanning walls (points $g$ and $h$) are conservative, but fairly close to the more accurate analysis. Although the wall bonded at the base is usually assigned a larger design capacity, this analysis indicates that they both have the same strength (point $d$). This is because initial cracking of the wall at the base transforms the bonded wall into an unbonded one with a small reserve of strength.

It is interesting to note that although the tensile strength of masonry is small, it greatly increases the strength of a wall compared to one with no bond strength. Even the strength of a cracked wall (point $e$) is about twice that of one with no bond strength (point $j$) because the tensile strength of the former wall forces the bed joint to open near mid-height rather than just below the top support.

Although walls normally have a much reduced strength when cracked at mid-height, this is not necessarily true when a compressive load is applied to the wall.

### 7.3.3 Effect of Superimposed Load

Most walls carry superimposed load from a roof or a floor. In the case of a roof, it is likely that the load is applied to the wall at a fixed position, usually on the center line of the wall thickness, and that the load continues to be applied at this location when the wall bends under wind load. On the other hand, the point of application of load from a concrete floor slab is not so definite and it varies as the wall bends. At the base of the wall, the reaction is likely to begin at the center but move to the edge with increasing wind load, as described earlier for the case with no applied compressive load. The general case to be considered is, therefore, one in which the eccentricity of the applied load at the top is different from the eccentricity of the reaction at the base. However, two particular cases will be considered here to illustrate the behavior.

Consider, first, the case of the wall shown in Fig. 7.4(a), and carrying a central compressive load at the top that produces an average compressive stress in the wall of $f_{as}$; see Fig. 7.4(b). The usual method of designing this wall is to equate the maximum bending moment at mid-height, for simple support conditions, to the section capacity based on the tensile strength of the masonry plus uniform axial compression. That is,

$$p = S \left( f'_{tn} + \frac{f_a}{2} + f_{as} \right) \frac{8}{h^2} \tag{7.11}$$

The actual behavior can be examined by adding the effects of Figs. 7.4(a) and 7.4(b). Not only does the additional axial load delay cracking at the base and elsewhere because of the precompressing effect, but, after cracking of the base, the counteracting moment due to eccentricity of the base reaction further increases the lateral load that must be applied to cause cracking near mid-height. (The usual design approach, based on cracking of the critical section under combined axial compression and bending, inherently incorporates the assumption of zero eccentricity of the reaction at the base. As can be seen from the previous calculations, there is potential for a significant increase in capacity above that given by the normal design approach.)

Secondly, consider the case of a wall between concrete floors. Not only will the base reaction relocate but the superimposed load will also move to an eccentric location at the top of the wall as the wall deflects. Tests[7.13] have shown that for average axial stresses less than 290 psi (2 MPa), the end eccentricities can be assumed to be e = t/2 (i.e., at the face of the wall).

Further discussion on the behavior and design of walls under axial load and vertical flexure is presented in Chap. 8.

### 7.3.4 Horizontal Flexure (Single-Wythe)

It is unusual for walls to be subjected to purely horizontal flexure from lateral forces because they usually rest on a base that provides some lateral support. The condition is approached, however, where a wall is free at the top, is simply supported along

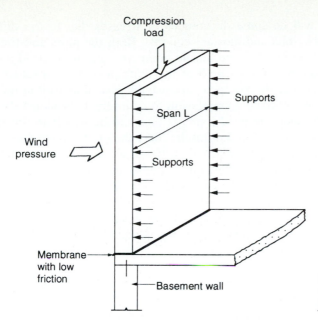

Compression load

Supports

Span L

Wind pressure

Supports

Membrane with low friction

Basement wall

**Figure 7.7** Unreinforced masonry wall (spanning horizontally) subjected to wind pressure.

its two vertical edges, and has flashing with a low coefficient of friction at its base, as shown in Fig. 7.7. Even when flashing is not used at the base, the top horizontal strip of such a wall is practically in pure horizontal flexure if the wall is high in comparison with its span. An understanding of the horizontal flexural strength of the masonry is also important when considering the two-way action of panels.

Panels in horizontal flexure have bending stresses parallel to the bed joints (that is, perpendicular to the head joints). The behavior of masonry in horizontal flexure is, however, more complicated than this simple description of bending stresses suggests. Near ultimate load, bending resistances of the head joints are diminished because they crack and stresses in the bed joints are of a torsional nature rather than bending.[7.14–7.17]

To illustrate behavior, the top strip of wall in Fig. 7.7 is considered to be in pure horizontal flexure. Applying normal bending theory, and initially ignoring any axial load, the uniformly distributed load to produce the ultimate tensile stress at the extreme fiber in horizontal bending at midspan is given by

$$p_h = f'_{tp}\left(\frac{8\,S}{L^2}\right) \tag{7.12}$$

where $f'_{tp}$ = flexural tensile strength parallel to the bed joints
    $L$ = horizontal span or length of the wall

Because of the changes in material and geometric properties along a vertical section, the stress $f'_{tp}$ is a hypothetical average value rather than an actual stress in the wall. This can be recognized by considering the behavior of a wall as lateral load is gradually applied.

When lateral load is first applied, the wall behaves elastically with more or less uniform bending strains occurring in the masonry units and the mortar elements of the masonry, but there will be some differences in stresses due to the different stiffnesses of the masonry units and mortar sections. Both the units and the head joints will be in bending, but because the tensile bond strength in the head joints is usually much less than the flexural strength of the units, a stage is reached when the head joints crack. Because the mortar is not usually placed as well in the head joints and there is no benefit of axial compression to counteract shrinkage stresses, the tensile bond strength along the head joints is likely to be less than the tensile bond strength along the bed joints, $f'_{tn}$. Research[7.18] has shown that once these joints have cracked they contribute very little to the stiffness of the masonry wall.

An idealized plot of the load-deflection response in Figs. 7.8(a) and 7.8(b) indicates a linear relationship up to point $b$ where cracking initiates in the head joints. Following this cracking, the sections at the head joints at alternate courses contribute very little moment resistance and the bulk of the applied moment is resisted by bending of the masonry units. As the moment cannot be transferred from unit to unit through the head joints, this is achieved mostly by torsion in the bed joints between adjacent units.[7.13,7.19]

For failure of the wall several possible modes exist. If the flexural strengths of the units are fairly low, they may not be able to resist the extra bending moment transferred from the cracked head joints and failure through the units can immediately follow cracking in the head joints. This line type of failure is illustrated by line $bc$ in Fig. 7.8(b). For stronger units, failure through the units might be delayed to point $g$ in Fig. 7.8(b). Additional axial compression will have negligible effect on this line type of failure.

If both the torsional resistance along the bed joints and the unit strength are adequate, lateral load can be increased prior to failure at, say, point $e$ in Fig. 7.8(a) at which point the toothed failure occurs. The slight slope of line $ed$ is due to friction along the bed joints from self-weight. If the bed joints are precompressed through the addition of superimposed axial load, the torsional shear capacity along the bed joints will increase and the toothed failure will be delayed to, say, point $f$ in Fig. 7.8(a). The added friction due to the superimposed axial load could result in the decreased slope for the unloading curve following failure.

Line $ah$ in Fig. 7.8(b) represents the observed failure mode for fully grouted

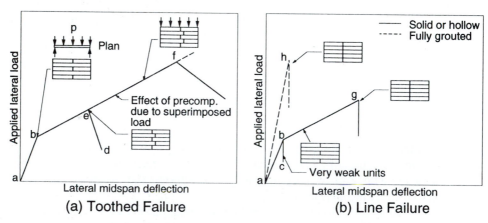

**Figure 7.8** Behavior of horizontally spanning walls.

hollow masonry that would normally fail in a toothed pattern if ungrouted. In this case, the grout cores crossing the bed joints stiffen and strengthen these joints thereby forcing failure through the units at alternate courses.

Research on brick masonry[7.15,7.19-7.21] has shown that the tensile strength parallel to the bed joint, $f'_{tp}$, can be related to the strength normal to the bed joint, $f'_{tn}$. It can be approximated by the lesser of Eq. 7.13 or 7.14.

$$f'_{tp} = C \sqrt{f'_{tn}} \left( 1 + \frac{f_a}{f'_{tn}} \right) \tag{7.13}$$

for toothed failure provided that $f'_{tn} > 22$ psi (0.15 MPa), where

$$C = 24 \text{ for stress in psi}$$

$$= 2 \text{ for stress in MPa}$$

or for failure through units,

$$f'_{tp} = 0.45 f'_{t \text{ unit}} + 0.55 f'_{tn} \tag{7.14}$$

where     $f'_{tn}$ = tensile strength normal to the bed joints
          $f_a$ = axial compressive stress (normal to bed joints)
    $f'_{t \text{ unit}}$ = modulus of rupture of units (tension parallel to the bed joints)

(A value in the order of 0.1 times the compressive strength of the unit may be used for $f'_{t \text{ unit}}$ if data are not available.)

### 7.3.5 Two-Way Flexure (Single-Wythe)

The previous two sections deal with one-way bending. In practice, most walls are supported on three or four sides so that there is a combination of vertical and horizontal bending. Such walls are statically indeterminate and their analysis is further complicated because the tensile strength in horizontal flexure $f'_{tp}$ can be between 2 to 5 times its strength in vertical flexure $f'_{tn}$ for most common ranges of material properties.

**Wall Supported on Three Sides.** A wall, such as shown in Fig. 7.7, may have sufficient friction at the base to provide lateral support against wind pressure. As wind pressure is applied, the wall deforms elastically as a plate supported on three sides. The joint at the bottom will open and the self-weight of the wall (and any superimposed loads) will have similar effects as considered previously. Even if the wall is initially bonded or continuous at the base, large bending stresses will be produced and a crack will finally form so that it then behaves as if simply supported.

As load is further increased, vertical and horizontal bending will provide principal stresses of varying magnitude and orientation throughout the wall. When the principal stresses at some position in the wall reach the strength of masonry at that location, a crack occurs. Tests[7.8,7.18,7.20] have shown that this first crack propagates immediately to form a mechanism, and the wall then has only a small residual strength due to the stabilizing effect of self-weight. The typical crack patterns at failure, shown in Fig. 7.9, depend on the ratio of height to length of the panel. The cracks approximately follow the maximum principal tensile stress locations in the wall. They

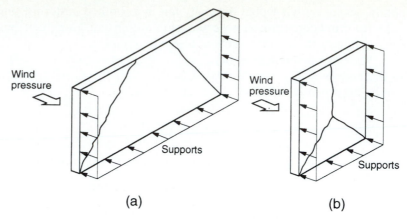

**Figure 7.9** Crack patterns at failure for walls supported on three edges.

also resemble the yield lines associated with the ultimate failure of reinforced concrete slabs.[7.22]

**Walls Supported on Four Sides.** Figure 7.10 is an illustrative plot of the load-deflection behavior of a panel simply supported on all four sides and about twice as long as it is high. As lateral load is applied, the wall deflects elastically, as shown by the line *ab*. A critical load, represented by point *b*, will be reached when the vertical bending moment at about mid-height of the wall reaches the resisting capacity of the bed joints. A crack initiates and the subsequent behavior of the wall depends upon the orthogonal strength properties of the masonry (see Chap. 5). If the strength in vertical flexure, $f'_{tn}$, is equal to the strength in horizontal flexure, $f'_{tp}$, then there is no reserve strength and the crack propagates along the bed joint (point

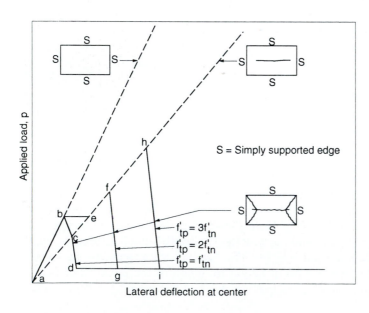

**Figure 7.10** Behavior of masonry wall with simple supports on all edges.

*c*) and the mechanism shown is immediately formed with only a small residual strength (point *d*) due to self-weight. However, in practice, most masonry has a horizontal flexural strength greater than its vertical flexural strength, so that when the crack propagates along the bed joint under constant load, a stable situation is reached at point *e*. The panel now (ideally) consists of two subpanels, each simply supported along three sides and free along the cracked bed joint. As load is further increased, each subpanel behaves as described previously until diagonal cracks initiate at point *f* for the case $f'_{tp} = 2f'_{tn}$ or at point *h* for the case $f'_{tp} = 3f'_{tn}$. These cracks immediately propagate to form a mechanism with a small residual strength (points *g* or *i*).

The foregoing description and the load-deflection plot in Fig. 7.10 are for an idealized case, but numerous tests[7.8,7.18] have confirmed this idealized behavior. Figure 7.11 shows a test wall loaded to failure using an air bag to apply uniform pressure.[7.23] The initial crack occurred in the bed joint at 53% of the failure load and gradually extended with each increment to effectively cover the width of the wall. The diagonal cracks formed at the failure load. Other tests[7.8,7.23] have confirmed that an uncracked wall and one with a preformed crack at mid-height both have the same failure mechanism. Once again, the final crack pattern resembles the yield line pattern in reinforced concrete slabs. Although it is recognized that yielding does not occur, this has led to development of design provisions[7.2] with moment coefficients similar to those from yield line analysis.

This same principle of staged cracking has been shown[7.8,7.11] to apply to other support conditions. For the case of a wall that is continuous past the supports so that its edges may be taken as *fixed* against rotation, the wall initially behaves elastically until the fixed end moments at the top and bottom supports reach the capacity of the masonry in vertical flexure. These joints then crack (line *bc*, Fig. 7.12) so that the panel then behaves as if it has simple supports at the top and

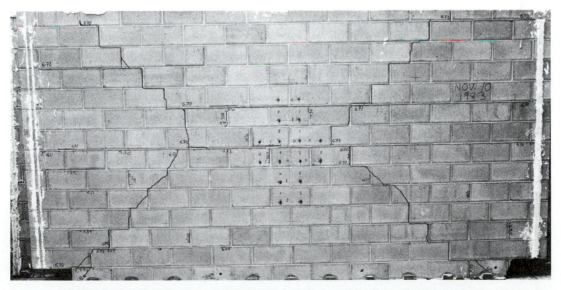

**Figure 7.11** Crack pattern at failure for a masonry wall simply supported on four edges.

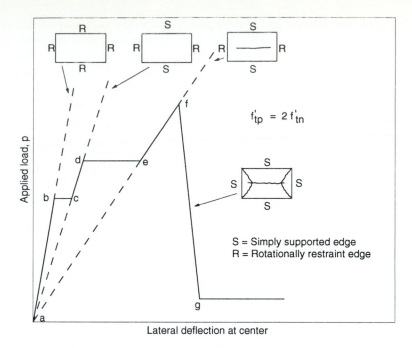

**Figure 7.12** Behavior of masonry wall with rotationally restrained edges.

bottom. A further increase in load causes a crack near mid-height (line *de*) forming two subpanels, *free* along the cracked joint. Further increase of load initiates additional cracking (point *f*) and the formation of a mechanism with a small residual strength (point *g*) due to self-weight.

### 7.3.6 Multiple-Wythe Walls

The sections dealing with vertical, horizontal, and two-way bending of single-wythe walls can be applied with appropriate adjustments to multiple-wythe walls. Composite walls may be treated in a similar manner to single wythe walls with the appropriate section properties and material strengths taken into account. A *diaphragm (utility) wall* composed of two wythes of masonry solidly connected by vertical masonry webs at regular spacing has composite action only in the vertical direction.

In a veneer wall, design practice has been to disregard the veneer in structural computations and to treat the backup as a single-wythe wall carrying all the applied lateral load. Design and construction of masonry veneers are covered in Chap. 12.

Cavity walls, as shown in Fig. 7.13, with two wythes connected by metal ties, exhibit complex behavior. The usual metal ties have negligible capacity to transfer shear between the two wythes, but should have sufficient axial strength and stiffness to cause the wythes to deflect similarly. Hence, the total lateral load is shared between the wythes, but each resists its share by bending about its own centroid.

Typically, a cavity wall, such as shown in Fig. 7.13(a), spans vertically between top and bottom supports and can be subjected to a combined external pressure $p_o$

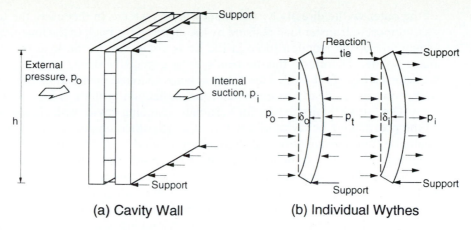

and internal pressure $p_i$. In this figure, a positive external pressure and an internal suction are shown. In general, some force will be transmitted through the ties and this can be considered a uniformly distributed load $p_t$ if the ties are closely spaced and the boundary conditions for the two wythes are the same.

By separating the wythes as free body diagrams in Fig. 7.13(b), it can be seen that the outer wythe carries a load $p_o - p_t$ and the inner wythe carries a load of $p_i + p_t$. Weep holes and vents in the outer wythe may permit pressure in the cavity to equal the external pressure $p_o$. In this case, the outer wythe carries the load from the ties, $p_t$, and the inner wythe carries the load $p_o + p_i - p_t$.

If both wythes have the same boundary conditions and the ties are very stiff axially, then the mid-height deflections are equal ($\delta_0 = \delta_i$) and the share of load resisted by each wythe is proportional to its flexural stiffness as are the resulting maximum bending moments. Hence,

$$M_0 = \left( \frac{E_0 I_0}{E_0 I_0 + E_i I_i} \right) M_T$$

$$(7.15)$$

$$M_i = \left( \frac{E_i I_i}{E_0 I_0 + E_i I_i} \right) M_T$$

where   $E_o$, $E_i$ = elastic modulus of outer and inner wythes, respectively
   $I_o$, $I_i$ = moments of inertia of outer and inner wythes, respectively
   $M_o$, $M_i$ = maximum bending moments in the outer and inner wythes, respectively
   $M_T$ = total bending moment (e.g., $M_T = (p_0 + p_i)h^2/8$ for simple support conditions)

The tie forces will be maximum when only one wythe is loaded directly, such as when cavity pressure equalization balances the pressure on the outer wythe or when pressure is only applied to one wythe (see rain screen principle in Chap. 12). For proportional bending moments in each wythe, the support conditions should be the same for both. However, in some cases, it is not possible to support the top of the outer wythe directly by the roof or floor system. In this case, the top tie acts as a support to transfer load resisted by the outer wythe back to the inner wythe. The magnitude of the reaction force in the tie is about half of the load resisted by the outer wythe. Therefore, this tie needs to be especially strong and stiff and should be near the top of the wall. This position helps to avoid overloading the inner wythe and produces, as nearly as possible, equal deflection profiles in both wythes.

The foregoing discussion for vertically spanning cavity walls is also applicable to the precracking behavior of cavity walls with other support conditions, including spanning both horizontally and vertically provided that the support conditions are identical. However, in many cases, it is not practical to provide similar support conditions. Also, some ties, particularly adjustable ties (see Chap. 14), are relatively flexible and, for very stiff masonry, do not force sufficiently similar deflections in both wythes. In addition, superimposed axial loads are normally applied to the inner wythe, which delays cracking in this wythe. Cracking of the outer wythe, therefore, often controls capacity. In fact, because the outer wythe is exposed to greater extremes of temperature, it may crack at low bending moment because of tensile stresses resulting from differential movement. For an equal total thickness of masonry, the most efficient resistance to both lateral and vertical load is to concentrate as much as possible of the total thickness in the axially loaded wythe. As discussed in Chap. 14, the implications of design to provide air, vapor and thermal barriers also affect the structural design approach.

Cavity walls with the wythes connected by masonry *bonding units*, often in the form of headers, behave very differently from those connected using metal ties. The two wythes are coupled together to produce a combined strength and stiffness greater than the sum of the individual values. However, the resistance to differential movement due to elastic deformations and to temperature and moisture deformation introduces stresses in the wythes that may result in premature cracking of one of the wythes or shear failure of the headers.

## 7.4 ANALYSIS AND DESIGN OF UNREINFORCED FLEXURAL WALLS

### 7.4.1 Introduction

The foregoing discussions give the designer an understanding of wall behavior under lateral loads normal to the face of the wall, but additional guidance is required for quantitative design. Various methods of analysis have been used to predict the ultimate strength of walls under two-way flexure. These include conventional and modified yield line approaches[7.24–7.28] and the fracture line approach.[7.29,7.30] Other methods attempt to simulate real material behavior based on the elastic performances of cracked panels.[7.23,7.31,7.32] They include variability of joint strength and load sharing between joints, principal stress criterion, and finite element methods of analyzing orthotropic plates using linear or nonlinear properties. Although all methods have been confirmed to some extent by limited experimental data, no one method has been able to closely predict the ultimate strengths of all the panels tested. Development of design methods based on finite element analyses results that were confirmed through comparison with test results seem to offer the best opportunity for reasonable predictions of capacity. Descriptions of various design approaches are included in this section.

### 7.4.2 Design from Basic Principles  *(WSD based)*

Masonry codes in the United States and Canada specify design stresses but do not give detailed design rules. The practice is to use linear elastic bending theory[7.33,7.34] or other proven analytical models. For either horizontal or vertical bending, beam theory is used. However, for two-way bending, practice is not standardized.

A simplistic conservative approach for two-way bending is to use the crossed strips method to apportion the lateral load to the two directions. For similar support conditions on all four sides, using equal midspan deflections and $p = p_h + p_v$:

$$p_h = p/(1 + K) \qquad \text{*Increase* } F_b \qquad (7.16)$$

$$p_v = pK/(1 + K) \qquad \text{*but } \tfrac{1}{3} \text{ for short term loads.*}$$

where
$$K = E_{m_v} I_v L^4/(E_{m_h} I_h h^4)$$

$E_{m_v}, E_{m_h}$ = modulus of elasticity in the vertical and horizontal directions, respectively

$I_v, I_h$ = moment of inertia for bending in the vertical and horizontal directions, respectively

$h, L$ = vertical height and horizontal length, respectively

$p, p_v, p_h$ = total lateral pressure, pressure resisted by a vertical strip, and pressure resisted by a horizontal strip, respectively

Alternately, for similar support conditions at the four sides, isotropic elastic plate analysis can be used by neglecting that $E_h I_h$ and $E_v I_v$ usually differ somewhat. Moment coefficients such as are available in many textbooks are reproduced in Table 7.1.

Walls may be initially supported on three sides or this may be effectively the case at the ultimate limit state, where a horizontal crack may occur first for walls supported on four sides. For such cases, simple analytical procedures are more difficult to formulate. Existing elastic plate analysis is normally not valid because of the orthotropic stiffness and strength characteristics of masonry. Also, maximum moments from textbook moment coefficients may not be the critical case because of the orthotropic strength characteristics. For specified strength and stiffness ratios, finite element analyses can be used to tabulate the critical lateral load. Complicated boundary conditions and wall geometry, including openings, can also be analyzed using this method.

### 7.4.3 Yield Line and Failure Mechanism Approaches

The Australian[7.6] and British[7.13] codes provide comprehensive design guidance expressed in terms of ultimate strength or limit states.

In the British code, the basic design expression is

$$\gamma_f p_k \leq \frac{F_{kh} S}{\alpha L^2 \gamma_m} \qquad (7.17)$$

## Table 7.1 - Moment Coefficients from Elastic Plate Analysis (from Ref. 7.33)

| b/a | β | β₁ |
|-----|------|------|
| 1.0 | 0.0479 | 0.0479 |
| 1.1 | 0.0554 | 0.0493 |
| 1.2 | 0.0627 | 0.0501 |
| 1.3 | 0.0694 | 0.0503 |
| 1.4 | 0.0755 | 0.0502 |
| 1.5 | 0.0812 | 0.0408 |
| 1.6 | 0.0862 | 0.0492 |
| 1.7 | 0.0908 | 0.0486 |
| 1.8 | 0.0948 | 0.0479 |
| 1.9 | 0.0985 | 0.0471 |
| 2.0 | 0.1017 | 0.0464 |
| 3.0 | 0.1189 | 0.0406 |
| 4.0 | 0.1235 | 0.0384 |
| 5.0 | 0.1246 | 0.0375 |
| ∞ | 0.1250 | 0.0375 |

$$(M_h)_{max} = \beta p a^2$$

$$(M_v)_{max} = \beta_1 p a^2$$

$\nu = 0.3$

$M_h$   Moment in the horizontal direction (produces stress parallel to bed joints).

$M_v$   Moment in the vertical direction (produces stress normal to bed joints).

Note: for b/a < 1 use a/b instead of b/a and $M_h = \beta_1 p b^2$ and $M_v = \beta p b^2$

where

$p_k$ = characteristic wind pressure

$\gamma_f$ = partial safety factor for loads (1.2 to 1.4)

$\gamma_m$ = partial safety factor for material stress (2.5 to 3.5)

$F_{kh}, F_{kv}$ = characteristic horizontal and vertical flexural strengths, respectively

$L$ = length of panel

$S$ = section modulus

$\alpha$ = moment coefficient based on yield line theory and test results and dependent upon the geometry of panel and support conditions (see Table 7.2)

$\mu$ = orthogonal strength ratio = $F_{kv}/F_{kh}$

Some caution is advised in use of this approach outside the limits of the British code where characteristic tensile strengths are normally taken to be much lower than in North America. In addition, the yield line basis for the moment coefficients is not theoretically justifiable for a material as brittle as unreinforced masonry and its validity depends on confirmation from test results. In particular, the use of full moment capacity at fixed bases may not be strictly valid because these locations tend to crack prior to ultimate load. However, a compensating feature not directly included in the design process is that the rigid supports required to provide the specified boundary conditions may introduce various degrees of two-way arching, as described in the next section. As will be shown, arching and the effects of superimposed vertical load can result in very large increases in lateral load resisting capacity.

The fracture line approach[7.29,7.30] does not utilize the idea of maintaining yield moments after cracking and is thus more fundamentally acceptable. For cases with edge moments and more than one stage of cracking, it predicts lower capacities than yield line capacities.[7.23]

# Table 7.2 - Moment Coefficients Based on Yield Line Theory (from Ref. 7.3)

| | | Values of α | | | | | | |
|---|---|---|---|---|---|---|---|---|
| | μ | h/L | | | | | | |
| | | 0.30 | 0.50 | 0.75 | 1.00 | 1.25 | 1.50 | 1.75 |
| **Wall Type A** | 1.00 | 0.031 | 0.045 | 0.059 | 0.071 | 0.079 | 0.085 | 0.090 |
| | 0.90 | 0.032 | 0.047 | 0.061 | 0.073 | 0.081 | 0.087 | 0.092 |
| | 0.80 | 0.034 | 0.049 | 0.064 | 0.075 | 0.083 | 0.089 | 0.093 |
| | 0.70 | 0.035 | 0.051 | 0.066 | 0.077 | 0.085 | 0.091 | 0.095 |
| | 0.60 | 0.038 | 0.053 | 0.069 | 0.080 | 0.088 | 0.093 | 0.097 |
| | 0.50 | 0.040 | 0.056 | 0.073 | 0.083 | 0.090 | 0.095 | 0.099 |
| | 0.40 | 0.043 | 0.061 | 0.077 | 0.087 | 0.093 | 0.098 | 0.101 |
| | 0.35 | 0.045 | 0.064 | 0.080 | 0.089 | 0.095 | 0.100 | 0.103 |
| | 0.30 | 0.048 | 0.067 | 0.082 | 0.091 | 0.097 | 0.101 | 0.104 |
| **Wall Type B** | 1.00 | 0.008 | 0.018 | 0.030 | 0.042 | 0.051 | 0.059 | 0.066 |
| | 0.90 | 0.009 | 0.019 | 0.032 | 0.044 | 0.054 | 0.062 | 0.068 |
| | 0.80 | 0.010 | 0.021 | 0.035 | 0.046 | 0.056 | 0.064 | 0.071 |
| | 0.70 | 0.011 | 0.023 | 0.037 | 0.049 | 0.059 | 0.067 | 0.073 |
| | 0.60 | 0.012 | 0.025 | 0.040 | 0.053 | 0.062 | 0.070 | 0.076 |
| | 0.50 | 0.014 | 0.028 | 0.044 | 0.057 | 0.066 | 0.074 | 0.080 |
| | 0.40 | 0.017 | 0.032 | 0.049 | 0.062 | 0.071 | 0.078 | 0.084 |
| | 0.35 | 0.018 | 0.035 | 0.052 | 0.064 | 0.074 | 0.081 | 0.086 |
| | 0.30 | 0.020 | 0.038 | 0.055 | 0.068 | 0.077 | 0.083 | 0.089 |
| **Wall Type C** | 1.00 | 0.009 | 0.023 | 0.046 | 0.071 | 0.096 | 0.122 | 0.151 |
| | 0.90 | 0.010 | 0.026 | 0.050 | 0.076 | 0.103 | 0.131 | 0.162 |
| | 0.80 | 0.012 | 0.028 | 0.054 | 0.083 | 0.111 | 0.142 | 0.175 |
| | 0.70 | 0.013 | 0.032 | 0.060 | 0.091 | 0.121 | 0.156 | 0.191 |
| | 0.60 | 0.015 | 0.036 | 0.067 | 0.100 | 0.135 | 0.173 | 0.211 |
| | 0.50 | 0.018 | 0.042 | 0.077 | 0.113 | 0.153 | 0.195 | 0.237 |
| | 0.40 | 0.021 | 0.050 | 0.090 | 0.131 | 0.177 | 0.225 | 0.272 |
| | 0.35 | 0.024 | 0.055 | 0.098 | 0.144 | 0.194 | 0.244 | 0.296 |
| | 0.30 | 0.027 | 0.062 | 0.108 | 0.160 | 0.214 | 0.269 | 0.325 |

Wall Type A — edges: top F (free), left S, right S, bottom S

Wall Type B — edges: top S, left S, right S, bottom S

Wall Type C — edges: top S, left S, right F (free), bottom S

NOTE 1. Linear interpolation of μ and h/L is permitted.

NOTE 2. When the dimensions of a wall are outside the range of h/L given in this table, it will usually be sufficient to calculate the moments on the basis of a simple span. For example, a panel of type A having h/L less than 0.3 will tend to act as a freestanding wall, whilst the same panel having h/L greater than 1.75 will tend to span horizontally.

The calculation of the design moment per unit height may be taken as either

$\alpha p \gamma_f L^2$, when the plane of failure is perpendicular to the bed joints; or

$\alpha p \gamma_f L^2$, when the plane of failure is parallel to the bed joints

where $\mu = F_{kv} / F_{kh}$ is the orthogonal strength ratio

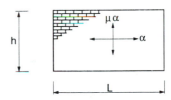

Key to support conditions
F - denotes free edge
S - simply supported edge

## 7.4.4 Crossed Strips Method

The Australian Code[7.6] uses an empirical approach in which the capacity of a panel is considered to be the sum of the capacities of a vertically spanning strip and a horizontally spanning strip without regard to compatibility of deflections. The

Australian code uses an empirical equation[7.36] that takes the form

$$p \le 10 \left( \frac{b_v M_v}{h^2} + \frac{b_h M_h}{L^2} \right) \tag{7.18}$$

where    $p$ = design wind pressure (factored load)
    $L, h$ = length and height of panel, respectively
    $b_v, b_h$ = vertical and horizontal bending coefficients, respectively
    $M_v$ = vertical bending moment capacity of a unit wide strip given as equal to the lesser of:

$$M_v = (C_m f'_{tn} + f_a) S \tag{7.19i}$$

or

$$M_v = 3.0 \, C_m f'_{tn} S \tag{7.19ii}$$

$M_h$ = horizontal bending moment capacity of a unit high strip given as equal to the lesser of

$$M_h = 24 C_m k_p \sqrt{f'_{tn}} \left( 1 + \frac{f_a}{f'_{tn}} \right) S, \tag{7.20i}$$

$$M_h = 48 C_m K_p \sqrt{f'_{tn}} \, S \tag{7.20ii}$$

or

$$M_h = C_m (0.44 f'_{tu} + 0.56 f'_{tn}) S \tag{7.20iii}$$

where  $C_m$ = capacity reduction factor = 0.6 for flexural design of unreinforced walls
    $f_a$ = axial compressive stress
    $f'_{tn}$ = tensile strength of masonry normal to the bed joints
    $f'_{tu}$ = characteristic modulus of rupture of units (taken equal to 260 psi (1.8 MPa) in the absence of test data)
    $K_p$ = bond pattern factor = 1.0 for running bond
    $S$ = section modulus based on net area

Coefficients $b_v$ and $b_h$ can be calculated from Table 7.3.
[Note: Equations (7.20i) and (7.20ii) are empirical and based on stresses expressed as psi. For use of metric units (MPa), replace the constants 24 and 48 by 2.0 and 4.0, respectively.]

    The strip method of design has been shown to generally give close predictions of strength or to be slightly conservative.[7.35] Its conservatism appears to increase for walls of hollow concrete blocks, but it may slightly overestimate the strengths of large walls. It can incorporate several of the features discussed in the previous sections and be expressed in both allowable stress or ultimate strength terms. Unfortunately, it is not readily adapted to wall panels with openings. It does not account for arching.

# Table 7.3- Moment Coefficients for Crossed Strip Method
## (from Ref. 7.6)

| Edge restraint to wall panels | Vertical bending coefficient, $b_v$ | Horizontal bending coefficient, $b_h$ | |
| --- | --- | --- | --- |
| | | Solid and cored units | Hollow units |
| S or R / S [ ] S / S or R | $b_v = 1.00$ | $b_h = 1.00$, or $= 2.00(3 - L/h)$, whichever is less | $b_h = 1.5$, or $= 3.0(3 - L/h)$, whichever is less |
| S or R / S [ ] R / S or R | $b_v = 1.00$ | $b_h = 1.25$, or $= 2.50(3 - L/h)$, whichever is less | $b_h = 1.8$, or $= 3.6(3 - L/h)$, whichever is less |
| S or R / R [ ] R / S or R | $b_v = 1.00$ | $b_h = 1.50$, or $= 3.00(3 - L/h)$, whichever is less | $b_h = 2.2$, or $= 4.4(3 - L/h)$, whichever is less |
| Free / S [ ] S / S or R | $b_v = 0.25$, or $= 0.25(6 - L/h)$, whichever is less | $b_h = 1.00$ | $b_h = 1.5$, or $= 1 + 0.5L/h$, whichever is less |
| Free / S [ ] R / S or R | $b_v = 0.25$, or $= 0.25(6 - L/h)$, whichever is less | $b_h = 1.25$ | $b_h = 1.8$, or $= 1.25 + 0.6 L/h$, whichever is less |
| Free / R [ ] R / S or R | $b_v = 0.25$, or $= 0.25(6 - L/h)$, whichever is less | $b_h = 1.50$ | $b_h = 2.2$, or $= 1.5 + 0.7 L/h$, whichever is less |

NOTES:
1. h = height of the wall panel
   L = length of the wall panel
   S = edge of wall is simply supported–that is, laterally supported but not rotationally fixed.
   R = edge of wall is rotationally restrained as well as laterally supported. (Note that structural continuity of wall construction at an edge does not necessarily provide rotational restraint at that edge).
   Free = edge of wall has neither lateral support nor rotational restraint.

2. A control joint in a wall is to be regarded as a free edge.

## 7.4.5 Effects of Axial Load

Walls built within and confined by a surrounding frame require detailed design not presented here. The Australian method, however, does incorporate the stabilizing effects of self-weight and superimposed load, and these can be included in North American design because they are based on principles of mechanics. To take advantage of these effects, it is necessary to ensure that the assumed eccentricities correspond to conditions in practice. Some guidance in this regard follows.

Elastic analysis generally shows eccentricities of the reactions due to self-weight and superimposed loads to be just a little less than half the wall thickness. It is a matter of engineering judgment as to what eccentricities should be reasonably assumed. Clearly, the maximum eccentricity can be limited by raked joints and the inclusion of bearing pad material between the wall and any slab it supports. For fully bedded units, the eccentricity is approximately one sixth the thickness when tension develops on one face and in practice eccentricities will be greater than this. For face shell bedded units, the eccentricity can be conservatively taken to correspond approximately with the inner edge of the face shell.

The deflection of a long span one-way slab is likely to cause the point of support to be near the inner face of the wall. Bending moment from external positive wind pressure further increases this eccentricity. Eccentricity toward the inner face of the wall is unfavorable for positive internal pressure, but as the wall bends, the point of support moves back toward the centerline. Where roof trusses, etc., have to be connected to the top of a wall, it is best to place them on the centerline of the wall so that an unfavorable eccentricity does not result when the wind blows from one direction.

Unless detailed information is known, the following values are suggested for eccentricities in bearing walls of thickness t where the direct axial stress does not exceed 200 psi (1.5 MPa).

**TRUSSES OR BEAMS CONNECTED TO TOP OF WALL:**

Use actual value but normally $\quad e_t = 0$

**SLAB RESTING ON TOP OF WALL:**

Stiff two-way slab $\qquad\qquad e_t = 0.25t$
Flexible one-way slab
    external positive pressure $\qquad e_t = 0.30t$
    external suction $\qquad\qquad e_t = 0$

**WALL CONTINUOUS PAST TOP SUPPORT:**

    fully bedded $\qquad\qquad e_t = 0.25t$
    face shell bedded $\qquad\quad e_t = 0.5t - \text{(face shell thickness)}$

**WALL SUPPORTED ON CONCRETE FOOTING OR SLAB:**

    assume fully bedded $\qquad\quad e_b = 0.25t$

**WALL CONTINUOUS PAST BOTTOM SUPPORT:**

    fully bedded $\qquad\qquad e_b = 0.25t$
    face shell bedded $\qquad\quad e_b = 0.5t - \text{(face shell thickness)}$

Use of soft bearing pad material in a strip along the edge of the bearing surface may be used to decrease the effective thickness, t, for eccentricity calculations.

For working stress design, the full benefit of the counteracting moments due to self-weight and superimposed load should not be used. To incorporate an adequate safety margin, the Australian code implies reduction of axial load to between half to two-thirds. For inclusion in North American practice, the use of half is recommended. Full details of the appropriate equations were provided in Sec. 7.3.2.

For two-way flexure, an alternative to the Australian approach is to have capacity controlled by development of the failure mechanism in one of the directions where equal deflections of the strips is required. This will result in a lower strength than obtained by the superposition of strengths.

## 7.4.6 Two-Way Bending Example

### Example 7.1

Calculate the design wind load (factored load) that can be carried by a rectangular brick masonry panel 8 ft (2.4 m) high by 20 ft (6 m) long. Flexural tensile strength normal to the bed joints, $f'_{tn}$ = 50 psi (0.35 MPa), and flexural tensile strength parallel to the bed joints, $f'_{tp}$ = 100 psi (0.70 MPa). Consider the panel to be simply supported along all four sides. The actual thickness of the brick panel is $3\frac{5}{8}$ in. (90 mm). Neglect any benefit due to axial compression, $f_a$.

**Solution A: Using Elastic Plate Theory:**[7.33] From Table 7.1, it is seen that b/a = 8/20 = 0.4 is not listed. Therefore, use a/b = 20/8 = 2.5 and interchange the moments for vertical and horizontal bending, $M_v$ and $M_h$, respectively.

Interpolating, $\beta$ = 0.1103 and $\beta_1$ = 0.0435:

$$M_v = \beta p b^2 = 0.1103 p b^2$$

$$M_h = \beta_1 p b^2 = 0.0435 p b^2$$

Considering linear elastic behavior up to ultimate conditions, if vertical bending controls,

$$M_v = f'_{tn} S = 0.1103 p(b)^2$$

$$0.1103 \, p(8)^2 12 = 50(12)(3.625)^2/6$$

$$p = 15.5 \text{ psf}$$

$$\left( p = \frac{0.35(1000)(90)^2/6}{0.1103(2.4)^2(1000)^2} = 0.74 \text{ kN/m}^2 \right)$$

If horizontal bending controls,

$$M_h = f'_{tp} S = 0.0435 \, p(b)^2$$

$$0.435 \, p(8)^2/2 = 100(12)(3.625)^2/6$$

$$p = 78.7 \text{ psf}$$

$$\left( p = \frac{0.7(1000)(90)^2/6}{0.0435(2.4)^2(1000)^2} = 3.77 \text{ kN/m}^2 \right)$$

Therefore, vertical bending controls and the predicted ultimate or factored load that can be resisted is 15.5 psf (0.74 kN/m²).

**Solution B: Using the British Code:** From Table 7.2, Case B, $\alpha$ = 0.021 for h/L = 0.4 and $\mu$ = 0.5. Using Eq. 7.17

$$\gamma_f p_k \leq \frac{\bar{F}_{kh} S}{\alpha L^2 \gamma_m}$$

where $\gamma_f = 1.3$
$\gamma_m = 3.0$
$F_{kh} = f'_{tp} = 100$ psi (0.7 MPa)
$L = 20$ ft (6 m)

$$1.3\,p_k = \frac{100(12)(3.625)^2/6}{0.021(20)^2(12)(3.0)} \qquad \text{and } p_k = 6.69 \text{ psf}$$

$$\left(1.3\,p_k = \frac{0.7(1000)(90)^2/6}{0.021(20)^2(12)(3.0)} \qquad \text{and } p_k = 0.32 \text{ kN/m}^2\right)$$

However, if the partial safety factors $\gamma_f$ and $\gamma_m$ are removed, a more meaningful comparison with solution A is obtained:

$$p = 6.69 \times 1.3 \times 3.0 = 26.1 \text{ psf } (1.25 \text{ kN/m}^2)$$

**Solution C: Using the Australian Code:** From Eq. 7.18

$$p \le 10\left(\frac{b_v M_v}{h^2} + \frac{b_h M_h}{L^2}\right)$$

From Table 7.3, Case 1,

$b_v = 1.0$

$b_h \le 1.0$

$\le 2.00\,(3 - L/h) = 2.00(3 - 2.5) = 1.0$ for solid units

From Eq. 7.19i,

$M_v \le (C_m\,f'_{tn} + f_a)\,S$
$= 0.6(50)12(3.625)^2/6 = 788$ in.-lb $= 65.7$ ft-lb/ft $\qquad$ (0.296 kN-m/m)

or from Eq. 7.19ii,

$$M_v \le 3.0\,C_m\,f'_{tn}\,S.$$

Because $f_a$ was taken as zero, Eq. 7.19i governs.
From Eq. 7.20i

$$M_h \le 24 C_m K_p \sqrt{f'_{tn}}\left(1 + \frac{f_a}{f'_{tn}}\right) S$$
$$\le 24(0.6)(1.0)\sqrt{(50)}\;12(3.625)^2/6 = 2676 \text{ in.-lb/ft} = 223 \text{ ft-lbs/ft}$$
$$\left(M_n = 2.0(0.6)1.0\sqrt{0.35}\,(1000)(90)^2/6(10)^6 = 0.96 \text{ kN-m/m}\right)$$

or from Eq. 7.20ii

$$M_h \le 48\,C_m\,K_p\,\sqrt{f'_{tu}}\,S > M_h \text{ from Eq. 7.20 i}$$

or from Eq. 7.20iii

$$M_h \le C_m(0.44 f'_{tu} + 0.56 f'_{tn})S$$
$$\le 0.6[0.44(260) + 0.56(50)]\,12\,(3.625)^2/6 = 2245 \text{ in.-lb/ft}$$
$$= 187.1 \text{ ft-lb/ft} \leftarrow \text{controls}$$
$$(M_h \le 0.6[(0.44)(1.80) + 0.56(0.35)](90)^2/6(10)^3 = 0.80 \text{ kN-m/m})$$

$$p = 10 \left[ \frac{1.0(65.7)}{8^2} + \frac{1.0(187.1)}{(20)^2} \right) = 14.9 \text{ psf}$$

$$\left( p = 10 \left( \frac{0.296}{(2.4)^2} + \frac{0.80}{(6)^2} \right) = 0.79 \text{ kN/m}^2 \right)$$

To compare this with the elastic plate analysis and the British method, the $C_m$ value should be removed. On this basis,

$$p = 14.9 \div 0.6 = 24.9 \text{ psf } (1.18 \text{ kN/m}^2)$$

**Discussion of Solutions.** It is difficult to draw general conclusions from a single example because the basis of computation for each method is different. For instance, the example used tensile strengths of 50 and 100 psi (0.35 and 0.70 MPa) in the elastic plate theory. However, the British method only uses the 100 psi (0.70 MPa) value assuming the other to be one third that value. Alternately, the Australian method used only the 50 psi (0.35 MPa) value calculating the other to be 142 psi (0.98 MPa). Nonetheless, use of an elastic plate analysis is generally the most conservative solution because it equates capacity with formation of the first crack. As mentioned earlier, tests have shown that significant reserve capacity exists after formation of the first horizontal crack for this case of simple support conditions on all four sides. On this basis, the extra 60 to 70% capacity indicated by the British and Australian analyses seem justified. Analysis of a wall panel 4 ft (1.2 m) high by 20 ft (6 m) long and free along one of the 20 ft (6 m) edges would also predict a similar load for development of the next set of cracks. In all cases, some small additional capacity would result from including the compressive stress due to self-weight, $f_a$, at mid-height of the wall.

The elastic plate analysis or finite element solutions can be used in a working stress design format using appropriate allowable flexural tensile stresses. Alternately, a strength method can be used when the specified lateral loads are factored up and the calculated capacities are factored down. For example, this is done in the British code, where the wind load is multiplied by $\gamma_f$ and the capacity is divided by $\gamma_m$. ∎

### 7.4.7 Diaphragm Walls

Diaphragm walls, sometimes referred to as utility walls (Fig. 2.7 in Chap. 2), are intended to be vertically spanning elements. As their primary behavior is similar to the vertically spanning single-wythe wall, they can be designed using the same methods, but using the properties of the appropriate section, which can be visualized as a wide flange section.

There is a limit to the portion of the inner and outer wythes that may be considered to act effectively as flanges with the connecting webs. Typical rules state that the effective flange beyond the face of the web shall not exceed:

- six times the thickness of the flange (wythe)
- one half the clear distance to the next web
- the distance to the end of the flange
- 1/25 the vertical span

Although the webs may be placed further apart than this, the additional masonry should not be regarded as effective in vertical bending.

The spacing of webs is limited by the ability of each unreinforced wythe to span horizontally between webs and resist bending due to lateral load without cracking. The shear strength of the web-to-wythe connection may also limit web spacing.

## 7.5 ARCHING

As discussed in Sec. 7.2.1, the resistance of unreinforced masonry walls to out-of-plane loads can be substantially increased by utilizing the large in-plane compressive forces that can be induced when the wall is butted up against rigid supports. It has been shown experimentally[7.27,7.37–7.40] that, under certain conditions, masonry walls can withstand much larger lateral loads than are predicted on the basis of conventional bending analysis. This can be explained by *arching* behavior.

### 7.5.1 Mechanics of Rigid Arching

When a wall is built between and in tight contact with supports that are restrained against outward movement, elongation of the tension face due to bending cannot occur without inducing a compressive force. Under lateral load, this induced in-plane compressive force results in arching, and analysis[7.10,7.41,7.42] has shown that the induced forces can increase the cracking load by a factor of about $2\frac{1}{2}$ if the end supports are completely rigid. With increased loading, flexural cracking occurs at the supports (location of maximum negative moments) followed by cracking at midspan (location of maximum positive moment); see Fig. 7.14(a). With the increased load, the wall is pushed against the unyielded supports creating clamping forces $P_u$ at the ends. A *three-hinged arch* is formed where the external moment is resisted by the internal couple $P_u r_u$ where $r_u$ is the height of the arch as shown.

 The thrust or clamping force $P_u$ is a function of the material properties (stress and strain at crushing), which is appropriate to relate to the mortar joint (rather than the masonry), and the contact area. The height of the arch $r_u$ is a function of the wall geometry, contact area $\alpha t/2$, and midspan deflection $\Delta_0$; see Fig. 7.14. McDowell et al.[7.37] proposed an arching theory for solid masonry walls assuming rigid supports and an idealized material stress-strain relationship, as shown in the inset in Fig. 7.15(a). Reasonable agreement was found between the proposed theory and experimental data. Based on their theory, curves presented in Fig. 7.15 were developed

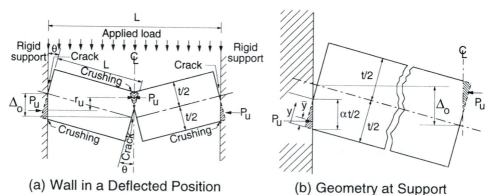

(a) Wall in a Deflected Position  (b) Geometry at Support

**Figure 7.14**  Rigid arching of masonry walls (from Ref. 7.37).

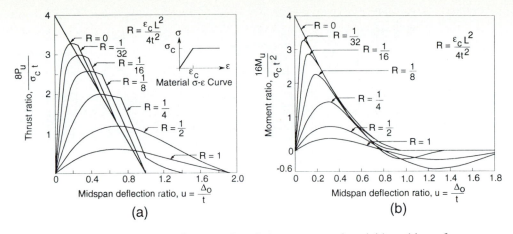

**Figure 7.15** Thrust force and resisting moment for rigid arching of masonry walls (from Ref. 7.37).

to express the clamping force and moment resistance in terms of wall deflection, material properties, and wall geometry. As can be seen, after the thrust reaches a peak value, the more the wall deflects, the lower the thrust is because of the decreased contact area. Also higher midspan deflections result in lower resisting moment because of the lower height of the arch.

Gabrielsen and his associates[7.38–7.40] conducted an extensive dynamic test program to study the resistance of masonry walls to uniform blast loading. Full scale walls mounted on a steel frame were subjected to blast waves in a large shock tunnel. These walls acted as simple plates and exhibited arching behavior once flexural cracking began. The steel frame became a perimeter restraining ring enabling arching thrusts to develop. The walls had a load carrying capacity approximately twice as strong as conventional simple plates and exhibited the ability to withstand mildly reversed cyclic loading.

### 7.5.2 Mechanism of Gapped Arching

If a masonry wall is separated from one support by a small gap due to poor construction, wall shrinkage, or an intentional movement joint, arching can still develop, but to a lesser extent. The wall must rotate about its base enough to close the gap and to push against the support at the gap so that clamping thrust forces can be generated. This results in an asymmetric arch, as shown in Fig. 7.16(b), compared to a symmetric arch in the case of rigid supports without a gap (Fig. 7.16(a)). A simple analysis based on equilibrium and neglecting wall deflection can be easily performed, as shown in Fig. 7.16, to show the difference in load resistance of a rigid arching wall versus a gapped arching wall. The analysis assumes a zero contact area (resultant force at the extreme fiber) for simplicity. The simplistic analysis presented in Fig. 7.16(b) shows a 50% reduction in load carrying capacity due to the gap, assuming that it is possible to develop the same thrusts, $P$.

It should be noted that in the upper segment of the gapped arch, the induced axial compression may actually weaken this part and there is danger of additional

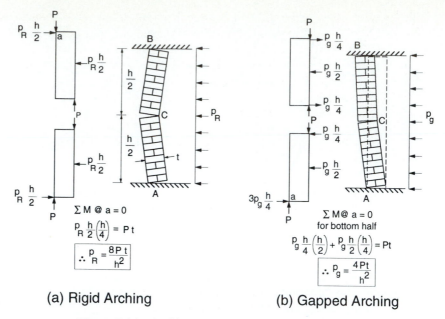

(a) Rigid Arching

$$\Sigma M @ a = 0$$

$$\frac{p}{R}\frac{h}{2}\left(\frac{h}{4}\right) = Pt$$

$$\therefore \frac{p}{R} = \frac{8Pt}{h^2}$$

(b) Gapped Arching

$$\Sigma M @ a = 0$$
for bottom half

$$\frac{p}{g}\frac{h}{4}\left(\frac{h}{2}\right) + \frac{p}{g}\frac{h}{2}\left(\frac{h}{4}\right) = Pt$$

$$\therefore \frac{p}{g} = \frac{4Pt}{h^2}$$

**Figure 7.16**  Arching mechanisms of masonry walls.

cracking and instability. Also if the gap is too large, the wall can simply rotate out of place without contacting the upper support. From simple geometry, the maximum gap $g_{max}$ to avoid this rotation for a wall of height $h$ and thickness $t$ is

$$g_{max} = \sqrt{h^2 + t^2} - h \qquad (7.21)$$

or for a 12 ft (3.6 m) high, 6 in. (150 mm) thick wall.

$$g_{max} = \sqrt{(144)^2 + (6)^2} - 144 = 0.125 \text{ in. (3.12 mm)}$$

There should be some safety, so only a very small gap can be accommodated.

Where a tight fit to the upper support is not ensured, it is good practice to provide clip angles or some other device to prevent rotation of the wall. In this case, lateral load will initially cause a crack to open at the base followed by cracking near

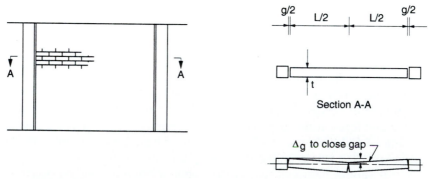

**Figure 7.17**  Gapped arching of horizontally spanning masonry wall.

mid-height. Then, provided that the gap is not too large, a mechanism such as shown in Fig. 7.16(a) will form and the capacity can be analyzed, as illustrated in what follows for a horizontally spanning wall with gaps $g/2$ at both ends.

Even in a situation where a horizontally spanning wall between rigid columns or cross walls has vertical gaps at both ends due to shrinkage or construction defects (Fig. 7.17), some arching can be developed. The reduction in the load carrying capacity, assuming rigid arching, can be assessed with certain assumptions to simplify the analysis.

Based on the arching theory proposed by McDowell et al.,[7.37] the resisting moment is expressed as

$$M_u = P_u \, r_u \tag{7.22}$$

The height of a rigid arch $r_u$ is defined by the following equation (refer to Fig. 7.14 for notation):

$$r_u = t \left( 1 - \frac{\Delta_0}{dt} - \frac{2\bar{y}}{t} \right) \tag{7.23}$$

where $\Delta_0$ = deflection due to axial shortening for a zero-gap condition.

Assuming $\Delta_0 = 0$ (which is allowed by the British Code[7.2] for walls with a length-to-thickness ratio less than 25) and taking the contact zone, $\alpha \, t/2$ equal to $(0.1)t$, the foregoing expression can be simplified to

$$r_u = t \left( 1 - 2 \left( \frac{1}{3} \right) (0.1) \right) = 0.933t \tag{7.24}$$

For a gap $g/2$ at each end of the wall, the midspan deflection of the wall to close these gaps can be approximated as

$$\Delta_g = \frac{gL}{4t} \tag{7.25}$$

The height of the arch can then be expressed as

$$r_{ug} = t \left[ 1 - \frac{gL}{4t(t)} - 0.067 \right]$$

$$= t \left( 0.933 - \frac{gL}{4t^2} \right) = 0.933t - \frac{gL}{4t} \tag{7.26}$$

The term $gL/4t$ is a measure of the reduction of arch height because of the gap, which in turn results in a reduced resisting moment. For example, a 0.04 in. (1 mm) gap at each end, total $g = 0.08$ in. (2 mm) of a 15 ft (4.5 m) long, $3\frac{5}{8}$ in. (90 mm) thick wall will result in a reduction in the height of the arch of $0.08(15 \times 12)/4(3\frac{5}{8}) = 0.99$ in. $(2(4500)/4(90) = 25$ mm), which is about 29% of the arch height for zero gap.

Assuming that the thrust force is the same, a reduction in lateral load resistance of 29% is expected due to the gap. Where the thrust causes movement of the supports and an increase in distance between them, the added distance can be divided into the equivalent of equal gaps at each end of the wall.

### 7.5.3 Design

When a nonloadbearing wall is built solidly between supports capable of resisting an arch thrust with no appreciable deformation or when walls are built continuously past vertical supports (horizontally spanning walls), the lateral load resistance of the wall can benefit from arching. The British code[7.2] contains specific provisions for design. It is important to realize that the equations for arching apply only to conditions after cracking. Therefore, design with these arching equations is a strength design, not a design for service loads.

For a wall with a length or height-to-thickness ratio of 25 or less, the following simple design method can be used. Referring to Fig. 7.18, the contact area/unit height of wall $= 0.1t$, as recommended by the British code. Therefore, the thrust force per unit height or length is $P_u = 0.85 f'_j (0.1t)$, where $f'_j$ is the compressive strength of the mortar joint at the contact surface and may be conservatively taken as the mortar cube strength.

As shown in Fig. 7.18, equilibrium results in the design force per unit area of the wall:

$$p = 0.61 f'_j \left(\frac{t}{L}\right)^2 \tag{7.27}$$

Figure 7.19 is a design chart for the lateral force capacity of arched walls using the foregoing approach.

For hollow walls, the foregoing approach can be used provided that premature

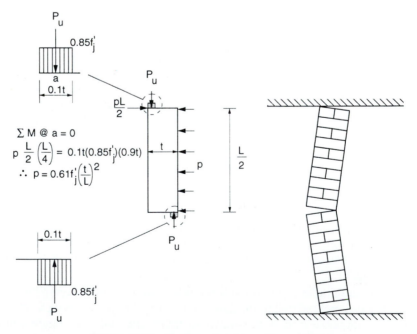

**Figure 7.18**  Design of rigid arching walls.

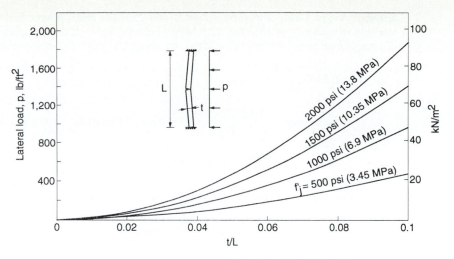

**Figure 7.19**  Resistance of walls with rigid arching.

web-shear failure is precluded.[7.39] This can be ensured by calculating the load to cause web-shear failure and checking that this load is higher than the design load calculated from arching. Limits on masonry shear stress, as defined in the relevant code, can be used. For working stress design, a safety factor of 3 can be used. For the ultimate limit state, a capacity reduction factor, $\Phi = 0.60$ similar to that for shear[7.5,7.6] is recommended for arching action.

### 7.5.4 Arching Action Example

**Example 7.2**

The masonry wall shown in Fig. 7.20 is built solid within a rigid reinforced concrete frame. Determine the ultimate lateral load that can be carried assuming the wall spans vertically and that rigid arching is developed in the vertical direction. The units are 6 in. (150 mm) hollow blocks with 2500 psi (17.3 MPa) strength and the mortar is type S. Consider shear strength of masonry $= 2\sqrt{f'_m}$ where $f'_m$ is in psi ($0.17\sqrt{f'_m}$ when $f'_m$ is in MPa).

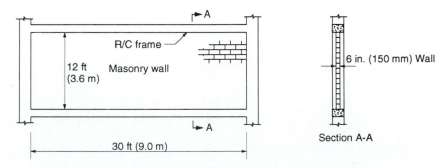

**Figure 7.20**  Rigid arching block masonry wall for Example 7.2.

**Solution A: No Gap:** For type S mortar and 2500 psi units, a specified compressive strength of masonry of 1833 psi (12.6 MPa)[7.1] can be used. If the compressive strength of type S mortar is 1800 psi (12.4 MPa), by using Eq. 7.27, for strength design,

$$p = 0.61 f_j' \left(\frac{t}{h}\right)^2 \Phi$$

$$= 0.61\,(1800) \left(\frac{5.625}{12 \times 12}\right)^2 \Phi = 1.68\,\Phi\,\text{psi} = 1.68\,(0.6) = 1.0\,\text{psi} = 144\,\text{psf}$$

$$\left(0.61(12.4) \left(\frac{140}{3600}\right)^2 (0.6)(10)^3 = 6.86\,\text{kN/m}^2\right)$$

By checking web-shear failure, the shear force/foot (meter) at the support is

$$144 \left(\frac{12}{2}\right) = 864\,\text{lb/ft}$$

$$\left(6.86 \left(\frac{3.6}{2}\right) = 12.35\,\text{kN/m}\right)$$

and at a distance t from the support,

$$V = 864 \left(\frac{72 - 5.63}{72}\right) = 797\,\text{lb/ft}$$

$$\left(V = 12.35 \left(\frac{1.8 - 0.14}{1.8}\right) = 11.4\,\text{kN/m}\right)$$

Shear stress can be calculated as

$$f_v = \frac{VQ}{Ib} \tag{7.28}$$

where the 1 in. (26 mm) face shells give $I$ = 130.3 in⁴/ft (1.72 × 10⁸ mm⁴/m), and the three 1 in. (26 mm) webs per block give $b$ = 3(1.0)(12/16) = 2.25 in./ft (195 mm/m), and $Q$ = 27.8 in.³/ft (1.48 × 10⁶ mm³/m).

$$f_v = \frac{797(27.8)}{130.3(2.25)} = 75.6\,\text{psi}$$

$$\left(f_v = \frac{11.4(1.48 \times 10^6)}{(1.72 \times 10^8)(195)}\,(1000) = 0.50\,\text{MPa}\right)$$

If the shear strength of masonry is taken as

$$F_v = 2.0\,\sqrt{f_m'}$$

$$= 2.0\,(\sqrt{1833}) = 85.6\,\text{psi} > 75.6\,\text{psi OK.}$$

(In SI units, shear strength is

$$0.17\,\sqrt{f_m'} = 0.17\,\sqrt{12.6} = 0.6\,\text{MPa} > 0.50\,\text{MPa} \qquad \text{OK.)}$$

**Solution B: Flexural Strength Comparison:** If arching is not utilized (or developed), the design load using conventional flexural strength can be found from

$$M_v = p\frac{h^2}{8} = f'_{tn} S$$

$$p = \frac{8M_v}{L^2} = \frac{8Sf'_{tn}}{h^2}$$

where $f'_{tn}$ is the tensile strength normal to the bed joint which is governed by the mortar bond strength at the unit-mortar interface. An experimentally determined value of $f'_{tn} = 60$ psi (0.41 MPa) based on net area is used.

For face shell mortared units, the section modulus is

$$S = \frac{12[(5.625)^3 - (5.625)^3]}{12}\left(\frac{2}{5.625}\right) = 46.3 \text{ in.}^3/\text{ft}$$

$$\left(S = \frac{1000[(140)^3 - (88)^3]}{12}\left(\frac{2}{140}\right) = 2.46 \times 10^6 \text{ mm}^3/\text{m}\right)$$

$$p = \frac{8(46.3)(60)}{(12)^2 12} = 12.9 \text{ psf}$$

$$\left(p = \frac{8(2.46 \times 10^6)(0.41)}{(3600)^2} = 0.62 \text{ kN/m}^2\right)$$

It is interesting to note that a more than tenfold increase in lateral load capacity is achieved by utilizing arching. The frame must be designed to resist the thrust forces due to arching. Also any gap at the top of the wall would dramatically reduce the calculated design load produced by the arching and must be taken into account in the design. The increase in span length due to deflection of the supporting beams can be treated as a gap.

**Solution C: With Gap:** As an illustration of the effect of a gap, if the blocks were partially wet when they were placed, a shrinkage strain of, say, 0.0002 might occur. Over the 12 ft. (3.6 m) height, this produces a gap of 0.029 in., say, 0.03 in. (0.75 mm).

By using Eq. 7.26

$$r_{ug} = t\left(0.933 - \frac{gh}{4t^2}\right)$$

$$= 5.625\left(0.933 - \frac{0.03(12 \times 12)}{4(5.625)^2}\right) = 5.06 \text{ in.}$$

$$\left(r_{ug} = 140\left(0.933 - \frac{0.75(3600)}{4(140)^2}\right) = 126 \text{ mm}\right)$$

Assuming the same thrust force, the reduction in height of the arch from 0.933 (5.625) = 5.25 in. (0.933(140) = 131 mm) to 5.06 in. (126 mm) represents a 4% loss in capacity.

Alternately, if the supporting beam at the base of the wall deflected h/720 after the wall was built (including the effects of thrust forces), the gap will reduce the capacity. A conservative approach would be to simply analyze a strip of wall at the point of maximum deflection. However, a reasonable and less conservative approach for walls without openings is to use a gap equal to the average deflection, which for a parabolic-shaped deflection profile is $\frac{2}{3}$ of the maximum deflection. In this case,

$$g_{\text{average}} = \frac{L}{720} \times \frac{2}{3} = \frac{30 \times 12}{720} \times \frac{2}{3} = 0.333 \text{ in.}$$

$$\left( g_{\text{average}} = \frac{9.0 \times 1000}{720} \times \frac{2}{3} = 8.33 \text{ mm} \right)$$

Again, by using Eq. 7.26

$$r_{ug} = t \left[ 0.933 - \frac{0.333(12 \times 12)}{4(5.625)^2} \right] = 0.55t$$

$$\left( r_{ug} = t \left( 0.933 - \frac{0.833(3.6 \times 1000)}{4(140)^2} \right) = 0.55t \right)$$

which, compared to 0.933t, represents 59% of the capacity without a gap. Clip angles or other support would be required to prevent the wall from initially starting to rotate out of the vertical. ■

## 7.6 REINFORCED FLEXURAL WALLS

### 7.6.1 Background

The use of reinforced masonry is a desirable and now common construction practice. In addition to increasing the flexural capacity, reinforcement can be included to carry tension and shear forces and to provide ductility and energy absorption characteristics in seismic areas. Minimum reinforcing is also used to control cracking due to temperature and shrinkage movements. Reinforced masonry can lead to thinner walls and substantial direct savings in the cost of the wall plus indirect savings through decreased mass (related to gravity and seismic forces) and increased available floor area.

The basic principles of flexural design of reinforced masonry are presented in Chap. 6 and are not repeated here. The main differences between flexural walls and beams relate to the direction of bending, the location of reinforcement, and the level of shear stress, which is usually low in walls.

Reinforced masonry walls designed to resist lateral loads are usually constructed of hollow units with reinforcement grouted into the cells or of solid or cored units with reinforcement grouted into a cavity between two wythes. In either case, the reinforcement is normally located at the center of the wall. Because of this, it is very difficult to provide effective lateral ties or to include shear reinforcement for out-of-plane shear. Consequently, reinforcement in compression is considered to be ineffective and the wall must be designed so that no stirrups are required for shear. This latter factor is rarely critical. Because reinforcement modifies the brittle behavior of masonry, increases its general robustness, and greatly reduces the variability of response, slenderness limits can often be relaxed.

### 7.6.2 Vertical One-Way Flexural Behavior

Reinforced masonry walls are commonly designed to span vertically between top and bottom supports because this is usually the shortest span and because the main reinforcement is often more readily incorporated in the vertical direction rather than the horizontal direction.

Test programs[7.43–7.47] conducted to study the behavior of reinforced masonry walls under out-of-plane loading showed that reinforced walls behave as unreinforced walls prior to cracking.

In an early test program,[7.44] 20 ft (6 m) high panels, as shown in Fig. 7.21(a), were constructed of conventional 8 in. and 6 in. (200 and 150 mm) blocks and tested under out-of-plane air bag loading with supports top and bottom. It was shown that for a given amount of reinforcement, an 8 ft (2.4 m) spacing of reinforcement in panels in running bond was as effective as a 2 ft (0.6 m) spacing. It was also noted that reinforced masonry deflects more without significant damage, and has greater earthquake damping characteristics than normally given credit.

Sample data in Fig. 7.21(b), for walls in running bond, indicated that the steel reinforcement contributed to a very significant increase in lateral load carrying capacity (more than a sixfold increase compared to unreinforced walls). After first cracking, wall stiffness dropped, but the wall continued to carry load up to and beyond yielding of the steel. These walls also exhibited a desirable ductile failure. The walls with a stack pattern apparently did not reach yield of the reinforcement before failure.

In New Zealand, Scrivener[7.45,7.46] tested thin reinforced brick walls under out-of-plane cyclic loading. The walls exhibited a ductile behavior characterized by large inelastic deformations. The unique pinched shape of the envelope of the hysteresis loops (Fig. 7.22) was related to the central location of the reinforcement in the brick walls.

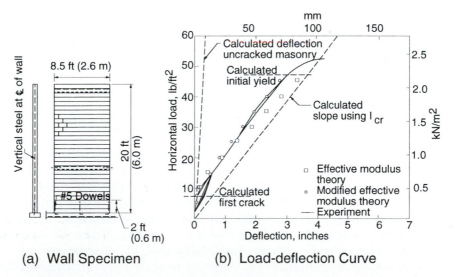

(a) Wall Specimen    (b) Load-deflection Curve

**Figure 7.21**  One-way flexural tests of reinforced masonry walls (from Ref. 7.44).

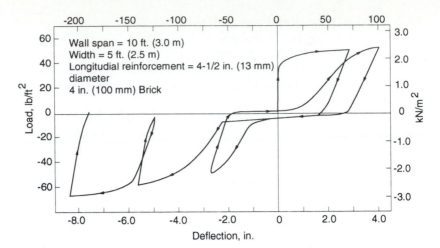

**Figure 7.22** Hysteresis curve of reinforced brick masonry wall (from Ref. 7.45).

As part of the U.S.–Japan coordinated program on Masonry Building Research (TCCMAR), a comprehensive test program[7.47–7.49] investigated the behavior of reinforced concrete masonry walls (Fig. 7.23) under monotonic and cyclic out-of-plane loading. The common mode of failure was a large open crack at the bed joints on the tension face and spalling of the mortar and face shells on the compression face (Fig. 7.24). At ultimate load, splitting of the face shell also occurred at the location of the vertical steel perhaps indicating the effect of localized stress concentration. Figure 7.25 shows load-deflection curves under monotonic loads. The sudden drops in the load coincide with horizontal cracking along bed joints.

The shape of the load-deflection curves for the wall panels are influenced by the percentage of vertical reinforcement with, as shown in Fig. 7.25(a), significant increases in capacity corresponding to increased reinforcing. The maximum strain decreases with increased percentage of reinforcement and indicates some loss of ductility. The extent of grouting significantly affected the cracking load and the wall deflection up to yielding of the reinforcement (see Fig. 7.25(b)). The extent of grouting had no significant effect on the flexural strength of the wall panels.

Cyclic tests revealed a ductile behavior of the walls with "pinched" shaped loops, Fig. 7.26(a), for centrally reinforced walls similar to those shown in Fig. 7.22. The large narrow region of zero stiffness in the hysteresis loops for load in the stages of cycles beyond yielding of the vertical steel is attributed to residual plastic deformations in the steel. Upon load reversal, "slack" exists in the system as cracks in the previous tension face close up and the previous compression face has cracks opening to produce strains compatible with the plastically strained reinforcement. This behavior differs considerably from the elastoplastic curves commonly used for ductile materials. On the other hand, walls with staggered depths of reinforcement (Fig. 7.23), where half of the bars were closer to the tension face for each direction of bending, did not show the pinching phenomenon, Fig. 7.26(b), and a higher energy absorption capacity was achieved.

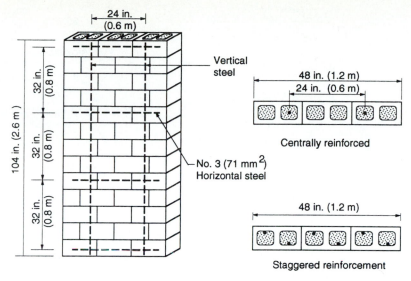

**Figure 7.23** Tests of vertically spanning reinforced masonry walls (from Ref. 7.47).

a) Crack pattern

b) Spalling and splitting of face shell

**Figure 7.24** Crack pattern and failure mode of reinforced walls spanning vertically.

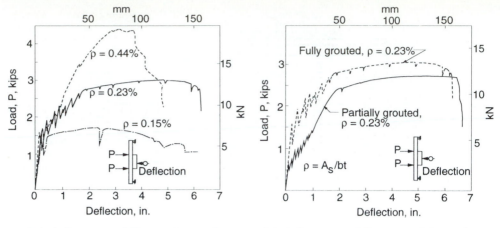

**(a) Influence of Percentage of Reinforcement, ρ**

**(b) Influence of Degree of Grouting**

**Figure 7.25** Effect of amount of reinforcement and degree of grouting on vertical bending of 6 in. (150 mm) block walls (from Ref. 7.47).

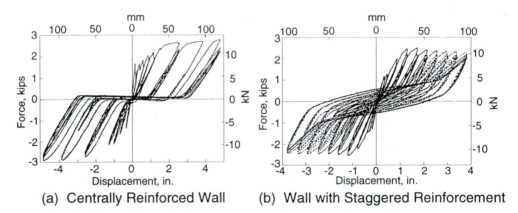

**(a) Centrally Reinforced Wall**

**(b) Wall with Staggered Reinforcement**

**Figure 7.26** Cyclic load-deflection curves of vertically spanning reinforced walls (from Ref. 7.47).

*Displacement ductility ratios*, measured by displacement at ultimate load to displacement at first yield of vertical steel, ranged from 1.79 for a wall with 0.44% of vertical steel to 29.4 for a wall with 0.15% of vertical steel. It was determined that a steel percentage less than or equal to 0.2% for centrally placed reinforcement would result in a ductility ratio greater than 4, which is accepted as adequate for energy absorption in seismic areas.[7.50] Staggering the reinforcement resulted in a higher displacement ductility.

### 7.6.3 Horizontal One-Way Flexural Behavior

Masonry walls may span horizontally between pilasters or cross walls in masonry buildings and between columns in concrete or steel buildings. Horizontal reinforcement may be steel reinforcing bars embedded in bond beams and/or joint reinforce-

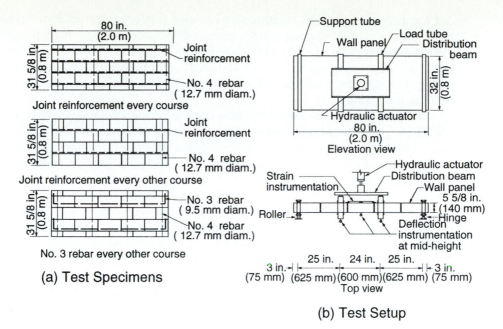

(a) Test Specimens

(b) Test Setup

**Figure 7.27** Test program on horizontally spanning reinforced walls (from Ref. 7.56).

ment. Numerous studies[7.27,7.51–7.56] have shown that joint reinforcement can be used as the principal steel to provide the bending resistance for laterally loaded walls. The test specimens shown in Fig. 7.27(a) were tested under monotonic and cyclic loading,[7.56–7.58] in a vertical position under two line loads, as shown in Fig. 7.27(b).

The common mode of failure was related to elongation of the joint reinforcement resulting in wide vertical cracks through combinations of head joints and face shells of blocks, as shown in Fig. 7.28. Crushing of the face shells in the compression zone

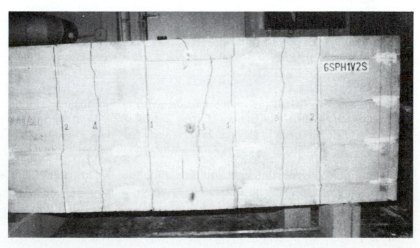

**Figure 7.28** Crack pattern of horizontally spanning reinforced wall.

did not take place. Walls reinforced with joint reinforcement in every course had more extensively developed crack patterns than walls with joint reinforcement in every other course. Grouting all cells reduced the number of cracks and, consistent with the results shown in Sec. 5.4, tended to prevent cracking along the bed joints. For walls built with a stack pattern, cracks occurred initially at the continuous vertical head joints.

In Fig. 7.29(a), the load-deflection curves for monotonic loading of walls with cold-drawn joint reinforcement exhibited very little post-peak inelastic deformation. The deflection of the wall with horizontal bar reinforcement continued to increase with only a gradual reduction in load; see Fig. 7.29(b). Grouting increased the flexural stiffnesses of the walls.

Interpolation of these results indicates that joint reinforcement corresponding to the normal minimum of $\frac{1}{3}$ of 0.2% (in Seismic Zones 3 and 4[7.1,7.5]) would increase the flexural strength by about 100%. Grouting also increased the flexural strength as is shown in Sec. 5.4.5. Partial grouting of the joint reinforced walls increased the ultimate load by 21% over the ultimate load of the ungrouted wall, and full grouting increased the ultimate load by 54%.

Figure 7.30 shows hysteresis curves for a joint reinforced wall and for a wall with the same percentage of steel, but with regular reinforcing bars in bond beams. The main differences noted are the shapes of the loops and the stability of the loops at post-peak load. The wall with bar reinforcement exhibited the "pinched" shape typical for centrally located reinforcement (similar to vertically reinforced walls), whereas joint-reinforced walls with rods nearer to the tension faces did not have this characteristic. However, the joint-reinforced walls exhibited unstable post-peak behavior, including fracture of the cold drawn reinforcement, whereas the walls with normal reinforcement showed a stable and ductile behavior, as clearly shown in Fig. 7.30(b). The ductility factors for the joint-reinforced walls ranged from 1.24 to 3.16, whereas the ductility factor for the bar-reinforced wall was 5.62. Use of

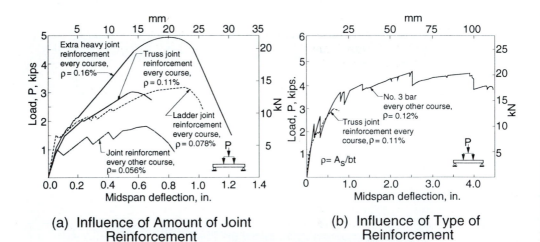

(a) Influence of Amount of Joint Reinforcement

(b) Influence of Type of Reinforcement

**Figure 7.29** Effect of type and amount of reinforcement for horizontal bending of 6 in. (150 mm) block walls (from Ref. 7.56).

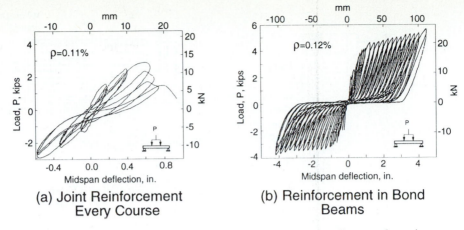

(a) Joint Reinforcement
Every Course

(b) Reinforcement in Bond
Beams

**Figure 7.30** Cyclic load-deflection curves of horizontally spanning reinforced walls (from Ref. 7.56).

stress-relieved hot-dipped galvanized joint reinforcement improves wall ductility and
provides a more ductile behavior suitable for use in high seismic risk areas.[7.59]

### 7.6.4 Two-Way Flexural Behavior

The support conditions for walls often result in both vertical and horizontal bending,
distributed in the two directions according to aspect ratio and boundary conditions. Because masonry is an anisotropic material, the strengths and flexural stiffnesses of the walls are different in the two directions. For fully grouted walls, the
masonry material is less anisotropic due to the continuity provided by the grouted
cores. For under-reinforced fully grouted panels, yield line theory, which has been
successfully used for reinforced concrete slabs[7.22,7.60] can be utilized to predict behavior and capacity at ultimate load.

Cajdert[7.27] tested masonry walls reinforced with two No. 3 reinforcing bars in
a joint (As = 0.22 in.$^2$ (142 mm$^2$); see Fig. 7.31(a). The walls were loaded with an
air bag and supported on either three or four sides. The crack patterns shown
in Fig. 7.31(b) indicate that the reinforcement produced greater distributions of
cracks. The crack patterns are consistent with predicted yield lines. A small amount
of reinforcement was shown to significantly increase the ultimate load. It was concluded that the ultimate load can be estimated within 20% using yield line theory.

## 7.7 ANALYSIS AND DESIGN OF REINFORCED FLEXURAL WALLS

### 7.7.1 Vertical Flexural Design

As discussed in Sec. 6.2, the elastic behavior for one-way flexure of reinforced
masonry has been modeled using procedures developed for working stress design
of reinforced concrete. For flexure of walls, the same procedures apply and are not
be repeated here.

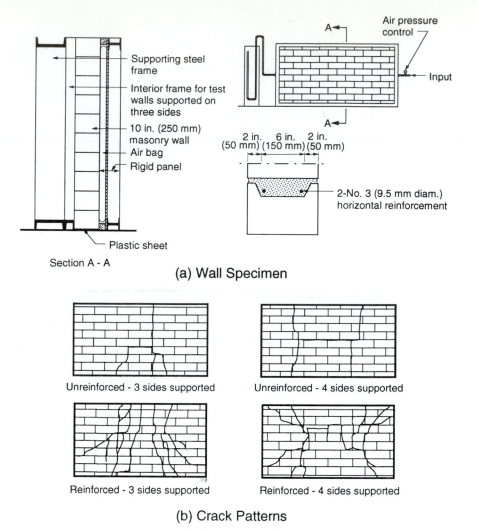

(a) Wall Specimen

Supporting steel frame

Interior frame for test walls supported on three sides

10 in. (250 mm) masonry wall

Air bag

Rigid panel

Plastic sheet

Section A - A

Air pressure control

Input

2 in. 6 in. 2 in.
(50 mm) (150 mm) (50 mm)

2-No. 3 (9.5 mm diam.) horizontal reinforcement

Unreinforced - 3 sides supported

Unreinforced - 4 sides supported

Reinforced - 3 sides supported

Reinforced - 4 sides supported

(b) Crack Patterns

**Figure 7.31**  Test program of reinforced walls—two-way flexure (from Ref. 7.27).

Masonry walls are usually lightly reinforced and ductile behavior under out-of-plane flexure is expected. For such cases, four stages are identified as shown in Fig. 7.32.

**Stage 1: All Materials Elastic, No Cracks.** The analysis is based on net section properties of the uncracked section and the minimal contribution of steel can be ignored. In this state, maximum tension stress in the masonry is less than the modulus of rupture of masonry. Methods of elastic analysis are covered in Sec. 7.2.4.

**Stage 2: All Materials Elastic, Masonry Cracks.** In the cracked region, the properties of the transformed section, neglecting the contribution of the masonry in tension, can be used (Fig. 7.32). Rectangular or T-sections result, depending on the

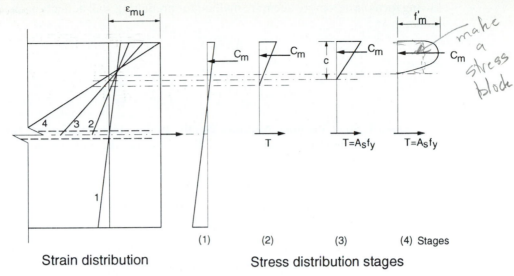

make a stress block

(1)    (2)    (3)    (4) Stages

Strain distribution      Stress distribution stages

**Figure 7.32**   Analysis of reinforced walls in flexure.

location of the neutral axis and whether the section is solid, hollow, or partially grouted. For light reinforcement, the neutral axis is normally within the face shell of partially grouted hollow masonry. This stage is a serviceability limit state when deflection is checked against allowable deflection. Calculations for effective moment of inertia similar to those for reinforced concrete have been proposed by Abboud[7.61] and Horton and Tadros[7.62] for deflection calculations. The UBC[7.5] contains the following formula for predicting midspan deflection of a reinforced masonry flexural wall simply supported at top and bottom:

$$\Delta = \frac{5\,M_{cr}h^2}{48E_mI_g} + \frac{5(M_s - M_{cr})h^2}{48E_mI_{cr}} \tag{7.29}$$

non cracked    $m \le M_n$

where    $h$ = height of the wall
       $M_s$ = service moment at the mid-height of the panel
   $I_g, I_{cr}$ = gross and cracked moments of inertia of the wall cross-section, respectively
     $M_{cr}$ = cracking moment strength of the masonry wall

The UBC specifies a maximum deflection under service load of $0.007h$. The foregoing equation is valid between $M_s$ greater than the cracking moment and less than the nominal flexural strength of the wall. The cracking moment can be calculated from the following formula:

$$M_{cr} = Sf_r \tag{7.30}$$

where   $S$ = section modulus of the uncracked section based on the net area
    $f_r$ = modulus of rupture, which is given in the range of $2.0\,\sqrt{f'_m}$ to $2.5\,\sqrt{f'_m}$ in the UBC code (based on psi) ($0.17\,\sqrt{f'_m}$ to $0.21\,\sqrt{f'_m}$ based on MPa)

Horton and Tadros[7.62] indicate that the UBC method tends to underestimate the deflection of walls with relatively low reinforcement ratios. They proposed the following formula as a more accurate method for predicting wall deflection under service moment $M_s$:

$$\Delta_s = \Delta_{cr}(1 - \alpha) \qquad (7.31)$$

where

$$\alpha = \left(\frac{M_{cr}}{M_s}\right)^2 \left(1 - \frac{I_{cr}}{I_g}\right)\left(2 - \frac{M_{cr}}{M_s}\right) \qquad (7.32)$$

**Stage 3: Steel Yielding, Masonry Elastic.** In this stage, cracked section analysis can be performed assuming linear stress distribution in the compression zone so that

$$M = A_s f_y \left(d - \frac{c}{3}\right) \qquad (7.33)$$

where $c$ is the depth to the neutral axis. As moment increases, the depth to the neutral axis decreases, as shown in Fig. 7.32.

**Stage 4: Steel Yielding, Masonry Inelastic.** This represents a strength limit state (Fig. 7.32). A strength approach using the Whitney Stress Block method from reinforced concrete is valid and slightly conservative for fully grouted reinforced masonry walls spanning vertically or horizontally for masonry compressive strengths less than 3000 psi (20 MPa).[7.63–7.65] The UBC[7.5] adopted a strength approach similar to the ACI 318[7.66] method for reinforced concrete with a maximum usable strain at the extreme masonry compression fiber of 0.003 (see Sec. 6.2.2). The moment capacity is

$$M_n = A_s f_y \, (d - a/2) \qquad (7.34)$$

where  $A_s$ = area of steel
$\quad f_y$ = yield strength of reinforcement
$\quad d$ = depth of steel, equal to $\frac{1}{2}$ wall thickness for centrally reinforced wall
$\quad a$ = depth of compression block = $A_s f_y/(0.85 f'_m \, b)$
$\quad b$ = width of compression zone

### 7.7.2 Horizontal Flexural Design

It is sometimes efficient for masonry to span horizontally between vertical supports. In this case, reinforcement is placed in a grouted cavity between wythes or in a specially shaped hollow unit (lintel blocks or blocks with knock-out webs). These special units often form bond beams. Design for horizontal flexure follows the same principles as for vertical flexure except that the effects of self-weight and other applied vertical load are neglected. The use of joint reinforcement was discussed earlier but it is important to note that for cold drawn reinforcement, a linear stress distribution with strain compatibility provides more accurate prediction[7.56] than Eq. 7.33. The reason is that cold-drawn joint reinforcement has limited yielding before rupture and therefore the maximum compression strain in the masonry may not be reached.

### 7.7.3 Two-Way Flexural Design

Reinforced walls supported on three or four sides behave as unreinforced walls until cracks penetrate to the reinforcement. The flexural strengths in the two orthogonal directions are related to the respective amounts of reinforcement in these directions and the panel behaves in much the same way as reinforced concrete slabs subjected to uniform load.

Such walls can be designed by allowable stress methods using elastic plate analysis to determine horizontal and vertical moments or by ultimate strength methods using a yield line approach. Alternatively, a strip method of design can be used in which the ultimate capacity is the sum of the capacities of a vertically spanning strip and a horizontally spanning strip. If high percentages of reinforcement or cold drawn joint reinforcement are used, compatibility of midspan deflections may have to be checked for the two crossing strips.

### 7.7.4 Walls with Openings

Openings can be dealt with in reinforced panels by making conservative simplifying assumptions regarding flexural behavior. This is commonly done by dividing the panels into strips that span either vertically or horizontally around the opening and then reinforce accordingly. It is good practice to reinforce along each edge of an opening and as close as is practical to the opening.

### 7.7.5 Reinforced Cavity and Veneer Walls

In cavity and veneer walls, only one wythe is generally reinforced to resist all of the applied lateral load. In these cases, the reinforced wythe is designed as a single wythe. If both wythes of a cavity wall are reinforced, each wythe is designed as a single wythe to resist the proportion of total load applied to it. The load sharing principles outlined for plain masonry apply here also except that if ultimate strength methods are used to design each reinforced wythe, the total load capacity can be taken as the sum of the individual capacities.

Because a reinforced backup wall can crack and allow the veneer to crack at service loads, this could have some impact on rain penetration and related serviceability considerations. The reader is referred to Chap. 12 for discussion of some of these factors.

### 7.7.6 Partially Reinforced Walls

Partial reinforcing of walls describes the situation where some parts of a wall are reinforced and other parts are not. In this case, the wall may be considered as bands or strips of reinforced masonry with unreinforced masonry spanning between the reinforced strips. The design of such walls is usually motivated by the desire to provide only very light reinforcing, for instance, to resist wind loads. In other cases, reinforcement at large spacing may exist as hold-down anchors for roofs. Partially reinforced masonry may not satisfy the minimum requirements for reinforcement or the maximum spacing limitations specified in some codes for reinforced masonry.

A common case is a hollow unit wall spanning vertically. If the maximum spacing of reinforcement is six times the thickness of the wall, $6t$, it is regarded as being fully reinforced because the whole of the compression face is assumed by most codes to be effective in compression. On the other hand, if the reinforcement is

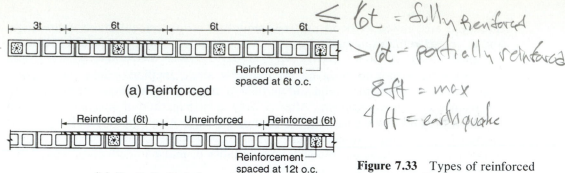

*handwritten annotations:*
≤ 6t = fully Reinforced
> 6t - partially reinforced
8ft = max
4 ft = earthquake

Reinforcement spaced at 6t o.c.

**(a) Reinforced**

Reinforced (6t) | Unreinforced | Reinforced (6t)

Reinforcement spaced at 12t o.c.

**(b) Partially Reinforced**

**Figure 7.33** Types of reinforced masonry walls.

spaced at greater distances, say, 12$t$, then the wall is regarded as consisting of reinforced strips 6$t$ wide with unreinforced strips in between, as shown in Fig. 7.33(b). The reinforced strips are designed to carry the full load and the unreinforced masonry must be capable of spanning a horizontal distance of 12$t$ between the reinforced cells. A limit of 8 ft (2.4 m) may be placed on the spacing of reinforcement.[7.4]

The same principles can be applied to horizontally spanning walls partially reinforced with bond beams and to panels in two-way bending with both vertically and horizontally reinforced strips.

Although not often thought of in these terms, reinforced pilasters supporting unreinforced wall panels is an example of partially reinforced masonry as defined before. However, pilasters may not be lightly reinforced. A very important consideration in the application of this design technique is consideration of the effects of consistent displacements. If reinforced pilasters must crack horizontally to resist the vertical bending moments, it is quite possible that the wall panel will also have to crack to have a consistent curvature. Therefore, unless vertical arching is possible, the wall panel will not be able to develop vertical bending resistance and must span horizontally between pilasters.

### 7.7.7 Shear Design

Shear stress, for walls under out-of-plane loading, is calculated using formulas presented in Sec. 6.3 for beams. Calculated shear stress should not exceed allowable shear stresses specified in building codes for working stress design. ACI 530/ASCE 5/TMS 402[7.1] specifies allowable shear for flexural members not to exceed $\sqrt{f'_m}$ psi (0.083 $\sqrt{f'_m}$ MPa) or 50 psi (0.35 MPa), whichever is less. It is seldom that shear stresses exceed code allowables. In such a case, increasing wall thickness and/or increasing compressive strength will be required because shear reinforcement is difficult to provide in flexural walls.

### 7.7.8 Anchorage of Reinforcement

The calculated tension in the reinforcement must be developed on each side of the section by an appropriate embedment length, a hook, mechanical device, or a combination of these. Requirements and details for development length discussed in Sec. 6.3 for beams are applicable for flexural walls.

Pull-out tests[7.67] of No. 4 and No. 7 bars (12.7 and 22.2 mm diameter) embedded in grouted concrete masonry prisms showed a longitudinal splitting failure, indicating that there was insufficient lateral resistance in the specimen to resist the wedging action of the reinforcement. Thus the bar size should be limited depending on wall thickness. ACI 530/ASCE 5/TMS 402[7.1] limits the diameter of the reinforcement to one-half the least clear dimension of the unit's cell in which it is placed. Providing confinement can help resist the splitting forces and larger bar sizes or smaller lengths of lap splices[7.68] may be accommodated.

Vertical steel is typically spliced at each floor level in multistory construction. The requirements for continuity of the reinforcement through proper splices depend on whether they are lapped, welded, mechanical, or end-bearing.

# 7.8 REINFORCED FLEXURAL WALL DESIGN EXAMPLES

## 7.8.1 Vertically Spanning Reinforced Block Wall

### Example 7.3

For the 20 ft (6 m) high, 6 in. (150 mm) block masonry wall with a No. 5 bar (15M) every 24 in. (600 mm) shown in Fig. 7.34, determine moments and corresponding deflections at (a) first crack, (b) service load and (c) initial yielding using allowable stress design, and (d) ultimate using strength design as per UBC.[7.5] The wall is fully grouted, $f'_m = 2500$ psi (17.3 MPa) and Grade 60 (400 MPa) steel is used. Plot the moment-deflection relationship for the wall. Ignore wall self-weight and consider the wall to be simply supported at the top and bottom.

**Solution.** Wall properties:

$I_g = bt^3/12 = 12(5.625)^3/12 = 178$ in.$^4$/ft

$(I_g = 1000(140)^3/12 = 2.29 \times 10^8$ mm$^4$/m$)$

$S = I_g/(t/2) = 178/(5.625/2) = 63.28$ in.$^3$/ft      $(2.29 \times 10^8/(140/2) = 3.27 \times 10^6$ mm$^3$/m$)$

$E_m = 750 f'_m = 750 \times 2500 = 1.875 \times 10^6$ psi (12,940 MPa)

$f_r = 2.5 \sqrt{f'_m} = 2.5 \sqrt{2500} = 125$ psi      (0.86 MPa)

(a) At first crack,

$$M_{cr} = f_r S = 125(63.28) = 7910 \text{ in.-lb/ft}$$

$$\left( M_{cr} = \frac{0.86(3.27 \times 10^6)}{(1000)(1000)} = 2.82 \text{ kN} \cdot \text{m/m} \right)$$

Just before cracking, the midspan deflection is

$$\Delta_{cr} = \frac{5 M_{cr} h^2}{48 E_m I_g} = \frac{5(7910)(20 \times 12)^2}{48(1.875 \times 10^6)(178)} = 0.14 \text{ in.}$$

$$\left( \Delta_{cr} = \frac{5(2.82 \times 10^6)(6000)^2}{48(12,940)(2.29 \times 10^6)} = 3.6 \text{ mm} \right)$$

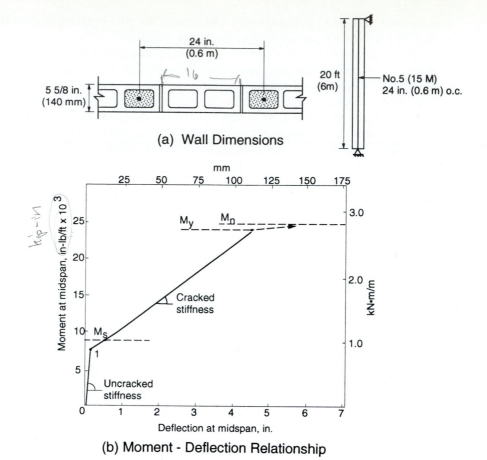

**(a) Wall Dimensions**

24 in.
(0.6 m)

5 5/8 in.
(140 mm)

20 ft
(6m)

No.5 (15 M)
24 in. (0.6 m) o.c.

**(b) Moment - Deflection Relationship**

**Figure 7.34** Reinforced concrete masonry wall spanning vertically
—Example 7.3.

(b) At allowable load,

$$F_b = 0.33 f'_m = 0.33 \times 2500 = 825 \text{ psi } (5.71 \text{ MPa})$$

$$F_s = 24,000 \text{ psi } (165 \text{ MPa})$$

$$n = E_s/E_m = 29,000/1875 = 15.5$$

Using Eqs. 6.6 and 6.10 from Chap. 6,

$$\rho = A_s/bd = 0.31/(24 \times 5.625/2) = 0.0046$$

$$(\rho = 200/(600 \times 140/2) = 0.0048)$$

$$n\rho = 15.5 \times 0.0046 = 0.071$$

$$(n\rho = 15.5 \times 0.0048 = 0.074)$$

$$(k = [(n\rho)^2 + 2n\rho]^{1/2} - n\rho = [(0.071)^2 + 2(0.071)]^{1/2} - 0.071 = 0.312$$

$$(k = [(0.074)2 + 2(0.074)]^{1/2} - 0.074 = 0.318)$$

Allowable service moment if compressive stress in masonry controls is calculated from Eq. 6.12,

$$F_b = \frac{M_s}{bd^2}\left(\frac{2}{jk}\right)$$

*(handwritten)* $M_b = \dfrac{F_b \, b d^2 j k}{2}$ "check masonry and steel i.n."

$$j = 1 - \frac{k}{3} = 1 - \frac{0.312}{3} = 0.90$$

$$(j = 1 - 0.318/3 = 0.89)$$

$$M_s = 825 \times 12 \times (2.813)^2 \, (0.9)(0.312)/2 = 10{,}995 \text{ in.-lb/ft}$$

$$\left(M_s = 5.71 \times 1000 \times (70)^2 \, (0.89)(0.318)/2(10^6) = 3.96 \text{ kN} \cdot \text{m/m}\right)$$

Allowable service moment if stress in steel controls is calculated from Eqs. 6.13 and 6.5,

$$F_s = \frac{M_s}{A_s j d}$$

*(handwritten)* $M_b = F_s A_s j d$

$$M_s = 24{,}000 \left(\frac{0.31}{2}\right)(0.9)(2.813) = 9416 \text{ in.-lb/ft controls.}$$

$$\left(M_s = 165(200(1000/600))(0.89)(70)/10^6 = 3.43 \text{ kN} \cdot \text{m/m controls}\right)$$

$$\boxed{I_{cr} = \frac{b(kd)^3}{3} + nA_s(d - kd)^2} \tag{7.34}$$

$$= \frac{12(0.878)^3}{3} + 15.5\left(\frac{0.31}{2}\right)(2.813 - 0.878)^2 = 11.65 \text{ in.}^4/\text{ft}$$

$$\left(I_{cr} = \frac{1000(22.3)^3}{3} + 15.5\left(\frac{200 \times 1000}{600}\right)(70 - 22.3)^2 = 1.54 \times 10^7 \text{ mm}^4/\text{m}\right)$$

At service load,

$$\Delta_s = \frac{5M_{cr}h^2}{48E_m I_g} + \frac{5(M_s - M_{cr})h^2}{48E_m I_{cr}}$$

*(handwritten)* lb-in $\left(\frac{lb}{12 \cdot in}\right)$

$$= 0.14 + \frac{5(9416 - 7910)(240)^2}{48(1.875 \times 10^6)(11.65)} = 0.55 \text{ in.}$$

$$\left(\Delta_s = 3.6 + \frac{5(3.43 \times 10^6 - 2.28 \times 10^6)(6000)^2}{49(12{,}938)(1.54 \times 10^7)} = 15.1 \text{ mm}\right)$$

(c) At initial yielding, assuming masonry behaves elastically, the location of the neutral axis and consequently $k$ and $j$ values are the same as before.

$$M_y = A_s f_y j d$$

$$M_y = \left(\frac{0.31}{2}\right) 60{,}000(0.9)(2.813) = 23{,}540 \text{ in.-lb/ft}$$

$$\left(M_y = \left(\frac{200 \times 1000}{600}\right) 400(0.89)(70) = 8.3 \times 10^6 \text{ N} \cdot \text{mm/m} = 8.3 \text{ kN} \cdot \text{m/m}\right)$$

At initial yielding,

$$\Delta_y = \frac{5M_{cr}h^2}{48E_mI_g} + \frac{5(M_y - M_{cr})h^2}{48E_mI_{cr}}$$

$$= 0.14 + \frac{5(23,540 - 7910)(240)^2}{48(1.875 \times 10^6)(11.65)} = 4.44 \text{ in.}$$

$$\left( \Delta_y = 3.6 + \frac{5(8.3 \times 10^6 - 2.82 \times 10^6)(6000)^2}{48(12,938)(1.54 \times 10^7)} = 106.7 \text{ mm} \right)$$

The moment-deflection relationship is shown in Fig. 7.34(b).

(d) At ultimate load,

$$M_n = A_s f_y \left( d - \frac{a}{2} \right)$$

$$a = \frac{A_s f_y}{0.85 f'_m b} = \frac{0.155(60,000)}{0.85(2500)(12)} = 0.36 \text{ in.}$$

$$\left( a = \frac{200(400)}{0.85(17.3)(600)} = 9.1 \text{ mm} \right)$$

$$M_n = \frac{0.31}{2} (60,000)(2.813 - 0.36/2) = 24,482 \text{ in.-lb/ft}$$

$$\left( M_n = \frac{200(1000)}{600} (400)(70 - 9.1/2) = 8.73 \times 10^6 \text{ N} \cdot \text{mm/m} = 8.73 \text{ kN} \cdot \text{m/m} \right)$$

For design, the ultimate capacity is $M_u \le \Phi M_n$ and for a $\Phi$ value of 0.80 for wind load,

$$\Phi M_n = 0.8(24,482) = 19,586 \text{ in.-lb/ft} = 1632 \text{ ft-lb/ft}$$

$$(\Phi M_n = 0.8(8.73) = 6.98 \text{ kN} \cdot \text{m/m})$$

The factored wind load $p_u$ corresponding to this moment capacity is

$$p_u = \frac{8M_u}{h^2} = \frac{8(1632)}{(20)^2} = 32.6 \text{ lb/ft}^2$$

$$\left( p_u = 8(6.98)(16)^2 = 1.55 \text{ kN/m}^2 \right)$$

Based on ultimate strength, a service wind load can be calculated as follows. Consider $M_u = 1.7 M_L$ (ignoring self-weight).

$$1.7M_L \le 0.8 M_n$$

$$M_L = 1632/1.7 = 960 \text{ ft-lb/ft}$$

$$(M_L = 6.98/1.7 = 4.11 \text{ kN} \cdot \text{m/m})$$

$$p = \frac{8M_L}{h^2} = \frac{8(960)}{(20)^2} = 19.2 \text{ lb/ft}^2$$

$$(p = 8(4.11)/(6)^2 = 0.91 \text{ kN/m}^2)$$

where the load factor of 1.7 and the capacity reduction factor of 0.8 were used to illustrate the process. ■

## 7.8.2 Two-Way Bending

**Example 7.4**

Consider that the wall of Example 7.3 is simply supported vertically and horizontally and spans 20 ft (6.0 m) in the two directions. If the same steel is used in the vertical and horizontal directions (No. 5 bar (15M) every 24 in. (0.6 m)), determine the increase in load carrying capacity due to two-way action.

**Solution A:**   Using yield line theory and considering yield lines as shown in Fig. 7.35(b):

From Example 7.3, $M_n$ = 24,482 in.-lb/ft = 2040 ft-lb/ft (8.73 kN · m/m). Allowing for the $\Phi$ factor, $M_u$ = 1632 ft-lb/ft (6.98 kN · m/m). This moment capacity is equal in the two directions.

From equilibrium of segment $abc$, Fig. 7.35(b), $\Sigma$ Moment about $ac$ = 0

$$M_u(10) + M_u(10) = p_u(0.5 \times 10 \times 20)(10/3)$$

$$(M_u(3) + M_u(3) = p_u(0.5 \times 3 \times 6)(3/3))$$

$$\therefore p_u - 98 \text{ lb/ft}^2 \ (4.65 \text{ kN/m}^2)$$

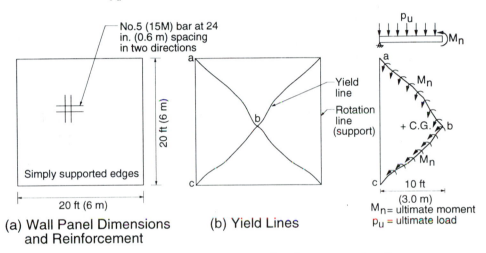

(a) **Wall Panel Dimensions and Reinforcement**     (b) **Yield Lines**

No.5 (15M) bar at 24 in. (0.6 m) spacing in two directions

20 ft (6 m)

Simply supported edges

20 ft (6 m)

Yield line

Rotation line (support)

a

b

c

$p_u$

$M_n$

$M_n$

+ C.G. b

$M_n$

10 ft

(3.0 m)

$M_n$ = ultimate moment

$p_u$ = ultimate load

**Figure 7.35**   Reinforced masonry wall under two-way bending —Example 7.4.

[Note that this load is three times the ultimate load for the wall spanning only vertically (see Example 7.3).]

**Solution B:**   Using the strip method

$$p_u = p_{uv} + p_{uh}$$

$$= \frac{8M_v}{H^2} + \frac{8M_h}{L^2}$$

and including $\Phi$ = 0.80,

$$p_u = \frac{8}{(20)^2}(2040 + 2040)(0.80) = 65 \text{ psf}$$

$$\left(p_u = \frac{8}{(6)^2}(8.73 + 8.73)(0.80) = 3.10 \text{ kN} \cdot \text{m}^2\right)$$

This load is 34% lower than that predicted from yield line analysis. Generally, the strip method will provide a more conservative estimate of wall capacity compared with the yield line theory. ∎

### 7.8.3  Horizontal Spanning Wall

**Example 7.5**

Determine the amount of joint reinforcement required for a 6 in. (150 mm) hollow concrete masonry wall (see Fig. 7.36(a)) spanning 8 ft (2.4 m) horizontally to carry 25 lb/ft² (1.2 kN/m²) unfactored wind load, $f_m' = 2000$ psi (13.8 MPa), and $E_m = 1,500$ ksi (10,350 MPa). Use ACI 530/ASCE 5/TMS 402.

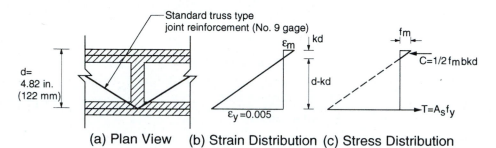

(a) Plan View    (b) Strain Distribution  (c) Stress Distribution

**Figure 7.36**  Reinforced masonry wall spanning horizontally —Example 7.5 and 7.6.

**Solution:**  Allowable masonry compressive stress,

$$F_b = \frac{1}{3}f_m' = \frac{1}{3}(2000) = 667 \text{ psi} \qquad (4.60 \text{ MPa})$$

Allowable steel stress,

$$F_s = 30,000 \text{ psi (200 MPa) for wire joint reinforcement}$$

$$n = E_s/E_m = 29,000/1500 = 19.3$$

From Eq. 6.19

$$\rho_b = \frac{nF_b}{2F_s(n + F_s/F_b)}$$

$$= \frac{19.3(667)}{2(30,000)(19.3 + 30,000/667)} = 0.0033$$

$$\left(\rho_b = \frac{19.3(4.6)}{2(200)(19.3 + 200/4.6)} = 0.0035\right)$$

Try No. 9 gage truss-type joint reinforcement every other course; $A_{s_{total}} = 0.05$ in.²/16 in. (32 mm²/400 mm)

$$\rho = \frac{A_s}{bd} = \frac{0.05/2}{16 \times 4.82} = 0.00036 < \rho_b$$

$$\left(\rho = \frac{32/2}{400 \times 122} = 0.00033 < \rho_b\right)$$

[Note that half the amount of steel is in tension.]

$$M = (A_s \, F_s) \, (jd)$$

From Eq. 6.10, $k = 0.105$ and, from Eq. 6.15, $j = 0.965$ resulting in an allowable moment

$$M = \frac{0.025}{16/12}(30{,}000)(0.965 \times 4.82) = 2616 \text{ in.-lb/ft} = 218 \text{ ft-lb/ft}$$

steel area per ft

$$\left( M = \frac{16}{400/1000}(200)(0.965 \times 122) = 0.94 \times 10^6 \text{ N} \cdot \text{mm/m} = 0.94 \text{ kN} \cdot \text{m/m} \right)$$

Moment from wind $= 25(8)^2/8 = 200 \text{ ft-lb/ft}$

$$(1.2(2.4)^2/8 = 0.86 \text{ kN} \cdot \text{m/m}^2)$$

which is less than the calculated allowable moment. Therefore, the design, with joint reinforcement every other course, is adequate. ∎

### 7.8.3 Factor of Safety in Allowable Stress Design

**Example 7.6**

Calculate the factor of safety for the wall designed in Example 7.5 using a strength design method. ($f_y = 98$ ksi (676 MPa), $\varepsilon_y = 0.005$, and $E_m = 1000$ ksi (6900 MPa).)
**Solution:** Because of the small area of joint reinforcement, the yield stress would be reached while the masonry is still elastic. Using the strain compatibility approach, the brittle nature of the joint reinforcement can be accounted for by limiting the steel strain to the nominal yield strain.
From similarity of triangles (Fig. 7.36),

$$\frac{\varepsilon_m}{kd} = \frac{0.005}{d - kd}$$

$$\therefore \varepsilon_m = 0.005 \, kd/(d - kd)$$

From equilibrium (Fig. 7.36(c)),

$$C = T$$

$$\frac{1}{2}f_m bkd = A_s f_y$$

$$\frac{1}{2}(\varepsilon_m E_m)bkd = A_s f_y$$

$$\frac{1}{2}\left(\frac{0.005 \, kd}{d - kd}\right)(1{,}000{,}000)(16)(kd) = 0.025(98{,}000) \qquad \text{and } kd = 0.51 \text{ in.}$$

$$\left( \frac{1}{2}\left(\frac{0.005 \, kd}{122 - kd}\right)(6900)(400)(kd) = 16(676) \qquad \text{and } kd = 13 \text{ mm} \right)$$

The moment capacity is

$$\therefore M_n = A_s f_y (d - kd/3)$$

$$= \frac{(0.025)}{16/12} (98,000) \left( 4.82 - \frac{0.51}{3} \right) = 8544 \text{ in.-lb/ft} = 712 \text{ ft-lb/ft}$$

$$\left( M_n = \frac{(16)}{400/1000} (676) \left( 122 - \frac{13}{3} \right) = 3.18 \times 10^6 \text{ N} \cdot \text{mm/m} = 3.18 \text{ kN} \cdot \text{m/m} \right)$$

$$\therefore \text{Factor of safety} = \frac{M_n}{\text{applied moment}}$$

$$= \frac{712}{200} = 3.56$$

$$\left( = \frac{3.18}{0.86} = 3.70 \right)$$

## 7.9 CLOSURE

Much of the masonry that is used goes into walls that are basically designed for flexure. Therefore, for the effective use of masonry, this topic deserves to be considered separately. Support conditions and two-way bending can have major effects on the design. Of particular note is the potential for very much increased capacities where support conditions produce an arching mechanism. For arching to be effective, the supports must be rigid and gaps at the support avoided or kept very small. The design details for flexural walls also affect the performance when these walls act as infill walls to resist in-plane forces as shear resisting elements. This topic is covered in Chap. 11. Another related topic is the design of veneer walls dealt with separately in Chap. 12.

The design of flexural walls is not specifically covered in current North American codes. As can be seen, accounting for two-way action, failure mechanisms, and arching can have dramatic effects on calculated capacities. In fact, these factors may be the real reason why many masonry walls, even some that are visibly cracked, have been able to carry high out-of-plane loads. Care must be taken in the application of these design approaches to ensure that the boundary conditions and other assumptions are valid.

The examples in this chapter have demonstrated the increased capacities that can be achieved through reinforcing. Therefore, thinner walls, resulting in savings in material and less mass, can be designed. Of particular interest is the potential for design of what has been defined as partially reinforced walls. In many cases, some reinforcement is required because of seismic loads, anchorage of roofs, or temperature and shrinkage crack control. It is useful to design this reinforcement for double duty as flexural reinforcement. The combination of unreinforced and reinforced parts of a wall or building requires careful consideration in design.

## 7.10 REFERENCES

7.1 Masonry Standards Joint Committee, "Building Code Requirements for Masonry Structures," ACI 530/ASCE 5/TMS 402, American Concrete Institute and American Society of Civil Engineers, Detroit and New York, 1992.

7.2 British Standards Institution, "Code of Practice for Use of Masonry: Part 1—Structural Use of Unreinforced Masonry," BS 5628, BSI, London, 1985.

7.3 British Standards Institution. "Code of Practice for Use of Masonry: Part 2—Structural Use of Reinforced and Prestressed Masonry," BS 5628, BSI, London, 1985.

7.4 Canadian Standards Association, "Masonry Design for Buildings," CAN3 S304-M84, CSA, Rexdale, Ontario, 1984.

7.5 International Conference of Building Officials, "Masonry Codes and Specifications," UBC, Chapter 24, ICBO, Whittier, CA. 1991.

7.6 Standard Association of Australia, "SAA Masonry Code," AS 3700, SAA, North Sydney, N.S.W., 1988.

7.7 L. R. Baker, "Masonry Walls to Resist Lateral Loads," in Proceedings of the Third International Seminar on Structural Masonry for Developing Countries, Mauritius, 1990, pp. 163–173.

7.8 L. R. Baker, "The Flexural Action of Masonry Structures Under Lateral Load," Ph.D. thesis, School of Engineering and Architecture, Deakin University, Geelong, Victoria, Australia, 1981.

7.9 L. R. Baker, and G. L. Franken, "Variability Aspects of the Flexural Strength of Brickwork," in Proceedings of the Fourth International Brick Masonry Conference, Brugge, Belgium, 1976, pp. 2b4.1–2b4.11.

7.10 L. R. Baker, "Precracking Behaviour of Laterally Loaded Brickwork Panels with In-Plane Restraints," in Proceedings of the British Ceramic Society, 1978, No. 27, 129–146.

7.11 L. R. Baker, "Lateral Loading of Masonry Panels: Structural Design of Masonry," Cement and Concrete Association of Australia, Sydney, 1980.

7.12 L. R. Baker, "The Failure Criterion of Brickwork in Vertical Flexure," in Proceedings of the Sixth International Symposium on Loadbearing Brickwork, London, 1977, pp. 203–216.

7.13 A. W. Hendry, "The Lateral Strength of Unreinforced Brickwork," Structural Engineer, Vol. 51, No. 2, 1973, 43–50.

7.14 L. R. Baker, "Flexural Strength of Brickwork Panels," in Proceedings of the Third International Brick Masonry Conference, Essen, West Germany, 1973, 377–383.

7.15 S. J. Lawrence, "Flexural Strength of Brickwork Normal to and Parallel to the Bed Joints," Journal of Australian Ceramic Society, Vol. 11, No. 1, 1975, 5–6.

7.16 S. J. Lawrence, "The Flexural Behaviour of Brickwork," in Proceedings of the First North American Masonry Conference, Boulder, CO, 1978, 20/1–20/18.

7.17 G. D. Base and L. R. Baker, "Fundamental Properties of Structural Brickwork," Journal of Australian Ceramic Society, Vol. 9, No. 1, 1973, pp. 1–6.

7.18 Lawrence, S. S. "Behaviour of Brick Masonry Walls Under Lateral Loading," Ph.D. thesis, University of New South Wales, 1983.

7.19 L. R. Baker, "A Failure Criterion for Brickwork in Bi-Axial Bending," in Proceedings of the Fifth International Brick Masonry Conference, Washington, D.C., 1979, pp. 41–43.

7.20 L. R. Baker, D. A. Gairns, S. J. Lawrence and J. C. Scrivener, "Flexural Behaviour of Masonry Panels–A State of the Art," in Proceedings of the Seventh International Brick Masonry Conference, Melbourne, Australia, 1985, 27–55.

7.21 D. A. Gairns, "Flexural Behaviour of Concrete Blockwork Panels," Master Thesis, University of Melbourne, Australia, 1983.

7.22 L. L. Jones and R. H. Wood, Yield Line Analysis of Slabs, American Elsevier, New York, 1967.

7.23 A. S. Essawy, "Strength of Block Masonry Walls Subject to Lateral (Out-of-Plane) Loading," Ph.D. thesis, McMaster University, Hamilton, Ontario, 1986.

7.24 A. Haseltine and R. Hodgkinson, "Wind Effects Upon Brick Panel Walls—Design Information," in Proc. of the Third International Brick Masonry Conference, Essen, West Germany, 1973, 399–406.

7.25 B. A. Haseltine, "Design of Laterally Loaded Wall Panels," Proc. of British Ceramic Society, Loadbearing Brickwork, Vol. 5, No. 24, 1975, 115–126.

7.26 B. A. Haseltine, H. W. H. West, and J. N. Tutt, "The Resistance of Brickwork to Lateral Loading: Part 2—Design of Walls to Resist Lateral Loads," Structural Engineer, Vol. 55, No. 10, 1977, 422–430.

7.27 A. Cajdert, "Laterally Loaded Masonry Walls," Publication 80:5, Chalmers University of Technology, Division of Concrete Structures, Goteberg, Sweden, 1980.

7.28 A. Cajdert, and A. Losberg, "Laterally Loaded Light Expanded Clay Block Masonry—The Effect of Reinforcement in Horizontal Joints," in Proceedings of the Third International Brick Masonry Conference, Essen, West Germany, 1973, 245–251.

7.29 B. P. Sinha, "A Simplified Ultimate Load Analysis of Laterally Loaded Model Orthotropic Brickwork Panels of Low Tensile Strength," Structural Engineer, Vol. 56B, No. 4, 1978, pp. 81–84.

7.30 B. P. Sinha, "An Ultimate Load Analysis of Laterally Loaded Brickwork Panels," International Journal of Masonry Construction, Vol. 1, No. 2, 1980, 57–60.

7.31 L. R. Baker, "An Elastic Principal Stress Theory for Brickwork Panels in Flexure," Proceedings of the Sixth International Brick Masonry Conference, Rome, 1982, 523–537.

7.32 D. W. Seward, "A Developed Elastic Analysis of Lightly Loaded Brickwork Walls With Lateral Loading," International Journal of Masonry Construction, Vol. 2, No. 3, 1982, 129–134.

7.33 S. Timoshenko, Theory of Plates and Shells, McGraw-Hill, New York, 1959.

7.34 C. T. Grimm, "Design of Non-Reinforced Masonry Panel Walls," in Proceedings of 2nd North American Masonry Conference, Maryland, 1982, Paper No. 11.

7.35 R. G. Drysdale and A. S. Essawy, "Out-of-Plane Bending of Concrete Block Walls," Journal of the Structural Division, Proceedings of ASCE, Vol. 114, No. 1, 1988, pp. 121–133.

7.36 L. R. Baker, S. J. Lawrence and A. P. Page, Australian Masonry Manual, Deakin University Press, Geelong, Victoria, 1991.

7.37 E. L. McDowall, K. E. McKee and E. Sevin, "Arching Action Theory of Masonry Walls," Journal Structural Division, Proceedings of ASCE, March 1956, pp. 915/1–915/18.

7.38 B. Gabrielsen, C. Wilton, and K. Kaplan, "Response of Arching Walls and Debris from Interior Walls Caused by Blast Loading," Report No. 7030-23, URS Research Company, San Mateo, CA, 1975.

7.39 B. Gabrielsen, and C. Wilton, "Shock Tunnel Tests of Arched Wall Panels," Report No. 7030-19, URS Research Company, San Mateo, CA, 1974.

7.40 B. Gabrielsen and K. Kaplan, "Arching in Masonry Walls Subjected to Out-of-Plane Forces," NBS Building Science Series 106, National Workshop on Earthquake Resistant Masonry Construction, 1977.

7.41 B. L. Gabrielsen, K. Kaplan and C. Wilton, "A Study of Arching in Non-Reinforced Masonry Walls," SSI 748-1, Scientific Services, Inc., Redwood City, CA, 1975.

7.42 A. W. Hendry, Structural Brickwork, pp. 128 and 141, The Macmillan Press Ltd., London, 1981.

7.43 ACI-SEASC Task Committee on Slender Walls, Test Report on Slender Walls, J. W. Athey, Ed., ACI, Detroit, 1982.

7.44 A. Mackintosh and W. L. Dickey, Results of Variation of "b" or Effective Width in Flexure in Concrete Block Panels, Masonry Institute of America, Los Angeles, CA, 1971.

7.45 J. C. Scrivener, "Face Load Tests on Reinforced Hollow Brick Non-Loadbearing Walls," New Zealand Engineering, Vol. 24, No. 7, 1969, 215–220.

7.46 J. C. Scrivener, "Reinforced Masonry—Seismic Behaviour and Design," Bulletin of New Zealand Society for Earthquake Engineering, Vol. 5, No. 4, 1972, 149–155.

7.47 A. A. Hamid, B. E. Abboud, M. W. Farah, M. K. Hatem and H. G. Harris, "Response of Reinforced Block Masonry Walls to Out-of-Plane Static Loads," Report No. 3.2(a), U.S.–Japan Coordinated Program for Masonry Building Research, Drexel University, Philadelphia, PA, 1989.

7.48 A. Hamid, M. Hatem, H. Harris and B. Abboud, "Hysteretic Response and Ductility of Reinforced Concrete Masonry Walls Under Out-of-Plane Loading," in Proceedings of the Fifth North American Masonry Conference, University of Illinois, Urbana-Champaign, 1990, 397–409.

7.49 A. Hamid, B. Abboud, M. Farah and H. Harris, "Flexural Behavior of Vertically Spanned Reinforced Block Masonry Walls," in Proceedings of the Fifth Canadian Masonry Symposium, University of British Columbia, Vancouver, 1989, 209–218.

7.50 M. Priestley, "Seismic Design of Concrete Masonry Shear Walls," ACI Journal, Vol. 83, January-February 1986, 58–68.

7.51 C. Anderson, "Lateral Loading Tests on Concrete Block Walls," Structural Engineer, Vol. 54, No. 7, 1976, 239–246.

7.52 W. L. Dickey, "Joint Reinforcement and Masonry," in Proceedings of the Second North American Masonry Conference, College Park, MD, 1982, pp. 15.1–15.15.

7.53 R. C. deVekey and H. W. H. West, "The Flexural Strength of Concrete Blockwork," Magazine of Concrete Research, Vol. 32, No. 113, 1980, 206–218.

7.54 Y. Omote, R. L. Mayes, S-W. J. Chen, and R. W. Clough, "A Literature Survey—Transverse Strength of Masonry Walls," Report to the Department of Housing, University of California, Berkeley, CA, 1977.

7.55 F. W. Cox and J. L. Ennenga, "Transverse Strength of Concrete Block Walls," ACI Journal, Vol. 54, 1958, 951–960.

7.56 A. A. Hamid, C. Chia and H. Harris, "Joint Reinforced Block Masonry Walls Under Out-of-Plane Lateral Loading," Technical Report No. STL-01/88, Structural Testing Laboratory, Drexel University, Philadelphia, 1988.

7.57 A. Hamid, C. Chia-Calabria and H. Harris, "Cyclic Behavior of Joint Reinforced Concrete Masonry Walls," TMS Journal, Vol. 10, No. 1, 1991, 31–42.

7.58 A. Hamid, C. Chia-Calabria and H. Harris, "Flexural Behavior of Joint Reinforced Concrete Masonry Walls," ACI Structural Journal, Vol. 89, No. 1, January-February 1992, 20–26.

7.59 A. A. Hamid and C. Chia-Calabria, "Effect of Properties of Steel on the Flexural Behavior of Joint Reinforced Concrete Masonry Walls," in Proceedings of the Masonry Society Journal, Vol. 8, No. 7, 1989, T1–T6.

7.60 E. Hognestad, "Yield Line Theory for the Ultimate Flexure Strength of Reinforced Concrete Slabs," Journal ACI, Vol. 24, No. 7, 1953, 637–656.

7.61 B. A. Abboud, "The Use of Small Scale Direct Models for Concrete Block Masonry Assemblages and Slender Reinforced Walls Under Out-of-Plane Loads," Ph.D. thesis, Drexel University, Philadelphia, 1987.

7.62 R. Horton and M. Tadros, "Deflection of Reinforced Masonry Members," ACI Structural Journal, Vol. 87, No. 4, 1990, 453–463.

7.63 A. A. Hamid, G. F. Assis and H. G. Harris, "Towards Developing a Flexural Strength Design Methodology for Concrete Masonry," ASTM Special Technical Publication 1063, J. Matthys, Ed., ASTM, Philadelphia, 1990.

7.64 J. M. Young and R. H. Brown, "Compressive Stress Distribution of Grouted Hollow Brick Masonry," Proceedings 8IBBMaC, Dublin, Ireland, September 1988.

7.65 J. M. Young and R. H. Brown, "Compressive Stress Distribution of Grouted Hollow Clay Masonry Under Strain Gradient," Report No. 1.2(b)-1, U.S.–Japan Coordinated Program for Masonry Building Research, Clemson University, Clemson, SC, 1988.

7.66 American Concrete Institute, "Building Code Requirements for Reinforced Concrete," ACI Standard 318-92, ACI, Detroit, 1992.

7.67 Z. Soric and L. Tulin, "Bond in Reinforced Concrete Masonry," in Proceedings of the Fourth North American Masonry Conference, Los Angeles, CA, 1987, 47/1–47/17.

7.68 T. Okada and F. Kumazawa, "Flexural Behavior of Reinforced Concrete Block Beams with Lap Joints and Spiral Reinforcement," in Proceedings of the Second Joint Technical Coordinating Committee on Masonry Research Meeting: U.S.–Japan Coordinating Program for Masonry Building Research, Keystone, CO, 1986.

## 7.11 PROBLEMS

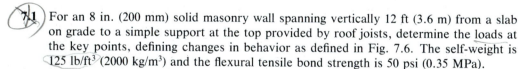

**7.1** For an 8 in. (200 mm) solid masonry wall spanning vertically 12 ft (3.6 m) from a slab on grade to a simple support at the top provided by roof joists, determine the loads at the key points, defining changes in behavior as defined in Fig. 7.6. The self-weight is 125 lb/ft$^3$ (2000 kg/m$^3$) and the flexural tensile bond strength is 50 psi (0.35 MPa).

**7.2** For the wall in Problem 7.1, determine the effect of an axial compression load at the roof level of 700 lb/ft (10.2 kN/m) or a tension load of 175 lb/ft (2.55 kN/m).

**7.3** A clay brick wall with a nominal thickness of 4 in. (100 mm) spans 12 ft (3.6 m) vertically. Data: Brick compression strength = 8000 psi (55 MPa) and type S mortar (compressive strength) = 2000 psi (13.8 Pa). Determine the maximum out-of-plane seismic load that can be carried using:
**(a)** Working stress design as per your local code.
**(b)** Arching between rigid supports (factor of safety = 3.0).
**(c)** Gapped arching with a $\frac{1}{32}$ in. (0.8 mm) gap at the top of the wall.
Comment on the results and the suitability of the methods.

**7.4** A 6 in. (150 m) thick (nominal) partition is 16 ft (4.8 m) high and is designed to carry its own weight plus lateral wind load. Consider it to be constructed of hollow concrete block and type S mortar ($f_m' = 2500$ psi (17.3 MPa)). Consider it to be simply supported top and bottom.

(a) Using allowable stresses, determine the allowable wind pressure.

(b) If a gap of $\frac{1}{4}$ in. (6.5 mm) exists at the top, what is the arching capacity of this wall. (Consider that clip angles prevent the wall from rotating out-of-vertical.)

**7.5** If the wall in Problem 7.3 also spans 12 ft (3.6 m) horizontally, what lateral load can be resisted if it is considered to be simply supported on all four edges? Data: $f'_{tn} = 100$ psi (0.69 MPa), and $f'_{tp} = 2.5f'_{tn}$.

(a) Use elastic plate theory.

(b) Use the British Code Approach.

(c) Use the Australian Code Approach.

Compare and comment on the three methods.

**7.6** An 8 in. (200 mm) nominal reinforced hollow clay brick masonry wall spans vertically 20 ft (6 m) and has No. 6 (20M) reinforcing bars spaced at 24 in. (600 mm) and located in the center of the wall. For out-of-plane wind loading, determine the following:

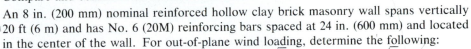

(a) Cracking load (modulus of rupture: $f_r = 2.5 \sqrt{f'_m}$ psi; ($f_r = 0.207 \sqrt{f'_m}$ MPa).

(b) Service wind load.

(c) Load at initial yielding of reinforcement.

(d) Deflection corresponding to the conditions for parts (a) to (c). Plot the load-deflection relationship.

(e) Load at failure.

(f) Repeat parts (a) to (e) for the same wall, but with No. 8 (25M) bars at 16 in. (400 mm) spacing.

Data: The wall is fully grouted; brick compressive strength = 8000 psi (56 MPa); type S mortar; $E_m = 750 f'_m$; and Grade 60 (400 MPa) steel.

**7.7** A 12 in. (300 mm) grouted brick wall spans 16 ft (4.8 m) horizontally between pilasters. If movement joints exist (simple support conditions), determine the amount and distribution of horizontal reinforcement required to resist a wind pressure of 25 lb/ft² (1.2 kN/m²). Brick compressive strength is 6000 psi (41 MPa), type S mortar is used, and the steel is Grade 40 (300 MPa).

a) Use allowable stresses from your local code.

b) Use strength design with a load factor of 1.5 and a Φ factor of 0.60.

**7.8** For the wall in Problem 7.3, if hollow units are used, determine the amount of reinforcement and maximum spacing that can be used to resist a wind load of 20 lb/ft² (0.96 kN/m²). Use working stress design. For a design that corresponds to a partially reinforced wall, calculate the stresses resulting from horizontal bending of the unreinforced zones.

**7.9** Review yield line analysis of reinforced concrete and design the wall in Problem 7.4 to carry a factored wind load of 30 lb/ft² (1.44 kN/m²) assuming that it is simply supported along its ends (horizontal span L = 20 ft (6 m)) and that the percentage of horizontal reinforcement is half the percentage of vertical reinforcement.

**7.10** Conduct a literature search for two-way arching of panels subject to out-of-plane loads over the face of the panel. For rigid edge supports, determine the ultimate capacity that the wall described in Problem 7.9 could carry by arching where reinforcement is omitted.

# CHAPTER 8

# Loadbearing Walls Under Axial Load and Out-of-Plane Bending

Diaphragm wall to carry gravity and wind loading.

## 8.1 INTRODUCTION

As was discussed in Chap. 3, loadbearing masonry buildings utilize the inherent strength and stiffness of masonry walls to transmit gravity loads (including the weight of the wall) to the foundation, to resist lateral loads, and to achieve overall stability of the structure. Therefore, as illustrated in Fig. 8.1, a wall can be required to carry concentric or eccentric vertical axial loads, out-of-plane loads applied normal to the face of the wall, and in-plane horizontal loads. The first two actions produce combined axial compression and bending forces which is the subject of this chapter. The last describes a shear wall function that is covered separately in Chap. 10. The behavior and design of walls, described as flexural walls, which carry only small axial loads is covered in Chap. 7.

Surprisingly, only a small amount of research has been done on masonry walls that carry significant axial compression and are subject to out-of-plane bending. Reflecting the small amount of background information, design codes differ markedly and have tended to incorporate simplifying empirical or semi-empirical relationships

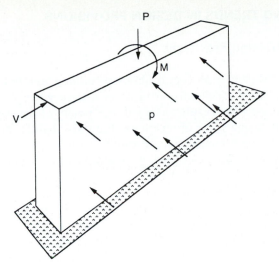

**Figure 8.1** Loading conditions on a wall.

or limitations. In some cases, the design provisions and limitations have some basis in traditional "rules-of-thumb" with the aim of producing designs consistent with experience of satisfactory performance.

Design loads calculated using various national codes,[8.1–8.6] different codes within the United States[8.1,8.2] or even different methods within the same code,[8.3] reveal large differences. It is fair to say that each of these methods is an attempt to provide sound and rational guidance within the objective of a reasonably easy-to-apply method. To some extent, the variation in level of safety within a design method and the relative conservativeness are related to the simplicity of the method. It should be remembered that different building traditions, economics, and different methods for assessing safety are also factors that affect the form and content of the design provisions. In the past, the inherently high load carrying capacity of massive masonry walls often led to large reserves of strength so that inconsistent levels of safety were not a major concern. However, in modern construction, market forces have provided greater incentive to develop rational design methods that are sufficiently comprehensive to provide satisfactory levels of safety without excessive reserves of capacity. A case in point is the provisions for Reinforced Masonry Slender Wall Design in Sec. 24.11 of the UBC.[8.2] Tests were conducted and a strength design method developed to permit slender masonry walls to compete effectively with precast concrete panel construction. Consequently, economic forces such as this and the simple existence of more extensive information on wall performance leads to continuing modification of existing design methods.

In this chapter, the behavior of laterally and axially loaded walls is presented to allow an understanding of future changes in design methods. The background and application of current design methods are covered to assist designers in the proper use of these methods.

## 8.2 OVERVIEW OF STATUS AND TRENDS IN DESIGN PROVISIONS

### 8.2.1 Current Status of Design Provisions

**Empirical Design.** Masonry is a very traditional building material long associated with tradesmen who were essentially artisans at a time when design was as much an art as it was a science. The result was the development of "rules-of-thumb" and other simple empirical design some of which exist in only slightly modified form even today. This form of design results in quite satisfactory levels of safety, when the details of construction, materials, and loading conditions are consistent with the experience used to establish the rules. However, extrapolation of these practices to situations not verified by long-standing experience can lead to unsafe conditions. Alternately, professional design based on engineering analysis ensures more consistent levels of safety and, in many cases, more economical construction.

Simple empirical design provisions are popular for small buildings where professional design involvement may be minimal. Thus, masonry codes[8.1,8.3] often include empirical design in a separate section as an alternative to more rigorous engineering calculations. The use of these methods is usually limited to certain seismic zones, maximum wind pressures, and, in some cases, to restrictions on building dimensions.

The minimal requirement for design input, and the restrictions on application of empirical design methods led to the decision not to include such designs in this book.

**Engineered Design.** Many design codes either have or are in the process of adopting strength design provisions. These strength design methods are typically incorporated into a *limit states* format consisting of *serviceability limit states* and *ultimate limit states*. The use of factored load combinations and mathematical representations of the nonlinear behavior of masonry at failure are essential to developing codes with reasonably consistent levels of safety. This is particularly important for the interactive effects of combined loads such as axial load and bending on wall sections.

In recognition of this situation and the likelihood that strength design will predominate in the not too distant future, special attention is given to discussion of fundamental aspects of behavior. Various design alternatives including the current design requirements are presented. This discussion reflects the transient nature of this topic area and the continuing efforts by code committees to achieve more consistent safety.

### 8.2.2 Types of Wall Construction

Traditional masonry and much of modern masonry construction are unreinforced. The time-tested success of this type of construction is related to development of designs that ensure that the walls remain in compression under the combined effects of axial load, bending, and shear. An outstanding case in point is the use of arching (see Sec. 7.5) to resist lateral loads. Although designers experienced in reinforced concrete design have an understandable hesitancy about using unreinforced or *plain* construction, an understanding of the inherent toughness and redundancy of well-designed masonry buildings will help develop a *comfort level* for using this type of construction.

Even when designed as unreinforced masonry, there are situations where it is wise to include some reinforcement. Many designers and contractors favor including joint reinforcement, as shown in Fig. 8.2(a), to control cracking due to moisture, thermal, and structural movements. (For masonry built in a stack pattern, joint reinforcement is required.) Similarly, reinforcement around openings controls cracking and bond beams at the tops of walls, ties the building together, and facilitates anchorage of roofs and floors. Also, reinforcement may be added to satisfy the minimum requirements for behavior under earthquake loads in moderately active seismic areas even though the wall may be designed as an unreinforced element.

For reinforced masonry design, minimum amounts of reinforcement distributed normal and parallel to the bed joints, as shown in Fig. 8.2.(b), are required by some codes.[8.3–8.6] The actual minimums may vary, depending on the seismic zone. For example, ACI 530/ASCE 5/TMS 402[8.1] only specifies minimum reinforcement ratios for Zones 3 and 4. Maximum spacing of vertical and horizontal reinforcement is also often limited by the design codes to ensure ductile performance at post-peak loads. However, for low risk seismic zones, masonry, lightly reinforced as required, can be highly efficient in providing adequate structural capacity and improving overall structural integrity. Partial reinforcing, as shown in Fig. 8.2(c), can be visualized as unreinforced wall segments between reinforced strips. As discussed in Sec. 7.6.8, design must account for the integral behavior of the reinforced and unreinforced parts.

Wythes of masonry can be *hollow*, constructed of hollow units, or they can be *solid*, constructed of solid units or of hollow units filled with grout. When only the cells containing reinforcement are grouted, this is partially grouted masonry.

Many masonry walls are designed to span vertically, as shown in Fig. 8.3(a). However, it is possible to take advantage of side supports to reduce the slenderness effects for vertical axial loads where horizontal bending, as shown in Fig. 8.3(b), limits deflection and increases stability.

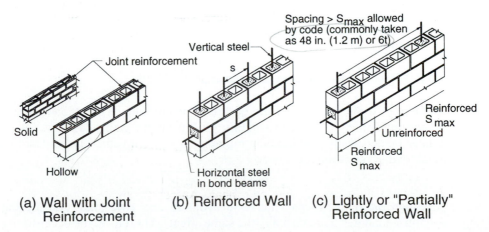

**Figure 8.2** Wall types.

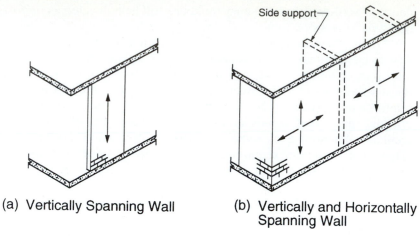

(a) Vertically Spanning Wall

(b) Vertically and Horizontally Spanning Wall

**Figure 8.3** Wall Support Conditions.

## 8.3 HISTORICAL INTERPRETATION OF BEHAVIOR OF MASONRY WALLS

### 8.3.1 General Discussion

The massive masonry walls of the past, where, for example, wall thickness increased by 4 in. (100 mm) for every story, led to relatively simplistic approaches to the design of masonry. Because walls have been traditionally thought of as carrying axial load, it was quite natural that the allowable axial load $P$ has been thought of in terms of section capacity for concentric axial load represented by

$$P = A_m F_m \tag{8.1}$$

where $A_m$ is the area of the masonry section, usually taken as the gross area even for hollow walls, and $F_m$ is the design compressive stress in the masonry. Although this approach is typically still used for empirical design of masonry, modern masonry construction with much thinner walls requires that bending and slenderness effects be considered. Before introducing existing design provisions for walls, it is worthwhile to briefly review masonry wall research which formed the basis for these provisions.

### 8.3.2 Effects of Bending on the Capacity of Walls

Bending in walls can result from eccentric positioning of axial loads, imperfections or nonuniformity of the construction (out-of-plumb walls), and lateral loads. For axially loaded walls, the bending moment $M$ can conveniently be described as the axial load $P$ times an eccentricity $e$ or,

$$e = M/P \tag{8.2}$$

Although the moment may not result from an eccentric load, considering it as an eccentric load will have virtually the same effect. Therefore, the eccentricity is called a *virtual eccentricity*.

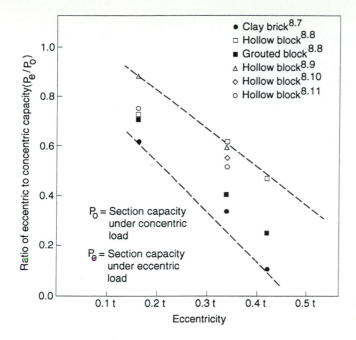

**Figure 8.4** Effect of eccentricity on axial load capacity of masonry.

Tests on short walls or prisms (see Sec. 5.3.2) with eccentrically applied loads show that increased eccentricity results in decreased axial load capacity, as shown in Fig. 8.4. It is possible to fit equations to the test data to derive empirical coefficients which account for bending within specified minimum and maximum limits of eccentricity. This is the approach taken by the Structural Clay Products Institute (now the Brick Institute of America) for their design criteria where an eccentricity coefficient $C_e$ is introduced in the right hand side of Eq. 8.1. Other design approaches involved limiting the eccentricity within prescribed limits (such as the middle third of the wall thickness) and limiting the compressive stress due to axial load to allow for the bending effect.

### 8.3.3 Section Capacity

Correlation of the capacity of prisms or short walls with the strengths of masonry units and type of mortar has normally been accomplished using prism tests, as described in Chap. 5. To test representative lengths of walls requires proportionally greater wall heights to minimize end-restraint effects of the load distribution system. The greater heights, in turn, introduce slenderness effects. As was discussed in Sec. 5.2.6, the section capacities of walls have been shown to be close to capacities determined by prism tests. Therefore, prism tests are accepted as a reasonable method for evaluating the section capacities of walls under compressive stress from the combined effects of axial load and bending.

Empirical design methods were originally based on the gross cross-sectional area, even when hollow masonry units were used. However, it is apparent that the cross-sectional geometry of the units and the effective mortar bedded area affect the stress distribution and the section capacity. As can be seen in Fig. 8.5, for eccentric

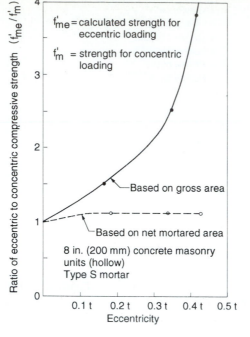

Figure 8.5 Effect of eccentricity on compressive strength of hollow masonry prisms (from Ref. 8.8).

loading, the extreme fiber stress at failure, calculated using a linear stress–strain relationship, is much more consistent for hollow prisms when the actual effective area is used. Use of the gross area distorts the relationship between axial load and moment resistance.

### 8.3.4 Effect of Wall Height

As discussed in Sec. 8.3.1, the idea of a wall as an axial load carrying member naturally led to methods that directly accounted for reduced capacity due to slenderness. For clay products, slenderness coefficients $C_s$ were developed[8.12] based on height-to-thickness ratio, $h/t$, and the ratio of end eccentricities, $-1 < e_1/e_2 < +1$. Combined with the eccentricity coefficient, Eq. 8.1 then became

$$P = C_e C_s A_m F_m \qquad (8.3)$$

Alternately, NCMA[8.13] introduced a reduction factor, designated here as $R$, so that

$$R = 1 - (h/40t)^3 \qquad (8.4)$$

which was applied independently to the axial load resistance (Eq. 8.1). This approach remains current in some building codes.[8.1,8.2]

Figure 8.6 illustrates the effect of slenderness on capacity for different end eccentricities for walls bent in symmetric single curvature ($e_1/e_2 = +1$). Both the slenderness coefficient $C_s$ and the reduction factor $R$ reflect the concept of buckling due to slenderness. The reduction factor $R$ is used to modify the allowable compres-

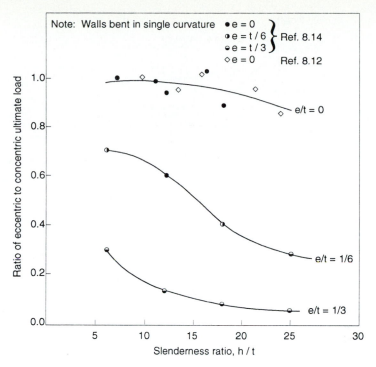

**Figure 8.6** Effect of slenderness on axial load capacity of masonry walls.

sive stress due to axial load $F_a$ used in the unity equation present in some masonry codes[8.1–8.3] for unreinforced masonry.

$$\frac{f_a}{F_a} + \frac{f_b}{F_b} = 1 \tag{8.5}$$

where $f_a, f_b$ = compressive stresses due to applied axial load and bending, respectively

$F_a, F_b$ = allowable axial and bending compressive stresses, respectively

The use of this relationship is schematically illustrated in Fig. 8.7. Bending due to an applied load is accounted for directly in the unity equation. The method of accounting for slenderness can be modified to include the effect of end eccentricities[8.12] or the effective height related to end restraints; see Fig. 8.8.

Relatively short walls and walls that do not carry heavy axial loads have fairly minor reductions in capacity due to slenderness. Therefore, in many cases, slenderness effects do not need to be considered in design. Further, because of existing slenderness design limits, few walls experience elastic buckling. Therefore, the main effect of slenderness in these cases is the development of additional bending due to deflection ($P–\Delta$ effect).

In most masonry codes, the stage of development for wall design for axial load and out-of-plane bending is part way between an historical approach and a more modern understanding of wall behavior. With this perspective as background, the

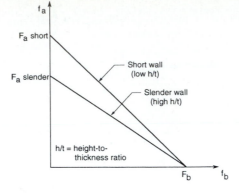

**Figure 8.7** Modifying the unity equation to account for slenderness.

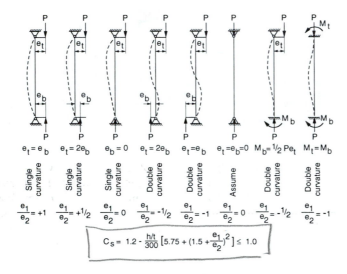

**Figure 8.8** Effect of end moment condition on the slenderness effect (from Ref. 8.12).

capacity of sections under combined axial load and bending is discussed in the next section and the effects of slenderness are covered in Sec. 8.5.

## 8.4 INTERACTION BETWEEN AXIAL LOAD AND BENDING

### 8.4.1 Introduction

The relationship between the combined effects of applied axial load and bending moment can be related to the virtual eccentricity, $e = M/P$. Regardless of whether the moment and axial force are from the same source (real eccentricity) or not, this is a convenient way of describing the combined load case.

If the axial load that a wall can carry is not affected by the applied bending moment or vice versa, there is no interaction and the relationship between the axial load and moment capacities can be represented by the diagram shown in Fig. 8.9(a). However, for linear elastic behavior of sections with equal limiting (allowable) stresses $F_m$ for combined axial compression and bending, the relationship between the maximum allowable load $P$ and moment $M$ that can be carried can be expressed as

$$F_m = \frac{P}{A} + \frac{M}{S} \tag{8.6}$$

where $A$ = area of section and $S$ = section modulus.

This equation can be used to define the linear interaction diagram shown in Fig. 8.9(b). If $P_0 = F_m A$ is the section capacity at zero eccentricity and $M_0 = F_m S$ is the moment that can be carried with zero axial load, this interaction diagram can be described by the unity equation written as

$$\frac{P}{P_0} + \frac{M}{M_0} = 1 \tag{8.7}$$

The unity equation in this form or in other variations such as in Eq. 8.5 is accurate for describing linear behavior of a section. Linear behavior requires that the section properties are not affected by the combined load and moment (no cracking) and that the material remains linearly elastic.

For masonry, the effects of tensile cracking, nonlinear stress–strain behavior of masonry, and varying contributions of reinforcement result in the unity equation, as written in Eq. 8.7, being a very conservative approximation of the true interaction. However, in the form of Eq. 8.5, the unity equation can be useful for working stress design of cracked sections where the limiting compressive stresses under axial compression and under bending are not equal. To account for the apparently higher

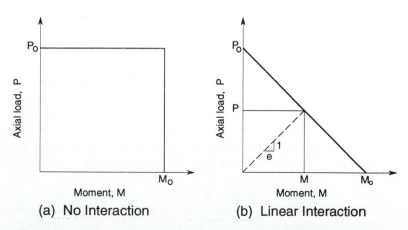

**(a) No Interaction**      **(b) Linear Interaction**

**Figure 8.9** Simple interaction diagrams for combined axial load and moment.

compressive strength due to the strain gradient (Sec. 5.3.4), the limiting compressive stress under bending $F_b$ has traditionally been greater than the limiting compressive stress under axial load $F_a$. The limiting (or allowable) stress under the combined action of axial load and bending can be defined as

$$F_m = f_a + F_b \left( 1 - \frac{f_a}{F_a} \right)$$

*For WSD*
*$F_b > F_a$*

(8.8)

where $f_a = P/A$.

### 8.4.2 Linear Elastic Analysis of Unreinforced Sections

**Solid Cross-Sections.**   Prior to cracking, the stress at the extreme fiber of a masonry section can be calculated as

$$f_m = \frac{-P}{A} \pm \frac{M}{S} = \frac{-P}{bt} \pm \frac{Pe}{bt^2/6}$$

where compression is negative and the virtual eccentricity e, defined as $M/P$, accounts for the total moment at the cross-section or, in absolute terms, the compressive stress is

$$f_m = \frac{P}{bt} \left( 1 + \frac{6e}{t} \right)$$

(8.9)

Then, for a limiting compressive stress $F_m$, the load that can be carried by the solid section is

*up to tension cracking*

*(a-b)* $\left( 0 \le e \le \frac{t}{6} \right)$

$$P = F_m bt \left( \frac{1}{1 + 6e/t} \right)$$

*$e \le t$ (Kearn)*

(8.10)

which is valid up to the point that tension cracking begins.  If tensile strength is not relied upon, then the limiting eccentricity for this condition is the kern distance $e_k$, which for a solid rectangular section is derived from

$$f_m = 0 = -\frac{P}{bt} + \frac{Pe_k}{bt^2/6}$$

*$e_{kearn}$*   *$+\frac{t}{6}$*

(8.11)

giving $e_k = t/6$.

As shown in Fig. 8.10(a), the allowable load under zero eccentricity is $P = btF_m$.  Therefore, the term $1/(1 + 6\,e/t)$ in Eq. 8.10 accounts for the effect of eccentricity for $0 < e < t/6$.  At an eccentricity, $e = t/6$ with the stress distribution as shown in Fig. 8.10(b), the allowable axial load is $P = (F_m/2)\,bt$.  The term $1/(1 + 6\,e/t)$ was the basis for the eccentricity coefficient, $C_e$ in some design standards[8.3,8.12] except that the values were multiplied by 1.3 to allow for the "strain gradient effect" (see Sec. 5.3).

For eccentricities greater than $t/6$, tensile stress develops at the extreme fiber.  If, for simplicity, masonry tensile strength is ignored, the depth of the crack can be specified as distance $x$ as shown in Fig. 8.11.  For this condition, the allowable axial load that can be carried is given as

$$P = b(t - x)\,F_m/2$$

(8.12)

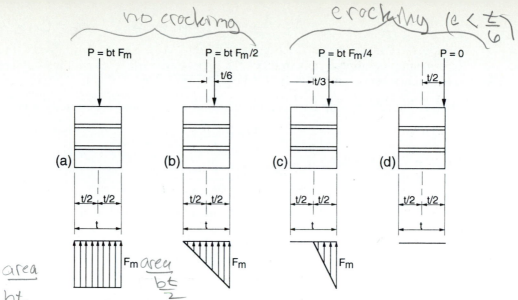

*(handwritten annotations)*
no cracking

cracking $(e < \frac{t}{6})$

area $\overline{bt}$

area $\frac{bt}{2}$

**Figure 8.10**  Effect of eccentricity on allowable axial load for solid masonry.

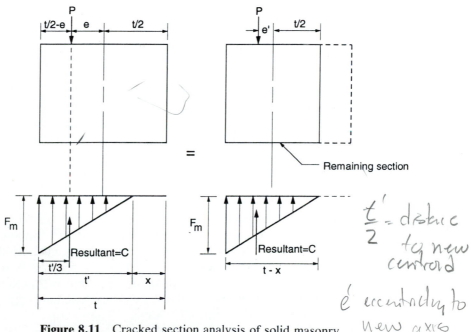

*(handwritten annotations)*

Remaining section

$\frac{t'}{2}$ = distance to new centroid

$e'$ eccentricity to new axis

**Figure 8.11**  Cracked section analysis of solid masonry.

where the value of x depends on the eccentricity. Assuming zero tensile strength, the load is located at the kern eccentricity, $e_k'$, of the uncracked thickness $t' = t - x$, so that

$$e_k' = (t - x)/6 \qquad \{ e' = e_k' \}$$

The distance between the centroids of the uncracked and the cracked sections is

$$D = \frac{t}{2} - \frac{t'}{2} = \frac{t}{2} - \frac{(t-x)}{2} = \frac{x}{2}$$

Hence, the relationship between the depth of crack $x$ and eccentricity $e$ is

$$e = \frac{x}{2} + \frac{(t-x)}{6} = \frac{x}{3} + \frac{t}{6} \tag{8.13}$$

Thus, the crack depth can be predicted by

$$\boxed{x = 3e - t/2} \tag{8.14}$$

and substituting Eq. 8.14 into Eq. 8.12, the allowable axial load is calculated as

*(b-c)* (handwritten)

or

$$P = b\left(t - \left(3e - \frac{t}{2}\right)\right)\frac{F_m}{2} \quad \left(\frac{t}{4} \le e \le \frac{t}{2}\right)$$

$$\boxed{P = btF_m\left(\frac{3}{4}\left(1 - 2\frac{e}{t}\right)\right)} \quad \tag{8.15}$$

*use this eq for e > t (cracking range)* (handwritten)

where the term $(\frac{3}{4}(1 - 2e/t)$ can be thought of as an eccentricity coefficient $C_e$ which is valid for the range $t/6 < e < t/2$. As can be seen in Fig. 8.10(c), at $e = t/3$, the crack progresses half way through the solid wall and the allowable load is reduced to 25% of the value when there is no bending moment, assuming that the allowable compressive stress $F_m$ is the same for both.

In the foregoing analysis, the combination of *largest axial load and largest bending moment* is critical for *compression controlled failure*. Alternately, for *tension failure* defined by eccentricity approaching $t/2$ and the crack opening up essentially across the entire wall cross-section (Fig. 8.10(d)), the combination of *largest moment* and *least axial load* is critical because it produces the highest virtual eccentricity.

Figure 8.12 graphically illustrates the reduction in allowable axial load with increasing eccentricity. For working stress design, some codes[8.1,8.2] do not allow cracking in unreinforced masonry in which case the eccentricity is limited by the allowable flexural tensile stress of the masonry. If the allowable tensile stress is taken equal to zero, the limiting eccentricity is $t/6$. Other codes[8.3,8.5] permit eccentricities up to $t/3$, which, as shown before, corresponds to cracking halfway through the wall, assuming zero flexural tensile strength. For this condition, it is important to note that the safety factor S.F. against tensile failure (instability) is actually 1.5 as defined by increasing the moment by 50% with the axial load held constant so that the virtual eccentricity approaches the limit, $t/2$. That is,

$$e = \frac{(Pt/3)\text{S.F.}}{P} = \frac{t}{2}$$

with the safety factor, S.F. = 1.5. Alternately, a one-third decrease in axial load would also produce tensile failure, where the moment, $P\, t/3$ remains constant so that

$$e = \frac{Pt/3}{2/3\,P} = \frac{t}{2}$$

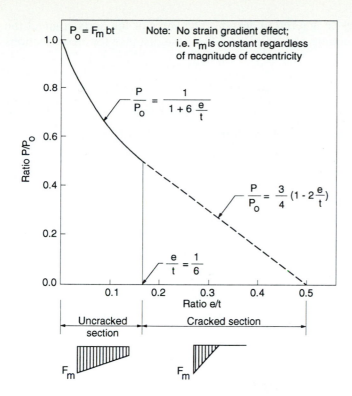

**Figure 8.12** Effect of eccentricity on allowable load for unreinforced solid masonry walls.

Therefore, where axial load and bending moment originate from independent loads, the margin of safety, achieved by limiting the eccentricity to $t/3$, may not be adequate and it is more appropriate to ensure that the maximum tensile stress in the wall is less than the design value.

**Hollow Cross-Sections.** Walls constructed with hollow unreinforced masonry units can be represented as having face shell mortared bed joints of average effective thickness $t_f$, as shown in Fig. 8.13(a). That is, the benefit of mortared webs

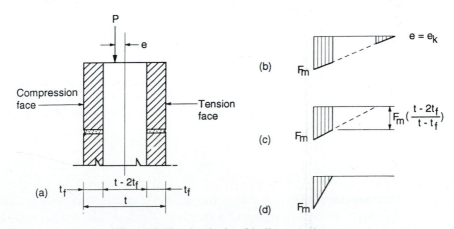

**Figure 8.13** Analysis of hollow sections.

which align vertically may be ignored in design. The webs of the units connect the face shells so that plane sections remain plane. The kern distance for cross-sections is then found from

$$e_k = \frac{S}{A_n}$$

where

$$A_n = 2t_f b$$

$$S = \frac{b[t^3 - (t - 2t_f)^3]}{12} \frac{2}{t}$$

For this analysis, the width $b$ is taken as unity.

For eccentricities up to or equal to the kern distance (see Fig. 8.13(b)), both face shell mortared areas are in compression and the allowable load that can be carried is

$$P = A_n F_m \Big/ \left(1 + \frac{e}{e_k}\right) \tag{8.16}$$

For $e > e_k$, cracking will gradually cross the tension face shell until it reaches the inside face of the tension face shell, as illustrated by the stress diagram in Fig. 8.13(c). At this point, the allowable axial load is

$$P = \frac{1}{4} A_n F_m \left(\frac{2t - 3t_f}{t - t_f}\right) \tag{8.17}$$

corresponding to an eccentricity

$$e = \left(t^2 - \frac{5}{2}tt_f + \frac{5}{3}t_f^2\right) \Big/ (2t - 3t_f) \tag{8.18}$$

which for $t_f = t/6$, gives an e value of $(34/81)\, t$.

After cracking of the tension face shell, the capacity of the compression face shell is

$$P = F_m \frac{A_n}{2} \left(\frac{t_f}{6\left(\frac{2}{3}t_f - \frac{t}{2} + e\right)}\right) \tag{8.19}$$

up to the eccentricity $e = (t/2) - (t_f/3)$ corresponding to zero tension at the tension side of the compression face shell, as shown in Fig. 8.13(d). For larger eccentricities up to $e = t/2$, Eq. 8.15 is applicable, which for a unit length of wall converts to

$$P = \frac{3}{2} F_m \left(\frac{t}{2} - e\right) \tag{8.20}$$

**Composite Walls.**  A *composite wall* is a multiwythe wall (often consisting of a clay brick outside wythe and a concrete block inside wythe) with a mortar or grout filled collar joint in between; see Fig. 8.14(a). Axial load from floor slabs or

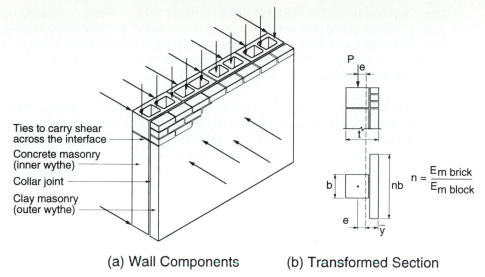

Ties to carry shear across the interface

Concrete masonry (inner wythe)

Collar joint

Clay masonry (outer wythe)

$$n = \frac{E_{m\ brick}}{E_{m\ block}}$$

(a) Wall Components    (b) Transformed Section

**Figure 8.14**  Composite walls.

the roof is typically applied on the inner wythe. If there is adequate shear-transfer capacity at the vertical interface between the wythes (at the collar joint), the wall will act as a single composite element.

Shear transfer can be accommodated by bond at the interface and by providing headers or horizontal ties to resist the shear. Due to the limited shear bond strength at the interface, current codes assign low allowable shear stresses. ACI 530/ASCE 5/TMS 402[8.1] specifies an allowable shear stress of 5 psi (0.035 MPa) for mortared collar joints and 10 psi (0.07 MPa) for grouted collar joints. For multiwythe walls connected by headers, a value equal to the square root of the unit compressive strength (based on net area) of headers is specified. Additional requirements for headers are specified regarding spacing and embedment; see Fig. 8.15(a). Because of

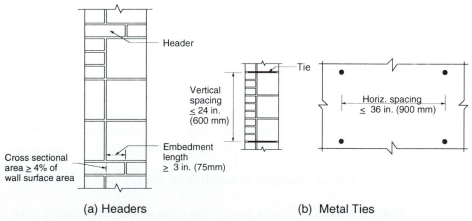

Header

Tie

Vertical spacing ≤ 24 in. (600 mm)

Horiz. spacing ≤ 36 in. (900 mm)

Embedment length ≥ 3 in. (75mm)

Cross sectional area ≥ 4% of wall surface area

(a) Headers    (b) Metal Ties

**Figure 8.15**  ACI 530/ASCE 5/TMS 402 requirements for headers and metal ties in composite walls.

possible cracking at the interface due to differential shrinkage, temperature changes, elastic deformations, and creep, a more positive shear-transfer mechanism is preferable. Metal ties and joint reinforcement have been traditionally used as a positive means of resisting interface shear in composite walls. The ACI 530/ASCE 5/TMS 402 minimum requirements for metal ties in composite sections are shown in Fig. 8.15(b).

Assuming no slip at the interface, elastic analysis of composite sections using transformed areas can be performed to calculate stresses in the section due to axial load and moment. As shown in Fig. 8.14(b), the cross-section area can be transformed into an equivalent section of concrete block masonry. The actual width of brick wythe is multiplied by $n$ to account for the difference in modulus of elasticity of clay brick masonry and concrete block masonry.

The analysis can be performed as follows:

1. Using a modular ratio, $n = E_{m\,brick}/E_{m\,block}$, the section is transformed to an equivalent concrete masonry section as shown in Fig. 8.14(b). The elastic moduli of the two wythes can be calculated from material properties using tabulated code values or from tests of masonry prisms. The properties of mortared or grouted collar joints are similar to concrete masonry and can be considered equal to simplify calculations.

2. The centroid of the transformed section can be calculated from simple mechanics and consequently the eccentricity $e$ of vertical load from the centroid of the transformed section may be calculated. Maximum stress in concrete masonry can be calculated as:

$$f_m = \frac{-P}{A_{tr}} \pm \frac{(P \cdot e \pm M)(t - \bar{y})}{I_{tr}}$$  (8.21)

where $A_{tr}, I_{tr}$ = area and moment of inertia of the transformed section, respectively

$M$ = moment from wind or seismic out-of-plane loads, if any

$\bar{y}$ = distance from the extreme fiber of the brick wythe to the centroid of the transformed section

3. The maximum stress in the clay brick masonry is

$$f_m = n \left[ \frac{-P}{A_{tr}} + \frac{(P \cdot e \pm M)(\bar{y})}{I_{tr}} \right]$$  (8.22)

If cracking is not allowed, tensile stresses in the clay brick masonry wythe can control the design. Compliance of maximum compressive stress in the masonry with code provisions should be checked (using the unity equation).

### 8.4.3 Linear Elastic Analysis of Reinforced Sections

**Solid Sections.**    To the point where the crack depth reaches the position of the reinforcement, the equations in the previous section are applicable, assuming that compressive forces in the reinforcing steel are ignored. As shown in Fig. 8.16,

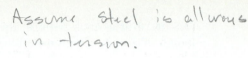

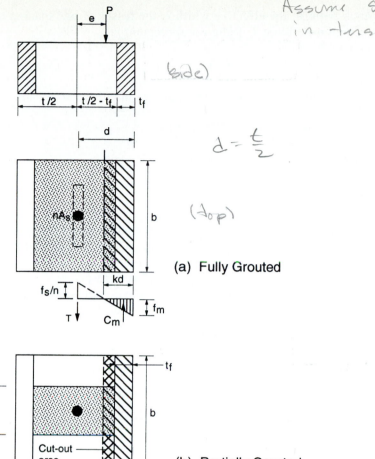

(side)

$$d = \frac{t}{2}$$

(top)

**(a) Fully Grouted**

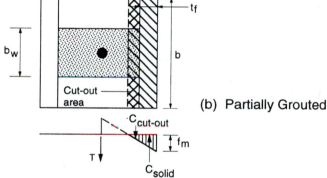

**(b) Partially Grouted**

**Figure 8.16** Combined axial load and bending with tension in the reinforcement.

cracked section analyses using a transformed area of steel is required when tension is developed in the reinforcement. The method of calculation is illustrated in what follows.

From equilibrium of vertical forces (Fig. 8.16(a)),

$$P = C_m - T \tag{8.23}$$

where

$$C_m = F_m \frac{kdb}{2} \tag{8.24}$$

The term $b$ is the effective width. ACI 530/ASCE 5/TMS 402[8.1] specifies $b$ for running bond masonry as the lesser of $s$ (bar spacing), $6t$ (wall thickness), or $= 72$ in.

Lesser of $S$, $6t$, $72$ in

(1.8 m). For a stack pattern, the effective width is the lesser of bar spacing or the length of a unit. The tensile force T in the reinforcement is

$f_m = F_m$

$$T = nA_s F_m \left( \frac{d - kd}{kd} \right)$$  (8.25)

From equilibrium of the internal and external moments about the centroid,

$$Pe = C_m \left( \frac{t}{2} - \frac{kd}{3} \right) + T \left( d - \frac{t}{2} \right)$$  (8.26)

By substituting for P from Eq. 8.23 into Eq. 8.26,

$$(C_m - T)e = C_m \left( \frac{t}{2} - \frac{kd}{3} \right) + T \left( d - \frac{t}{2} \right)$$  (8.27)

The position of the neutral axis, $kd$, can now be determined either by solving the resulting cubic equation or by trial-and-error substitution of $kd$ values until the correct value of eccentricity $e$ is obtained. [Note that if the reinforcement is concentric, as is commonly the case, the term $T(d - t/2)$ in Eq. 8.27 is zero.]

The previous formulation is based on the allowable compressive stress in the masonry, $F_m$, controlling, but the allowable stress in the steel, $F_s$, may be critical. To check this, the steel stress may be calculated as

$F_s =$

$$f_s = nF_m \left( \frac{d - kd}{kd} \right)$$  (8.28)

If the allowable steel stress $F_s$ is less than the calculated steel stress, $f_s$, then the steel is critical and Eq. 8.24 must be expressed in terms of $F_s$. Thus

$$C_m = \frac{F_s}{n} \left( \frac{kd}{d - kd} \right) \frac{kdb}{2}$$  (8.29)

and Eq. 8.25 is simply

$$T = A_s F_s$$  (8.30)

Substituting these values into Eq. 8.27, allows $kd$ to be determined as before. The allowable load for the section can be determined by calculating the values of $T$ and $C_m$ for Eq. 8.23.

**Partially Grouted Sections.**    For the section shown in Fig. 8.16(b), the grouted cells in the wall produce a T-beam cross-section. However, if the value of $kd$ is less than the thickness of the face shell, then the analysis illustrated in Eqs. 8.23 to 8.30 is applicable. If $kd$ is greater than the thickness of the face shell, a flanged section analysis is required as follows. From equilibrium of internal and external axial forces,

$$P = C_{\text{solid}} - C_{\text{cut-out}} - T$$  (8.31)

where, from Eq. 8.24,

$$C_{\text{solid}} = \frac{1}{2} F_m bkd,$$

and from Eq. 8.25,

$$T = nA_sF_m\left(\frac{d - kd}{kd}\right)$$

and

$$C_{\text{cut-out}} = \frac{1}{2}F_m\left(\frac{kd - t_f}{kd}\right)(kd - t_f)(b - b_w) \tag{8.32}$$

For equilibrium of internal and external moments

$$Pe = C_{\text{solid}}\left(\frac{t}{2} - \frac{kd}{3}\right) - C_{\text{cut-out}}\left(\frac{t}{2} - \frac{kd}{3} + \frac{2}{3}t_f\right) + T\left(d - \frac{t}{2}\right) \tag{8.33}$$

Thus, for a particular eccentricity, the value of $kd$ can be found from Eqs. 8.31 and 8.33. Again, because the derivation was based on the allowable compressive stress in the masonry controlling the design, it is necessary to check if the limiting stress in the reinforcement controls. If

$$F_s < nF_m\left(\frac{d - kd}{kd}\right) \tag{8.34}$$

Then Eq. 8.29 is used in place of Eq. 8.24 and Eq. 8.32 becomes

$$C_{\text{cut-out}} = \frac{1}{2}\frac{F_s}{n}\left(\frac{kd}{d - kd}\right)\left(\frac{kd - t_f}{kd}\right)(kd - t_f)(b - b_w) \tag{8.35}$$

Equation 8.30 is used in place of Eq. 8.25.

In this manner, for a given eccentricity, the maximum allowable axial load can by calculated for specified allowable masonry and steel stresses. [Note that the last term in Eq. 8.33 is zero for a centrally reinforced wall.]

### 8.4.4 Strength Analysis of Masonry (Ultimate Limit State)

Various forms of strength design have been incorporated in design codes in several countries. Factored loads and material properties allow for more rational consideration of load combinations and material variations leading to a more consistent level of safety. Thus, it seems certain that strength design of masonry structures will become the preferred method of design with the addition of pertinent serviceability limit states. With this in mind, an introduction to strength design follows.

**Stress–Strain Relationships for Masonry and Steel.** A reasonably accurate stress–strain relationship for reinforcing steel is a perfectly elastoplastic stress–strain relationship as shown in Fig. 8.17(a). For masonry, a singular stress-strain relationship is harder to establish because the shape of the curve is more variable and measured values of ultimate compressive strain vary over a large range. In general, as is indicated in the New Zealand code,[8.6] the objective is to replace the actual nonlinear stress–strain relationship with a mathematically well defined and easy-to-use shape that provides a reasonable estimate of the magnitude and location of the

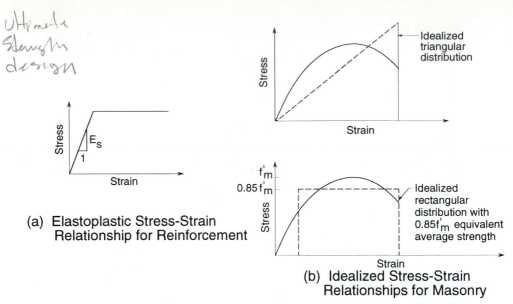

**Figure 8.17** Stress–strain relationships for steel and masonry.

(a) Elastoplastic Stress-Strain Relationship for Reinforcement

(b) Idealized Stress-Strain Relationships for Masonry

resultant compressive force. Although parabolic or trapezoidal shapes may most accurately model actual behaviour, triangular or rectangular shapes, as discussed in Sec. 6.2.2 and shown again in Fig. 8.17(b), are more commonly adopted for their ease of use.

A rectangular representation similar to the Whitney Stress block in reinforced concrete can be used for reinforced masonry.[8.15] Most tests have indicated that the ultimate strain varies between 0.0025 and 0.0035 (see Secs. 5.3.5 and 6.2.2).

Regardless of the chosen geometric shape of stress block used to calculate the magnitude and location of the resultant force in the compression zone, the design procedure is quite similar. Of more importance is the assumed maximum extreme fiber compression strain, because this assumption affects the calculations of strain in the reinforcement which is proportional to the distance from the neutral axis.

**Unreinforced Masonry.** Assuming zero tensile strength of masonry, from equilibrium, the internal compression force resultant must be coincident with the eccentricity or virtual eccentricity of the external force. As shown in Sec. 8.4.2 for a linear stress–strain relationship and a solid cross section, substitution of the specified masonry compressive strength $f'_m$ for the allowable compressive stress $F_m$ into Eq. 8.10 results in a nominal load-carrying capacity $P_n$ in the form

$$P_n = btf'_m\left(\frac{1}{1 + 6\,e/t}\right) \tag{8.36}$$

for cases where $e < t/6$ whereas for $e > t/6$, Eq. 8.15 becomes

$$P_n = btf'_m\left(\frac{3}{4}\left(1 - 2\frac{e}{t}\right)\right) \tag{8.37}$$

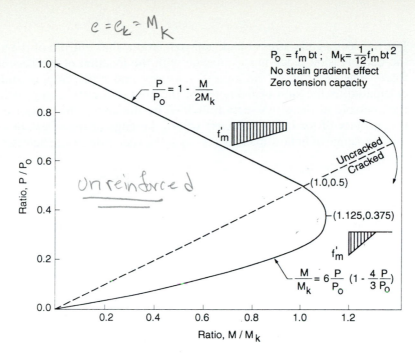

Figure 8.18  Load–moment interaction curve for unreinforced solid masonry.

An interaction curve for axial load and moment is shown in Fig. 8.18. For hollow masonry, the more complex calculations follow those in Sec. 8.4.2 assuming linear stress–strain relationship with $f_m'$ substituted for $F_m$ in Eqs. 8.16, 8.17, 8.19 and 8.20 corresponding to different degrees of cracking through the bed joint.

Alternately, as shown in Fig. 8.19(a) for a rectangular stress block, the nominal axial load capacity is given as

$$P_n = 0.85 f_m' b(t - 2e) \tag{8.38}$$

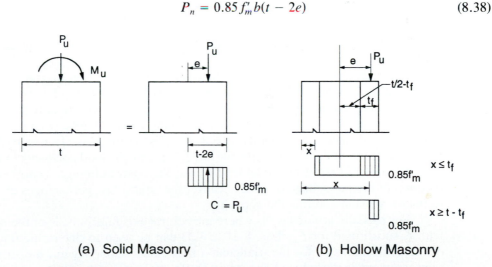

(a)  Solid Masonry          (b)  Hollow Masonry

Figure 8.19  Strength design of unreinforced masonry using rectangular stress block.

For the rectangular stress block approximation, the centroid of the area covered by the compression block must coincide with the location of the eccentrically applied force. The centroid of the area of masonry under the compression block can be calculated as the sum of moments of the area under the compression block about the centroid of the cross-section divided by the area. This distance is equal to the eccentricity of load. For the case where the edge of the stress block is within the tension flange of hollow masonry ($x \le t_f$), the eccentricity can be calculated as shown below

$$e = \frac{(t - x)x}{2(2t_f - x)} \tag{8.39}$$

and for a particular eccentricity, $e$, the depth of the compression block is $(2t_f - x)$ (Fig. 8.19) where

$$x = \left(\frac{t}{2} + e\right) - \frac{1}{2}\sqrt{t^2 + 4te + 4e^2 - 16et_f} \tag{8.40}$$

and

$$P_n = 0.85 f'_m b(2t_f - x) \tag{8.41}$$

For $e \ge (t/2 - t_f/2)$, Eq. 8.36 can be used and $x = 2e$.

**Reinforced Masonry.** For solid sections, the analyses of reinforced masonry using a *triangular stress block* follow Eq. 8.23 to 8.27 with $f'_m$ substituted for $F_m$ except that Eq. 8.25 is not applicable and can be replaced by:

$$T = A_s f_s \tag{8.42}$$

where

$$f_s = \left(\frac{d - kd}{kd}\right) \varepsilon_u E_s \tag{8.43}$$

but $f_s$ is not greater than $f_y$. The ultimate strain in the masonry $\varepsilon_u$ can be taken as 0.003.

Figure 8.20 shows typical axial load–moment interaction diagrams for a centrally reinforced masonry wall where the contribution of the reinforcement in compression is neglected. The concentric reinforcement is effective only when located on the tension side of the neutral axis. Cracking to mid-depth, where zero tensile strength is assumed, corresponds to $e = t/3$. Therefore, it is clear from the figure that reinforcement is only effective for low levels of axial load corresponding to high eccentricities. At these low axial load levels, the contribution of reinforcement is a function of the percentage of steel because, for typical low percentages of steel (0.1 to 0.4%), tension controls the capacity.

For partially grouted walls where the neutral axis falls outside of the face shell, with $f'_m$ substituted for $F_m$, Eqs. 8.31 to 8.33 can be used to determine the position of the neutral axis $kd$ for the triangular stress distribution. Again, the stress in the reinforcement is determined according to the ultimate compression strain and the position of the neutral axis as defined by Eqs. 8.42 and 8.43.

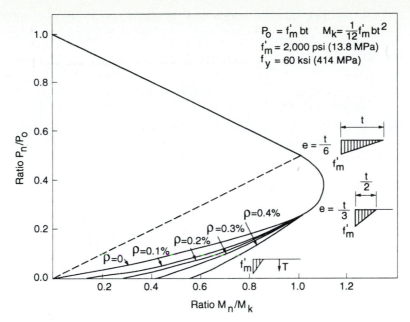

**Figure 8.20** Load–moment interaction curves for centrally reinforced masonry.

For a *rectangular stress block* (Fig. 8.21), Eq. 8.24 becomes

$$C_m = 0.85 f'_m d \beta_1 c \tag{8.44}$$

where   $c$ = distance from the extreme compression fiber to the neutral axis
$\beta_1$ = ratio of depth of the rectangular stress block to depth of the

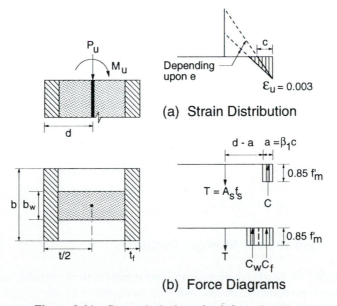

(a)  Strain Distribution

(b)  Force Diagrams

**Figure 8.21**  Strength design of reinforced walls.

neutral axis, which may be taken as 0.85 for $f'_m < 3000$ psi (20 MPa)

The tension force is $T = A_s f_s$ where

$$f_s = \left(\frac{d-c}{c}\right) \varepsilon_u E_s \leq f_y \qquad (8.45)$$

Therefore, by using Eq. 8.23 and substituting for $T$ and $C_m$ from Eqs. 8.42, 8.43 and 8.45, the position of the neutral axis $c$ and, consequently, the capacity can be calculated.

In the case shown in Fig. 8.21(b) for partially grouted construction when $\beta_1 c > t_f$, force equilibrium is given as

$$P = C_f + C_w - T \qquad (8.46)$$

where the compression forces are

$$C_f = 0.85 f'_m b\, t_f \qquad (8.47)$$

$$C_w = 0.85 f'_m b_w (\beta_1 c - t_f) \qquad (8.48)$$

and $T$ is calculated with $f_s$ determined by Eq. 8.45 depending on the value of $c$, but not higher than $f_y$.

Moment equilibrium about the centroid of the cross-section is given as

$$Pe = C_f \left(\frac{t}{2} - \frac{t_f}{2}\right) + C_w \left(\frac{t}{2} - \frac{\beta_1 c - t_f}{2}\right) + T\left(d - \frac{t}{2}\right) \qquad (8.49)$$

Thus from Eqs. 8.46 and 8.49, $c$ is determined for a particular eccentricity. When the grouted area is small, a simplification is to neglect the contribution of the web so that the compressive force resultant is given as

$$C = C_f = 0.85 f'_m b t_f \qquad (8.50)$$

The corresponding force equilibrium is

$$P = C_f - T \qquad (8.51)$$

and the moment equilibrium about the centroid of the cross-section is

$$Pe = C_f \left(\frac{t}{2} - \frac{t_f}{2}\right) + T\left(d - \frac{t}{2}\right) \qquad (8.52)$$

which permits $c$ to be found for a specified eccentricity. Knowing $c$ permits the calculation of capacity. [Note that in Eqs. 8.49 and 8.52, the term $T(d - t/2)$ is zero for centrally reinforced walls.]

## 8.4.5 Application of Strength Design

Variability in material properties and construction variabilities such as positioning of the reinforcement are normally accounted for in the ultimate strength provisions in limit states codes by using reduction factors applied to the individual material properties. Therefore, rather than use the full values of $f'_m$ and $f_s$, $\Phi_m f'_m$ and $\Phi_s f_s$ values are substituted. Statistical analysis and calibrations with existing satisfactory designs are used to determine appropriate reduction factor values. In addition, where the accuracy of the design equation is questionable or where the consequences of a particular type of failure are considered to be more critical, an overall reduction of the capacity of the element, $\Phi e$, can be introduced.

The more traditional strength design approach, which parallels the ACI-318[8.16] method for reinforced concrete, introduces a single reduction factor $\Phi$ applied to the nominal capacity. This factor varies for different types of members and failures. The Australian masonry code[8.5] suggests strength reduction factors of 0.70 or lower and the UBC[8.2] contains factors ranging from 0.4 to 0.8, depending on size, inspection, and construction details.

## 8.5 EFFECTS OF SLENDERNESS

### 8.5.1 Introduction

The axial capacity of a masonry wall decreases with increasing wall height. For very slender walls, this decrease in capacity can be associated with elastic buckling, but for most practical wall heights, lower capacity is the result of material failure under the combination of axial load and bending. For all walls, additional bending moments are caused by deflection of the walls. As was the case for reinforced concrete columns many years ago, because the main function of walls was to carry axial compression, methods were developed to directly account for the decreased capacity due to slenderness. Typically, the slenderness effect was introduced by multiplying the section capacity by the reduction factor $R$ so that

$$P_{\text{slender}} = A_m F_m R \qquad (8.53)$$

Initially, this type of factor was expressed in terms of height-to-thickness ratios, $h/t$, but to more correctly include both hollow and solid masonry, the slenderness ratio has generally been changed to height-to-radius of gyration ratio, $h/r$, where

$$r = \sqrt{I/A} \qquad (8.54)$$

From the Euler buckling equation,

$$P_c = \frac{\pi^2 EI}{h^2} \qquad (8.55)$$

substituting $Ar^2 = I$ and dividing by $A$ gives

$$f_{m\ \text{critical}} = \frac{\pi^2 E}{(h/r)^2} \qquad (8.56)$$

As a result, the ratio $h/t$ or $h/r$ has been related to the reduction in capacity due to slenderness.

Alternately, wall tests[8.17] and computer modeling[8.18] have shown that bending moments in the wall significantly influence the effect of slenderness. Other factors such as the presence, location, and amount of reinforcement, rotational restraint at the top and bottom of the wall, application of forces normal to the surface of the wall, tensile strength, variation in modulus of elasticity $E_m$, and long term creep and shrinkage all contribute to making the rational prediction of the effects of slenderness very complex. Various international codes have adopted different approaches to this problem.

### 8.5.2 Moment Magnification

Secondary bending moment due to axial load and member deflection must be included when calculating the critical combination of axial force and total bending moment at any section. Although it is possible to calculate this secondary moment directly as the product of the axial load $P$ and the displacement of the section, a moment magnifier approach is typically more convenient. In reinforced concrete,[8.16] a moment magnifier equation of the following form is used:

$$\delta = \frac{C_m}{1 - \dfrac{P_u}{\Phi P_c}} \tag{8.57}$$

where $\delta$ is the moment magnification factor for load $P_u$, $C_m$ is a factor relating the actual moment diagram to an equivalent uniform moment diagram, $P_c$ is the Euler buckling load, and $\Phi$ is a capacity reduction factor. This equation has been adapted to masonry in the Canadian code.[8.3] However, as is the case for reinforced concrete, the main problem is to derive a simple but sufficiently accurate formula for the EI term used in the equation for the Euler buckling load, which, by restating Eq. 8.55, becomes

$$P_c = \frac{\pi^2 E I_{\text{eff}}}{(kh)^2} \tag{8.58}$$

where  $E$ = modulus of elasticity
$I_{\text{eff}}$ = effective moment of inertia of the wall section
$kh$ = effective height of the wall

In the application of this method in the Canadian code, lack of adequate data led to several conservative assumptions being introduced. The first was that the term, $C_m$, which for concrete is given as

$$C_m = 0.6 + 0.4 M_1/M_2 \geq 0.4 \tag{8.59}$$

was set equal to 1.0 for masonry.

The effect of the ratio of end moments was allowed for to a limited extent by the equations for $I_{\text{eff}}$ that follow.

For unreinforced masonry

$$I_{\text{eff}} = (I_1 + I_2)/4 \qquad \text{for } 0 \leq e_1/e_2 \leq 1 \tag{8.60}$$

$$I_{\text{eff}} = (I_2 + I_0)/4 \qquad \text{for } -1 \leq e_1/e_2 < 0 \tag{8.61}$$

For reinforced masonry

$$I_{eff} = (I_1 + 2I_{cr} + I_2)/4 \qquad \text{for } 0 \le e_1/e_2 \le 1 \qquad (8.62)$$

$$I_{eff} = (I_2 + 2I_{cr} + I_0)/4 \qquad \text{for } -1 \le e_1/e_2 < 0 \qquad (8.63)$$

where $I_1$ and $I_2$ are the cracked or uncracked moments of inertia at ends 1 and 2, respectively, where $e_2$ and $I_2$ refer to the end with the highest eccentricity. $I_0$ is the moment of inertia of the uncracked section. $I_{cr}$ is the moment of inertia of the transformed section subject to pure moment, $M_0$

To allow for creep and shrinkage of the masonry, the nonlinear stress–strain behavior near failure, and the use of the unfactored working stress load $P$ instead of $P_u$, a value of

$$EI_{eff} = E_m I_{eff}/4 \qquad (8.64)$$

was chosen. Also, the effective height of the wall $kh$ was taken as the clear height, $h$ to reflect uncertainty regarding the effectiveness of rotational restraint at the top and bottom of walls. The $\Phi$ factor in Eq. 8.57 was taken as 1.0 for working stress design. Although the moment magnifier method represents a logical way of accounting for slenderness effects, additional research is required to reduce some of the conservative features of the CSA[8.3] version.

### 8.5.3 Use of the Unity Equation: UBC Method

For the basic unity equation as written in Eq. 8.5, the slenderness for unreinforced and reinforced walls is accounted for in the bracketed part of the following equation for the allowable axial compressive stress, $F_a$,

$$F_a = 0.2 f'_m \left[ 1 - \left( \frac{kh}{42t} \right)^3 \right] \qquad (8.65)$$

The effective height, $kh$, can be set as less than the clear height to reflect top and bottom restraint or the deflected shape of the wall.

Because the allowable stresses for axial load, $F_a$, and for bending, $F_b$, are different, the allowable compressive stress due to the combined loading is defined by Eq. 8.8. Analysis of the capacity of a section follows the methods presented in Secs. 8.4.2 and 8.4.3 for linear elastic analysis of solid or hollow unreinforced or reinforced walls.

### 8.5.4 Use of the ACI 530/ASCE 5/TMS 402 Provisions for Slenderness

WSD

**Unreinforced Masonry.** For unreinforced masonry, the ACI 530/ASCE 5/TMS 402 provisions require that the unity equation, Eq. 8.5, be satisfied where the allowable compressive stresses, $F_a$ and $F_b$ are given as follows:

$$F_b = \frac{1}{3} f'_m \qquad (8.66)$$

$$F_a = \frac{f'_m}{4} \left( \frac{70r}{kh} \right)^2 \qquad \text{for } \frac{kh}{r} > 99 \qquad (8.67)$$

or

$$F_a = \frac{f'_m}{4}\left[1 - \left(\frac{kh}{140r}\right)^2\right] \qquad \text{for } \frac{kh}{r} \le 99 \tag{8.68}$$

Also, the applied axial load must be less than 25% of the Euler buckling load, or

$$P \le P_e/4 \tag{8.69}$$

where the Euler buckling load is given as

$$P_e = \frac{\pi^2 E_m I}{(kh)^2}\left(1 - 0.577\frac{e}{r}\right)^3 \tag{8.70}$$

For this case, another design restriction is that under service loads, the allowable tensile stress cannot be exceeded. $f_q < f_q$

**Reinforced Masonry.** For reinforced masonry walls, the compressive stress due to axial load, $f_a = P/A$, must be less than $F_a$ as defined by Eqs. 8.67 or 8.68. Also, the combined stresses due to axial load and bending must be less than $F_b = \frac{1}{3}f'_m$.

The benefit of compression reinforcement can be included if lateral reinforcement is provided that effectively encloses and provides lateral support to the longitudinal bars. In walls with one layer of reinforcement, it is not possible to satisfy this requirement using standard ties.

### 8.5.5 Special Provisions for Slender Reinforced Walls: UBC Method

Masonry codes have traditionally incorporated restrictions on the maximum height-to-thickness ratio of walls to limit designs to the range of application of empirical or semi-empirical provisions for slenderness. In addition, there is an element of common sense to staying within ranges where experience has verified satisfactory performance. However, through development of more accurate design provisions confirmed by test results, it is possible to relax these arbitrary limits.

As the result of a test program and corresponding analysis of very slender walls,[8.19] Sec. 2411 of UBC[8.2] was developed to facilitate the design of slender reinforced masonry walls. The developed design procedure is based on strength design for factored loads with serviceability deflection limits for unfactored loads. It is intended to be used for walls subject to relatively low axial loads. Therefore, the axial stress due to unfactored loads at the location of maximum bending is limited to

$$f_a = P/A \le 0.04 f'_m \tag{8.71}$$

In addition to providing at least the minimum specified reinforcement, the design moment capacity, $\Phi M_n$, must be greater than or equal to the factored applied moment which, at mid-height of the wall, is calculated as

$$M_u = \frac{p_u h^2}{8} + P_{ou}\frac{e}{2} + (P_{wu} + P_{ou})\Delta_u \tag{8.72}$$

Simply supported wall

where $p_u$ = factored uniformly distributed lateral load
$h$ = height of wall between supports
$P_{wu}$ = factored weight of wall at the section under consideration
$P_{ou}$ = factored load from floor or roof loads
$e$ = eccentricity of load $P_{ou}$
$\Delta_u$ = horizontal deflection at mid-height under factored load with the $P$–$\Delta$ effects included in the deflection calculation

For serviceability, the mid-height deflection for unfactored loads must be limited to

$$\Delta_s = 0.007\,h \tag{8.73}$$

$$\boxed{\Delta_s @ \tfrac{h}{2} \le .007h}$$

except that the limit is reduced to

$$\Delta_s = 0.005\,h \tag{8.74}$$

for hollow brick masonry of 5 in. (127 mm) or less in thickness when the vertical reinforcement is not held in position during placement of the grout.

Assuming symmetric single curvature, the UBC specifies the mid-height deflection of simply supported walls at service loads, $\Delta_s$, to be computed by

$$\boxed{\Delta_s = \frac{5M_s h^2}{48\,E_m I_g} \qquad \text{for } M_s < M_{cr}} \tag{8.75}$$

or

$$\boxed{\Delta_s = \frac{5M_{cr} h^2}{48\,E_m I_g} + \frac{5(M_s - M_{cr})h^2}{48\,E_m I_{cr}} \qquad \text{for } M_{cr} < M_s < M_n} \tag{8.76}$$

where $M_s$ = service moment at mid-height of the wall
$I_g, I_{cr}$ = gross and cracked moment of inertia, respectively of the wall cross-section
$M_{cr}$ = cracking moment strength of the wall = $S f_r$
$S$ = section modulus
$f_r$ = modulus of rupture ranging from $2\sqrt{f'_m}$ to $4\sqrt{f'_m}$ for $f'_m$ in psi ($0.166\sqrt{f'_m}$ to $0.33\sqrt{f'_m}$ (MPa)) depending on full or partial grouting of hollow units or two-wythe walls
$M_n$ = nominal moment capacity of the masonry wall = $A_{se} f_y\,(d - a/2)$
$A_{se}$ = effective area of steel = $(A_s f_y + P_u)/f_y$
$a$ = depth of stress block due to factored loads = $(P_u + A_s f_y)/0.85\,f'_m b$
$\Phi$ = strength reduction factor equal to 0.8 for flexure where the requirement for special inspection must be satisfied.

Using established principles of mechanics to calculate the deflection at the critical mid-height section for use in Eq. 8.72, the deflection due to primary moment from Eq. 8.77 can be used.

without P$\Delta$ effects.

$$\Delta_0 = \frac{5}{384}\frac{p_u h^4}{E_m I_{\text{eff}}} + \frac{1}{16}\frac{P_{ou} e h^2}{E_m I_{\text{eff}}} \tag{8.77}$$

The additional deflection due to additional moments caused by this deflection can be calculated directly or the deflection, $\Delta_0$, can be magnified as shown in Eq. 8.78

$$A_{se} = \frac{(A_s f_y + P_{ou} + P_{wu})}{f_y} \quad \text{factored loads}$$

to produce the total deflection

$$\Delta_u = \left( \frac{5}{384} \frac{p_u h^4}{E_m I_{eff}} + \frac{1}{16} \frac{P_{ou} e h^2}{E_m I_{eff}} \right) \frac{1}{1 - P_u/P_c}$$ (8.78)

where $P_c = \pi^2 E_m I_{eff}/h^2$, where for service loads,

$p_u$ = factored uniformly distributed lateral load

$P_{ou}$ = factored load from floor or roof loads

$P_u$ = factored axial load at mid-height of the wall = $P_{wu} + P_{ou}$

$I_{eff}$ = effective moment of inertia using a cracked section (use $A_{se}$)

For other than symmetric single curvature, formulations similar to Eqs. 8.77 and 8.78 can be used in place of Eqs. 8.75 or 8.76 to calculate deflection under service loading, $\Delta_s$.

### 8.5.6 Discussion

For design of masonry walls subject to combined axial load and bending, it is useful to define slenderness limits below which buckling is not likely and the secondary bending moments due to deflection have relatively minor effects. Also, intersecting walls and flanged walls tend to be self-braced against out-of-plane deflection. For slender walls, the need to have simple design procedures is likely to mean that some variability in levels of safety will exist.

Four different design methods to account for slenderness were reviewed to illustrate the variety of approaches that have been adopted in various codes. However, in comparing these different approaches, it should be considered that, in some cases, greater apparent levels of safety may have been intentionally introduced for what were felt to be critical loading conditions, slendernesses, or types of construction. Design methods tend to be formulated into integrated packages, including loading, definition of material strengths and design stresses, as well as the mechanics for calculation of section capacities and slenderness effects. Therefore, it is recommended that slenderness and other design provisions from different codes or different design methods not be arbitrarily mixed.

## 8.6 DESIGN EXAMPLES

### 8.6.1 Wall Under Concentric Axial Compression

**Example 8.1**

For the wall shown in Fig. 8.22, determine the allowable design load.

**Solution A:** By using ACI 530/ASCE 5/TMS 402,[8.1] consider a design compressive strength of 2400 psi (16.6 MPa) and $E_m = 750 f_m' = 1.8 \times 10^6$ psi (12,420 MPa). For the end conditions shown, the effective height may be taken as

$kh = 0.8h = 0.8 (15 \text{ ft–9 in.}) = 151.2 \text{ in.}$

$(kh = 0.8(4800) = 3840 \text{ mm})$

$r = \sqrt{I/A} = 0.29t = 2.79 \text{ in.}$

$(r = 0.29(240) = 69.1 \text{ mm})$

*for class, Assume $k = 0.8$*

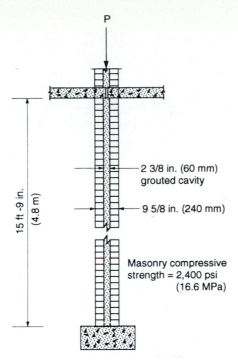

2 3/8 in. (60 mm) grouted cavity

9 5/8 in. (240 mm)

Masonry compressive
strength = 2,400 psi
(16.6 MPa)

15 ft -9 in. (4.8 m)

**Figure 8.22** Wall dimensions for
Example 8.1.

$h/r < 99$, therefore, from Eq. 8.68

$$F_a = \frac{f'_m}{4}\left[1 - \left(\frac{kh}{140r}\right)^2\right] = \frac{2400}{4}(0.85) = 510 \text{ psi}$$

$$\left(F_a = \frac{16.6}{4}(1 - 0.155) = 3.51 \text{ MPa}\right)$$

For concentric loading,

$$f_a = F_a$$

$$P = A_n F_a = (12 \times 9.625)\,510 = 58{,}910 \text{ lb/ft} \quad \nearrow \text{linear foot of wall}$$

$$(P = (1000 \times 240)\,3.51/1000 = 842 \text{ kN/m})$$

but the condition, $P \le (1/4)\,P_e$, must be satisfied where from Eq. 8.70

$$P_e = \pi^2 \frac{1.8 \times 10^6}{(151.2)^2} \frac{12(9.625)^3}{12}\left[1 - 0.577\,(0)\right]^3 = 692.2 \times 10^3 \text{ lb/ft} = 692.2 \text{ kips/ft.}$$

$$\left(P_e = \pi^2 \frac{12420}{(3840)^2} \frac{1000(240)^3}{12} \frac{1}{1000} = 9567 \text{ kN/m}\right)$$

Checking $P < P_e/4$, $P = 58.9$ kips/ft (842 kN/m) is allowed because

$$\frac{P_e}{4} = \frac{692.2}{4} = 173.1 \text{ kips/ft} \left(\frac{9567}{4} = 2392 \text{ kN/m}\right)$$

If, rather than a concentric load, a minimum accidental eccentricity of $0.1t$ is assumed,
$e = 0.963$ in. (24 mm).

By using the unity equation,

$$\frac{f_a}{F_a} + \frac{f_b}{F_b} = \frac{P/A}{F_a} + \frac{M/S}{F_b} = 1$$

$$\frac{P}{\dfrac{12(9.625)}{510}} + \frac{P(0.963)}{\dfrac{12(9.625)^2/6}{2400/3}} = 1$$

$$\left( \frac{P}{\dfrac{1000(240)}{3.51}} + \frac{P}{\dfrac{1000(240)^2/6}{16.6/3}} = 1 \right)$$

(handwritten note: $S = \dfrac{bh^2}{6}$)

For $F_b = f'_m/3$, the solution gives $P = 42.6$ kips/ft (620 kN/m).

$$P_e = \frac{\pi^2 1.8 \times 10^6}{(151.2)^2} \frac{12(9.625)^3}{12} [1 - 0.577\,(0.963/2.79)]^3$$

$$= 355.9 \text{ kips/ft} > 4\,(42.6) = 170.4 \text{ kips/ft} \qquad \text{O.K.}$$

$$\left( P_e = \frac{\pi^2 (12,420)}{(3840)^2} \frac{1000(240)^3}{12} (1 - 0.577(24/69.6))^3/1000 \right.$$

$$\left. = 4922 \text{ kN/m} > 4(620) = 2480 \text{ kN/m} \qquad \text{O.K.} \right)$$

**Solution B:** Using the UBC,[8.2] consider a design compressive strength of 2400 psi (16.6 MPa):

$$kh = 0.8h = 151.2 \text{ in. (3840 mm)}$$

$$F_a = 0.2f'_m \left[ 1 - \left( \frac{kh}{42t} \right)^3 \right] = 0.2(2400)(0.948) = 455 \text{ psi}$$

$$(F_a = 0.2(16.6)(0.945) = 3.14 \text{ MPa})$$

and the wall capacity is

$$P = A_n F_a = 12(9.625)(455) = 52,550 \text{ lb/ft} = 52.6 \text{ kips/ft}$$

$$(P = 1000(240)(3.14)/1000 = 754 \text{ kN/m})$$

Again, if we assume an accidental eccentricity of $0.1t$, using the unity equation and substituting $F_b = 0.33 f'_m = 800$ psi (5.53 MPa), and $F_a = 455$ psi (3.14 MPa) results in $P = 39.2$ kips/ft (575 kN/m).

**Solution C:** Using CAN3-S304-M84:[8.3]

$$e_{\min} = 0.1t = 0.963 \text{ in. (24 mm) and } e_1/e_2 = +1$$

For uncracked unreinforced masonry, $I_{\text{eff}} = 2I_o/4$, from Eq. 8.60:

$$I_{\text{eff}} = \frac{1}{2} \frac{12(9.63)^3}{12} = 447 \text{ in.}^4/\text{ft}$$

$$\left( I_{\text{eff}} = \frac{1}{2} \frac{1000(240)^3}{12} = 5.76 \times 10^8 \text{ mm}^4/\text{m} \right)$$

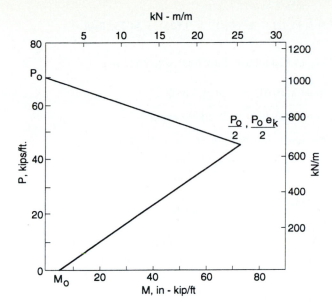

**Figure 8.23** Simplified interaction curve for Example 8.1.

From Eqs. 8.58 and 8.64,

$$P_c = \frac{\pi^2 E_m I_{\text{eff}}}{4h^2} = \frac{\pi^2 1.8 \times 10^6 (447)}{4(189)^2 10^3} = 55.5 \text{ kips/ft}$$

$$\left( P_c = \frac{\pi^2 12{,}420(5.76 \times 10^8)}{4(4800)^2(1000)} = 765 \text{ kN/m} \right)$$

and magnifying the moment using Eq. 8.57 gives a design moment of

$$M = Pe/(1 - P/P_c)$$

A simplified interaction diagram can be drawn by joining the lines between the three points on the horizontal and vertical axes of Fig. 8.23 defined by coordinates (0, $P_o$) and ($P_o$ $e_k/2$, $P_o/2$) and ($M_o$, 0) where $P_o = F_m A_n$ and $F_m = F_a$ for point (0, $P_o$), $F_m = F_b$ for the other points, $e_k = S/A_n$, and $M_o = SF_t$.

For

$$F_m = F_a = 0.25 f_m' = 0.25(2400) = 600 \text{ psi } (4.15 \text{ MPa})$$

$$P_0 = 600(12)(9.625) = 69.3 \text{ kips/ft}$$

$$(P_0 = 4.15(1000)(240)/1000 = 996 \text{ kN/m})$$

For

$$F_m = F_b = 0.32 f_m' = 0.32 (2400) = 768 \text{ psi } (5.31 \text{ MPa})$$

$$e_k = S/A_n = \frac{12(9.63)^2/6}{12(9.63)} = 1.61 \text{ in.}$$

$$\left( e_k = \frac{1000(240)^2/6}{1000(240)} = 40 \text{ mm} \right)$$

$$\frac{P_0}{2} = \tfrac{1}{2}(F_b A_n) = \tfrac{1}{2}(768)(12 \times 9.625) = 44{,}352 \text{ lb/ft}$$

$$\left(\frac{P_0}{2} = \frac{1}{2}(5.31)(1000 \times 240)/1000 = 637 \text{ kN/m}\right)$$

$$P_0 e_k/2 = 44.35(1.61) = 71.4 \text{ in. kip/ft}$$

$$(P_0 e_k/2 = 637(40)/1000 = 25.5 \text{ kN} \cdot \text{m/m})$$

For an allowable flexural tensile stress $F_t$ of 36 psi (0.25 MPa)

$$M_0 = (bt^2/6)(F_t) = \frac{12(9.625)^2}{6}\frac{(36)}{1000} = 6.67 \text{ in. kips/ft}$$

$$\left(M_0 = \frac{1000(240)^2}{6}\frac{(0.25)}{(1000)^2} = 2.4 \text{ kN} \cdot \text{m/m}\right)$$

By using these results, the simplified interaction curve shown in Fig. 8.23 is drawn. By using iteration, an initial value of $P$ is guessed and $\delta$ can be calculated from the equation $\delta = 1/(1 - P/Pc)$ so that

$$M_{\text{mag}} = \delta(P \cdot e)$$

With the magnified moment $M_{\text{mag}}$, a $P$ value can be calculated from the interaction diagram and compared with the initial value of $P$. If it is not close, another value is used and the iteration continues. After iteration, the given example results in a value of $P = 27.0$ kips/ft (395 kN/m) for a moment magnification, $\delta = 1.60$. Comparing the results from the different codes indicates that the CSA slenderness provisions are much more conservative than the American codes. This is due to the value of $P_c$ used. If $P_c = \pi^2 E_m I_{\text{eff}}/(0.8h)^2$ is used, the value of $P_c = 463$ kips/ft (6380 kN/m) results in $P = 52$ kips/ft (760 kN/m).

## 8.6.2 Wall Under Eccentric Axial Load

**Example 8.2**

Design a concrete block wall to carry a vertical load of 300 lb/ft (4.4 kN/m) with a 4 in. (100mm) eccentricity in addition to a lateral load of 20 lb/ft² (1.0 kN/m²). The wall is 16 ft (4.8 m) high and is simply supported at top and bottom; see Fig. 8.24(a).

Consider a design compressive strength of masonry = 2000 psi (13.8 MPa).

(a) Use unreinforced masonry construction and design the wall in accordance with the ACI 530/ASCE 5/TMS 402 working stress design method.

(b) Repeat part (a) but with reinforced masonry.

(c) Use reinforced masonry and design the wall in accordance with the UBC strength design of slender walls.

**Solution A:** *Unreinforced.* Try an 8 in. (200 mm) thick wall
Face shell area, $A_n = (1.25 \times 2)\, 12 = 30$ in.²/ft of wall
$(A_n = (32 \times 2)(1000) = 6.4 \times 10^4$ mm²/m)
$I = 12((7.625)^3/12 - 12(5.125)^3)/12 = 308.7$ in.⁴/ft of wall
$(I = 1000\,((190)10^3 - (126)^3)/12 = 4.08 \times 10^8$ mm⁴/m)
Wall weight $= 45 \times 16 = 720$ lb/ft $(2.2 \times 4.8 = 10.56$ kN/m)
Moment at mid-height $= 20(16)^2/8 + 300\,(4/12)/2 = 690$ ft-lb/ft;
$(1.0(4.8)^2/8 + 4.4\,(100/1000)/2 = 3.1$ kN.m/m)
(See Fig. 8.24(b))

*must use ACI for hollow block*    *("it is undeterminate)"*
*for hollow walls)*

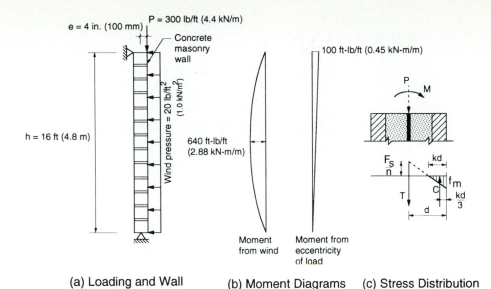

Figure 8.24 Single wythe reinforced masonry wall for Example 8.2.

Checking the flexural tension stress:

$$P \text{ at mid-height} = 720/2 + 300 = 660 \text{ lb/ft}$$

$$(10.56/2 + 4.4 = 9.68 \text{ kN/m})$$

$$f_t = -P/A_n + My/I = -\frac{660}{30} + \frac{690(12)(7.625/2)}{308.7} = +80.3 \text{ psi}$$

$$\left(f_t = -\frac{9.68 \times 10^3}{6.4 \times 10^4} + \frac{3.1 \times 10^6(190/2)}{4.08 \times 10^8} = +0.57 \text{ MPa}\right)$$

The code allows 25 psi (0.16 MPa) of flexural tension, which, with the $\frac{1}{3}$ increase allowed for wind loading, amounts to 33 psi (0.21 MPa). This is much less than the calculated 80.3 psi (0.57 MPa). Therefore, this design is not adequate.

Try a 12 in. (300 mm) thick hollow masonry unit weighing 64 lb/ft² (3.1 kN/m²): Considering face shell areas

$$A_n = (1.5 \times 2)(12) = 36 \text{ in.}^2/\text{ft}$$

$$(A_n = (38 \times 2)(1000) = 7.6 \times 10^4 \text{ mm}^2/\text{m})$$

$$I = 12((11.625)^3 - (8.625)^3)/12 = 929.4 \text{ in.}^4/\text{ft}$$

$$(I = 1000((290)^3 - (214)^3)/12 = 1.22 \times 10^9 \text{ mm}^4/\text{m})$$

$$P \text{ at mid-height} = 64 \times 8 + 300 = 820 \text{ lb/ft}$$

$$(3.1 \times 2.4 + 4.4 = 11.84 \text{ kN/m})$$

$$\therefore f_t = -820/36 + 690(12)(5.8125)/929.4 = 29 \text{ psi}$$

$$\left(f_t = -\frac{11.84 \times 10^3}{7.6 \times 10^4} + \frac{3.1(10)^6(290/2)}{1.22 \times 10^9} = 0.21 \text{ MPa}\right)$$

This tensile stress is allowed. Therefore, the 12 in. (300 mm) unreinforced wall is adequate for tension.

For this section

$$r = \sqrt{\frac{I}{A}} = \sqrt{\frac{929.4}{36}} = 5.08 \text{ in.}$$

$$\left( r = \sqrt{\frac{1.22 \times 10^9}{7.6 \times 10^4}} = 126.7 \text{ mm} \right)$$

Thus $kh/r = 1.0 \times 16 \times 12/5.08 = 37.8$ ($4800/126.7 = 37.9$) which is less than 99 so that,

$$F_a = \frac{1}{4} f'_m \left[ 1 - \left( \frac{kh}{140r} \right)^2 \right] = \frac{2000}{4} \left[ 1 - \left( \frac{37.8}{140} \right)^2 \right] = 464 \text{ psi}$$

$$\left( F_a = \frac{13.8}{4} \left[ 1 - \left( \frac{37.9}{140} \right)^2 \right] = 3.20 \text{ MPa} \right)$$

Using $F_b = f'_m/3 = 667$ psi (4.6 MPa), the unity equation gives

$$\frac{f_a}{F_a} + \frac{f_b}{F_b} = \frac{P/A}{F_a} + \frac{M/S}{F_b} = \frac{820/36}{464} + \frac{690(12)(5.813)/929.4}{667} = 0.127$$

$$\left( \frac{11.84 \times 10^3/(7.6 \times 10^4)}{3.20} + \frac{3.1 \times 10^6(290/2)/1.22 \times 10^9}{4.6} = 0.129 \right)$$

which is well below the 1.33 value permitted where the one third increase in stress for wind loading is included.

It is also required that $P \le 1/4(P_e)$ to provide safety against buckling

$$P_e = \frac{\pi^2 E_m I}{h^2} \left( 1 - 0.577 \frac{e}{r} \right)^3$$

Consider

$$E_m = 1000 f'_m = 1000(2000) = 2.0 \times 10^6 \text{ psi} \qquad (13,800 \text{ MPa})$$

$$r = \sqrt{I/A} = \sqrt{929.4/36} = 5.08 \text{ in.}$$

$$(r = \sqrt{1.22 \times 10^9/7.6 \times 10^4} = 127 \text{ mm})$$

$$P_e = \frac{\pi^2(2.0 \times 10^3)(929.4)}{(16 \times 12)^2} (1 - 0.577(4/5.08))^3 = 82.0 \text{ kips/ft}$$

$$\left( P_e = \frac{\pi^2(13,800)(1.22 \times 10^9)}{(4800)^2(1000)} (1 - 0.577(100/127))^3 = 1189 \text{ kN/m} \right)$$

[Note that the eccentricity used in the above equation is the actual eccentricity of the applied load at the top of the wall.]

Since $P < P_e/4$, the design satisfies code requirements for compression and the 12 in. (300 mm) unreinforced wall is adequate.

**Solution B:** *Reinforced.* Try an 8 in. (200 mm) wall, reinforced with Grade 60 (400 MPa) steel. The wall is partially grouted and reinforced at a 4 ft (1.2 m) spacing.

For a balanced design:

$$\text{Allowable tension in steel} = 24{,}000 \text{ psi (165 MPa)}$$
$$\text{Allowable masonry compression} = \tfrac{1}{3} f_m' = \tfrac{1}{3}(2000)$$
$$= 667 \text{ psi (4.6 MPa)}$$

both of which can be increased by $\tfrac{1}{3}$ for wind loading.
Equilibrium of forces is checked on a 4 ft (1.2m) length of wall (see Fig. 8.24(c)).

Design for tension controlled condition: ACI 530/ASCE 5/TMS 402 limits the effective width to 6t, where t is the actual wall thickness. Therefore, $b = 6(7.625) = 45.75$ in. $(b = 6(190) = 1140$ mm).

$$P \text{ at mid-height} = 55 \times 8 + 300 = 740 \text{ lb/ft}$$
$$(2.7 \times 2.4 + 4.4 = 10.88 \text{ kN/m})$$
$$T = A_s F_s$$
$$C = \frac{F_s}{n}\left(\frac{kd}{d - kd}\right)\frac{bkd}{2} \quad \text{and } n = 15$$

Summing the forces vertically,

$$P = C - T \tag{8.80}$$

$$4 \times 740 = \frac{24{,}000(4/3)}{15}\left(\frac{kd}{7.625/2 - kd}\right)\frac{(45.75)kd}{2} - A_s(24{,}000)(4/3)$$

$$\left(1.2(10.88 \times 10^3) = \frac{165(4/3)}{15}\left(\frac{kd}{95 - kd}\right)\frac{(1140)kd}{2} - A_s(165)(4/3)\right)$$

The sum of the moments about the center of the wall which, in this case, is also about the reinforcement, is

$$M = C(d - kd/3) \tag{8.81}$$

$$4(690)(12) = \frac{24{,}000(4/3)}{15}\left(\frac{kd}{7.625/2 - kd}\right)\frac{45.75\,kd}{2}\left(\frac{7.625}{2} - \frac{kd}{3}\right)$$

$$\left(1.2(3.1 \times 10^6) = \frac{165(4/3)}{15}\left(\frac{kd}{95 - kd}\right)\left(\frac{(1140)kd}{2}\right)\left(95 - \frac{kd}{3}\right)\right)$$

Solving Eq. 8.81 results in $kd = 0.76$ in. $< 1.25$ in. $(kd = 19.5$ mm $< 32$ mm). Therefore, the neutral axis is inside the face shell. From Eq. 8.80

$$A_s = 0.20 \text{ in.}^2/4 \text{ ft } (132 \text{ mm}^2/1.2\text{m})$$

Using one No. 4 bar every 4 ft (15M every 1.2 m, which provides more than required) will be adequate.

Check the compressive stress to determine if the initial assumption of tension controlled design is valid. From similar triangles in Fig. 8.24 (c),

$$\frac{f_m}{F_s/n} = \frac{kd}{d - kd}$$

$$\therefore f_m = \frac{24{,}000(4/3)}{15}\left(\frac{0.76}{3.81 - 0.76}\right) = 532 \text{ psi} < F_b = 667 \text{ psi} \times 1.33 \quad \therefore \text{OK}$$

$$\left(f_m = \frac{165(4/3)}{15}\left(\frac{19.5}{95 - 19.5}\right) = 3.79 \text{ MPa} < 4.6 \text{ MPa} \times 1.33 \quad \therefore \text{OK}\right)$$

Compression does not control the section capacity and the previous design is adequate. If compression did control, Eqs. 8.24 and 8.25 would be used for $C$ and $T$ in Eqs. 8.80 and 8.81.

[Note: In some cases neither the allowable tensile stress in the steel nor the allowable compressive stress in the masonry control. That is, there is a potential for excess capacity. In the analysis of such a case, an unknown value of $f_s$ must be used in place of $F_s$ in Eqs. 8.80 and 8.81. With three unknown values and only two equations, it is necessary to select an amount of reinforcement and check the design. Alternately, interaction diagrams could be used.]

The slenderness effect of reinforced walls is satisfied by checking the axial stress $f_a = P/A = 660/30 = 22$ psi (9680/(1000)(64) = 0.15 MPa) against Eq. 8.68 for

$$\frac{kh}{r} = \frac{16 \times 12}{3.21} = 59.8 \qquad \left(\frac{kh}{r} = \frac{4800}{79.8} = 60.1\right)$$

where $r = \sqrt{I/A} = 3.21$ in. (79.8 mm) from Solution A, Eq. 8.68 gives

$$F_a = \frac{f'_m}{4}\left(1 - \left(\frac{kh}{140r}\right)^2\right) = \frac{2000}{4}\left(1 - \left(\frac{59.8}{140}\right)^2\right) = 409 \text{ psi} > 22 \text{ psi}$$

$$\left(F_a = \frac{13.8}{4}\left(1 - \left(\frac{60.1}{140}\right)^2\right) = 2.81 \text{ MPa} > 0.15 \text{ MPa}\right)$$

Therefore, slenderness does not control.

**Solution C:** *UBC Slender Wall Design.* Use a 6 in. (150 mm) block with the same properties as in Solution A but fully grouted. To use the UBC method, the axial stress due to the unfactored axial load must be $< 0.04 f'_m = 0.04(2000) = 80$ psi (0.04 (13.8) = 0.55MPa)

$$P_0 = 300 \text{ lb/ft} (4.4 \text{ kN/m})$$

$$P_w \text{ at mid span} = 63 \text{ lb/ft}^2 (8 \text{ ft}) = 504 \text{ lb/ft}$$

$$(P_w = 3020 \text{ kN/m}^2 (2.4 \text{ m}) = 7.25 \text{ kN/m})$$

$$f_a = \frac{P_w + P_0}{A_n} = \frac{300 + 540}{(5.625)12} = 12 \text{ psi} \qquad \left(f_a = \frac{4400 + 7250}{1000(140)} = 0.083 \text{ MPa}\right)$$

which satisfies the above condition by being less than $0.04 f'_m$.

The UBC slender wall design method (Sec. 2411) is based on ultimate loads calculated as

$$p_u = 1.7 \times 20 = 34 \text{ lb/ft}^2 (1.7 \text{ kN/m}^2)$$

$$P_{wu} = 1.4 \times 504 = 706 \text{ lb/ft} (10.15 \text{ kN/m})$$

$$P_{ou} = 1.4 \times 300 = 420 \text{ lb/ft} (5.6 \text{ kN/m})$$

$$P_u = P_{wu} + P_{ou} = 706 + 420 = 1126 \text{ lb/ft}$$

$$(P_u = 10.15 + 5.6 = 15.75 \text{ kN/m})$$

To calculate deflections and the nominal bending strength required in the design, it is necessary to choose a trial amount of reinforcement. Try No. 5 bars at 24 in. spacing (15M at 600 mm).

$$\therefore A_s = \frac{0.31 \times 12}{24} = 0.155 \text{ in}^2/\text{ft} \qquad \left( A_s = \frac{200}{0.6} = 333 \text{ mm}^2/\text{m} \right)$$

$$\rho_s = \frac{A_s}{bt} = \frac{0.155}{12(5.625)} = 0.0023$$

$$\left( \rho_s = \frac{333}{1000(140)} = 0.00238 \right)$$

In calculations done according to the UBC, properties of the section are calculated using an effective area of steel $A_{se}$ which includes the effect of axial compression load so that

$$A_{se} = \frac{P_u + A_s f_y}{f_y} = \frac{1126 + 0.155(60,000)}{60,000} = 0.174 \text{ in.}^2/\text{ft}$$

$$\left( A_{se} = \frac{15750 + 333(400)}{400} = 372 \text{ mm}^2/\text{m} \right)$$

Then, the depth of the area under the rectangular stress block is

$$a = \frac{A_{se} f_y}{0.85 f'_m b} = \frac{0.174(60,000)}{0.85(2000)12} = 0.51 \text{ in.}$$

$$\left( a = \frac{372(400)}{0.85(13.8)(1000)} = 12.7 \text{ mm} \right)$$

The distance from the extreme compression fiber to the neutral axis is

$$c = a/0.85 = 0.51/0.85 = 0.60 \text{ in.}$$

$$(c = 12.7/0.85 = 14.9 \text{ mm})$$

Thus, the nominal moment capacity is

$$M_n = A_{se} f_y (d - a/2) = 0.174(60,000)(2.81 - 0.51/2) = 26,675 \text{ in-lb/ft}$$

$$= 2223 \text{ ft-lb/ft}$$

$$(M_n = 372(400)(70 - 12.7/2)/10^6 = 9.471 \text{ kN} \cdot \text{m/m})$$

$$\Phi M_n = 0.8(2223) = 1778 \text{ in.lb/ft} \quad \text{true capacity}$$

$$(\Phi M_n = 0.8(9.471) = 7.58 \text{ kN} \cdot \text{m/m})$$

To determine the factored ultimate moment, the effect of the deflection under the factored loads must be included. Therefore taking the elastic modulus of the masonry as

$$E_m = 750 f'_m = 750 \times 2000 = 1,500,000 \text{ psi}$$

$$(E_m = 750(13.8) = 10,350 \text{ MPa})$$

$$n = E_s/E_m = 29,000,000/1,500,000 = 19.3 \qquad (n = 200,000/10,350 = 19.3)$$

The research used to formulate the UBC slender wall design method[8.19] led to calculation of the moment of inertia of the cracked section $I_{cr}$ using the effective steel area $A_{se}$.

The member is treated as if it is under bending moment without axial load, and the depth to the neutral axis $c$ calculated at ultimate conditions is used. This approach provides an acceptable approximation of the $I_{cr}$ value. Therefore, using the transformed area of steel $nA_{se}$.

$$I_{cr} = nA_{se}(d - c)^2 + \frac{bc^3}{3} = 19.3(0.174)(2.81 - 0.60)^2 + 12(0.60)^3/3 = 17.3 \text{ in}^4/\text{ft}.$$

$$(I_{cr} = 19.3(372)(70 - 14.9)^2 + 1000(14.9)^3/3 = 22.9 \times 10^6 \text{ mm}^4/\text{m})$$

Conservatively, deflections could be calculated using the moment of inertia for the cracked section. However, to judge how conservative this might be, the moment required to produce cracking, $M_{cr}$, can be calculated where the UBC specified modulus of rupture $f_r$ for solidly grouted hollow unit masonry is

$$f_r = 4 \sqrt{f'_m} (\text{psi}) \leq 235 \text{ psi}$$

$$= 4 \sqrt{2000} = 179 \text{ psi} (1.23 \text{ MPa})$$

$$M_{cr} = f_r S_{\text{solid}} = 179(12)(5.625)^2/6 = 11,330 \text{ in.lb/ft} = 994 \text{ ft-lb/ft}.$$

$$(M_{cr} = 1.23(1000)(140)^2/6(10^6) = 4.02 \text{ kN.m/m})$$

To calculate the moment under factored load considering wall deflection $\Delta_u$, an estimate of $\Delta_u$ is needed. Using the cracked moment of inertia and nominal moment, an estimate can be expressed as

$$\Delta_u = \frac{5 M_n h^2}{48 E_m I_{cr}}$$

$$= \frac{5}{48} \frac{26675(16 \times 12)^2}{(1.5 \times 10^6)(17.3)} = 3.9 \text{ in.}$$

$$\left( \Delta_u = \frac{5}{48} \frac{9.471 \times 10^6 (4800)^2}{10,350(22.9 \times 10^6)} = 96 \text{ mm} \right)$$

At mid-height, the estimated moment due to factored loads, including an estimated mid-height deflection of 3.9 in. (75 mm) can be calculated from Eq. 8.72

$$M_{u_{\text{estimated}}} = \frac{p_u h^2}{8} + P_{ou} \frac{e}{2} + (P_{wu} + P_{ou}) \Delta_{u_{\text{estimated}}}$$

$$= \frac{34(16)^2(12)}{8} + 420 \left( \frac{4}{2} \right) + (706 + 420)(3.9) = 18,287 \text{ in.-lb/ft}$$

$$= 1524 \text{ ft-lb/ft}$$

$$\left( M_{u_{\text{estimated}}} = \frac{1.7(4.8)^2}{8} + 5.6 \left( \frac{100}{2 \times 1000} \right) \right.$$

$$\left. + (10.15 + 5.6) \left( \frac{96}{1000} \right) = 6.69 \text{ kN} \cdot \text{m/m} \right)$$

Because these moments are not several times more than the cracking moment, it is worthwhile weighting the effects of the uncracked and cracked moments of inertia which is in fact what is done in Eq. 8.76. Conservatively, because the cracking moment is more than half of the ultimate moment for this problem, they can be weighted equally and the effective moment of inertia $I_{\text{eff}}$ can be calculated where the moment of inertia of the uncracked section $I_g$ is

$$I_g = \frac{bt^3}{12} = \frac{12(5.625)^3}{12} = 178 \text{ in.}^4/\text{ft}$$

$$\left(I_g = \frac{1000(140)^3}{12} = 2.29 \times 10^8 \text{ mm}^4/\text{m}\right)$$

so that,

$$\frac{1}{I_{\text{eff}}} = \frac{1}{2}\left(\frac{I}{I_{cr}} + \frac{1}{I_g}\right) = \frac{1}{2}\left(\frac{1}{17.3} + \frac{1}{178}\right)$$

$$I_{\text{eff}} = 31.5 \text{ in.}^4/\text{ft}$$

$$(I_{\text{eff}} = 41.8 \times 10^6 \text{ mm}^4/\text{m})$$

The total deflection can be calculated using Eq. 8.78 where

$$P_u = P_{wu} + P_{ou} = 706 + 420 = 1126 \text{ lb/ft}$$

$$(P_u = 10.15 + 5.6 = 15.75 \text{ kN/m})$$

$$P_c = \frac{\pi^2 E_m I_{\text{eff}}}{h^2} = \frac{\pi^2(1.5 \times 10^6)(31.5)}{(16 \times 12)^2} = 12{,}560 \text{ lb/ft}$$

$$\left(P_c = \frac{\pi^2(10{,}350)(41.8 \times 10^6)}{(4800)^2(10)^3} = 185.3 \text{ kN/m}\right)$$

$$\Delta_u = \left(\frac{5}{384}\frac{p_u h^4}{E_m I_{\text{eff}}} + \frac{1}{16}\frac{P_{ou} e h^2}{E_m I_{\text{eff}}}\right)\frac{1}{1 - P_u/P_c}$$

$$= \left(\frac{5}{384}\frac{(34)(16)^4(12)^3}{(1.5 \times 10^6)(31.5)} + \frac{1}{16}\frac{(420)(4)(16)^2(12)^2}{(1.5 \times 10^6)(31.5)}\right)\frac{1}{1 - (1126/12{,}650)}$$

$$= 1.14 \text{ in. } (1.10) = 1.25 \text{ in.}$$

$$\left(\Delta_u = \left(\frac{5}{384}\frac{(1.7)(4.8)^4 10^{12}}{10{,}350(41.8 \times 10^6)} + \frac{1}{16}\frac{5600(100)(4800)^2}{(10{,}350)(41.8 \times 10^6)}\right)\frac{1}{1 - (15.75/185.3)}\right)$$

$$= 29.0 \text{ mm } (1.09) = 31.6 \text{ mm}\bigg)$$

This is considerably less than the estimated value of 3.9 in. (96 mm) used above and therefore the estimate of $I_{\text{eff}}$ is even more conservative.

The factored ultimate moment from Eq. 8.72 is

$$M_u = \frac{p_u h^2}{8} + P_{ou}\frac{e}{2} + (P_{wu} + P_{ou})\Delta_u$$

$$= \frac{34(16)^2 \times 12}{8} + 420(4/2) + (706 + 420)(1.25) = 15304 \text{ in.-lb/ft}$$

$$= 1293 \text{ ft.-lb/ft}$$

$$\left(M_u = \frac{1.7(4.8)^2}{8} + 5.6\left(\frac{100}{2 \times 10^3}\right) + (10.15 + 5.6)\frac{31.6}{1000} = 5.70 \text{ kN.m/m}\right)$$

Because $M_u$ is less than $\Phi M_n$, the strength criteria is satisfied and, in fact, a slightly larger spacing of the reinforcement could be considered.

At service loads, the maximum deflection permitted is

$$\Delta_{s_{\text{limit}}} = 0.007h = 0.007(16 \times 12) = 1.34 \text{ in.}$$

$$(\Delta_{s_{\text{limit}}} = 0.007(4800) = 33.6 \text{ mm})$$

The moment at service load $M_s$ can be calculated using Eq. 8.72 but using unfactored loads and the deflection at service loads so that

$$M_s = \frac{ph^2}{8} + P_0 e/2 + (P_w + P_0)\Delta_s$$

where we may conservatively take

$$\Delta_s = \Delta_{s_{\text{limit}}} = 0.007h = 1.34 \text{ in. (33.6 mm)}$$

$$\therefore M_s = \frac{20(16)^2}{8} \times 12 + 300(4/2) + (504 + 300)(1.34) = 9357 \text{ in.lb/ft} = 780 \text{ ft.lb/ft.}$$

$$\left( M_s = \frac{1(4.8)^2}{8} + 4.4 \left(\frac{0.10}{2}\right) + (7.25 + 4.4)(0.0336) = 3.49 \text{ kN.m/m} \right)$$

Since $M_s$ is less than $M_{cr}$ previously calculated, the wall will not crack at service loads. From Eq. 8.75

$$\Delta_s = \frac{5}{48} \frac{M_s h^2}{E_m I_g} = \frac{5}{48} \frac{(9357)(16 \times 12)^2}{(1.5 \times 10^6)(178)} = 0.13 \text{ in.} < 1.34 \text{ in.}$$

$$\left( \Delta_s = \frac{5}{48} \frac{(3.49 \times 10^6)(4800)^2}{(10,350)(2.29 \times 10^8)} = 3.5 \text{ mm} < 33.6 \text{ mm} \right)$$

Therefore, the deflection criterion is satisfied.

It is interesting to note that the UBC strength design method allows use of a much thinner wall which does not crack under the service load because of the high modulus of rupture specified for fully grouted walls. ∎

### 8.6.3 Allowable Load for a Composite Wall

**Example 8.3**

Determine the maximum allowable load $P$ for the composite wall shown in Fig. 8.25. The wall is 15 ft (4.5 m) high and is simply supported at top and bottom. Type S mortar is used. The brick has a compressive strength of 10,000 psi (70 MPa) and the block has a compressive strength of 3,000 psi (20 MPa). The axial load is applied at the centroid of the ungrouted block. Ignore the self-weight of the wall. Use the ACI 530/ASCE 5/ TMS 402 design method and consider the allowable flexural bond equal to 36 psi (0.25 MPa).

**Solution.** For 10,000 psi (70 MPa) clay brick and type S mortar, consider a design compressive strength of 2900 psi (20 MPa) with $E_{m \text{ brick}} = 2.2 \times 10^6$ psi (15,000 MPa).

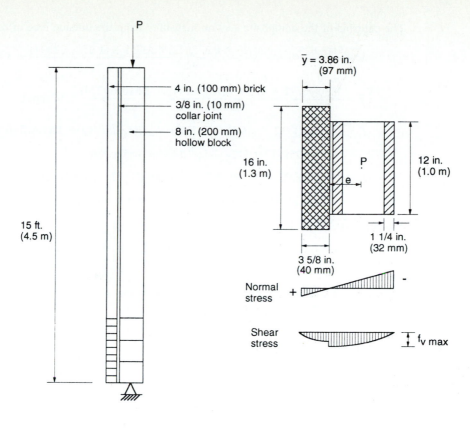

**(a) Wall Geometry**

**(b) Stress distribution**

**Figure 8.25** Composite masonry wall for Example 8.3.

For 3,000 psi (20 MPa) concrete block and type S mortar, consider a design compressive strength of 2200 psi (15 MPa) with $E_{m\ block} = 1.65 \times 10^6$ psi (11,250 MPa). The modular ratio for the brick wythe compared to the block is

$$n = E_{m\ brick}/E_{m\ block} = 1.33$$

Therefore, transform the brick into an equivalent length of block and calculate the section properties. Consider the vertical mortar joint to have the same properties as the block masonry.

The transformed length of the brick is

$$1.33 \times 12 = 16.0 \text{ in./ft (1330 mm/m)}$$

The transformed area of the section is

$$A_t = 3.625 \times 16.0 + (1.25 \times 2 + 0.375)(12) = 92.5 \text{ in.}^2/\text{ft}$$

$$(A_t = 90\,(1330) + (32 \times 2 + 10)\,1000 = 1.94 \times 10^5 \text{ mm}^2/\text{m})$$

The centroid of the composite section measured from the outside face of the brick is

$$\bar{y} = \frac{16(3.625)^2/2 + (1.25 \times 0.375)(12)(4.4375) + (1.25 \times 12)(11)}{92.5} = 3.86 \text{ in.}$$

$$\left( \bar{y} = \frac{1330(90)^2/2 + (32 + 10)1000(111) + 32(1000)(274)}{1.94 \times 10^5} = 97 \text{ mm} \right)$$

$$I_t = 16(3.625)^3/12 + (16)(3.625)(2.05)^2 + 12(1.625)^3/12 + 12(1.625)(0.58)^2$$
$$+ 12(1.25)^3/12 + (12)(1.25)(7.14)^2 = 1085 \text{ in.}^4/\text{ft}$$

$$(I_t = 1330(90)^3/12 + 1330(90)(52)^2 + 1000(42)^3/12 + 1000(42)(14)^2$$
$$+ 1000(32)^3/12 + 1000(32)(177)^2 = 1.424 \times 10^9 \text{ mm}^4/\text{m})$$

Eccentricity of load is

$$e = (3.625 + 0.375 + 7.625/2) - 3.86 = 3.95 \text{ in.}$$

$$(e = 90 + 10 + 190/2 - 97 = 98 \text{ mm})$$

If the allowable tension $F_t$ controls in the brick wythe, $F_t = 36/n = 27$ psi (0.188 MPa) for the transformed brick. Hence, from

$$F_t = -P/A + My/I = -P/A + Pey/I$$

$$27 = -P/92.5 + P(3.95)(3.86)/1085$$

$$\therefore P = 8333 \text{ lb/ft}$$

$$(0.188 = -P/1.94 \times 10^5 + P(98)(97)/1.424 \times 10^9 \qquad \therefore P = 123.6 \text{ kN/m})$$

If compression controls in the block wythe

$$F_a = \frac{1}{4}f_m' \left[ 1 - \left( \frac{h}{140r} \right)^2 \right]$$

where

$$r = \sqrt{I/A} = \sqrt{\frac{1085}{92.5}} = 3.42 \text{ in.}$$

$$\left( r = \sqrt{\frac{1.424 \times 10^9}{1.94 \times 10^5}} = 85.7 \text{ mm} \right)$$

and

$$F_a = \frac{1}{4}(2200) \left[ 1 - \left( \frac{15 \times 12}{140 \times 3.42} \right)^2 \right] = 472 \text{ psi}$$

$$\left( F_a = \frac{1}{4}(15) \left( 1 - \left( \frac{4500}{140(85.7)} \right)^2 \right) = 3.22 \text{ MPa} \right)$$

$$F_b = \frac{1}{3}f_m' = \frac{1}{3}(2200) = 733 \text{ psi}$$

$$\left( F_b = \frac{15}{3} = 5 \text{ MPa} \right)$$

Using the unity equation

$$\frac{f_a}{F_a} + \frac{f_b}{F_b} = \frac{P/92.5}{472} + \frac{P(3.95)7.77/1085}{733} = 1.0$$

$$\therefore P = 16,260 \text{ lb/ft}$$

$$\left( \frac{P/1.94 \times 10^5}{3.22} + \frac{P(98)(193)/1.42 \times 10^9}{5.0} = 1.0 \quad \therefore P = 234.5 \text{ kN/m} \right)$$

The wall capacity is 8333 lb/ft (123.6 kN/m) and is controlled by the maximum allowable tensile stress developed in the brick wythe due to the eccentricity of the load measured from the centroid of the transformed section of the composite wall. ∎

### 8.6.4 Axial Load and Bending of a Composite Wall

**Example 8.4**

For the composite wall described in Example 8.3, calculate the maximum design wind pressure that the wall can carry in addition to an axial load $P$ of 3.0 kips/ft (44 kN/m) acting in the center of the block wythe. Consider the wall simply supported at the top and bottom and ignore self-weight. Use the ACI 530/ASCE 5/TMS 402 design method.
**Solution:** Use the transformed section properties calculated in Example 8.3.

$$A_t = 92.5 \text{ in.}^2/\text{ft} \quad (1.94 \times 10^5 \text{ mm}^4/\text{m})$$

$$I_t = 1085 \text{ in.}^2/\text{ft} \quad (1.424 \times 10^9 \text{ mm}^4/\text{m})$$

$$\bar{y} = 3.86 \text{ in.} \quad (97 \text{ mm})$$

$$e = 3.95 \text{ in.} \quad (98 \text{ mm})$$

Maximum moment at mid-height (critical for wind loading) with $p$ in lb/ft$^2$ (kN/m$^2$)

$$M = P \cdot \frac{e}{2} + \frac{ph^2}{8} = 3000 \left( \frac{3.95}{2} \right) \frac{1}{12} + \frac{p(15)^2}{8} = (493.8 + 28.1 \, p) \text{ ft-lb/ft}$$

$$\left( M = 44 \frac{(98)}{2} \frac{1}{1000} + p \frac{(4.5)^2}{8} = (2.16 + 2.53 \, p) \text{ kN.m/m} \right)$$

The addition of the two moments implies a suction or outward wind pressure which is the governing case if tension controls on the brick face. If the allowable tensile stress is increased by one third for wind loading,

$$\frac{4}{3} F_t/n = \frac{48}{n} = \frac{48}{1.33} = -\frac{P}{A_t} + \frac{My}{I_t}$$

$$36 = -\frac{3000}{92.5} + \frac{[(493.8 + 28.1p)12](3.86)}{1085}$$

$$\therefore p = 39.5 \text{ lb/ft}^2$$

$$\left( \frac{0.25(4/3)}{1.33} = -\frac{44,000}{1.94 \times 10^5} + \frac{(2.16 + 2.53p)10^6(97)}{1.424 \times 10^9} \quad \therefore p = 1.91 \text{ kN/m}^2 \right)$$

For positive wind pressure on the wall, the maximum moment to cause tension on the inside face of the concrete masonry wythe is

$$M = -\frac{P_e}{2} + \frac{ph^2}{8} = -493.8 + 28.1p$$

$$(M = -2.16 + 2.53p)$$

and recalculating as above but with $y = 7.77$ in. (193 mm).

$$36 = -\frac{P}{A_t} + \frac{M_y}{I} = -\frac{3000}{92.5} + \frac{(-493.8 + 28.1p)12(7.77)}{1085}$$

$$p = 45.9 \text{ lb/ft}^2 > 39.5 \text{ lb/ft}^2$$

$$\left( 0.25 = -\frac{44,000}{1.94 \times 10^5} + \frac{(-2.16 + 2.53p)10^6(193)}{1.424 \times 10^9} \right.$$

$$\left. p = 2.24 \text{ kN/m}^2 > 1.91 \text{ kN/m}^2 \right)$$

Therefore tension on the inside face of the block wythe does not control the capacity.

If compression in the concrete masonry controls, the unity equation, multiplied by 4/3 to allow for wind, can be used.

$$\frac{f_a}{F_a} + \frac{f_b}{F_b} = 1.33$$

The following values are substituted

$$f_a = P/A_t = 3000/92.5 = 32.4 \text{ psi}$$

$$(f_a = 44000/1.94 \times 10^5 = 0.23 \text{ MPa})$$

$$f_b = \frac{My}{I} = \frac{(493.8 + 28.1p)12(11.63 - 3.86)}{1085}$$

$$\left( f_b = \frac{(2.16 + 2.53p)10^6(290 - 97)}{1.424 \times 10^9} \right)$$

$$F_a = 472 \text{ psi (3.22 MPa)} \qquad F_b = 733 \text{ psi (5 MPa) from Example 8.3}$$

$$\therefore p = 365 \text{ lb/ft}^2 \text{ (15.7 kN/m}^2\text{) which clearly does not control}$$

$\therefore$ Tension controls and the maximum allowable design wind pressure is 39.5 lb/ft$^2$ (1.91 kN/m$^2$).

Check the maximum shear stress, $f_{max}$, at the interface between wythes (Fig. 8.25)

$$f_{v,max} = \frac{VQ}{I_t b}$$

$$V_{\text{shear at support}} = ph/2 = 39.5 (15/2) = 296 \text{ lb/ft}$$

$$\left( V = 1.91 \left( \frac{4.5}{2} \right) = 4.30 \text{ kN/m} \right)$$

$$f_v = \frac{296(3.625)(16.0)(2.05) + (0.23(12)(0.12))}{1085(12)} = 2.71 \text{ psi}$$

$$\left( f_v = \frac{4300((90)(1330)(52) + 7(1000)(3.5))}{1.424 \times 10^9 \times 1000} = 0.019 \text{ MPa} \right)$$

which is less than the code allowable stress of 5 psi (0.034 MPa) for a mortar collar joint. Therefore, the design is adequate for shear. ∎

## 8.7 CLOSURE

In this chapter, the fundamental mechanics of section behavior for unreinforced and reinforced sections are presented for the combined actions of axial load and bending moment. Analyses for both the working stress and the ultimate strength approaches are included. An important observation is the differences in the approaches taken by various codes to account for slenderness. This factor plus the degree to which tensile stress or cracking is allowed can have significant effects on the design of masonry walls. Use of the strength (ultimate strength) method of design should result in more consistent designs for the full range of load combinations.

The basic approach to design is applicable to combined axial load and bending moment for either in-plane or out-of-plane bending. Special features of design for in-plane bending, such as occurs in shear walls, are covered in Chap. 10. The design of columns and pilasters covered in Chap. 9 also follows the basic principles presented in this chapter.

## 8.8 REFERENCES

8.1  The Masonry Joint Committee, "Building Code Requirements for Masonry Structures," ACI 530/ASCE 5/TMS 402, American Concrete Institute and American Society of Civil Engineers, Detroit/New York, 1992.

8.2  International Conference on Building Officials, "Masonry Codes and Specifications," 1991 UBC, Chapter 24, ICBO, Whittier, CA, 1991.

8.3  Canadian Standards Association, "Masonry Design and Construction for Buildings," CSA Standard S 304-1984, CSA, Rexdale, Ontario, 1984.

8.4  British Standards Institution, "Code of Practice for Structural Use of Masonry: BS 5628," BSI, London, 1978 (confirmed April 1985).

8.5  Standard Association of Australia, "Australian Standard 3700-1988: Masonry in Buildings," SAA, Sydney, NSW, 1988.

8.6  Standard Association of New Zealand, "Code of Practice for the Design of Masonry Structures," NZS 4230 Part 1: 1990, SANZ, Wellington.

8.7  R.G. Drysdale and A.A. Hamid, "Effect of Eccentricity on the Compressive Strength of Brickwork," Journal of the British Ceramic Society, No. 30, 1982, 140-149.

8.8  R.G. Drysdale and A.A. Hamid, "Capacity of Concrete Block Masonry Prisms Under Eccentric Compressive Loading," ACI Journal, Vol. 80, No. 11, March–April 1983, 102–108.

8.9  C. Yao, "Failure Mechanism of Concrete Masonry," Ph.D. Thesis, the University of British Columbia, Vancouver, Canada, Jan. 1989.

8.10 P. Guo, "Investigation and Modelling of the Mechanical Properties of Masonry," Ph.D. thesis, McMaster University, Ontario, Canada, Feb. 1991.

8.11 A.H.P. Maurenbrecher, "Compressive Strength of Eccentrically Loaded Masonry Prisms," in Proceedings of the Third Canadian Masonry Symposium, Edmonton, Canada, 1983, pp. 10-1–10-13.

8.12 Brick Institute of America, "Recommended Practice for Engineering Brick Masonry," BIA, McLean, VA, Nov. 1969.

8.13 National Concrete Masonry Association, "Specification for the Design and Construction of Load-Bearing Concrete Masonry," NCMA, Herndon, VA, February, 1987.

8.14 S. Hasan and A. Hendry, "Effect of Slenderness and Eccentricity of the Compressive Strength of Walls," in Proceedings of The Fourth International Brick Masonry Conference, Brugge, Belgium, April 1976, pp. 4d.3–4d3.9.

8.15 A. Hamid, G. Assis, and H. Harris, "Towards Developing A Flexural Strength Design Methodology for Concrete Masonry," ASTM STP 1063, J. Matthys, ed., ASTM, Philadelphia, 1991.

8.16 J. MacGregor, Reinforced Concrete–Mechanics and Design, Prentice Hall, Englewood Cliffs, NJ, 1988.

8.17 M. Hatzinikolas, J. Longworth and J. Warwaruk, "Concrete Masonry Walls," Report No. 70, Department of Civil Engineering, University of Alberta, Edmonton, 1978.

8.18 P. Suwalski and R. Drysdale, "Influence of Slenderness on the Capacity of Concrete Block Walls," in Proceedings of the Fourth Canadian Masonry Symposium, Fredericton, New Brunswick, 1986, pp. 122–136.

8.19 J. Amrhein, L. Selna, W. Simpson, and R. Tobin, "Slender Masonry Wall Test," in Proceedings of the Third Canadian Masonry Symposium, Edmonton, Canada, 1983, pp. 11/1–11/14.

## 8.9 PROBLEMS

**8.1** Review the different methods used to determine design loads for combined axial load and bending on masonry walls. Comment on the differences and on the relative conservativeness of the methods.

**8.2 (a)** For an 8 in. (200 mm) nominal concrete block wall reinforced with No. 5 bars (15M) at 24 in. (600 mm) spacing, develop the interaction diagram for allowable moment and axial load using ACI 530/ASCE 5/TMS 402 or the code applicable to your area. Change the percentage of steel and study the effect on the interaction diagram by using the same reinforcement, but change the spacing to 8 in. (200 mm) and then to 72 in. (1.8 m). Consider the compressive strength to be 2000 psi (15 MPa) and $E_m = 1.5 \times 10^6$ psi (11,250 MPa). Use Grade 60 (400 MPa) steel. Consider the wall to be fully grouted.

**(b)** Using an ultimate strength approach, develop a new interaction diagram for the previous case with reinforcement spaced at 24 in. (600 mm).

**8.3 (a)** Develop the interaction diagram for working stress design for a 6 in. (150 mm) solid brick wall section under axial load and bending moment (similar to the plot in Fig. 8.18). Assume a specified compressive strength $f'_m$ of 3000 psi (20 MPa). Show the effect of changing the allowable tension from 10 psi (0.07 MPa) to 36 psi (0.25 MPa). Use the ACI 530/ASCE 5/TMS 402 design method.

**(b)** Using a rectangular stress block and a tensile strength of 100 psi (0.7 MPa), develop the interaction diagram for ultimate strength conditions for the wall section described in part (a). Compare and comment on the safety factor for the working stress design.

**(c)** Repeat part (a) using the local masonry code if it is different from ACI 530/ASCE 5/TMS 402.

**8.4 (a)** Design an unreinforced composite block-brick masonry wall to carry a superimposed axial load of 4 kips/ft (58 kN/m) and a lateral wind load of 25 lb/ft² (1.25 kN/m²). The wall is 20 ft. (6.1 m) high and is considered to have pin supports at top and bottom. Use ACI 530/ASCE 5/TMS 402. The brick and block compressive strengths are 8000 psi (55 MPa) and 2000 psi (14 MPa), respectively. Type S mortar is used and the collar joint between the brick and block wythes is filled with mortar.

**(b)** Complete part a) using the local masonry code.

**8.5** If the section described in Prob. 8.3(a) is made with hollow clay units and fully grouted with No. 6 bars at 32 in. (20 M at 1.0m) using Grade 60 (400 MPa) steel, determine what lateral wind pressure can be carried for an 18 ft (5.5 m) high simply supported wall that carries a concentric axial load of 800 lb/ft (12 kN/m). If the axial load is placed at an eccentricity of 1.5 in. (38 mm), how much less wind pressure can be resisted. Use the design procedure for tall walls specified in the UBC code.

**8.6** A wall made up of two wythes of 4 in. (100 mm) (nominal) brick is made into a solid wall by filling the $2\frac{3}{4}$ in. (70 mm) cavity with grout. Ladder-type joint reinforcing at 16 in. (400 mm) vertical spacing is used to tie the wythes together. If 10,000 psi (70 MPa) brick, type S mortar, and Grade 60 (400 MPa) reinforcement are used determine the following:

**(a)** For a ground floor 14 ft. (4.2 m) high wall of a multistory building, determine the allowable axial load if the eccentricity of loading is 0.1t.

**(b)** If this wall is 24 ft (7.2 m) high, pinned at the top, and rests on a strip footing, determine what reinforcement is required for a wind load of 20 lb/ft² (1 kN/m) and an axial load of 2000 lb/ft (30 kN/m) applied at an eccentricity of 2.5 in. (63 mm) at the top of the wall.

**8.7** Repeat the design for the case described in Prob. 8.6(b) using concrete block walls with a block strength of 3000 psi and type S mortar. Try the following options using ACI 530/ASCE 5/TMS 402 and/or your local code.

**(a)** Unreinforced masonry (consider single or multiwythe, hollow or solid).

**(b)** Unreinforced 8 in. (200 mm) blocks in combination with pilasters used to transfer load vertically.

**(c)** Reinforced masonry.

**(d)** Reinforced masonry designed by the ultimate strength method using the UBC code.

**(e)** Compare and comment on the different designs and prepare a cost comparison using local prices.

# CHAPTER 9

# Columns and Pilasters

Reinforced pilaster (*Courtesy of National Concrete Masonry Association*).

## 9.1 INTRODUCTION

A *masonry column* is a vertical compression member with a width-to-thickness ratio (*b*/*t*) less than or equal to three[9.1] usually built as a separate supporting member. Where it is contained within a wall, it is commonly referred to as a *pier*; see Fig. 9.1(a). This pier can be enlarged in the out-of-plane direction of the wall for added strength and stability. If a column is built integrally with the wall and interacts with the wall to resist an out-of-plane lateral load, it is called a *pilaster*. A *buttress* is a pilaster with a cross-section that increases from top to bottom.[9.2] The section is enlarged toward the base to resist an increasing moment due to cantilever action. Traditionally, buttresses have been made large enough so that the thrust line remains within the kern distance (middle third of the section). This keeps the entire section in compression so that, typically, there is no need to reinforce buttresses. Information on analysis of thrust lines is covered in Chaps. 1 and 2.

A pilaster within a wall can project from either or both faces; see Fig. 9.1(b). It serves to carry vertical concentrated load and to stiffen the wall against lateral buckling. In addition, the wall can span horizontally between pilasters to carry lateral

**396**

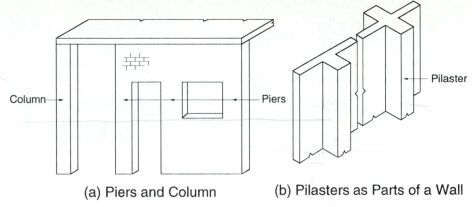

| (a) Piers and Column | (b) Pilasters as Parts of a Wall |

**Figure 9.1**  Columns and Pilasters.

loads. In this case, the pilaster must be designed to span vertically and carry these loads by bending in the vertical direction.

It is important to note that, because pilasters are usually defined by a thickened section of the wall, they constitute a comparatively stiff region for out-of-plane bending. Therefore, they tend to attract bending moments from the attached wall segments which are less stiff. For vertical bending, the wall can share in carrying the load to the extent that it forms part of the pilaster as flanges.

Because a pilaster is a change in wall section, cracking due to thermal and moisture movement will tend to occur at the intersection with the wall. Hence, depending on pilaster spacing, it is common to locate movement joints at or near the intersection of the wall and pilaster. This has obvious implications for horizontal spanning of the wall and for the wall acting as a flange of the pilaster. Usually some connectors are required to transfer out-of-plane shear from the wall to the pilaster over its height.

Columns support vertical gravity loads and can have added axial forces due to overturning effects. Pilasters may be required to perform similar functions. Bending from lateral loads, possible frame action, and eccentricity of vertical loads must also be resisted. Bending in columns can occur about their two principal axes. In such cases, they should be designed to carry biaxial bending combined with axial load.

Columns and pilasters can be unreinforced or reinforced. However, due to the high vulnerability of columns and their structural significance to overall building stability, inclusion of reinforcement is recommended, and ACI 530/ASCE 5/TMS 402[9.1] requires that all masonry columns be reinforced with at least four bars with a minimum percentage of steel of 0.25% and a maximum of 4%. Most codes have minimum limits in the 0.25-to-0.5% range as compared to the 1% minimum for reinforced concrete.

Figure 9.2 shows typical details and reinforcement for masonry columns and pilasters. Columns and pilasters can be constructed using standard blocks and bricks or by using special pilaster or chimney units. Whereas conventional practice has been to allow ties to be placed in the mortar joints away from direct contact with vertical reinforcement, the more recent trend is to require ties to be directly in

$\rho_{min} = .25\%$

$\rho_{max} = 4\%$

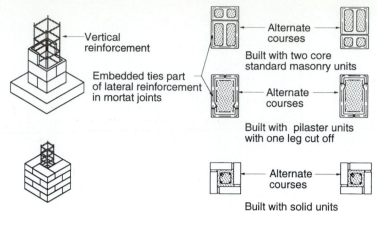

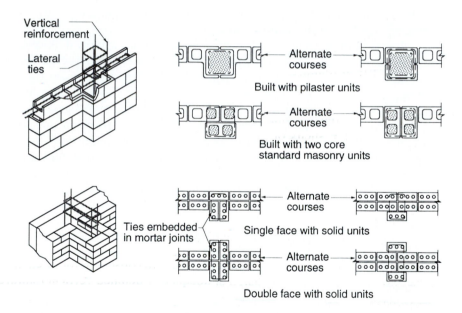

(a) Columns

(b) Pilasters

**Figure 9.2** Column and Pilaster Details.

contact with this steel. This helps prevent buckling of longitudinal steel and results in a more ductile behavior.

For both pilasters and columns, the units in successive courses should be arranged in a running bond to avoid continuous head joints.[9.2–9.4] For pilasters where the adjacent wall sections are intended to form part of the pilaster, connection by maintaining this overlapping at the intersection is preferred. Typical arrangements for alternate courses are illustrated in Fig. 9.2(b).

## 9.2 COLUMN BEHAVIOR

### 9.2.1 Introduction

Compared to reinforced concrete, relatively few experimental studies have been conducted to document the behavior of masonry columns under axial load and combined axial load and bending. Therefore, where information is lacking, behavior of masonry walls and, to a lesser extent, reinforced concrete columns can be used as background for the development of rational column design methods.

### 9.2.2 Failure Modes and Compressive Strength

Experimental research[9.5,9.6] shows that masonry columns fail in one of three modes:

- overall vertical splitting and crushing of the masonry shell and the grout core, which is characteristic of unreinforced columns,
- simultaneous splitting of the masonry shell, crushing of the grout core, and buckling of vertical reinforcement between ties, and
- same as the second mode, but with pulling out of lateral tie hooks and buckling of vertical reinforcement over two or more courses.

The mode of failure is different for unreinforced and reinforced columns and is influenced by the detailing of the lateral reinforcement. As was discussed in Chap. 5 for prisms, the shells of grouted masonry columns fail before they attain their ungrouted strength. The type of embedment of lateral ties significantly affects the failure modes and strengths of masonry columns.[9.5–9.7] The failure mode changed from vertical splitting to spalling of the outer masonry shell when lateral ties were embedded in the grout. Figure 9.3 shows typical failure of reinforced masonry columns.

Experimental work[9.7] on concentrically loaded reinforced block masonry columns shows that column behavior was essentially elastic for loads up to about 75 percent of the ultimate load. Similar behavior was observed for brick masonry under eccentric loading.[9.8] The inclusion of adequately detailed vertical reinforcement and

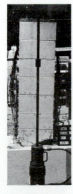

a) Test under cyclic lateral load

b) Full capacity retained at large lateral displacement

**Figure 9.3**  Failure of axially loaded reinforced masonry columns. (*Courtesy N. Cook*).

lateral ties helped to increase the amount of post-ultimate vertical deformation that the columns could accommodate before failure.

### 9.2.3 Slenderness Effect

As discussed in Chap. 8, slenderness can affect capacity either as a result of buckling or because of additional bending moments due to the deflection (i.e., $P–\Delta$ effect). In the latter case, material failure occurs at a lower axial load because of the increased bending moment. As discussed in Chap. 8, the complexity of the factors affecting reductions in capacity due to slenderness have led to quite different approaches being adopted by various design codes. Traditionally, tabulated or calculated capacity reduction factors have been applied to the axial load carrying part of the capacity calculation.

For slender columns, secondary moments due to the $P–\Delta$ effect can be directly accounted for in determining section capacity. The magnitude of the secondary moment is a function of the level of axial load in relation to the critical buckling load, which is dependent upon the modulus of elasticity, the effective moment of inertia, and the effective height of the column. The CSA code[9.9] adopted a moment magnification procedure similar to that for reinforced concrete as the basis for calculation of secondary moment effects. This method is demonstrated in Section 8.6 for walls.

In ACI 530/ASCE 5/TMS 402,[9.1] slenderness effects have been included in the calculation of the allowable compressive stress and the possibility of significant effects of slenderness on unreinforced masonry have been guarded against by limiting the slenderness and stipulating a maximum axial load based on the Euler buckling equation. Slenderness effects can be expressed as a function of $kh/r$. The value $k$ is a coefficient defining effective height and depends upon the end conditions, or shape of the moment diagram over the clear height of the column, $h$. The radius of gyration of the cross-section is $r = \sqrt{I/A}$. Figure 9.4 illustrates the load-carrying

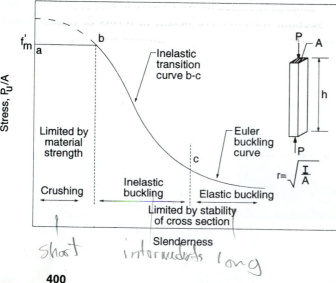

**Figure 9.4** Effect of slenderness on compression capacity of masonry columns.

capacity of a column versus $kh/r$ as expressed by the inelastic and elastic buckling relationships. As shown, there are three limit stages:

- for low $kh/r$ (short columns), capacity is limited by material strength,
- for high $kh/r$ (long, slender columns), failure or column stability is controlled by elastic buckling, and
- for intermediate $kh/r$, inelastic buckling controls column capacity.

## 9.3 COLUMN DESIGN

### 9.3.1 Design Considerations

Column design requires consideration of several limiting factors, which are briefly discussed in what follows.

**Minimum Eccentricity.**  Columns are rarely subjected to perfectly concentric axial loads. Eccentricity due to imperfections, being out of plumb, continuity of the structure, lateral loads, and eccentrically applied axial loads must be considered in design. Many masonry codes[9.1,9.9] require a $0.1t$ minimum eccentricity about each axis of the column, where $t$ is the side dimension. Minimum values as low as $0.05t$ have been specified for unreinforced masonry.[9.10,9.11] Figure 9.5 shows the minimum design eccentricity for a column as required by ACI 530/ASCE 5/TMS 402.[9.1] When the actual eccentricity exceeds the minimum eccentricity required by the code, the actual eccentricity is used.

**Effective Height.**  The *effective height* of a column is the distance between points of inflection. Because slenderness, which is proportional to the effective column height, can have a large effect on column capacity, correct choice of this value is very important. Effective height depends on the boundary conditions at the top and bottom of the column. Figure 9.6 illustrates effective height recommendations for various restraint conditions as specified by ACI 530/ASCE 5/TMS 402.[9.1] Alternately, other codes[9.10,9.11] account for effective height in terms of ratios of end eccentricities. In cases where minimum eccentricities control or where there is doubt

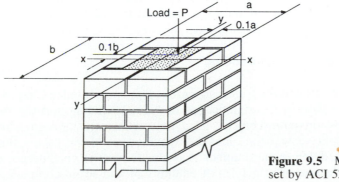

**Figure 9.5**  Minimum design eccentricity set by ACI 530/ASCE 5/TMS 402.

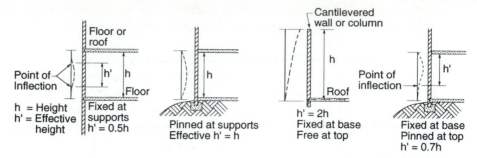

**Figure 9.6**  Effective height of masonry columns (from Ref. 9.1).

concerning the restraint at the ends of columns, use of an effective height equal to the clear height between floors is usually recommended.

**Minimum Dimension.**  Because of the structural importance of columns and their vulnerability as isolated members, many codes specify an 8 in. (200 mm) nominal minimum dimension.  In high seismic risk areas, a minimum dimension of 12 in. (300 mm) is specified in ACI 530/ASCE 5/TMS 402.[9.1]

**Maximum _h/t_ ratio.**  Some codes set maximum limits of effective height to minimum thickness, _h/t_. The ACI 530/ASCE 5/TMS 402[9.1] and CSA[9.9] limit is 25, whereas the Australian[9.10] and British[9.11] codes limit _h/t_ to 27.  If the _h/t_ ratio exceeds the set limit, the design provisions are deemed not to be applicable and lacking provisions similar to the UBC formulations discussed in Sec. 8.5.5 for slender walls, design of more slender columns may not be possible.  Alternately, where permitted, design based on fundamental principles of structural mechanics could be used to directly account for secondary moment due to deflections.

**Reinforcement.**  ACI 530/ASCE 5/TMS 402[9.1] requires that all masonry columns be reinforced with at least four bars having an area between 0.0025 and 0.04 times the cross-sectional area of the column. Lateral ties at least $\frac{1}{4}$ in. (6.3mm) diameter should be provided at a spacing not exceeding 16 longitudinal bar diameters, 48 tie bar diameters, or the least cross-sectional dimension of the column.  Where the percentage of vertical steel used is small, its contribution to the axial compression capacity is small and can be ignored in design.  Unreinforced columns are permitted in some codes[9.9-9.11] in low seismic risk areas.

### 9.3.2  Design for Axial Load and Bending

Column design closely parallels the wall design procedures presented in Chap. 8.  The principal differences relate to the significance of biaxial bending and the relative importance of compression reinforcement.  Also, for working stress design, allowable compressive stresses in some codes may be specified as lower values than for walls. Various codes differ in the manner that slenderness is considered and in the extent to which the use of principles of equilibrium and strain compatibility is recommended to consider the interaction between axial compression and bending moment.

A full discussion of the background for design and the calculation of member capacities under axial load and bending is found in Chap. 8. Therefore, only a brief summary of the main points is included here. Basically, two cases exist, depending on the magnitude of the virtual eccentricity of the load defined by $e = M/P$. The division between the two cases is the eccentricity that produces zero stress at the extreme fiber as defined by the kern distance,

$$e_k = S/A \qquad (9.1)$$

where  $e_k$ = kern distance
  $S$ = section modulus for the direction of bending
  $A$ = area of the column cross-section

For a solid rectangular section, $e_k = t/6$.

The design approaches are illustrated using the column shown in Fig. 9.7(a).

**Case 1: Entire Section in Compression ($e \leq e_k$).** In this case, the vertical reinforcement can be ignored and the area $A$ and section modulus $S$ can be calculated for the masonry section. Alternately, the effect of the reinforcement can be included by transforming the areas of the steel and adding $(n - 1)A_s$ to the cross-sections, as shown in Fig. 9.7(b), where $n$ is the modular ratio, $n = E_s/E_m$. The corresponding properties of the transformed section, $A_t$ and $S_t$, can be calculated.

To incorporate different allowable stresses for axial compression ($F_a$) and flexural compression ($F_b$), the unity equation can be used:

$$\frac{f_a}{F_a} + \frac{f_b}{F_b} \leq 1.0 \qquad (9.2)$$

where  $f_a$ = $P/A$
  $f_b$ = $M/S$

Alternately, where moment magnification is used to account for slenderness,[9.9] the bending stress is evaluated for the increased moment and $f_a + f_b$ must be less than the allowable compressive stress.

The combination of stresses due to axial load and bending is shown in Fig. 9.7(c). Calculations, using this approach, produce the upper linear part of the interaction relationships shown in the interaction diagram illustrated schematically in Fig. 9.8.

**Case 2: Cracked Section ($e > e_k$).** The tensile strength for the masonry is neglected after cracking, as is done for reinforced concrete. Once cracking occurs, there are potentially three subcategories of masonry behaviour within this case. In the equations to follow, $F_s$ is the allowable steel stress, and $F_m$ is the allowable masonry compressive stress under combined axial load and bending. In some codes, the allowable compressive stress in the masonry $F_m$ is equal to the allowable flexural compressive stress $F_b$. When different allowable stresses exist for axial load and for bending, the unity equation (Eq. 9.2), can be used to define the allowable compressive stress under the combined loading. This gives

$$f_b = \left(1 - \frac{f_a}{F_a}\right) F_b$$

Hence, for

$d = 5''$

$$f_m = f_a + f_b \tag{9.3}$$

$$F_m = f_a + \left(1 - \frac{f_a}{F_a}\right) F_b \quad \text{allowable stress in masonry}$$

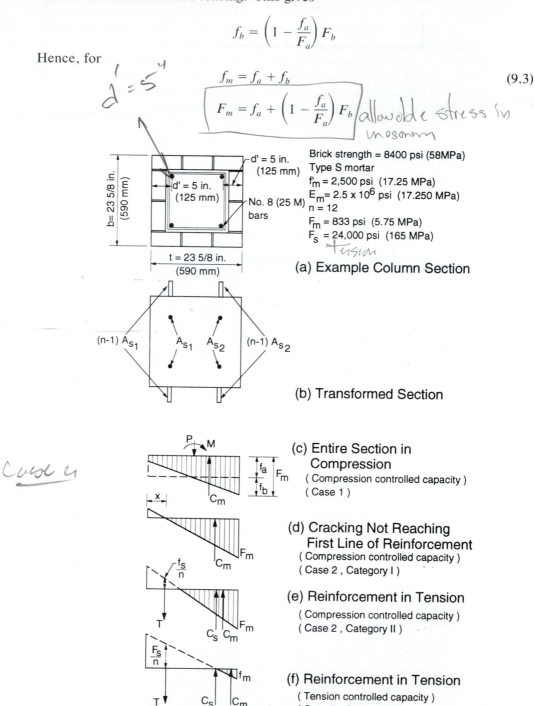

d' = 5 in.
(125 mm)

d' = 5 in.
(125 mm)

b = 23 5/8 in.
(590 mm)

No. 8 (25 M)
bars

Brick strength = 8400 psi (58MPa)
Type S mortar
$f'_m$ = 2,500 psi (17.25 MPa)
$E_m$= 2.5 x 10⁶ psi (17.250 MPa)
n = 12
$F_m$ = 833 psi (5.75 MPa)
$F_s$ = 24,000 psi (165 MPa)
tension

t = 23 5/8 in.
(590 mm)

**(a) Example Column Section**

(n-1) $A_{s1}$   $A_{s1}$   $A_{s2}$   (n-1) $A_{s2}$

**(b) Transformed Section**

Case 4

P   M

$f_a$   $F_m$
$f_b$

x

$C_m$

**(c) Entire Section in Compression**
( Compression controlled capacity )
( Case 1 )

$F_m$

$C_m$

**(d) Cracking Not Reaching First Line of Reinforcement**
( Compression controlled capacity )
( Case 2 , Category I )

$\frac{f_s}{n}$

T

$C_s$  $C_m$  $F_m$

**(e) Reinforcement in Tension**
( Compression controlled capacity )
( Case 2 , Category II )

$\frac{F_s}{n}$

T

$C_s$  $C_m$  $f_m$

**(f) Reinforcement in Tension**
( Tension controlled capacity )
( Case 2 , Category III )

**Figure 9.7**   Column under combined axial load and bending.

**_Category I._** In this case, the crack depth x has not reached the reinforcement $(x \le d')$, as shown in Fig. 9.7(d). From equilibrium, the applied eccentric load must coincide with the compressive stress resultant, $C_m$. If we neglect the reinforcement, the crack depth for a solid section is given as

$$x = 3e - t/2 \qquad (9.4)$$

Thus, from force equilibrium, the maximum load is derived as

$$P = F_m A \left[ \frac{3}{4}\left(1 - 2\frac{e}{t}\right) \right] \qquad (9.5)$$

This equation can be used to draw the next small segment of the interaction curve shown in Fig. 9.8.

**_Category II._** In this case, the crack has passed the first layer of steel and it is in tension, as shown in Fig. 9.7(e). In this case, the allowable design loads are controlled by the allowable masonry compressive stress. This condition is described by the stress relationship:

$$\frac{f_s}{n} = \left(\frac{x - d'}{t - x}\right)F_m \le \frac{F_s}{n} \qquad \text{tensial} \qquad (9.6)$$
$$\text{Stress}$$

The axial load is equal to the sum of internal forces:

$$P = C_m + C_s - T = \frac{F_m}{2}(t - x)b + (n - 1)A_{s2}\left(\frac{t - x - d'}{t - x}\right)F_m - nA_{s1}\left(\frac{x - d'}{t - x}\right)F_m \qquad (9.7)$$

$$e_{min} = .1t$$

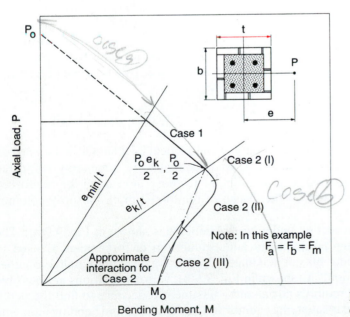

**Figure 9.8** Schematic interaction diagram for columns.

and from moment equilibrium about the centroid of the cross-section

$$M = (C_s + T)\left(\frac{t}{2} - d'\right) + C_m\left(\frac{t}{6} + \frac{x}{3}\right)$$ (9.8)

Hence, for values of x, P and M can be calculated and plotted in nondimensional form as shown in Fig. 9.8.

*Category III.* In this case, the tension steel has reached its allowable tensile stress and this controls the maximum permissible load, as shown in Fig. 9.7(f). It should be noted that, except for low percentages of reinforcement, this category may not exist for some columns. This condition is described by the stress relationship:

$$f_m = \frac{F_s}{n}\left(\frac{t - x}{x - d'}\right) \leq F_m$$ (9.9)

From force equilibrium

$$P = C_m + C_s - T = \frac{F_s}{n}\left(\frac{t - x}{x - d'}\right)\left(\frac{t - x}{2}\right)b + (n - 1)A_{s2}\frac{F_s}{n}\left(\frac{t - x - d'}{x - d'}\right) - A_{s1}F_s$$ (9.10)

and from moment equilibrium about the centroid of the cross-section

$$M = (C_s + T)\left(\frac{t}{2} - d'\right) + C_m\left(\frac{t}{6} + \frac{x}{3}\right).$$ (9.11)

Hence, for various values of x, P and M can be calculated and plotted, as shown in the lower part of the normalized interaction diagram in Fig. 9.8.

An option to the foregoing, specified in the Canadian code,[9.9] is to draw the interaction diagram by joining the points corresponding to (M, P) coordinates of (0, $P_o$) to ($e_k P_o/2$, $P_o/2$) to ($M_o$, 0) by straight lines as shown in Fig. 9.8. This approximate interaction diagram is generally slightly conservative. Construction of this simplified interaction diagram is illustrated in Example 8.1(c) and is not repeated here.

In most codes, the maximum design axial load is limited by a minimum eccentricity, as shown by the horizontal solid line in Fig. 9.8.

### 9.3.3 Biaxial Bending

Biaxial bending originates from eccentricity of the vertical load about the two principal axes of the member cross-section. Corner columns are commonly subjected to biaxial bending. Interaction diagrams similar to those developed for reinforced concrete columns can be used for the design of reinforced brick masonry columns under biaxial bending.[9.8]

A masonry column section under biaxial bending is shown in Fig. 9.9(a). The equations given by strain compatibility and equilibrium of forces can be used to analyze the section. Under biaxial bending, the neutral axis is not parallel to either principal axis of the section. As shown in Fig. 9.9(b), various positions are possible and the solution usually requires a trial-and-adjustment procedure to find the depth and angle of the neutral axis satisfying equilibrium of axial force and bending moments

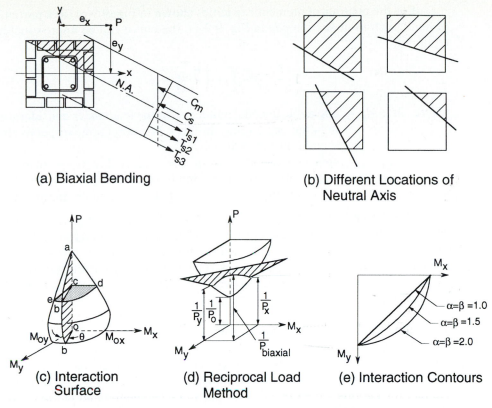

**(a) Biaxial Bending**

**(b) Different Locations of Neutral Axis**

**(c) Interaction Surface**

**(d) Reciprocal Load Method**

**(e) Interaction Contours**

**Figure 9.9** Analysis of columns under biaxial bending.

about the two principal axes of the section. The capacities over the full range of combinations form an interaction surface, as shown in Fig. 9.9(c).

For a given $e_x/e_y$ ratio, an interaction curve such as $ab$ in Fig. 9.9(c) can be established. By varying this ratio, expressed in terms of the angle $\theta$, a series of interaction curves similar to $ab$ can be obtained. If a horizontal section $cde$ is taken through the interaction surface at a particular load, line $de$ represents the possible combinations of maximum moments, $M_x$ and $M_y$, for that axial load.

By using reinforced concrete design procedures for guidance,[9.12] the allowable load on a section under biaxial bending, $P_{\text{biaxial}}$, can be approximated by a plane passing through points $1/P_0$, $1/P_x$, and $1/P_y$ on an inverted interaction diagram, as shown in Fig. 9.9(d), resulting in the equation

$$\frac{1}{P_{\text{biaxial}}} = \frac{1}{P_x} + \frac{1}{P_y} - \frac{1}{P_0} \tag{9.12}$$

where  $P_x$, $P_y$ = maximum allowable design loads for the specified eccentricities in the $x$ and $y$ directions for uniaxial bending

$P_0$ = maximum allowable design load for zero eccentricity

However, for cases with low axial loads ($P < 0.1P_0$), it is recommended that the contour method outlined in what follows be used.

For the contour of moment capacities shown in Fig. 9.9(e) for a particular level of load, such as at section *cde* in Fig. 9.9(c), the curve can be approximated by the equation

$$\left(\frac{P_{\text{biaxial}}e_x}{M_x}\right)^\alpha + \left(\frac{P_{\text{biaxial}}e_y}{M_y}\right)^\beta \leq 1.0 \qquad (9.13)$$

where $M_x$, $M_y$ = allowable uniaxial moments for the load under consideration when bending occurs only in the $x$ or $y$ direction, respectively, points $d$ and $e$ in Fig. 9.9(c).

$e_x$, $e_y$ = specified eccentricities in the $x$ and $y$ directions, respectively.

$\alpha$, $\beta$ = exponentials that affect the fit of the equation. (Use of $\alpha = \beta = 1.5$ is slightly conservative, but a reasonable approximation.)

This latter procedure involves iteration. A value of $P_{\text{biaxial}}$ is estimated for which $M_x$ and $M_y$ are calculated or found from interaction diagrams. Then $P_{\text{biaxial}}$ is calculated using Eq. 9.13. If this value is not close to the estimated load, a new estimate is made and the procedure is repeated. The speed of convergence is increased by using a new estimate between the previous estimate and the calculated value of $P_{\text{biaxial}}$.

### 9.3.4 Seismic Design Considerations

For columns subjected to high axial compression and bending moments resulting from the effects of seismic loads, building codes have more stringent requirements for tie spacing and tie location. Ties must act as shear reinforcement to prevent premature shear failure prior to plastic hinging in the column. Also ties must support vertical bars to improve strength and ductility over the length of the column and, particularly near the ends of the column, more closely spaced ties are required to ensure adequate ductility by allowing plastic hinges to occur. A maximum spacing of 8 in. (200 mm) is specified[9.1,9.13] over the full height of columns stressed by tensile or compressive axial overturning forces due to seismic loads. In high seismic zones, lateral ties for compression reinforcement should be embedded in grout.[9.1] This provision prohibits the use of lateral ties embedded in mortar joints because ties in this location do not support the vertical steel and do not confine the enclosed volume of masonry as effectively as ties placed in contact with the vertical steel and embedded in grout.

ACI 530/ASCE 5/TMS 402[9.1] specifies that the least nominal dimensions of a masonry column in Seismic Zones 3 and 4 not be less than 12 in. (300 mm).

## 9.4 COLUMN DESIGN EXAMPLES

### 9.4.1 Eccentrically Loaded Column

#### Example 9.1

For the column shown in Fig. 9.7, the normalized interaction diagram shown in Fig. 9.10 was produced using the detailed procedure described in Sect. 9.3.2. For the develop-

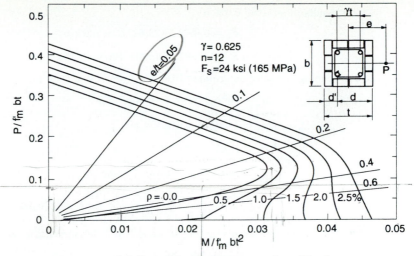

(a) Considering Compression Steel

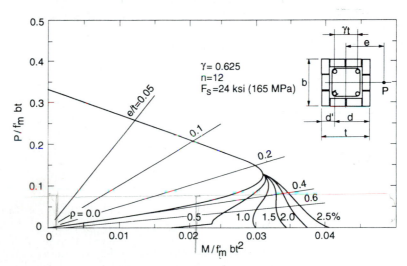

(b) Ignoring Steel in Compression

**Figure 9.10** Interaction curves for masonry columns (working stress design).

ment of these diagrams, it is assumed that the allowable compressive stress under combined axial load and bending, $F_m$, is equal to $f'_m/3$.

**a) Eccentric Axial Load.** For an 8 in. (200 mm) eccentricity of load in the x-direction, determine the capacity of the column. The effective height is 18 ft (5.4m).

**Solution:** From Fig. 9.10, for $e/t = 0.34$ and considering compression steel with

$$\rho = 4(0.79)/(23.625 \times 23.625) = 0.0057 = 0.57\%$$

$$(\rho = 4(500)/(590)^2 = 0.0057),$$

$$P/(f'_m bt) = 0.09$$

which results in an axial capacity

$$P_x = 0.09(2.5)(23.625)^2 = 126 \text{ kips}$$

$$(P_x = 0.09(7.25)(590)^2/1000 = 540 \text{ kN})$$

and a bending capacity determined from $Pe/f'_m bt^2$ in Fig. 9.10 or

$$M_x = Pe = 126(8) = 1008 \text{ in.–kips}$$

$$(M_x = 540(0.2) = 108 \text{ kN.m})$$

$$f_a = P_x/A = 126(1000)/(23.625)^2 = 226 \text{ psi}$$

$$(f_a = 540(1000)/(590)^2 = 1.55 \text{ MPa})$$

$$r = \sqrt{I/A} = \sqrt{\frac{(23.625)^4/12}{(23.625)^2}} = 6.82 \text{ in. (170 mm)}$$

Checking for slenderness using the ACI 530/ASCE 5/TMS 402 approach,[9.1]

$$F_a = \frac{1}{4}f'_m\left[1 - \left(\frac{h}{140r}\right)^2\right] = \frac{1}{4}(2500)\left[1 - \left(\frac{18 \times 12}{140 \times 6.82}\right)^2\right]$$

$$= 593 \text{ psi} > 226 \text{ psi} \qquad \therefore \text{OK}$$

$$\left(F_a = \frac{1}{4}(17.25)\left[1 - \left(\frac{5.4 \times 1000}{140 \times 170}\right)^2\right] = 4.09 \text{ MPa} > 1.55 \text{ MPa} \qquad \therefore \text{OK}\right)$$

$\therefore$ The design is satisfactory.

Alternately, by using the moment-magnifier approach[9.9] the eccentricity would be magnified and the allowable load determined for this higher eccentricity.

**b) Biaxial Bending.** For the previous example, determine the allowable load if there is an additional eccentricity of 4 in. (100 mm) in the y-direction.

**Solution:** Using the same procedure as before, the $P_0/f'_m bt$ value in Fig. 9.10, corresponding to $e/t = 4/23.625 = 0.17$, is 0.17 and

$$P_y = 0.17 \times 2.5 \times 23.625 \times 23.625 = 237 \text{ kips}$$

$$(P_y = 0.17 \times 17.25 \times 590 \times 590/1000 = 1021 \text{ kN})$$

$$M_y = P_y(4) = 948 \text{ in.–kips}$$

$$(M_y = P_y(100/1000) = 102 \text{ kN.m})$$

Also $P_0$ for $e/t = 0$ is:

$$P_0 = F_a(A_n + (n - 1)A_s)$$

$$= \frac{2.5}{3}[23.625 \times 23.625 + (12 - 1)(3.16)] = 494 \text{ kips}$$

$$\left(P_0 = \frac{17.25}{3}[590 \times 590 + (12 - 1)(2000)]1000 = 2128 \text{ kN}\right)$$

Then

$$\frac{1}{P_{\text{biaxial}}} = \frac{1}{P_x} + \frac{1}{P_y} - \frac{1}{P_0} = \frac{1}{126} + \frac{1}{237} - \frac{1}{494} = \frac{1}{99}$$

$$\left(\frac{1}{P_{\text{biaxial}}} = \frac{1}{540} + \frac{1}{1021} - \frac{1}{2128} = \frac{1}{423}\right)$$

gives

$$P_{\text{biaxial}} = 99 \text{ kips (423 kN)}$$

**Alternate Solution:**   Estimating $P_{\text{biaxial}} = 99$ kips (423 kN), $P/f'_m bt = 0.071$. The value of $Pe/f'_m bt^2$ corresponding to this is 0.031 for uniaxial bending, which converts to

$$M_x = M_y = 0.031 f'_m bt^2 = 0.031 \times 2.5 \times 23.625 \times (23.625)^2 = 1022 \text{in.–kips}$$

$$(M_x = M_y = 0.031 \times 17.25 \times 590 \times (590)^2/(1000)^2 = 110 \text{ kN.m})$$

Using

$$\left(\frac{P_{\text{biaxial}} e_y}{M_y}\right)^{1.5} + \left(\frac{P_{\text{biaxial}} e_x}{M_x}\right)^{1.5} = \left(\frac{4P_{\text{biaxial}}}{1022}\right)^{1.5} + \left(\frac{8P_{\text{biaxial}}}{1022}\right)^{1.5} = 1$$

$$\left(\left(\frac{0.1\,P_{\text{biaxial}}}{110}\right)^{1.5} + \left(\frac{0.2\,P_{\text{biaxial}}}{110}\right)^{1.5} = 1\right)$$

$$P_{\text{biaxial}} = 105 \text{ kips (450 kN)}$$

Conservatively, the estimated axial load capacity of 99 kips (423 kN) can be taken as the capacity because it is close to, but lower than, the calculated value. For greater precision, a new estimate of $P_{\text{biaxial}}$ of, say 102 kips (440 kN), could be tried to determine new values of $M_y$ and $M_x$ for use in the foregoing equation. This should reduce the difference between the estimated and calculated values of $P_{\text{biaxial}}$. ■

### 9.4.2  Concentric Axial Compression

**Example 9.2**

Design a block masonry column composed of two standard 8 in. (200 mm) blocks, as shown in Fig. 9.11, to carry an axial load of 80 kips (356 kN) for a column with a height of 16 ft (4.8 m). For illustrative purposes, values conforming with ACI 530/ASCE 5/TMS 402 are used. (No design aids are used.)

**Solution:**   Lacking other information on boundary conditions, the effective height is taken as $kh = h = 16$ ft $= 192$ in. (4800 mm). The radius of gyration is

$$r = \sqrt{\frac{I}{A}} = \sqrt{\frac{bt^3/12}{bt}} = 0.29t = 0.29(15.625) = 4.53 \text{ in. (113 mm)}$$

Then, the allowable axial stress:

$$F_a = 1/4 f'_m \left[1 - \left(\frac{h}{140r}\right)^2\right] = 0.227 f'_m$$

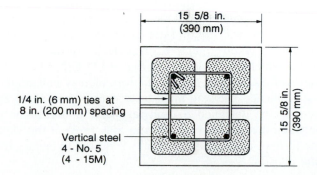

1/4 in. (6 mm) ties at 8 in. (200 mm) spacing

Vertical steel 4 - No. 5 (4 - 15M)

15 5/8 in. (390 mm)

15 5/8 in. (390 mm)

**Figure 9.11**   Column cross-section for Example 9.2.

Even for concentrically applied loads, the minimum eccentricity is $e = 0.1t = 1.56$ in. (39 mm), which, because it is less than the kern distance $e_k = t/6$, means that the section is uncracked. Using the unity equation for combined axial load and bending (ignoring the contribution of any reinforcement that may be added).:

$$f_a = P/A = 80 \times 1000/(15.625)^2 = 328 \text{ psi}$$

$$(f_a = 356 \times 1000/(390)^2 = 2.34 \text{ MPa})$$

$$f_b = \frac{M}{S} = \frac{Pe}{bt^2/6} = \frac{80(1000)(1.56)}{(15.625)^3/6} = 196 \text{ psi}$$

$$\left(f_b = \frac{356(1000)(39)}{(390)^3/6} = 1.40 \text{ MPa}\right)$$

For $F_b = f'_m/3$

$$\frac{f_a}{F_a} + \frac{f_b}{F_b} = \frac{328}{0.227f'_m} + \frac{196}{0.33f'_m} = 1$$

$$\left(\frac{2.34}{0.227f'_m} + \frac{1.40}{0.33f'_m} = 1\right)$$

Therefore, the required $f'_m = 2039$ psi (14.55 MPa).

For $f'_m = 2039$ psi (14.55 MPa) and type $S$ mortar, a block with a compressive strength of 2870 psi (19.9 MPa) is required according to the ACI 530/ASCE 5/TMS 402 specifications. [*Note*: Grout to fill the cores of the blocks must be at least equal to $f'_m$ but not less than 2000 psi (13.8 MPa).]

For good practice or as required by ACI 530/ASCE 5/TMS 402,[9.1] the minimum amount of reinforcement is $A_{s,min} = 0.0025 A_n = 0.0025 (15.625)^2 = 0.61$ in.$^2$ (0.0025 $(390)^2 = 380$ mm$^2$). Therefore, the minimum of four No. 4 bars (4-10M) giving $A_s = 0.80$ in.$^2$ (400 mm$^2$) is adequate. Lateral ties of $\frac{1}{4}$ in. (6.3 mm) diameter at 8 in. (200 mm) spacing on center are required.[9.1] ∎

## 9.5 PILASTER DESIGN

### 9.5.1 Introduction

A pilaster is a stiffening element within a wall that is laterally supported in the plane of the wall. Therefore, the critical buckling dimension is perpendicular to the wall. A pilaster can be designed as a column with a flanged section, as shown in Fig. 9.12. For cases where the wall and pilaster are built integrally, criteria to determine the effective width of the flanges created by the wall section are given in building codes. The effective flange width is commonly taken equal to $6t$ on each side or the actual center line-to-center line dimension, whichever is less; see Fig. 9.13. Other codes[9.9] may also limit the effective flange width according to other factors such as wall height. For a wall to act as a flange for a pilaster, the wall must be effective in resisting applied load and the intersection must be designed to transfer shear in the plane of the wall. Movement joints between the wall and the pilaster may prevent the wall from acting as a flange but some form of connection is usually required to ensure transfer of out-of-plane loads to the pilaster.

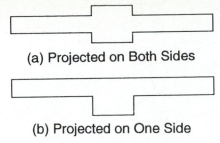

(a) Projected on Both Sides

(b) Projected on One Side

**Figure 9.12**  Possible geometries of pilasters.

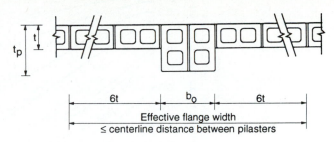
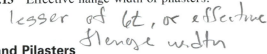

6t  |  $b_0$  |  6t

Effective flange width
≤ centerline distance between pilasters

**Figure 9.13**  Effective flange width of pilasters.

*lesser of 6t, or effective flange width*

## 9.5.2 Load Sharing Between Walls and Pilasters

**Effective Flanges.**  To consider the wall as part of the pilaster, the ACI 530/ASCE 5/TMS 402[9.1] requirements are the following:

* The wall should be in running bond.
* The connection (at the intersection) should conform to <u>one</u> of the following requirements:
  1. Fifty percent of the masonry units at the interface shall interlock, or
  2. Walls shall be regularly toothed with 8 in. (200 mm) maximum offsets and anchored by steel connectors meeting the following requirements:
     (a) Minimum size: $\frac{1}{4}$ in. (6.3 mm) by $1\frac{1}{2}$ in. (38 mm) by 28 in. (700 mm) long, including 2 in. (50 mm) long 90° bent legs at each end to form a U or Z shape.
     (b) Maximum spacing: 4 ft (1.2 m).
* Intersecting bond beams shall be provided in intersecting walls at a maximum spacing of 4 ft (1.2 m) on center.  Bond beams shall be reinforced and the area of reinforcement shall not be less than 0.1 in.$^2$/ft (210 mm$^2$/m) of wall.  Anchorage of the reinforcement should be developed on each side of the intersection.

**Transfer of Load to Pilasters.**  Masonry walls are commonly designed to span horizontally between pilasters, particularly for unreinforced walls. In such cases, pilasters are usually sufficiently closely spaced so that the unreinforced wall can span horizontally in one-way bending. All of the lateral load is transferred to the pilasters, which in turn transfer the load to the top and bottom supports by vertical bending.

In other situations, when the pilasters are spaced further apart, the masonry wall can span vertically and horizontally in two-way bending for load transfer.  Design of the wall for such situations was covered in Chap. 7.  Depending upon the design of the wall for two-way bending for either reinforced or unreinforced construction, the amount of lateral load transmitted in the two directions is a function of its edge restraint and the horizontal-to-vertical span ratio of the wall. Curves[9.4] presented in Fig. 9.14 provide a simplified method for determining the lateral load transferred to

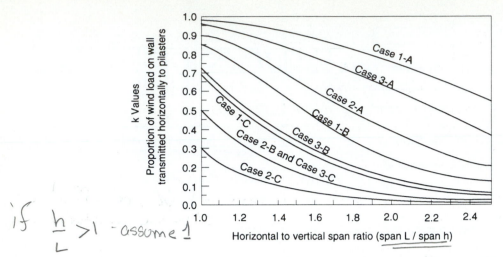

if $\dfrac{h}{L} > 1$ - assume 1

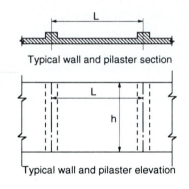

Typical wall and pilaster section

Typical wall and pilaster elevation

Case 1: Walls fixed at pilasters

A.  Fixed at bottom, free at top
B.  Supported top and bottom
C.  Fixed at bottom, supported at top
    "Pinned"

Case 2: Walls supported at pilasters

A.  Fixed at bottom, free at top
B.  Supported top and bottom
C.  Fixed at bottom, supported at top

Case 3: Walls fixed at one end,
        supported at the other

A.  Fixed at bottom, free at top
B.  Supported top and bottom
C.  Fixed at bottom, supported at top

**Figure 9.14**  Calculation of load transmission to masonry pilasters (from Ref. 9.4).

$P_L$ is horizontal load (lb/ft)

a pilaster $p_L$, per unit of height,

$$p_L = kpL \tag{9.14}$$

where  $k$ = coefficient given in Fig. 9.14 for different aspect ratios and support conditions of the wall
$p$ = lateral design load
$L$ = center-to-center horizontal span between pilasters.

The bending moment from the distributed lateral load on the pilaster is usually taken as $p_L(h^2)/8$ located at mid-height, assuming that pinned conditions are used to model the actual supports. Alternately, this bending moment may be located at the base of the wall if sufficient restraint exists to provide a fixed end condition. Loads from beams supported on the pilaster are likely to cause additional bending moments due to eccentric loading.

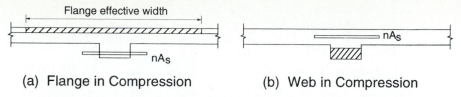

Flange effective width

nAs

(a) Flange in Compression       (b) Web in Compression

**Figure 9.15** Change in effective section of pilaster due to direction of loading after cracking.

**Influence of Reversed Loading.** Wind loading, causing positive pressure and suction, can result in roughly equal bending of the pilaster in either direction. This has a significant effect on the capacity of asymmetric pilaster sections. As shown in Fig. 9.15, cracked flanges on the tension side of the pilaster under the combination of axial load and bending do not add to the resistance. Because of the reversal of bending due to wind or seismic loads, flanges located at one face of the pilaster will not have much effect on the allowable load unless high axial load or the moment due to eccentric axial load dominates. In these cases, flanges located on the compression side of the pilaster can be effective. Similarly, flanges located at the center of a pilaster projecting from both sides of the wall (Fig. 9.12(a)) will only have significant effects for high axial compression loads. In cases where the pilaster mainly resists bending due to lateral loading, locating vertical movement joints at the intersection between the wall and the pilaster does not significantly reduce the capacity of the pilaster.

**Slenderness.** Slenderness effects for pilasters can be evaluated using procedures similar to columns except that biaxial bending is not a concern. For unreinforced pilasters, ACI 530/ASCE 5/TMS 402[9.1] specifies allowable compressive stress in terms of the h/r ratio and limits the axial load to not more than 25% of the buckling load given by

$$P_e = \frac{\pi^2 E_m I}{h^2}\left(1 - 0.577\frac{e}{r}\right)^3 \tag{9.15}$$

This equation includes the effect of eccentricity of applied load. The calculated limiting axial load is sensitive to the value of the modulus of elasticity, $E_m$.

### 9.5.3 Pilaster Design Example

**Example 9.3**

A warehouse 30 ft (9 m) wide by 160 ft (48 m) long has the roof supported by trusses at 8 ft (2.4 m) on center. The 22 ft (6.6 m) high external wall is built with 8 in. (200 mm) thick ungrouted hollow masonry units and 16 × 16 in. (400 × 400 mm) pilasters at the truss locations as shown in Fig. 9.16.

Roof Dead Load = 10 lb/ft² (0.5 kN/m²)
Live Load = 20 lb/ft² (1 kN/m²)
Wind Load = 25 lb/ft² (1.25 kN/m²)

aproximate $M = \frac{P_L h^2}{8}$ - Symple supported beam.

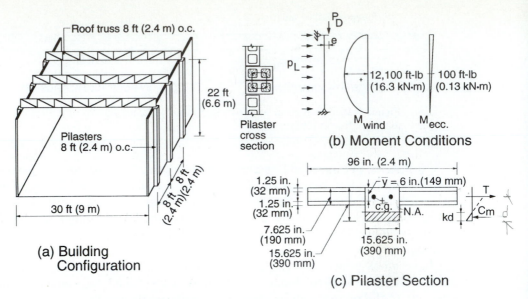

**Figure 9.16** Pilaster design: Example 9.3.

Design a typical pilaster to carry the vertical load from the roof trusses and the lateral wind load. The axial load from the truss is applied at 4 in. (100 mm) from the inside face of the wall. The bearing plate for the trusses is $4 \times 4$ in. ($100 \times 100$ mm). Use ACI 530/ASCE 5/TMS 402.[9.1]

**Solution:**

### 1. Section Properties:

The possible overall flange width of

$$12t + b_o = 12(7.625) + (15.625)$$
$$= 107.13 \text{ in.} > 8 \text{ ft} \qquad (12(190) + 390 = 2.67 \text{ m} > 2.4 \text{ m})$$

Therefore use a flange width of 96 in. (2.4 m) equal to the spacing of pilasters. The net section is shown in Fig. 9.16(c).

$$A_g = (15.625 \times 15.625 + 2(1.25)(96 - 15.625) = 445 \text{ in.}^2$$

$$(A_g = (390)^2 + 2(32)(2400 - 390) = 280{,}740 \text{ mm}^2)$$

$$\bar{y} = [15.625(15.625)^2/2 + 80.375(2.5)(7.625/2)]/445 = 6.0 \text{ in.}$$

$$(\bar{y} = (390)^3/2 + 2010(32)2(190/2))/280{,}740 = 149 \text{ mm})$$

$$I = (15.625)^4/12 + (15.625)^2((15.625/2) - 6.0)^2$$
$$\quad + 80.375[(7.625)^3/12 - (5.125)^3/12] + 80.375(2.5)(6 - 7.625/2)^2$$
$$\quad = 8800 \text{ in.}^4$$

$$(I = 390)^4/12 + (390)^2((390/2) - 149)^2 + 2010((190)^3/12 - (126)^3/12)$$
$$\quad + 2010(64)(149 - (190/2)^2 = 3.44 \times 10^9 \text{ mm}^4)$$

$$r = \sqrt{I/A} = 4.45 \text{ in.} \qquad (111 \text{ mm})$$

**2. Loads:**

Axial load on pilasters:

$$P_{Dead} = 10 \times 8 \times 30/2 = 1200 \text{ lb}$$
$$(P_{Dead} = 0.5\,(2.4)\,(9/2) = 5.4 \text{ kN})$$
$$P_{Live} = 20 \times 8 \times 30/2 = 2400 \text{ lb}$$
$$(P_{Live} = 1.0\,(2.4)\,(9/2) = 10.8 \text{ kN})$$
$$\text{Eccentricity of Load} = 6.0 - 4.00 = 2.0 \text{ in. } (49 \text{ mm})$$

*use trib areas*

Wind load is distributed in two-way action. The panel is considered free at the top and fixed at the other three sides. By using the chart in Fig. 9.14, the lateral load acting on a pilaster can be calculated using Eq. 9.14:

$$p_L = kpL$$

For Case 1-A in Fig. 9.14, for the aspect ratio $L/h = 8/22 = 0.36$, this is off the chart and $k$ is essentially equal to 1.0. Thus the lateral load on a pilaster is

$$p_L = (1.00)\,(25)\,(8) = 200 \text{ lb/ft}$$

$$(p_L = 1.0\,(1.25)\,(2.4) = 3.0 \text{ kN/m})$$

The moment at mid-height due to wind, assuming a simply supported condition:

$$M = \frac{p_L h^2}{8} = 200\,\frac{22^2}{8} = 12{,}100 \text{ ft–lb.} \checkmark$$

$$\left(M = 3.0\,\frac{(6.6)^2}{8} = 16.3 \text{ kN.m}\right)$$

Moment at the top of the wall due to the eccentricity of the vertical load:

$$M = Pe = (1200)\,(2.0/12) = 200 \text{ ft-lb}$$

$$(Pe = 5.4\,(49/1000) = 0.26 \text{ kN.m})$$

Critical moment at mid-height, Fig. 9.16(b):

$$M = 12{,}100 - (200/2) = 12{,}000 \text{ ft-lb}$$

$$(M = 16.3 - (0.26/2) = 16.17 \text{ kN.m})$$

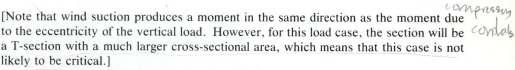

*this compressny contals*

[Note that wind suction produces a moment in the same direction as the moment due to the eccentricity of the vertical load. However, for this load case, the section will be a T-section with a much larger cross-sectional area, which means that this case is not likely to be critical.]

**3. Allowable Stresses:**

For a masonry compressive strength of 2,000 psi (13.8 MPa).
Allowable axial compression stress for $h/r = 59.4 < 99$:

$$F_a = \frac{1}{4}f'_m[1 - (h/140r)^2] = 440 \text{ psi} \qquad (3.03 \text{ MPa})$$

Allowable bending compression stress:

$$F_b = \frac{1}{3}f'_m = 667 \text{ psi} \qquad (4.6 \text{ MPa})$$

*for wind*

Allowable tension in steel:

$$F_s = 24,000 \text{ psi (165 MPa) for Grade 60 steel}$$

The above stresses can be increased by $\frac{1}{3}$ for wind loading.

### 4. Axial Compression:

Consider the dead plus the live load where the weight of the pilaster is 130 lb/ft³ (2090 kg/m³) and the weight of the 8 in. (200 mm) wall is 45 lb/ft² (222 kg/m²)

$$\text{Total } P = P_{\text{Dead}} + P_{\text{Live}} + P_{\text{self-weight at mid-height}} = 1200 + 2400$$

$$+ [(15.625 \times 15.625/12 \times 12) \times 11 \times 130$$

$$+ (80.375/12) \times 11 \times 45] = 9340 \text{ lb}$$

$$(\text{Total } P = 5.4 + 10.8 + (0.39)^2 (3.3) (2090 \times 9.8/1000)$$

$$+ (2.01) (3.3) (228) 9.8/1000 = 40.9 \text{ kN})$$

Axial compressive stress:

$$f_a = P/A = 9340/445 = 21 \text{ psi} < 440 \text{ psi} \qquad \text{O.K.}$$

$$(f_a = 40,900/280,740 = 0.15 \text{ MPa} < 4.6 \text{ MPa} \qquad \text{O.K.})$$

### 5. Combined Axial Load and Bending:

The allowable compressive stress for combined axial load and bending is $F_m = F_b$. In ACI530/ASCE 5/TMS 402, this can be increased by $\frac{1}{3}$ due to wind loading.

Consider the critical case of wind plus the dead load (including self-weight): At mid-height

$$e = M/P_{\text{Dead}} = 12,000 \times 12/6940 = 18.75 \text{ in.}$$

$$(e = 16.17 (1000)/30.1 = 537 \text{ mm})$$

For such a large eccentricity ($e > t$), the section is cracked. Therefore, tension reinforcement is required. Use the strain and equilibrium approach and ignore steel in compression. Refer to Fig. 9.16(c).

For tension control,

$$f_m = (F_s/n) (kd/(d - kd))$$

Consider

$$E_m = 1.5 \times 10^6 \text{ psi} (10,350 \text{ MPa})$$

$$n = E_s/E_m = 19.3$$

$$f_m = ((4/3) F_s/19.3) [kd/(d - kd)] = 1658 [kd/(d - kd)]$$

$$(f_m = 11.4 (kd/(d - kd))$$

$$C_m = \tfrac{1}{2} f_m bkd = 12,960 (kd)^2/(d - kd)$$

$$(C_m = 2223 (kd)^2/(d - kd))$$

$$T = A_s F_s = (4/3) 24,000 A_s$$

$$(T = (4/3) 165 A_s)$$

From equilibrium of axial forces

$$P = C_m - T \tag{9.16}$$

Equilibrium of moments, $\Sigma M_{centroid} = 0$,

$$M = C_m (t - \bar{y} - kd/3) + T_s(\bar{y} - t_{wall}/2) \tag{9.17}$$

From Eqs. 9.16 and 9.17,

$$kd = 3.15 \text{ in. (80 mm)}$$

and substituting into Eq. 9.16,

$$A_s = 0.25 \text{ in.}^2 (164 \text{ mm}^2)$$

Checking for tension controlled capacity

$$f_m = 1658 \, (kd/(d - kd)) = 603 \text{ psi} < 667 \, (4/3) \text{ psi} \qquad \therefore \text{O.K.}$$

$$(f_m = 4.24 \text{ MPa} < 4.6 \, (4/3) \text{ MPa} \qquad \therefore \text{O.K.})$$

Use two No. 4 (10M): $A_s = 2 \times 0.20 = 0.40 \text{ in.}^2 (200 \text{ mm}^2)$
Provide a total of four bars as shown in Fig. 9.16(a).
[Note that for the reversed load case where the flange is in compression, the lower kd value and the larger lever arm result in a higher moment resistance for the cross section. Therefore, this case is not likely to be critical but should be checked to make sure.]

Use $\frac{1}{4}$ in. (6.3 mm) diameter lateral ties at 8 in. (200 mm) o.c. at the mortar joints as shown in Fig. 9.16(a). Check other load combinations for stress compliance. ∎

## 9.6 CLOSURE

The reader is referred to Chap. 8 for a more thorough treatment of the behavior of masonry under combined axial load and bending. In addition, the background for various design methods for wall design including strength design (ultimate limit state) are discussed. The principal differences between walls and the columns discussed in this chapter relate to the added vulnerability of columns due to their isolation from other parts of the structure, the effect of compression reinforcement, and the potential for biaxial bending. For pilasters, the interaction with adjoining walls is the main difference. As discussed in Chap. 7, the support conditions for wall panels subject to out-of-plane loads affect the loading conditions on pilasters. Columns and pilasters may have to be reinforced to resist shear forces, in which case the design follows the provisions of Chap. 6 for beams.

## 9.7 REFERENCES

9.1   The Masonry Standards Joint Committee, "Building Code Requirements for Masonry Structures," ACI 530/ASCE 5/TMS 402, American Concrete Institute and American Society of Civil Engineers, Detroit and New York, 1992.

9.2   A. Elmiger, "Architectural and Engineering Concrete Masonry Details for Building Construction," National Concrete Masonry Association, Herndon, 1976.

9.3   Brick Institute of America, "Columns, Piers, Pilasters and Buttresses," BIA Technical Note 12, BIA, Reston, VA, April 1962.

**9.4** National Concrete Masonry Association, "Reinforced Concrete Masonry Columns and Pilasters," TEK-Notes 30, Herndon, VA, 1971.

**9.5** G.R. Sturgeon, J. Longworth, and J. Warwaruk, "Reinforced Concrete Block Masonry Columns," in *Proceedings of the Third North American Masonry Conference,* Arlington, TX, 1985, pp. 20.1–20.16.

**9.6** G.J. Edgell, "Reinforced Brickwork Columns," in *Proceedings of the 7th International Brick Masonry Conference*, Melbourne, Australia, 1985, pp. 20-1, 20-16.

**9.7** J. Warwaruk and J. Longworth, "Response of Masonry Columns Using Standard Wall Units," in *Proceedings of the Fourth Canadian Masonry Symposium*, Vol. 2, Fredericton, New Brunswick, 1986, pp. 897–912.

**9.8** S.R. Davis and E.A. Elfraify, "Biaxial Bending of Reinforced Brick Masonry Columns," in *Proceedings of the British Ceramic Society*, No. 30, Stoke-on-Trent, 1982, pp. 355–359.

**9.9** Canadian Standards Association, "Masonry Design for Buildings," CAN3 S304-M84, CSA, Rexdale, Ontario, 1984.

**9.10** Standards Association of Australia, "SAA Masonry Code," AS3700 SAA, North Sydney, New South Wales, 1988.

**9.11** British Standards Institution, "Code of Practice for Use of Masonry," BS 5628, BSI, London, 1985.

**9.12** C. Wang, and C. Salmon, *Reinforced Concrete Design*, 4th ed., Harper & Row, New York, 1985.

**9.13** International Conference of Building Officials, *Uniform Building Code*, UBC-91, Chap. 24: Masonry, ICBO, Whittier, CA, 1991.

## 9.8 PROBLEMS

**9.1** Develop an interaction diagram for a 16 × 16 in. (400 × 400 mm) concrete masonry column reinforced with four No. 8 (25M) bars. $f'_m = 2500$ psi (17.3 MPa), $f_y = 60$ ksi (400 MPa), and $d' = 4$ in. (100 mm). Use ACI 530/ASCE 5/TMS 402 or your local code if it is different. Ignore the slenderness effect and consider equal allowable compressive stress of $f'_m/3$ for an axial load and for a combined axial load and moment.

**9.2** Given a 24 × 24 in. (600 × 600 mm) brick masonry column which is 24 ft (7.2 m) high, design the column to carry a 120 kip (535 kN) axial load with 16 in. (400 mm) eccentricity. Use the interaction diagrams presented in Fig. 9.10. Ignore steel in compression.

**9.3** For the column designed in Prob. 9.2, what is the reduction in allowable axial load due to an 8 in. (200 mm) eccentricity in the other direction [in addition to the 16 in. (400 mm) eccentricity]?
   **(a)** Use the $1/P = 1/P_x + 1/P_y - 1/P_o$ method.
   **(b)** Use the contour method.

**9.4** A masonry wall consists of 4 in. (100 mm) nominal thick brick spanning between 12 in. (300 mm) square pilasters spaced 5 ft (1.5 m) on center, as shown in Fig. P9.4. The wall is 16 ft (4.8 m) high and, for the type of construction, can be most closely modeled as being simply supported (no resistance to rotation) top and bottom. The wall supports a roof load of 2 kips/ft (30 kN/m) applied at the center of the 4 in. (100 mm) (nominal) wall.

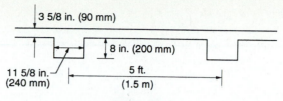

3 5/8 in. (90 mm)

8 in. (200 mm)

11 5/8 in.
(240 mm)

5 ft.
(1.5 m)

**Figure P9.4**

**(a)** If the wall and pilasters are not reinforced, determine the maximum allowable wind pressure for the allowable tension of 36 psi (0.25 MPa) and allowable compression of 1200 psi (8.3 MPa).

**(b)** If the pilaster is reinforced with a No. 8 bar (25M), located 3 in. (75mm) from each face, and the allowable tensile stress in this reinforcement is 24,000 psi (165 MPa), determine the maximum allowable wind pressure.

**9.5** A design calls for an 8 in. (200 mm) reinforced concrete block wall which is 24 ft (7.2 m) high to support a concentric vertical roof load of 1.5 kips/ft (22 kN/m) and a 20 lb/ft² (1 kN/m²) horizontal wind load. An alternate design calls for pilasters and an unreinforced wall. Design the thickness of wall and the size and spacing of pilasters using ACI 530/ASCE 5/TMS 402 or your local code for the following conditions. Use $f'_m = 2200$ psi (15 MPa), Grade 60 (400 MPa) reinforcement.

**(a)** If unreinforced construction is used.

**(b)** If reinforced pilasters are used. Also determine the amount of reinforcement required in the pilaster. Show the reinforcing details.

**(c)** What is the impact of introducing vertical movement joints at the intersections between pilasters and the wall sections?

**9.6** Using the strength design approach outlined in Chap. 8, solve Prob. 9.1.

# CHAPTER 10

# Shear Walls

Concrete masonry shear wall
building (*Courtesy of National
Concrete Masonry Association*).

## 10.1 INTRODUCTION

Loadbearing masonry buildings usually have walls arranged at a fairly uniform spacing to carry gravity loads to the foundation without the addition of frames or columns. These bearing walls also act as shear walls to resist lateral loads due to wind or seismic forces. The long walls required to provide support for the floor system produce a building that is inherently very stiff and strong in the predominant direction of these loadbearing walls. In some cases, however, it is necessary to introduce additional walls to provide lateral load resistance in the perpendicular direction even though they are not required to support gravity loads. The lateral load-carrying capacity of shear wall structures is mainly dependent on the in-plane resistances of the shear walls because the in-plane stiffness of a shear wall is far greater than its out-of-plane stiffness. Thus, shear walls are usually arranged to correspond with the two major axes of the building.

In multistory buildings, concrete floors and roofs act as rigid diaphragms to distribute the shear forces to walls according to their stiffnesses, taking into account both shear and flexural deformations. As a result, the design of shear wall buildings tends to be self-fulfilling, that is, larger walls are stiffer and thereby attract more of the lateral load.

Although reinforced concrete shear walls and masonry shear walls fulfill the same function, there are some unique considerations for masonry shear walls that are identified in this chapter.

Infill masonry panels confined between the columns and beams of reinforced concrete or steel frame buildings can be designed to resist shear forces resulting from lateral load. Because of the special considerations involved, the behavior and design of these infill masonry shear walls are covered separately in Chap. 11.

## 10.2 INFLUENCE OF TYPES AND LAYOUT OF SHEAR WALLS

Masonry shear walls can be described not only in terms of the type of masonry unit used (solid or hollow, brick or block, and clay or concrete), but also as loadbearing or nonloadbearing, reinforced or unreinforced, single-wythe or multiwythe, solid or perforated, rectangular or flanged, and cantilevered or coupled. Shear wall behavior is affected by the shapes of shear walls in plan, sizes and distribution of multiple openings, and boundary elements such as cross walls, returns, and columns. As a general rule, structural designers tend to favor arrangements of openings that result in solid or nearly solid walls as the main lateral load-resisting elements. The distribution of in-plane shear and bending moment in walls with openings as shown in Fig. 10.1(a) is much more complex than for solid walls.

Compared to solid walls, walls having large openings can be considered to be made up of systems of piers and spandrels to produce what is sometimes called a perforated shear wall, as shown in Fig. 10.1(b). This wall can be analyzed and designed as a frame composed of short stiff members and is suitable for low to medium rise construction (less than about 5 to 6 storys). In this case, both flexural and shear deformations should be included in the analysis of the frame. Alternately, large wall sections joined across openings by spandrels produce what is recognized as coupled shear walls, in which case, the design of the spandrels for shear capacity and ductility becomes a main feature of the design.

The behavior of perforated shear walls and coupled shear walls is much more complicated than solid walls and, because of this, they are normally not built of unreinforced masonry. Instead, a recommended practice is to create individual cantilever walls supporting strips of openings, such as shown in Fig. 10.2. The behavior of these cantilever walls is relatively simple to predict. The thin floor slabs that cross the openings are usually sufficiently flexible in their out-of-plane direction to

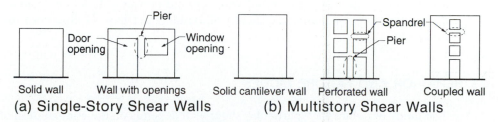

**Figure 10.1** Types of masonry shear walls.

**Figure 10.2** Isolation of individual shear walls using strips of openings.

avoid introducing appreciable coupling. Where beams span the opening between walls, uncoupled cantilever action can be ensured if the beams are isolated by movement joints and are designed to permit relative displacements at the ends without developing significant bending moments. For reinforced masonry, the reinforcement can be used to resist the shear and bending forces and, by providing continuity, the capacity and energy absorbing properties will be improved.

Flanged walls, created by connecting intersecting walls or by adding returns at the ends of walls, substantially increase the moment capacity and axial load capacity, but the shear capacity is relatively unaffected. Therefore, as is illustrated in Sec. 10.5, shear design often can be the limiting factor. Also, where extra ductility or extra flexural capacity is required, built-in boundary elements such as pilasters or thickened wall sections can be used. These elements more readily accommodate lateral ties for lateral support of compression steel and confinement of part of the masonry in the compression zone. This improves ductility and helps to avoid failures due to shear–compression and toe crushing.

Multistory loadbearing shear walls are suited to floor plans that are divided into a large number of compartments with the identical plan repeated at each floor. As is discussed in Chap.3 regarding planning of wall layouts to provide stability, the rigid diaphragm action of the floor system permits all the walls to participate fully in resisting the lateral load. They can consist of walls in the two orthogonal directions arranged in a rational manner and walls around elevator shafts and stairwells. The design example of a multistory shear wall building in Chap. 17 illustrates this feature of design.

## 10.3 BEHAVIOR AND FAILURE MODES

The mode of failure for a particular shear wall depends on the combination of applied loads, wall geometry, properties of the materials, and details of reinforcement, if any. It is important for the designer to understand the effects of design decisions on behavior. The following sections include a brief review and discussion of some of the important aspects of behavior.

### 10.3.1 Unreinforced Shear Walls

Typical modes of failure of unreinforced masonry shear walls are illustrated in Fig. 10.3. All are characterized by brittle behavior with rapid decreases in capacity and very limited deformations after reaching the ultimate load.

As shown in Fig. 10.3(a), predominance of the axial load at ultimate load leads to vertical cracking failure, which, as discussed in Sec.5.2, is attributed to the incompatibility between the deformational characteristics of the materials (unit, mortar, and grout) and geometric changes in horizontal cross-section over the height of the wall. With increased influence of the overturning moment due to lateral load, more localized compression failure can occur at the toe of the wall, as shown in Fig. 10.3(b). The failure mechanism is similar to Fig. 10.3(a), but confined to a more localized area.

Sliding or shear slip failure along a bed joint is a failure mode that results from lateral shear forces exceeding the adhesion and shear-friction resistance between the mortar and the units or the mortar and the floor, as shown in Fig. 10.3(c).[10.1–10.4] Because of the reduced effect of friction, this type of failure is most likely to occur where low axial loads are combined with high shear forces. For high shear and high axial loads, diagonal tension failures such as shown in Fig. 10.3(d) have been observed. In this case, the shear strength associated with diagonal tension cracking depends on the tensile strength of the units and the mortar bond strength.[10.5,10.6]

### 10.3.2 Reinforced Shear Walls

Design and detailing of reinforced masonry shear walls can ensure sufficient ductility to permit redistribution of lateral load and to provide good energy dissipation characteristics for seismic loading. It is desirable that reinforced masonry shear walls be able to develop their full resistance to axial load and bending and to withstand substantial inelastic deformation without either significant loss of strength or severe

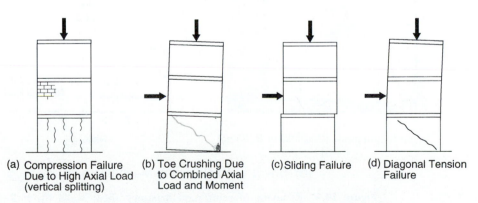

(a) Compression Failure Due to High Axial Load (vertical splitting)

(b) Toe Crushing Due to Combined Axial Load and Moment

(c) Sliding Failure

(d) Diagonal Tension Failure

**Figure 10.3**   Failure modes of unreinforced masonry shear walls.

stiffness degradation. This is particularly important in high seismic risk regions. Thus, brittle modes of failure in the masonry such as diagonal tension failure, bed joint slip, premature crushing at the toe of the wall, or loss of anchorage of reinforcement must be prevented. Under the combined effects of vertical and lateral loads, the strength and deformation characteristics primarily depend on wall geometry, level of axial load, and the amount and distribution of vertical and horizontal steel.[10.7–10.17] The two distinct failure modes associated with shear walls are:

- flexural failure, Fig. 10.4(a), characterized by bed-joint cracking, yielding of vertical reinforcement, and toe crushing, and
- shear failure, Fig. 10.4(b), characterized by diagonal tension cracking.

As shown by the idealized load–deflection curves in Fig. 10.4, axial load affects wall behavior by delaying initial cracking of the masonry and yielding of the reinforce-

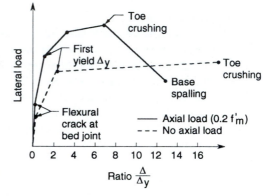

Flexural failure mode (*Courtesy of B. Shing*)

### (a) Flexural Load Resisting Mechanism

Reinforced failure modes.

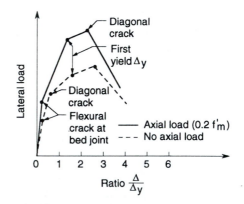

Shear failure mode (*Courtesy of B. Shing*)

### (b) Shear Load Resisting Mechanism

**Figure 10.4** Behavior of reinforced masonry shear walls (from Ref. 10.10).

ment. Higher axial compression loads result in increased shear capacities and can increase flexural capacities except for compression-controlled failures.

In the early 1970s, the effects of the amount and distribution of vertical and horizontal reinforcement were investigated at the University of Canterbury.[10.7,10.8] The racking test results showed that horizontal and vertical reinforcement were equally effective for crack control and development of shear strength. Also, evenly distributed reinforcement delayed the onset of severe cracking compared to walls with reinforcement concentrated near the edges. Failure occurred soon after the onset of severe cracking at low percentages of uniformly distributed reinforcement, whereas, at higher percentages of reinforcement, the failure load was much greater than the load that caused severe cracking.

Because test conditions affect the interpretation and presentation of results, it is difficult to quantify optimum amounts of horizontal steel. However, it is worth noting that in the previous tests, higher failure loads coincided with higher percentages of distributed reinforcement, but that additional reinforcement, over 0.3% of the gross area, had little effect on the failure load. At the 0.3% or higher amounts of reinforcement, the average ultimate shear stress was 170 psi (1.17 MPa). This general finding, relating optimum shear reinforcement to relatively low quantities of steel, agrees with research by Schneider,[10.9] who found that 0.2% was the maximum effective quantity. Recent tests on reinforced masonry shear walls, conducted at the University of Colorado,[10.10–10.12] show that the amount of horizontal reinforcement has a significant effect on the cracking pattern, shear strength, and ductility of walls. Specimens with adequate quantities of vertical and horizontal reinforcement (in the range of 0.25 to 0.40%) exhibited ductile behavior with large deformations after yielding of the reinforcement. Conversely, specimens that failed in shear had limited ductility, depending on the horizontal steel reinforcement ratio; see Fig. 10.5(a). Thus, it appears that the maximum effective amount of horizontal reinforcement is not a fixed quantity, but, because of the shear–compression mode of failure, depends on the compressive strength of the wall, its geometry, and special features, such as lateral confinement at the ends of walls and lateral tie support for compression reinforcement.

Pier tests conducted at the University of California–Berkeley[10.13] showed three possible modes of failure: flexural mode, shear mode, and sliding mode, depending primarily on the aspect ratio of the pier and the level of axial load. The test results revealed higher shear strengths for lower aspect ratios. The results also showed a positive influence of the horizontal reinforcement on the inelastic behavior of the piers. Increasing amounts of horizontal reinforcement resulted in a more uniform crack pattern, increased the ultimate strength of the piers, and increased their deformation capacity. Test results[10.14] also showed that partially grouted walls exhibited ductility which compared well with fully grouted walls. There were, however, significant reductions in strength and stiffness.

Cyclic tests conducted in the 1970s by Williams[10.8] and more recently by Shing et al.[10.10–10.12] showed that strength and stiffness degradation characteristics depend on the primary mode of failure. For walls that exhibit yielding in the reinforcement or flexure failure, the hysteresis loops were stable (Wall A in Fig. 10.5(a)) and a relatively minor stiffness degradation was observed. In contrast, degradation for walls failing predominantly in shear was relatively large (Wall B in Fig. 10.5(b)).

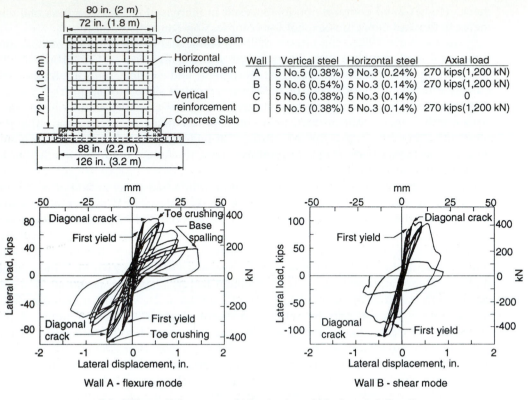

| Wall | Vertical steel | Horizontal steel | Axial load |
|---|---|---|---|
| A | 5 No.5 (0.38%) | 9 No.3 (0.24%) | 270 kips(1,200 kN) |
| B | 5 No.6 (0.54%) | 5 No.3 (0.14%) | 270 kips(1,200 kN) |
| C | 5 No.5 (0.38%) | 5 No.3 (0.14%) | 0 |
| D | 5 No.5 (0.38%) | 5 No.3 (0.14%) | 270 kips(1,200 kN) |

(a)  Effect of Amount of Vertical and Horizontal Steel

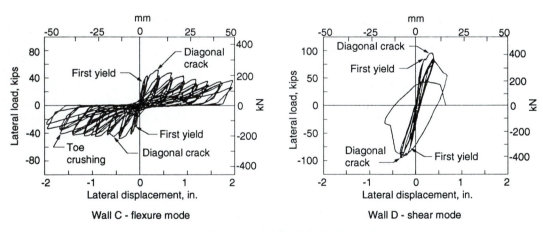

(b)  Effect of Level of Axial Stress

**Figure 10.5**  Load-deflection curves for reinforced concrete masonry shear walls (from Ref. 10.10).

Tests results[10.10] indicated the significance of the level of axial load on the response of single-story shear walls under cyclic loads, as shown in Fig. 10.5(b). The presence of an axial load of 270 kips, (1200 kN) which results in a compressive stress of about 20% of the prism compressive strength, changed the mode of deformation from a ductile flexure (Wall C) to a brittle shear mode (Wall D) with much reduced inelastic deformation capability beyond the peak load.

Tests[10.10,10.15,10.16] show that the performance of cantilever shear walls under cyclic reversed loading can be improved by adding confinement in the critical compression zones where plastic hinges occur near the base of the wall. The confinement allows large inelastic stains to develop, resulting in significant rotations under flexure. The curvatures plotted in Fig. 10.6 illustrate the very large rotations that can occur when sufficient ductility is present to permit the plastic hinge to form over the high moment region of the wall height. As shown in Fig. 10.7, higher ductility levels were obtained by introducing confinement to increase the ultimate compression strain. Placing steel plates in critical mortar courses[10.17] and providing continuous helical steel wires in the grouted cavity around the vertical end bars[10.15,10.16] are two methods that have been used to confine the ends of masonry walls. In Japan, spiral reinforcement has been successfully used in critical plastic hinge regions to enhance the ductility of masonry shear walls.[10.16]

Pseudo-static cyclic load tests of wide flange T-section masonry shear walls[10.18] showed a strongly asymmetric response when subjected to reversed loading in the direction parallel to the web. As shown in Fig. 10.8, the walls loaded in the direction causing compression in the web were stronger, stiffer, and less ductile compared with the reversed loading direction, causing compression in the flange.

For multistory shear walls, the mode of failure and ductility depend on the wall geometry and the curvature capacity of the wall.[10.19] Fig. 10.9 shows different types of reinforced masonry shear walls. The simple cantilever shear wall shown in Fig. 10.9(a) assures a ductile response and good energy dissipation capacity when properly designed and constructed to provide for plastic hinges at the base of each wall. The response of perforated walls such as shown in Figs. 10.9(b) and 10.9(c), is different and is much more complex than the response of cantilever walls. Under inelastic response to lateral loading, hinging may initiate in the piers, Fig. 10.9(b), or in the spandrels, Fig. 10.9(c). In the former, which is the more common case, there are high ductility demands on the piers. In this case, designers should give special

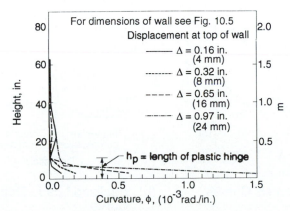

**Figure 10.6** Curvature variation along wall height (from Ref. 10.11).

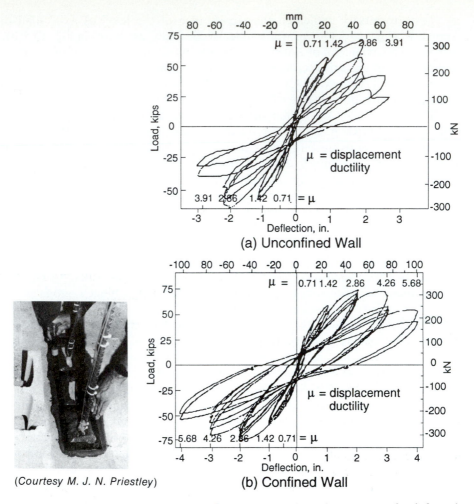

(Courtesy M. J. N. Priestley)

(a) Unconfined Wall

(b) Confined Wall

**Figure 10.7** Effect of confinement on lateral response of reinforced masonry walls (from Ref. 10.17).

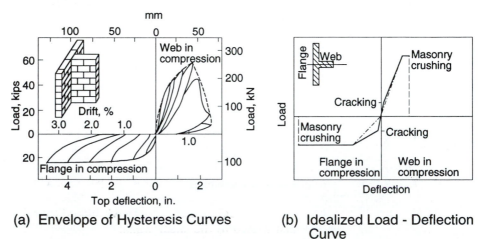

(a) Envelope of Hysteresis Curves

(b) Idealized Load - Deflection Curve

**Figure 10.8** Response of flanged walls to lateral load reversal (from Ref. 10.19).

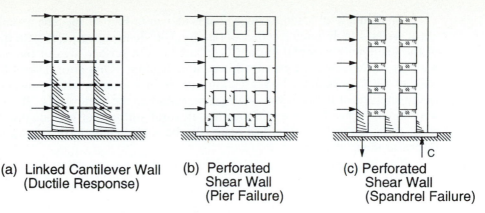

(a) Linked Cantilever Wall
    (Ductile Response)

(b) Perforated
    Shear Wall
    (Pier Failure)

(c) Perforated
    Shear Wall
    (Spandrel Failure)

**Figure 10.9**  Response of reinforced masonry multistory shear walls to seismic loading (from Ref. 10.20).

attention to the shear design and reinforcing details of the piers. Plastic hinging in the spandrels is consistent with coupled shear wall behavior.

## 10.4  DISTRIBUTION OF LOADS TO SHEAR WALLS

### 10.4.1  Gravity Loads

Calculated dead and live loads due to gravity forces are normally based on the tributary areas for each wall. In this regard, depending upon the type of occupancy, use of live load reduction factors based on tributary area can significantly reduce the axial compression load on shear walls, particularly near the base of a multistory building. In addition to other dead loads, the self-weights of walls can comprise a significant part of the mass of the building and, therefore, cannot be ignored in the analysis of forces.

The presence of an axial compression force can prevent the development of flexural tensile stresses for unreinforced masonry or it can reduce the amount of tensile reinforcement required for reinforced masonry. For this reason, it is important not to overestimate the minimum axial load. In strength design, this situation is accounted for by reducing the load factor for dead load to less than 1.0 when dead load enhances capacity.

### 10.4.2  Lateral Forces

Concrete floors in multistory buildings have in-plane stiffnesses that are considered to be very rigid compared to the flexural stiffnesses of the masonry shear walls in their in-plane direction. The rigid diaphragm action of the floors that connect the cantilever shear walls ensures that the positions of the walls with respect to each other do not change as a result of lateral loading. This characteristic behavior of buildings with rigid diaphragms affects the manner in which the lateral loads due to wind or earthquake are distributed to the shear walls.

An example of a simple floor plan is shown in Fig. 10.10. The arrangement of shear walls is such that the center of rigidity, CR, does not coincide with the location of the resultant horizontal force, CG, related to the surface area of the building for wind or to the center of gravity, for seismic loading. Therefore, lateral loads from wind or earthquake in the N-S direction will produce torsion, which must be included in the analysis of the distribution of lateral forces to the shear walls. Because of the very much higher in-plane stiffness, the flexural stiffness of rectangular walls with respect to their weak axis (out-of-plane) can be neglected in the distribution of lateral forces to these elements.

The location of the center of rigidity of the shear walls at a particular floor level can be calculated as

$$\bar{x} = \frac{\sum R_{yi} x_{\text{ref},i}}{\sum R_{yi}} \qquad (10.1)$$

$$\bar{y} = \frac{\sum R_{xi} y_{\text{ref},i}}{\sum R_{xi}}$$

where  $R_{xi}, R_{yi}$ = rigidity of shear wall $i$ to bending in the $x$ and $y$ directions, respectively

$x_{\text{ref},i}, y_{\text{ref},i}$ = distance in the $x$ and $y$ directions, respectively, from a reference point to shear wall $i$

$\bar{x}, \bar{y}$ = distances of center of rigidity of the building from the reference point

The rigidity $R$ of a shear wall is dependent on its dimensions, modulus of elasticity $E_m$, modulus of rigidity $G_m$, and the support conditions. For a cantilever wall subject to a horizontal load $V$ at the top, as shown in Fig. 10.11(a), the deflection due to bending and shear is

Shape factor for rectangular Shapes

$$\Delta_c = \frac{Vh^3}{3E_m I} + \frac{1.2 Vh}{G_m A} \qquad (10.2)$$

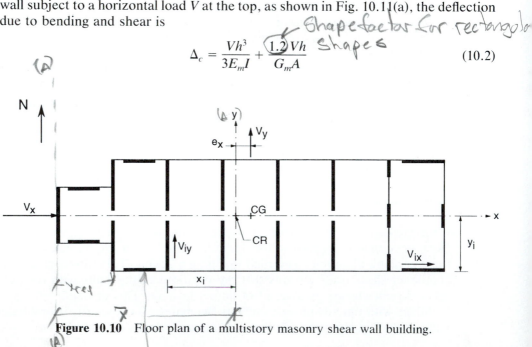

**Figure 10.10**  Floor plan of a multistory masonry shear wall building.

neglect out of plane stiffness

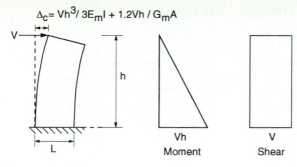

(a) Cantilever Wall

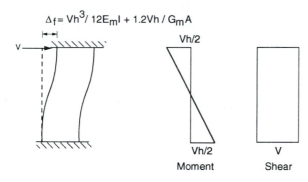

**Figure 10.11** Deflection of walls due to bending and shear deformations.

(b) Wall Fixed Against Rotation

where, in the last term, the factor 1.2 is a shape factor for rectangular sections, and area $A$ is the length of the wall times the effective thickness $t$. (The effective thickness can be calculated as the minimum mortared and grouted area of the wall section divided by the wall length.)

For $G_m = 0.4E_m$, and considering a rectangular wall with $A = Lt$ and $I = (L^3t)/12$, Eq. 10.2 can be simplified to

$$\Delta_c = \frac{V}{Et}\left[4\left(\frac{h}{L}\right)^3 + 3\left(\frac{h}{L}\right)\right] \tag{10.3}$$

In a similar way, a wall with ends fixed against rotation, such as shown in Fig. 10.11(b) has a deflection due to bending and shear of

$$\Delta_f = \frac{Vh^3}{12E_mI} + \frac{1.2\,Vh}{G_mA} \quad \text{Fixed against} \atop \text{Rotation.} \tag{10.4}$$

which, for $G_m = 0.4\,E_m$ and a rectangular section, simplifies to

$$\Delta_f = \frac{V}{Et}\left[\left(\frac{h}{L}\right)^3 + 3\left(\frac{h}{L}\right)\right] \tag{10.5}$$

As an indication of the relative contribution of shear deformation to the total deflection, the values in Table 10.1 are for rectangular walls with various aspect

Sec. 10.4    Distribution of Loads to Shear Walls

**TABLE 10.1** EFFECT OF ASPECT RATIO ON DEFLECTION DUE TO SHEAR

| Aspect ratio, h/L | Percentage deflection due to shear | |
|---|---|---|
| | Cantilever wall | Fixed-end wall |
| 0.25 | 92 | 98 |
| 1 | 43 | 75 |
| 2 | 16 | 43 |
| 4 | 5 | 16 |
| 8 | 1 | 4.5 |

ratios $h$ to $L$. As can be seen, for cantilever walls with high $h/L$ ratios, the component of shear deformation is small and can be ignored in calculating wall rigidity.

The rigidity of a wall is proportional to the inverse of the deflections. Therefore, the calculations of rigidities to be used in Eq. 10.1 is as follows:
For cantilever walls,

$$R = \frac{V}{\Delta}$$

$$R_c = \frac{Et}{4\left(\dfrac{h}{L}\right)^3 + 3\left(\dfrac{h}{L}\right)} \tag{10.6}$$

and for fixed-end walls,

$$R_f = \frac{Et}{\left(\dfrac{h}{L}\right)^3 + 3\left(\dfrac{h}{L}\right)} \tag{10.7}$$

These relationships are valid only for loads at the top of the wall and change for other load distributions. In addition, the relative contributions of the bending and shear deformations depend on the wall aspect ratio ($h/L$) and, therefore, the rigidity varies over the height of a building. However, for high $h/L$ ratios, the effect of shear deformation is very small and the calculation of wall rigidities based on flexural stiffness is reasonably accurate, although it should be recognized that shear distortions will be more significant for flanged walls with large increases in flexural rigidity but little change in shear deformation. For very squat shear walls (with $h/L \le 0.25$), rigidities based on shear deformation are reasonably accurate, but for intermediate cantilever walls with $h/L$ ratios ranging roughly between 0.25 and 4, it is very important to include both components of deflection in the calculation of relative rigidities. As an aid to carrying out such analyses, Fig. 10.12 is a plot of the relative rigidities, $R/Et$, for rectangular wall sections.

Walls connected by floors that do not provide significant coupling may be assumed to act as cantilevers from the base. Then, with certain simplifications including linear elastic behavior, the total lateral load can be distributed among all cantilever walls according to their combined flexural and shear rigidities. With reference to Fig. 10.10, the distribution of the total lateral load in the N–S direction, $V_y$, among all cantilever walls can be approximated by the following expressions:

$$V_{yi} = V_{yit} + V_{yir} \tag{10.8}$$

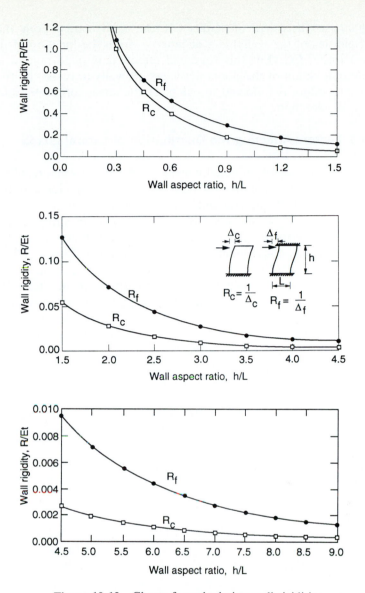

**Figure 10.12** Charts for calculating wall rigidities.

where $V_{yi}$ is the shear resisted by wall i for external load in the y-direction. The translation mode or direct shear $V_{yit}$ is calculated as

$$V_{yit} = \frac{R_{yi}}{\sum R_{yi}} V_y \qquad (10.9)$$

and the rotational mode or torsion shear $V_{yir}$ is calculated as

$$V_{yir} = \frac{x_i R_{yi}}{\sum (x_i^2 R_{yi} + y_i^2 R_{xi})} e_x V_y \qquad (10.10)$$

Sec. 10.4    Distribution of Loads to Shear Walls    $x_i = x_{ref,i}$    **435**

In this case, the center of rigidity CR (the mathematical point that concentrates all the rigidities of the walls) is a distance $e_x$ from the location of the resultant of the lateral force, CG. For this torsional effect, it is important to note that the rigid diaphragm action of the floors activates all walls to resist the twisting action. The same procedure is followed to calculate the shear forces $V_{xi}$ due to lateral load in the E–W direction.

### 10.4.3 Factors Affecting the Distribution of Lateral Loads

Neglecting the added torsional resistance of flanged walls results in errors in lateral load distribution calculations that are likely to be smaller than those resulting from ignoring the effects of cracking on stiffness.[10.20] Studies[10.10,10.11,10.21] of reinforced masonry shear walls show significant reductions in stiffness due to cracking and inelastic deformations. Also, wall uplift and base sliding result in apparent increases in deflection and, consequently, a reduction in wall stiffness. The potential for cracking and subsequent loss of stiffness is affected by the magnitude of the axial compressive load. Figure 10.13 shows the shear stiffness of square walls ($h/L = 1.0$) as a ratio of the secant stiffness measured just prior to formation of the first major diagonal crack. As shown, shear stiffness measured at lateral loads greater than or equal to 50% of the ultimate load increases significantly with increase in axial stress.

For shear walls that are expected to experience significant amounts of inelastic deformations under earthquake loads, use of the secant stiffnesses at first yield was suggested[10.19] because this has more relevance in the inelastic phase of response. The secant stiffness is a function of the percentage and grade of steel reinforcement and the level of axial load. The stiffness of cantilever shear walls with predominantly flexural deformation can be estimated using the effective moment of inertia of the wall cross-section at first yield.[10.22]

For flanged walls, an effective flange width of not more than six times the flange thickness on each side of the web is commonly used for strength calculations. This

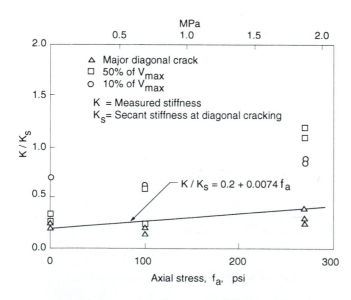

**Figure 10.13** Shear stiffness of single-story reinforced masonry shear walls (from Ref. 10.10).

may be conservative for combined flexural and axial load capacity calculations. However, additional unaccounted for stiffness due to longer flanges could attract a larger share of the lateral load to such walls and result in them being underdesigned for shear. The usual practice is to use the same flange width for both the stiffness and strength calculations, but, particularly for unreinforced walls, it is wise to ensure that there is ample reserve strength to accommodate an increased share of the lateral load. For reinforced shear walls, the limit on effective flange width has less significance because highly stressed walls will crack, resulting in a large decrease in stiffness that will tend to redistribute the load to other uncracked walls.

The assumption of rigid diaphragm action is an approximation because all floor and roof diaphragms deform to some extent. For reinforced shear walls this approximation is not too critical because inaccuracies in the elastic distribution of lateral loads are compensated for by redistribution following cracking and inelastic deformation in the shear walls. Unreinforced shear walls do not have this capability for redistribution and therefore it is more important to ensure very rigid behavior or to leave some margin of safety in the design. For example, it is reported[10.22] that in-plane deformations of precast concrete floors can be significant, particularly in buildings of five stories or less. Variations of 20 to 40% from loads calculated using rigid diaphragm behavior can be expected.

## 10.5 EFFECTS OF OPENINGS IN WALLS

### 10.5.1 Multistory Shear Walls

As is discussed in Chap. 3 and earlier in this chapter, the placement of openings in walls can have marked effects on both the appearance of the building and the structural response. Therefore, it is very important that locations for windows, doors, and services be rationally planned at an early stage of design. Depending upon the size and location of openings, large variations in the stiffness of a shear wall can result.[10.23] This effect is illustrated in Fig. 10.14 where staggered openings allow the

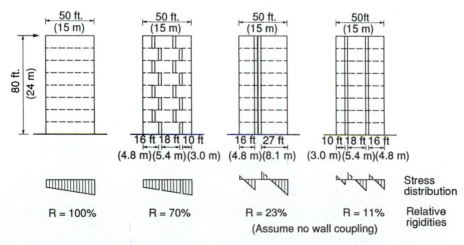

**Figure 10.14**   Relative rigidities of walls with openings (from Ref. 10.23).

wall to act essentially as a single section and retain much of the stiffness of a similar solid wall. Walls with vertically aligned openings tend to be divided into separate strips with greatly reduced total stiffness. In the case of aligned openings, the separate wall strips can be interconnected either just by the floor slab alone or by specially designed short, deep coupling beams. This *coupling* of the shear walls introduces a complexity in masonry construction because of the difficulty of providing continuity at the intersection of the walls and horizontal beams.

Several different methods of analysis have been used for coupled shear walls to determine the design requirements and to predict displacements. Sinha et al.[10.24] reported that standard methods of analysis do not provide reliable predictions of either stresses or displacements for coupled unreinforced masonry walls. The cantilever method, which ignores any coupling, was found to overestimate the extreme fiber stresses and the deflections. The continuum method, assuming wall interaction, provided a reasonable estimate of stress, but underestimated deflection. The conclusion is that analysis based on an *equivalent frame* or *wide column frame* for larger shear wall strips is a reasonable approach. However, it can produce widely varying results, depending on whether centerline span lengths are used or rigid arms are introduced at the ends of members. The effects of shear deformation can also be significantly underestimated. Unlike the other methods, finite element analyses can model nonlinear strain and stress distributions along the wall and produces results consistent with test results, which also showed decreasing rigidity at higher shear loads.

For shear walls with openings such as shown in Fig. 10.15, the very rigid wall area above the openings causes the cross-section defined by the isolated piers to act as a composite section for which the centroid and section properties can be calculated. Then, the distribution of stresses due to the overturning moment $M_{ovt}$

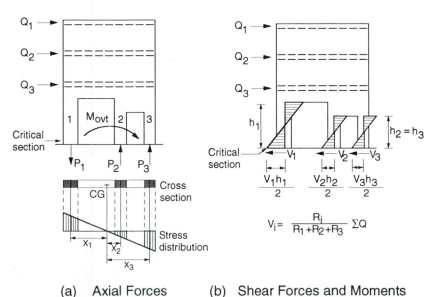

(a)   Axial Forces        (b)   Shear Forces and Moments

**Figure 10.15**   Forces acting on the bottom story piers of a multistory building.

can be calculated from $\sigma_i = M_{\text{ovt}}x_i/I$ for various points $i$ at distance $x_i$ from the centroid of these piers. As can be seen in the figure, the effect is that the sum of the axial stresses on each pier produces an axial force equal to the pier area times the average stress on the pier. Each of these resultant forces $P_i$ will be located just slightly outside the center of each pier because of the stress gradient. Therefore, a slightly conservative approximation is to assume that the axial force developed in each pier due to overturning moment is at the center of the pier. Then the axial force in any pier due to overturning, Fig. 10.15(a), can be calculated from

$$P_i = \frac{M_{\text{ovt}}x_iA_i}{\sum A_ix_i^2} \tag{10.11}$$

In addition to the axial forces developed by overturning, the shear force $V_i$ resisted by each pier will cause additional end moments $M_i$ in that pier of

$$M_i = V_ih_i/2 \tag{10.12}$$

where $h_i$ is the clear height of that pier. In this analysis, the ends are considered to be fixed; see Fig. 10.15(b). The method for distributing the total shear to the piers in the wall is discussed in the next section.

If the area of openings in a wall is only 5 to 10% of the total wall area and the openings are relatively small and well distributed, the effect of the discontinuities in strain distribution over the length of the wall is usually ignored and the analysis of rigidity can be based on the stiffness of the gross cross-sectional area. Stresses due to axial load, bending, and shear can be based on the minimum section accounting for the openings in the wall but assuming a linear strain distribution over the length of the wall.

### 10.5.2 Horizontal and Vertical Combinations of Shear Wall Segments

As was mentioned before, where large openings occur, effective coupling of the wall segments may be difficult to achieve. Therefore, an acceptable approach is to minimize the amount of coupling and to analyze the wall as a horizontal combination of cantilever strips such as shown in Fig. 10.16(a). In this case, the combined rigidity $R$ is

$$R = R_{c1} + R_{c2} + R_{c3} \tag{10.13}$$

where $R_{c1}$, $R_{c2}$, and $R_{c3}$ are the rigidities of cantilever strips 1, 2, and 3, respectively.

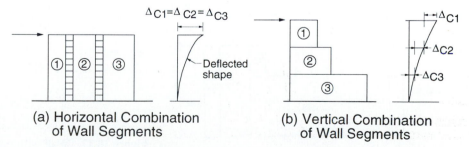

(a) Horizontal Combination of Wall Segments

(b) Vertical Combination of Wall Segments

**Figure 10.16** Wall combinations for calculating rigidities of walls with openings.

If the wall segments are combined vertically, as shown in Fig. 10.16(b), the combined rigidity $R$ can be calculated as

$$R = \frac{1}{\Delta_{c1} + \Delta_{c2} + \Delta_{c3}} = \frac{1}{\frac{1}{R_{c1}} + \frac{1}{R_{c2}} + \frac{1}{R_{c3}}} \tag{10.14}$$

where $R_{c1}$, $R_{c2}$, and $R_{c3}$ are the rigidities of wall segments 1, 2 and 3, respectively, where they are treated as cantilevers. This expression ignores the rotations that occur at the tops of segments 2 and 3 and therefore overestimates the rigidity of the wall. In addition, it is valid only for the application of loads at the top level of the building. When used in combination with other walls, this results in the conservative assignment of a larger share of the lateral load to this type of wall. An alternate approach is to calculate the deflection accurately for a cantilever with changes in cross-section. The more usual case of change in section over the height of walls in low-rise buildings is covered in the next section.

### 10.5.3 Rigidity of Walls with Openings in Low-Rise Buildings

For a wall with openings where the shear force is applied at the top of the wall as in a single-story building, two different approximate methods are used. The calculated total deflections and thus rigidities differ because of the different approaches adopted in each method for adding up the deflections attributed to the wall segments that incorporate openings.

**Method 1.** In this method, the wall is divided into segments stacked on top of each other in vertical series and the deflections of all segments are added to obtain the total wall deflection. The deflection of a segment containing openings is based on assumed fixed-end conditions top and bottom. The deflections of the solid segments are calculated assuming cantilever action. Because this method does not account for the rotations at the tops and bottoms of successive parts, it is more applicable for squat walls ($h/L \leq 0.5$), where the dominant component of deflection is shear, which does not produce rotation.

For each vertical segment of the wall, defined by a horizontal strip, the deflection of the piers connected in horizontal series is the reciprocal of the sum of the rigidities. This method is illustrated in what follows for the wall shown in Fig. 10.17,

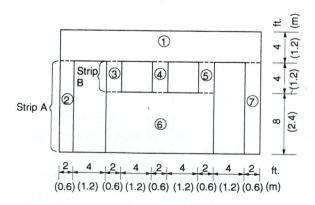

**Figure 10.17** Wall with openings (calculation of rigidity and load distribution).

where the overall rigidity is given by

$$R_{\text{wall}} = \frac{1}{\Delta_{1c} + \Delta_{2,3,4,5,6,7(f)}} \qquad (10.15)$$

$$= \frac{1}{\dfrac{1}{R_{1c}} + \dfrac{1}{R_{2,3,4,5,6,7(f)}}} \qquad \text{(vertical combination)} \qquad (10.16)$$

where $R_{2,3,4,5,6,7(f)} = R_{2f} + R_{3,4,5,6(f)} + R_{7f}$  (horizontal combination)

and $\qquad R_{3,4,5,6(f)} = \dfrac{1}{\dfrac{1}{R_{3,4,5(f)}} + \dfrac{1}{R_{6f}}} \qquad$ (vertical combination) $\qquad (10.17)$

$$R_{3,4,5(f)} = R_{3f} + R_{4f} + R_{5f} \qquad \text{(horizontal combination)} \qquad (10.18)$$

For the dimensions shown,

$$R_{1c} = 2.10\, E_m t$$

$$R_{2f} = R_{7f} = 0.004\, E_m t$$

$$R_{3f} = R_{4f} = R_{5f} = 0.071\, E_m t$$

$$R_{6f} = 0.526\, E_m t$$

resulting in $R_{\text{wall}} = 0.149\, E_m t$

**Method 2.**  Rather than adding up the deflections for the individual parts, this method involves calculation of the deflection at the top of the wall considering the wall to be solid. Then, for the strip containing openings, the deflection as a solid section is subtracted and, finally, the deflection of the system of piers between openings with fixed ends is added. The calculations that follow illustrate the calculation of the rigidity for the wall in Fig. 10.17.

$$\Delta_{\text{wall}} = \Delta_{\text{solid wall}(c)} - \Delta_{\text{stripA}(c)}$$
$$+ \Delta_{2,3,4,5,6,7(f)} \qquad (10.19)$$

where $\qquad \Delta_{2,3,4,5,6,7(f)} = \dfrac{1}{R_{2,3,4,5,6,7(f)}} \qquad (10.20)$

and $\qquad R_{2,3,4,5,6,7(f)} = R_{2f} + R_{3,4,5,6(f)} + R_{7f}$  (horizontal combination) $\qquad (10.21)$

$$R_{3,4,5,6(f)} = \dfrac{1}{\Delta_{3,4,5,6(f)}} \qquad (10.22)$$

for which $\qquad \Delta_{3,4,5,6(f)} = \Delta_{\text{solid},3,4,5,6(f)} - \Delta_{\text{stripB}(f)} + \Delta_{3,4,5(f)}$

In this case, the fixed-end condition is assumed for all parts because of the effect of the rotational restraint provided by Part 1. For the horizontal combination of segments 3, 4, and 5,

$$\Delta_{3,4,5(f)} = \frac{1}{R_{3f} + R_{4f} + R_{5f}} \qquad (10.23)$$

For the dimensions shown,

$$\Delta_{\text{solid } c} = 2.778/E_m t \text{ and, for later comparison, } R_{\text{solid wall}} = 0.36 \, E_m t$$

$$\Delta_{\text{stripA(c)}} = 1.778/E_m t$$

$$\Delta_{\text{solid3,4,5,6(f)}} = 3.201/E_m t$$

$$\Delta_{\text{stripB (f)}} = 0.880/E_m t$$

$$\Delta_{3,4,5(f)} = 4.667/E_m t$$

$$\Delta_{3,4,5,6(f)} = 3.201/E_m t - 0.880/E_m t + 4.667/E_m t$$

$$= 6.987/E_m t, \text{ so that } R_{3,4,5,6(f)} = 0.143 \, E_m t$$

$$R_{2f} = R_{7f} = 0.004 \, E_m t$$

$$\Delta_{2,3,4,5,6,7(f)} = 6.623/E_m t$$

resulting in
$$\Delta_{\text{wall}} = 7.623/E_m t$$

or
$$R_{\text{wall}} = 1/\Delta_{\text{wall}} = 0.131 \, E_m t$$

For the wall in Fig. 10.17, compared to a solid wall, the openings result in rigidities of 41.4% and 36.4%, respectively, for Methods 1 and 2. Although there is no clear evidence regarding which method is more accurate, current practice is to use Method 2.

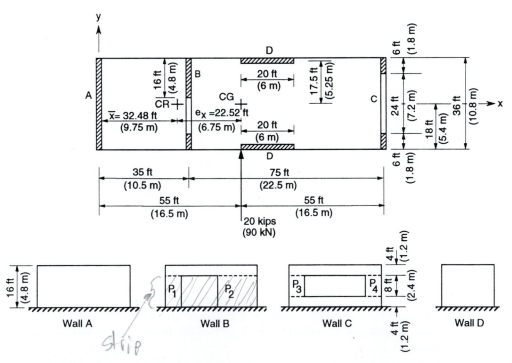

**Figure 10.18** Single-story building with rigid roof diaphragm (Distribution of lateral load): Example 10.1.

Shear Walls    Chap. 10

### 10.5.4 Example of Lateral Load Distribution to a System of Walls

**Example 10.1**

Distribute the 20 kip (90 kN) lateral load shown in Fig. 10.18 to walls A, B, and C in the single-story building for the case where the roof acts as a rigid diaphragm. Consider all walls to have the same material properties and thickness.

**Calculation of Wall Rigidities.** For walls constructed with similar materials of equal thickness, the term $Et$ is constant and is omitted in the following calculations.

**Wall A.** By using Eq. 10.6, $h = 16$ ft (4.8 m) and $L = 36$ ft (10.8 m),

$$R_A = \frac{1}{4\left(\dfrac{h}{L}\right)^3 + 3\left(\dfrac{h}{L}\right)} = \frac{1}{4\left(\dfrac{16}{36}\right)^3 + 3\left(\dfrac{16}{36}\right)} = 0.594$$

**Wall B.** By using Method 2 outlined in Sec.10.5.3,

$$R_B = \frac{1}{\Delta_B}$$

and

$$\Delta_B = \Delta_{solid(e)} - \Delta_{strip(c)} + \Delta_{1,2(c)}$$

where

$$\Delta_{solid(e)} = \frac{1}{R_A} = \frac{1}{0.594} = 1.68$$

and for the 12 ft (3.6 m) high strip

$$\Delta_{strip(c)} = \frac{1}{R_{strip(c)}} = 4\left(\frac{12}{36}\right)^3 + 3\left(\frac{12}{36}\right) = 1.15$$

In this case, Piers 1 and 2 are considered to be cantilevers because the shallow section above the opening does not provide much fixity at the top of the much stiffer piers.

$$\text{Pier 1, } R_{1c} = \frac{1}{4\left(\dfrac{12}{6}\right)^3 + 3\left(\dfrac{12}{6}\right)} = \frac{1}{38.0} = 0.026$$

$$\text{Pier 2, } R_{2c} = \frac{1}{4\left(\dfrac{12}{16}\right)^3 + 3\left(\dfrac{12}{16}\right)} = \frac{1}{3.94} = 0.254$$

$$R_{1,2(c)} = R_{1c} + R_{2c} = 0.280 \text{ and } \Delta_{1,2} = \frac{1}{R_{1,2}} = 3.57$$

Therefore, $\Delta_B = 1.68 - 1.15 + 3.57 = 4.10$ and $R_B = 1/4.10 = 0.244$ or 41% of $R_A$.

**Wall C.**

$$R_C = \frac{1}{\Delta_C}$$

and

$$\Delta_C = \Delta_{solid(c)} - \Delta_{strip(c)} + \Delta_{3,4(f)}$$

where

$$\Delta_{solid(c)} = 1.68$$

$$\Delta_{strip(c)} = 4\left(\frac{8}{36}\right)^3 + 3\left(\frac{8}{36}\right) = 0.711$$

Pier 3 is the same as Pier 4, and by using Eq. 10.7, for fixed-end conditions

$$R_{3f} = \frac{1}{\left(\dfrac{h}{L}\right)^3 + 3\left(\dfrac{h}{L}\right)} = \frac{1}{\left(\dfrac{8}{6}\right)^3 + 3\left(\dfrac{8}{6}\right)} = \frac{1}{6.37} = 0.157$$

$$R_{3,4(f)} = 2(0.157) = 0.314$$

and

$$\Delta_{3,4(f)} = 3.185$$

$$\Delta_C = 1.68 - 0.711 + 3.185 = 4.15$$

$$R_C = \frac{1}{\Delta_C} = 0.241 \text{ or } 41\% \text{ of } R_A$$

**Wall D.**

$$R_{D(c)} = \frac{1}{4\left(\dfrac{16}{20}\right)^3 + 3\left(\dfrac{16}{20}\right)} = 0.224$$

**Load Distribution.** From Eq. 10.1, the location of the center of rigidity, CR, can be calculated as

$$\bar{x} = \frac{\sum R_{yi}x_i}{\sum R_{yi}} = \frac{R_A(0) + R_B(35) + R_C(110)}{R_A + R_B + R_C}$$

$$= \frac{0.594(0) + 0.244(35) + 0.241(110)}{0.594 + 0.244 + 0.241} = 32.48 \text{ ft}$$

$$\left(\bar{x} = \frac{0.594(0) + 0.244(10.5) + 0.241(33)}{0.594 + 0.244 + 0.241} = 9.75 \text{ m}\right)$$

from the $y$-axis reference line. From the symmetry of walls $D$ oriented in the $x$ direction, $\bar{y} = 0$.

The eccentricity of load in the $y$-direction is

$$e_x = 55 - 32.48 = 22.52 \text{ ft.}$$

$$(e_x = 16.5 - 9.75 = 6.75\text{m})$$

For direct shear due to loading in the $y$-direction, Eq. 10.9 gives

$$V_{yit} = \frac{R_{yi}}{\sum R_{yi}} V_y$$

$$V_A = \left(\frac{0.594}{0.594 + 0.244 + 0.241}\right) 20 = 11.01 \text{ kips} \uparrow (49.55 \text{ kN})$$

$$V_B = \left(\frac{(0.244)}{(0.594 + 0.173 + 0.241)}\right) 20 = 4.52 \text{ kips} \uparrow (20.34 \text{ kN})$$

$$V_C = \left(\frac{(0.241)}{(0.594 + 0.244 + 0.241)}\right) 20 = 4.47 \text{ kips} \uparrow (20.11 \text{ kN})$$

Shear Walls    Chap. 10

For shear due to torsion, EQ. 10.10 gives

$$V_{yir} = \frac{R_{yi}x_i}{\sum R_{yi}x_i^2 + \sum R_{xi}y_i^2} V_y e_x$$

$$\sum R_{yi}x_i^2 = 0.594(32.48)^2 + 0.244(2.52)^2 + 0.241(77.52)^2 = 2076.4$$

$$\left( \sum R_{yi}x_i^2 = 0.594(9.75)^2 + 0.244(0.75)^2 + 0.241(23.25)^2 = 186.88 \right)$$

$$\sum R_{xi}y_i^2 = 0.224(17.5)^2 + 0.224(17.5)^2 = 137.20$$

$$(R_{xi}y_i^2 = 0.244(5.25)^2 + 0.224(5.25)^2 = 12.35)$$

$$V_A = \frac{(0.594)(32.48)}{2076.4 + 137.2} 20(22.52) = 3.93 \text{ kips } (17.9 \text{ kN}) \downarrow$$

Similarly,

$$V_B = 0.13 \text{ kips } (0.58 \text{ kN}) \uparrow$$

$$V_C = 3.80 \text{ kips } (17.02 \text{ kN}) \uparrow$$

and substituting $R_{xi}y_i$ for $R_{yi}x_i$ in the above equation,

$$V_D = 0.80 \text{ kips } (3.58 \text{ kN}) \text{ in the } x-\text{direction}$$

Because most building codes require that negative torsional shear be neglected [i.e., from Eq. 10.8, $V_{yi} \geq V_{yit}$], the design values for shear are

$$V_A = 11.01 \text{ kips } (49.55 \text{ kN})$$

$$V_B = 4.52 + 0.13 = 4.65 \text{ kips } (20.92 \text{ kN})$$

$$V_C = 4.47 + 3.80 = 8.27 \text{ kips } (37.13 \text{ kN})$$

$$V_D = 0.80 \text{ kips } (3.58 \text{ kN})$$

It is likely that lateral loading in the $x$-direction will govern for Wall D.

As can be seen, the torsional moment has a very significant effect on the distribution of shear forces and, in particular, it almost doubled the shear force for wall C containing the two 6 ft (1.83 m) long piers. ∎

## 10.6 DESIGN OF SHEAR WALLS

Although wall design for axial load and bending has been covered in Chap.8, some aspects of this part of design that are specifically related to shear walls are included in this section.

### 10.6.1 Unreinforced Shear Walls

The capacity of unreinforced masonry shear walls is sensitive to the weaker planes along the mortar bed and head joints and, typically, the shear mode of failure will be either shear slip along a bed joint or a stepped diagonal crack by breaking the bond along a combination of head and bed joints. In addition, the shear capacity is significantly affected by the level of axial compression, which has the effect of delaying tensile cracking and enhancing the shear-friction component of the shear strength. The recommended use of unreinforced masonry shear walls is limited to conditions that do not produce tensile stresses normal to the bed joints at working loads.

Currently, design codes allow for the interaction between shear and axial compression and between bending and axial compression. This design approach is consistent with failures that essentially relate to bond strength between the units and mortar.

**Axial Load.** As opposed to walls subject to out-of-plane bending, the slenderness effect on shear walls under high axial compression is not related to the slenderness in the in-plane direction of bending. It is a normal practice to determine the reduction in axial compression capacity as functions of height-to-thickness ratio, $h/t$, or height-to-radius of gyration ratio, $h/r$, where $t$ and $r$ relate to the out-of-plane direction and $h$ is the wall height between horizontal supports usually existing at each floor level. For typical ranges of floor height and wall thickness, the reduction of axial capacity is usually governed by material failure rather than by buckling.

If the part of the cross-section of the shear wall under high compressive stresses is flanged or composed of intersecting walls tied together, the web and flange areas tend to brace each other and minimize the effects of slenderness. In such cases, it may be justifiable to use a reduced slenderness or even ignore slenderness effects.

For working stress design, a common approach in masonry codes[10.25–10.27] has been to define the allowable axial stress by

$$F_a = \alpha f'_m R \tag{10.24}$$

where $\alpha$ defines the basic level of safety desired and is usually taken between 0.2 and 0.3 so that the nominal factor of safety $(1/\alpha)$ is between 3 and 5. $R$ is the capacity reduction factor for slenderness, and $f'_m$ is the specified masonry compressive strength.

A variation of this for design based on the ultimate limit state[10.28] is

$$P_n = \beta A_n f_k / \gamma_m \tag{10.25}$$

where $\beta$ = a capacity reduction factor to account for slenderness and accidental eccentricity in the out-of-plane direction

$A_n$ = net cross-sectional area

$f_k$ = characteristic compressive strength

$\gamma_m$ = material partial safety factor, which varies depending on the quality of material manufacturing control and the category of construction control

**Combined Axial Load and Bending.** Because it is good practice not to allow in-plane flexural tensile stresses on bed joints of unreinforced shear walls at working loads, the allowable bending moment may be limited by the level of axial load as indicated by the relationship,

$$P/A_n \geq M/S \qquad (10.26)$$

Flexural tension in shear walls due to in-plane bending should not be allowed at working loads because the shear walls are critical elements in the overall stability of a building. The uncracked section is also relied upon to resist in-plane shear.

To check for compression-controlled capacity, the unity equation can be used:[10.25,10.26]

$$\frac{f_a}{F_a} + \frac{f_b}{F_b} = 1.0 \qquad (10.27)$$

where  $f_a = P/A_n$ (axial load divided by cross-sectional area)
$F_a$ = allowable compression stress for axial load including capacity reduction for slenderness
$A_n$ = net cross-sectional area of the wall
$f_b = M/S$ (bending moment divided by the section modulus)
$F_b$ = allowable compression stress for bending

[Note, In USA Codes, a one-third increase in allowable stress is normally allowed for load combinations that include wind or earthquake loads.]

For intersecting walls, flanges can be considered to be effective in resisting the applied axial load and bending moment provided that the connection is capable of transferring the shear. This can be accomplished using one of the following construction details:[10.25]

1. Fifty percent of the units interlocked at the web-flange interface; see Fig. 10.19(a).
2. Mechanical connection with steel anchors at a maximum spacing of 4 ft (1.2 m); see Fig. 10.19(b).
3. Provide reinforced bond beams at a maximum spacing of 4 ft (1.2 m); see Fig. 10.19(c).   $A_g = .1 \text{ in}^2$

The width of the flange considered effective on each side of the web can be taken equal to six times the flange thickness, equal to the actual flange width, or equal to half the distance to the next cross wall, whichever is least.

**Shear.**    For unreinforced masonry shear walls, even though the mode of failure and deformation behavior are affected by the interactions between bending moment and shear (as represented by the height-to-length ratio), design provisions tend to be based on material properties and the level of axial load. For North American codes, the allowable shear stresses have been dependent on the type of material (concrete or clay)[10.26,10.27] but this difference has been eliminated in ACI 530/ASCE 5/TMS 402.[10.25] Limits on maximum values also exist either as fixed values, in terms of the square root of the specified compressive strength, or indirectly through limits on compressive stress.

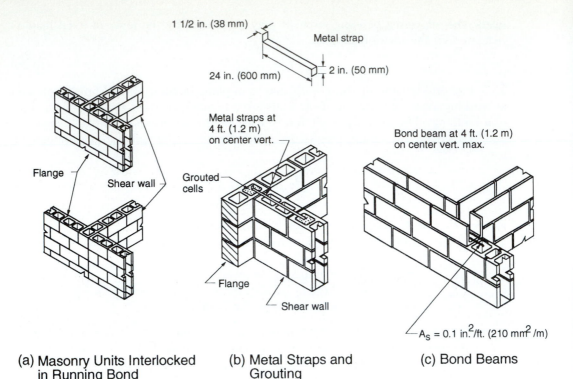

**Figure 10.19**  Masonry wall intersections.

The beneficial effect of axial compressive stress is generally accounted for using the following equation:[10.25,10.27]

$$F_v = F_{vo} + \mu f_{a,\text{dead}} \qquad (10.28)$$

where  $F_v$ = allowable shear stress
  $F_{vo}$ = allowable shear bond stress at zero axial compression
  $\mu$ = coefficient resembling a coefficient of friction and normally taken as a value between 0.3 and 0.45
  $f_{a,\text{dead}}$ = axial compressive stress due to the dead load that gives the minimum benefit

In terms of the overall shear capacity of the wall, the allowable shear force has traditionally been calculated using the relationship

$$V = F_v L t \qquad (10.29)$$

where  $F_v$ = allowable shear stress
  $L$ = length of web of the wall
  $t$ = effective thickness of the web of the wall

However, for unreinforced masonry, the brittle nature of cracking has led to dropping the average stress concept[10.25] and, instead, the shear stress is calculated at a point in a section using

$$\bar{\tau}_v = \text{min of} \begin{cases} \text{max specified} \\ 1.5\sqrt{f_m} \\ f_{vo} + \mu \frac{t_a}{dead} \end{cases}$$

$$f_v = \frac{VQ}{It} \qquad (10.30)$$

where  $V$ = design shear force

$Q$ = first moment of the area of the section beyond the point in question about the centroid of the section

$I$ = moment of inertia

$t$ = thickness of the wall at the point in question

For rectangular sections, the 50% increase in maximum shear using Eq. 10.30 compared to Eq. 10.29 has been offset by adopting higher allowable stresses.

### 10.6.2 Reinforced Shear Walls

The behavior of reinforced masonry shear walls is quite different from unreinforced walls because the reinforcement and grout contribute significantly to both the load-carrying and the deformation properties. Allowable stresses, based on elastic analysis of working loads, are actually determined by factoring observed ultimate strength results. Therefore, the more modern design provisions[10.26,10.28] tend to have forms similar to the ultimate strength provisions for reinforced concrete.

Tension reinforcement uniformly distributed along the length of the wall results in some improvements in shear wall behavior and this has become a recent requirement of masonry codes.[10.29,10.30] Benefits include better resistance to sliding, a higher compression-splitting resistance at the toe of the wall, and more effective control of diagonal cracking. Another advantage is the reduced congestion of reinforcement near the ends of the wall. Similarly, horizontal reinforcement uniformly distributed over the wall height has proven to be very effective in resisting shear and ensuring ductile flexural failure.

**Axial Load.**   Design codes based on working stresses[10.25-10.27] specify allowable axial stress in reinforced masonry similar to unreinforced masonry shear walls. Because of the typically low percentage of vertical reinforcement in masonry shear walls and because the steel stress is generally quite low, the contribution of compression reinforcement to axial load-carrying capacity is often neglected. If it is included, special provisions for lateral support of the reinforcement by ties must be satisfied. For walls designed with reinforcement located in the center of the wall, the efficiency of any such tying scheme is questionable. In zones of high seismic risk, an exception is the use of special nonstandard details, such as spiral reinforcement at the toe of the wall, to help develop the strength and ductility contributions of this reinforcement.

For strength design, adequate tying to permit use of the full yield stress of compression steel can add a significant component to the axial capacity as shown in the following equation:[10.29]

$$P_o = 0.85f'_m(A_n - A_s) + A_s F_y \qquad (10.31)$$

where $f'_m$ is the masonry compressive strength, $A_n$ is the masonry net area, and $A_s$ and $F_y$ are area and yield strength of vertical steel, respectively. Axial design strength

$P_u$, provided by the shear wall cross-section is defined by the UBC code[10.26] as

$$P_u \leq \Phi \, (0.80) \, P_o \qquad (10.32)$$

where $\Phi = 0.65$. The 0.80 factor is included to provide additional reserve strength for this critical loading condition.

**Combined Axial Load and Bending.**   For working stress design, calculation of stresses in the reinforcement and determination of the required amount of reinforcement are based on the same type of linear elastic analysis as presented in Chap. 8 for walls subject to out-of-plane bending. The allowable tension in the reinforcement must not be exceeded and the total compressive stress must be limited to the allowable value normally in the range of $0.23f'_m$ to $0.33f'_m$. Certain codes[10.25,10.26] permit a one-third increase in allowable stress for load combinations that include short term duration loads due to wind or earthquake. Other codes,[10.27] based on probability of simultaneous occurrence of maximum loads, allow a similar one-third increase for the combination of wind or seismic load plus dead and live loads. All flexural tension should be carried by the vertical reinforcing steel assuming zero tensile strength for masonry.

For intersecting walls, the effective flange width defined in masonry codes can be considered in stress calculations. However, the designer should check to ensure that the intersection has adequate shear capacity. The presence of unsymmetric flanges at the ends of the wall result in unsymmetric response of the walls under lateral load reversal.

As discussed in Chap. 8 (Sec. 8.4), for codes in which the allowable axial compressive and flexural compression stresses differ or where the slenderness is accounted for in the allowable axial stress $F_a$, the unity equation can again be used to combine the effects of axial load and bending moment. In its simplest form, the compressive stresses $f_a$ and $f_b$ due to the axial load and bending moment, respectively, can be calculated independently and then introduced into Eq. 10.27:

$$\frac{f_a}{F_a} + \frac{f_b}{F_b} \leq 1.0 \quad \text{or } 1.33$$

However, a less conservative approach that allows calculation of stresses under the combined loading is to use the allowable compression stress defined by Eq. 8.8.

Where slenderness of the shear wall in the weak axis (out-of-plane) direction is significant or where significant out-of-plane bending occurs simultaneously with in-plane bending, codes that use the moment-magnifier approach[10.27] can incorporate the out-of-plane effect. Considering biaxial bending using the contour method described by the relationship in Eq. 9.14 but using $\alpha = \beta = 1.0$, because of the aspect ratio of the wall section,

$$\frac{M_{x,\text{design}}}{M_x} + \frac{M_{y,\text{design}}}{M_y} = 1 \qquad (10.33)$$

where $M_{x,design}$, $M_{y,design}$ = design moments in the $x$ and $y$ directions, respectively, with the moment in the out-of-plane direction magnified to account for slenderness

$M_x$, $M_y$ = allowable moment capacities in the x and y directions corresponding to the design axial load

It should be noted that, unless out-of-plane bending is used as the means for assessing slenderness effects, the effects of minimum out-of-plane eccentricity usually are not included in the design of masonry (or, for that matter, reinforced concrete) shear walls. Fortunately, as has been found for reinforced concrete shear walls, slenderness is usually not a major factor and this extra complexity can usually be avoided.

The ACI 530/ASCE 5/TMS 402[10.25] approach to slenderness effects for reinforced masonry specifies a separate limit on the magnitude of compressive stress due to axial load in addition to the allowable compressive stress for combined axial loads and bending. This guards against significant effects of slenderness and is a simple and efficient method of design. Section 8.5.4 and example calculations in Chap. 8 illustrate this approach.

With strength design[10.26] and limit states design[10.30] methods, the capacities at critical sections can be calculated using assumptions similar to those for reinforced concrete.

Based on the assumptions of the UBC code, which are similar to those in ACI 318-89 for reinforced concrete members,[10.29] Shing et al.[10.12] proposed the following equations for axial capacity $P_n$ and moment capacity $M_n$ of reinforced masonry shear walls with symmetrically distributed flexural steel:

$$P_n = 0.72f'_m cb - \sum_{i=1}^{n} A_{si}f_{si} \qquad (10.34)$$

$$M_n = 0.72f'_m cb \left( \frac{L}{2} - \frac{0.85c}{2} \right) + \sum_{i=1}^{n} A_{si}f_{si} \left( d_i - \frac{L}{2} \right) \qquad (10.35)$$

where
$$f_{si} = 0.003E_s \frac{d_i - c}{c} \le f_y \qquad (10.36)$$

For $c$, $b$, $d_i$, $L$, and other terms refer to Fig. 10.20(a).

For cases where the axial load is relatively small compared to bending and where relatively light vertical reinforcing is used, the small compression zone results in a strain distribution that ensures yielding of virtually all of the tension steel.[10.19] In such situations, as illustrated in Fig. 10.20(b), the nominal moment capacity, $M_n$, can be approximated using an equation of the form[10.13,10.19]

$$M_n = A_{st}f_y \frac{d}{2} + P_n \frac{L}{2} \qquad (10.37)$$

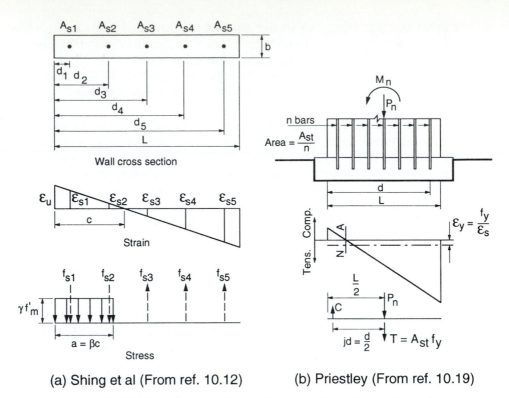

Wall cross section

Strain

Stress

(a) Shing et al (From ref. 10.12)

(b) Priestley (From ref. 10.19)

**Figure 10.20** Contribution of vertical steel in flexural strength design.

where $A_{st}$ = total area of vertical reinforcement
$f_y$ = yield strength of reinforcement
$d$ = distance from the extreme compression fiber to the extreme layer of reinforcement
$L$ = length of wall
$P_n$ = nominal axial compression capacity

Strength reduction factors ranging from 0.65 to 0.85, according to the influence of axial load, have been recommended[10.19,10.29] to account for statistical variations in material properties and dimensions of elements plus inaccuracies in the analysis.

It has been shown[10.31] that little penalty (generally less than 5%) results from uniformly distributing the vertical reinforcement compared to placing equal amounts of steel concentrated toward the wall ends. Examples of dimensionless interaction diagrams at low levels of axial load for unconfined shear walls with distributed flexural reinforcement are presented in Fig. 10.21. Such design aids are relatively easy to construct and are very useful.

**Shear.** If the allowable shear stresses for reinforced masonry are exceeded, current practice in North American codes is to provide horizontal shear reinforcement to resist the entire shear force for working stress design. However, to guard against the possibility of shear-compression failures, even though the horizontal steel is

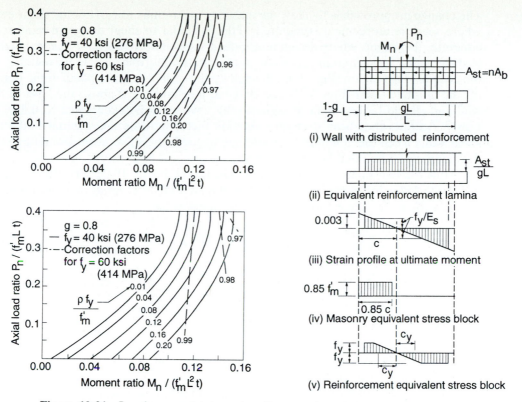

**Figure 10.21** Load-moment interaction diagrams for reinforced masonry shear walls with distributed flexural steel (from Ref. 10.31).

designed to carry the entire shear force, the nominal shear stress in the masonry, calculated using $f_v = V/bjd$, must be limited. The limits on nominal shear stress in the masonry have typically been expressed as some function of the specified compressive strength of masonry, $f'_m$, and more recently[10.25,10.26] as a function of the aspect ratio of the wall represented by the $M/Vd$ ratio. Higher limits on these shear stresses correspond to lower $M/Vd$ values.

In the strength design provisions of the Uniform Building Code,[10.26] nominal shear capacity $V_n$ is specified as

$$V_n = V_m + V_s \tag{10.38}$$

where

$$V_m = C_d A_{mv} \sqrt{f'_m} \tag{10.39}$$
$$V_s = A_{mv} \rho_n f_y \tag{10.40}$$

$A_{mv}$ = net area of horizontal section of the wall

$C_d$ = shear strength coefficient, which, for $f'_m$ in psi, is equal to 2.4 for $M/Vd \le 0.25$ and 1.2 for $M/Vd \ge 1.0$ (0.20 and 0.10, respectively for $f'_m$ in MPa). (Use straight-line interpolation for $M/Vd$ values between 0.25 and 1.0.)

$\rho_n$ = reinforcement ratio (of distributed steel) on a vertical plane

The change in approach is to allow the masonry to continue to carry some shear even where shear reinforcement is required. This is similar to shear design in reinforced concrete. However, where design requirements specify provision for plastic hinging in shear walls, the shear reinforcement must carry the entire shear in these regions. In the critical hinge region (Fig. 10.6), defined by the wall base and a plane at the height $h_p$, the ultimate shear force $V_u$ is calculated at a distance $h_p/2$ above the base of the wall and should be less than or equal to $\Phi V_n$, where $\Phi$ is taken equal to 0.6. The UBC[10.26] allows the shear strength reduction factor $\Phi$ to be taken equal to 0.8 for shear walls when the nominal shear strength exceeds the shear corresponding to the development of nominal flexural strength.

For the two ranges of aspect ratios illustrated in Fig. 10.22, Priestley[10.19] proposed the following equations to calculate the area of shear reinforcement required to resist the entire shear force in cantilever shear walls:

$$A_v = \frac{V_u s}{\Phi_s f_y d} \qquad \text{for } M/Vd > 1.0 \qquad (10.41)$$

and

$$A_v = \frac{V_u s}{\Phi_s f_y L} \qquad \text{for } M/Vd < 1.0 \qquad (10.42)$$

where  $V_u$ = ultimate shear force
  $s$ = vertical spacing of horizontal reinforcement
  $f_y$ = yield strength of horizontal reinforcement
  $d$ = effective depth of vertical tension reinforcement
  $L$ = wall length
  $\Phi_s$ = strength reduction factor = 0.80

### 10.6.3  Special Seismic Design Considerations

In high seismic risk areas, special design considerations and reinforcement details should be implemented in order to achieve good performance.

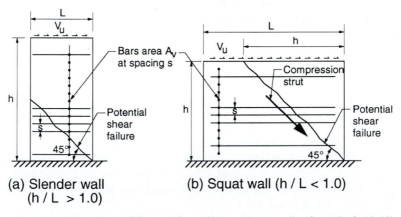

**Figure 10.22**   Shear failures of cantilever shear walls (from Ref. 10.19).

**Layout.** Shear walls should be adequately distributed in the two orthogonal directions of the building and they should be placed as symmetrically as possible to minimize torsional effects. The designer should try to avoid complicated geometry and wall coupling and rely on the simple predicted inelastic response of cantilever walls for energy dissipation.[10.22] Height-to-thickness ratios of shear walls should be checked against lateral instability. Further discussion of wall layout is presented in Sec. 3.6.4.

**Ductility.** Because masonry structures are very stiff, response acceleration during an earthquake can be very high; see Fig. 10.23(a). For 5% critical damping, a peak elastic response could be as high as 0.8g.[10.19] It is not economically feasible to design masonry structures for such high seismic loads using elastic response. According to current building code provisions, the design loads are typically reduced by a factor of 3 to 4, Fig. 10.23(b), implying a certain ductility demand.[10.19] Ductility of shear walls is an indicator of capability for inelastic deformation beyond the peak load without appreciable loss in strength and with minimal stiffness degradation.

Ductility can be calculated based on an elastoplastic approximation of load-deflection response of cantilever shear walls using an equivalent compressive stress block. Plots of ductility capacity of unconfined and confined shear walls are presented in Fig. 10.24. As can be seen, the percentage of steel, the level of axial load, and the confinement are the critical parameters affecting wall ductility. Lower percentages of flexural (vertical) steel and lower levels of axial load result in higher ductility capacities.

**Story Drift.** Horizontal displacement or *interstory drift* should be limited to control the $P$–$\Delta$ effect and ensure stable behavior in the inelastic range of response. A drift limit of 0.005 times the wall height under unfactored loads is specified by UBC[10.26] for design of shear walls.

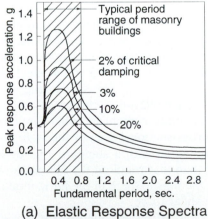

(a) Elastic Response Spectra
(El Centro 1940 N-S)

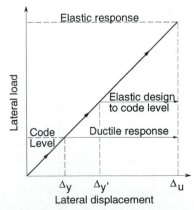

(b) Peak Lateral Load Using "Equal
Displacement Principle"

**Figure 10.23** Seismic loads on masonry buildings (from Ref. 10.19).

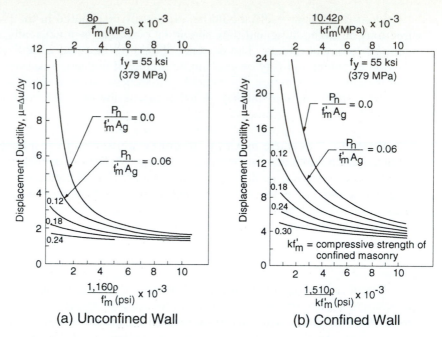

**Figure 10.24** Ductility of reinforced masonry shear walls with an aspect ratio of three (from Ref. 10.19).

**Anchorage.** Masonry shear walls should be adequately anchored to all floors and the roof. This anchorage should provide direct connection capable of resisting the design horizontal forces due to seismic loads. United States codes[10.25,10.26] require a minimum horizontal design force of 200 lb/ft (2.92 kN/m). Anchors should be fully embedded in reinforced bond beams or reinforced vertical cells.

**Reinforcement.** Horizontal reinforcement required to resist shear should be uniformly distributed and embedded in mortar or grout. The maximum spacing in each direction, according to North American code provisions, should not exceed the smaller of one-half of the length or height of the wall nor more than 48 in. (1.2 m) within the plastic hinge region. In Seismic Zones 3 and 4, steel is required in both the horizontal and vertical directions.[10.25–10.27] The sum of the areas of the distributed reinforcement should be at least 0.002 times the gross cross-sectional area of the wall and the minimum area of reinforcement in either direction should not be less than 0.0007 times the gross cross-sectional area of the wall.

Shear reinforcement should be terminated with a standard 135° or 180° hook or with an extension equal to or greater than that required for full anchorage. This requirement is essential to fully develop the tensile force in the horizontal reinforcement so that it can effectively resist the shear force.

Sufficient horizontal reinforcement should be provided in shear walls laid in other than running bond to provide full shear-transfer capability at the head joints. In this situation, ACI 530/ASCE 5/TMS 402[10.25] requires a minimum amount of horizontal steel of 0.0015 times the wall cross-section. The New Zealand Code does not allow stack pattern construction in the critical plastic hinge regions of shear walls.[10.30]

Reinforcement should be continuous around wall corners and through intersections. The designer should avoid locating lap splices in the critical hinge regions.

**Boundary Members.** Boundary members consisting of thickened and specially reinforced sections should be provided at the end of the shear wall when the failure mode is flexural and when the maximum compressive fiber stress from factored loads exceeds $0.2f_m'$.[10.26] Boundary members should also be reinforced to confine the vertical steel at the ends of walls when the failure mode is flexural and the maximum compressive stress exceeds $0.4f_m'$. Confinement of boundary members using horizontal steel or confining plates results in significantly increased deformation capacity and ductility of the wall as described earlier in Sec. 10.3.2.

### 10.6.4 Wall Connections

Connections between walls and floors and between intersecting walls play an important role in the structural response of multistory masonry buildings and in their resistance to lateral load. Typical connection locations are shown in Fig. 10.25. Floor-to-wall connections must be capable of transferring lateral forces from the floor and roof diaphragms to the shear walls. The pre- and post-fracture behavior of a masonry connection depends on the type of floor slab (precast concrete, cast in-place concrete, and hollow core planks), the level of precompression due to gravity loads, and whether or not the wall is reinforced.[10.33] In unreinforced masonry construction, the connection strength is mainly dependent upon the shear-friction resistance due to compression loads. In reinforced masonry construction, the pre-cracking strength is also a function of the shear-friction resistance, whereas the post-cracking stiffness and strength are dependent on the dowel action of the vertical reinforcing bars crossing the wall-floor interface and the corresponding shear-friction

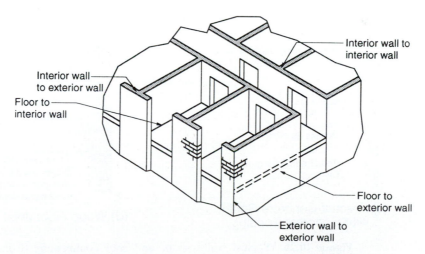

Interior wall to interior wall

Interior wall to exterior wall

Floor to interior wall

Floor to exterior wall

Exterior wall to exterior wall

**Figure 10.25**  Wall connections in multistory masonry building.

capacity. Tests of full-scale wall-floor connections[10.33] showed that load cycling causes stiffness degradation of the interface. Typical construction details of floor-wall connections are shown in Fig. 10.26.

If the wall designs utilize flanges to resist loads, proper details should be used to ensure adequate shear transfer between the intersecting walls. This can be accomplished by interlocks of units, metal anchors, or bond beams as discussed in Sec. 10.6.1.

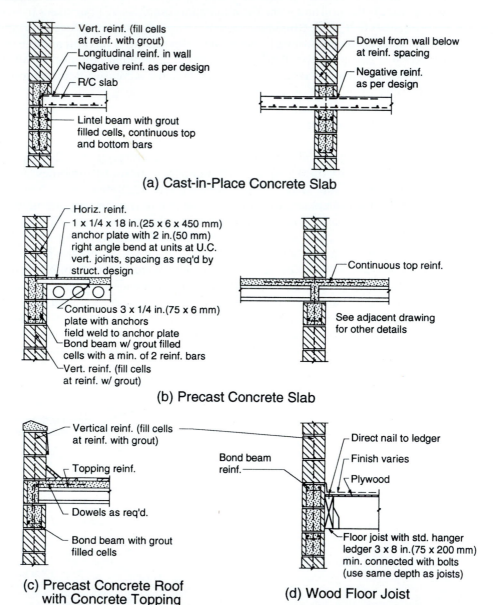

(a) Cast-in-Place Concrete Slab

(b) Precast Concrete Slab

(c) Precast Concrete Roof with Concrete Topping

(d) Wood Floor Joist

**Figure 10.26** Typical wall-to-floor and roof connections (from Ref. 10.32).

# 10.7 DESIGN EXAMPLES

### 10.7.1 Unreinforced Shear Wall

#### Example 10.2

Design the solid brick masonry wall shown in Fig. 10.27(a) to carry the prescribed vertical and lateral loads. The moment indicated is due to the overturning moment from the floors above.

Given:

6 in. (150 mm) (nominal) clay brick
$f'_m = 4000$ psi (27.6 MPa)
Allowable shear stress = 120 psi (0.83 MPa)
Allowable axial compressive stress = 1000 psi (6.9 MPa) (ignoring slenderness effect)
Allowable flexural compressive stress = 1330 psi (9.18 MPa)

[These allowable stresses are typical North American design values.[10.25]]
**Solution A:** Assume that the intersecting walls are not connected (ignore the effect of cross walls).

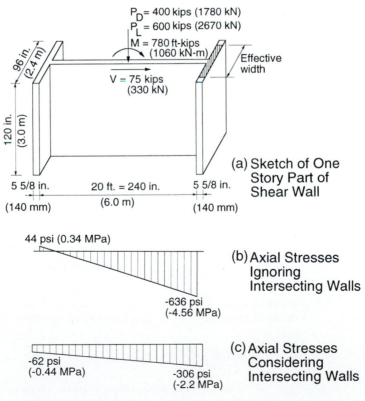

$P_D$ = 400 kips (1780 kN)
$P_L$ = 600 kips (2670 kN)
M = 780 ft-kips (1060 kN-m)
V = 75 kips (330 kN)

96 in. (2.4 m)

120 in. (3.0 m)

Effective width

5 5/8 in. (140 mm)    20 ft. = 240 in. (6.0 m)    5 5/8 in. (140 mm)

**(a) Sketch of One Story Part of Shear Wall**

44 psi (0.34 MPa)

-636 psi (-4.56 MPa)

**(b) Axial Stresses Ignoring Intersecting Walls**

-62 psi (-0.44 MPa)

-306 psi (-2.2 MPa)

**(c) Axial Stresses Considering Intersecting Walls**

**Figure 10.27** Influence of shear wall flanges: Example 10.2.

**1.** Section properties

$$A = (5.625)(240) = 1350 \text{ in.}^2$$

$$(A = (140)(6000) = 8.4 \times 10^5 \text{ mm}^2)$$

$$I = 5.675(240)^3/12 = 6,480,000 \text{ in.}^4$$

$$(I = 140(6000)^3/12 = 2.52 \times 10^{12} \text{ mm}^4)$$

**2.** Normal stresses (consider the bottom section of the wall): For tension controlled capacity, $P_D = 400$ kips (1780 kN) (ignoring self-weight).

$$M = 780 + 75 \times \frac{120}{12} = 1530 \text{ ft-kips}$$

$$(M = 1060 + 330(3) = 2050 \text{ kN-m})$$

$$f_m = -\frac{P}{A} \pm \frac{My}{I} = -\frac{400 \times 1000}{1350} \pm \frac{1530 \times 12000 \times 120}{6480000}$$

$$= 44 \text{ psi tension or } 636 \text{ psi compression}$$

$$\left( f_m = \frac{-1780 \times 10^3}{8.4 \times 10^5} \pm \frac{2050 \times 10^6 \times 3000}{2.52 \times 10^{12}} \right.$$

$$\left. = 0.34 \text{ MPa tension or } 4.56 \text{ MPa compression} \right)$$

The resulting stress distribution is shown in Fig. 10.27(b). Because tensile stresses occur, the section should be reinforced (see Sec. 10.6.1).

**Solution B:** Assume that the intersecting walls are connected. For an effective flange width of 6t on either side of the web:

$$b_{\text{eff}} = 5.625 + 2 (6 \times 5.625) = 73.1 \text{ in.}$$

$$(b_{\text{eff}} = 140 + 2(6 \times 140) = 1820 \text{ mm})$$

**1.** Section properties:

$$A = 1350 + (73.1)(5.625) = 2172 \text{ in.}^2$$

$$(A = 8.4 \times 10^5 + 2(1820)(140) = 1.35 \times 10^6 \text{ mm}^2)$$

$$I = \frac{5.625(240)^3}{12} + 2\left[ \frac{73.1(5.625)^3}{12} + 73.1 \times 5.625 \times (122.8)^2 \right] = 18,883,452 \text{ in.}^4$$

$$\left( I = \frac{140(6000)^3}{12} + 2\left[ \frac{1820(140)^3}{12} + 1820(140)(3070)^2 \right] = 7.33 \times 10^{12} \text{ mm}^4 \right)$$

**2.** Normal stresses:

For the distance from the centroid to the extreme fiber of $y = 125.62$ in. (3140 mm),

$$f_m = -\frac{P}{A} \pm \frac{My}{I}$$

$$= 62 \text{ psi (0.44 MPa) compression or } 306 \text{ psi (2.20 MPa) compression}$$

The resulting stress distribution is shown in Fig. 10.27(c).

Therefore, no tension stresses are developed and flexural steel is not required. Compressive stress compliance can be checked using the unity equation. Using full dead and live loads,

$$\text{Axial stress } f_a = \frac{P}{A} = 460 \text{ psi (3.30 MPa)}$$

$$\text{Bending stress } f_b = \frac{My}{I} = 122 \text{ psi (0.88 MPa)}$$

$$\frac{f_a}{F_a} + \frac{f_b}{F_b} = \frac{460}{1000} + \frac{122}{1330} = 0.55 < 1 \quad \therefore \text{O.K.}$$

$$\left( \frac{3.30}{6.9} + \frac{0.88}{9.18} = 0.57 < 1 \quad \therefore \text{O.K.} \right)$$

The significance of including the flanges in reducing the normal stresses due to axial load and bending moments is evident.

**Shear Design:** Shear stresses (for either Solution A or B). Considering only the web to carry the shear,

$$f_v = \frac{V}{A_{\text{web}}} = \frac{75,000}{5.625(240)} = 56 \text{ psi}$$

$$\left( f_v = \frac{330(1000)}{140(6000)} = 0.39 \text{ MPa} \right)$$

[Note: For Solution B, the web length could include the thicknesses of the two flanges.] Typical[10.25] allowable shear stress $F_v$ without shear reinforcement can be taken as the least of:

$F_v = 120 \text{ psi (0.83 MPa)}$

$F_v = 1.5 \sqrt{f_m'} = 1.5 \sqrt{4000} = 95 \text{ psi (0.66 MPa)} \leftarrow \text{controls}$

$F_v = 37 + 0.45 P_D/A = 37 + 0.45(400 \times 1000/2172) = 120 \text{ psi (0.83 MPa)}$

Therefore, because the applied shear stress $f_v = 56$ psi (0.39 MPa) is less than the allowable shear stress, $F_v = 95$ psi (0.66 MPa), no shear reinforcement is required.

For Solution B, the intersection between the web and the flange should be capable of carrying the induced shear. This can be accomplished by interlocking of the units at alternate courses or by mechanical anchors (Fig. 10.19). ■

## 10.7.2 Reinforced Shear Wall

### Example 10.3

Design the reinforced concrete block wall shown in Fig. 10.28(a) to carry the prescribed loads using working stress design (Solution A) and strength design (Solution B). The wall is in a moderate seismic area. The design is for minimum dead loads and full seismic loads.

The following are given:

Fully grouted wall construction, $f_m' = 3000$ psi (20.7 MPa), $E_m = 2500$ ksi (17,250 MPa)

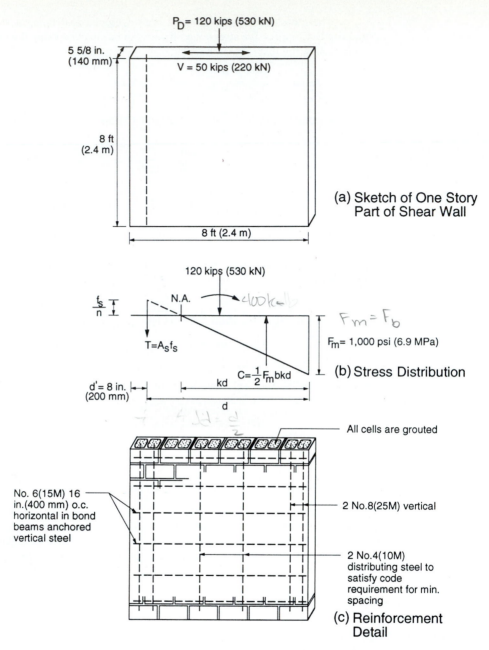

**Figure 10.28** Working-stress design of masonry shear wall: Example 10.3.

Allowable axial compression stress $F_a$ = 750 psi (5.18 MPa)
Allowable shear stress $F_v$ = 60 psi (0.41 MPa)
Allowable flexural stress $F_b$ = 1000 psi (6.9 MPa)
Grade 60 steel (400 MPa), allowable stress, $F_s$ = 24 ksi (165 MPa), $E_s$ = 29,000 ksi (200,000 MPa)
Allowable bond stress for reinforcement = 160 psi (1.1 MPa)

**Solution A: Working stress design:** The critical load combination for calculating the required amount of vertical steel is $0.9D + E$.

1. Combined axial load and bending:

$$f_a = \frac{P}{A} = \frac{0.9(120)(1,000)}{(5.625)(96)} = 200 \text{ psi}$$

(8×12 in)

$$\left( f_a = \frac{0.9(530)(1,000)}{140(2400)} = 142 \text{ MPa} \right)$$

$$f_b = \frac{M}{S} = \frac{(50 \times 8)(12,000)}{(5.625)(96)^2/6} = 556 \text{ psi}$$

$$\left( f_b = \frac{220(2.4)(10^6)}{140(2400)^2/6} = 3.93 \text{ MPa} \right)$$

Because $f_b > f_a$, tension occurs and tension steel is required.

Using the equilibrium and strain compatibility approach as illustrated in Chap. 8, the locations of the neutral axis and forces, shown in Fig. 10.28(b), can be calculated as follows, where $F_m = F_b$ for combined stresses due to axial load and bending.

$\Sigma M_a$ about the tension steel = 0.

$$\frac{1}{2} F_m t k d \left[ (L - d') - \left( \frac{kd}{3} \right) \right] = P \left( \frac{L}{2} - d' \right) + M$$

(50×6)

$$\frac{1}{2}(1000)(5.625)(kd) \left[ (96 - 8) - \left( \frac{kd}{3} \right) \right] = 0.9(120)\,1000(48 - 8) + 400(12,000)$$

$$\left( \frac{1}{2}(6.9)(140)(kd) \left[ (2400 - 200) - \frac{kd}{3} \right] = 0.9(530)(1000)(1200 - 200) + 528(10^6) \right)$$

Solving for kd:

$kd$ = 44.5 in. (1,130 mm) and $k$ = 44.5/88 = 0.51 (1130/2200 = 0.51)

Sum of vertical forces = 0.

$$T = C - P$$

$$C = \frac{1}{2}(F_m)(t)(kd) = 125 \text{ kips (544 kN)}$$

$$T = 125 - 0.9(120) = 17 \text{ kips}$$

$$(T = 544 - 0.9(530) = 67 \text{ kN})$$

From strain compatibility, assuming that plane sections remain plane after deformation,

$$f_s = \left(\frac{1-k}{k}\right) nF_m$$

$$= \left(\frac{1-0.51}{0.51}\right) \left(\frac{29000}{2500}\right) (1000) = 11{,}145 \text{ psi}$$

$$\left( f_s = \left(\frac{1-0.51}{0.51}\right) \left(\frac{200{,}000}{17{,}250}\right) 6.9 = 76.9 \text{ MPa} \right)$$

$$< 24 \text{ ksi (165 MPa)} \qquad \therefore \text{O.K.}$$

$$A_s = T/f_s = \frac{17}{11.45} = 1.53 \text{ in.}^2$$

$$\left( A_g = \frac{67(1000)}{76.9} = 872 \text{ mm}^2 \right)$$

Provide two No. 8 bars ($A_s = 2 (0.79) = 1.58$ in.$^2$; (two 25M bars gives $A_s = 2(500) = 1000$ mm$^2$) in the two cells at each end of the wall. See the details in Fig. 10.28(c).

2. Shear design for reinforced masonry:

$$f_v = \frac{V}{bd} = \frac{50 \times 1{,}000}{(5.625)(96 - 8)} = 101 \text{ psi}$$

$$\left( f_v = \frac{220(1000)}{140(2400 - 200)} = 0.71 \text{ MPa} \right)$$

Therefore, because $f_v > F_v = 60$ psi (0.41 MPa) shear reinforcement must be provided to carry all shear.

Consider bars in bond beams for every other course, $s = 16$ in. (400 mm),

$$A_v = \frac{Vs}{F_s d} = \frac{50{,}000(16)}{24{,}000(88)} = 0.38 \text{ in.}^2$$

$$\left( A_v = \frac{220 \times 1000(400)}{165(2200)} = 242 \text{ mm}^2 \right)$$

Use No. 6 bars (20M) ($A_s = 0.44$ in.$^2$ (300 mm$^2$) spaced 16 in. (400 mm) o.c. for a total of six bars. See details in Fig. 10.28(c). This horizontal reinforcement should be adequately anchored around the vertical steel at the ends.

3. Bond stress: For two No. 8 (two 25M) bars, $\Sigma_0 = 2(3.14) = 6.28$ in. (157mm)

$$u = \frac{V}{\Sigma_0 jd}$$

$$= \frac{50 \times 1{,}000}{(6.28)(0.87)(88)} = 104 \text{ psi} < 160 \text{ psi} \qquad \therefore \text{ bond stress O.K.}$$

$$\left( u = \frac{220 \times 1000}{(157)(0.87)(88)} = 0.73 \text{ MPa} < 1.1 \text{ MPa} \qquad \therefore \text{ bond stress O.K.} \right)$$

[Note that some codes specify a ⅓ increase in allowable stresses for design under wind or seismic forces. This may reduce the amounts of flexure and shear reinforcement required.]

**Solution B:  Strength Design:**

1. Combined axial and bending: The critical load combination for calculating the required amount of vertical steel, for the case where axial load is beneficial, is $0.9D + 1.4E$:

   Factored loads at base section:

   $$P_u = 0.9(120) = 108 \text{ kips}$$

   $$(P_u = 0.9(530) = 477 \text{ kN})$$

   $$M_u = 1.4(400) = 560 \text{ ft-kips}$$

   $$(M_u = 1.4(528) = 739 \text{ kN.m})$$

Consider uniformly distributed vertical steel, as shown in Fig. 10.29, and assume that all bars will yield because of the low percentage of steel.

By using Eq. 10.37 with a capacity reduction factor of 0.85

$$M_u = \Phi\left\{ A_{st}f_y\frac{d}{2} + P_u\frac{L}{2} \right\}$$

$\Phi = .65 - .85$

$$\frac{560 \times 12}{0.85} = A_s(60)\left(\frac{96 - 4}{2}\right) + 108\left(\frac{96}{2}\right)$$

$$\left(\frac{739 \times 10^6}{0.85} = A_s(400)\left(\frac{2400 - 100}{2}\right) + 477 \times 1000\left(\frac{2400}{2}\right)\right)$$

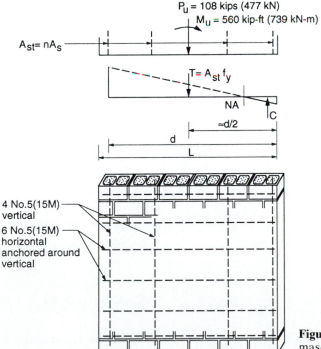

$P_u$ = 108 kips (477 kN)
$M_u$ = 560 kip-ft (739 kN-m)

$A_{st} = nA_s$

$T = A_{st}f_y$

NA

C

$\approx d/2$

d

L

4 No.5(15M) vertical

6 No.5(15M) horizontal anchored around vertical

**Figure 10.29**  Strength design of masonry shear wall: Example 10.3.

Therefore, the total area of vertical steel $A_{st} = 0.99$ in.$^2$ (646 mm$^2$). Use four No. 5 (4-15M) bars, $A_s = 4 \times 0.31 = 1.24$ in.$^2$ ($4 \times 200 = 800$ mm$^2$).

2. Shear: Load reversal will cause cracking and reduction in compression-shear transfer, aggregate interlock, and dowel action. Therefore, it may not be too conservative to carry all shear by shear reinforcement.[10.19]

By using Eq. 10.41, with a factored shear $V_u = 1.7(50)$kips ($1.7 \times 220$ kN), where the load factor 1.7 is applicable to the lateral load,

$$A_v = \frac{V_u s}{\Phi_s F_y d}$$

and by assuming a vertical spacing of 16 in. (400 mm),

$$A_v = \frac{1.7(50)(16)}{(0.8)(60)(88)} = 0.32 \text{ in}^2/16 \text{ in. } (208 \text{ mm}^2/400 \text{ mm})$$

Therefore use No. 5 (15M) bar at 16 in. (400 mm) spacing o.c.

For the details of reinforcement, see Fig. 10.29.

[Note that the strength design method results in lower amounts of vertical and horizontal reinforcing steel compared to working stress design. In high seismic areas, some codes[10.25,10.26] require that shear walls be designed to resist 1.5 times the calculated seismic shear forces.] ∎

### 10.7.3 Pier in a Perforated Wall

**Example 10.4**

Perform the shear design of pier A of the perforated shear wall shown in Fig. 10.30(a). The wall is an 8 in. (200 mm) reinforced fully grouted concrete masonry wall. Use $f'_m = 2500$ psi (17.3 MPa).

1. Load Distribution: The approach described in Sec. 10.4.4 for lateral load distribution is used. The lintel above the door opening is assumed to cause Segments 1 and 2 to deflect equally at the top of the wall, but not to contribute to the rigidity of the wall. The relative rigidities of the parts shown in Fig. 10.30(b) can be calculated as

$$R_1 = \frac{1}{\Delta_1}$$

$$R_2 = \frac{1}{\Delta_2} = \frac{1}{\Delta_{solid} - \Delta_3 + \dfrac{1}{R_4 + R_5}}$$

By referring to Fig. 10.12 for the calculation of wall rigidities,

$$R_1 = 0.0007Et \qquad \text{(Cantilever pier, } h/L = 7.0)$$

For Wall 2,

$$\Delta_{solid} = \frac{1}{R_{solid}} = \frac{1}{0.1Et} = \frac{10.0}{Et} \qquad \text{(Cantilever pier, } h/L = 1.167)$$

$$\Delta_3 = \frac{1}{R_3} = \frac{1}{0.6 \, Et} = \frac{1.03}{Et} \qquad \text{(Fixed pier, } h/L = 0.334)$$

note:
Vertical bars
are used @ ends
of wall

$$R_4 = R_5 = 0.25\,Et \qquad\qquad \text{(Fixed pier, } h/L = 1.0\text{)}$$

$$\therefore R_2 = \frac{1}{\Delta_2} = \cfrac{1}{\cfrac{10.0}{Et} - \cfrac{1.03}{Et} + \cfrac{1}{0.25\,Et + 0.25\,Et}} = 0.091\,Et$$

By referring to Fig. 10.30(c),

$$V_2 = V\left(\frac{R_2}{R_1 + R_2}\right) = 40\left(\frac{0.091\,Et}{0.0007\,Et + 0.091\,Et}\right) = 39.6 \text{ kips (178.6 kN)}$$

[Note that the majority of the lateral load is resisted by Wall 2 because Segment 1 is very flexible.]

2. **Maximum Shear Stress:** Shear on Pier 4 $= \frac{1}{2}(39.6) = 19.8$ kips (89.3 kN) (see Fig. 10.30(c))

Maximum moment in Pier 4, $M = Vh/2 = 19.8(2) = 39.6$ ft-kips
($M = 89.3 \times 0.6 = 53.6$ kN.m)
Minimum axial load at section a-a is equal to the dead load $P_D$ plus the self-weight from the tributary area shown in Fig. 10.30(a)
$$P_{min} = (4(8) + (6(6) + 2(2))\,70/1000 = 34.8 \text{ kips}$$
$$(P_{min} = (60(2.4) + (1.8(1.8) + 0.6(0.6))3.4 = 156 \text{ kN})$$

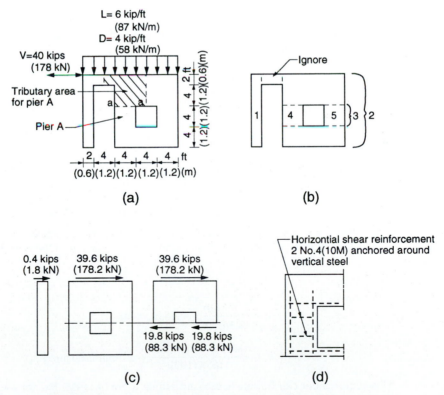

Figure 10.30  Design of shear wall with openings: Example 10.4.

The combined effect of axial load and bending is calculated from

$$f_m = -\frac{P}{A} \pm \frac{M}{S}$$

$$f_m = -\frac{34.80 \times 1,000}{7.625 \times 48} \pm \frac{39.6 \times 12,000}{7.625 \times (48^2)/6} = -95.1 \pm 162.3 \text{ psi}$$

$$\left( f_m = -\frac{156 \times 1000}{190 \times 1200} \pm \frac{53.6 \times 10^6}{190 \times 1200^2/6} = -0.68 \pm 1.18 \text{ MPa} \right)$$

This results in extreme fiber stresses of $-257$ psi ($-1.86$ MPa) (compression) and $+67$ psi ($+0.50$ MPa) (tension). Therefore, the section has tensile stress and should be reinforced vertically using an analysis based on the cracked section.

For reinforced masonry, the shear stress is,

$$f_v = \frac{V}{bjd}$$

$$f_v = \frac{19.8 \times 1000}{(7.625)\left(\frac{7}{8}\right)(44)} = 67 \text{ psi}$$

$$\left( f_v = \frac{88.3 \times 1000}{(190)\left(\frac{7}{8}\right)(1100)} = 0.48 \text{ MPa} \right)$$

**Solution A: Shear Design Using ACI 530/ASCE 5/TMS 402 (Allowable Stress Design).** For the relationship

$$\frac{M}{Vd} = \frac{39.6 \times 12}{(19.8)(44)} = 0.55 < 1$$

the allowable masonry shear stress is

$$F_v = \frac{1}{3}\left[\left(4 - \left(\frac{M}{Vd}\right)\right)\right]\sqrt{f'_m}$$

$$= \frac{1}{3}\left[\left(4 - 0.55\right)\right]\sqrt{2500}$$

$$= 58 \text{ psi } (0.40 \text{ MPa}) < 67 \text{ psi } (0.48 \text{ MPa})$$

Therefore, shear reinforcement must be provided to carry all shear, as shown in Fig. 10.30(d). The area of horizontal shear reinforcement per 8 in. (200 mm) height is

$$A_v = \frac{Vs}{F_s d}$$

$$A_v = \frac{(19.8)(8)}{(24)(44)} = 0.15 \text{ in.}^2/8 \text{ in.}$$

$$\left( A_v = \frac{(88.3 \times 1000)(200)}{(165)(1100)} = 97 \text{ mm}^2/200 \text{ mm} \right)$$

This requirement can be satisfied by providing a No. 4 (10M) bar for every course or 2 No. 4 (10 M) bars every second course, as shown in Fig. 10.30(d).

$P_0 = .85 f'_m (A_n - A_s) + A_s f_y$

$P_u \leq \phi(.8)(P_0), \phi = .65$

The maximum shear stress allowed by the code when reinforcement carries all the shear is

$$F_v = \frac{1}{2}\left[\left(4 - \frac{M}{Vd}\right)\right]\sqrt{f_m'}$$

$$= \frac{1}{2}\left[4 - 0.55\right]\sqrt{2500} = 86 \text{ psi } 5 \text{ (0.60 MPa)} > 67 \text{ psi (0.49 MPa)} \qquad \therefore \text{OK}$$

**Solution B:  Shear Design Using UBC-91 (Strength Design).**

$$V_u \leq \Phi V_n$$

$$\leq 0.6\,(V_m + V_s)$$

where
$$V_m = C_d A_n \sqrt{f_m'}$$

From Eq. 10.39, $C_d = 1.92$ (0.16) for $M/Vd = 0.55$ (Using linear interpolation)

$$\therefore V_m = 1.92(7.625 \times 48)\sqrt{2500}/1000 = 35.14 \text{ kips}$$

$$(V_m = 0.16(190 \times 1200)\sqrt{13.8}/1000 = 152 \text{ kN})$$

$$V_u = 1.7 \times 19.8 = 33.7 \text{ kips}$$

$$(V_u = 1.7 \times 88.3 = 150 \text{ kN})$$

$$\therefore V_s = (V_u - 0.6\,V_m)/0.6 = (33.7 - 0.6(35.14))/0.6 = 21.0 \text{ kips}$$

$$(V_s = (150 - 0.6(152))/0.6 = 98 \text{ kN})$$

Then from $V_s = A_{mv}\rho_n f_y$

$$\rho_n = \frac{21.0 \times 1000}{(7.625 \times 48)(60{,}000)} = 0.096\%$$

$$\left(\rho_n = \frac{98 \times 1000}{(190 \times 1200)(400)} = 0.107\%\right)$$

Therefore the required area of shear reinforcement is

$$A_v = \rho_n ts = \frac{0.096}{100}(7.625 \times 12) = 0.088 \text{ in.}^2/\text{ft}$$

$$\left(A_v = \frac{0.107}{100}(190 \times 1000) = 203 \text{ mm}^2/\text{m}\right)$$

Total area of steel required $= 0.088(4) = 0.35 \text{ in}^2$ (203 × 1.2 = 244 mm²) Use two No. 4 bars (2-15M) in the place of those shown in Fig. 10.30(d).

[Note that strength design yields a lower amount of shear reinforcement when masonry is allowed to carry shear. Other codes such as the New Zealand Code[10.30] do not allow masonry to carry shear from seismic forces.] ∎


## 10.8 CLOSURE

Masonry buildings are typically loadbearing shear wall buildings. In most cases, the walls required to carry gravity loads and to enclose or sub-divide the interior space can be easily designed to also provide ample resistance to lateral loads. In low seismic risk regions, reinforcement is often not required. Where reinforcement is required, good detailing can result in structures that perform very well under extreme loading conditions such as may be experienced in strong earthquakes.

The reader is also directed to Chap. 3 regarding layout of masonry wall systems, Chap. 8 for a more in-depth coverage of design for axial load and bending, Chap. 15 for details of construction, and Chaps. 16 and 17 for examples of shear wall design in buildings.

## 10.9 REFERENCES

10.1  F. Yokel and G. Fatal, "Failure Hypothesis for Masonry Shear Walls," *Journal of the Structural Division, Proceedings of ASCE*, Vol. 102, ST3, March 1976, pp. 515–532.

10.2  R. Drysdale, R. Vanderkyle, and A. Hamid, "Shear Strength of Brick Masonry Joints," in *Proceedings of the Fifth International Brick Masonry Conference*, Washington, D.C. 1979, pp. 106–113.

10.3  A. Hamid, R. Drysdale and A. Heidebrecht,, "Shear Strength of Concrete Masonry Joints," *Journal of the Structural Division, Proceeding of ASCE*, Vol. 105, ST7, July 1979, 1227–1240.

10.4  A. Hendry, B. Sinha and S. Davis, *An Introduction to Loadbearing Brickwork Design*, Ellis Horwood Ltd., London, 1981.

10.5  R. Drysdale and A. Hamid, "Tensile Strength of Concrete Masonry," *Journal of the Structural Division, Proceeding of ASCE*, Vol. 105, ST7, July 1979, 1261–1276.

10.6  R.G. Drysdale and A.A. Hamid, "Anisotropic Tensile Strength Characteristics of Brick Masonry," in *Proceedings of the Sixth International Brick Masonry Conference*, Rome, May, 1982, pp. 143–153.

10.7  J. Scrivener, "Static Racking Tests on Concrete Masonry Walls," in *Proceedings of Conference on Masonry Structural System*, Gulf Publishing, Houston, Texas, 1967, pp. 185–191.

10.8  D. Williams, "Seismic Behaviour of Reinforced Masonry Shear Walls," Ph.D. thesis, University of Canterbury, Christchurch, New Zealand, 1971.

10.9  R. Schneider, "Lateral Load Tests on Reinforced Grouted Masonry Shear Walls," Report No. 70-101, University of Southern California Engineering Center, September 1959.

10.10 P. Shing, J. Noland, H. Spaeh, E. Klamerus and M. Schuller, "Response of Single-Story Reinforced Masonry Shear Walls to In-Plane Lateral Loads," U.S.–Japan Coordinated Program for Masonry Building Research, Report No. 3.1(a)-2, Department of Civil and Architectural Engineering, University of Colorado at Boulder, January 1991.

10.11 P. Shing, M. Schuller, V. Hoskere and E. Carter, "Flexural and Shear Response of Reinforced Masonry Walls," *ACI Structural Journal*, Vol. 87, No. 6, November-December 1990, pp. 646–656.

10.12 P.B. Shing, M. Schuller and V. Hoskere, "In-Plane Resistance of Reinforced Masonry Shear Walls," *Journal of the Structural Division, Proceedings of ASCE*, Vol. 116, No. 3, March 1990.

10.13 B. Sveinsson, R. Mayes and H. McNiven, "Evaluation of Seismic Design Provisions for Masonry in the United States," Report No. UCB/EERC–81/10, College of Engineering, University of California, Berkeley, California, August 1981.

10.14 S. Thurston, and D. Hutchinson, "Reinforced Masonry Shear Walls Cyclic Load Tests in Contraflexure," *Bulletin of the New Zealand Society for Earthquake Engineering*, Vol. 15, No. 1, March 1982.

10.15 A. Salim, "Experimental Investigation of the Behaviour of Masonry Walls With Confinement Reinforcement," in *Proceedings of the Second North American Masonry Conference*, College Park, MD, August 1982, pp. 32/1–32/14.

10.16 T. Okada, S. Okamoto, Y. Tamazaki, A. Baba, T. Kaminosono, M. Teshigawara and O. Senbu, "U.S.-Japan Coordinated Earthquake Research Program on Masonry Buildings–Japanese Side Research Outline," *Proceedings of the Fourth North American Masonry Conference*, University of California at Los Angeles, August 1987, 23/1–23/14.

10.17 M.J.N. Priestley, "Ductility of Confined and Unconfined Concrete Masonry Shear Walls," *The Masonry Society Journal*, Vol. 1, No. 2, July–December 1982.

10.18 M.J.N. Priestley and H. Limin, "Seismic Response of T-Section Masonry Shear Walls," in *Proceedings of the Fifth North American Masonry Conference*, University of Illinois at Urbana–Champaign, June 1990, 359–372.

10.19 M.J.N. Priestley, "Seismic Design of Concrete Masonry Shear Walls," *ACI Journal*, Vol. 83, No. 1, January–February 1986, 58–68.

10.20 R. Park, and T. Pauley, *Reinforced Concrete Structures*, Chap. 12, John Wiley, New York, 1975.

10.21 G. Hart, R. Englekirk, and W. Hong, "Structural Component Models of Flexural Walls", in *Proceedings of the Fourth Meeting of the Joint Technical Coordinating Committee on Masonry Research*, U.S.–Japan Coordinated Program, San Diego, CA, October 1988.

10.22 T. Paulay and M.J.N. Priestley, "Seismic Design of Reinforced Concrete and Masonry Buildings", Chapter 5, John Wiley & Sons, New York, 1992.

10.23 J. Amrhein, *Reinforced Masonry Engineering Handbook*, 5th Ed., Masonry Institute of American, Los Angeles, 1992.

10.24 B. Sinha, A. Maurenbrecher and A. Hendry, "Model and Full Scale Test on a Five-Storey Cross Wall Structure under Lateral Loading," in Proceedings of the Second International Brick Masonry Conference, Stoke-on-Trent, April 1975, pp. 201–208.

10.25 The Masonry Standards Joint Committee, "Building Code Requirements for Concrete Masonry Structures," ACI 530/ASCE 5 TMS 402, American Concrete Institute and American Society of Civil Engineers, Detroit and New York, 1992.

10.26 International Conference of Building Officials, "Masonry Codes and Specifications," in *UBC-91, Chap. 24*, ICBO, Whittier, CA, 1991.

10.27 Canadian Standards Association, CSA Standard S 304-1984, "Masonry Design for Buildings," CSA, Rexdale, Ontario, 1984.

10.28 British Standards Institution, "Code of Practice for Use of Masonry—Part 1: Structural Use of Unreinforced Masonry," BS 5628, BSI, London, 1978 (confirmed April 1985).

10.29 G. Hart, R. Englekirk, R. Mayes and T. Kelly, "1988 Uniform Building Code Limit State Design Criteria For Shear Walls," in *Proceedings of the Fourth North American Masonry Conference*, University of California at Los Angeles, August 1987, pp. 4/1–4/18.

10.30 M.J.N. Priestley, "Ultimate Strength Design of Masonry Structures—The New Zealand Masonry Design Code," in *Proceedings of the Seventh International Brick Masonry Conference*, Melbourne, Australia, February 1985, pp. 1449–1462.

10.31 M.J.N. Priestley, "Flexural Strength of Rectangular Unconfined Shear Walls with Distributed Reinforcement," *The Masonry Society Journal Proceedings*, Vol. 5, No. 2, July–December 1986, pp. T1–T15.

10.32 A. Elmiger, *Architectural and Engineering Concrete Masonry Details for Building Construction*, National Concrete Masonry Association, Herndon, VA, 1976.

10.33 A. Anvar, S. Arya and Hegemier, G., "Behaviour of Floor-to-Wall Connections in Concrete Masonry Buildings," *The Masonry Society Journal Proceedings*, Vol. 2, No. 3, July–December 1983, pp. T11–T25.

## 10.10 PROBLEMS

**10.1** For the shear walls shown in Fig. P10.1, determine rigidity in terms of $Et$ using Methods 1 and 2 outlined in Sec. 10.4.

**10.2** Figure P10.2 shows a shear wall in a 5-story building. The wall is nonloadbearing. Determine the axial load, shear, and bending moments acting on the bottom story piers due to self-weight and the lateral loads shown. Consider the effect of the overturning moment on the axial load on the piers. The wall self-weight is 70 lb/ft² (350 kg/m²).

**10.3** Figure P10.3 shows a plan of a single story building. Distribute the 20 kip (90 kN) lateral load between the shear walls A, B, C, and D assuming rigid diaphragm action.

**10.4** Check to see if the building described in Prob. 10.3 can be constructed with 8 in. (200 mm) unreinforced hollow concrete block masonry. (Neglect axial load). Use either ACI 530/ASCE 5/TMS 402 or your local code. Perform the design for only in-plane loading.

**10.5** For the shear wall shown in Fig. P 10.5;

**(a)** Determine the required thickness if it is to be constructed with solid clay bricks; ($f'_m = 6000$ psi (41.4 MPa).

**(b)** Determine the required amount of vertical and horizontal reinforcement if it is constructed with 6 in. (150 mm) hollow clay masonry. Use the working stress design method of ACI 530/ASCE 5/TMS 402 or your local code.

**(c)** Repeat part (b) using the UBC-91 strength design method for shear walls. Compare the amount of reinforcment required using the two design methods.

**(d)** Show the reinforcement details for the steel calculated in part (c).

**(e)** Determine wall displacement ductility using the charts in Fig. 10.25.

[*Note*: For parts (b), (c), and (e), use $f'_m = 3000$ psi (20.7 MPa) and $f_y = 60$ ksi (400 MPa) with fully grouted construction.]

**10.6** For the shear wall of Prob. 10.5, if the axial load is doubled, how will this increase affect wall reinforcement requirements and ductility?

**10.7** For the shear wall of Prob. 10.5 and using the steel calculated in part (c), develop the load-moment interaction diagram for this wall using the graphs presented in Fig. 10.21.

**10.8** For advanced students: For the walls of Prob. 10.1(c), perform an elastic finite element analysis to determine the wall deflection under a top lateral load $P$ and then calculate the wall rigidity. Compare the results with your answer to Prob. 10.1 and comment on the accuracy of the approximate methods used in Prob. 10.1.

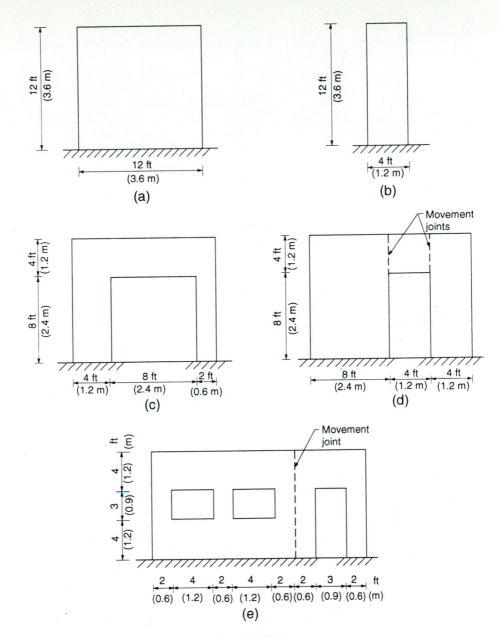

**Figure P10.1**

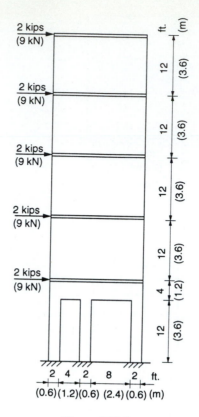

**Figure P10.2**

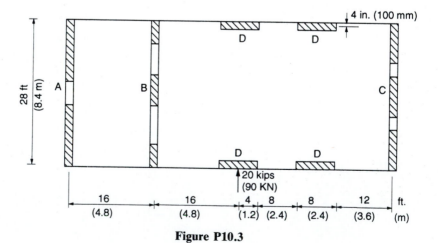

**Figure P10.3**

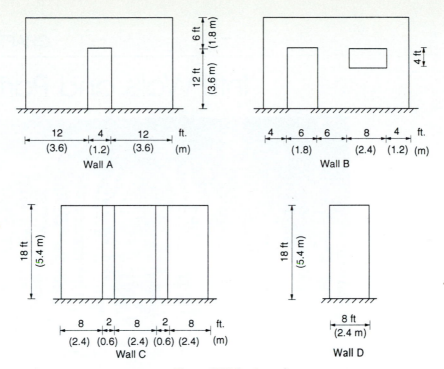

Wall A

Wall B

Wall C

Wall D

**Figure P10.3** *(cont.)*

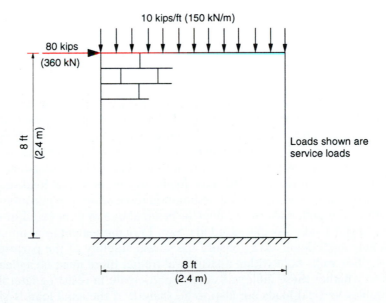

**Figure P10.5**

# CHAPTER 11

# Infill Walls and Partitions

Reinforced concrete frame building with masonry infill walls (*Courtesy of National Concrete Masonry Association*).

## 11.1 INTRODUCTION

Masonry walls are widely used as interior partitions and as exterior walls to form part of the building envelope in steel and reinforced concrete frame structures. In most cases, interior partition walls serve as barriers and their design is governed by the fire rating and sound transmission classification specified in local building codes (see Chap. 14). Where these walls are intended to be nonloadbearing, they are not designed to contribute to the axial load-carrying or lateral load-resisting capacity of the structure. For this design assumption to be valid, the partition must not be built tight to the underside of the floor or beam above or connected in such a manner as to transfer loads. As discussed in Chap. 15, a movement joint must be provided and include such details as to ensure the lateral stability of the partition.

For walls built within a structural frame, there must be a clear design decision as to whether these infill walls are to participate in resisting lateral loads and, if so, whether vertical loads are also to be shared. If the axial load is shared by building an infill wall tight to the underside of the floor system, the infill wall will also resist lateral load transferred to it by relative displacement of successive floors. Friction or mechanical anchorage along the top will transfer lateral load to the wall regardless of whether or not it is built tight to the surrounding columns. Conversely, walls built tight to the columns will share in resisting lateral load but may not share the vertical load, depending on the provision of a movement joint at the top of the wall. To avoid having the infill wall interact with the frame, movement joints must be provided on the sides and top and must be of sufficient thickness to isolate the wall from the effects of interstory drift, floor deflections, and differential deformations. Such infill walls behave very much as partitions and only interact with the supporting beam or floor, as is discussed in Sec. 11.3.

Infill walls and partitions must be designed for out-of-plane loads from wind or earthquake inertia forces, where the connections to the structure must be able to transfer reactions. As shown in Chap. 7, where these walls are nonloadbearing through provisions of movement joints, flexural behavior will control. When the walls are built tight to the frame, arching action can be the dominant mechanism for resisting out-of-plane loading (see Chap. 7).

The potential for interaction of infill walls and partitions with the structural frame has often been ignored to simplify the design or because the lack of design information has made it difficult to assess the extent of composite action. There are two very important reasons why this practice is not satisfactory. First, in today's competitive market, the choice of the structural system for a building may be largely determined by the efficiency of transmitting lateral loads to the foundations, particularly for multistory buildings in high wind or seismic areas. In most cases, the efficiency is related to the ability of the structural system to limit the interstory drift of the building under lateral loads. Ignoring the substantial stiffening effect of infill walls can lead to an inefficient and uneconomical design of the structural frame, where both the strength and stiffness requirements for the frame could be substantially reduced.

Second, and perhaps more important, ignoring the contribution of the infill walls does not always lead to a conservative design. Infill walls can greatly stiffen a flexible frame and significantly affect the distribution of lateral loads to various parts of the building. Thus, higher loads than expected may be attracted to an infilled section, possibly leading to cracking of the wall and overstressing of the frame. Therefore, the interaction of an infill wall and the frame should be considered in order to achieve an efficient design and to ensure that neither the wall nor frame is overstressed. Similarly, ignoring the interaction of a partition wall with its supporting beam may lead to an inefficient design of the beam and/or cracking of the partition wall.

In this chapter, the in-plane interactions of infill walls with their surrounding frames and of partition walls with their supporting beams are considered. Out-of-plane behavior and design are covered in Chap. 7.

## 11.2 INFILL WALLS

Infill walls can be totally enclosed in a surrounding frame of beams and columns, as is usually the case in a multistory building, Fig. 11.1(a), or simply be built between the columns of a single-story frame building, Fig. 11.1(b). In the former case, interac-

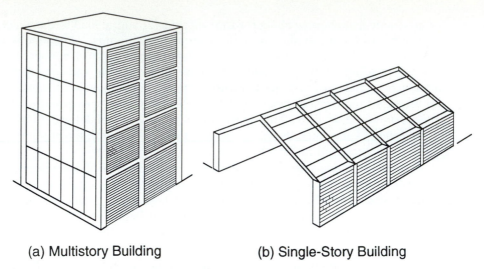

(a) Multistory Building        (b) Single-Story Building

**Figure 11.1**   Infill masonry walls.

tion with the frame under in-plane loads depends on the infill being constructed snugly inside the surrounding frame. Care is needed in construction to ensure a tight fit in the frame, but it is also necessary to ensure that temperature and moisture movements do not lead to overstressing of either the wall or the frame. Thermal and moisture expansions of the infill wall and shrinkage of the frame have been shown[11.1] to cause structural distress and, in such cases, a movement joint must be provided. As is discussed in Sec. 11.2.4, the presence of movement joints has significant implications on load transfer between the frame and the wall. Similar effects occur due to separations (gaps) created by differential horizontal or vertical shortening between the infill wall and the frame.

For masonry infill between steel frames, as shown in Fig. 11.1(b), there is rarely a substantial connecting member for anchorage at the eave level between frames and the masonry must rely on anchorage to the columns for its bracing action in resisting lateral loads. In this case, the steel frames bend about their weak axes and essentially transmit all shear to the infill panels.

The design of shear-resisting infill panels is similar to the shear wall design covered in Chap. 10 for the case when the only vertical load is self-weight of the wall. The resistance of a totally enclosed wall to in-plane lateral loads is more complex and is discussed in the following sections.

### 11.2.1 Interaction Mechanism and Failure Modes

The interaction mechanism between an infill wall and the surrounding structural frame depends on the area of contact at the interfaces of the two components. Thus, the extent of composite action will depend on the level of lateral load, degree of bond or anchorage at the interfaces, and geometric and stiffness characteristics of the two components.[11.2]

At very low levels of lateral load, there is full composite action between the infill wall and the frame as long as bond or anchorage at the interfaces ensures full

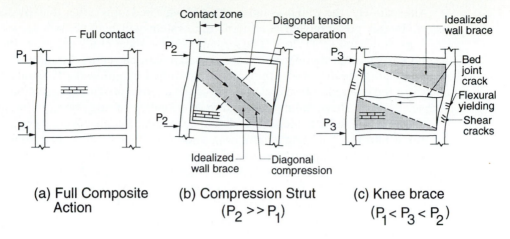

**Figure 11.2** Deformations and failure modes of infilled frames.

contact; Fig. 11.2(a). At this stage, the contribution of the infill wall to the total stiffness of the system is maximum; see Fig. 11.3(b). As the load increases, deformations increase and separation between the wall and the frame takes place except in the vicinity of the two corners where compression forces are transmitted through the wall; see Fig. 11.2(b). At this stage, the masonry wall develops a diagonal compression strut within the frame that converts the structural system to a type of *truss*. Lateral stiffness decreases as cracks are developed within the wall; see Fig. 11.3(a). With increasing lateral load, stiffness is reduced as additional cracks develop and the effectiveness of the diagonal compression brace or strut in the wall is also reduced. As the load increases, further separation and cracking of the wall occur until the masonry wall fails in shear (diagonal tension) and the columns in the frame yield in flexure.

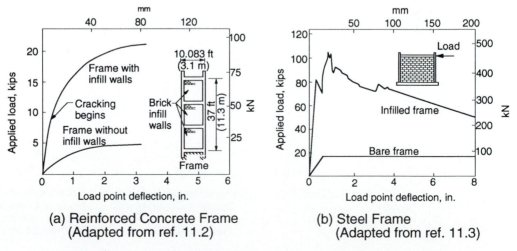

**Figure 11.3** Load-deflection curves of frames with and without infill walls.

Alternatively, the wall may fail in shear along a bed joint rather than by diagonal tension. In this post-cracked condition, the system will behave as a *knee-braced system*, as shown in Fig. 11.2(c). The formation of the shear crack separates the panel into two parts so that the effective column height can be reduced to approximately half. The failure mechanism in this case is controlled by either the flexural or shear capacity of the columns. Flexural yielding of the column provides protection against collapse, whereas shear failure can lead to overloading of neighboring columns.

The joint-slip failure mechanism should be avoided because it creates a situation very much like the half-height infill wall shown in Fig. 11.4(a). For such cases, because the bending moment capacity of the columns is unchanged, designs to avoid shear failure of the column can be negated as a result of halving the distance between the plastic moments (i.e., plastic moment now formed at mid-height), thus doubling the corresponding shears. In addition, as shown in Fig. 11.4(c), the transfer of the plastic moment to mid-height of the column can place it in a region where inadequate confining reinforcement is provided.

### 11.2.2 Analysis of Infilled Frames

The analysis of a framed building with masonry infill walls has two major parts, namely; the analysis to distribute forces to the frames of the building and the analysis of each frame under the action of the distributed force. These two interdependent parts are considered in turn here.

**Distribution of Lateral Load.** Methods of structural analysis commonly used to distribute wind and seismic forces to the lateral load-resisting components of a building apply equally to buildings with infilled frames provided that the increased stiffnesses of the infilled frames are considered. This point is illustrated by consider-

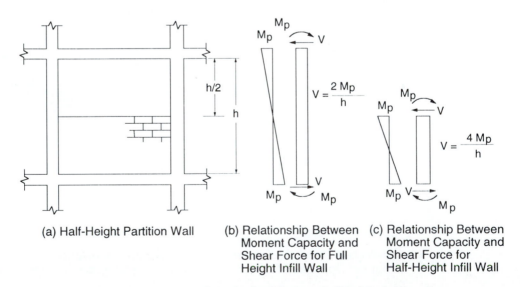

(a) Half-Height Partition Wall

(b) Relationship Between Moment Capacity and Shear Force for Full Height Infill Wall

(c) Relationship Between Moment Capacity and Shear Force for Half-Height Infill Wall

**Figure 11.4** Influence of shear-slip failure or half-height infill walls on column loading.

ing the effects of partially infilling a building frame, as shown in plan in Fig. 11.5. The symmetrically arranged infilled frames in Figs. 11.5(a) and 11.5(b) have ideal geometry for resisting lateral loads. They have equal strength about both major axes, good torsional resistance, and do not induce torsional forces as their centers of rigidity are centrally located and coincide approximately with the resultant of applied lateral loads. Thus, distribution of lateral loads to the component frames is a straightforward matter.

If, however, the frames are infilled with masonry on adjacent property lines but have glazing on the street frontages, as shown in Figs. 11.5(c) and 11.5(d), then a completely different response to lateral loads occurs. For the case in Fig. 11.5(c), the center of rigidity of the combined structural system is displaced from the central location to near the corner because the infilled frames are much stiffer. Thus, very large torsional forces from the now eccentric lateral loads are introduced. In addition, the natural period of vibration of the combined system is decreased, leading to increased seismic forces. This building plan has very little resistance to torsion because the center of rigidity almost coincides with the line of action of these main shear-resisting infilled frames.

The need to consider the stiffening effects of infill panels was evident in the aftermath of the 1985 earthquake in Mexico City, where buildings on corner sites with infills, such as in Fig. 11.5(c), constituted a large portion of building failures.[11.4] The configuration of infill walls shown in Fig. 11.5(d) has essentially the same large torsional effect as in Fig. 11.5(c), but has significantly improved resistance to torsion because two of the infilled frames are located away from the center of rigidity of the combined system.

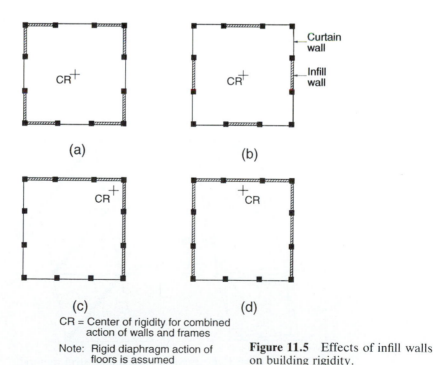

(a)

(b)

(c)        (d)

CR = Center of rigidity for combined action of walls and frames

Note: Rigid diaphragm action of floors is assumed

**Figure 11.5** Effects of infill walls on building rigidity.

**Behavior of Infill Walls.** For individual infilled frames, significant experimental and analytical research has been conducted to develop analysis and design procedures.[11.2,11.3,11.5–11.11] Sophisticated analyses using finite elements or theory of elasticity may not be warranted due to the uncertainty of defining the practical boundary conditions. Therefore, approximate analyses are acceptable for this type of problem where the behavior is dependent upon a multitude of parameters that are highly variable. In this regard, it is important to avoid shear-slip failure to ensure that the *knee-brace* action does not occur and thereby control the lateral load-carrying capacity.

Various approximate methods have been proposed by researchers, the most simple and highly developed being that based on the concept of *equivalent diagonal struts*, originally proposed by Polyakov[11.6] and subsequently developed by Stafford-Smith.[11.7] In this method, the system is modeled as a braced frame where the infill walls provide the web elements (equivalent diagonal struts), as shown in Fig. 11.6. The geometric properties of the diagonal struts are functions of the lengths of contact between the wall and the columns, $\alpha_h$, and between the wall and the beams, $\alpha_L$, shown in Fig. 11.7.[11.5] A range of contact length between one-fourth and one-tenth of the length of the panel is expected. Holmes[11.12] recommended a width of the diagonal strut equal to one-third of the diagonal length of the panel, whereas the New Zealand Code[11.13] specifies a width equal to one quarter of its length.

Using a beam on elastic foundation formulation, the following equations are proposed[11.7] to determine $\alpha_h$ and $\alpha_L$, which are in turn dependent on the relative stiffnesses of beams, columns, and infill.

$$\alpha_h = \frac{\pi}{2} \sqrt[4]{\frac{4\,E_f I_c h}{E_m t \sin 2\theta}} \tag{11.1}$$

and

$$\alpha_L = \pi \sqrt[4]{\frac{4\,E_f I_b L}{E_m t \sin 2\theta}} \tag{11.2}$$

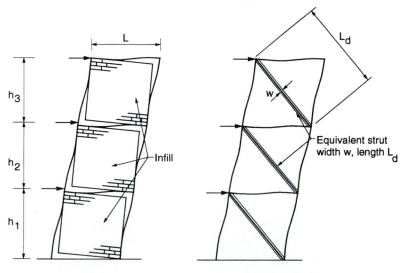

**Figure 11.6** Equivalent diagonal strut method.

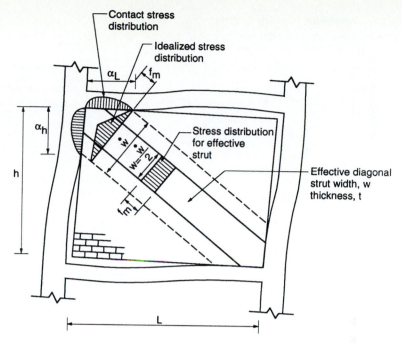

**Figure 11.7** Equivalent diagonal strut.

where: $E_m$, $E_f$ = elastic moduli of the masonry wall and frame material, respectively

$t$, $h$, $L$ = thickness, height, and length of the infill wall, respectively

$I_c$, $I_b$ = moments of inertia of the column and the beam of the frame, respectively

$\theta = \tan^{-1} (h/L)$

Assuming a triangular stress distribution along the width $\dot{w}$ of the strut (Fig. 11.7), the force in the strut is $\frac{1}{2} f_m \dot{w} t$, where the average compressive stress is one-half of the maximum stress $f_m$. Hendry[11.5] proposed the following equation to determine the equivalent or effective strut width $w$, where the strut is assumed to be subject to uniform stress $f_m$:

$$w = \tfrac{1}{2} \sqrt{\alpha_h^2 + \alpha_L^2} \tag{11.3}$$

Once the geometric and material properties of the struts are calculated, conventional braced frame analysis can be used to determine the stiffness of the infilled frames, the internal forces, and the deflections.

### 11.2.3 Strength of Infill Walls

The strut force obtained from the truss analysis is used to check the capacity of masonry infill walls. Although there is the possibility of crushing at the end of the diagonal strut, Dhanasekar and Page[11.11] showed that the tensile and shear bond strengths of the masonry are critical to the behavior and ultimate load capacity of infill frames (see Fig. 11.8).

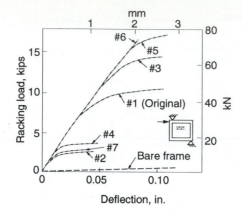

| Analysis | Tensile bond strength psi (MPa) | Shear bond strength psi (MPa) | Ultimate load, kips (kN) | Mode of failure* |
|---|---|---|---|---|
| 1† | 58 (0.40) | 43 (0.30) | 9.7 (43) | 1 |
| 2 | 0 (0.00) | 43 (0.30) | 2.2 (10) | 1 |
| 3 | 116 (0.80) | 43 (0.30) | 14.1 (63) | 1 |
| 4 | 58 (0.40) | 22 (0.15) | 3.6 (16) | 1 |
| 5 | 58 (0.40) | 86 (0.60) | 16.9 (75) | 1 |
| 6 | 116 (0.80) | 86 (0.60) | 16.9 (75) | 2 |
| 7 | 29 (0.20) | 22 (0.15) | 2.7 (12) | 1 |

*1. Diagonal cracking; 2. Corner crushing.
†Original material model.

**Figure 11.8** Effect of mortar bond strength on the behavior of infill frames (from Ref. 11.11).

A simple and conservative means of assessing the capacity of the masonry is to assume that the horizontal component of the strut force $F_h$ is resisted by bed joint shear on the horizontal cross-sectional area $A_n$ of the infill wall. The shear stress at the center of the infill wall $f_v$ can be calculated as

$$f_v = 1.5 \, F_h / A_n \tag{11.4}$$

The above expression assumes a parabolic distribution of shear stresses over the length of the infill. By neglecting shear friction in the bed joint resulting from the vertical component of the strut force, typical shear strengths are 50 to 100 psi (0.35 to 0.70 MPa) for brick masonry and 25 to 50 psi (0.17 to 0.35 MPa) for hollow block masonry.

A more accurate approach is to check the state of stress in the strut at the center of the panel using an appropriate failure envelope that accounts for the inherent anisotropy of masonry, the planes of weakness along bed and head joints, and the orientation of the joints.[11.14]

Because a shear-slip failure along a bed joint can double the shear in the column and result in a brittle failure with overstressing of adjacent columns, the design must ensure that the shear bond strength along these joints will not be exceeded. However, where masonry has a low shear strength along the bed joints or the designer has reason to believe that the masonry may fail prematurely along a bed joint, an analysis based on the *knee-braced frame* concept[11.13,11.15] should be performed to assess the impact that this premature failure could have on the performance of the infilled frame structure. Alternately, especially for high seismic risk areas, the infill should be reinforced to prevent shear-slip failure (see Sec. 11.2.6).

### 11.2.4 Infill Walls With Openings

Infill walls frequently contain openings of different sizes and locations. The effects of small openings for running conduits, cables, and ducts and of openings outside the diagonal struts, may be negligible. However, in assessing the effects of openings, it should be noted that when the load reverses, the other diagonal of the panel becomes a *strut*.

Dawe and Seah[11.3] studied the effects of openings in masonry infill in steel frames. The load-deflection curves in Fig. 11.9 indicate that the openings greatly reduced stiffnesses and load-carrying capacities compared to infill walls without openings. This is attributed to the opening interfering with the diagonal bracing action, thereby causing premature shear failure of the sections on either side of the

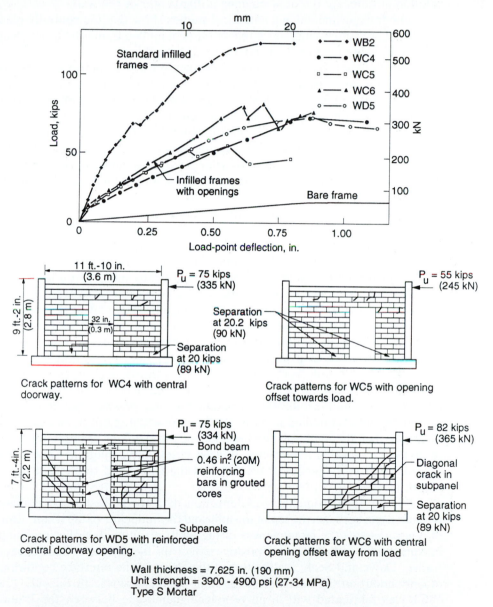

**Figure 11.9**  Behavior of infilled steel frames with openings (from Ref. 11.3).

opening. Failure in many cases was initiated by separation of the wall from the frame on each side of the opening. For specimen WD5, the inclusion of steel reinforcement around the opening increased the initial stiffness, but did not increase the ultimate load. For specimen WC6 with the opening located away from the load, diagonal cracks developed and the partial strut that developed resulted in a slightly higher ultimate load than for openings located nearer to the load. For this reason, and because lateral load is usually applied in both directions, it appears that the best solution is to locate the door opening at the center of the wall.

Other experimental and analytical studies show that the centrally placed openings in square panels can reduce the stiffness of infilled frames by as much as 75%[11.12] and the ultimate load of the panel by up to 40%.[11.15]

Development of simple analytical procedures for infill frames with openings is not easy to accomplish due to the multiple parameters affecting behavior, including wall geometry and the location, shape, and size of openings. Finite element analysis can be used to check stresses in the panels due to composite action under lateral loads.[11.16] Otherwise, it is recommended that masonry infill with openings be designed in the same way as shear walls in a loadbearing masonry building with shear distributed to the parts in accordance with their rigidities, as illustrated in Sec. 10.4.3. Then each part is designed for the appropriate self-weight, shear, and bending moment.

### 11.2.5 Infill Walls With Movement Joints at the Top

In reinforced concrete buildings with infill walls built tight to the surrounding frames, initial elastic shortening and long-term creep and shrinkage of the concrete can result in most of the vertical load being transferred to the infill. (To a lesser extent, this can also happen in steel frame buildings due to elastic shortening of the columns, deflection of the beams, and moisture expansion when clay masonry is used.) The distribution of vertical load between the frame and the infill walls depends on the short and long-term deformations of both elements. Unless an analysis is done to apportion the axial load to the frame and infill wall, a conservative approach for design of the infill walls is to check the axial compressive stress in the wall assuming that it will eventually be required to carry a large share of the load. In cases where the axial stress on the infill walls would be too large or where nonuniform shortening of infilled and non-infilled parts of the frames would result in sloped floors, use of a gap between the infill wall and the underside of the floor or beam above is required to allow free vertical movement. In such cases, the conditions shown in Fig. 11.7 do not exist until significant deflection of the frame and rotation of the wall have occurred.

Figure 11.10 contains load-deflection data for infilled steel frames with and without a gap between the top of the infill and the underside of the beam. Introduction of a gap reduces the effectiveness of the infill, resulting in earlier cracking, but the masonry infill wall still has a positive effect on the ultimate shear capacity of the frame. Dawe and Seah[11.3] also studied the effect of the interface condition (bonded or unbonded) on the behavior of the infill in steel frames. In Fig. 11.11, specimen WC1 was fabricated with a polyethylene membrane between the frame and the masonry infill wall to act as a bond breaker. It exhibited lower initial stiffness and lower ultimate strength compared to specimen WB2 with mortar bond at the

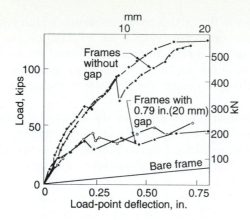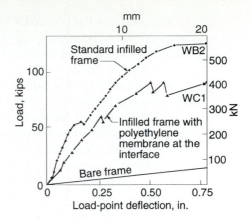

**Figure 11.10** Effect of gap at the top of the infill on behavior of infilled steel frames (from Ref. 11.3).

**Figure 11.11** Effect of interface condition on response of infilled steel frames (from Ref. 11.3).

interface. This difference in behavior can be attributed to the elimination of the interface shear that affects the distribution of cracks and the effective width of the diagonal strut.

Where vertical compression in the infill wall is not a problem, a gap or unbonded surface at the top of the infill wall should be avoided by using proper coursing and sizing of masonry units for the clear floor heights and by packing mortar into the top joint to achieve tight contact at the frame–wall interface. Such mortaring of the top wall should be uniform to minimize the possibility of eccentric vertical loading of the wall. Where top gaps are required but shear-resisting capacity of the infill is also desirable, it is recommended that the infill walls be designed as independent cantilever shear walls carrying their share of the lateral load from the floors above.

Anchoring of the sides of the infill wall to the frame columns restricts rotation of the wall and should help avoid local crushing failure by increasing the contact zone for load transfer from the frames to the infill. However, with rigid anchors, differential vertical deformations can introduce either vertical tension or compression in the infill.

### 11.2.6 Seismic Design Considerations

Masonry infill walls have been identified as a major cause of poor performance of reinforced concrete frames under high seismic loading. However, this is not due to the inadvisability of infilling such frames, but rather due to the traditional practice of considering the infill walls as nonstructural elements and ignoring their effects on the structural behavior of the building. This practice has continued despite the fact that the significant effect of the infill on altering the structural response of the frame has been recognized and documented.

When infill panels are constructed without full separation from the frame, structural stiffness is greatly increased and the natural period reduced, resulting in increased seismic forces. When separation is provided at the tops of the panels, the infill panels may still tend to stiffen the supporting beams and force plastic hinges to migrate from the preferred location in the beams to the columns, which causes a breakdown of the *weak-beam strong-column* concept. If the infill panels are separated also from the columns, the gap and connection details should be such that hammering of the frame against the infill during an earthquake is prevented. When masonry infill does not extend the full story height, shear failure of the columns can be induced, as shown in Fig. 11.4.

As mentioned previously, it is important to consider the effect that stiffening of the frame has on the distribution of lateral load to the other component frames of the building. The effect on the location of the torsional center of rigidity and the consequent torsional effect of lateral loads should be included. Designers should pay attention to the effects of concentrated loads from the diagonal compression strut acting on the beam–column joints. For reinforced concrete frames, adequate shear reinforcement details must be provided in the joints and in the beams and columns near the joints to resist the loads.

Shear failure of masonry infill panels at a particular elevation, perhaps due to openings, can cause a *soft-story* effect by reducing the interstory stiffness and increasing the ductility demands on the columns. This could also cause asymmetry of load application, resulting in increased torsional forces and changes in the distribution of shear forces between lateral load-resisting elements. Therefore, it is particularly important that the potential for these undesirable behaviors be minimized by prudent choices of infill wall plans and the locations of openings.

Under reversed loading in seismic areas, the effectiveness of the strut and contact area is questionable unless the infill is anchored to the frame. Anchoring is seen as a reasonably effective way to ensure composite action under large interstory drift and under load reversals. Horizontal and vertical reinforcement can be used to improve the strength, stiffness, and deformability of masonry infill walls.

In view of the complex behavior of infill frames under seismic loads, it is recommended[11.13] that elastic analysis be used unless a comprehensive rational analysis is carried out to determine the available structural ductility.

### 11.2.7 Example Analysis of the Stiffness of an Infilled Frame

**Example 11.1**

Determine the increase in stiffness of the frame shown in Fig. 11.12(a) when infilled with masonry.

**Properties**:

Infill wall:
  Thickness $t = 5\frac{5}{8}$ in. (140 mm)
  Elastic modulus $E_w = 2000$ ksi (13,800 MPa)
Frame:
  Areas of beams and columns, $A_b = A_c = 144$ in.$^2$ (90,000 mm$^2$)
  Moments of inertia of beams and columns, $I_b = I_c = 1728$ in.$^4$ (6.75 × 10$^8$ mm$^4$)
  Elastic modulus $E_f = 3000$ ksi (20,700 MPa)
**Solution:**   (a) Calculate the width of the equivalent diagonal strut

$$\theta = \tan^{-1}\frac{h}{L} = 38.65°$$

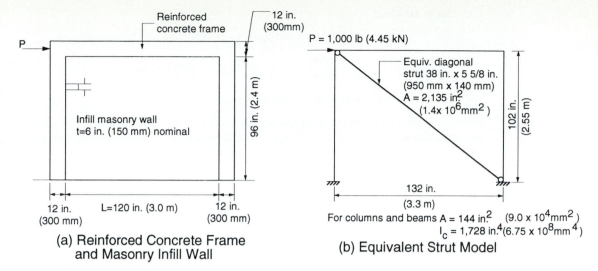

**Figure 11.12** Analysis of the stiffness of infilled frames: Example 11.1.

From Eq. 11.1,

$$\alpha_h = \frac{\pi}{2} \sqrt[4]{\frac{E_f I_c h}{2 E_m t \sin 77.3°}} = 32.7 \text{ in. } (0.811 \text{ m})$$

and from Eq. 11.2,

$$\alpha_L = \pi \sqrt[4]{\frac{E_f I_b L}{E_m t \sin 77.3°}} = 68.5 \text{ in. } (1.716 \text{ m})$$

By using Eq. 11.3,

$$w = \tfrac{1}{2} \sqrt{\alpha_h^2 + \alpha_L^2} = 38 \text{ in.}^* (949 \text{ mm})$$

Area of diagonal strut, $A_d = 38 \times 5.625 = 213.8 \text{ in.}^2$

$$(A_d = 949(140) = 132,860 \text{ mm}^2).$$

(b) Analyze the frame using classical methods of structural analysis or a plane frame computer program with and without the diagonal pin-jointed strut, as illustrated in Fig. 11.12(b). [Note: It is convenient and slightly conservative to model the strut as connecting the center lines of the surrounding frame, as shown in Fig. 11.12(b).]
   Without the strut:

$$\Delta_{\text{horiz}} = 0.013 \text{ in. } (0.33 \text{ mm}) \text{ due to } 1000 \text{ lb } (4.45 \text{ kN}) \text{ lateral force}$$

$$\therefore \text{ stiffness, } k = \frac{P}{\Delta} = \frac{1000}{1.3 \times 10^{-2}} = 77000 \text{ lb/in. } = 77 \text{ kips/in.}$$

$$(k = 4.45/0.33 = 13.5 \text{ kN/mm})$$

*The length of the diagonal strut is $L_d = \sqrt{h^2 + (L)^2} = 154 \text{ in. } (3.84 \text{ m})$ and the ratio $w/L_d = 0.247$, which is very close to the ratio ($\tfrac{1}{4}$) specified in the New Zealand code.[11.13]

With the strut:

$$\Delta_{\text{horiz}} = 0.0002 \text{ in. } (0.005 \text{ mm}) \text{ due to } 1000 \text{ lb } (4.45 \text{ kN}) \text{ lateral force}$$

$$\therefore \text{ stiffness, } k = \frac{P}{\Delta} = \frac{1000}{0.0002} = 5 \times 10^6 \text{ lb/in.} = 5000 \text{ kips/in.}$$

$$(k = 4.45/0.005 = 890 \text{ kN/mm})$$

Hence, the infill wall increases the stiffness of the frame by a factor of 65. This clearly indicates the significance of the infill walls in attracting most of the lateral loads in frame structures and shows why infill walls should be accounted for in design. ∎

## 11.3 WALLS SUPPORTED ON BEAMS

### 11.3.1 Introduction

A masonry partition or exterior infill wall with movement joints along the sides and top does not interact with any part of the frame of a building other than the supporting beam or slab. It has long been recognized that a masonry wall acts compositely with a supporting steel or concrete beam, but prior to about 1950, this was simply allowed for by assuming that the beam carried only the load from a triangular section of masonry above it. The remaining load was assumed to be carried over the span by arching action in the wall; see Fig. 11.13(a). More recent experimental and theoretical studies have resulted in a better understanding of the interaction mechanism and have allowed more realistic theoretical models to be developed for analysis and design.[11.17–11.20]

### 11.3.2 Interaction Mechanism and Failure Modes

The partition wall resting on the simply supported beam in Fig. 11.13(a) acts compositely with it to resist the weight of the wall and beam and any applied vertical

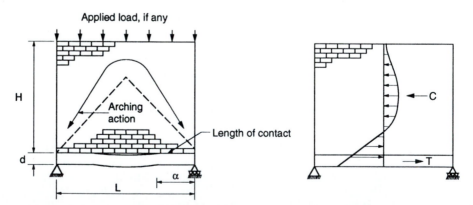

(a) Triangular Load Distribution and Arching    (b) Stress Distribution at Midspan Due to Deep Beam Behavior

**Figure 11.13**  Composite action of a masonry wall on a supporting beam.

loads. Deep beam action of the wall and beam system, Fig. 11.13(b), produces tension in the beam, which acts as a tie and facilitates arching in the masonry.

The weight of the wall is transferred to the beam near the supports and the beam separates or tends to separate from the wall near midspan. Detailed behavior of the system is largely influenced by the relative stiffnesses of the wall and the beam. At one extreme, a very flexible beam can bend completely away from the wall and will not attract load so that it acts purely as a tie for the wall that spans the opening by arching action alone. In this case, there would be a very high concentration of compressive stress in the wall near the supports. At the other extreme, when the beam is extremely stiff, the entire weight of the wall is transferred as a uniform load to the beam and no arching occurs.

Between these limits there are varying contact lengths between the wall and the beam with resulting variations of compressive stress in the wall and shear stress at the beam–wall interface (Fig. 11.14). Correspondingly, the beam will be subjected to varying combinations of axial tension force and bending. A more flexible beam results in a greater concentration of compressive stress near the ends of the wall, a greater tendency of the beam to bend away from the wall at midspan, smaller bending moment in the beam, and a greater axial tension force in the beam.

The wall can fail by crushing of the masonry near the supports, by slip at the interface with the beam, or the beam can fail under combined bending and axial forces.

### 11.3.3 Analysis

Several approximate methods of analysis have been proposed for design. In 1952, Wood[11.17] proposed moment coefficients and a tension coefficient to enable the beam to be designed for combined tension and bending. Wood and Simms[11.18] later suggested a method for determining the maximum compressive stress in the wall.

Based on experimental and analytical studies, Stafford-Smith and Riddington[11.19] introduced a relative wall-to-beam stiffness parameter

$$K_S = \sqrt[4]{\frac{E_w t L^3}{E_b I_b}} \tag{11.5}$$

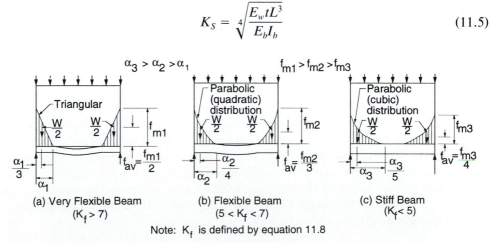

**Figure 11.14** Vertical compressive stress distribution at contact zone between wall and beam (from Ref. 11.20).

where $E_w$, $E_b$ = moduli of elasticity of the wall and the beam, respectively
$I_b$ = moment of inertia of the beam
$t, L$ = wall thickness and length of wall, respectively

Expressions based on this parameter were given for maximum bending moment in the beam and maximum compressive stress in the wall. The axial force in the beam was taken as approximately 30% of the vertical load. Also the contact length between the infill wall and the beam was defined as

$$\alpha = L/K_S \qquad (11.6)$$

It is noted that this approach does not reflect the effect of the wall aspect ratio (height-to-length ratio) on the wall–beam relative stiffness. A comparative study[11.5] indicates that this approach gives reasonable results for bending in the beam, but is overly conservative for axial force in the beam and for compressive stress in the wall. Also it does not include calculation of shear at the beam-wall interface. Therefore, the following method is recommended.

Based on a finite element study, Davies and Ahmed[11.20] proposed the following two expressions for relative stiffness parameters

$$K_F = \sqrt[4]{\frac{E_w t h^3}{E_b I_b}} \qquad (11.7)$$

$$K_A = \frac{E_w t h}{E_b A_b} \qquad (11.8)$$

where $A_b$ = cross-sectional area of the beam
$h$ = height of the wall

and other terms are the same as those in Eq. 11.5.

$K_F$ is similar to $K_S$ but because it accounts for the height of the wall, it is a better representation of the relative wall-to-beam flexural stiffness. Because the wall and the beam have the same length, the $K_A$ term is a measure of the relative wall-to-beam axial stiffness. Based on the finite element study and experimental data, they proposed the following relationship for contact length:

$$\alpha = \frac{L}{1 + \beta K_F} \qquad (11.9)$$

where $\beta$ is based on wall height-to-length ratio, $h/L$, as shown in Fig. 11.15.

By determining $\alpha$ and assuming a triangular stress distribution for simplicity, the moments and stresses on the beam and the wall can be determined using conventional structural analysis. The controlling criteria is either stresses in the wall and the beam or the limiting deflection of the beam. The previous equations indicate that increasing the beam stiffness relative to the infill wall increases the length of contact $\alpha$, and this in turn increases the bending moment in the beam and reduces the wall stresses.

Alternately, design can be carried out directly using the following:
Maximum vertical stress in wall,

$$f_m = C_1 \frac{W}{Lt} \qquad (11.10)$$

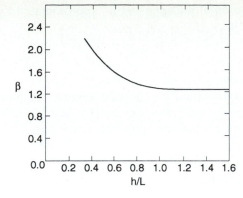

**Figure 11.15** $\beta$ factor versus $h/L$ ratio for analysis of a composite wall-beam assembly (from Ref. 11.20).

Axial tension in beam,

$$T = C_2 W \tag{11.11}$$

Maximum shear stress along interface,

$$\tau_m = C_1 C_2 \frac{W}{Lt} \tag{11.12}$$

Maximum bending moment in beam,

$$M = \frac{1}{C_1}\left(C_4 - C_2 C_3 \frac{d}{L}\right) WL \tag{11.13}$$

where   $d$ = depth of beam
$W$ = weight of wall plus other superimposed load

The values of $C_1$ and $C_2$ are shown in Fig. 11.16.
For $K_F \le 5$, $C_3 = 2.0$ and $C_4 = 0.20$.
For $K_F$ between 5 and 7, $C_3 = 1.5$ and $C_4 = 0.19$.
For $K_F \ge 7$, $C_3 = 1.0$ and $C_4 = 0.17$.

    These three sets of values for $C_3$ and $C_4$ reflect the assumption that the distributions of compressive stresses over the contact length between the wall and the beam change from linear to parabolic to cubic as the relative flexural stiffness of wall to beam decreases. (See Fig. 11.14.)

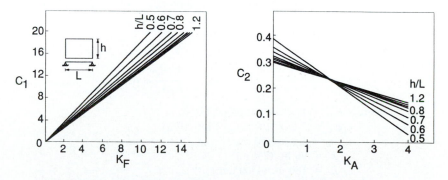

**Figure 11.16** Values of coefficients $C_1$ and $C_2$ for analysis of a composite wall-beam assembly (from Ref. 11.20).

The maximum deflection at midspan can be approximated by

$$\delta = \frac{WL^3}{240E_bI_b}\left[\frac{5C_1 - 2}{C_1^3} + \frac{25W_b}{8W} - \frac{10C_2d}{L}\right] + \frac{3WL}{10E_wht} \qquad (11.14)$$

where $W_b$ = weight of beam.

The resulting deflection should be checked against maximum allowable values. For unreinforced masonry walls, it is specified by ACI 530/ASCE 5/TMS 402[11.21] that the beam deflection not exceed the span/600, nor 0.3 in. (7.6 mm), whichever is less.

## 11.3.4 Limitations of Analysis

The analyses presented before assume that the beams are shored during construction of the wall so that the whole of the weight of the wall is resisted by composite action. If the beam is not shored during construction, the degree to which the partially built wall will share in resisting the wall weight by composite action depends on the rate of construction and generally cannot be relied upon. Loads superimposed later can be resisted by the composite construction.

Research[11.20] on wall–beam structures with door or window openings shows that centrally placed openings have only a small effect on behavior. This is because arching can still develop over the opening. Openings close to the supports, however, destroy complete arching action over the full span. In this case, it may be advisable to design the beam to resist the full weight of the wall.

The previous analyses are not readily applied to continuous spans.

## 11.3.5 Example of a Masonry Wall Supported on a Beam

### Example 11.2

A 15 ft (4.5 m) long reinforced concrete beam, 10 in. wide × 16 in. deep (250 mm × 400 mm) supports a brickwork partition wall 8 ft high × 3 5/8 in. thick (2.4 m × 90 mm), which weighs 4000 lb (17.8 kN); see Fig. 11.17. Calculate the maximum compressive stress in the wall, the maximum shear between the wall and the beam, and the maximum bending moment and axial force in the beam.

Using the Davies and Ahmed approach, from Eqs. 11.7 and 11.8,

$$K_F = \sqrt[4]{\frac{E_wth^3}{E_bA_b}} = 4.40$$

$$K_A = \frac{E_wth}{E_bA_b} = 0.87$$

where
$E_w = 1 \times 10^6$ psi (6900 MPa)
$E_b = 2.5 \times 10^6$ psi (17,250 MPa)
$t = 3\frac{5}{8}$ in. (90 mm)
$h = 96$ in. (2.4 m)
$L = 180$ in. (4500 mm)
$I_b = 10(16)^3/12 = 3413$ in.$^4$ $(250(400)^3/12 = 13.3 \times 10^8$ mm$^4)$
$A_b = 10(16) = 160$ in.$^2$ $(250(400) = 1.0 \times 10^5$ mm$^2)$

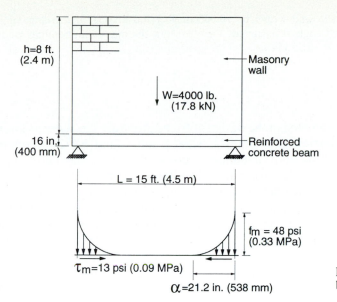

For $(h/L) = (8/15) = 0.533$, Fig. 11.16 gives

$$C_1 = 7.8 \text{ for } K_F = 4.4 \text{ and } C_2 = 0.27 \text{ for } K_A = 0.87.$$

$$\text{For } K_F = 4.4 < 5, C_3 = 2.0 \text{ and } C_4 = 0.2.$$

From Fig. 11.15,

$$\beta = 1.7 \text{ for } h/L = 0.533$$

From Eq. 11.9,

$$\alpha = 21.2 \text{ in. (531 mm)}$$

From Eq. 11.10, the maximum compressive stress,

$$f_m = C_1 \frac{W}{Lt} = 48 \text{ psi (0.33 MPa)}$$

From Eq. 11.12, the maximum interface shear,

$$\tau_m = C_1 C_2 \frac{W}{Lt} = 13 \text{ psi (0.09 MPa)}$$

From Eq. 11.11, the axial tension in the beam

$$T = C_2 W = 1080 \text{ lb (4.8 kN)}$$

From Eq. 11.13, the maximum bending moment in the beam with $d/L = 0.089$,

$$M = \frac{1}{C_1}\left(C_4 - C_2 \frac{d}{L}\right) WL = 14{,}025 \text{ in.–lb (1.56 kN} \cdot \text{m)}$$

It should be noted that if there is no composite action and the beam is assumed to carry the full weight of the wall as a uniform load, the maximum bending moment from the weight of the wall would be 90,000 in.-lb (10.0 kN-m). If a triangular load distribution with a 45° angle is assumed, Fig. 11.13(a), the maximum bending moment would be 56, 250 in.-lb (6.25 kN-m). ∎

## 11.4 CLOSURE

Masonry walls are often included in the design of frame structures to serve as interior partitions because of their excellent fire resistance and sound barrier properties and because of low finishing and maintenance costs. Similarly, aesthetic appeal, durability and low maintenance, strength and stiffness against out-of-plane wind loading, and the facility for accommodating thermal, air, and moisture barriers lead to extensive use of masonry as part of the exterior building envelope. Inclusion of these walls requires that the designer make a conscious decision regarding their structural interaction with the frame.

If participation in carrying lateral load is not intended, adequate movement joints at the ends and tops of the infill panels must be provided to permit the frame to deflect without transferring load to the walls. Alternately, advantage can be taken of the very significant strength and stiffness characteristics of such walls by incorporating them into the design for lateral load resistance. Provided that openings in the infill can be avoided, and that the walls are properly connected to the frame, the design of the structure to resist lateral load can be greatly simplified and considerable savings can be achieved in the cost of the structural frame.

For beams supporting masonry walls, the interaction of the masonry wall can significantly reduce the bending moments in the beams. However, high compression near the ends of the infill walls and cracking due to arching of the wall should be checked.

Readers are directed to Chap. 7 containing information on the design of wall panels for out-of-plane bending, Chap. 14 for information on design of walls for environmental loads, and to Chap. 15 for construction details.

## 11.5 REFERENCES

11.1 G.T. Suter, and J. Hall, "How Safe Are Our Cladding Connections," in *Proceedings of the First Canadian Masonry Conference*, Calgary, Alberta, 1976, pp. 95–109.

11.2 G.M. Sabnis, "Interaction Between Masonry Walls and Frames in Multistory Structures," in *Proceedings of the First Canadian Masonry Symposium*, Calgary, Alberta, 1976, pp. 324–337.

11.3 J.L. Dawe and C-K. Seah, "Behavior of Masonry Infill Steel Frames," in *Canadian Journal of Civil Engineering*, Vol. 16, December 1989, 865–876.

11.4 J.E. Amrhein, J. Anderson and V. Robles, "Mexico Earthquake—September 1985," *The Masonry Society Journal*, Vol. 4, No. 2, July–December 1985, G.12–G.17.

11.5 A. Hendry, *Structural Brickwork*, Macmillan, London, 1981.

11.6 S.V. Polyakov, Masonry in Framed Buildings, Gosudalst-Vennoe Izdatel' stvo Literature po Straitel'stvu i Arkitecture, Moskva, 1956, Trans. G.L. Cairns, Building Research Station, Watford, Herts, 1963.

11.7  B. Stafford-Smith, "Behaviour of Square Infilled Frames," *Journal of the Structural Division, Proceedings of ASCE*, Vol. 91, No. ST1, 1966, 381–403.

11.8  L. Esteva, "Behaviour Under Alternating Loads of Masonry Diaphragms Framed by Reinforced Concrete Members," *Symposium on the Effects of Repeated Loading on Materials and Structural Elements*, RILEM, Mexico City, 1966.

11.9  A.E. Fiorato, M.A. Sozen and W.L. Gamble, "An Investigation of the Interaction of Reinforced Concrete Frames with Masonry Filler Walls," University of Illinois, Urbana-Champaign, November 1970.

11.10 J.M. Leuchars and J.C. Scrivener, "Masonry Infill Panels Subjected to Cyclic In-Plane Loading," *South Pacific Regional Earthquake Engineering Conference*, Wellington, New Zealand, May 1975.

11.11 M. Dhanasekar and A. Page, "The Influence of Brick Masonry Infill Properties on the Behaviour of Infilled Frames," in *Proceedings of the Institution of Civil Engineers*, Part 2, Paper 9061, December 1986.

11.12 M. Holmes, "Combined Loading on Infilled Frames," *Proceedings of the Institute of Civil Engineers*, Vol. 25, 1963, pp. 31–38.

11.13 Standard Association of New Zealand, "Code of Practice for the Design of Masonry Structures," NZS 4230: Part 1, Wellington, 1990.

11.14 M. Dhanasekar, A. Page, and P. Kleeman, "The Failure of Brick Masonry Under Biaxial Stresses," *Proceedings of the Institute of Civil Engineers*, Part 2, Paper 8871, June 1985.

11.15 M.R.A. Kadir, "The Structural Behaviour of Masonry Infill Panels in Framed Structures," Ph.D. thesis, University of Edinburgh, 1974.

11.16 D.J. Male and P.F. Arbon, "A Finite Element Study of Composite Action in Walls," *Second Australasian Conference on Mechanics of Structures and Materials*, University of Adelaide, August 1969.

11.17 R.H. Wood, "Studies in Composite Construction. Part 1, The Composite Action of Brick Panels Supported on Reinforced Concrete Beams," National Building Studies Research Paper 13, HMSO, London, 1952.

11.18 R.H. Wood and L.G. Simms, "A Tentative Design Method for the Composite Action of Heavily Loaded Brick Walls Supported on Reinforced Concrete Beams," BRS CP 26/69, Building Research Station, Watford, Herts, 1969.

11.19 B. Stafford-Smith and J.R. Riddington, "The Composite Behaviour of Masonry Wall on Steel Beam Structures," in *Proceedings of the First Canadian Masonry Symposium*, Calgary, Alberta, 1976, pp. 292–303.

11.20 S.R. Davies and A.F. Ahmed, "An Approximate Method for Analyzing Composite Wall-Beams," in *Proceedings of the British Ceramic Society*, No. 27, 1978, pp. 305–320.

11.21 The Masonry Standards Joint Committee, "Building Code Requirements for Masonry Structures," ACI 530/ASCE 5/TMS 402, American Concrete Institute and American Society of Civil Engineers, Detroit and New York, 1992.

## 11.6 PROBLEMS

11.1 Identify framed buildings in your area that employ infill masonry walls. If possible determine whether the walls are built tight to the frame or whether movement joints have been provided to avoid transfer of lateral load from the frame to the wall. Prepare sketches of the wall layout and discuss the effect of this layout on the response of the building to lateral load for the following:

**(a)** A single-story building.

**(b)** A building with two or more stories.

**11.2** If an exterior infill wall or an interior partition is not intended to be part of the lateral load resistance of a frame building;

**(a)** Indicate how the size of movement joints required at each end of the wall and at the top should be determined. A typical panel size or one identified in Problem 11.1 can be used as an example.

**(b)** How can the wall be separated from the frame and yet be supported by the frame against out-of-plane loads? Provide sketches of connection details.

**11.3** A single-story steel frame building has 24 ft (7.2 m) long columns and beams across a 20 ft (6 m) bay (clear span) as shown in Fig. P11.3. Determine the resistance of an unreinforced masonry infill wall to an in-plane shear load applied at the top for the following:

**(a)** If frame–wall interaction is ignored and the wall is modeled as carrying the shear load as a shear wall. (This could represent the case with either no top beam or a very flexible beam.)

**(b)** Utilizing the diagonal strut compressive action of infill walls.

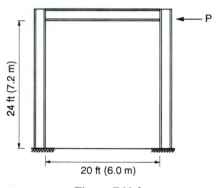

**Figure P11.3**

Masonry wall properties: The infill wall is 8 in. (200 mm) hollow blocks constructed in running bond.

Allowable shear stress = 60 psi (0.40 MPa)
Allowable tensile stress = 40 psi (0.28 MPa)
Allowable compressive stress = 400 psi (2.75 MPa)
$E_m$ = 1.45 × $10^6$ psi (10,000 MPa)

Steel frame properties: $E_s$ = 29 × $10^6$ psi (200,000 MPa), $I_c$ = 400 in.$^4$ (1.7 × $10^8$ mm$^4$), $I_b$ = 518 in.$^4$ (2.2 × $10^8$ mm$^4$).

**11.4** Calculate the increase in the stiffness of the frame shown in Fig. P11.4 when the clay brick infill walls are included in the analysis of the frame. Use a standard plane frame computer analysis to determine the deflection at the top under lateral load $P$.

Infill wall properties: $f'_m$ = 4000 psi (27.6 MPa), $E_m$ = 2000 ksi (13,800 MPa), $t$ = $3\frac{5}{8}$ in. (90 mm).

Frame properties: $f'_c$ = 5000 psi (34.5 MPa), $E_c$ = 4000 ksi (27,600 MPa), $b$ = 8 in. (200 mm) (width of beams and columns).

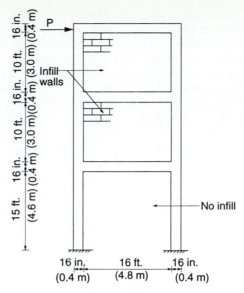

**Figure P11.4**

**11.5 (a)** If a gap (movement joint) of $\frac{1}{4}$ in. (6.3 mm) is left at the top of the wall to avoid load transfer due to column shortening and/or expansion of the masonry infill, determine the effect that this has on lateral load resistance for the building frames described in

  (i) Prob. 11.3
  (ii) Prob. 11.4.

**(b)** If for Prob. 11.3, the gap was 1/16 in. (1.6 mm) due to shrinkage of the concrete masonry (because blocks were allowed to get wet prior to being placed in the wall), would the reduction in capacity be 25% of that which occurred for Prob. 11.5(a) (i)? Explain.

**11.6** For the single-story reinforced concrete infilled frame shown in Fig. P11.6, determine the following:

**(a)** What is the best position for a 10 ft (3 m) long by 8 ft (2.4 m) high opening? Explain.

**(b)** What is the shear capacity of this wall for the chosen location?

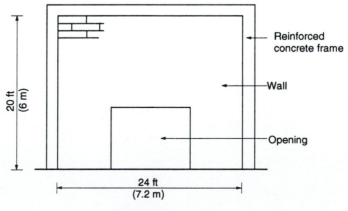

**Figure P11.6**

**11.7** Determine whether a 16 in. (400 mm) deep by 8 in. (200 mm) wide reinforced concrete foundation wall beam spanning between piles at 24 ft (7.2 m) spacing is adequate when it supports a 16 ft (4.8 m) high, 8 in. (200 mm) hollow block wall that carries a roof load of 1000 lb/ft (14.6 kN/m). The soft soil beneath the foundation beam is capable of supporting the beam only during construction of the wall. Use the Davies and Ahmed approach considering the foundation beam to be simply supported. Base your answer on the requirements of ACI 530/ASCE 5/TMS 402 or your local code. Consider the deflection limits. $f'_m = 2000$ psi (13.8 MPa), $E_m = 1500$ ksi (10,350 MPa), and $E_c = 3000$ ksi (20,700 MPa).

**11.8** Determine the axial force, shear, and bending moments at the critical section for the foundation beam designed in Prob. 11.7.

**11.9 (a)** For a steel beam supporting a 4 in. (100 mm) brick veneer with a 6 in. (150 mm) concrete block backup over a span of 16 ft (4.8 m) with a 10 ft (3 m) wall height, investigate the effect of relative beam-to-wall stiffness on the intensity of the vertical compressive stress in the masonry and the bending moment requirements for the beam.

**(b)** For part (a), what is the influence of reducing the height of the wall to 5 ft (1.5 m)?

# CHAPTER 12

# MASONRY VENEER
# AND CAVITY WALLS

Multistory residential buildings with brick veneer cladding (*Courtesy of Brick Institute of America*).

## 12.1 INTRODUCTION

### 12.1.1 Background

The change from the very thick walls of the past to the present use of thin masonry walls not only affected structural design, but also resulted in increased rain penetration problems. One solution to the rain penetration problem is to seal the wall with rendering, stucco, paint, or other sealants or a combination of these. The need for frequent maintenance and because such coverings mask the natural appearance of masonry have made it an unpopular solution in North America.

Another solution is to build a wall with two major elements, namely, an outer wythe of masonry connected to but separated by an air space or cavity from an inner wall. Here the outer masonry wythe serves as the first line of resistance to rain penetration and as the architectural or aesthetic finish. In this way, the traditional benefits of durability, low maintenance, fire resistance, and user acceptance of masonry have been incorporated into modern buildings. Such a wall system can be either a veneer wall or a cavity wall.

*Masonry veneer* is an exterior facing of masonry applied to a structural backing. The veneer provides ornamentation but is not counted on to contribute to the structural performance of the wall. It is anchored to its backing by a series of metal ties. [Other thin masonry facings are adhered to backing with an adhesive or portland cement matrix. These veneer systems have historically been applied by following prescriptive requirements and, because of the different methods of attachment, are not included in this book.] Structurally, the masonry veneer must be able to carry its own weight and transfer face loads, from wind or earthquakes, through the ties to the backup wall. Therefore, the backup wall is designed to carry all of the face load and, if loadbearing, the floor and roof gravity loads. It can be masonry, concrete, wood studs, or steel studs.

A *cavity wall* is similar to a veneer wall with two wythes of masonry and has many of the same rain resistance and building envelope characteristics. The principal difference is in the structural design concept employed. For cavity walls, both the interior and exterior wythes are designed to share in resisting vertical and/or lateral loads. Aside from the design philosophy employed, many of the characteristics and construction details are similar and, indeed, the distinctions given before can vary from place to place. However, the foregoing terminology is used throughout this book. The structural design of cavity walls is discussed in Chaps. 7 and 8, and, therefore, is not treated in detail in this chapter.

### 12.1.2 Components of Masonry Veneer and Cavity Walls

A typical masonry veneer wall is shown in Fig. 12.1. The main components of the masonry veneer wall and the corresponding cavity wall components are discussed in what follows.

**Masonry Veneer (or Outer Wythe).** In addition to serving as the architectural finish on the building, the veneer or outer wythe protects the other wall compo-

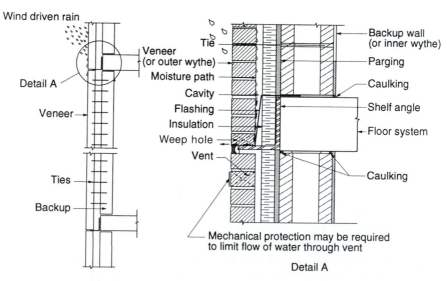

**Figure 12.1** Masonry veneer wall components (from Ref. 12.1).

nents from direct exposure to sunlight, moisture, and wind. It is the outermost barrier in the building envelope. In design of cavity walls, the outer wythe is designed to carry a share of the face load and any axial load directly applied to it.

**Shelf Angle.** Use of shelf angles is the preferred method for support of masonry veneer on high-rise structures at levels above the foundation. The alternative of supporting the veneer directly on the floor results in the exposed edges of the floor slabs acting as thermal bridges. This construction tends to be more susceptible to rain penetration.

**Cavity.** The cavity is the total distance between the outer and inner wythes or between the veneer and backup wall. If insulation is included in the cavity, then the *air space* will be smaller than the overall cavity width. The air space is also a barrier to migration of water across the cavity.

**Backup Wall (or Inner Wythe).** The backup or backing wall can be another wythe of loadbearing or nonloadbearing masonry or it can be some other structural system. The backup wall is intended to provide lateral support for the veneer and is designed to resist the full lateral load. In the case of a cavity wall, the inner and outer wythes are designed to share in resisting the lateral load.

**Ties and Anchors.** Ties are used to connect the two wythes and transfer lateral load from one to the other. For veneer walls, the ties must be able to transfer the total load from the veneer to the backup. Where the veneer (or outer wythe) is attached to other structural elements such as steel or concrete columns, a different type of connector known as an anchor is often used (see Chap. 13).

**Weep Holes and Vents.** Weep holes are intentional openings in the veneer (or outer wythe) located just above the support to permit water in the cavity to drain out of the wall. Vents are intentional openings near the top of the veneer that are intended to allow air circulation to dry out the cavity. Both allow air flow into the cavity, tending to equalize the air pressure in the cavity with the external air pressure.

**Flashing.** Flashing directly under the weep holes is used to direct water in the cavity out through the weep holes.

**Movement Joints.** A horizontal space at the top of the veneer is necessary to avoid transfer of vertical load from the structure to the veneer because of differential movement. This horizontal space is called a horizontal movement joint. Vertical movement joints are necessary to allow the veneer to expand or contract horizontally without buildup of excessive stresses. Proper sealing of movement joints is necessary to minimize rain penetration.

**Air Barrier.** Depending on the location of insulation and other design considerations, the air barrier can be located on either face of the backup wall (or inner wythe). It minimizes air flow through the backup wall (or inner wythe). The air barrier helps develop air pressure equalization in the cavity and it minimizes condensation in the wall by controlling movement of moisture laden air through the wall.

**Compartment Divider.** To equalize the air pressure in the cavity with the external air pressure, the cavity must be comparted into localized pressure areas. Whereas the shelf angles and flashings serve as horizontal air barriers across the cavity, vertical air barriers must be included as separate components. As shown later in Sec. 12.2.1, metal or plastic flashings in vertical movement joints can serve this purpose.

### 12.1.3 Critical Features in Design and Construction

The basic requirements for proper design and construction of masonry veneer on wood stud backup walls have been well defined[12.2,12.3] and are relatively simple to implement on the site. The requirements for brick veneer and cavity walls on high-rise structures are also well documented.[12.1,12.4] However, lack of formal education of designers on masonry veneers and inadequate training of builders have combined to produce serious problems in some walls. In many cases, the more severe environmental loading conditions encountered in high-rise construction have led to an increased amount of distress. In particular, in high-rise buildings, roof overhangs and other shielding elements are usually not available to minimize the exposure of walls to rain. In addition, the greater tendency for air flow through the wall due to wind pressure, stack effect,[12.5] and mechanical ventilation[12.6] increases the potential for condensation of airborne moisture in the wall. Also the differential vertical movement between the veneer (or outer wythe) and the structure due to shortening of the structure and expansion of the outer wythe is much larger for high-rise construction.[12.7–12.9] Therefore, the quality of design and construction is more severely tested in high-rise construction. The remainder of this chapter emphasizes high-rise construction, but most concepts also apply to low-rise buildings.

Many design and construction mistakes are usually easily avoided at minimal, if any, extra cost. On the other hand, ignorance of proper procedures or ill-conceived minor savings, such as using ties with poor corrosion resistance, can result in remedial costs far in excess of the original cost of the veneer. Therefore, designers and builders should take the small additional care needed to produce the quality of masonry wall that has proven to provide a superior exterior finish for a wide variety of buildings.

**Differential Movement.** One of the most frequent problems in masonry constructed during the past two decades is lack of horizontal movement joints to accommodate vertical movement.[12.10,12.11] In some cases, where the veneer is in contact with the shelf angle above it, as shown in Fig. 12.2(a), the veneer may have sufficient capacity to carry the load transferred to it through the shelf angle and may not show any signs of distress. However, such conditions with the shelf angle in tight contact with the veneer below are not recommended, as they impose undesigned stresses in the veneer. Also, bending of the shelf angle due to differential vertical movement can cause eccentric loading and, thus, out-of-plane bending in the veneer. If the joint at a shelf angle is only partially filled with mortar, Fig. 12.2(b), the

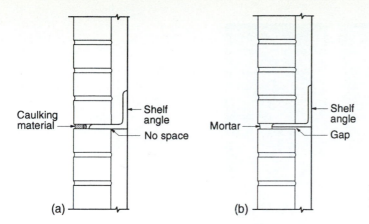

Caulking material — Shelf angle — No space

Mortar — Shelf angle — Gap

(a)　　　　　(b)

**Figure 12.2** Veneer on shelf angle with no movement joint.

localized concentration of compression force can cause either local spalling such as shown in Fig. 12.3, or a buckling of the masonry veneer, as shown in Fig. 12.4. This buckling, or bending, indicates that eccentric forces are present, tending to cause a very dangerous failure condition. Both of these effects can occur in relatively low buildings.

Differential movement between the inner and outer wythes of cavity walls can affect the distribution of axial load when the outer wythe is designed to carry some of the vertical load. Differential movement can invalidate the load-sharing assumptions made and lead to distress by overloading the outer wythe. It is, therefore, common to carry vertical load on the inner wythe only and allow relative movement of the outer wythe to occur. Relative differential movement can be accommodated by using adjustable ties or ties that are sufficiently flexible in bending. Design, detailing, and inspection to ensure adequate movement joints should be top priorities.

For low-rise buildings, the entire height of the masonry veneer can be supported on the foundation wall provided that differential vertical movement is allowed for

**Figure 12.3** Spalling of brick veneer at shelf angle.

**Figure 12.4** Buckling of brick veneer (*Courtesy of C. T. Grimm*).

in elements projecting through the veneer, such as balconies and window frames. The top of the wall must also be detailed to permit relative movement.

**Rain Penetration.** The design of masonry veneer or cavity walls must consider rain penetration. Provision of an adequate air space, an airtight backup wall, clean cavity with open weep holes, properly positioned flashing, and use of good quality materials will help ensure good performance. Omission of any part of the design can result in water related damage.

Excessive water in the cavity air space, besides being more likely to penetrate the backup wall, can result in increased rates of corrosion of ties or other metal components, including "eating away" of galvanizing. Where weep holes do not function properly to drain water, and vents do not allow air circulation to dry out the cavity, the damp conditions can lead to growth of mold and fungi and to efflorescence due to migration of the moisture through the veneer. In cold climates, saturation of the brick can lead to freeze-thaw deterioration of the veneer and water retained in the cavity can freeze and push the veneer off the shelf angle.

For masonry backup walls, although water penetration does not usually harm the masonry, wetting of insulation, staining or flaking of interior finishes, and wetting of carpets and ceiling areas are all unacceptable, but possible, consequences. For other backup wall systems, the integrity of the backup itself can be compromised as a result of wetting. In addition to the wetting of insulation, softening of sheathing, corrosion of metal components or connectors, rotting of wood, and growth of molds or fungi are all serious concerns.

The vast majority of masonry veneer walls provide excellent rain protection. Problems arise when some (usually more than one) aspects of design or construction are not properly carried out. Investigation of such problems usually reveals missed details either at the design or the construction stage.

## 12.2 MASONRY RAIN SCREEN WALLS

### 12.2.1 Design of Masonry Veneer Rain Screens

The masonry veneer and cavity wall systems should be thought of as three-stage barriers to rain penetration, as shown in Fig. 12.5. As the outermost element, the outer wythe is the first barrier and should be designed and constructed to minimize the amount of water that can penetrate through it. This is the first stage of the masonry rain screen. Control of air flow in the cavity and drainage of the cavity are the second and third stages of the rain screen.

**Stage 1.** For rain to reach the cavity, it first must hit the outer wall surface. Therefore, design of roof overhangs, drips under window sills, and rain gutters

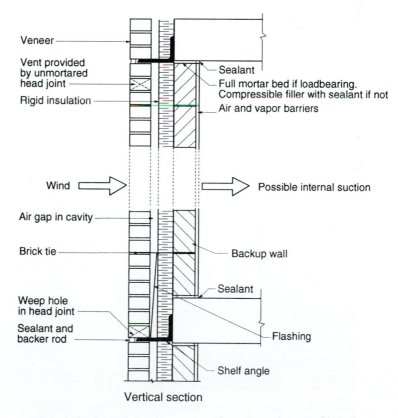

Veneer

Vent provided by unmortared head joint

Rigid insulation

Sealant

Full mortar bed if loadbearing. Compressible filler with sealant if not

Air and vapor barriers

Wind

Possible internal suction

Air gap in cavity

Brick tie

Backup wall

Sealant

Weep hole in head joint

Sealant and backer rod

Flashing

Shelf angle

Vertical section

**Figure 12.5** Brick veneer as a rain screen (from Ref. 12.1).

can all help to reduce the amount of water available on the exterior surface of the wall. Water on the surface of the wall can be carried into and through the wall by one or more of the following:

- gravity drainage
- capillary action
- kinetic energy
- air flow

Provided that relatively impermeable and uncracked masonry units and good quality mortar are used, the main avenues for rain penetration are through partially filled, poorly compacted or debonded head and bed joints.[12.12–12.16] Therefore it is very important that mortar joints be completely filled and that the mortar, units and construction conditions combine to produce good bond to resist cracking.[12.17–12.19] Also it is strongly recommended that the exterior face of the mortar joint be tooled to produce a concave profile (see Fig. 12.6). The cylindrical jointer (tooling instrument) produces a denser compacted surface and forces the mortar into tight contact with the masonry unit. This joint best resists water penetration. As shown in Fig. 12.7, alternative joints that may be satisfactory but are not as good are the V-joint and the weathered joint. Flush, grape vine, struck, raked, or extruded joints tend to allow easy access of water into the joint and are not recommended for rain resistance.

Although the design concept for open rain screens incorporates provisions for dealing with water that reaches the cavity, it is logical to minimize the amount of water that does penetrate the outer wythe so that the second and third stages of resistance are not severely tested. The fewer components put to the ultimate test, the fewer are the walls that will experience distress due to rain penetration.

**Figure 12.6** Tooling of mortar joints (concave type) (*Courtesy of Brick Institute of America*).

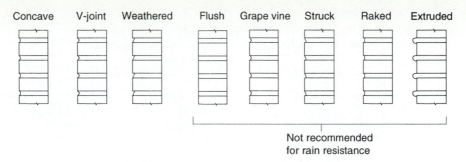

**Figure 12.7**  Types of mortar joints.

A well-built veneer (or outer wythe) constructed with high quality materials will allow relatively small amounts of moisture to penetrate to the cavity due to gravity flow through the wall or the kinetic energy of rain hitting the surface of the wall.[12.14] Similarly, although capillary action can result in gradual wetting through parts of the veneer (or outer wythe), by itself it is not a source of free-flowing water in the cavity.

**Stage 2.**    By far the largest potential for rain penetration is from water carried into the wall by air flow. The best defense is to use the open rain screen principle where, ideally, the air pressure within the cavity is equalized with the external air pressure, thereby eliminating air pressure as a driving force. To accomplish this, sufficient weep holes and vents must be present to allow the cavity pressure to react almost instantaneously with variations of external air pressure due to wind gusts. For this to be effective, the cavity enclosure consisting of an air barrier and compartment dividers must also be sufficiently airtight to minimize the amount of air that flows into the cavity. It is not just the pressure equalization itself, but also the lack of air flow due to pressure equalization that minimizes the amount of water that will be carried into the wall. The provision of an adequate air space and limitation of air flow into that space is the second stage of the open rain screen.

As is illustrated in Fig. 12.8(b), air flowing into the cavity carries water with it and even minor leakage through the backup wall can transport this moisture into areas where it can cause damage. Particularly at corners of buildings, where suction on side walls (see Fig. 12.8(a)) can significantly increase air flow through the cavity, it is important that vertical barriers be introduced across the cavity to create pressure equalization compartments. As mentioned in Sec. 12.1.2, shelf angles and flashing can serve as the horizontal barriers.

For vertical barriers or dividers, use of closed-cell compressible filler material such as foamed rubber or urethane and the alternative of positioning a vertical flashing in vertical movement joints are illustrated in Fig. 12.8(c). Although there are no absolute rules regarding size of compartment or spacing of dividers, it is very important that dividers be located near corners and any other areas where high local differences in air pressure are likely to occur. For other areas subject to less local variation in air pressure, spacing dividers to coincide with movement joints seems to be a reasonable approach.

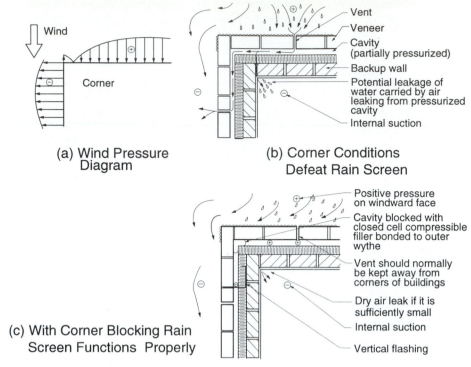

**Figure 12.8** Pressure equalization at corners of buildings (from Ref. 12.1).

**Stage 3.** Regardless of the quality of the veneer construction and the effectiveness of air barriers and compartments to prevent air flow, it must be assumed that some water will reach the cavity. The drainage system is provided for this reason and it forms the third stage of the rain screen.

The primary component of the drainage system is the air space in the cavity. It must be free of any materials that would impede the drainage of water from the cavity or direct water onto the backup wall. Furthermore, the flashing must be properly installed and weep holes must be clear to drain water to the exterior of the veneer. Although drawings may show water draining freely down the cavity, the effect of mortar droppings adhering to wall ties and collecting at the bottom of the cavity, Fig. 12.9, must be considered. In general, to minimize the potential for mortar bridging across the cavity, an air gap of $1\frac{1}{2}$ to 2 in. (40 to 50 mm) is required. Otherwise, water may be directed onto the backup wall by flowing across these bridges.

Care must be taken to ensure that mortar does not collect at the bottom of the cavity and plug weep holes or otherwise restrict drainage of the cavity. Mortar droppings at the bottom of the cavity should be reduced by using one or more of the techniques illustrated in Fig. 12.10. These are discussed as part of workmanship in Sec. 15.3. Investigations of existing buildings[12.20] have shown cases where mortar droppings have formed a solid layer some distance up the wall, as shown in Fig. 12.11. It is important that drainage channels be opened, even through small layers

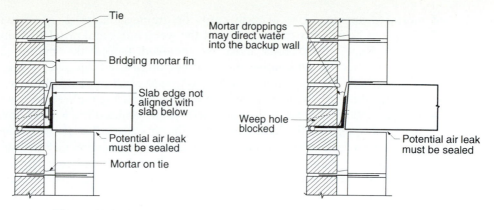

**Figure 12.9** Transportation of water across a cavity wall (from Ref. 12.1).

a) Wood strip to remove mortar droppings

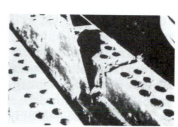

b) Smearing the back with mortar

c) Leaving units out at the bottom for cleaning

**Figure 12.10** Methods for elimination of mortar droppings (*Courtesy of National Concrete Masonry Association*).

**Figure 12.11** Mortar droppings blocking drainage path in cavity wall (*Courtesy H. Keller*).

of mortar droppings. Use of sash cords or drainage material, as shown in Fig. 12.12(a), can help to prevent plugging of the weep holes and provide drainage. For proper drainage, sash cords must be removed before the mortar droppings harden. An effective alternative, but one not often used because of the work involved, is to initially leave out a brick on one side of all or some of the weep holes; see Fig. 12.10(c). At the maximum recommended spacing for weep holes[12.21,12.22] of 24 to 32 in. (600 to 800 mm), the space left by the omitted brick will allow most of the mortar droppings to be cleaned out when the air space is 2 in. (50 mm) wide. When the omitted brick is mortared into place, the weep hole and cavity in that area will be clear and free-draining. Another alternative is to lay a rope lengthwise along the bottom of the cavity and to shake this to break up the mortar droppings as they are hardening. Coarse aggregate at the bottom of the cavity or a screen to catch the mortar will help keep weep holes clear of mortar, Figs. 12.12(b) and 12.12(c). A drainage fabric can be placed on the outside face of the backing to aid in drainage; see Fig. 12.12(d).

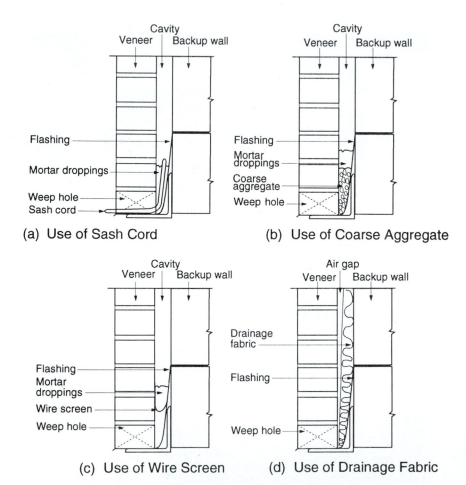

**Figure 12.12** Methods for keeping drainage path open in cavity walls (from Ref. 12.1).

The flashing that directs the draining water out of the weep holes must be continuous or adequately lapped and must be turned up at the ends to form a dam. It must be firmly attached in the backup wall and of sufficient height to prevent flow of water into the backup.

Ties crossing the cavity should not be shaped or sloped in a way that would cause them to act as pathways for water to cross the cavity. As is discussed in Chap. 13, some ties incorporate a localized bend that acts as a drip to eliminate flow across the cavity. However, in some cases, the decreased strength and stiffness resulting from inclusion of this drip may be more significant. Whether a drip is included or not, ties that tend to act as platforms for mortar droppings can produce bridges for transport of water across the air space.

The backup wall is also part of the third stage of the open rain screen. Although it is intended that using good quality masonry construction and minimizing the air flow through the outer wythe will eliminate most of the rain penetration and that the air space will allow drainage of the water that does reach the cavity, the backup wall must be designed to shed water that may cross the cavity. Use of insulation, sheathing, or other membranes on this surface that shed water is recommended. If tar impregnated building paper or other loose forms of moisture barriers are used, special care is required to ensure that these materials remain attached to the backup or inner wythe and do not fall across the cavity and act as funnels for transporting water across the cavity. Similarly, flashing must be firmly attached, so that water draining down the outside of the backup (or inner wythe) will be directed out the weep holes.

Air flow through the backup wall (or inner wythe) can be a major mechanism for transporting water across the air gap and through the wall. Hence, the airtightness of the backup wall is very important to the overall rain resistance of a wall.

### 12.2.2 Additional Precautions to Avoid Rain Penetration

A major source of rain penetration involves situations where water is directed into the wall. Examples are balconies not properly drained away from the building and ineffective sealing around window frames. Balcony slabs should be sloped and preferably built with a curb, as shown in Fig. 12.13(a).

In regions susceptible to rain penetration, it is not good practice to support the veneer directly on the slab, as shown in Fig. 12.13(b). Even with the flashing properly run up and fastened into the backup wall,there is a high probability of cavity water entering the building either under the flashing or through accidental holes in the flashing. Using a recess in the edge of the slab to support the veneer, shown as an alternative in Fig. 12.13(b), will help prevent cavity water from entering the building by lowering the level at which water will accumulate. In cold climates, thermal bridging by the slab makes the practice of supporting a veneer directly on the slab questionable.

Flashings at roof or parapet levels and at openings in the wall should extend over the exterior face of the masonry a sufficient distance, as shown in Fig. 12.14, to prevent the wind from carrying water into the cavity. Although caulking can be effective, building movements and aging of the caulking can render it ineffective and allow large amounts of water into the cavity. In addition, the need to replace caulking

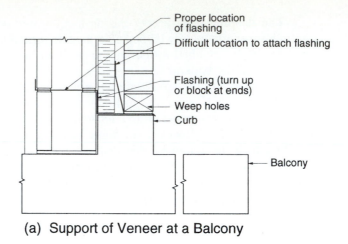

Proper location of flashing

Difficult location to attach flashing

Flashing (turn up or block at ends)

Weep holes

Curb

Balcony

### (a)  Support of Veneer at a Balcony

Joint raked out to permit positive anchorage of flashing.

Continuous flashing

Weep hole @ 32 in. (800mm) spacing

Drip

Recess alternate detail

Sealant
Movement joint
Vent

Cavity wall tie

Mortar parging air barrier

Vapor barrier

Gypsum board

Flexible sealant at perimeter of air barrier

Wrong flashing location allows water in cavity  to flow into the building.

Sealant

### (b)  Veneer Supported on the Floor Slab

**Figure 12.13**   Flashing details for brick veneer walls (from Ref. 12.1).

at regular time intervals and the possibility that the caulking may become concealed within the wall or otherwise inaccessible should be considered.

Parapets are exposed to the extremes of climate on both sides including very high air pressure differences. Therefore they are susceptible to rain penetration and subsequent damage due to free-thaw cycles in cold climates. Also, differential movement of the parapet compared to the wall below and rotation of the roof tend to cause cracking at the parapet when it is built integral with the wall below. For parapets which extend significantly above the roof level, separate support, as shown in Fig. 12.14(a), is recommended for the veneer covering the parapet.  Where parapets are not insulated, for cold climates, solid grouting of the parapet, Fig. 12.14(b),

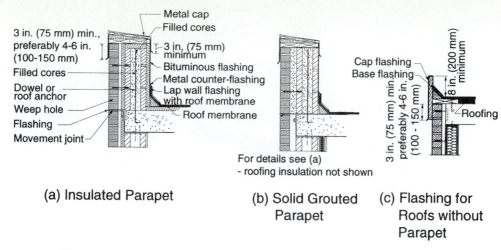

**Figure 12.14**  Parapet flashing and construction details (from Ref. 12.1).

can lessen the potential for freeze-thaw damage by removing spaces for water to accumulate and by reducing the amount of water carried into the wall by air flow.

Special design details and attention to these details in construction are very important. For example, air barriers that are discontinuous at openings or at tops of walls tend to allow the moisture from exfiltrating air to be carried through into backup walls at these points. The importance of the air barrier in prevention of rain penetration should not be underestimated and its structural resistance to air pressure in either direction should be part of the design. Unparged masonry, particularly hollow masonry, is not airtight. Thus a separate component is needed. Air barriers should be continuous and fully supported over the height of the wall and across all joints. Additional discussion on detailing of air barriers is included in Sec. 14.5.

Vents to permit pressure equalization in the cavity and circulation of air to aid in drying the wall should be detailed. However, if significant air leakage occurs through the backup walls or if the comparting of cavities is not effective, air flowing into the cavities will carry in water. In cases where significant air movement through the wall or along the cavity occurs, the vents may act as large openings for penetration of airborne moisture. Therefore, in addition to reducing air flow through the veneer by comparting and using airtight backup walls, vents should be located away from corners and away from potential drainage paths for water running down the veneer. Positions directly beneath weep holes and at the ends of window sills are examples of obvious locations to be avoided. Studies have shown[12.14] that vents created by omitting the mortar in head joints can be more widely spaced than weep holes and still satisfy the requirements for equalization of cavity pressure.

### 12.2.3  Detailing Requirements for Rain Screen Performance

The flashing at the base of the cavity must be continuous through proper lapping and sealing, as shown in Fig. 12.15. It must be fastened to structural components such as columns as well as the backup. Where columns extend beyond the face of the

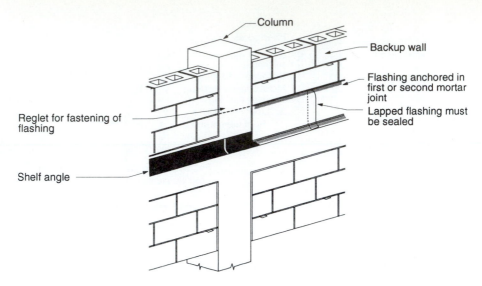

**Figure 12.15**  Continuity of flashing (from Ref. 12.1).

backup wall, special details are required for fitting the flashing around the corners. In addition, the flashing should pass completely through the veneer, have a positive outward slope, and form a drip at the exterior face, as shown in Fig. 12.16(a). Sealant should be located as shown to provide for a continuous drainage surface. (Application of the sealant may be difficult unless rigid flashing is used.) Unless the shelf angle extends to the face of the veneer, the alternative detail, shown in Fig. 12.16(b), is deficient because it allows water to sit on the top of the masonry and therefore produces a much greater chance for rain penetration. As is illustrated in Fig. 12.17, for veneer supported on the slab, tolerances in alignment of slab edges will often leave the top of veneer exposed to water running down the face of the building. This is another reason for not supporting the veneer (or outer wythe) directly on the floor slab. Although similar problems can be encountered for veneer supported by shelf angles, adjustments to the positions of the shelf angles are possible.

Because of possible interference with shelf angle bolts and other logistics of placing the flashing in the confined cavity space, some designs have incorporated

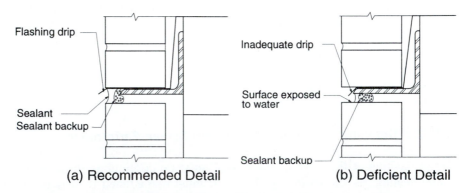

**Figure 12.16**  Flashing and sealant detail at shelf angle (from Ref. 12.1).

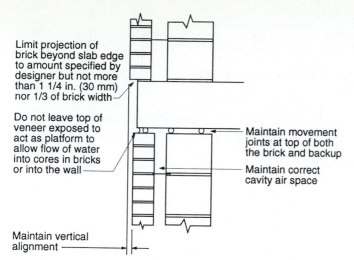

Limit projection of brick beyond slab edge to amount specified by designer but not more than 1 1/4 in. (30 mm) nor 1/3 of brick width

Do not leave top of veneer exposed to act as platform to allow flow of water into cores in bricks or into the wall

Maintain vertical alignment

Maintain movement joints at top of both the brick and backup

Maintain correct cavity air space

**Figure 12.17** Requirements for veneer supported on concrete slab (from Ref. 12.1).

the flashing at the bottom of the second course of brick, as shown in Fig. 12.18. However, this practice is not recommended because any small punctures in the flashing can allow water to accumulate on the shelf angle. This can result in corrosion, efflorescence, or freezing damage. The specification of an adequate cavity width to allow for minor dimensional tolerances will ensure that the flashing can be placed directly on the shelf angle where it can be continuously supported. The use of fixed and adjustable shelf angles is discussed in Sec. 12.3.1.

To the extent possible, sealant should be avoided as the main barrier to rain penetration. Even properly installed, very good quality sealant will have a limited service life and must be checked for replacement at regular intervals. Where sealant is used, a backing material is usually necessary. This should be a compressible material such as closed-cell polystyrene foam rod, as shown in Fig. 12.19(a). In

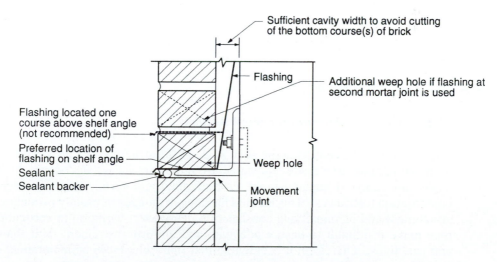

Sufficient cavity width to avoid cutting of the bottom course(s) of brick

Flashing

Additional weep hole if flashing at second mortar joint is used

Flashing located one course above shelf angle (not recommended)

Preferred location of flashing on shelf angle

Sealant

Sealant backer

Weep hole

Movement joint

**Figure 12.18** Flashing detail to avoid shelf angle bolt (from Ref. 12.1).

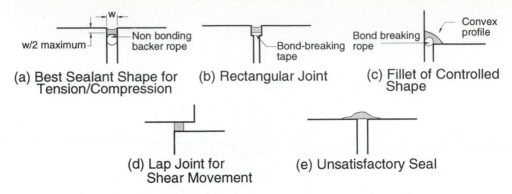

**Figure 12.19**  Sealant shapes and bond breakers (from Ref. 12.1).

addition to serving as a bond breaker and support, the *backer rod* or rope can help produce the desirable "dog bone" shape. This shape reduces the stresses in the areas where the sealant adheres to the sides of the joint and thus reduces the likelihood of debonding during opening of the joint. When the joint is compressed, there will be less tendency for the sealant to bulge out beyond the face of the veneer.

For the sealant to deform without rupturing or peeling away from the sides of the joint under repeated cycles of extension and compression, the amount of deformation should be limited to the manufacturers' recommendations. The acceptable strain is generally less than 25%. For this reason, it is sometimes necessary to create a large joint, as shown in Fig. 12.19(b), to reduce the strain for a given movement. [A $\frac{1}{4}$ in. (6.3 mm) movement would require a 1 in. (25 mm) wide joint.] Figures 12.19(c) and 12.19(d) illustrate other acceptable types of joints, but Fig. 12.19(e) shows a smear type of caulking technique that will generally either rupture at the smaller section near the edge or debond from the surface due to a gradual peeling action.

In some cases, it may be possible to use commercially available water-stop types of seals or other flashing arrangements to provide additional protection from water entering the wall at movement joints and at the junctions with other elements such as windows.

## 12.3 SHELF ANGLE SUPPORTS

### 12.3.1 Design of Anchors for Shelf Angles

The method of attachment of the shelf angle to the structure and the degree to which its position should be made adjustable are important design decisions. If the shelf angle connection is nonadjustable, its horizontal and vertical positions depend on construction tolerances. This method has the advantages of usually providing secure attachment and of simplifying construction. However, variation in vertical position may make it difficult to ensure adequate and uniform movement joint thicknesses and true lines. Therefore, it is essential that tight tolerances be maintained on con-

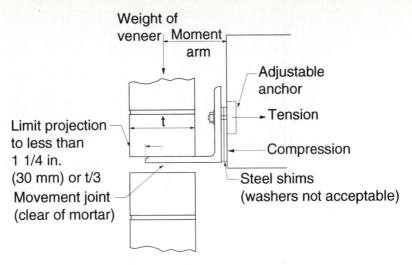

Weight of veneer

Moment arm

Adjustable anchor

Tension

Compression

Limit projection to less than 1 1/4 in. (30 mm) or t/3

Movement joint (clear of mortar)

Steel shims (washers not acceptable)

t

**Figure 12.20** Adjustable shelf angle (from Ref. 12.1).

struction with fixed shelf angles. Placement of the top course of veneer below the fixed shelf angle may also be difficult.

Use of an adjustable shelf angle, such as shown in Fig. 12.20, employs cast-in or drilled-in anchors to permit vertical adjustment of the shelf angle using slots in the shelf angle. Horizontal out-of-plane adjustments can be made using shim plates, as shown. Where anchor bolt locations do not match the slots in the shelf angle, correction by burning out a larger hole is not recommended. It destroys the corrosion protection locally and touch up painting may not be as effective as the factory-applied protection. The enlarged slot also may not leave sufficient bearing surface for the nut and washer. Drilling new anchors may be a preferable solution.

For adjustable shelf angles, greater care and inspection effort are required to ensure proper positioning and proper tension on the bolts for the friction connection.

Typically, bolts or welded anchors should be not spaced farther apart than 4 ft (1.2 m) and should be sized to carry the shear and tensile forces resulting from loading the outstanding leg of the angle. The shear is taken directly by the bolt or welded anchor. The required tensile force is calculated as the resisting moment divided by the distance from the compression component of the couple. Therefore, for adjustable anchors, washers should not be used as shims for horizontal adjustment of the shell angle position because the moment arm between the compressive and tensile forces will be very small. Also more rotation of the shelf angle can occur. Shims, such as shown in Fig. 12.21, should be used.

### 12.3.2 Example Calculation of Anchor Requirement

#### Example 12.1

A 9 ft (2.7 m) height of brick veneer is supported by a shelf angle anchored at 4 ft (1.2 m) intervals (see Fig. 12.21(a)) and the distance from the center of the brick veneer

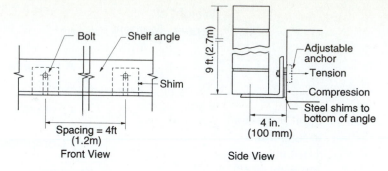

**Figure 12.21** Anchor bolt design: Example 12.1.

to the floor slab is 4 in. (100 mm). The applied moment per anchor bolt is

$$M = \text{(veneer weight per unit length)} \times \text{(anchor bolt spacing)}$$

$$\times \text{(moment arm between center of veneer and edge of slab)}.$$

$$= (36 \text{ lb/ft}^2)(9 \text{ ft})(4 \text{ ft})(4 \text{ in.}) = 5184 \text{ in.-lb}$$

$$(M = (176 \text{ kg/m}^2)\, 9.8\, (2.7)\, (1.2)\, (100)/10^6 = 0.559 \text{ kN} \cdot \text{m})$$

where the weight of the veneer is 36 lb/ft$^2$ (176 kg/m$^2$).

For an allowable tensile stress in the anchor bolt of $F_s = 44{,}000$ psi (300 MPa) for A325 high strength bolts, the required area of the bolt can be found from

$$M = A_s F_s jd$$

giving

$$A_s = M/F_s jd \qquad (12.1)$$

where $jd$ is the distance between tension and compression force resultants and, as a trial value, $j = \frac{7}{8}$ is approximated. Then, if the distance from the anchor bolt to the bottom of the shim is $d$ [taken to be 4 in. (100 mm) in this case], the estimated value of $A_s = 0.034$ in.$^2$ (21.3 mm$^2$). A minimum $\frac{1}{2}$ in. (13 mm) diameter bolt is recommended.

To check the bearing force on the concrete and the value of $j$, a conservative approximation using the reinforced concrete analogy with working stress equations similar to those in Chap. 6 can be performed as follows where $b = 5$ in. (125 mm) is the width of the shim.

$$n = E_s/E_c = 8.06 \qquad \text{(from Eq. 6.1)}$$

$$\text{[for 4000 psi (27.5 MPa) concrete]}$$

$$\rho = A_s/bd = 0.01 \qquad \text{(from Eq. 6.6)}$$

$$k = \sqrt{(n\rho)^2 + 2n\rho} - n\rho = 0.32 \qquad \text{(from Eq. 6.10)}$$

$$j = 1 - k/3 = 0.89 \qquad \text{(from Eq. 6.15)}$$

where

$$E_s = 29 \times 10^6 \text{ (200,000 MPa)}$$

$$E_c = 3.6 \times 10^6 \text{ psi (24,800 MPa) for 4000 psi (27.5 MPa) concrete}$$

$$A_s = 0.20 \text{ in.}^2 \text{ (129 mm}^2\text{)}.$$

Checking the maximum compressive bearing stress on the concrete using an equation of the same form as Eq. 6.12,

$$f_c = 2M/jkbd^2 = 455 \text{ psi (3.14 MPa)}$$

which is adequate for 4000 psi (27.5 MPa) concrete.

The foregoing calculation could also be done in a similar manner using factored loads and equivalent ultimate strength theory for reinforced concrete. [Note that the capacity of the anchorage in the concrete, including possible pull out, should be checked in a complete design.] ∎

### 12.3.3 Design of Shelf Angles

A rigorous design of shelf angles should take into account the influence of supporting a story height of rigid veneer on a relatively flexible shelf angle. The shelf angle is subject to large torsional forces and the extended leg bends due to the eccentricity of the veneer weight from the anchorages.[12.23] It is important that the shelf angle be properly designed, particularly for modern construction where cavity insulation and up to 2 in. (50 mm) air spaces result in larger moment arms and correspondingly higher bending moments.

A simplistic approach has been to design the thickness of the shelf angle based on simple cantilever bending. The moment arm is usually taken as the distance from the center of the veneer to the center of the upright leg of the angle. The moment due to the weight of the veneer will actually be less than that calculated because, as the angle deflects, the force from the veneer will be transmitted to the angle at a contact point closer and closer to the back face of the veneer, as shown in Fig. 12.22. On the other hand, the assumption that the entire length of the angle will resist bending of the outstanding leg requires that plastic behavior be relied upon. In many cases, the required angle thickness is relatively small and some extra capacity can be built in using slightly greater thicknesses than required. In other cases, shimming the shelf angle out beyond the insulation can reduce the bending on the extended leg of the shelf angle. Closer spacing of anchors can also be used to ensure that the entire length of angle is effective in the elastic range of behavior.

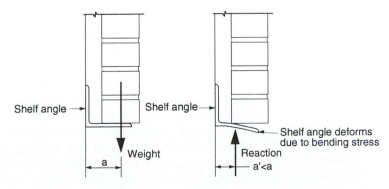

**Figure 12.22**  Loads on shelf angle due to weight of veneer.

An alternative to the foregoing is for design firms to prepare in-house design aids using finite element analyses or other methods[12.23] for a range of veneer heights, moment arms and anchor bolt spacings.

Corrosion of shelf angles is another area of concern for designers.[12.24] Corrosion reduces the section capacity and the products of corrosion can take up to $2\frac{1}{2}$ times the original volume of the steel, thus reducing the size of the movement joint. Except in very dry environments, painted shelf angles typically do not have sufficient corrosion resistance and thus hot-dipped galvanizing should be the minimum level of corrosion resistance.

### 12.3.4 Example Design of the Thickness of a Shelf Angle

**Example 12.2**

Using Example 12.1, shown in Fig. 12.21, but with a moment arm of $3\frac{1}{2}$ in. (90 mm), from the center of the veneer to the centre of the upright leg of the angle, the applied moment is

$$M = \text{veneer weight per unit length} \times \text{moment arm}$$

$$= 36 \text{ lb/ft}^2 \times 9 \text{ ft} \times 3.5 \text{ in.} = 1134 \text{ in.-lb/ft}$$

$$(176 \text{ kg/m}^2 \times 9.8 \,(2.7)(90)/10^6 = 0.42 \text{ kN.m/m})$$

and from

$$M = F_s S$$

where  $S = bt^2/6$
  $b = $ unit length of the shelf angle $=$ unit length of veneer
  $F_s = 0.6\,(36) = 21.6$ ksi (149 MPa) for A36 steel

Hence, the required thickness $t$ of the shelf angle is

$$t = 6M/bF_s \tag{12.2}$$

which in this case indicates that a 0.16 in. (4.1 mm) thick shelf angle is required. Normally a minimum thickness of $\frac{1}{4}$ in. to $\frac{5}{16}$ in. (6 to 8 mm) is used. ∎

### 12.3.5 Construction Details for Shelf Angles

Stability of the free-standing initial courses of veneer during construction dictates that the masonry veneer cannot extend more than half the thickness of the veneer beyond the edge of the shelf angle. However, due to the slope at the toe of the shelf angle and to avoid stress concentrations, the projection is normally limited to one-third of the veneer thickness, as shown in Fig. 12.20. Extending the leg of the shelf angle out to near the exterior face of the veneer has the advantage of providing support for the flashing, but requires that the sealant and backer material be positioned between the top of the veneer and the underside of the shelf angle. In this case, a larger movement joint is normally required to satisfy the sealant's strain limits.

At corners of the building, it is important that the shelf angle be detailed to provide continuous support for the veneer. As shown in Fig. 12.23, mitering of the

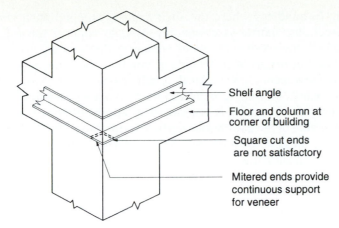

Shelf angle

Floor and column at corner of building

Square cut ends are not satisfactory

Mitered ends provide continuous support for veneer

**Figure 12.23** Shelf angle detail at corners of buildings.

**Figure 12.24** Cracking of brick veneer at corners of buildings.

shelf angle is appropriate. In buildings where the shelf angle is stopped flush with the corner in the two directions, mortar is commonly used to support the brick around the corner. Therefore, the small sections of veneer at the building corners become continuous over the full height of the building, whereas the other parts of the veneer are separated by movement joints under the shelf angles. As the structure shortens and the veneer expands, extremely high localized compression forces are introduced into these corner areas, resulting in compressive failure either as spalling or vertical cracking on either side of the continuous vertical strip of veneer, as shown in Fig. 12.24. In some cases, these vertical cracks extend nearly the full heights of the buildings. In addition to exposing the walls to excessive rain penetration, there is a potential safety hazard from falling pieces of masonry. This type of problem should not occur. The correct detail for supporting veneer around corners is simple to install. However, this cracking does graphically point out the need for a completely thought-out support system for the veneer so that unintentional forces are not built up.

Where brick veneer spans over small openings, it is usual to support it with a *lintel angle* built into the veneer at each side of the opening, as shown in Fig. 12.25(a). In this case, the lintel angle is not attached directly to the structure and is sometimes referred to as a "loose lintel." For larger openings, rather than design the lintel to span the entire length of the opening, it is often more practical to support the lintel on the backup wall or structure, in which case it acts as a shelf angle. In such designs, it is necessary to introduce vertical movement joints, such as shown in Fig. 12.25(b), to allow for the varying differential movements associated with support at different elevations.

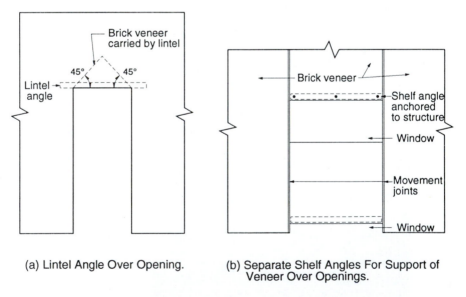

(a) Lintel Angle Over Opening.

(b) Separate Shelf Angles For Support of Veneer Over Openings.

**Figure 12.25** Support of veneer over openings (from Ref. 12.1).

## 12.4 STRUCTURAL DESIGN CONSIDERATIONS

### 12.4.1 Structural Requirements for Masonry Veneer Walls

The design concept for masonry veneer walls is that no axial force other than self-weight should be imposed on the veneer and the backup wall should be designed to resist the full lateral load. Therefore, in theory, the only strength requirement for a veneer is for it to be able to transmit wind or seismic loads to the backup wall and structure through the ties, anchors, and shelf angles. In this regard, sufficient flexural strength of the veneer must exist for it to at least span between ties so that the potential for rotation and instability of veneer sections, such as shown in Fig. 12.26, cannot occur. This function has traditionally been met by prescriptive requirements found in most building codes. As shown in the Fig. 12.26, stability requires that potential crack spacing be at least twice the spacing between ties. Therefore, in addition to adequate bond between the units and the mortar, limits on maximum vertical spacing of ties must be observed.

Regardless of the design philosophy, uncracked brick veneer shares in carrying part of the lateral load where stiff ties force the veneer to follow the same curvature as the backup wall. If the veneer and backup walls have the same height and support conditions and are forced to deflect equally because of rigid connectors, the share of the lateral load taken by the veneer can be calculated from

$$p_v = p_{\text{total}}(EI)_v/[(EI)_v + (EI)_{\text{backup}}] \qquad (12.3)$$

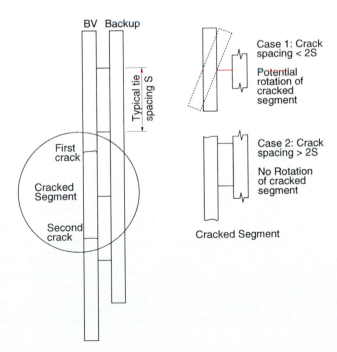

BV  Backup

Typical tie spacing S

First crack

Cracked Segment

Second crack

Case 1: Crack spacing < 2S

Potential rotation of cracked segment

Case 2: Crack spacing > 2S

No Rotation of cracked segment

Cracked Segment

**Figure 12.26**  Potential instability of cracked veneer for large tie spacing (from Ref. 12.1).

where

$$p_{\text{total}} = \text{total uniformly distributed lateral pressure}$$

$$p_v = \text{uniformly distributed lateral pressure resisted by the veneer}$$

$$(EI)_v, (EI)_{\text{backup}} = \text{section rigidity of the veneer and backup, respectively.}$$

If the veneer and the backup are not the same height, are not supported in the same manner, and are not forced to deflect equally, plane frame analysis computer programs with ties modeled as springs can be used to predict the tie forces and load sharing (see Sec. 13.2.5 for additional discussion of such an analysis). However, the foregoing equation can still be used as a rough estimate. When the backup wall is much stiffer than the veneer, the backup wall resists most of the lateral load and the deflection is quite small. Alternately, when the backup is relatively flexible, such as with wood or steel studs, the veneer resists most of the load up to the point of cracking. If avoidance of veneer cracking under service load is a design limit, the fact that the structural contribution of the veneer is taken into account violates the strict definition of veneer. Nonetheless, evaluation of the potential for cracking may be considered in some designs.

The potential for cracking of the veneer, either through exceeding the flexural design strength or because of workmanship (see Chap. 15), should be considered in terms of serviceability requirements. For a stiff backup wall, the potential for cracking is relatively small, and if a crack does occur, the relatively small deflection of the backup prevents it from opening significantly. On the other hand, the likelihood of cracking is much greater where flexible backup walls are present and because of their flexible nature, a crack will tend to open wider. It is generally not economically nor practically feasible to stiffen the backup systems sufficiently to ensure that cracking will not occur. Therefore, the potential for development of cracks should be an accepted part of the design.

Two measures should be taken to help achieve satisfactory serviceability performance. First, because cracking can allow increased water penetration where the open rain screen is not effectively implemented, increased attention to design detailing and construction supervision must be provided to ensure that this water does not compromise the serviceability or long-term capacity of the wall. Related to this approach, the second measure is to use deflection design criteria for the backup wall to limit the size of cracks that may occur in the veneer and thus limit the extra amount of water that may enter a wall.

By using simple geometry and rigid uncracked sections, as shown in Fig. 12.27, a deflection limit of height divided by 360 ($h/360$) for a $3\frac{5}{8}$ in. (90 mm) thick veneer cracked at mid-height produces a maximum crack width of about 0.04 in. (1 mm), which is quite large. BIA[12.4] and CMHC[12.1] recommend that the maximum veneer deflection be limited to $h/720$, which produces a maximum crack width of 0.02 in. (0.5 mm) and an average crack width (at mid-thickness) of 0.01 in. (0.25 mm). Although still large, this latter deflection limit will help to limit the amount of water that can enter the veneer through the crack.

Masonry backup walls can be designed to resist the lateral load using the information provided in either Chap. 7 or Chap. 8, depending on whether the wall

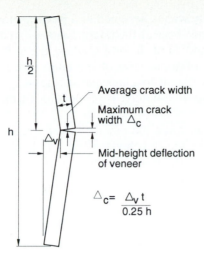

Average crack width

Maximum crack width $\triangle_c$

Mid-height deflection of veneer

$$\triangle_c = \frac{\triangle_v\, t}{0.25\, h}$$

**Figure 12.27** Calculation of crack width in brick veneer (from Ref. 12.1).

is nonloadbearing or loadbearing. Other types of backup walls should be designed to satisfy the deflection criteria for the veneer.

### 12.4.2 Structural Requirements for Masonry Cavity Walls

The structural design concept for cavity walls is that both wythes share in resisting lateral wind or seismic loads and that one or both wythes can carry superimposed vertical loads. If only one wythe supports the superimposed axial load, then that wythe should be designed to independently resist the entire compression force. No mutual lateral support from the other wythe should be assumed. This is because the lightly loaded wythe has much less resistance to flexural cracking. The more heavily loaded wythe benefits from the precompression caused by the axial load. Cracking of the unloaded wythe at lateral loads below the ultimate capacity of the loaded wythe results in it having very little stiffening effect on the loaded wythe.

Where both wythes of a cavity wall support compressive loads, each wythe should be designed as an independent compression member subject to the applied compression force and the calculated share of any lateral load. In these cases, some degree of mutual lateral stiffening is likely, but this is difficult to determine. The simple assumption of independent action is recommended. Alternately, the stronger of the two wythes can be designed to carry the total lateral load provided that the other wythe has sufficient section capacity to resist the axial load actually applied to it.

If axial load is shared between wythes, it is of great importance to account for all factors likely to affect the distribution of load between them. These factors include different material properties and capacities of the wythes, differential movements, and rotational effects from floors and beams. Because the effects of these factors are difficult to evaluate with any certainty, cavity walls are commonly constructed so that the total superimposed vertical load is applied to the inner wythe.

Each wythe of a cavity wall is designed as a flexural element resisting a portion of the total lateral load. The share of the lateral load resisted by one wythe depends on the relative flexural rigidities of the two wythes, the stiffness and spacing of the wall ties, and the plane(s) of application of the lateral load (i.e., on the interior or exterior wythe or both). If the tie spacings and stiffnesses are adequate, the wythes will share the total load approximately in proportion to their flexural rigidities as represented by Eq. 12.3.

In some cases, such as when the inner wythe is reinforced or carries an axial compression force, it is possible to achieve greater total capacity by assuming that the inner wythe resists all of the lateral load. In this case, the outer wythe should be designed as a masonry veneer.

The design procedures discussed in Chaps. 7 and 8 are applicable to the structural design of the individual wythes of cavity walls.

### 12.4.3 Requirements for Ties and Anchors

Chapter 13 contains information on both the behavior and the requirements for masonry ties and anchors and should be referred to for a more complete discussion. The information that follows is limited to a few specific points.

Traditionally, tie spacing provisions in building codes have corresponded to a tributary area concept of tie loading. Thus, ties spaced, for instance, at 3 ft (0.9 m) horizontally and 16 in. (400 mm) vertically could be considered to transfer the load from a 4 ft² (0.36 m²) area. If the backup wall is very stiff compared to the veneer, this is only a slightly unconservative approximation. However, because the top tie tends to act as the top support for the share of the load carried by the veneer (or outer wythe), it is required to transfer higher forces than the other ties. Hence, the greater the share of load carried by the veneer, the higher this force will be. Similarly, if the veneer cracks near the mid-height of the wall, the line of ties nearest the crack will also be required to carry up to nearly half of the load transferred from the veneer (or outer wythe) to the inner wall. The strength requirements for such ties are much greater than would be estimated using the tributary area approach. Rational approaches to the design of ties are discussed in Chap. 13.

In addition to strength criteria, tie stiffness can be important to the overall performance of a masonry veneer wall. To limit the deflection of the veneer, maximum mechanical play and minimum stiffness limits should be satisfied. For adjustable ties, a lower maximum spacing may be required[12.25] to achieve a total stiffness equivalent to rigid (nonadjustable) ties. Suggested minimum strength and stiffness limits for ties are presented in Chap. 13.

It is good practice to use additional ties at support locations and around openings in masonry veneer. For flexible backup walls, the added load transferred from windows or other elements attached to the backup creates a greater demand for load transfer between the veneer and the backup around the opening. the greater demand occurs because the ties force the veneer and backup to deflect equally. In the absence of a more thorough analysis, increasing the number of ties (i.e., using reduced spacing) immediately around the opening in proportion to the increased load is recommended.

## 12.5 REQUIREMENTS FOR MOVEMENT JOINTS

### 12.5.1 Horizontal Movement Joints

As was mentioned in Sec. 12.1.3, omission of movement joints is one of the principal causes of distress in masonry walls. Loadbearing frames and walls of buildings tend to shorten in the vertical direction with time due to elastic compression, creep, and shrinkage. Conversely, the veneer (or outer wythe) may tend to increase in length due to moisture expansion and increases in temperature. If there are no horizontal movement joints, the occurrence of this vertical differential movement results in transfer of load from the structure to the veneer, which can lead to the types of failures shown in Figs. 12.3 and 12.4.

It is relatively easy to estimate the amount of differential movement that must be accommodated. For example, a reinforced concrete frame of 4000 psi (27.5 MPa) concrete might have a compressive stress of 1600 psi (11 MPa) under service loads, which results in an elastic strain of about 0.04%. The long term creep strain for a reinforced concrete column would normally be in the range of $1\frac{1}{2}$ times the elastic deformation, say, 0.06%, and shrinkage could be up to 0.05%. For clay brick, the moisture expansion strain might be 0.02% and for a temperature rise of as much as 80°F (45°C) (taking into account solar radiation), the thermal expansion could be about 0.03%. This gives a total differential strain of 0.04 + 0.06 + 0.05 + 0.02 + 0.03 = 0.20%. For a 9 ft (2.7 m) story height of veneer, this converts to a 0.22 in. (5.4 mm) differential movement. However, if it is assumed that the elastic deformation and one-third of the shrinkage and creep will have occurred prior to installation of the veneer, the differential movement could reasonably be reduced to 0.13 in. (3.4 mm).

For steel structures, creep and shrinkage will not be factors, but elastic strain will be higher. For loadbearing concrete block walls, the elastic strain, shrinkage, and creep will normally be less than for reinforced concrete. Calculations of differential movement can be performed using the material properties discussed in Chaps. 4 and 5.

The required width of a movement joint depends on the characteristics of the sealant being used in the joint. If $\varepsilon_{seal}$ is the maximum cyclic deformation strain rating for the sealant and $\Delta h$ is the calculated differential movement, the joint must have a thickness $t_j$ at the location of the sealant of

$$t_j = \Delta h / \varepsilon_{seal} \tag{12.4}$$

which for a 25% limit of strain in the sealant and the 0.13 in. (3.4 mm) differential movement calculated before would require a joint thickness of 0.52 in (13 mm). For greater heights of veneer, this requirement increases proportionally. Allowance for deflections of floors would require larger joints.

If the sealant is placed beyond the tip of the shelf angle, the joint width dimension applies to the distance between the flashing and the course of bricks below the shelf angle, as shown in Fig. 12.28(a). However the 0.13 in. (3.4 mm) clearance plus a tolerance of 50 to 100% of this distance must be maintained between the bottom of the shelf angle and the course of bricks below. An inescapable result of these provi-

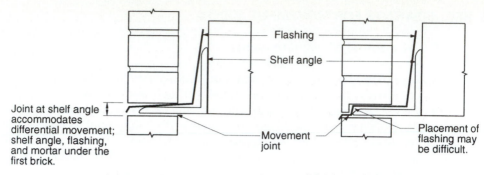

**(a) Requirement for Thicker Mortar Joints**

**(b) Use of Notched Brick to Maintain Standard Mortar Joint Thickness**

**Figure 12.28** Movement joint at the shelf angle (from Ref. 12.1).

sions is that the sum of the thickness of the shelf angle, and flashing and mortar on top of the shelf angle, and the required clearance below the shelf angle add up to more than a standard 3/8 in. (10 mm) joint. Attempts by the builder or designer to maintain the standard joint are a principal cause of omission of the movement joint or of failing to provide a sufficient space for movement. Designers should recognize that thicker joints will be visible at the shelf angles. In special circumstances, where the thicker joint detracts from the appearance of the building, specially cut bricks, such as shown in Fig. 12.28(b), can be used. However, this detail increases the construction costs, and the small pieces of brick over the shelf angle may not be durable. In addition, it may not provide sufficient width for placing the sealant.

### 12.5.2 Vertical Movement Joints

Vertical movement joints are required to relieve the stresses resulting from the tendency for horizontal deformation in masonry walls. Assuming similar values to

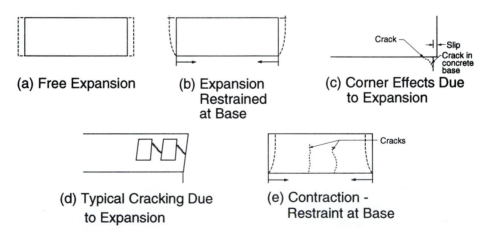

**(a) Free Expansion**

**(b) Expansion Restrained at Base**

**(c) Corner Effects Due to Expansion**

**(d) Typical Cracking Due to Expansion**

**(e) Contraction - Restraint at Base**

**Figure 12.29** Effects of expansion and contraction on long masonry walls (from Ref. 12.1).

those before, for thermal and moisture expansion, a 45 ft (13.5 m) long clay brick wall laid up at 70°F (20°C) would undergo a free expansion of about a 1/4 in. (7 mm), as illustrated in Fig. 12.29(a). Decreases in temperature and shrinkage of initially wet concrete units could lead to a similar magnitude of contraction. Base support for the wall will offer some restraint to movement, see Fig. 12.29(b), and typical types of distress shown in Figs. 12.29(c) to 12.29(e) can occur.

Requirements for location and spacing of vertical movement joints are covered in more detail in Sec. 15.6.

## 12.6 CLOSURE

Masonry veneer and cavity walls combine the traditional benefits of masonry with excellent resistance to rain penetration. The inclusion of an air space in the cavity utilizes the open rain screen principle as part of the system for minimizing rain penetration. With proper attention to a relatively few details, this wall system has proven to be very effective and one that typically requires very little maintenance. Past problems and the need for structural repairs have arisen mainly from lack of proper detailing and insufficient attention to accommodating differential movement.

This chapter identifies recurring problems and the simple steps necessary to avoid them. In addition to understanding the design requirements, appropriate corresponding construction details are required to ensure an attractive durable wall with superior resistance to rain penetration. Topics covered in Chap. 14 (building science) and Chap. 15 (construction) also relate to these subjects.

## 12.7 REFERENCES

12.1   R.G. Drysdale and G.T. Suter, "Exterior Wall Construction in High-Rise Buildings: Brick Veneer on Concrete Masonry or Steel Stud Wall Systems," Canada Mortgage and Housing Corporation, Public Affairs Centre, Ottawa, Ontario, 1991.

12.2   Brick Institute of America, "Brick Veneer New Construction," BIA Technical Notes on Brick Construction 28, BIA, Reston, VA, 1986.

12.3   Brick Institute of America, "Brick Veneer Existing Construction," BIA Technical Notes on Brick Construction 28A, BIA, Reston, VA, 1987.

12.4   Brick Institute of America, "Brick Veneer Steel Stud Wall Panels," BIA Technical Notes on Brick Construction 28B, BIA, Reston, VA, 1987.

12.5   A.G. Wilson and G.T. Tamura, "Stack Effect in Buildings," Canadian Building Digest, No. 104, Division of Building Research, National Research Council, Ottawa, Ontario, 1968.

12.6   G.O. Handegord, "Air Leakage, Ventilation and Moisture Control in Buildings," DBR Paper No. 1063, National Research Council of Canada, Ottawa, Ontario, 1982.

12.7   Brick Institute of America, "Differential Movements," Technical Notes on Brick Construction, No. 18, BIA, Reston, VA, 1991.

12.8   C.T. Grimm, "Design for Differential Movement in Brick Walls," *Journal of the Structural Division, Proceedings of ASCE*, November 1975, pp. 2385–2403.

12.9   Experimental Building Station, "Differential Movements in Buildings Clad with Clay Bricks, 2—The Design of Expansion Gaps," Notes on the Science of Building, NSB 85, Department of Housing and Construction, Australia, 1975.

12.10  W.G. Plewes, "Failure of Brick Facings on High-Rise Buildings," Canadian Building Digest, No. 185, Division of Building Research, National Research Council, Ottawa, Ontario, 1976.

12.11  J.S. Hall, and G.T. Suter, "How Safe are Our Masonry Cladding Connections?" in *Proceedings of First Canadian Masonry Symposium*, University of Calgary, Calgary, Aberta, 1976, pp. 95–109.

12.12  Brick Institute of America, "Water Resistance of Brick Masonry—Construction and Workmanship," BIA Technical Notes on Brick Construction 7B, BIA, Reston, VA, 1985.

12.13  C.T. Grimm, "Water Permeance of Masonry Walls: A Review of the Literature," *Masonry: Material Properties and Performance*, ASTM STP 778, ASTM, Philadelphia, 1982, pp. 178–199.

12.14  R.G. Drysdale and M.J. Wilson, "A Report on Lateral Load and Rain Penetration Tests of Full Scale Brick Veneer/Steel Stud Walls," Research Report, Project Implementation Division, Canada Mortgage and Housing Corporation, Ottawa, Ontario, July 1990.

12.15  T. Ritchie, "Rain Penetration of Walls of Unit Masonry," Canadian Building Digest, No. 6, Division of Building Research, National Research Council, Ottawa, Ontario, 1960.

12.16  L.A. Palmer and D.A. Parsons, "Permeability Tests of 8-in. Brick Wallettes," in *Proceedings, American Society for Testing Materials*, Vol. 34, Part II, 1934, pp. 419–431.

12.17  J.I. Davison, "Rain Penetration and Masonry Wall Systems," Building Practice Note No. 12, Division of Building Research, National Research Council, Ottawa, Ontario, 1979.

12.18  J.I. Davison, "Rain Penetration and Design Detail for Masonry Walls," Building Practice Note No. 13, Division of Building Research, National Research Council, Ottawa, Ontario, 1979.

12.19  J.I. Davison, "Workmanship and Rain Penetration of Masonry Walls," Building Practice Note No. 16, Division of Building Research, National Research Council, Ottawa, Ontario, 1980.

12.20  Suter Keller Inc., "Field Investigation of Brick Veneer/Steel Stud Wall Systems," Project Implementation Division, Canada Mortgage and Housing Corporation, Ottawa, Nov. 1989.

12.21  Canadian Standards Association, CAN3-A371-M84, "Masonry Construction for Buildings," CSA, Rexdale, Ontario, 1984.

12.22  W.C. Panarese, S.H. Kosmatka and F.A. Randall, *Concrete Masonry Handbook for Architects, Engineers, Builders*, Portland Cement Association, Skokie, Illinois, 1991.

12.23  C.T. Grimm and J.A. Yura, "Shelf Angles for Masonry Veneer," *Journal of Structural Division, Proceedings of ASCE*, Vol. 115, No. 3, 1989, pp. 509–525.

12.24  C.T. Grimm, "Corrosion of Steel in Brick Masonry," ASTM STP 871, ASTM, Philadelphia, PA, 1985, pp. 67–87.

12.25  The Masonry Standards Joint Committee, "Building Code Requirements for Masonry

Structures," ACI 530/ASCE 5/TMS 402 American Concrete Institute and American Society of Civil Engineers, Detroit and New York, 1992.

## 12.8 PROBLEMS

**12.1** Discuss the different mechanisms for water penetration through brick veneer walls. How is the open rain screen designed to avoid rain penetration problems?

**12.2** What are the critical design features for the control of rain penetration through veneer walls?

**12.3** **(a)** List and discuss the functions of flashing material for the cavity of masonry veneer and cavity walls.

**(b)** Sketch appropriate details required to ensure that the flashing will function satisfactorily.

**(c)** Obtain three different types of flashing material. Evaluate the anticipated performance of each of these materials, including construction considerations such as bending, lapping and sealing.

**12.4** Identify two examples of existing multiwythe masonry walls. Are these masonry veneer or cavity walls? Present your analysis and arguments.

**12.5** Design a shelf angle to carry a two-story 20 ft (6.0 m) high brick veneer. The shelf angle is anchored at 3 ft. 4 in. (1 m) intervals. The distance from the center of the brick veneer to the edge of the floor slab is 5 in. (125 mm). Check bearing stresses. Masonry compressive strength = 3000 psi (20 MPa) and concrete strength = 4000 psi (27.5 MPa). Consider the fixed shelf angle to be anchored in the concrete with Grade 40 (300 MPa) concrete reinforcing bars. (*Hint*: The compression zone of contact between the shelf angle and the concrete floor can be conservatively approximated as having a width $b$ of two times the moment arm distance $d$ to the anchor.)

**12.6** Calculate the average crack width for a 10 ft (3 m) high brick veneer wall, 4 in. (100 mm) thick, assuming that deflection is limited to height $h$ divided by 720. Comment on the calculated crack width from a rain penetration standpoint. Considering the brick to be simply supported top and bottom, what wind pressure would be required to cause cracking if the flexural tensile bond strength is 75 psi (0.52 MPa)?

**12.7** For a cavity wall composed of a 4 in. (100 mm) brick outer wythe and an 8 in. (200 mm) hollow concrete block inner wythe, determine the flexural stresses in both wythes due to a lateral wind load of 20 lb/ft² (1.0 kN/m²). Assume that both wythes have the same height and boundary conditions.

Brick compressive strength = 8000 psi (55 MPa)

Block compressive strength = 2000 psi (13.8 MPa)

Type S mortar.

**12.8** A 36 ft (10.8 m) high, three-story masonry building has a wall system composed of an 8 in. (200 mm) solidly grouted and reinforced concrete block loadbearing wall and a 4 in. (100 mm) clay brick masonry veneer. If the veneer is supported vertically at the foundation and is continuous over the full height of the building, determine the following:

**(a)** Calculate the anticipated magnitude of differential vertical movement that must be designed for at the top of the building.

**(b)** What other aspects of design will be affected? Discuss.

**12.9** Construct a 16 in. (400 mm) square segment of brick veneer.

**(a)** Observe the time required for water to penetrate to the unwetted side under the action of a gentle water spray. (Seal polyethylene sheet around the perimeter of the wetted face to ensure that water reaching the unwetted face must pass through the veneer.)

**(b)** Repeat part (a), but spray the water on the polyethylene above the specimen so that surface water runs down and covers the face of the veneer. Use a fan to create air pressure on the wet face of the specimen. Observe rain leakage and compare with part (a).

**(c)** Carefully load the specimen in bending to cause a horizontal crack along a mortar bed joint. Repeat tests (a) and (b) and comment.

# CONNECTORS

Rectangular wire tie in a brick cavity wall (*Courtesy of Brick Institute of America*).

## 13.1 INTRODUCTION

The term "connector" is a generic name for mechanical devices used to attach two or more elements together, and for masonry construction includes wall ties, anchors, and fasteners. This classification of masonry connectors is convenient in discussing behavior, design functions, and design approach, although in practice, the distinctions can be blurred.

*Wall ties* are used to connect two or more wythes of masonry together either as cavity walls, veneer walls, or composite walls, as shown in Fig. 13.1(a). The wall tie also can be used to connect a masonry veneer to a nonmasonry backup wall. They are spaced uniformly over the face area of the wall, except at edges. *Anchors*, on the other hand, are devices used to connect masonry walls at their intersections, to attach masonry to supports and other structural members (e.g., floors, columns), or to anchor roof and floor systems into the wall. Examples are shown in Fig. 13.1(b). They are spaced along the line of support. Whereas wall ties generally transmit lateral loads by axial forces, anchors must often transmit the larger axial forces and shear forces at supports. Finally, a *fastener* is a device used to attach

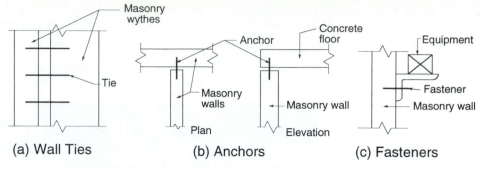

**Figure 13.1**  Types of connectors in masonry construction.

various items such as equipment and services to masonry. An example is shown in Fig. 13.1(c).

The development of design codes for engineered masonry and the various cladding failures that have occurred from the 1960s onward have served to focus attention on the design of connectors. Although many of the connector problems were associated with precast concrete or stone facing units, they are applicable to masonry cladding. Some of the failures created hazardous situations, and nearly all entailed very expensive repairs. Engineering analysis revealed deficiencies in knowledge and in the requirements of relevant codes and standards. Many connectors had not been appropriately tested, no doubt due to the lack of standard evaluation procedures. Although this situation is gradually being remedied as various codes incorporate performance criteria and design requirements for connectors, test data for particular connectors are still sparse. As the performance requirements are becoming more clearly defined, it is evident that some existing products are unsatisfactory and new connectors are being developed.

## 13.2  WALL TIES

### 13.2.1  Basic Functions

**Masonry Veneer and Cavity Walls.** The basic function of wall ties is to transfer force between parallel wythes of masonry or between masonry veneer and backup walls without excessive relative movement. Where there is a cavity (air space) between parallel walls, ties are usually required to transfer lateral load by both axial compression and tension. The tie forces inevitably cause the wythes to move slightly closer together or further apart, depending upon the stiffness of the ties and the effectiveness of their anchorage in the mortar. To the degree that equal deflections of the two or more wythes are achieved, the wythes will have similar curvatures and will therefore share the lateral load in proportion to their section rigidities, $EI$. In this case, the wythe that reaches its limiting stress or strain first will control the capacity unless the remaining wythe has sufficient reserve capacity to carry the total load. In the case of a masonry veneer wall system, even though the backup wall is designed to resist the entire lateral load (see Chap. 12), ties must support the veneer and transfer load from the veneer to the backup wall. If the backup wall is sufficiently stiff, providing stiff ties will minimize bending in the veneer, which otherwise could result in flexural cracking or the opening of cracks already present.

The outer wythe of a cavity wall or a masonry veneer wall usually experiences more thermal and moisture movements but less elastic or inelastic deformations than a loaded interior wythe or the backup structure. Normally, there is significant differential movement between the outer wythe and the inner wythe or backup wall. The present design approach is to permit unrestrained relative movements to occur in directions parallel to the plane of the wall. This design concept requires the ties to be sufficiently flexible to allow the differential displacement of the ends of the tie, as shown in Fig. 13.2(a). To avoid significant coupling between wythes, many ties are made with relatively thin plates or rods that have low flexural stiffnesses. Alternately, flexurally stiff sections can be used with narrow sections incorporated to accommodate the foregoing movement. Adjustable ties, as shown in Fig. 13.2(b), can be designed to provide adequate axial strength and stiffness, but allow free differential movement in the plane of the wall. For convenience of laying, when the exterior wythe is constructed last, the trend in North America has been to use adjustable ties.

**Coupled Cavity Walls.** Tying wythes together with flexurally stiff connectors will couple the wall to some extent so that the resulting bending strength and stiffness are more than the sum of the individual values. For this type of coupling, it is important to determine whether the increased resistance to external loads outweighs the effects of internal stresses produced by the tendency for differential movement. This concept is used in the design of some cavity wall systems. For example, a shear connector tie such as shown in Fig. 13.3 has been used in cavity wall construction. This tie is designed to transfer shear between the the exterior brick wythe and the backup wall. In this system, the exterior wythe or veneer is utilized as a structural component and the coupling action may result in increased capacity. However, unless the effects of differential movements are taken into account, it is advisable to avoid coupling.

**Composite Walls.** For composite walls, the mortar in the collar joint or the grout-filled cavity space acts as the shear transfer mechanism between the wythes of masonry. However, typical allowable shear stresses along the interface are small; 5 and 10 psi (0.034 and 0.069 MPa), respectively, for mortared or grouted collar

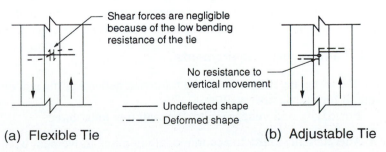

(a) Flexible Tie    (b) Adjustable Tie

**Figure 13.2** Deformation of ties due to relative movement of masonry wythes.

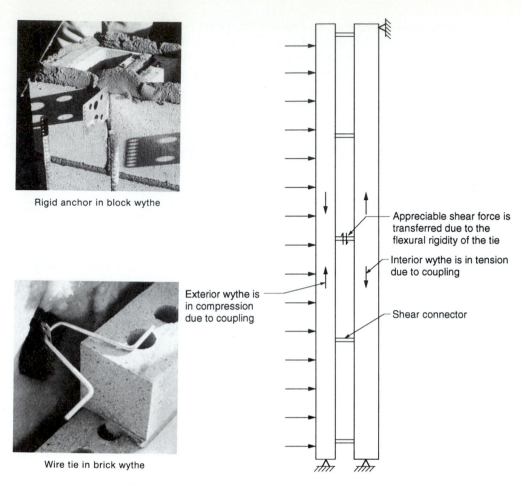

Rigid anchor in block wythe

Wire tie in brick wythe

Appreciable shear force is transferred due to the flexural rigidity of the tie

Interior wythe is in tension due to coupling

Exterior wythe is in compression due to coupling

Shear connector

**Figure 13.3** Shear connector tie. (*Courtesy of G. R. Sturgeon*)

joints.[13.1] In the case where the shear bond strength along this interface is exceeded, the ties act in tension to develop a shear-friction resistance to relative movement along the debonded interface. Ties used in composite walls hold the two wythes together during construction. In the case of a grouted cavity, large tensile forces can be induced during grouting. The fluid grout exerts outward pressure on each of the two wythes, which must be prevented from bursting apart by tension in the ties. This tensile force can be very large and should be considered in design.

### 13.2.2 Performance Requirements

From the foregoing discussions, the following general performance requirements are essential for satisfactory wall ties.

For cavity and veneer walls, the wall ties must have:

- Adequate flexibility to accommodate in-plane movement between wythes without anchorage failure. (This is still a requirement even where some coupling of wythes is desired.)

- Adequate tensile and compressive capacity (including embedment requirements) to transmit lateral loads between wythes even after displacements due to relative movement have occurred.
- Acceptable axial stiffnesses and deformation characteristics in both tension and compression (after displacements due to relative movement have occurred).
- Adequate resistance to transfer of moisture across the cavity.
- Adequate corrosion resistance.

- For composite walls, the wall ties must have:
- Adequate tensile and anchorage strengths to hold the wythes together under the action of lateral loads. (In the case of grouted double wythe walls, the bursting pressure of the fluid grout can to be more critical than applied loads.)
- Adequate shear strength to maintain composite action after debonding of the collar joint or grouted cavity. (Shear results from both out-of-plane bending and in-plane differential movements.)
- Adequate corrosion resistance.

Tie selection in design has been based traditionally on prescriptive specifications expressed in terms of horizontal and vertical spacing and minimum size. Although setting out of the foregoing qualitative requirements is an essential first step to the rational design of wall ties and their incorporation as structural elements in wall systems, quantitative performance criteria are required. To provide guidance for this second stage is more difficult as the performance requirements vary from country to country and even within countries. The design of wall ties to provide coupling in cavity walls is an example where the forces to be resisted are strongly dependent upon the potential differential movements in each particular case and the extent of coupling required. Quantitative criteria are simpler for uncoupled cavity and veneer walls. For example, the Australian standard[13.2] specifies criteria to satisfy the foregoing performance requirements for cavity and veneer ties to include the following:

- The test specimen shown in Fig. 13.4 is subjected to 3/8 in. (10 mm) horizontal and vertical in-plane differential displacements with the wythes held parallel

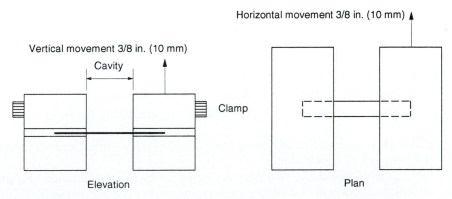

**Figure 13.4**   Test specimen for Australian tie test (from Ref. 13.2).

to each other. To ensure that adequate in-plane movement is permitted, these differential movements must not cause failure of the tie itself or the bond at the mortar joints.

- After the foregoing 3/8 in. (10 mm) displacements have been introduced, a tensile or compressive force is applied to ensure adequate axial strength. The ultimate strength is taken at 0.06 in. (1.5 mm) axial displacement unless the maximum load is reached at a smaller displacement.
- The stiffness under working load is defined as the slope of the load versus the displacement line joining zero load and half of the ultimate load. Anchorage slip and mechanical play in adjustable ties are included in the displacement measurement.
- Ties are placed in a standard device with the far end displaced $\frac{3}{8}$ in. (10 mm) downward to check that water cannot travel across the tie.
- Corrosion resistant materials or coating are specified for various exposure conditions. The measured strength and stiffness are used to classify the tie as light, medium, or heavy duty according to tabulated criteria.

In North America, prescriptive requirements are most widely used in design, but increased use of adjustable ties has resulted in some recognition of differences in performance. ACI 530/ASCE 5/TMS 402[13.1] introduced more limited spacing and tributary areas for adjustable ties, whereas CSA CAN3 A370[13.3] requires equivalence to standard nonadjustable ties. Specific stiffness requirements and limits on mechanical play similar to BIA recommendations[13.4] are to be introduced in the next edition of Ref. 13.3.

### 13.2.3 Types of Ties

There are many different types of ties available. Because of the proprietary nature of many of these, the manufacturer must take the responsibility for providing information on their performance characteristics. In addition to designating whether ties are standard (common) or proprietary, they can be classified based on type of material used (galvanized steel, stainless steel, aluminum, plastics), geometry (rectangular, Z-shape, corrugated), shape of material used (wire or sheet), stiffness (rigid or flexible), adjustability (adjustable or nonadjustable), continuity (discrete or continuous), and corrosion resistance (corrosion resistant or noncorroding).

The following are examples of common or standard nonadjustable ties.[13.3]

1. corrugated strip tie (Fig. 13.5(a))
2. rectangular wire tie and Z-wire tie (Fig. 13.5(b))
3. continuous ladder and truss joint reinforcement (Fig. 13.5(c))

Other ties, such as the combined ladder joint reinforcement and rectangular tie, Fig. 13.5(d), are also available.

Use of adjustable two-part ties helps avoid the interference of a projecting tie during construction and allows for some adjustment to match the levels of the mortar

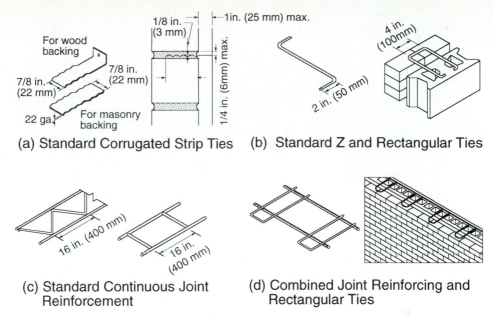

(a) Standard Corrugated Strip Ties   (b) Standard Z and Rectangular Ties

(c) Standard Continuous Joint Reinforcement   (d) Combined Joint Reinforcing and Rectangular Ties

**Figure 13.5**   Typical nonadjustable wall ties.

bed joints. Most of the proprietary ties fall into the adjustable category and can be described according to the adjustability mechanism as follows:

1. *Pintle*. Pintles made with bent wire, Fig. 13.6(a), have been used to provide vertical adjustability, created by the pintle passing through a restraining eye or other opening in the receiving unit. The receiving unit can be a plate with a hole.

2. *Slot*. Many adjustable ties employ a triangular wire tie in a vertical slot, such as shown in Fig. 13.6(b). The length of the slot governs the range of adjustment. Here a backing plate is used to limit tie movement perpendicular to the wall because it is normally mounted over exterior sheathing as shown.

3. *Fastener Adjustment*. Several types of ties allow for vertical adjustment at the fastener. The example wire tie shown in Fig. 13.6(c) has a slot in the tie to allow the tie to be repositioned relative to the fastener during construction of the outer wythe.

4. *Fixed Position*. Nonadjustable ties that can be attached to the backup after the required position has been determined provide a form of adjustability. Self-drilling ties, Fig. 13.6(d), or ties attached by other post-construction techniques fall into this category.

The available types of ties encompass a wide range of strengths and stiffnesses that vary depending on the tie configuration in its adjustable range and whether the applied force is tensile or compressive. These properties are discussed in the following sections.

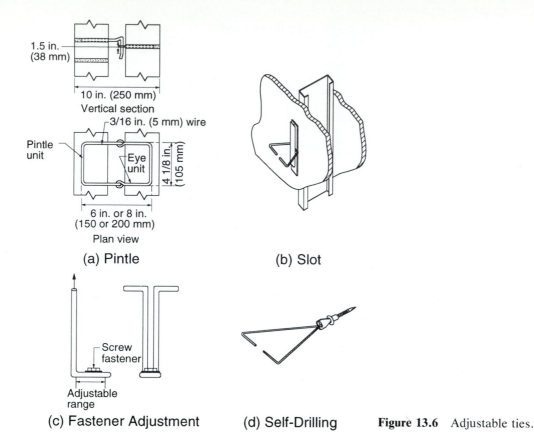

**1.5 in.**
**(38 mm)**

**10 in. (250 mm)**
Vertical section

3/16 in. (5 mm) wire

Pintle
unit

Eye
unit

4 1/8 in.
(105 mm)

**6 in. or 8 in.**
**(150 or 200 mm)**
Plan view

(a) Pintle

(b) Slot

Screw
fastener

Adjustable
range

(c) Fastener Adjustment

(d) Self-Drilling

**Figure 13.6**  Adjustable ties.

### 13.2.4 Strength

**Requirements.**   Wind pressures (positive or negative) and seismic forces result in both tensile and compressive forces in ties.  Traditionally, tie loads have been approximated according to tributary area around the tie, resulting in prescriptive limits on tie spacing.  This practice is based on development of equal forces in all ties, which would be the case if rigid body motion of the inner and outer wythes occurred.  However, a real wall system will deflect, and the force in each tie will depend on the rigidities of the wythes and the tie stiffness and location.

If the backup wall is much stiffer than the veneer or exterior wythe, adopting tie forces proportional to tributary areas is a reasonable approximation except that, when the top of the exterior wythe of masonry is not directly supported by the roof or floor, the top tie will act as the top support or reaction point for the share of the load carried by this exterior wythe. The simplified two-wythe wall in Fig. 13.7(a) can be used to illustrate the tie force distribution, assuming the ties to be infinitely stiff. For the total load applied to the outer wythe, the ties transfer some load to the inner wythe so that they share load in proportion to their stiffnesses.  For a case when the outer wythe has a rigidity $(EI)_v$ of 25% of the backup wall rigidity $(EI)_b$, 80% of the load will be transferred by the ties to the inner wythe $((EI)_b/[(EI)_v$ +

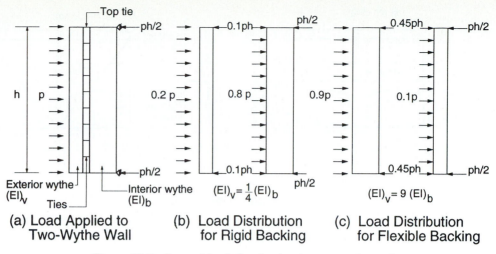

**Figure 13.7**  Lateral load distribution in two-wythe walls.

$(EI)_b] = 0.8$). From simple equilibrium, the top and bottom lateral supports for the outer wythe must each transfer half of the load on the outer wythe back to the structure. Thus, the top tie would act as a support and be required to transfer an additional force equal to 10% of the total load applied to the wall system (one half of the 20% resisted by the outer wythe), as shown in Fig. 13.7(b). Alternately, where the exterior wythe is much stiffer than the backup (say, nine times), only 10% of the total load will be transferred to the backup by the intermediate ties, and as illustrated in Fig. 13.7(c), the additional reaction force at the top tie will be 45% of the total load. In each case, the top tie will also transfer some additional load according to its tributary area. The method of calculation and relative magnitude of the approximate capacity requirements for top ties is valid for most similar systems, but will be affected by different heights and support conditions for the outer wythe and backup wall, by the location of the top tie, and by tie stiffness.

**Performance.**    Types of behavior and modes of failure under both tensile and compressive loading have been determined[13.5] for the common ties shown in Fig. 13.8. These data are reproduced in Table 13.1 for the tie configurations shown. Compression failures were predominantly buckling of the tie in the cavity, although push out of the mortar joint in the concrete block masonry did occur for the stiffer Z-tie. Pullout of the tie from the mortar joint was the main tension failure mode. As might be expected, the amount of embedment in the mortar joint had a significant influence. The higher capacity of the ladder joint reinforcement used as a tie compared to the truss reinforcement is to be expected because the diagonal wires in the truss reinforcement cross the cavity at an angle and, therefore, have significantly longer lengths, greater axial forces, and some induced bending. Corrugated ties exhibit very limited compressive capacity. The tabulated test values indicate the relative effects of cavity width and embedment conditions. Only test results for the specific grade of steel and manufacturing process should be used in design.

To prevent water crossing cavities on the ties, some ties have a small crimp or *drip* built into them at the position of the wall cavity (Fig. 13.9). Tests of such

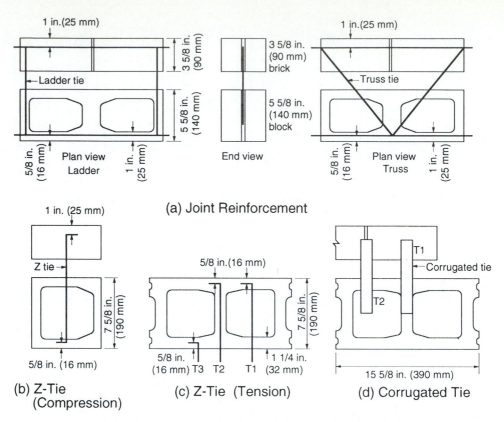

**Figure 13.8** Test specimens for determining capacity of wall ties (from Ref. 13.5).

ties show significant reduction in compressive strength due to buckling. As shown in Table 13.2, for a 3.5 in. (89 mm) cavity width, the crimp reduced the compressive strength of wire ties by about 50%. The crimp also reduced the horizontal in-plane shear transfer capacity of truss reinforcement. This may be due to shear introducing axial tension and compression forces in alternate diagonal wires where those wires under compression have a lower buckling capacity. ACI 530/ASCE 5/TMS 402[13.1] specifies halving the normal tie spacing when ties with drips are used unless evidence is provided to demonstrate equivalent strength.

### 13.2.5 Stiffness

Analysis of a cavity wall based on the assumption of equal deflections of the wythes implies that the ties are infinitely rigid or, in practical terms, that the tie deformations are very small compared to the deflection of the wall system. Some ties, such as the Z-tie, may satisfy this condition, but many others, such as the corrugated tie and some adjustable ties, do not because they are relatively flexible and may allow unrestrained movement due to mechanical play.

For cases where wythes are supported independently at the top and bottom, as shown in Fig. 13.10(a), the stiffness of the ties has a marked effect on the load

Table 13.1 Capacity of Cavity Wall Ties (from Ref. 13.5)

| Specimen type and loading | Cavity width, in. (mm) | Tie placement | Mean failure load, lb (kN) | C.O.V. (%) | Failure mode |
|---|---|---|---|---|---|
| Truss reinforcement (compression) | 2-$\frac{3}{8}$ (60) <br> 4-$\frac{3}{8}$ (110) | See Fig. 13.8(a) | 900 (4.01) <br> 680 (3.03) | 10.8 <br> 13.0 | Buckling of diagonal rod |
| Ladder reinforcement (compression) | 2-$\frac{3}{8}$ (60) <br> 4-$\frac{3}{8}$ (110) | See Fig. 13.8(a) | 1470 (6.56) <br> 1080 (4.81) | 11.0 <br> 7.3 | Buckling of cross wire |
| Z-tie compression | 2-$\frac{3}{8}$ (60) <br><br> 4-$\frac{3}{8}$ (110) | See Fig. 13.8(b) | 740 (3.28) <br> 660 (2.94) | 16.7 <br> 8.9 | Push through of tie at block |
| Corrugated tie (compression) See Fig. 13.8(d): | $\frac{3}{4}$ (19) <br> $\frac{3}{4}$ (19) <br> 2-$\frac{3}{8}$ (60) <br><br> 4-$\frac{3}{8}$ (110) | tie on web, mortar two sides <br> tie crossing face shell <br> tie on web <br> tie on web | 550 (2.43) <br> 610 (2.70) <br> 240 (1.05) <br> 88 (0.39) | 21.4 <br> 25.5 <br> 14.0 <br> 14.8 | Buckling of tie |
| Truss reinforcement (tension) | 2-$\frac{3}{8}$ (60) | See Fig. 13.8(a) | 1330 (5.92) | 10.8 | Pull out at brick and wire failure |
| Ladder reinforcement (tension) | 2-$\frac{3}{8}$ (60) | See Fig. 13.8(a) | 1320 (5.89) | 16.8 | Pull out at block |
| Z-tie (tension) See Fig. 13.8(c): | N/A | $T_1$ <br> $T_2$ <br> $T_3$ | 1170 (5.20) <br> 1780 (7.90) <br> 950 (1.54) | 8.6 <br> 16.5 <br> 19.0 | Pull out of tie at block |
| Corrugated tie (tension) See Fig. 13.8(d): | N/A | $T_1$, 4 in. (102 mm) depth <br> $T_1^*$, 2 in. (51 mm) depth <br> $T_2$ tie crossing face shell | 1240 (5.52) <br> 1170 (5.21) <br> 750 (3.34) | 12.3 <br> 11.9 <br> 13.6 | Pull out of tie at block |

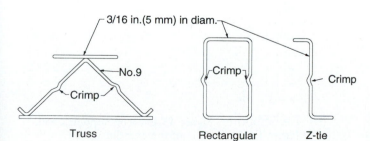

**Figure 13.9** Crimp in wall ties (from Ref. 13.6).

**Table 13.2** Effect of Crimp on Tie Capacity (from Ref. 13.6)

| Type of tie | Yield Loads, lb (kN) | |
| --- | --- | --- |
| | Compression (P) | Shear (V) |
| Straight ties (no crimp) | | |
| 3/16 in. (4.76 mm) Z | 1970 (8.77) | 45 (0.20) |
| 3/16 in. (4.76 mm) rectangular | 3640 (16.2) | 100 (0.45) |
| 3/16 in. (4.76 mm) truss | 2310 (10.3) | 1830 (8.14) |
| No. 8 gage truss | 1370 (6.10) | – |
| No. 9 gage truss | 1080 (4.81) | 890 (3.96) |
| Crimped ties | | |
| 3/16 in. (4.76 mm) Z | 920 (4.10) | 38 (0.17) |
| 3/16 in. (4.76 mm) rectangular | 1830 (8.14) | 98 (0.44) |
| 3/16 in. (4.76 mm) truss | 1250 (5.56) | 1230 (5.47) |
| No. 8 gage truss | 810 (3.60) | – |
| No. 9 gage truss | 510 (2.27) | 450 (2.00) |

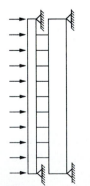

**(a) Independent Support at Top**

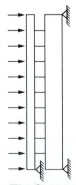

**(b) Tie Support Near Top**

**Figure 13.10** Top support conditions of the exterior wythe.

sharing. Very flexible ties will not transfer much load between the wythes, whereas very rigid ties will produce load sharing proportional to the stiffnesses of the wythes.[13.7]

The situation is less easily visualized for cases such as shown in Fig. 13.10(b), where the veneer is supported only by ties near the top. The distribution of tie forces to transfer lateral load is affected by tie stiffness, the relative stiffness of the backup wall, and the stiffness ratio between wythes.[13.7,13.8] Tests and analyses show that the maximum tie force occurs at the top tie or at the tie nearest a crack in the veneer. In each case, the tie acts as the support for the flexural behavior of the uncracked veneer height(s).[13.9] Although tie stiffness does affect tie forces, the stiffness criteria should be primarily based on practical considerations such as the acceptable amount of movement of the veneer and fatigue effects on other wall components such as caulking. At relatively rigid support areas, such as at columns and corners, overloading of stiff connectors should be investigated. The possibility of unexpected veneer cracking because of unforeseen load-carrying mechanisms, such as horizontal versus vertical spanning, should also be considered.

### 13.2.6 Adjustability

If both wythes of masonry are built at the same time, it is usually not difficult to match the levels of the mortar bed joints at regular intervals so that nonadjustable ties can be placed during construction. However, if one of the wythes is to be constructed later, nonadjustable ties must be left protruding from the constructed wythe. These ties tend to interfere with construction of the second wythe. Also, if the ties become bent or if the bed joints are out of level so that the ties have to be bent during construction, fitting the ties in the $\frac{3}{8}$ in. (10 mm) mortar joint can be a problem. Bending is also likely to weaken the tie or even make it ineffective.

For convenience of construction, a wide variety of adjustable ties (see Sec. 13.2.3) has been developed. Unfortunately, the lack of recognized performance requirements has resulted in a wide range of strength and stiffness properties. Research[13.10] on a variety of adjustable ties designed to transfer load from masonry veneer to stud backup walls shows that there is a wide range in strength and stiffness between the different ties and between the different adjustment positions for each type of tie. Typical test results are reproduced in Table 13.3 for the ties selected to represent different forms of adjustment (see Fig. 13.11).

As an example, test results[13.11,13.12] presented in Table 13.4 indicate that a pintle type of tie (Fig. 13.12) has greatly varying capacities depending on the adjusted position of the tie from the aligned position (zero eccentricity). These test results indicate that the effect of bending of the pintle and shank on the stiffness of the tie is roughly inversely proportional to the cube of the eccentricity. For example, a 1.5 in. (38 mm) eccentricity will result in a displacement of more than 27 times that for a 0.5 in. (12.7 mm) eccentricity.

The Brick Institute of America[13.13] suggests a minimum stiffness for veneer ties defined by a displacement of 0.05 in. (1.2 mm) at a load of 100 lb (445 N). In addition, a maximum *mechanical play* of 0.05 in. (1.2 mm) is suggested. Building codes will likely introduce provisions that are variations of this type of criteria to provide a uniform basis for judging acceptability of various adjustable ties. The allowance for

**Table 13.3** Strength and Stiffness of Adjustable Ties in Veneer/Steel Stud System (from Ref. 13.10)

| Tie type | Adjusted position | Ultmate load, lb (kN) | C.O.V. (%) | Load @ 0.05 in. (1.2 mm), lb (kN) |
|----------|-------------------|-----------------------|------------|-----------------------------------|
| I | Middle<br>Quarter<br>End | 590 (2.64)<br>610 (2.73)<br>780 (3.45) | 4.6 | 190 (0.83)<br>240 (1.06)<br>290 (1.28) |
| II | Minimum<br>15 mm<br>35 mm | 380 (1.70)<br>260 (1.14)<br>130 (0.57) | 10.9 | 98 (0.44)<br>53 (0.24)<br>18 (0.08) |
| III | 15 mm<br>25 mm<br>30 mm | 290 (1.29)<br>270 (1.20)<br>250 (1.10) | 4.6 | 110 (0.48)<br>115 (0.50)<br>64 (0.28) |

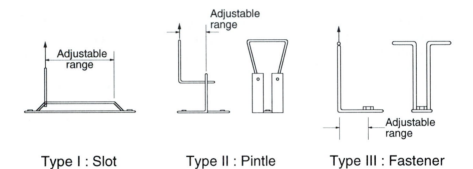

Type I : Slot          Type II : Pintle          Type III : Fastener

**Figure 13.11**  Adjustable ties for masonry veneer supported on flexible backup.

**Table 13.4** Effect of Eccentricity on Capacity of Pintle Tiles (from Ref. 13.11)

| Unit (3/16 in. dia. wire) | Load position | y (Exposed shank), in (mm) | e (Eccentricity) in (mm) | Yield load, lb (kN) |
|---------------------------|---------------|----------------------------|--------------------------|---------------------|
| Single eye | – | – | – | 550 (2.45) |
| Single pintle | 1 | 1 3/4 (45) | 0 | 490 (2.18) |
| Single pintle | 2 | 1    (25) | 1 1/2 (38) | 50 (0.22) |
| Single pintle | 2 | 1 3/4 (45) | 1 1/2 (38) | 40 (0.18) |

Note:  For distances y and e, refer to Fig. 13.12.

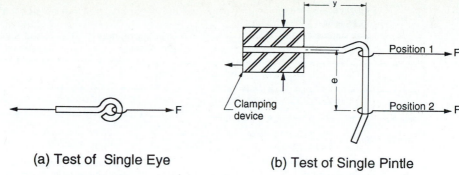

(a) Test of Single Eye          (b) Test of Single Pintle

**Figure 13.12** Effect of adjustability on capacity of pintle ties (from Ref. 13.11).

mechanical play between the joining parts is required for manufacturing tolerances, including varying thicknesses of hot dipped galvanizing.

The potential effects of mechanical play on tie force and load transfer can be illustrated by the following example. A simply supported 8 ft (2.4 m) high wall of $3\frac{5}{8}$ in. (90 mm) thick brick veneer subject to a uniform pressure of 20 lb/ft$^2$ (1 kN/m$^2$) will have a mid-height elastic deflection of something in the order of 0.02 in. (0.5 mm), depending on the modulus of elasticity of the masonry. For a large mechanical play of, say, 0.05 in. (1.2 mm), some ties will be positioned within the range of mechanical play so that they do not become engaged and thus cannot contribute to the load transfer. Sharing of the load by these ties will only occur if the loaded ties are not very stiff and deform significantly or if the veneer cracks allowing greater deflection. However, there are many ties along a particular bed joint and at several levels over the height of the wall. On average, the ties will be located midway in the range of mechanical play, but some ties at each level will be effectively engaged over the complete loading range. Therefore, it is suggested that much tighter tolerances on mechanical play should be adopted to avoid overloading individual ties and prematurely cracking the veneer.

### 13.2.7 Design Considerations

The design of ties is usually a process of selecting from the ties available. In addition to strength and stiffness properties (including variability), considerations such as susceptibility to damage on the job site, potential for improper installation, interaction with other wall components (i.e., insulation and air barriers), and potential vulnerability to corrosion should influence the choice.

Design guidance is available for specific conditions in the form of maximum spacings for limited ranges of cavity widths and types of wall systems. As an example, one of several recommendations by BIA[13.13] is a 3/16 in. (4.76 mm) diameter tie for each 4.5 ft$^2$ (0.42 m$^2$) of wall area, but not spaced more than 24 in. (600 m) vertically or 36 in. (900 mm) horizontally for cavity widths not exceeding 3.5 in. (89 mm). For multiwythe masonry walls, ACI 530/ASCE 5/TMS 402[13.1] has similar requirements, as well as the requirement for maximum spacing of cross-wires of joint reinforcement, as shown in Fig. 13.13. The ACI 530/ASCE 5/TMS 402 requirements for adjustable

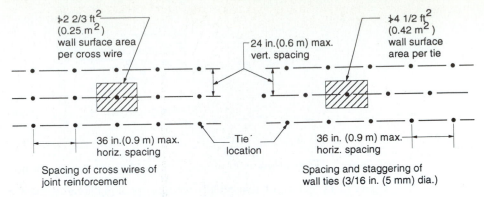

Figure 13.13 shows (per the figure):
2 2/3 ft² (0.25 m²) wall surface area per cross wire
24 in. (0.6 m) max. vert. spacing
4 1/2 ft² (0.42 m²) wall surface area per tie
36 in. (0.9 m) max. horiz. spacing
Tie location
36 in. (0.9 m) max. horiz. spacing
Spacing of cross wires of joint reinforcement
Spacing and staggering of wall ties (3/16 in. (5 mm) dia.)

**Figure 13.13** ACI 530/ASCE 5/TMS 402 code requirements for wall ties (from Ref. 13.1).

ties are more stringent. The spacing requirements are shown in Fig. 13.14 for the double pintle adjustable tie shown in Fig. 13.6. The maximum tributary area is 1.77 ft² (0.16 m²) compared to the 4.5 ft² (0.42 m²) limit for nonadjustable ties.

Computer analyses were run by the authors for the wall configuration shown in Fig. 13.15(a). For stiff ties and a very stiff backup wall, the distribution of the tie forces shown in Fig. 13.15(b) is reasonably uniform, except for the top tie. Very low tie forces over most of the wall height and much higher forces in the top ties correspond to a decreased relative stiffness of the backup. As discussed in Sec. 13.2.4, the top tie has the greatest increase in force because it acts as the reaction for the part of the load carried by the outer wythe. For a rigid backup wall, the effects of large changes in the tie stiffness (Fig. 13.15(c)) were not as significant as large changes in stiffness of the backup wall. For the case of flexible ties with a flexible backup, if the stiffness of the ties is reduced to correspond to one of the more flexible pintle ties, the tie forces were shown to be less uniform, with very little load being transferred to the backup. The implications of this type of behavior on wall design are discussed in Chap. 12.

In addition to tie stiffness and relative stiffnesses of the wythes, the support conditions also affect the magnitude and distribution of tie forces. As is shown in Fig. 13.16, if fixed end conditions (restraint against rotation) are introduced in the

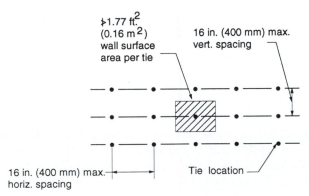

1.77 ft² (0.16 m²) wall surface area per tie
16 in. (400 mm) max. vert. spacing
16 in. (400 mm) max. horiz. spacing
Tie location

**Figure 13.14** ACI 530/ASCE 5/ TMS 402 tie spacing requirements for adjustable ties (from Ref. 13.1).

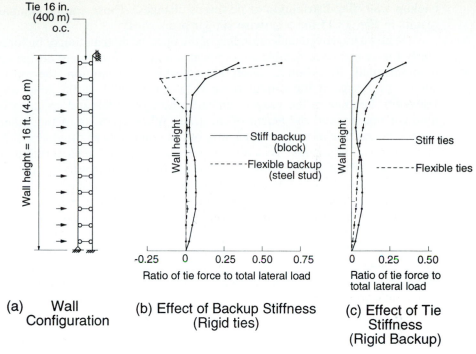

**Figure 13.15** Rigorous analysis of two-wythe wall (backup is simply supported).

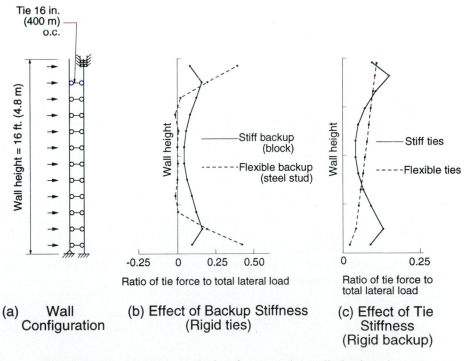

**Figure 13.16** Rigorous analysis of two-wythe wall (backup is continuous).

backup wall, the distribution of tie forces changes. These can be compared to the results in Fig. 13.15 for otherwise similar properties.

With an improved understanding of the unequal distribution of tie forces, particularly for veneer over flexible backup walls, a new approximate design approach is recommended. Based on a conservative tie force assumption, the top line of ties should be designed for a total force between 40–50% of the load on the wall. However, bending in the veneer can cause horizontal flexural cracking in the high moment region near mid-height of the wall. When cracking of the veneer occurs, the horizontal line of ties in the vicinity of the crack acts as the reaction for the two uncracked wall segments. As a result, these ties also need to have sufficient capacity to resist 40–50% of the lateral load and the practical solution is to specify this capacity requirement for all ties. It is apparent that much greater capacity requirements apply to ties used to connect veneer to flexible backup, and that the requirements are a function not only of the lateral pressure, but also of the span of the veneer.

To facilitate proper design of ties used with masonry veneer over flexible backup walls, manufacturers of proprietary ties should provide verified information regarding strength and stiffness characteristics. Such information should cover the normal conditions of use, including the ranges of adjustability (if any) and the method of attachment to the walls.

### 13.2.8 Design

**Rigorous Analysis.** Where support conditions result primarily in vertical bending, a strip of wall can be analyzed using standard frame analysis. The veneer and backup can be modeled as continuous vertical beams with appropriate pinned, fixed, or free end conditions. Axial force springs can be used to represent the ties connecting the wythes. This type of analysis was used to generate the results shown in Figs. 13.15 and 13.16. The influence of cracking at a location (say, mid-height) can also be investigated by introducing a hinge in the veneer at that location. Wall systems that develop significant two-way bending because of edge supports or large openings can be analyzed by finite element modeling of the two wythes. Representative boundary conditions and connection of wythes with axial springs to represent the ties can be accommodated with most commercially available finite element programs. [As a warning, the effort required to define the wall and input the appropriate properties can be significant. Therefore, except in critical situations, such an analysis is normally not justifiable.]

**Simplified Analysis.** For multiwythe walls of equal wythe span and similar support conditions, the nominal load on each tie can be calculated as the net load transferred between the veneer and the backup for the tributary area around a tie. [Note that the load can be applied to the inner wythe and transferred back to the outer wythe by tie forces.] The load carried by the outer wythe can be estimated using the ratio of flexural rigidities as follows:

$$p_v = p\,\frac{(EI)_{\text{veneer}}}{(EI)_{\text{backup}} + (EI)_{\text{veneer}}} \tag{13.1}$$

and

$$p_b = p - p_v \tag{13.2}$$

where  $(EI)_{\text{veneer}}$ = flexural rigidity of the veneer
$(EI)_{\text{backup}}$ = flexural rigidity of the backup wall
$p$ = total applied pressure
$p_v$ = net pressure carried by the veneer
$p_b$ = net pressure carried by the backup wall

### 13.2.9 Example Calculation of Tie Design Forces for Masonry Veneer

**Example 13.1**

For the brick veneer / steel stud wall configuration shown in Fig. 13.15, assuming 16 in. (400 mm) horizontal spacing of ties, representative relative rigidities are:

$$(EI)_{\text{veneer}} = 9\,(EI)_{\text{steel stud}}$$

From Eqs. 13.1 and 13.2, $p_v = 0.9p$ and $p_{ss} = 0.1p$.

For $p = 20$ lb/ft$^2$ (1.0 kN/m$^2$), applied on the veneer, and assuming that the ties are rigid, the tie force due to a tributary area of 1.77 ft$^2$ (0.16 m$^2$) can be approximated as

$$\text{Tie force} = 0.1(20)(1.77) = 3.5 \text{ lb} \qquad (0.1(1)(0.16) = 0.016 \text{ kN})$$

For the 16 in. (400 mm) wide vertical strip of veneer resisting a total load of 0.9(20 lb/ft$^2$)(1.33 ft $\times$ 16 ft) = 384 lb (0.9(1.0)(0.4)(4.8) = 1.728 kN), the additional load on the top tie due to it acting as a reaction can be approximated as half of the load carried by the veneer, or

$$\text{Top tie force} = 3.5 \text{ lb} + \tfrac{1}{2}(384) = 195.5 \text{ lb} \qquad (0.016 + \tfrac{1}{2}(1.728) = 0.88 \text{ kN})$$

If the fact that the top tie is 8 in. (200 mm) below the top is included in the simplified analysis, the recalculated top tie force is

$$\text{Tie force} = 3.5 \text{ lb} + 0.9(20)(1.33 \times 16)(16 \text{ ft}/2)/(16 \text{ ft} - 0.67 \text{ ft})$$

$$= 203.8 \text{ lb } (0.92 \text{ kN})$$

which is not much different from the 195.5 lb (0.88kN) force.

Compared to the value of 0.66(20)(16 $\times \frac{16}{12}$) = 282 lb (1.27 kN) determined using the results of the more rigorous analysis plotted in Fig. 13.15(b), the design force obtained using the simplified analysis is about 27% unconservative but is much closer than predicted from tributary area. Some load sharing with the next level of ties or an arbitrary inclusion of an additional 30% tie capacity can provide satisfactory performance.

For the case of a rigid wall with $(EI)_{\text{veneer}} = \tfrac{1}{3}(EI)_{\text{backup}}$, the corresponding top tie forces are 80 lb (0.36 kN) from the simplified method compared to 109 lb (0.49 kN) from the computer analysis (Fig. 13.15(b)). The percentage differences are similar to the case with the flexible backup wall.

Alternately, if the lateral load $p$ is applied on the steel stud backup, based on tributary area, the tie force is

$$0.9(20)(1.77) = 31.86 \text{ lb } (0.9(1.0)(0.16) = 0.144 \text{ kN})$$

This force can be either tension or compression, depending on the direction of the lateral load $p$. The top tie reaction force is calculated as in the foregoing case with load applied directly on the veneer. However, for load applied on the backup, the algebraic sum of the top tie reaction force and the tie force per tributary area is less than when load is applied on the veneer. Load applied on the backup wall is a less critical case for maximum tie force.

When the spans of the veneer and backup are significantly different or when relatively flexible ties and backup are used, the foregoing procedure, which is based on relative wall rigidities, will be less accurate. ■

### 13.2.10 Construction Considerations

Some performance problems with ties can be directly related to either poor detailing or poor construction practice. It is important that ties appropriate for the cavity width be used to ensure adequate compressive capacity and to permit the tie to be properly positioned in the mortar joint. Nonadjustable ties that are bent to accommodate misalignment of mortar joints, Fig. 13.17(a), or adjustable ties that are installed at an angle, Fig. 13.17(b), are considerably weakened and have lower stiffnesses.

Selection of ties should be made, noting that some ties are more sensitive than others to the method of installation. A wall tie in this category is the corrugated strip tie. To provide strength and stiffness, this type of tie must span the cavity in a straight line, free of loops, bends, and other forms of slack.[13.14] Slack can result in very little resistance to movement, as illustrated in Fig. 13.18(c).

Design standards and product literature generally provide information on the required positioning of the parts of the tie in the mortar joints. For solid units with full mortar beds, the anchor portion is typically placed in the middle third of the thickness of the unit. This allows some adjustment for small variations in the relative position of the two wythes; see Fig. 13.19. On the other hand, for anchorage into hollow masonry construction, there is very little leeway for placement location. The building code requirements and/or manufacturer's recommendations must be followed.

The interaction of ties with other wall components is a critical consideration for construction. Installation of sheathing, air barriers, cavity insulation, and even

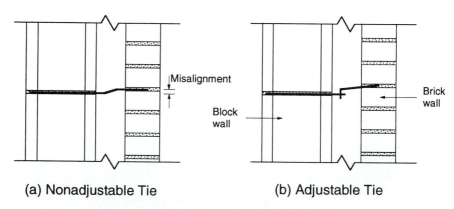

(a) Nonadjustable Tie          (b) Adjustable Tie

**Figure 13.17**   Improper installation of wall ties.

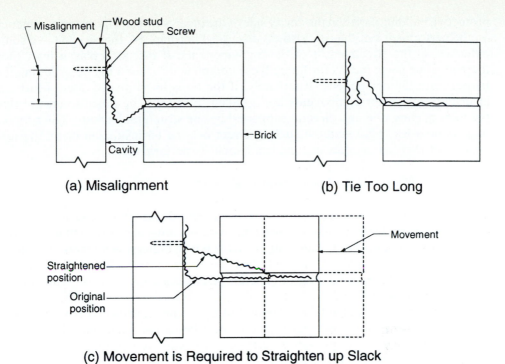

(a) Misalignment　　　　　(b) Tie Too Long

(c) Movement is Required to Straighten up Slack

**Figure 13.18**　Improper installation of corrugated ties (from Ref. 13.14).

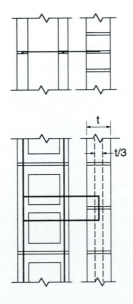

**Figure 13.19**　Embedment of ties in mortar joints.

protective building paper on the cavity side of the inner wythe, all have some influence on the performance of the tie. Also, as discussed in Chap. 14, the tie can have some influence on the effectiveness of these components. If the tie is placed first, then there must be some means for fitting these components around the tie or letting the tie puncture the component. Alternately, if the tie is to be placed after sheathing, insulation, and other components are installed, care must be taken to ensure that the tie is anchored to and directly supported by the structural backup. The support of a tie on a less rigid material such as direct bearing on insulation or on gypsum board that may soften after wetting can impair its performance.

### 13.2.11 Tie Materials and Corrosion Resistance

The majority of ties are currently made using cold-drawn steel wire and/or cold-rolled carbon steel sheets. Cold-drawn steel wire is a low-carbon wire that conforms to ASTM A82 requirements and does not have a well defined yield stress nor much ductility. Similarly, cold-rolled sheet steel, conforming to ASTM A366, is not as ductile as normal structural steel or concrete reinforcement. Stainless steel components are also used.

All steel tie components are required to have some corrosion resistance. Use of stainless steel ties or zinc-coated carbon steel ties is recommended. Galvanizing is the standard method of providing corrosion protection to carbon steel ties. Electroplating with zinc is not an acceptable coating; instead, hot dipped galvanizing must be used. The required weight (or thickness) of zinc coating depends on the expected life of the building and the atmospheric or other conditions of use.[13.15] As is shown in Table 13.5, there are different recommended levels of galvanizing for wire and for sheet steel.

Mill galvanizing (ASTM A641) is generally allowed for steel components completely embedded in mortar for interior use. This method of galvanizing supplies 0.1 ounces of zinc per square foot (30 g/m$^2$) of area of wire and 0.3 ounces per square

**Table 13.5** ACI 530/ASCE 5 Specifications for Corrosion Resistance of Ties (from Ref. 13.1)

| Type | Application | Classification |
|------|-------------|----------------|
| Wires | Joint reinforcement, wire ties, or anchors in exterior walls or interior walls exposed to moist environments (e.g., swimming pools and food processing) | ASTM A 153 Class B2 (1.50 oz/ft$^2$) |
| Sheet metal | Sheet metal ties or anchors in exterior walls or interior walls exposed to moist environments | ASTM A 153 Class B2 (1.50 oz/ft$^2$) |
| | Sheet metal ties or anchors in interior walls | ASTM A 525 Class G60 (0.60 oz/ft$^2$) |

Note: 1 oz/ft$^2$ = 300 g/m$^2$.

foot (90 g/m$^2$) on each side of sheet metal. These thin zinc coatings will normally accommodate later bending or cutting without spalling or cracking. However, ties in other than interior walls require heavier weights of hot dipped galvanizing, which must be applied after the parts are formed. Applying the galvanizing after the parts are formed allows weights of 1.5 ounces per square foot (450 gm/m$^2$) or greater to be applied to all parts including cut edges. These greater thicknesses of zinc are much more than were normally supplied by industry during the 1970s and 1980s and should provide significantly improved corrosion resistance.

For a given amount of galvanizing, the service life of the tie will depend on the atmospheric conditions. A general guide to service life for various environments published by the American Galvanizers Association[13.16] is reproduced in Fig. 13.20. For a 1.5 ounce per square foot coating (450 g/m$^2$), service life differs by a factor of 3 between industrial and rural environments. An alternative way of looking at the life of the galvanizing is to evaluate the sacrificial rate of zinc loss. Table 13.6 contains data from several locations in North America. For example, the rate of loss per square foot per year in Bethlehem, Pennsylvania, is $(0.02/2)/(4/12)(6/12) = 0.06$ oz/ft$^2$. Therefore, the life expectancy for hot dipped galvanizing of 1.5 ounces per square foot in accordance with ASTM A153 is $1.5/0.06 = 25$ years.

The information available and the development of standards for galvanizing have greatly improved. However, in terms of predicting life, it is important to recognize that local environments can vary significantly and that the environment within a masonry wall is not necessarily the same as the outdoor climate. Therefore, the service life can be quite different.

For buildings designed for extended life or where potential corrosion is seen as a critical aspect of design, there has been a gradual increase in the use of noncorroding stainless steel ties. Although the cost of stainless steel material is considerably more than galvanized steel, the actual cost of the tie, particularly when judged in terms

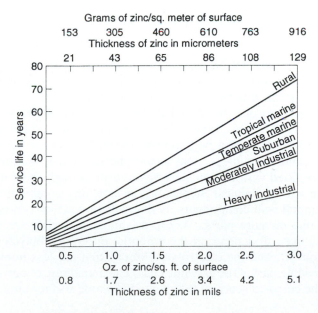

**Figure 13.20** Service life of metal for different levels of galvanization (from Ref. 13.16).

**Table 13.6** Effect of Climate on Corrosion of Zinc Galvanizing
(from Ref. 13.17)

| Location | Zinc lost, oz(g), after 2-year outdoor exposure* |
|---|---|
| Phoenix, AZ | 0.005 (0.14) |
| Cape Kennedy, FL, 1/2 mile from ocean | 0.018 (0.50) |
| Bethlehem, PA | 0.020 (0.56) |
| Point Reyes, CA | 0.023 (0.64) |
| East Chicago, IN | 0.028 (0.78) |
| Columbus, OH | 0.033 (0.92) |
| Pittsburgh, PA | 0.040 (1.12) |
| Cleveland, OH | 0.042 (1.18) |
| Newark, NJ | 0.057 (1.60) |
| Cape Kennedy, FL<br> 60 miles (96 km) from ocean, elevation 30 ft (9.1 m) | 0.62 (1.74) |
| 60 miles (96 km) from ocean, ground level | 0.064 (1.79) |
| 60 miles (96 km) from ocean, elevation 60 ft (18.3 m) | 0.168 (1.90) |
| Halifax, Nova Scotia | 0.114 (3.19) |

\*  The weight loss shown here is for 4×6 in. (100×150 mm) test specimens fully exposed to weathering. Tests have not been done on zinc lost from galvanized metals embedded in masonry walls at different locations.

of the overall cost of the wall, may not be prohibitively greater. In fact, when the financial impact of premature tie corrosion is weighed against the very small increase in overall cost of the wall system, it is very inexpensive insurance.

Not all stainless steels are appropriate. A nickel–chromium type of steel satisfying the requirements of the ASTM A167, Types 304 or 316, offers the best corrosion resistance. These austenitic steels cannot be hardened thermally and are normally not magnetic. Because austenite stainless steel cannot be hardened, self-drilling sheet metal screws must be fitted with carbon steel tips to drill through sheet metal.

A concern regarding use of stainless steel, and, in fact, any combination of metals for ties and other metal wall components, is that galvanic action can occur, resulting in corrosion of one or more parts. Where dissimilar materials are used, the amount of galvanic action will depend on the potential difference between the metals, the intimacy of contact, and the electrolyte. Metals that are less noble are always the anode, the corroding part of the galvanic cell. The actual rate of corrosion is affected by the size of the corrosion current relative to the anode surface area. For

**Table 13.7** Corrosion of Dissimilar Metals (from Ref.13.17)

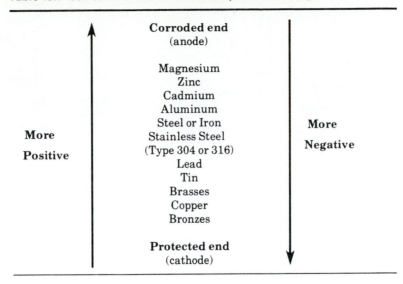

```
                    ↑   Corroded end                    ↓
                        (anode)

                        Magnesium
                        Zinc
                        Cadmium
                        Aluminum
                        Steel or Iron          More
   More                 Stainless Steel        Negative
   Positive             (Type 304 or 316)
                        Lead
                        Tin
                        Brasses
                        Copper
                        Bronzes

                        Protected end
                        (cathode)
```

example, if a fastener with a small surface area relative to the surrounding metal is the anode, the current density will be high and rapid corrosion of the fastener will occur. Table 13.7 lists common tie materials arranged in order of nobility. From this table and the previous discussion, it can be concluded that the small fasteners (screws) used as parts of some ties should be of more noble material than the attached metal. For this reason, use of stainless steel screws in galvanized carbon sheet steel is usually a satisfactory design.

## 13.3 ANCHORS

### 13.3.1 Basic Functions

As briefly introduced in Sec. 13.1, anchors are employed to attach various building components together.[13.18] In many cases, there is not much difference in the basic functions of ties and anchors except that anchors tend to have higher load-carrying capacity. As shown in Fig. 13.21(a), various types of anchors that allow vertical movement can be used to laterally support nonloadbearing walls at floor levels. Anchors that permit similar degrees of movement at columns are shown in Fig. 13.21(b). Where both horizontal and vertical load transfer are required, normal reinforcement can be used as the anchor device.

Anchors for both fixed and adjustable shelf angles are discussed in Sec. 12.3.

On single-story buildings, a critical type of anchor is that required to attach roofs to walls so that uplift due to wind forces can be resisted (see Sec. 16.3.2 in Chap. 16). Also, where it is desired that intersecting loadbearing walls be fixed together to form flanged sections, use of anchors, as shown in Fig. 13.21(c), is common practice.

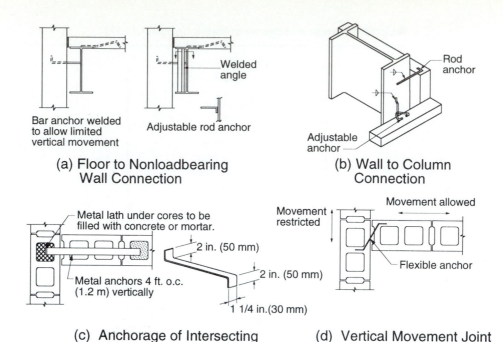

(a) Floor to Nonloadbearing Wall Connection

(b) Wall to Column Connection

(c) Anchorage of Intersecting Walls to Form Flanges

(d) Vertical Movement Joint

**Figure 13.21** Typical anchors in masonry construction.

The function of an anchor must be clearly defined. For instance, as shown in Fig. 13.21(d), at a vertical movement joint, transfer of out-of-plane shear across the movement joint may be needed to provide extra support for lateral loading. However, unrestrained in-plane horizontal movement of the two walls must be allowed so that the movement joint functions properly. Rods in greased sleeves or anchors that have stiff shear and flexible axial load resistance properties, such as shown, are appropriate.

### 13.3.2 Design Considerations

As mentioned before, it is important to define the functions of anchors. It can be just as critical in the design of the anchor to allow some types of movement as it is to provide the intended restraint. Clip angles at the tops of nonloadbearing partition walls (Chap. 11) are an example of this type of anchor. The design of metal anchors follows the principles of structural steel design, whereas minimum requirements for spacings and size of anchors relate to design limitations for stress and displacement in the masonry. Although the minimum requirements often can provide strengths in excess of those required, this is not always the case. For example, prescriptive design requirements for strap anchors, such as shown in Fig. 13.21(c), to be spaced at a maximum of 4 ft (1.2 m) along the joint between intersecting walls, will not necessarily satisfy the shear transfer requirements for bending in the flanged wall section. Example designs of a roof anchor and of the anchors at interesting walls are included in Chaps. 16 and 17, respectively.

### 13.3.3 Construction Considerations

Where parts of anchors are not completely protected by adequate embedded in mortar or grout, corrosion protection similar to the requirements for ties should be provided. Anchors, like ties, require proper installation to ensure expected performance.

Anchors depending on proper embedment in mortar or grout for their strength must be solidly encompassed in these materials. Alignment and edge distance are also important in terms of both the shear and tensile strengths of the anchor.

Anchors, such as a rod in a greased sleeve, designed to permit movement in one direction while restraining displacement in the other directions require accurate alignment in order to function properly. Similarly, when one part of an anchor is installed in advance of another part, some adjustability to accommodate misplacement is an advantage. The potential interference or interaction of parts of an anchors with other elements of the construction (i.e., water stops, flashing, air barriers, and interior finishes) should be considered.

## 13.4 FASTENERS

### 13.4.1 Basic Function

As discussed in Sec. 13.1, a fastener is a device used to fix elements such as equipment, fixtures, pipes, and even ties and anchors to masonry elements. They are normally required to transfer forces to masonry through shear and/or tension. Therefore, the performance of the fastener depends both on its strength and deformation properties and on its anchorage into the masonry. Load ratings for various fasteners should include the effects of repeated and cyclic loading.

### 13.4.2 Types of Fasteners

The most common type of fastener is the anchor bolt. A variety are shown in Fig. 13.22. These can be anchored in mortar joints, in grout, or in concrete during construction. If bond alone is not sufficient, some form of mechanical anchorage

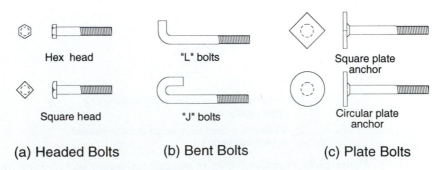

| Hex head | "L" bolts | Square plate anchor |
| Square head | "J" bolts | Circular plate anchor |
| (a) Headed Bolts | (b) Bent Bolts | (c) Plate Bolts |

**Figure 13.22**   Anchor bolts for masonry (from Ref. 13.1).

such as an enlarged or bent end is usually required. Alternately, a commercially available expansion anchor can be inserted into predrilled holes to receive the threaded end of a bolt or screw. The ratings for pullout and shear should include allowance for construction conditions.

A wide variety of shot, wedged, screwed, and nailed fasteners also exists. In general, these types of fasteners exhibit very large scatter in strength.

### 13.4.3 Strength of Anchor Bolts

Anchor bolt fasteners can be subjected to loads resulting from wind, earthquake, gravity forces, vibration, or differential movement. These loads can introduce various combinations of shear and tension forces, depending on the orientation of the fastener.

Tests results[13.19] indicate that the failure mode and strength of anchor bolts embedded in mortar are affected by the diameter and the geometry of the bolt, embedment length, strength of masonry, yield strength of the bolt, edge distance, bolt spacing, and bolt grouping. Other tests on "J" bolts embedded in concrete and clay masonry[13.20,13.21] revealed three failure modes:

1. Failure of the masonry (typical conical shear failure, as shown in Fig. 13.23).
2. Fracture of the bolt.
3. Straightening of the J-bolt.

Failure modes 1 and 3 prevailed under axial tension load, whereas modes 1 and 2 were observed for combined tensile and shear loads.

Test results for the specimens shown in Fig. 13.24(a) indicate some interaction for combined shear and axial tension loading of J-bolts. This is particularly evident for larger diameter bolts, as shown in Figs. 13.24(b) and 13.24(c). Shear tests on bolts anchored through the face shells of concrete block, as shown in Fig. 13.25, indicated that shear capacity increased with increased bolt diameter. Increased bolt

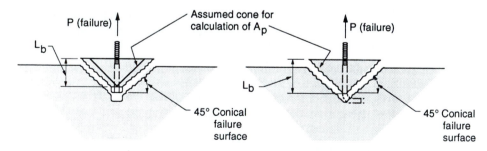

$A_p$ = Projected area on the masonry surface, of a right circular cone

$L_b$ = Effective embedment length of plate, headed or bent anchor bolt

**Figure 13.23** Tension failure of masonry at anchor bolt location (from Ref. 13.1).

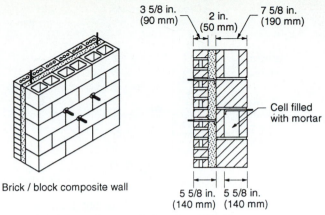

Cell filled with mortar

Brick / block composite wall

**(a) Locations of Anchor Bolts**

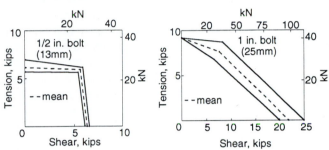

**(b) Interaction Diagrams for Brick Masonry Embedments**

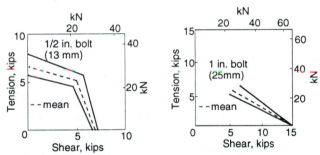

**(c) Interaction Diagrams for Concrete Masonry Embedments**

**Figure 13.24** Strength of anchor bolts in masonry under combined shear and tension (from Ref. 13.20).

size was not significant for direct tension where the size of the conical failure was basically controlled by the washer anchor at the end of the bolt.

When information for masonry fasteners is lacking, conservative use of information available for fasteners in similar concrete applications is common practice. However, the influence of edge distance, particularly where the anchor bolt is embedded vertically in a grout core, can be much more significant for hollow grouted masonry than for monolithic concrete. A conservative approach would be to base strengths on edge distance within the grout.

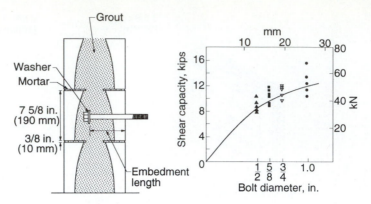

**Figure 13.25** Shear capacity of anchor bolts in masonry (from Ref. 13.19).

Except for anchor bolts, which are generic fasteners, most other fasteners are proprietary products. It should be ensured that data or design information provided for these products has been prepared in accordance with the relevant standards. In addition to determining that the strengths satisfy the design requirements, consideration should be given to durability, likelihood and consequences of faulty installation, repeated loading, and the mode of failure. In practice, fasteners designed for shear may be subject to some tension as a result of prying action. The potential for such a situation should be considered.

### 13.4.4 Design of Fasteners

Because of the multitude of parameters affecting the strength of anchor bolts in masonry, it is not surprising that test results are highly variable[13.19] with coefficients of variation of 35% not uncommon. This phenomena is often reflected in manufacturers' design information where a safety factor of 5 relative to the mean strength is often used. For other types of fasteners, even greater variabilities have been observed and factors of safety up to 10 may be required, depending on the number of fasteners acting together. Development of strength and limit states design codes should lead to a more consistent rationalization of the level of safety.

For plate, headed, and bent bar anchor bolts (Fig. 13.22), ACI 530/ASCE 5/ TMS 402[13.1] specifies the allowable load as the lesser of

$$B_a = 0.5A_p \sqrt{f'_m} \text{ in lbs. } \text{ or } B_a = 0.042 A_p \sqrt{f'_m} \text{ in } N \qquad (13.3)$$

$$B_a = 0.2 A_b f_y \qquad (13.4)$$

where: $A_b$ = bolt cross-sectional area (in.$^2$ or mm$^2$)
$A_p$ = $\pi L_b^2$ (in.$^2$ or mm$^2$) (area of failure cone, see Fig. 13.23)
$f'_m$ = specified masonry compressive strength (psi or MPa)
$f_y$ = yield strength of the bolt (psi or MPa)
$L_b$ = embedment length of anchor bolt or edge distance measured from the surfaces of an anchor bolt to the nearest free edge, whichever is least.

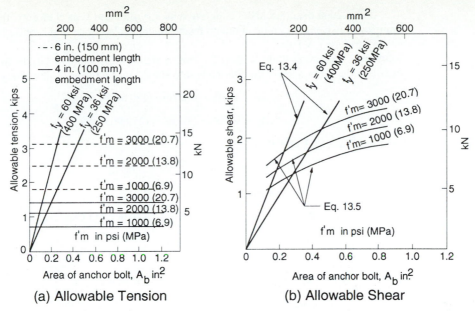

**Figure 13.26** Allowable tension and shear loads of anchor bolts (from Ref. 13.1).

Equation 13.3 represents masonry controlled capacity, whereas Eq. 13.4 represents capacity controlled by bolt failure. These two equations are shown graphically in Fig. 13.26(a).

The allowable shear load for anchor bolts with edge distances of at least 12 bolt diameters is specified as

$$B_v = 350 \sqrt[4]{f'_m A_b} \quad \text{in lbs.} \quad \text{or } B_v = 1070 \sqrt[4]{f'_m A_b} \quad \text{in N} \tag{13.5}$$

or

$$B_v = 0.12 A_b f_y \tag{13.6}$$

If the edge distance is less than 12 bolt diameters, the value of $B_v$ in Eq. 13.5 is *linearly* reduced to zero at an edge distance of 1 in. (25 mm). The foregoing two equations are plotted in Fig. 13.26(b) for different values of the design parameters.

For combined axial tension force $b_a$ and shear force $b_v$, the following unity equation can be used to approximate the interaction:

$$\frac{b_a}{B_a} + \frac{b_v}{B_v} \le 1 \tag{13.7}$$

For other types of anchor bolts, allowable loads[13.1] can be based on 20% of the average of five samples tested in accordance with ASTM E488.

### 13.4.5 Anchor Bolt Design Example

**Example 13.2**

A "J" shaped anchor bolt required to resist 500 lb (2.23 kN) in tension and 300 lb (1.34 kN) in shear can be designed in accordance with the provisions discussed before.

For a 2000 psi (13.8 MPa) concrete masonry unit and type S mortar, the specified compressive strength can be obtained from the applicable code and in this case is taken as 1555 psi (10.7 MPa).

**Solution:** Assuming that a $\frac{1}{2}$ in. (12.7 mm) diameter bolt made of Grade 40 (276 MPa) steel is to be embedded 8 in. (200 mm) into the 12 in. (300 mm) thick block wall,

$$A_b = \pi(\tfrac{1}{4})^2 = 0.20 \text{ in.}^2 (129 \text{ mm}^2)$$

$$A_p = \pi(8)^2 = 201 \text{ in.}^2 (125,600 \text{ mm}^2)$$

and from Eqs. 13.3 and 13.4,

$$B_a = 0.5 A_p \sqrt{f'_m} = 0.5\,(201)\,\sqrt{1555} = 3963 \text{ lb}$$

$$(B_a = 0.042(125,600)\,\sqrt{10.7}/10^3 = 17.3 \text{ kN})$$

or $\qquad B_a = 0.2 A_b f_y = 0.2(0.20)(40,000) = 1600 \text{ lb } (7.34 \text{ kN}) - \text{controls}$

Therefore, $B_a = 1600$ lbs. (7.34 kN).
From Eqs. 13.5 and 13.6,

$$B_v = 350 \sqrt[4]{f'_m A_b} = 350\,\sqrt[4]{1555(0.20)} = 1470 \text{ lbs}$$

$$(B_v = 1070 \sqrt[4]{10.7(129)} = 6520 \text{ N})$$

or $\qquad B_v = 0.12 A_b f_y = 0.12(0.20)(40,000) = 960 \text{ lbs } (4.27 \text{ kN}) - \text{controls}$

Therefore, $B_v = 960$ lbs (4.27 kN).
Then from Eq. 13.7,

$$\frac{b_a}{B_a} + \frac{b_v}{B_v} = \frac{500}{1600} + \frac{300}{960} = 0.63 < 1$$

$$\left( \frac{2.23}{7.34} + \frac{1.34}{4.27} = 0.62 < 1 \right)$$

Therefore, the use of the $\frac{1}{2}$ in. (12.7 mm) diameter bolt is satisfactory. ∎

## 13.5 CLOSURE

Ties, anchors, and fasteners have very little impact on the overall cost of masonry buildings but are key elements for the long term satisfactory performance of the building. It is important that designers properly design and clearly specify the type of connector to be used and include reference details for their proper installation. Similarly, it is important that builders follow proper installation procedures and avoid damaging or otherwise altering the properties of the connectors.

Experience with remedial work has clearly shown the folly of cost cutting, resulting in poor connector performance. The potential savings of cost cutting of ties are trivial compared to the consequences of inadequate service performance. In particular, provision of adequate resistance to corrosion is essential.

Some information and discussion regarding the influence of connectors on the behavior and design of masonry are also included in Chaps. 10, 12, and 14.

## 13.6 REFERENCES

13.1 The Masonry Standards Joint Committee, "Building Code Requirements for Masonry Structures," ACI 530/ASCE 53/TMS 402, American Concrete Institute and American Society of Civil Engineers, Detroit and New York, 1992.

13.2 Standards Association of Australia, AS 2699-1984: Wall Ties for Masonry Construction SAA, North Sydney, N.S.W., 1984.

13.3 Canadian Standards Association, "Connectors for Masonry," CAN3-A370-M84, CSA, Rexdale, Ontario, March 1984.

13.4 Brick Institute of America, "Brick Veneer Panel and Curtain Walls," BIA Technical Notes on Brick Construction 28B, BIA, Reston, VA, 1987.

13.5 B. Pitoni, R. Drysdale, E. Gazzola and A. Hamid, "Capacity of Cavity Wall Tie System," in *Proceedings of the Fourth Canadian Masonry Symposium*, Fredericton, New Brunswick, June 1986, pp. 81–93.

13.6 Dur-O-Wal, "Investigation of Masonry Wall Ties," Technical Bulletin No. 64-3, Northbrook, IL, 1985.

13.7 R. Brown and R. Elling, "Lateral Load Distribution in Cavity Walls," in *Proceedings of the Fifth International Brick Masonry Conference*, Washington, D.C., 1979, pp. 351–359.

13.8 W. McGinley, J. Warwaruk, J. Longworth and M. Hatzinikolas, "Masonry Veneer and Steel Stud Curtain Walls," in *Proceedings of the Fourth Canadian Masonry Symposium*, Fredericton, New Brunswick, June 1986, pp. 733–746.

13.9 R.G. Drysdale, "Defining Better Wall Systems," Research Report, Canada Mortgage and Housing Corporation, Ottawa, Ontario, 1991.

13.10 R.G. Drysdale and M.J. Wilson, "A Report on Behaviour of Brick Veneer/Steel Stud Tie Systems," Research Report, Project Implementation Division, Canadian Mortgage and Housing Corporation, Ottawa, Ontario, 1989.

13.11 Dur-O-Wal, "Dur-O-Wal Adjustable Wall Ties, Their Structural Properties and Recommended Use," Technical Bulletin No. 64-4, Northbrook, IL, 1986.

13.12 M. Wilson and R.G. Drysdale, "Influence of Adjustability on the Behaviour of Brick Veneer/Steel Stud Wall Ties," in *Proceedings of the Fifth Canadian Masonry Symposium*, Vancouver, British Columbia, 1989, pp. 521–530.

13.13 Brick Institute of America, "Wall Ties for Brick Masonry," BIA Technical Notes on Brick Construction 44B, BIA, Reston, VA, 1987.

13.14 G. Bell and W. Gumpertz, "Engineering Evaluation of Brick Veneer/Steel Stud Walls Part 2—Structural Design. Structural Behavior and Durability," in *Proceedings of the Third North American Masonry Conference*, Arlington, TX, 1985, pp. 5.1–5.20.

13.15 R. Heidersbach and J. Lloyd, "Corrosion of Metals in Concrete and Masonry Buildings," Paper 258, Corrosion/85, NACE, Houston, March 1985.

13.16 American Hot Dip Galvanizers Association, "Hot Dip Galvanizing for Corrosion Protection of Steel Products," Technical Report, Washington, D.C., 1984.

13.17 M. Catani, "Preventing Corrosion," *Aberdeen's Magazine of Masonry Construction*, Vol. 4, No. 4, 1991.

13.18 C.T. Grimm, "Metal Ties and Anchors for Brick Walls," *Journal of the Structural Division*, Proceedings of ASCE, Vol. 102, No. ST4, 1976, pp. 839–858.

13.19 M. Hatzinikolas, J. Longworth and J. Warwaruk, "Strength and Behavior of Anchor

Bolts Embedded in Concrete Masonry," in *Proceedings of the Second Canadian Masonry Conference*, Carleton University, Ottawa, Ontario, 1980, pp. 549–563.

13.20 R.H. Brown and A.R. Whitlock, "Strength of Anchor Bolts in Grouted Concrete Masonry," *Journal of the Structural Division*, Proceedings of ASCE, Vol. 109, No. 6, 1983, pp. 1362–1374.

13.21 Whitlock, A.R. and Brown, R.H., "Cyclic and Monotonic Strength of Anchor Bolts in Concrete Masonry," in *Proceedings of the Third Canadian Masonry Symposium*, University of Alberta, Edmonton, Alberta, 1983, pp. 20.1–20.18.

## 13.7 PROBLEMS

**13.1** Obtain samples of three different types of brick veneer ties. Comment on their suitability for use in brick veneer or cavity wall construction.

**13.2** From manufacturing literature, building codes or other sources, determine the design load for tension for the following fasteners:

(a) A standard $2\frac{1}{2}$ in. (63 mm) spiral nail in a spruce board.

(b) A screw in 18 gage cold formed steel.

(c) A $\frac{5}{8}$ in. (15 mm) diameter bolt with a $1\frac{3}{4}$ in. (45 mm) washer at the end, embedded 5 in. (127 mm) through the face shell in grouted 2000 psi (13.8 MPa) concrete block masonry (Grade 60 (400 MPa) steel).

**13.3** How can the thickness of galvanizing be determined? (*Hint*: Weights and thicknesses before and after stripping off of the zinc can be measured.)

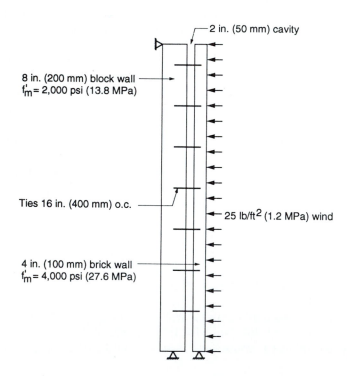

8 in. (200 mm) block wall
$f'_m$ = 2,000 psi (13.8 MPa)

2 in. (50 mm) cavity

Ties 16 in. (400 mm) o.c.

25 lb/ft$^2$ (1.2 MPa) wind

4 in. (100 mm) brick wall
$f'_m$ = 4,000 psi (27.6 MPa)

**Figure P13.4**

**13.4** For a 10 ft (3 m) high clay brick / concrete block cavity wall, as shown in Fig. P13.4, determine the critical design load for ties as follows:

**(a)** Use the simplified analysis.

**(b)** For senior students, use a plane frame structural analysis program to determine this value.

**(c)** If the veneer is cracked at the location of the fourth tie from the base, determine the maximum tie force using the analyses suggested in parts (a) and/or (b).

**(d)** If half of the load is positive pressure on the outer wythe and half is suction on the inner wythe, what difference does this make to the tie forces?

# CHAPTER 14

# Application of Building Science for Environmental Loads

Hose steam test during fire test of a masonry wall (*Courtesy of National Concrete Masonry Association*).

## 14.1 INTRODUCTION

For buildings to perform satisfactorily, it is increasingly evident that knowledge crossing the boundaries of the traditional sciences is required. Defined as *building science,* the resulting interdisciplinary subject area "draws on the knowledge and experience of almost every branch of engineering science."[14.1] The application of building science to the design and construction of buildings for good performance under a wide range of environmental loads is the main focus of this chapter.

A major part of this book is devoted to topics related to the structural design of masonry buildings. However, it is important to realize that the success and long term satisfactory performance of buildings are equally dependent on factors that can be broadly defined as relating to *environmental loads*. Although building science is a comparatively new term, the basis for this developing discipline, in terms of environmental considerations, is as old as mankind. In their search for safe and comfortable living environments, humans have gradually developed buildings suitable to their most important needs. These needs are directly related to the desire to separate the internal environment from the external environment and, as such, are climatically, geographically, and culturally dependent.

For example, the shelters built by North American Indians varied greatly, depending on their culture (nomadic versus agrarian) and on their geographic location and climate. Hence, the hunting tribes on the central plains tended to build tent or teepee type structures with materials that were readily available and easily transportable to new hunting grounds. Tribes further south relied on agriculture as their main food source and remained in the same location. As a result, some became cliff dwellers in caves or built massive permanent shelters which provided them with protection from their enemies and from the effects of climatic extremes.

The exterior walls of buildings separate the internal and external environments and are thus subjected to environmental loads (both man-made and from nature). Therefore, in addition to structural safety and aesthetic considerations, the wall must act as part of the building envelope to

- resist the initiation and spread of fire
- control heat flow
- control air flow, including the effects of internal and external air pressures
- control the movement of water either as vapor or liquid
- control light and other solar radiation
- control sound and vibration
- provide privacy and security, including limiting entry by unwanted intruders

The foregoing requirements must be included in the design and integration of all components of the building envelope. Durability, economical construction, operating and long term maintenance costs, as well as accommodation of features such as mechanical and electrical components are other factors that affect the suitability of a particular building envelope design.

To achieve an effective building envelope that satisfies all of these requirements, the application of building science must account for individual environmental requirements and their interactions. It is quite possible that wall designs that work well in cold dry climates may not be effective in warm humid climates and vice versa. In addition, traditionally satisfactory designs may start to exhibit problems as the environmental conditions change (e.g., increased interior humidity) or as new materials or construction processes are introduced (e.g., relatively impermeable rigid plastic insulation that acts as an air and vapor barrier).

## 14.2 APPLICATIONS OF BUILDING SCIENCE TO MASONRY CONSTRUCTION

Masonry has proven to provide cost-effective, aesthetically pleasing, and durable walls for the full range of climatic conditions and for countries covering the full range of economic and technological development. However, in this chapter, the range of applications is focused on nontropical climates.

Whether masonry walls are loadbearing or simply wind bearing, they have many of the inherent characteristics necessary to satisfy the building science requirements of environmental barriers. For features such as thermal insulation and air barriers, masonry alone may not satisfy the requirements, but does provide an ideal base upon which to incorporate other materials. Therefore, the discussions in this chapter include other materials required to achieve the desired performance.

Many of the books written on environmental design have been devoted to individual building science topics such as sound control or fire resistance. Only basic backgrounds in some of the building science areas are addressed in this chapter. To assist the reader who wishes to explore certain areas in depth, references to other sources of information are provided. Available information is quite diverse; however, valuable single sources of practical advice and technical information are available from the Brick Institute of America[14.2] and the National Concrete Masonry Association.[14.3] These publications are readily available and are subject to periodic review and revision. In the broad range of building science areas, publications of the Division of Building Research (now the Institute for Research in Construction) of the National Research Council of Canada in Ottawa, Ontario are excellent sources of information. The *Canadian Building Digests*[14.4] and the American Society of Heating, Refrigeration and Air Conditioning Engineers (ASHRAE)[14.5] provide very focused treatments of specific topics. Of English language sources, publications from the National Institute for Standards and Technology (USA), Building Research Establishment (UK) and CSIRO (Australia) also provide comprehensive background in many of the identified topics.

Building codes have for some time incorporated well defined requirements for certain aspects of building science such as fire protection and sound transmission. Other aspects, such as those relating to rain penetration, are not well defined and specifications for some requirements, such as air barriers, tend to be general performance statements. As a result of the lack of prescriptive code requirements in many building science areas and the lack of standardization in many facets of construction, the designer bears a large responsibility for exercising judgment to develop an integrated design to satisfy the relevant building science requirements for environmental loads.

Traditionally, the architect had the responsibility for ensuring the environmental adequacy of the building. This included design of the walls and coordination and integration of the structural, mechanical, and electrical designs with each other and with the architectural aspects. Today, however, because of the wide variety of technical topics involved, the application of building science to the environmental design of a building often requires the cooperation of many specialists. Usually, the individual architect, engineer, or builder having overall responsibility must have a broad expertise developed through experience and on-going study. Alternately, building science specialists may perform an advisory or review function. The introduction of new materials and new methods of construction coupled with changing environmental requirements may mean that repeating previous practices simply repeats previous problems or creates new problems because of changes in other aspects of the design.

The successful application of building science principles to the design of an environmental barrier depends on providing details and specifications that are easily understood and readily built. Communication with the builder, including proper inspection, is very important. The suitability of the design and the details should be evaluated in terms of maintenance requirements and remedial measures. Designs that depend upon construction perfection for successful performance should be avoided. It is suggested that any design should be assessed for vulnerability[14.6] to unsatisfactory performance resulting from minor flaws in construction which, despite good quality control, may be unavoidable. The use of highly vulnerable designs should be reconsidered, whereas, in other cases, the potential for problems can be reduced to an acceptable level by employing additional quality control.

## 14.3 FIRE RESISTANCE

### 14.3.1 Introduction

Fire is one of the major causes of loss of life and property. Because it has been shown to have such a large impact on safety, most building codes have well developed regulations relating to both *fire prevention* and *fire protection*. Whether fires are started by elements of the building (mechanical or electrical systems), by carelessness (grease fires or smoking in bed), or by external sources (radiation from or direct contact with adjacent burning buildings), many of the same general methods for resisting ignition and limiting the spread of fire are applicable. These measures essentially fall into the following categories:

- limitation of combustible content (fuel load)
- limitation of flame spread for building components
- limitation of the development of smoke and other toxic gases
- containment of fire within the area of fire origin
- provision of sufficient residual structural capacity during and after the fire to allow occupants to escape and the opportunity for the fire to be extinguished without building collapse
- provision for safe areas of refuge or protected means of egress
- provision for fire extinguishing systems such as sprinklers
- inspection to reduce hazards
- education of occupants

The degree to which these types of provisions have been implemented in practice depends on the type of occupancy (e.g., a single family house versus a high-rise apartment versus a factory) and the type of construction (combustible versus noncombustible).

Masonry walls themselves do not contribute fuel to the fire, are not subject to flame spread, and do not produce smoke or toxic gases in the presence of fire. They do provide solid noncombustible barriers to the spread of fire from the original fire area and can be equally useful in creating safe compartments or escape routes from the building. Therefore, masonry construction is ideally suited to minimizing the potential for ignition of fires and the consequences of fires. The main fire resistance aspects of the design of masonry walls relate to the following:

- ability to maintain sufficient load-carrying capacity to support floors and roofs in loadbearing construction
- ability of nonloadbearing firewalls to maintain sufficient strength during and after specified durations of fire to avoid collapse
- thermal characteristics of the wall so as to prevent temperature rise that could possibly cause new ignition
- impact of the failure of other structural elements on the stability of the masonry
- impact of the use of other materials as part of the wall system

### 14.3.2 Design Practice

Fire records in the United States indicate that deaths, injuries, and property losses due to fire have remained unacceptably high during the past decade.[14.7] Except for the substantial reduction in residential fire deaths associated with the increased use of smoke detectors during the 1970s and early 1980s, there has been very little recent progress toward improving this poor record. Of concern is the recent trend in some building codes to permit trade-offs between active and passive fire-safety design. Automatic detection and alarms are active aspects of design aimed principally at life safety as are automatic sprinkler systems, which may also provide property protection, depending on the design of the system. Alternatively, masonry construction is a noncombustible form of passive fire safety aimed at protection of life and property by providing fire resistance to ensure stability during the fire and secure escape routes.

It is suggested[14.7,14.8] that balanced design incorporating both active and passive protection is the most reliable practice and that the use of masonry walls as the passive system provides an overall economic benefit. The discussion that follows focuses on passive fire-safety design, which is one of the major benefits of masonry construction.

### 14.3.3 Fire Resistance Rating

Provisions to ensure reasonable fire resistance of wall systems have been developed based on standardized test procedures. The standardized tests do not accurately represent the true behavior of walls in buildings, in actual fire situations, but do give basic measures of fire resistance. The behavior of a wall in a building is affected by factors such as boundary conditions, size, shape, and load combinations. For example, intersecting members can either brace the wall or introduce additional loads, parts of larger walls exposed to less intense heat can share load from more heat affected zones, walls with flanges will be less sensitive to the effects of bowing due to heat on one side, and walls will react differently to different combinations of axial load and bending. Interactions with other parts of the structure such as an expanding roof system can also affect the performance of the wall.

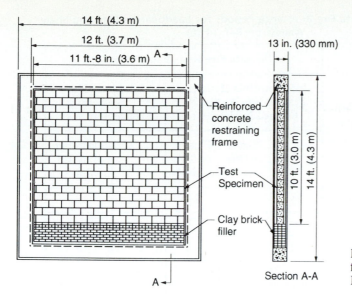

14 ft. (4.3 m)

12 ft. (3.7 m)

11 ft.-8 in. (3.6 m)

A

13 in. (330 mm)

Reinforced
concrete
restraining
frame

Test
Specimen

Clay brick
filler

10 ft. (3.0 m)

14 ft. (4.3 m)

Section A-A

**Figure 14.1** Typical test specimen for standard fire test (from ASTM E119-1985).

Although varying slightly from country to country, the test conditions of ASTM E119[14.9] are typical of those used to determine fire resistance periods of walls. A wall panels of at least 100 ft² (9.3m²) surface area with no dimension less than 9 ft (2.7 m), built to represent the actual construction is mounted in a test frame and positioned on the face of a fire chamber. Figure 14.1 is an illustration of a typical test wall. Nonloadbearing walls are required to be restrained at all four sides whereas loadbearing walls are not restrained at the vertical edges. For loadbearing walls, the design level of axial load is applied on the wall. Then, fire loading following the temperature curve shown in Fig. 14.2 is applied. After the time for the desired fire resisting rating has elapsed, the wall is subjected to a hose stream test to establish that the wall can withstand the thermal shock and the pressure from the cold water without either collapsing or allowing water through the wall assembly. [Note: An option allows a second wall exposed for half of the fire rating time to be used for the hose stream test.]

Prior to expiration of the test period, a rise of temperature of 325°F (163°C) at any one of nine points on the face of the wall not exposed to fire or an average rise

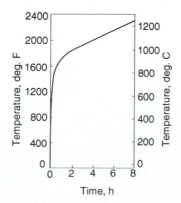

**Figure 14.2** Standard time–temperature curve for ASTM E119 fire test (from Ref. 14.9).

**Table 14.1** Ultimate Fire Resistance Periods for Loadbearing Clay and Shale Walls (from Ref. 14.10)

| Nominal Wall Thickness, in. (mm) | Wall Type | Ultimate Fire Resistance Period (h) | | | | |
|---|---|---|---|---|---|---|
| | | Incombustible Members Framed Into Wall or No Framed-In Members | | | Combustible Members Framed Into Wall | |
| | | No Plaster | Plaster on One Side* | Plaster on Two Sides* | No Plaster | Plaster on Exposed Side* |
| 4 (100) | Solid | $1\frac{1}{4}$ | $1\frac{3}{4}$ | $2\frac{1}{2}$ | – | – |
| 8 (200) | Solid | 5 | 6 | 7 | 2 | $2\frac{1}{2}$ |
| 12 (300) | Solid** | 10 | 10 | 12 | 8 | 9 |
| 12 (300) | Solid† | 12 | 13 | 15 | – | – |
| 9 to 10 (230 to 250) | Cavity | 5 | 6 | 7 | 2 | $2\frac{1}{2}$ |

\*  To achieve these ratings, each plastered wall face must have at least $\frac{1}{2}$-inch (13 mm) 1:3 gypsum-sand plaster.
\*\*  Based on load failure (for loadbearing walls).
†  Based on temperature rise (for nonloadbearing walls).

of over 250°F (121°C) constitutes failure of the wall as a fire barrier. Also, partial collapse of the wall or fire penetration through the wall is deemed to signify failure of integrity. The fire endurance (resistance) of the wall is determined by the time to reach the first of fire penetration through the wall, temperatures rise on the unexposed side, collapse, or termination of the test. The fire resistance rating is the fire endurance rounded down to the nearest whole hour.

Building code requirements for fire ratings typically vary from 1 to 4 hours depending on type of building and occupancy. For clay brick walls,[14.10] Table 14.1 contains design guidance in the United States for brick walls of various thicknesses with and without plaster on either side. In this table, "walls with combustible members framed in," refers to ends of wood joists, as shown in Fig. 14.3, and the fire

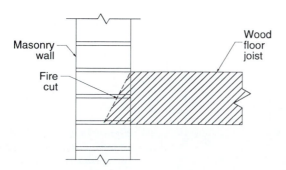

**Figure 14.3** Framing of combustible members into masonry wall.

resistances shown are less than the ultimate fire resistance periods as determined by standard tests.

For hollow masonry units, fire ratings are typically expressed in terms of equivalent wall thicknesses where the volume of the masonry unit is divided by the area of the exposed face to give the equivalent solid thickness. Table 14.2 contains minimum equivalent thicknesses for various North American fire resistance ratings of loadbearing concrete masonry. As can be seen, the aggregate used to manufacture the ma-

**Table 14.2** Minimum Equivalent Thickness, in. (mm), of Loadbearing Concrete
Masonry Unit Walls for Fire Resistance Ratings
(from Ref. 14.11)

| A. SBC**‡ | 4 h | 3 h | 2 h | 1 h |
|---|---|---|---|---|
| Pumice or expanded slag aggregates | 4.7 (119) | 4.0 (102) | 3.2 (81) | 2.1 (53) |
| Expanded shale, clay or slate aggregates | 5.1 (130) | 4.4 (112) | 3.6 (91) | 2.6 (66) |
| Limestone, cinders, or unexpanded slag aggregates | 5.9 (130) | 5.0 (127) | 4.0 (102) | 2.7 (69) |
| Calcareous gravel aggregates | 6.2 (157) | 5.3 (135) | 4.2 (107) | 2.8 (71) |
| Siliceous gravel aggregates | 6.7 (170) | 5.7 (145) | 4.5 (114) | 3.0 (76) |
| **B. UBC**‡ and BOCA/NBC†‡** | 4 h | 3 h | 2 h | 1 h |
| Pumice or expanded slag | 4.7 (119) | 4.0 (102) | 3.2 (81) | 2.1 (53) |
| Expanded clay, shale or slate | 5.1 (130) | 4.4 (118) | 3.6 (91) | 2.6 (66) |
| Limestone, cinders, or air-cooled slag | 5.9 (150) | 5.0 (127) | 4.0 (102) | 2.7 (69) |
| Calcareous or siliceous gravel | 6.2 (157) | 5.3 (135) | 4.2 (107) | 2.8 (71) |

**    Where all of the core spaces of hollow-core wall panels are filled with loose-fill material, such as expanded shale, clay or slag, or vermiculite or perlite, the fire-resistance rating of the wall is the same as that of a solid wall of the same aggregate type and of the same overall thickness.

†    Two fire-resistance rated walls composed of hollow concrete masonry units having a nominal thickness of 8 in. (200 mm) or greater are permitted to be classified as having 4 h of fire resistance when all the core spaces of the units are filled. Grout, insulation or dry granular materials as described in the footnote above are considered as acceptable fill.

‡    SBC: Southern Building Code, US. UBC: Uniform Building Code, US. NBC: National Building Code of Canada. BOCA: The BOCA National Building Code

sonry units has a significant effect on the fire resistance rating. For a particular fire resistance rating, a smaller equivalent thickness is required for hollow units made with lightweight aggregates.

The National Bureau of Standards, Technical Report BMS 92[14.12] and BIA TEK Note 16B[14.13] contain similar methods for estimating fire resistance R based on insulation criteria represented by the formula

$$R = (C_t V_t)^n \tag{14.1}$$

where, for $m$ layers of material

$$V_t = V_1 + V_2 + \ldots V_m$$

$$C_t = (C_1 V_1 + C_2 V_2 + \ldots C_m V_m)/V_t$$

where  $R$ = fire resistance period in hours as determined by heat transmission
$C$ = coefficient depending on type of material, wall design, and units of measurement of $R$ and $V$
$V$ = volume of solid material per unit area of wall surface
$n$ = an exponent, generally taken as 1.7.

Equation 14.1 can be modified to

$$R = (C_1 V_1 + C_2 V_2 + \ldots + C_m V_m)^n \tag{14.2}$$

From Eqs. 14.1 and 14.2, total resistance can be expressed as

$$R = (R_1^{1/n} + R_2^{1/n} + \ldots + R_m^{1/n})^n \tag{14.3}$$

where the subscripts relate to each layer of dissimilar material.

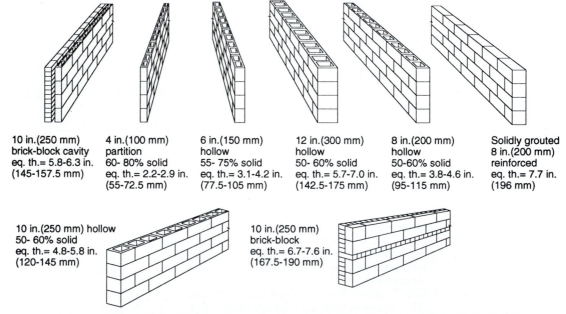

10 in.(250 mm) brick-block cavity eq. th.= 5.8-6.3 in. (145-157.5 mm)

4 in.(100 mm) partition 60- 80% solid eq. th.= 2.2-2.9 in. (55-72.5 mm)

6 in.(150 mm) hollow 55- 75% solid eq. th.= 3.1-4.2 in. (77.5-105 mm)

12 in.(300 mm) hollow 50- 60% solid eq. th.= 5.7-7.0 in. (142.5-175 mm)

8 in.(200 mm) hollow 50-60% solid eq. th.= 3.8-4.6 in. (95-115 mm)

Solidly grouted 8 in.(200 mm) reinforced eq. th.= 7.7 in. (196 mm)

10 in.(250 mm) hollow 50- 60% solid eq. th.= 4.8-5.8 in. (120-145 mm)

10 in.(250 mm) brick-block eq. th.= 6.7-7.6 in. (167.5-190 mm)

**Figure 14.4**   Wall sections showing equivalent thicknesses (from Ref. 14.15).

In the Australian code,[14.14] the effects of the air space in cavity walls can be neglected, allowing the insulation resistance to be calculated on the basis of the thickness of the two wythes. Alternately, the BIA[14.13] recommended inclusion of an additional term for each continuous air space, results in greater fire resistance ratings for such assemblies.

Figure 14.4 contains equivalent thicknesses for typical single wythe, composite, and cavity walls.

### 14.3.4 Methods for Increasing Fire Resistance

**Insulation.** For heat transmission controlled ratings, the fire resistance of a masonry wall can be increased by increasing the equivalent thickness of the wall. When hollow units are used, the cells can be filled with grout, which delays the temperature rise on the unexposed side of the wall. The Uniform Building Code[14.16] allows fully grouted walls to be equated to solid walls. Considering the grout as one layer in a multilayered wall, Eq. 14.3 can be used to calculate total wall resistance. For example, a lightweight 6 in. (150 mm) concrete block made with expanded clay aggregate can be filled with grout containing siliceous aggregate, to increase the fire resistance. If the grout occupies a volume equal to 42% of the gross volume, the equivalent thicknesses of the hollow masonry and the grout are $0.58 \times 5.62 = 3.26$ in. (83 mm) and $0.42 \times 5.62 = 2.36$ in. (60 mm), respectively. As an approximation of the resistance periods of each layer, linear interpolation of the information in Part A of Table 14.2 gives, $R_{block} = 106$ min. and linear extrapolation gives $R_{grout} = 36$ min. Then $R = [(106/60)^{1/1.7} + (36/60)^{1/1.7}]^{1.7} = 3.65$ h fire resistance period. Therefore, a 3 h rating is achieved.

Similar increases can be achieved by dry rodding the cell spaces full of loose expanded slag, shale, or clay aggregate or water repellent vermiculite masonry fill insulation. Equivalent thicknesses greater than the overall thickness of the wall are not normally allowed.

The fire resistance period for either hollow or solid masonry can be increased by the addition of exterior plaster or wallboard finishes. In North American codes,[14.16,14.17] this is incorporated directly by adding the listed endurance times assigned to various layers. For plaster or wallboard applied to the fire-exposed side of a wall, the contribution of the finish is limited by its ability to stay in place during the fire test, which is related to the method of attachment. Alternately, as shown in Table 14.3 for plaster finishes on the unexposed side of concrete masonry, multiplication factors are available to modify the thickness of the plaster to give the total equivalent wall thickness as

$$t_t = t_e + n t_p \tag{14.4}$$

where  $t_t$ = total equivalent wall thickness
$t_e$ = equivalent thickness of block wall
$t_p$ = actual thickness of the plaster
$n$ = multiplier in Table 14.3, which depends on the type of block and type of plaster.

Alternatively, the BIA[14.13] provides the *thickness coefficients,* pl, listed in Table 14.4

**Table 14.3** Multiplying Factor for Finishes on the Unexposed Side of the Wall (from Ref. 14.11)

| Type of Finish Applied to Wall | Type of Aggregate Used in Concrete Masonry | | | |
|---|---|---|---|---|
| | Siliceous or Calcareous Gravel | Limestone, Cinders, or Unexpanded Slag | Expanded Clay, Shale, or Slate | Pumice or Expanded Slag |
| Portland cement–sand plaster | 1.00 | 0.75* | 0.75* | 0.50* |
| Gypsum–sand plaster or gypsum wallboard | 1.25 | 1.00 | 1.00 | 1.00 |
| Gypsum-vermiculite or perlite plaster | 1.75 | 1.50 | 1.25 | 1.25 |

\* For portland cement–sand plaster 5/8 in. (16 mm) or less in thickness and applied directly to concrete masonry on the non fire-exposed side of the wall, the multiplying factor is 1.00.

to be added to Eq. 14.3 so that

$$R = \left( \sum_1^m (R_i)^{1/1.7} + \text{pl} \right)^{1.7} \tag{14.5}$$

By referring back to the previous discussion regarding the insulating effect of air space, the fire resistance equation can then be written as:

$$R = (R_1^{1/1.7} + R_2^{1/1.7} + \ldots + R_m^{1/1.7} + \text{pl} + as)^{1.7} \tag{14.6}$$

where *as* represents the resistance of a 0.5 to 3.5 in. (13-89 mm) wide air space and is estimated as 0.3 for resistances expressed in hours.[14.13]

**Structural Adequacy and Integrity.** In terms of design for structural capacity and integrity, fire resistance is improved by reducing the effective slenderness of the

**Table 14.4** Coefficients for Plaster, pl* (from Ref. 14.13)

| Thickness of Plaster, in. (mm) | One-Side | Two-Side |
|---|---|---|
| 1/2 (12.7) | 0.30 | 0.60 |
| 5/8 (16) | 0.37 | 0.75 |
| 3/4 (19) | 0.45 | 0.90 |

Values listed are for 1:3 sanded gypsum plaster.

wall thus increasing the buckling capacity and reducing secondary bending moments due to deflection. This can be achieved by increasing the thickness, by providing supports along the vertical edges of the wall, by providing pilasters, or by introducing other forms of redundancy.

### 14.3.5 Other Aspects of Fire Protection

If the floor or roof system has lower fire endurance, design of the connection so that it can fail without pulling down the wall is appropriate. Use of fire cut wood joists is a simple example of this practice. For protection of firefighters and neighboring buildings, it is important that the walls not collapse prematurely. Figure 14.5 is a not uncommon example where the steel roof system collapsed entirely, but the masonry walls remained intact.

A common form of fire protection is the use of fire walls to separate the building into fire resistive compartments, thus reducing the risk of fire sweeping through large areas. Although definitions and requirements differ somewhat for different building codes, fire walls are generally constructed of materials that do not contribute to the combustible content and that have restricted or protected openings. Although some codes in the United States permit lower ratings, masonry fire walls with up to a 4 h fire rating are easily incorporated into the structure and provide superior protection against property loss due to spread of fire. Fire walls should have sufficient structural stability under fire conditions to allow collapse of construction on either side without collapse of the fire wall itself.

Examples of single and double masonry fire walls are shown in Fig. 14.6. In these cases, the walls are not loadbearing and are detailed so that collapse on the fire side will not affect the integrity of the fire wall. For single walls, this can be achieved by the use of cantilever walls, walls that span horizontally between pilasters or protected columns, or combinations of these. The double fire wall is designed so that the building collapse on one side can pull over the wythe on that side, but the other wythe will still function to prevent the spread of fire. It is important that fire walls extend above the roof level as shown to prevent flame spread along the roof area.

The excellent fire resisting properties for masonry have resulted in their common specification for elevator shafts, stairwells and other exit routes, and for safe areas of refuge.

**Figure 14.5** Masonry walls stand intact after fire, whereas the steel frame and roof collapsed (*Courtesy of Canadian Concrete Masonry Producers Association*).

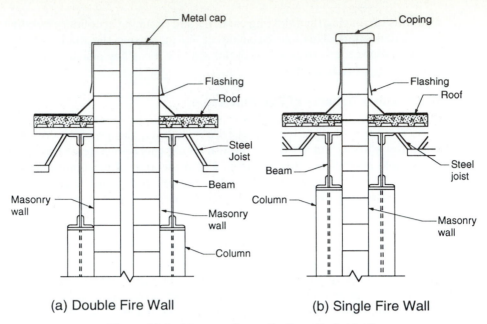

(a) Double Fire Wall          (b) Single Fire Wall

**Figure 14.6**   Masonry fire walls (from Ref. 14.18).

## 14.4 THERMAL PERFORMANCE

### 14.4.1 Introduction

Section 3.3.1 contains a brief introduction to some of the thermal factors affecting the design of masonry walls. For the thick wall construction used in the past, the modifying effects of the mass of masonry and the accumulated thermal resistance often provided a satisfactory separator in terms of maintaining a reasonably comfortable interior thermal environment. Even today, in locations where the effect of variation in the daily temperature can be modified by the mass of the masonry, or where the energy requirements for cooling or heating are modest, masonry walls without additional insulation can be used. For greater extremes in climate, the requirement for addition of insulation depends on economics (the initial cost versus savings in energy), the degree of comfort desired (avoidance of cold surfaces and the ability to humidify), and, finally, the impact of insulation on structural or other building science related performance requirements of the wall (avoidance of condensation and thermal stresses in the wall).

Even when climatic conditions justify the use of insulation materials, the mass of masonry walls can be used effectively for passive solar heating. Heat stored during daylight hours will be gradually released at night.

### 14.4.2 Introduction to Heat Transfer at Building Surfaces

A heat balance, such as illustrated in Fig. 14.7, exists at the surface of any opaque material. For a wall exposed to solar radiation, the temperature at the surface will be affected by the amount of *solar radiation* and the angle of incidence. Thus, clouds

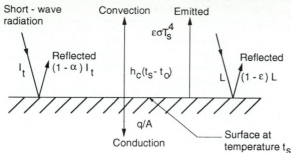

Long - wave radiation

**Figure 14.7**  Components of heat balance at an opaque surface.

can reduce the amount of radiation, and the time of year, time of day, and orientation of the wall surface affect the amount received on the wall surface. The heat balance equation is then[14.1,14.19]

$$q/A = \alpha I_t + h_c(t_o - t_s) + (\varepsilon L - \varepsilon \sigma T_s^4) \tag{14.7}$$

where

$q/A$ = heat transferred through the wall by conduction, Btu/ft² h (watts/m²)

$\alpha I_t$ = absorbed short-wave solar radiation as the product of the absorptivity for solar radiation, $\alpha$, and the short-wave radiation incident to the surface, $I_t$

$h_c(t_o - t_s)$ = heat transfer by convection, where $h_c$ is the surface coefficient for heat transfer by convection, and $(t_0 - t_s)$ is the temperature difference between the outside air and the surface of the wall

$(\varepsilon L - \varepsilon \sigma T_s^4)$ = net long-wave radiation with emissivity $\varepsilon$ of the surface applied to the long-wave radiation $L$ incident to the surface and the long-wave energy emitted from the surface as described by $\sigma T_s^4$

$\sigma$ = Stefan–Boltzmann constant for energy emission from a surface

$T_s$ = absolute temperature at the surface

The short-wave radiation absorbed depends on the *absorptivity* of the surface, which, as indicated in Table 14.5, is principally dependent on color. For example, nonmetallic black surfaces have an absorptivity of about 0.9 compared to 0.4 for a white surface. The radiation not absorbed is reflected. *Convection loss* is the transfer of heat between the surface of the material and the air. For moving air, this is dependent on the temperature difference between the outside air and the surface $(t_0 - t_s)$ and the *surface conductance,* which depends on the velocity of air movement. A rough surface can trap a thin layer of air along the surface and reduce the heat transfer due to convection. An additional heat loss not shown in the equation is evaporation loss when energy is used to evaporate moisture from the surface.

Net long-wave radiation gain is calculated from the absorbed long-wave radiation from other terrestrial objects minus the emitted long-wave radiation. The amounts of both the absorbed and emitted long-wave radiation depend upon the *emissivity* properties of the surface.

| Surface | Emissivity (Low–Temperature Radiation) | Absorptivity (Solar Radiation) |
|---|---|---|
| Small hole in an enclosure | 0.97 – 0.99 | 0.97 –0.99 |
| Black, nonmetallic surfaces | 0.90 – 0.98 | 0.85 – 0.98 |
| Red brick and tile, stone and concrete, rusted iron and dark paints | 0.85 – 0.95 | 0.65 – 0.80 |
| Yellow and buff building materials | 0.85 – 0.95 | 0.50 – 0.70 |
| White or light cream surfaces | 0.85 – 0.95 | 0.30 – 0.50 |
| Glass | 0.90 – 0.95 | Transparent (8% reflected) |
| Bright aluminum paint | 0.40 – 0.60 | 0.30 – 0.50 |
| Dull brass, copper, aluminum, polished iron | 0.20 – 0.30 | 0.40 – 0.65 |
| Polished brass, copper | 0.02 – 0.05 | 0.30 – 0.50 |
| Highly polished tin, aluminum, nickel, chrome | 0.02 – 0.04 | 0.10 – 0.40 |
| Aluminum | 0.05 | 0.2 |
| Asphalt | 0.95 | 0.9 |
| Brick (dark) | 0.9 | 0.6 |
| Paint: white | 0.9 | 0.3 |
| Paint: black | 0.9 | 0.9 |
| Slate | 0.9 | 0.9 |

Because of the differences in wave length distributions between solar radiation and radiation emitted by bodies at the earth's surface, as shown in Table 14.5, emissivity and absorptivity values can be very different. For example, a bright metallic surface can have an emittance value of about 0.04 with an absorptance of as much as 0.4 to solar radiation. As a result, such surfaces can become quite hot. The expression that you could "fry eggs on the sidewalk" describes this type of phenomenon, where surface temperatures of 100°F (56°C) above air temperature can occur. Conversely, in the absence of sunlight, energy loss usually exceeds energy gain and the surface temperature can be below the outside air temperature.

In calculations of heating or cooling loads as well as temperatures of the layers within the wall, it is necessary to account for the *heat-balance* effect described before. This is usually done by assigning a fictitious temperature, known as the *sol–air temperature* (S.A.T.), to the outside air. Chapters 26 and 27 of the 1985 ASHRAE *Handbook of Fundamentals*[14.20] provide the information to complete calculations for a variety of situations, including time of year, time of day, and orientation of the face of the wall. As a general guide, the maximum effective temperature, or S.A.T., for a vertical wall is usually in the range of 50 to 100°F (28 to 56°C) above the air temperature, whereas the minimum is in the range of about 10°F (6°C) colder than the ambient air temperature.[14.1]

### 14.4.3 Heat Transfer and Thermal Resistance

For steady-state heat flow in one direction, the rate of heat flow $q$ can be written as

$$q = A\frac{k}{l}(t_1 - t_2) \qquad \text{Btu/h (watts)} \qquad (14.8)$$

where

$A$ = surface area transverse to direction of flow (ft$^2$ or m$^2$)
$k$ = coefficient of thermal *conductivity* (Btu $\cdot$ in./ft$^2$ $\cdot$ °F $\cdot$ h or W/m$^2$ °C)
$l$ = length of flow path (or thickness of material) (in. or m)
$t_1 - t_2$ = temperature difference producing flow (°F or °C)

Two or three-dimensional heat flow, nonsteady-state conditions, nonuniform flow path, and the use of material with variable conductivity as a function of temperature are beyond the scope of the simplified treatment provided here.

For some materials with nonuniform cross-section (such as hollow masonry units) or where standard thicknesses are available (such as gypsum board), the term *thermal conductance, C,* is often applied to the specific layer of material ($C = k/l$). In these cases,

$$q = AC(t_1 - t_2) \qquad (14.9)$$

where $C$ is expressed in terms of Btu/ft$^2$ °F h (W/m$^2$ °C). An alternate convenient expression for heat flow is

$$q = \frac{A(t_1 - t_2)}{R} \qquad (14.10)$$

where $R$ is the *thermal resistance* equal to $1/C$ for the building material or combination of materials.

The total resistance of a building element to heat flow is the sum of the resistances of the individual layers, so that for layers 1 to $n$,

$$R_{\text{total}} = R_1 + R_2 \ldots + R_n \qquad (14.11)$$

Thus, the relative heat flow through composite sections can be evaluated using a term $U = 1/R_{\text{total}}$ to describe the overall *coefficient of heat transmission* (Btu/ft$^2$ °F h or W/m$^2$ °C).

It is also useful to be able to predict the temperature at various points through a section. Calculation of thermal movements and determination of the potential for condensation of water vapor are example uses of this information.

Table 14.6 is a partial listing of conductivity and/or conductance values for materials commonly associated with masonry construction. This information can be used to calculate the thermal profile through the wall where the temperature difference $\Delta t_i$ across a layer $i$ is proportional to the thermal resistance $R_i$ of that layer divided by the total thermal resistance $R_{\text{total}}$ of the wall

$$\Delta t_i = |t_e - t_i| \frac{R_i}{R_{\text{total}}} \tag{14.12}$$

where $|t_e - t_i|$ is the difference between the indoor air temperature $t_i$ and the effective outdoor temperature $t_e$. The latter term should account for the sol–air temperature effect and, therefore, may not be the outdoor air temperature.

As shown in Table 14.6, the air layers on both sides of the wall add to the overall resistance to heat flow through the wall. For the air layer on the outside of the wall, it is common practice to include its effect for the condition of either a 7.5 mph or 15 mph (12 km/h or 24 km/h) wind, which obviously significantly decreases the resistance compared to the still indoor air conditions. Air layers in the wall (i.e., air gap in cavity walls) also increase the overall thermal resistance of the wall, but the amount of resistance depends on the thickness of this space and the extent of air exchange with the exterior. Thicker air spaces permit increased convection thereby decreasing thermal resistance. Large weep holes and vents at the bottom and top of the exterior wythe may result in air exchange with the exterior, further reducing the effectiveness of this layer.

### 14.4.4 Thermal Resistance of Masonry Wall Systems

Table 14.7 contains the information required to calculate heat flow and thermal profiles for the very common wall cross-section shown in Fig. 14.8. For an indoor air temperature of 70°F (21°C), effective outdoor temperatures (S.A.T.'s) of 170°F (77°C) and 0°F (−18°C) are used as representative summer and winter conditions, respectively. The results are plotted in Fig. 14.8. As can be seen, the brick wythe will cycle through a large temperature range whereas use of an insulation layer on the outside of the concrete block wall significantly reduces the temperature variation in this element. As discussed in Chap. 12 and as illustrated in Chap. 15, provision of details to accommodate the total movement and the differential movement between these layers is an important part of the design. If the insulation is omitted, the temperature on the inside surface of the wall would range from 88°F (31°C) in summer to 57°F (14°C) in winter.

Typical thermal resistance values for brick veneer / wood stud walls and for single-wythe concrete block walls are provided in Fig.14.9(a). Filling the cells of hollow units with insulation and the use of lightweight units can significantly increase the thermal resistance of the masonry as indicated in Fig. 14.9(b). Figure 14.10 contains information on the thermal resistance effect of insulation in cavity walls.

For walls with nonuniform cross-sections, two- and three-dimensional analyses for multidirectional heat flow can be performed using electric or hydraulic analo-

**Table 14.6** Heat Transmission Coefficients of Building Materials (from Ref. 14.21)

| Materials Description | Density* (lb/ft³) | Conductivity, k, or Conductance, C** | | Resistance (R) 1/Btu/(ft²·°F·h) | |
|---|---|---|---|---|---|
| | | k Btu·in./ (ft²·°F·h) | C Btu/ (ft²·°F·h) | Per Inch (25 mm) Thickness 1/k | For Thickness Listed 1/C |
| **Masonry Units** | | | | | |
| Face brick | 130 | 9.00 | | 0.11 | |
| Common brick | 120 | 5.00 | | 0.20 | |
| Hollow brick: | | | | | |
| 4 in. (100 mm) (62.9% solid) | 81 | | 1.36 | | 0.74 |
| 6 in. (150 mm) (67.3% solid) | 86 | | 1.07 | | 0.93 |
| 8 in. (200 mm) (61.2% solid) | 78 | | 0.94 | | 1.06 |
| 10 in. (250 mm) (60.9% solid) | 78 | | 0.83 | | 1.20 |
| Hollow brick, vermiculite fill: | | | | | |
| 4 in. (100 mm) (62.9% solid) | 83 | | 0.91 | | 1.10 |
| 6 in. (150 mm) (67.3% solid) | 88 | | 0.66 | | 1.52 |
| 8 in. (200 mm) (61.2% solid) | 80 | | 0.52 | | 1.92 |
| 10 in. (250 mm) (60.9% solid) | 80 | | 0.42 | | 2.38 |
| Hollow concrete block (100 lb/ft³): | | | | | |
| 4 in. (100 mm) | 78 | | 0.71 | | 1.40 |
| 6 in. (150 mm) | 66 | | 0.65 | | 1.53 |
| 8 in. (200 mm) | 60 | | 0.57 | | 1.75 |
| 10 in. (250 mm) | 58 | | 0.51 | | 1.97 |
| 12 in. (300 mm) | 55 | | 0.47 | | 2.14 |
| Hollow concrete block (100 lb/ft³), vermiculite fill: | | | | | |
| 4 in. (100 mm) | 79 | | 0.43 | | 2.33 |
| 6 in. (150 mm) | 68 | | 0.27 | | 3.72 |
| 8 in. (200 mm) | 62 | | 0.21 | | 4.85 |
| 10 in. (250 mm) | 61 | | 0.17 | | 5.92 |
| 12 in. (300 mm) | 58 | | 0.15 | | 6.80 |
| Hollow concrete block (125 lb/ft³): | | | | | |
| 4 in. (100 mm) | 98 | | 0.93 | | 1.07 |
| 6 in. (150 mm) | 83 | | 0.83 | | 1.21 |
| 8 in. (200 mm) | 75 | | 0.74 | | 1.35 |
| 10 in. (250 mm) | 73 | | 0.69 | | 1.45 |
| 12 in. (300 mm) | 69 | | 0.65 | | 1.54 |
| **Building Board** | | | | | |
| 3/8 (10 mm) drywall (gypsum) | 50 | | 3.10 | | 0.32 |
| 1/2 in. (13 mm) drywall (gypsum) | 50 | | 2.25 | | 0.45 |
| Plywood | 34 | 0.80 | | 1.25 | |
| 1/2 in. (13 mm) fiberboard heathing | 18 | | 0.76 | | 1.32 |

TABLE 14.6 (cont.)

| | | | | | |
|---|---|---|---|---|---|
| **Siding** | | | | | |
| 7/16 in. (11 mm) hardboard | 40 | | 1.49 | | 0.67 |
| 1/2 × 8 in. (13 × 20 mm) wood bevel | 32 | | 1.23 | | 0.81 |
| Aluminum or steel over sheathing | – | | 1.61 | | 0.61 |
| **Insulating Materials** | | | | | |
| Batt or blanket: | | | | | |
| 2 to $2\frac{3}{4}$ in. (50–70 mm) | | | | | 7.00 |
| 3 to $3\frac{1}{2}$ in. (75–90 mm) | 1.2 | | | | 11.00 |
| $5\frac{1}{2}$ to $6\frac{1}{2}$ in. (140–165 mm) | | | | | 19.00 |
| Boards: | | | | | |
| Expanded polystyrene: | | | | | |
| Cut cell surface | 1.8 | 0.25 | | 4.00 | |
| Smooth skin surface | 1.8 | 0.20 | | 5.00 | |
| Expanded polyurethane | 1.5 | 0.16 | | 6.24 | |
| Polyisocyanurate | 2 | 0.14 | | 7.14 | |
| Loose fill: | | | | | |
| vermiculite | 4–6 | 0.44 | | 2.27 | |
| perlite | 5–8 | 0.37 | | 2.70 | |
| **Woods** | | | | | |
| Hard woods | 45 | 1.10 | | 0.91 | |
| Soft woods | 32 | 0.80 | | 1.25 | |
| **Metals** | | | | | |
| Steel | | 312.0 | | 0.003 | |
| Aluminum | | 1416.0 | | 0.0007 | |
| Cooper | | 2640.0 | | 0.0004 | |
| **Air Space** | | | | | |
| 3/4 to 4 in. (20 to 100 mm), winter | | | 1.03 | | 0.97 |
| 3/4 to 4 in. (20 to 100 mm), summer | | | 1.16 | | 0.86 |
| **Air Surfaces** | | | | | |
| Inside: still air | | | 1.47 | | 0.68 |
| Outside: | | | | | |
| 15 mph (24 km/h) wind, Winter | | | 6.00 | | 0.17 |
| 7.5 mph (12 km/h) wind, Summer | | | 4.00 | | 0.25 |

\*       To convert density in lb/ft$^3$ to kg/m$^3$, multiply by 16.02.

\*\*      To convert conductivity values k given in Btu in/ft$^2$°F h to W/m°C, multiply by 0.142.

†       To convert conductance values C given in Btu/ft$^2$°F h to W/m$^2$ °C, multiply by 5.678.

**Table 14.7** Tabulated Winter Temperature and Vapor Pressure Calculations for the Wall Section in Fig. 14.8

| | | Air Film | | Brick | | Cavity | | Insulation | | Block (115 lb/ft³) | | Air Film | | Total |
|---|---|---|---|---|---|---|---|---|---|---|---|---|---|---|
| Thickness, n, in. | | | | 3-5/8 | | 2 | | 2 | | 7-5/8 | | | | |
| Thermal conductance, c, Btu/ft² °F h* | | 6.00** | | 9.00 | | 1.03 | | | | 0.68 | | 1.47 | | |
| Thermal conductivity, k, Btu-in/ft² °F h† | | | | | | | | 0.25 | | | | | | |
| Thermal resistance, $R_t = 1/C$ or n/k | | 0.17 | | 0.40 | | 0.97 | | 8.00 | | 1.47 | | 0.68 | | 11.69 |
| Temperature drop, °F | | 1 | | 2 | | 6 | | 48 | | 9 | | 4 | | 70 |
| Temperature, °F | 0 | | 1 | | 3 | | 9 | | 57 | | 66 | | 70 | 70 |
| Saturated vapor pressure, $P_B$, in. -Hg‡ | 0.038 | | 0.038 | | 0.044 | | 0.060 | | 0.468 | | 0.644 | | 0.74 | |
| Permeance, M | | | | 0.8 | | | | | | 2.4 | | | | |
| Permeability, μ | | | | | | | | 1.6 | | | | | | |
| Vapor resistance, $R_v = 1/M$ or n/μ | | 0 | | 1.25 | | 0 | | 1.25 | | 0.42 | | 0 | | 2.92 |
| Vapor pressure drop for continuity, in. Hg‡ | | 0 | | 0.098 | | 0 | | 0.098 | | 0.033 | | 0 | | 0.229 |
| Vapor pressure for continuity, $P_c$, in. Hg‡ | 0.03§ | | 0.03 | | 0.128 | | 0.128 | | 0.226 | | 0.259 | | 0.259‖ | 0.259 |
| Actual vapor pressure, $P_a$, in. Hg | 0.03 | | 0.03 | | 0.044 | | 0.060 | | 0.206 | | 0.259 | | 0.259 | 0.259 |

\*   To convert thermal conductance values given in Btu/(ft²·°F·h) to W/(m²·°C), multiply by 5.678.
†   To convert thermal conductivity values given in Btu·in/(ft²·°F·h) to W/(m·°C), multiply by 0.1442.
‡   To convert vapour pressure values given in in·Hg to Pa, multiply by 3386.
\*\*   For 15 mph (24 km/hr) wind.
§   For 80% relative humidity of outside air.
‖   For 35% relative humidity of inside air.

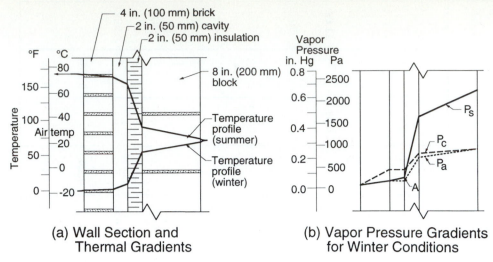

(a) Wall Section and Thermal Gradients

(b) Vapor Pressure Gradients for Winter Conditions

**Figure 14.8** Temperature and vapor pressure gradients for a cavity wall.

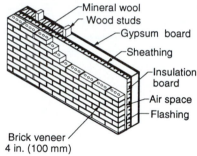

| Wall type | R* |
|---|---|
| No Insulation | 4.2 |
| 2 in. (50 mm)  Mineral wool insulation | 11.1 |
| 3 in. (75 mm)  Mineral wool insulation | 12.5 |
| 2 in. (50 mm)  Polyurethane board | 12.5 |
| 2 in. (50 mm)  Polyurethane board | 16.6 |
| 3 in. (75 mm)  Polyurethane board | 16.6 |
| 3 in. (75 mm)  Polyurethane board | 20.0 |

*R = 1/C - To convert conductance values, C, from Btu/ (hr. °F. ft²) to W/m² °K multiply by 5.678

(a)  Brick Veneer/ Wood Stud Walls (From ref. 14.21)

| Nominal wall thickness in. (mm) | R* value based on concrete unit weight, lb/ft³ (kg/m³) | | | | | |
|---|---|---|---|---|---|---|
| | Insulation cells | 60 (960) | 80 (1280) | 100 (1600) | 120 (1920) | 140 (2240) |
| 4 (100) | Filled | 3.36 | 2.79 | 2.33 | 1.92 | 1.14 |
| | Empty | 2.07 | 1.68 | 1.40 | 1.17 | 0.77 |
| 6 (150) | Filled | 5.59 | 4.59 | 3.72 | 2.95 | 1.59 |
| | Empty | 2.25 | 1.83 | 1.53 | 1.29 | 0.86 |
| 8 (200) | Filled | 7.46 | 6.06 | 4.85 | 3.79 | 1.98 |
| | Empty | 2.30 | 2.12 | 1.75 | 1.46 | 0.98 |
| 10 (250) | Filled | 9.35 | 7.45 | 5.92 | 4.59 | 2.35 |
| | Empty | 3.00 | 2.40 | 1.97 | 1.63 | 1.08 |
| 12 (300) | Filled | 10.98 | 8.70 | 6.80 | 5.18 | 2.59 |
| | Empty | 3.29 | 2.62 | 2.14 | 1.81 | 1.16 |

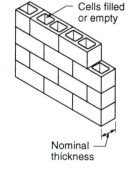

*R = 1/C - To convert conductance values, C, from Btu/ (ft²h °F) to W/m² °K multiply by 5.678

(b)  Concrete Masonry Walls (From ref. 14.22)

**Figure 14.9** Thermal resistance of masonry walls.

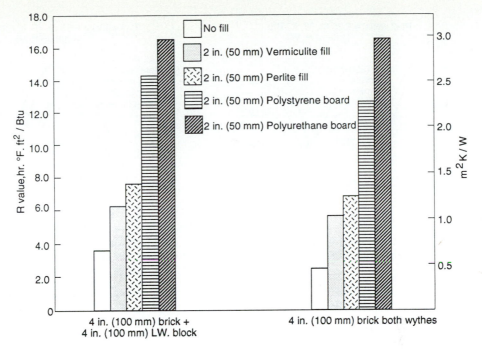

**Figure 14.10** Influence of various insulation methods in the cavity of 10 in. (250 mm) thick cavity walls (from Ref. 14.21).

gies. Computer programs to perform such calculations are relatively easily written and commercial versions are available.[14.23] As described in the next section, for more accurate predictions of thermal performance, it may be necessary to perform dynamic analyses to include variation in outdoor air temperature and radiation and the effects of *thermal inertia* relating to the mass of the construction.

For human occupancy, requirements for air changes and the inclusion of windows and other elements with low thermal resistances means that there is some rational limit to the benefits of increasing the amount of insulation. For example, doubling the original insulation from an $R$ value of, say, 10/Btu/ft$^2$ °F · h to a value of 20 halves the amount of energy lost through the wall. However, doubling this again to $R = 40$/Btu/ft$^2$ °F h only results in savings of a further one-fourth of the original heat loss through the wall. Because the heat loss through the walls only accounts for a part of the heating energy costs, it is fairly easy to determine the optimum amount of insulation given the relative costs of adding insulation versus the difference in energy costs. Some energy codes allow trade-offs of heat loss through different components as long as the total heat loss does not exceed the maximum specified value.

### 14.4.5 Thermal Inertia

The effect of mass on slowing the thermal response of building elements by absorption and retention of heat is a well documented fact of masonry construction.[14.24–14.27] The time lag in thermal response is due to the heat storage capacity of the massive

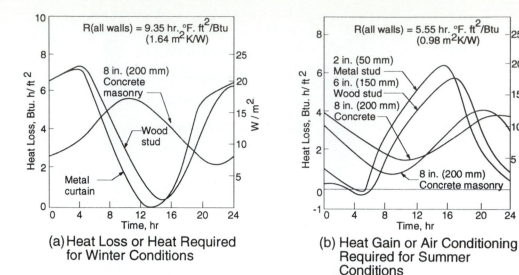

**Figure 14.11** Comparisons of heat loss and heat gain for buildings with walls of different masses (from Ref. 14.26).

element. This phenomenon is referred to as *thermal inertia* or *thermal storage capacity*.[14.24]

The effects of mass on heat loss and heat gain of different wall elements having the same $U$ or $R$ values (calculated using Eq. 14.11) are illustrated in Fig. 14.11. The results shown are for dynamic analyses to predict peak heating or cooling loads for identical climatic conditions. As can be seen in Fig. 14.11(a), the peak heating load for the more massive masonry walls is about 20% less than for the lighter walls. The actual heat loss for the masonry walls is 18% lower than predicted by steady-state calculations. In Fig. 14.11(b), the differences for heat gain and peak air conditioning loads are more dramatic where the peak requirements for the masonry walls are between about 28 to 40% less than for the lighter wall systems.

Figure 14.12 is a plot of data from a dynamic analysis[14.27] to illustrate the effect of mass on the temperature of the inside surface of the wall for summer conditions but without the effects of solar radiation. As can be seen, the mass of the masonry wall resulted in a more uniform and, therefore, more comfortable interior temperature.

The effect of the mass of masonry walls on thermal response is considerable and should be accounted for in the design loads for both heating and cooling. Appropriate allowances can be introduced using one of the following two approaches:

1. Perform a dynamic thermal analysis using commercially available computer programs such as those developed by ASHRAE and the NBS.[14.23]
2. Modify the steady-state formula to account for mass using the $M$ factor. Application of this $M$ factor to account for the effect of thermal inertia results in equivalent $U$ or $R$ values equal to $U_s M$ or $R_s/M$ where $U_s$ and $R_s$ are values calculated from steady-state formulas. The value of the $M$ factor depends on the density of the material and the annual heating degree days. Typical values

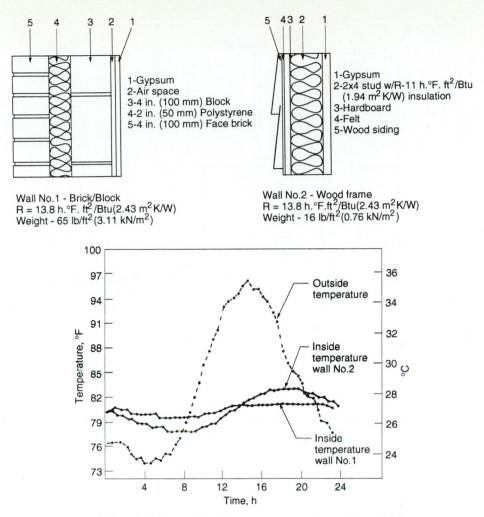

Wall No.1 - Brick/Block
R = 13.8 h.°F. ft$^2$/Btu(2.43 m$^2$K/W)
Weight - 65 lb/ft$^2$(3.11 kN/m$^2$)

1-Gypsum
2-Air space
3-4 in. (100 mm) Block
4-2 in. (50 mm) Polystyrene
5-4 in. (100 mm) Face brick

Wall No.2 - Wood frame
R = 13.8 h.°F.ft$^2$/Btu(2.43 m$^2$K/W)
Weight - 16 lb/ft$^2$(0.76 kN/m$^2$)

1-Gypsum
2-2x4 stud w/R-11 h.°F. ft$^2$/Btu
(1.94 m$^2$K/W) insulation
3-Hardboard
4-Felt
5-Wood siding

**Figure 14.12** Variation of inside temperature for walls of different masses (from Ref. 14.27).

range from 0.6 to 1.0.[14.25] This is a simple and easy-to-use method that enables the designer to directly account for the effect of wall mass. For details of the application of the *M* factor to masonry walls, refer to BIA Technical Note 4B.[14.25]

## 14.4.6 Other Considerations

**Thermal Bridging.** Thermal bridging through the main insulating layer in any wall can substantially increase the amount of heat transfer compared to calculations that ignore this effect. To the extent possible, thermal bridging should be minimized by appropriate design and detailing. Because floor slabs are a major region for heat transfer, it is important that the edge of the slab be insulated, as shown in Fig. 14.13. In the case shown, the shelf angle is still a source of heat loss, but the

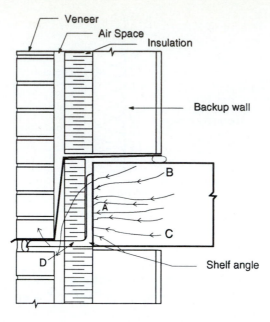

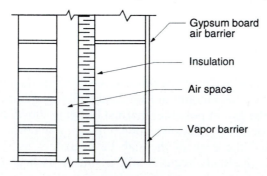

**Figure 14.13** Insulating the edge of floor slabs to minimize thermal bridging (from Ref. 14.6).

temperature variation in the floor itself would be relatively small. For winter conditions, temperatures at A, B, and C remain warm because the steel angle's heat pickup area A is large and conductive while its heat releasing area D is relatively restricted.

Ties and other elements that penetrate through the insulating layer are other sources of heat loss. In addition to increased costs for energy use, such thermal bridging can result in localized cold areas in the wall and the potential for condensation in these areas (see Sec. 14.5). For heat loss calculations, information in BIA Technical Note 4 (Revised)[14.21] can be used to modify steady-state calculations to account for thermal bridging effects.

**Effectiveness of Insulation.** The design decision regarding the location of the insulation should include practical as well as theoretical considerations. The insulation design should be an integral part of the design to control condensation due to air flow and water vapor transmission (see Sec. 14.5). All other things being equal, it is best to locate the insulation as close to the exterior of the building as practical. For walls with an air space, the insulation should be placed on the outside of the inside wythe or backup wall, as shown in Fig. 14.14, rather than on the interior

**Figure 14.14** Location of insulation in double wythe walls.

face of the wall. This location minimizes the thermal fluctuations in the inside wythe and limits the amount of thermal bridging. It has the disadvantage of being more difficult to inspect and, because of weather conditions and interference from ties and other components, attachment and proper fitting of the insulation may be difficult. This location also has the very large handicap of being essentially inaccessible for repair or maintenance.

For insulation to be fully effective, air must be prevented from circulating through and around the insulation and, for this reason, some porous types of insulation have a "breathing" membrane attached to the exterior side. However, the potential of such sheets to form a vapor trap for condensation from exfiltrating moist air must be considered. Insulation must also fit tightly to the surface of the wall to prevent air from circulating behind the insulation. Insulation held in place by daubs of adhesive, such as shown in Fig. 14.15(a), is likely to have some gaps between it and the wall and is not recommended. A grid of adhesive, as shown in Fig. 14.15(b), will reduce air circulation by comparting any gaps that exist, but a full bed of adhesive, as shown in Fig. 14.15(c), is the best method of application.

Mechanical anchorage can be used both to hold the insulation layer in place and, if the insulation is sufficiently compressible, force the insulation to conform and fit tightly to slightly irregular wall surfaces. It is also important that insulation be anchored sufficiently to the backup wall to resist differential air pressure.

To avoid air gaps, such as shown in Fig. 14.16, insulation between furring strips must be cut to fit snugly between these strips without folds or compressed areas.

In certain cases, insulation can be placed on the inside of the wall. This is unavoidable in single-wythe wall construction, where the masonry is also the exterior architectural finish. When both wythes of a veneer or cavity wall are constructed at the same time to avoid high scaffold costs on high-rise buildings, installation of insulation in the cavity may be difficult. In such cases, the disadvantages of extra energy consumption and temperature variation in the floors near the walls, due to placing insulation on the interior, may be outweighed by the savings in construction costs. Locating the insulation on the inside of the wall typically allows for more complete inspection, and installation conditions are generally conducive to better workmanship. However, the potential for problems (e.g., condensation) arising from such a design decision should be carefully considered.

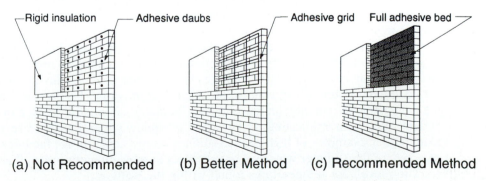

(a) Not Recommended    (b) Better Method    (c) Recommended Method

**Figure 14.15**   Application of rigid insulation with adhesives.

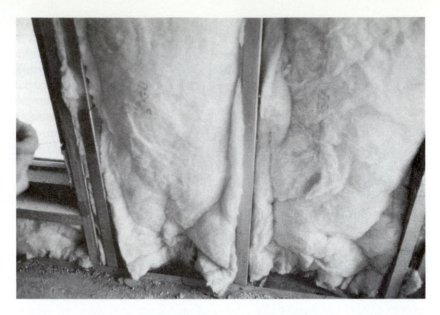

**Figure 14.16**  Compressed insulation.

## 14.5 CONDENSATION CONSIDERATIONS

### 14.5.1 Introduction

Moisture movement through a wall system results in condensation if the vapor pressure in the wall reaches the saturation or *dew point*. The accumulation of water as *concealed condensation* can affect the performance of some materials (e.g., wetting changes the insulating value of insulation). It can have dramatic effects on durability (e.g., freeze–thaw damage, corrosion, rotting, softening of moisture-sensitive materials), requires maintenance (e.g., straining and blistering), and results in growth of mold and fungi, some of which may create health problems in the indoor air.[14.28] The mechanisms for condensation of water are by water vapor diffusion through the wall section and by moist air passing through the wall. Limitation of condensation from these sources is the topic of this section.

Condensation can be either a summer or winter phenomenon, depending on the climatic conditions and the type of building. In cold climates, humidification of warm inside air in buildings in winter results in vapor pressure levels, which will lead to condensation if air exfiltration occurs or if vapor barriers are not adequate. Conversely, in hot humid climates, summer air conditioning can lead to condensation from infiltrating air or when adequate vapor barriers are not provided. Other examples of high condensation risks are cold storage facilities and ice rinks, where the summer water vapor pressures in most climates are well above the saturation levels for temperatures found within the walls.

## 14.5.2 Water Vapor Condensation

*Relative humidity* (RH) is the common term used to describe the amount of water vapor present in air. It is expressed as a percentage of the amount of water in saturated air at that temperature. However, for condensation considerations, a more directly useful measure of water content of air is the *vapor pressure*. Vapor pressure is independent of other gases in the air and, therefore, even when the air pressures on either side of a membrane are equal, any vapor pressure differential tends to move water vapor from the higher pressure zone to the lower pressure zone (an example of Dalton's Law of Partial Pressures).

Air that cannot hold more water vapor is said to be saturated with the corresponding vapor pressure defined as the *saturated vapor pressure* (SVP). Because the weight of water vapor that air can hold increases with increased temperature, the SVP also increases, as shown in Table 14.8. Therefore, the vapor pressure for 80% RH at 0°F (18°C) of 0.8(0.0185 psi) = 0.015 psi (103 Pa) is considerably lower than for air at 40% RH and 70°F (21°C) which is 0.4(0.363 psi) = 0.145 psi (1000 Pa). These conditions might represent winter conditions and indoor climate in a temperate climate with a vapor pressure difference of 0.130 psi (897 Pa). Alternatively, for a hot humid day of 90°F (32°C) and 70% RH, where air conditioning provides an interior environment of 70°F (21°C) and 50% RH, the vapor pressure difference would be 0.7 (0.698) − 0.5 (0.363) = 0.307 psi (2120 Pa). These two conditions illustrate the potential for vapor flow through a wall system from the interior to the exterior in winter and from the exterior to the interior in summer.

Except for the minor extra demands for humidification or dehumidification, the main concern regarding vapor flow through a wall system relates to the potential for condensation of some of that moisture at a plane where the dew point is reached. Use of a simplified psychometric chart, such as shown in Fig. 14.17[14.20] is a convenient method of determining the dew point temperature or saturation temperature for the existing vapor pressure. From the foregoing example, after locating the point corresponding to 70°F (21°C) (on the horizontal axis) and 40% RH (curved lines sloping upward to the right), a horizontal line drawn to the left intersects the saturation curve at 45°F (7.2°C), which is the dew point. If the interior air is cooled to below this temperature, condensation will occur on surfaces below that temperature.

Condensation within a wall can reduce the effectiveness of some materials such as batt insulation. Moisture will reduce its thermal resistance and the extra weight of water can cause it to sag, leaving uninsulated zones. Also, this moisture can lead to deterioration of materials (e.g., corrosion of steel, decay of wood, softening of gypsum board) and possibly lead to loosening of interior finishes. Movement of this moisture can produce internal staining, damp areas on the interior, and possibly cause efflorescence on the exterior face of a masonry wall.

Icicles, forming on walls, such as shown in Fig. 14.18(a), or the buildup of ice within the walls can lead to dangerous conditions. Moist air, exfiltrating near the top of the wall, can cause ice lensing and lift-off of the top course of the outer wythe of masonry, as shown in Fig. 14.18(b).

Design to restrict the amount of condensation is very important, particularly in cold climates.

Table 14.8 Thermodynamic Properties of Water at Saturation

| Temp., t (°F)* | Absolute Pressure $p_s \times 10^{-2}$ | | Temp., t (°F)* | Absolute Pressure $p_s$ | | Temp., t (°F)* | Absolute Pressure $p_s$ | |
|---|---|---|---|---|---|---|---|---|
| | psi** | in. Hg | | psi** | in. Hg | | psi** | in. Hg |
| −30 | 0.344 | 0.700 | 32 | 0.08859 | 0.18036 | 92 | 0.7434 | 1.5136 |
| −26 | 0.436 | 0.888 | 34 | 0.09600 | 0.19546 | 94 | 0.7909 | 1.6103 |
| −24 | 0.491 | 0.999 | 36 | 0.10396 | 0.21166 | 96 | 0.8410 | 1.7124 |
| −22 | 0.551 | 1.122 | 38 | 0.11249 | 0.22904 | 98 | 0.8937 | 1.8200 |
| −20 | 0.618 | 1.259 | 40 | 0.12164 | 0.24767 | 100 | 0.9496 | 1.9334 |
| −18 | 0.693 | 1.410 | 42 | 0.13145 | 0.26763 | 102 | 1.0083 | 2.0529 |
| −16 | 0.776 | 1.579 | 44 | 0.14191 | 0.28899 | 104 | 1.0700 | 2.1786 |
| −14 | 0.868 | 1.766 | 46 | 0.15317 | 0.31185 | 106 | 1.1351 | 2.3110 |
| −12 | 0.969 | 1.974 | 48 | 0.16517 | 0.33629 | 108 | 1.2035 | 2.4503 |
| −10 | 1.082 | 2.203 | 50 | 0.17799 | 0.36240 | 110 | 1.2754 | 2.5968 |
| −8 | 1.207 | 2.457 | 52 | 0.19169 | 0.39028 | 112 | 1.3510 | 2.7507 |
| −6 | 1.344 | 2.737 | 54 | 0.20630 | 0.42003 | 114 | 1.4305 | 2.9125 |
| −4 | 1.496 | 3.047 | 56 | 0.22188 | 0.45176 | 116 | 1.5139 | 3.0823 |
| −2 | 1.664 | 3.388 | 58 | 0.23819 | 0.48558 | 118 | 1.6014 | 3.2606 |
| 0 | 1.849 | 3.764 | 60 | 0.25618 | 0.52160 | 120 | 1.6933 | 3.4477 |
| 2 | 2.052 | 4.178 | 62 | 0.27502 | 0.55994 | 122 | 1.7897 | 3.6439 |
| 4 | 2.276 | 4.633 | 64 | 0.29505 | 0.60073 | 124 | 1.8907 | 3.8196 |
| 6 | 2.521 | 5.134 | 66 | 0.31636 | 0.64411 | 126 | 1.9966 | 4.0651 |
| 8 | 2.791 | 5.683 | 68 | 0.33900 | 0.69021 | 128 | 2.1075 | 4.2910 |
| 10 | 3.087 | 6.286 | 70 | 0.36304 | 0.73916 | 130 | 2.2237 | 4.5274 |
| 12 | 3.412 | 6.946 | 72 | 0.38856 | 0.79113 | 132 | 2.3452 | 4.7750 |
| 14 | 3.767 | 7.669 | 74 | 0.41564 | 0.84626 | 134 | 2.4725 | 5.0340 |
| 16 | 4.156 | 8.461 | 76 | 0.44435 | 0.90472 | 136 | 2.6055 | 5.3049 |
| 18 | 4.581 | 9.326 | 78 | 0.47478 | 0.96668 | 138 | 2.7446 | 5.5881 |
| 20 | 5.045 | 10.27 | 80 | 0.50701 | 1.0323 | 140 | 2.8900 | 5.8842 |
| 22 | 5.552 | 11.30 | 82 | 0.54112 | 1.1017 | 142 | 3.0419 | 6.1934 |
| 24 | 6.105 | 12.43 | 84 | 0.57722 | 1.1752 | 144 | 3.2006 | 6.5164 |
| 26 | 6.708 | 13.66 | 86 | 0.61510 | 1.2530 | 146 | 3.3662 | 6.8536 |
| 28 | 7.365 | 15.00 | 88 | 0.65575 | 1.3351 | 148 | 3.5390 | 7.2056 |
| 30 | 8.080 | 16.45 | 90 | 0.69838 | 1.4219 | 150 | 3.7194 | 7.5727 |

\*     °C = (°F − 32) 5/9.

\*\*    1 psi = 6900 Pa.

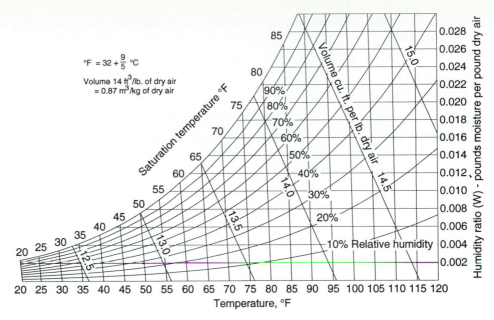

**Figure 14.17** Simplified psychrometric chart.

a) Icicles on wall

b) Ice lensing causing lifting of top course.

**Figure 14.18** Problems caused by formation of ice from exfiltrating moist air.

### 14.5.3 Diffusion of Water Vapor and Potential for Condensation

The various layers of material in a wall section have characteristic resistances to transmission of water vapor. The *water vapor transmission coefficient* for a material is expressed as a *permeance M* in units of perm (1 perm = 1 grain of water passing through 1 ft² of wall in 1 hour under a vapor pressure differential of 1 inch of mercury) or in units of nanograms per second per square meter for 1 pascal vapor pressure difference (ng/s-m²-Pa). These values apply to specific thicknesses of material, whereas the more general measure of *permeability* $\mu$ is expressed in terms of a standard reference thickness of 1 in. (1m) (perm-in. or ng/Pa-s-m).

The resistance to vapor flow through a layer of material is

$$R = \frac{1}{M} = \frac{l}{\mu} \tag{14.13}$$

where $l$ is the thickness of the layer.

Therefore, the total resistance of a wall assembly consisting of $n$ layers of material is

$$R_{\text{total}} = R_1 + R_2 + R_3 + \ldots + R_n \tag{14.14}$$

and for steady-state conditions and uniform wall sections, the vapor flow can be calculated as

$$W = A \, (\Delta t) \, \Delta P / R_{\text{total}} \tag{14.15}$$

where  $W$ = mass of water vapor transmitted, grains (ng)
   $A$ = area of cross-section of flow path, ft$^2$ (m$^2$)
   $\Delta t$ = time interval, hours (seconds)
   $\Delta P$ = vapor pressure difference, in. of mercury (Pa)

The foregoing equation also assumes that there is no storage of moisture in the wall assembly. Table 14.9 provides representative permeance or permeability values for materials typically associated with masonry construction. Higher values indicate that more water vapor will move through the material.

For diffusion of water vapor through a wall, the vapor pressure decreases in proportion to the resistance to vapor flow thereby resulting in a *vapor pressure gradient*. For the cavity wall shown in Fig. 14.8(a), the vapor pressure gradient $P_c$ is plotted in Fig. 14.8(b) for winter conditions. The values of the saturated vapor pressure $P_s$ corresponding to the temperatures at each layer are also plotted. Table 14.7 contains the values relevant to these calculations. Condensation will occur wherever the vapor pressure is greater than the saturated vapor pressure. When condensation occurs, the amount of condensation can be calculated by using the saturated vapor pressure at that plane and independently calculating the amount of water transmitted to the plane and the amount transmitted away from that plane. The difference provides an estimate of the accumulation of condensation. Such simplified calculations are estimates because they ignore the influence of latent heat released by the water vapor and the influence that wetting can have on the thermal profile through the wall.

By using the example wall section and environmental conditions previously studied in Sec. 14.4 for thermal effects, the potential for condensation under winter conditions can be assessed by comparing plots for vapor pressure versus saturated vapor pressure. As can be seen, despite the reasonable vapor resistance of the insulation, the vapor pressure for continuity of flow, $P_c$, is greater than the saturated vapor pressure, $P_s$, near the exterior of the insulation. Therefore, condensation occurs at this location and the actual pressure gradient, $P_a$, shown at the bottom of Table 14.7, must be calculated using the saturated vapor pressure at the point of condensation.

As an indication of the potential accumulation of moisture, the calculated flow to the exterior of the insulation at point A is obtained from Eq. 14.15 where

$$\frac{\Delta P}{\sum R} = \frac{0.259 - 0.060}{1.25 + 0.42} = 0.119 \text{ grains/ft}^2/\text{hr} \ (22,800 \text{ ng/s} \cdot \text{m}^2)$$

whereas, for saturated air in the cavity, the flow out through the brick is

$$= \frac{0.044 - 0.03}{1.25} = 0.011 \text{ grains/ft}^2/\text{h} \ (2200 \text{ ng/s} \cdot \text{m}^2)$$

**Table 14.9** Water Vapor Transmission Coefficients

| Material | Permeance | | | | | |
|---|---|---|---|---|---|---|
| | Dry Cup 50–0% | | Wet Cup 100–50% | | Inverted Wet Cup | |
| | grains/ (h ft²×in. Hg) | ng/Pa-s-m² | grains/ (h ft²×in. Hg) | ng/Pa-s-m² | grains/ (h ft²×in. Hg) | ng/Pa-s-m² |
| Foamed polyurethane insulation 1 in. (25 mm): | | | | | | |
| 1.75 lb/ft³ (28 kg/m³) | 1.30 | 75 | 1.31 | 75 | – | – |
| 1.93 lb/ft³ (31 kg/m³) | 1.10 | 63 | 1.10 | 63 | – | – |
| Foamed polystyrene insulation, 1 in. (25 mm): | | | | | | |
| Extruded, 1.81 lb/ft³ (29 kg/m³) | 1.60 | 92 | 1.61 | 92 | – | – |
| Extruded, 2.19 lb/ft³ (35 kg/m³) | 0.77 | 44 | 0.74 | 42 | – | – |
| Polyethylene film | | | | | | |
| (0.02 in. 0.5 mm) | 0.16 | 9 | 0.14 | 8 | – | – |
| 0.004 in. (0.10 mm) | 0.09 | 5 | 0.07 | 4 | – | – |
| 0.006 in. (0.15 mm) | 0.05 | 3 | 0.04 | 2 | – | – |
| Nylon film, 0.001 in. (0.025 mm) | 0.68 | 39 | 0.70 | 40 | – | – |
| Vinyl film, 0.002 in. (0.05 mm) | 0.33 | 19 | 0.33 | 19 | – | – |
| Cellulose acetate film, 0.001 in. (0.25 mm) | 4.73 | 270 | 11.20 | 640 | – | – |
| Waxed building paper: | | | | | | |
| medium weight | 0.09 | 5 | 0.16 | 9 | – | – |
| heavyweight | 0.11 | 6 | 0.89 | 51 | – | – |
| Asphalt-saturated sheathing paper | | | | | | |
| 0.16 lb/ft² (0.75 kg/m²) | 4.73 | 270 | 8.40 | 480 | 12.69 | 725 |
| 0.26 lb/ft² (1.25 kg/m²) | 3.33 | 190 | 6.48 | 370 | – | – |
| heavy weight | 0.82 | 47 | 6.30 | 360 | 8.75 | 500 |
| Asphalt-saturated roofing felt 0.75 kg/m³ (0.05 lb/ft³) | 1.93 | 110 | 11.90 | 680 | 15.92 | 910 |

**Table 14.9** (cont.)

| Material | | | | | | |
|---|---|---|---|---|---|---|
| Tar-infused sheathing paper | 6.56 | 375 | 30.98 | 1770 | 70.88 | 4050 |
| Asphalt-infused sheathing paper | 6.39 | 365 | 18.90 | 1080 | 42.00 | 2400 |
| Asphalt-coated building paper | 0.82 | 47 | 1.10 | 63 | 2.01 | 115 |
| Perforated asphalt-coated sheathing paper | 11.02 | 630 | 14.00 | 800 | 15.05 | 860 |
| Structural clay tile, 0.25 in. (6 mm) | – | – | 11.55 | 660 | – | – |
| Vitreous ceramic tile, 0.35 in. (9 mm) | 0.01 | 0.6 | 0.40 | 23 | – | – |
| Fiberboard, untreated, 0.5 in. (12.5 mm) | 43.22 | 2470 | 44.10 | 2520 | – | – |
| Fiberboard, sheathing grade, 0.5 in. (12.5 mm) | 30.10 | 1720 | 31.15 | 1780 | – | – |
| Asbestos cement board | 4.99 | 285 | 8.40 | 480 | – | – |
| Brick masonry, 4 in. (100 mm) | – | – | – | – | 0.81 | 46 |
| Concrete block, 8 in. (200 mm), cored limestone aggregate | – | – | – | – | 2.42 | 138 |
| Tile masonry, glazed, 4 in. (100 mm) | – | – | – | – | 0.12 | 7 |
| Asbestos cement board, 0.2 in. (5 mm) | 0.54 | 31 | – | – | – | – |
| Plaster on wood lath | – | – | 11.03 | 630 | – | – |
| Plaster on plain gypsum lath on studs | – | – | – | – | 20.13 | 1150 |
| Gypsum wallboard, 3/8 in. (9.5 mm), plain | – | – | – | – | 50.22 | 2870 |
| Hardboard, 0.1 in. (3 mm), tempered | – | – | – | – | 5.08 | 290 |
| Plywood, douglas fir, exterior glue, 1/4 in. (6.5 mm) | – | – | – | – | 0.70 | 40 |

**Table 14.9** (cont.)

| | | | | | | |
|---|---|---|---|---|---|---|
| Enamels, 2 coats on smooth plaster | | | – | – | 0.51-1.51 | 29-86 |
| Primers, sealers, 2 coats on insulation board | | | – | – | 0.91-2.10 | 52-120 |
| Various primers, 2 coats + 1 coat flat oil paint on plaster | | | – | – | 1.61-3.01 | 92-172 |
| Flat paint, 2 coats on insulation board | | | – | – | 4.03 | 230 |
| Water emulsion, 2 coats on insulation board | | | – | – | 30.1 to 85.75 | 1720 to 4900 |
| Exterior paint, 3 coats white lead and oil on wood siding | 0.3-1.0 | 17-57 | – | – | | – |
| Styrene butadiene latex coating, 0.13 $lb/ft^2$ (0.62 $kg/m^2$) | 11.03 | 630 | – | – | | – |
| Polyvinyl actetate latex coating, 0.26 $lb/ft^2$ (1.25 $kg/m^2$) | 5.60 | 320 | – | – | | – |

giving a condensation rate of $0.119 - 0.011 = 0.108$ grains/ft$^2$/h (20,600 ng/s · m$^2$), which for a 30 day time period produces 0.011 lb of water/ft$^2$ [1 lb = 7000 grains], which is equivalent to a film of water 0.002 in. (0.05 mm) thick. To avoid condensation for the specified conditions, it would be necessary to add a layer of material on the warm side of the insulation that would have a much higher resistance to water vapor transmission. In this case, a permeance of 0.09 perm or lower would be required.

In some cases, the simple step of painting the interior of the wall may be all that is required. Some types of thermal insulation also have low permeability. In other situations, a specific sheet of relatively impermeable material is required to reduce the amount of vapor flow.

The need for an identified layer of material to control vapor transmission has led to such layers being referred to as vapor barriers. An alternate term used in some building codes is vapor retarders, emphasizing the fact that most of these layers do allow some vapor transmission. The vapor barrier must be located on the warm side of the insulation to prevent condensation. At this location, the saturated vapor pressure will be well above the actual vapor pressure, whereas at colder locations, the two are much closer.

### 14.5.4 Condensation from Airborne Moisture

If air flows through a wall, condensation of airborne moisture can occur within the wall at the point where the air temperature reaches the dew point, as shown in Fig. 14.19. The amount of condensation that will occur depends on the rate of air leakage. For example, air at 70°F (21°C) and 40% RH contains 0.0062 pounds of water per pound of dry air (obtained from Fig. 14.17) becomes saturated when cooled to 20°F ($-6.6$°C) and at 100% RH contains 0.0024 pounds of water per pound of dry air (from Fig. 14.17). Therefore, the amount of condensation per unit volume of air is $(0.0062 - 0.0024)(0.075) = 0.0003$ lb/ft$^3$ (0.005 kg/m$^3$) for the density of air taken as 0.075 lb/ft$^3$ (1.2 kg/m$^3$). For an air leakage rate of 0.02 ft$^3$/ft$^2$ min. (0.1 liter/m$^2$ · s), the moisture accumulated over a 30 day period would be 0.26 lbs per square foot (1.30 kg/m$^2$) of wall area. This corresponds to a 0.05 in. (1.3 mm) thick sheet of water.

The potential for air flow through a wall arises from air pressure differences on opposite sides of the wall. As illustrated in Fig. 14.20, air pressure differences are caused by external forces such as:

1. the stack effect, which accounts for differences in mass of air at different temperatures
2. positive pressure or suction from wind
3. internal forces such as from mechanical air distribution systems

Therefore, in winter, mechanically pressurized buildings, prevailing winds causing suction on certain walls, and warm interior (lower density) air can all combine to provide significant air pressure differences. This differential air pressure results in exfiltration of the warm moist air through the walls, particularly in the upper parts of the building.

To avoid problems due to condensation of moisture from exfiltrating air, it is necessary that the wall (including intersections and connections with other compo-

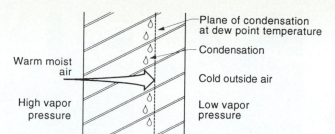

Plane of condensation
at dew point temperature

Condensation

Warm moist
air

Cold outside air

High vapor
pressure

Low vapor
pressure

**Figure 14.19**  Concealed condensation.

nents) be very airtight.  In some cases, components of the wall may have sufficient
resistance to air flow, but this should not be taken for granted.  Despite the solid
appearance of masonry, significant air leakage can occur through such walls.  Parging
of one face of the masonry or several coats of paint can increase sufficiently the
resistance to air flow, but this initially improved resistance to air flow will decrease
if mortar joints crack or if previous cracks open and close and extend through the
parging or paint.  Otherwise, trowel on, torch on, or stick on membranes, as shown
in Fig. 14.21, painted and sealed drywall, or jointed and sealed rigid insulation can
be used to provide the main resistance to air flow through a wall.  Any of these layers
could be designated as the air barrier.

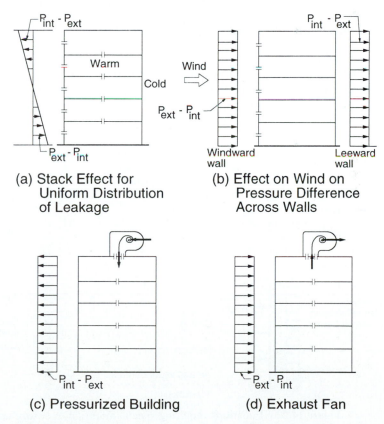

(a) Stack Effect for
Uniform Distribution
of Leakage

(b) Effect on Wind on
Pressure Difference
Across Walls

(c) Pressurized Building

(d) Exhaust Fan

**Figure 14.20**  Forces creating air pressure differences across walls.

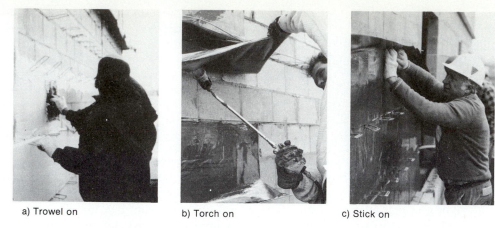

a) Trowel on          b) Torch on          c) Stick on

**Figure 14.21**   Air barrier membranes. (*Courtesy of Bakor Inc.*)

Because air barriers are subject to air pressure differences, they must be structurally able to resist such forces without failing. Areas requiring particular attention to ensure continuity of the air barrier are at the intersections of the wall with other components of the building and at openings in the wall, as illustrated in Fig. 14.22.

### 14.5.5 Interaction Between Air and Vapor Barriers

Virtually all materials have some resistance to air flow and some resistance to vapor flow. Therefore, when we speak of an air or vapor barrier, we are usually identifying the layer that is designed for that purpose. However, it must be recognized that some vapor barriers such as polyethylene sheet also offer resistance to air flow and,

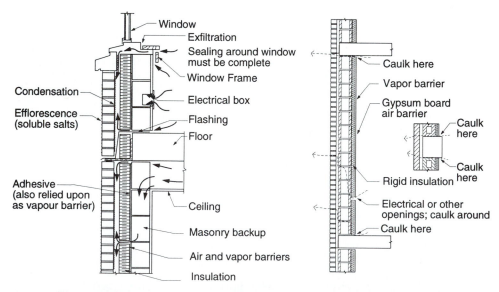

**Figure 14.22**   Locations requiring special attention to ensure continuity of air barriers.

therefore, must be supported to prevent tearing or other types of failure under the air pressure load. On the other hand, most types of air barriers have significant resistance to vapor transmission. If condensation from airborne moisture occurs on the cold side of the vapor barrier, it can become trapped between the intended vapor barrier and another unintended vapor barrier consisting of the air barrier located nearer to the cold side of the wall. Because the amount of condensation due to air leakage can be much greater than due to vapor transmission, drying out of the *vapor trap* may take a long time.

Design of wall assemblies to have the desired thermal, air leakage, and vapor transmission resistance characteristics requires that the interactions of these components be studied. The vapor barrier must be located on the warm side of the insulation. However, the air barrier can be located anywhere from the interior finish to the outside face of the interior wythe or backup wall. Although many designers favor the exterior location because it keeps this barrier away from damage by inhabitants of the buildings, the requirements for inspection, repair, and maintenance of the air barrier should be considered when making this decision. Also, where there is potential for formation of a vapor trap, the vulnerability of the wall system to deterioration as a result of condensation should be considered. Even with very good inspection and good quality construction, perfection is rarely achieved. Therefore, it is wise to anticipate the existence of minor flaws in the air barrier and to choose a wall design that is not sensitive to the resulting condensation.

## 14.6 RAIN PENETRATION

### 14.6.1 Background

Water penetration from rain through a wall can have a variety of unacceptable consequences ranging from staining or damage to interior finishes or contents to growth of mold or fungi within or on the surface of the wall. As was the case with condensation, water in the wall can lead to deterioration such as corrosion of metal parts, softening of sheathing, and reduced effectiveness of insulation. In cold climates, freezing of water can cause damage either from freeze–thaw action within materials or formation of ice in spaces within the wall. A primary requirement of design is that water not penetrate to parts of the wall that will be susceptible to these types of problems.

The concept of the open rain screen is discussed in detail in Sec. 12.2 and will not be repeated here. It is generally acknowledged that superior resistance to rain penetration will result from use of masonry cavity or veneer walls designed as open rain screens. The discussion in this section applies to these walls and to single-wythe walls.

### 14.6.2 Factors Affecting Rain Penetration

For rain penetration to occur, there must be a source of water, a driving force or mechanism to carry water through the wall and some pathway through the material. Simply put, if water is kept off the wall by roof overhangs, drips, and rain gutters, rain leakage should not occur. Conversely, if the wall surface is covered with water, there is obviously a greater probability of rain penetration.

Measurements reported by Lacy[14.29] using rain gages set into walls indicated that the amount of rain driven onto a wall is directly proportional to the product of rainfall and the wind speed during the rain. In the absence of wind, rain hardly wets a wall. Wind speed during rain is not normally recorded separately, but the average wind speed, over a given period, is approximately proportional to the average wind speed occurring during rain over that same period.[14.30] Hence, a valid parameter for indicating the average annual amount of rainfall driven onto a wall is the product of the annual rainfall and the average annual wind speed. This product, known as the *Driving Rain Index* (DRI), can be simply calculated from data that are usually readily available. Contour maps of such information are available in Britain[14.29] and the United States.[14.31]

Water on the surface of the wall can penetrate the wall by:[14.32]

- gravity flow
- capillary action
- kinetic energy
- air flow

For both cavity and single-wythe walls, transport of water through the wall by air flow is potentially the largest source of rain penetration. Therefore, it is important that both the wall and the joints with other components such as floors and windows be airtight. For hollow masonry, an interior air barrier will not prevent air flow in and out of the outer parts of the wall. Different air pressures on the same face or adjoining faces of the building act to force this air flow. Therefore, it is good practice to ensure that hollow walls are also comparted into pressure equalized chambers (see Sec. 12.2). Bond beams, flashing and movement joint material can serve as separators for these compartments.

Except for very minor flow through voids hit directly by rain drops, transport of water by kinetic energy is usually not a significant factor. However, gravity flow along cracks and openings and even through porous material can be a significant source of moisture penetration. In addition, migration of water through the wall by capillary action can result in rapid wetting if the materials have high absorption characteristics.

For single-wythe masonry construction, the amount of rain penetration can be greatly reduced by using walls that are:

- constructed with relatively impermeable units
- constructed with full head and bed joints with mortar compacted in weathertight joints (see Fig. 12.7 in Chap. 12)
- free of cracks

Use of coatings on walls can cause problems by trapping moisture in the wall, in which case, the wall may be saturated for a long period of time. Coating the wall is generally not recommended for brick masonry construction. Successful applications of particular products in the same geographic region should be used as a guide as to whether or not to use a coating. Test results of full-scale block walls[14.33] show that the use of a clear coating resulted in a significant reduction in the amount of leakage.

### 14.6.3 Design Provisions to Minimize the Potential for Rain Penetration

One of the major causes of rain penetration is the direct funneling of water into the wall. This can occur at tops of walls, roof intersections, and at intersections with projecting elements. Water from the roof and other horizontal surfaces should be kept off the wall. Properly detailed roof flashings and a sufficient height of roof curbs or parapets should be provided on flat roofs to force water to drain through drain pipes. On sloped roofs, use of adequate rain gutters and leaders is essential to avoid overflow. Window sills and other projections beyond the exterior face of the wall should be sloped away from the wall and be provided with a drip to enable the water to fall free from the wall. In addition, through-the-wall flashings should protrude beyond the wall to provide a drip for water draining from the wall.

Movement joints should be carefully detailed and, in many cases, a double seal such as shown in Fig. 14.23 should be used to apply the rain screen principle (Sec. 12.2) to the joint to prevent water from passing through the wall. In addition, movement joints should be positioned away from exposed corners and away from likely drainage paths on the surface of the wall.

Where hollow units are used, the likelihood of water penetrating to the cells of the units dictates that provisions must be implemented to direct the water back out. Flashings with weepholes through the exterior face shell should be included. Two alternatives for placing the flashing at the base of a wall, are shown in Fig. 14.24. Where floors penetrate into the masonry, water may tend to collect on the floor and run into the building or else migrate around the edge of the floor and under at the ceiling level below. Use of through-the-wall flashing and weep holes are again recommended.

Inadequate flashing details around windows, doors, and at intersections of building columns with infill walls are common causes of leakage. Use of inferior

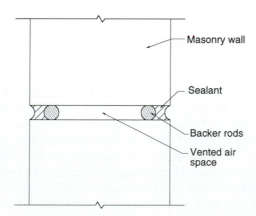

Masonry wall

Sealant

Backer rods

Vented air space

**Figure 14.23** Vertical movement joint (plan view; not to scale).

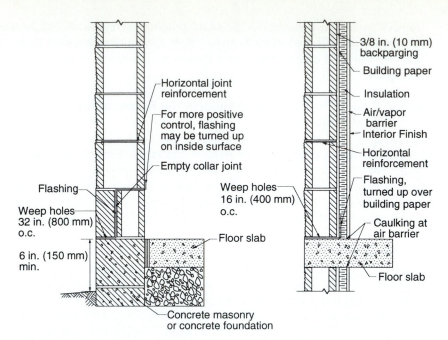

**Figure 14.24**  Details for water resistant construction of single-wythe masonry walls.

quality flashing materials including those subject to damage during construction, can result in leakage.

Cracks in masonry are potential pathways for rain penetration.[14.34] Therefore, distances between movement joints should be limited, as suggested in Sec. 15.6.2. Joints should generally be positioned near openings and changes in wall direction where concentration of stress from restrained movement would otherwise likely cause cracking.

### 14.6.4  Measurement of Water Permeance

The choice of materials and construction details should include evaluation of the water permeance. ASTM E514[14.35] is a laboratory test commonly used in North America to evaluate the extent of moisture passing through the wall under a specified head of water and wind speed for a four hour test period. Previous versions of the test provided ratings that depended upon the extent of dampness, visibility of moisture, and the rate of leakage collected at the inside face of the wall. Although ratings are no longer used, an indication of relative water permeance can be obtained from observation of the time of first dampness on the back of the wall, time of first visible water on the back of the wall, and the area of dampness and total water collected at the end of the test. This test method has also been modified and adapted for field conditions.[14.36] This test is not appropriate for masonry veneer and cavity walls, which rely on the previously discussed multistage rain screen to avoid problems due to rain penetration. Even though the test has been criticized for the severe conditions imposed, it is a means to compare performance of different types of walls on a relative scale. Some attempts have been made to relate time of rain penetration in the standard test to the suitability of the wall for locations having a specified Driving Rain Index.[14.30]

## 14.7 SOUND CONTROL

### 14.7.1 Introduction

Unwanted sound is classified as *noise,* and the elimination or reduction of noise is an important feature of building design. What is considered to be noise is affected by the listener and the circumstances. Loud music at a concert might not be thought of as noise by a teenager, but even barely audible conversation might be considered as noise by someone trying to get to sleep. Standards have been established defining the minimum acceptable acoustic properties for building elements in various types of buildings.[14.17] These standards principally relate to the minimum acceptable sound insulation for separations between adjoining residential dwellings. Noise invades many parts of modern day living. With the high density of residential buildings, use of lightweight building envelopes and partitions, and generally higher sound levels, the concern for effective sound control has resulted in this being one of the major areas of complaint in modern apartments and condominiums. Building regulations governing acoustic performance are sometimes only recommendations and not usually enforceable by law. In addition to the requirements for design of the building elements, the overall design of the building and the quality of construction can have marked effects on the actual sound control within a building.

The area of acoustics is much too broad to be covered in depth here, however, some background information is provided in order that the acoustic properties of masonry construction can be discussed within a rational quantitative framework.

### 14.7.2 Sound Transmission

Sound is the sensation perceived by the human ear resulting from rapid fluctuations in air pressure usually created by a vibrating object. Longitudinal sound waves are elastic waves that can occur in any media with properties of mass and elasticity. If a particle is displaced, then the elastic forces present tend to pull the particle back to its original position. However, through the inertia of the displaced particle, the initial disturbance will propagate throughout the media. Ripples from a pebble dropped into a pond are an analogy except that sound waves travel in three dimensions.

Using a loudspeaker diaphragm as an example, the series of *compressions* and *rarefactions* produced by its movement constitute *sound waves.* The frequency is determined by the rate of oscillation (vibration) of the diaphragm. As is illustrated in Fig. 14.25(a), one complete oscillation, including both the compression and rarefaction motions, is defined as one *wavelength.* One cycle in Fig. 14.25(b) is the movement of one particle through one complete sound wave. The *frequency f* is defined as the number of cycles per second, commonly referred to as hertz. Alternatively,

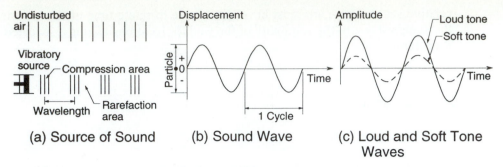

(a) Source of Sound      (b) Sound Wave      (c) Loud and Soft Tone Waves

**Figure 14.25**   Properties of sound.

the time taken for one cycle is the period $T$ given by,

$$T = 1/f \tag{14.16}$$

The speed of sound $c$ is dependent on the mass and elasticity of the medium. For air, this has been determined to be

$$c = \sqrt{kP_0/\rho} \tag{14.17}$$

where   $c$ = velocity of sound in medium, ft/sec (m/s)
       $P_0$ = atmospheric pressure, psi (Pa)
       $\rho$ = density of medium, lb/ft$^3$ (kg/m$^3$)
       $k$ = $6.5 \times 10^3$ for U.S. Customary Units (1.4 for metric units)

With the wavelength $\lambda$ being the distance between successive pressure maxima or minima in a plane wave,

$$c = \lambda f \tag{14.18}$$

The normal *amplitudes* of airborne sound waves are very small, ranging from $10^{-7}$ mm to a few mm. As indicated in Fig. 14.25(c), smaller amplitudes correspond to sound just perceptible to the human ear, whereas the larger amplitudes represent the limiting range beyond which the ear would suffer damage. Thus, the strength or loudness of sound depends on the energy causing the displacement of air particles.

Pressure fluctuations above and below normal atmospheric pressure, about 14.5 psi (100 kPa), is the common method for quantifying sound. The pressure fluctuations of interest range from about $3 \times 10^{-9}$ psi ($20 \times 10^{-6}$ Pa), which is roughly the threshold for hearing, to approximately 0.029 psi (200 Pa), which represents the threshold of pain. Because the latter pressure is $10^7$ larger than the first, it was convenient to convert this rather unwieldly range of values using a logarithmic scale known as the *decibel scale,* with units of *decibels* (dB).

The decibel is a relative measure determined from the pressure ratio with respect to the standard value for the threshold of hearing. Table 14.10 contains a listing of typical decibel levels for sound sources over the range of hearing. The logarithmic calculation of the decibel measure of sound is appropriate because the perception of loudness by the human ear is also logarithmic. Therefore, a 10 dB increase in sound pressure is perceived as a doubling of the loudness.

**Table 14.10** Sound Levels (from Ref. 14.37)

| Loudness | dB | Source | Comments |
|---|---|---|---|
| Deafening | 150 | | Short exposure can cause hearing loss |
| | 140 | Jet plane takeoff | |
| | 130 | Artillery fire<br>Machine gun<br>Riveting | |
| | 120 | Siren at 100 ft. (30 m)<br>Jet plane (passenger ramp)<br>Thunder, sonic boom | Threshold of pain |
| | 110 | Woodworking shop<br>Accelerating motorcycle<br>Hard rock band | Threshold of discomfort |
| Very Loud | 100 | Subway (steel wheels)<br>Loud street noise<br>Power lawnmower<br>Outboard motor | |
| | 90 | Truck unmuffled<br>Train whistle<br>Kitchen blender<br>Pneumatic jackhammer | |
| Loud | 80 | Printing press<br>Subway (rubber wheels)<br>Noisy office<br>Average factory | Intolerable for phone use |
| | 70 | Average street noise<br>Quiet typewriter<br>Freight train at 100 ft (30 m)<br>Average radio | |
| Moderate | 60 | Noisy home<br>Average home<br>Normal conversation | |
| | 50 | General office<br>Quiet radio<br>Average home<br>Quiet street | |
| Faint | 40 | Private office<br>Quiet home | |
| | 30 | Quiet conversation<br>Broadcast studio | |
| Very Faint | 20 | Empty auditorium<br>Whisper | |
| | 10 | Rustling leaves<br>Sound proof room<br>Human breathing | |
| | 0 | | Threshold of audibility |

As a sound wave propagates from its source, the amplitude and sound level decrease. This is called *attenuation* of sound. As the area of the sound wave front increases, the pressure and hence the amplitude decrease. Megaphones and directional speakers are devices that limit the spreading out of sound waves and hence permit sound to carry further.

For different frequencies at the same pressure level, sounds will not be perceived as being equally loud. As an example, sound at 3 kHz at a level of 54 dB will sound as loud as one at 50 Hz at a 79 dB level.[14.38] Therefore, it is necessary to weigh sound according to frequency in order to get a more meaningful measure of the effects of sound. In general, the ear is most sensitive in the 2 to 5 kHz frequency range.

The tolerance of noise is a function of hearing sensitivity and the environment. In addition to the type of environment, the frequency structure and duration of the noise affect its acceptance.

### 14.7.3 Sound Transmission Loss

One of the significant benefits of masonry walls is the property of sound transmission loss. Masonry walls can be very effective sound transmission barriers and are used to ensure desired levels of privacy and living or working comfort. For the most part, it is the mass of the masonry that provides an economical method of achieving an effective *sound barrier* otherwise often referred to as *sound insulation*.

ASTM E90[14.39] is a standard test method to determine sound transmission loss of airborne noise through walls. The data from the laboratory tests are interpreted using ASTM E413[14.40] to define the *sound transmission class,* STC. The STC is a single number rating that provides a convenient way of describing *transmission loss* (TL). It is determined by fitting the sound transmission loss contour to a reference contour. As shown by the dashed line in Fig. 14.26,[14.38] the reference contour extends from 125 to 4000 Hz. The STC rating is based only on the 16 TL values measured between these frequencies, but, as shown by the solid line in Fig. 14.26, laboratories often test at higher and lower frequencies to get additional information.

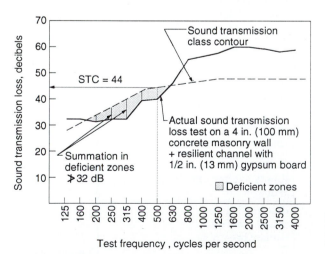

**Figure 14.26** Sound-transmission loss for concrete block masonry walls (from Ref. 14.43).

The reference contour is fitted to the TL curve so that the sum of the differences at the reference frequencies in the deficient zones is less than or equal to 32 dB, as indicated by the shaded area. A second constraint is that, at any frequency, the deficiency in TL cannot be greater than 8 dB. The application of this constraint is shown in Fig. 14.26 at the 500 Hz frequency. This requirement limits the effects of such a *coincidence dip,* sometimes called an *acoustic hole,*[14.41] as has been measured in some non-masonry wall systems. For the fitted position of the reference contour, the TL value of the reference contour at the 500 Hz frequency is the STC.

For effective sound insulation, a key requirement is mass. The greater the mass, the lower is the amplitude of reradiated sound waves. The *mass law* predicts a 6 dB increase in insulating value for each doubling of mass for single-wythe walls. However, practical construction details limit this increase to about 5 dB, so that, for example, a 4 in. (100 mm) thick brick wall with an STC of 45 dB has to be increased to an 8 in. (100 mm) thickness to achieve an STC of 50 dB. This concept applies to composite single-wythe construction. Where other layers of materials such as gypsum board are bonded directly to the masonry wall, there is little change in mass. Hence, the main sound insulating effect of directly applying such layers is to cover small holes or cracks that otherwise might allow direct transmission of sound waves.

Figure 14.27 is an example of the traditional approach to estimating STC for concrete masonry walls, where the range of values for each wall weight is indicative

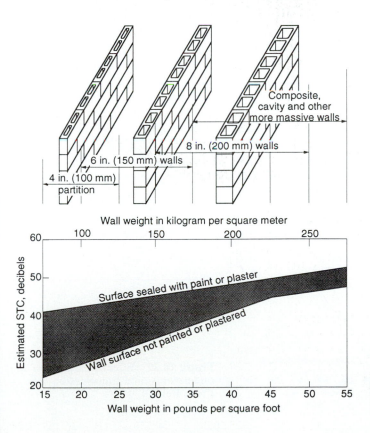

**Figure 14.27** Estimation of STC for concrete masonry walls (from Ref. 14.42).

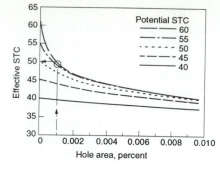

**Figure 14.28** Effect of small gaps on sound transmission (from Ref. 14.38).

of the effects of sealing the wall. A plaster coated or painted surface covers over or seals openings, porous surfaces, and flaws that otherwise permit freer passage of sound waves.

Small gaps in a wall can have very significant effects on transmission loss and, therefore, airtightness and uniformity of walls are very important characteristics. Figure 14.28 shows the calculated influence of very small leaks, where a hole representing 0.001% of the wall area can result in a reduction in STC from 55 or 60 dB down to about 50 dB.[14.38] Figure 14.29 is a practical illustration showing the effect (3 dB decrease) of improperly sealing back-to-back electrical outlets. Good detailing would require the outlets to be offset from each other as well as sealed.

Inclusion of doors or windows can result in substantially lower STC values. It is logical that the element with the lowest STC should be improved before putting effort into other areas. Additions of gaskets around doors and use of windows with specially designed frames and glass mounting is usually necessary to reliably achieve STC values above 45 dB.

Stiffness and isolation are other factors that affect sound transmission loss. In general, the combination of high mass and flexible materials provides the best sound insulation. Although not generally practical, separation of elements of the wall can be effective in reducing sound transmission as additional energy is lost as sound waves are converted to different wave motions at each layer. Isolation by use of cavities, resilient mountings, and materials that have a damping effect on vibrations all can help reduce sound transmission.

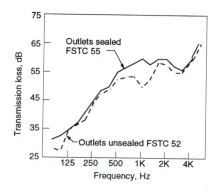

**Figure 14.29** Effect of sealing of outlets on sound transmission loss (from Ref. 14.38).

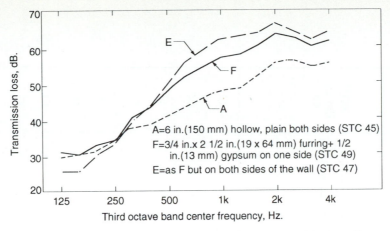

**Figure 14.30**  Transmission losses for 6 in. concrete block walls.

For porous blocks or where fine cracks may exist along mortar joints, applications of a sealant such as latex paint can be effective by ensuring that sound transmission is due to vibration rather than due to airborne passage of sound waves.

Recent tests[14.43],[14.44] on concrete block walls demonstrate the influence of block thickness and surface treatments on measured STC values. Figure 14.30 is a plot of transmission loss data for a 6 in. (150 mm) concrete block wall. Covering one side of the wall with gypsum board mounted on wood furring produces a significant increase in STC. As can be seen in Table 14.11, separation of the gypsum board covering by mounting it on wood furring strips, Z-bars or resilient channels can greatly increase the transmission loss (TL), especially where fiberglass batt insulation is placed between the gypsum board and the block wall. For example, an 8 in. (200 mm) hollow concrete block wall covered on both sides with $\frac{5}{8}$ in. (15 mm) gypsum board attached to $2\frac{1}{2}$ in. (63 mm) steel studs and with glass fiber batts in the stud spaces produced an STC value of 72[14.44] (not shown).

Cavity wall systems can also be very effective as sound barriers. As can be seen in Table 14.12, the size of the air space is very important and the change from a 1 in. (25 mm) air space for wall *A* to a $2\frac{3}{8}$ in. (61 mm) air space for wall *B* resulted in an STC increase from 62 to 77 dB.[14.44] Comparison of wall *C* with wall *B* indicates the influence of the gypsum board on covering wall leaks. The result for wall *D* without cavity insulation compared to wall *B* illustrates the benefit of a combination of air space and insulation in the cavity. It has been shown that sound transmission loss through a masonry cavity wall is frequently about 8 dB better than a solid wall of equal weight.[14.45]

For design purposes, housing authorities and other regulatory bodies normally establish minimum requirements for STC values. The requirements may depend on building location and location within the building. Table 14.13 is an example of such minimum requirements. However, it is not uncommon to see STC values of 55 or higher specified for a wide variety of walls.

**Table 14.11** Sound Transmission Classes for Single-Wythe Concrete Block Walls (from Refs. 14.43–14.45)

| Block Size (in.) | Wall Construction (thicknesses in inches*) | | | | | | | | | | | | STC Rating (dB) |
| --- | --- | --- | --- | --- | --- | --- | --- | --- | --- | --- | --- | --- | --- |
| | Interior | | | | | | | Exterior | | | | | |
| | Gypsum Board (in.) | | 1½ in. Wood Furring | Glass Fiber Batts (in.) | | | | 1½ in. Wood Furring | Glass Fiber Batts (in.) | | Gypsum Board (in.) | | |
| | ½ | ⅝ | | 1.0 | 1.5 | 2.0 | 2.5 | | 1.5 | 2.0 | ½ | ⅝ | |
| 4† | | | | | | | | | | | | | 46 |
| 4 | X** | | | X | | | | | | | X | | 52 |
| 6† | X | | | | | | | | | | | | 45 |
| 6 | | X | X | | X | | | X | X | | X | | 57 |
| 8 | | X | X | | X | | | | | | | | 50 |
| 8 | | X | X | | | X | | | | | | | 57 |
| 8 | | X† | | | | | | | | X | | X† | 64 |
| 8† | | | | | | | | X | X | | | X | 55 |
| 8 | | X | X | | X | | | X | X | | | X | 59 |

\*\* Metal channels are used instead of the wood furring strips.

† Gypsum boards are attached to the block by Z-bars.

‡ Thicknesses can be converted using 1 in. = 25.4 mm.

Note: To illustrate the use of the table, the wall construction represented by the last line consists of an 8 in. block with 5/8 in. gypsum board on 1 1/2 in. furring with glass fiber batts covering both sides.

**Table 14.12** STC Results for Concrete Block Cavity Walls (from Ref. 14.44)

| Designa-tion* | Cavity | | Surface Covering on Interior Wall | STC Rating (dB) |
|---|---|---|---|---|
| | Air Space | Insulation | | |
| A | 1 in. (25 mm) | 2.5 in. (65 mm) glass fiber | $\frac{5}{8}$ in. (16mm) gypsum board | 62 |
| B | $2\frac{3}{8}$ in. (60 mm) | 2.5 in. (65 mm) glass fiber | $\frac{5}{8}$ in. (16mm) gypsum board | 77 |
| C | $2\frac{3}{8}$ in. (60 mm) | 2.5 in. (65 mm) SM styrofoam | none | 73 |
| D | 5 in. (125 mm) | none | $\frac{5}{8}$ in. (16mm) gypsum board | 69 |

\* Both the interior and exterior wythe are $3\frac{5}{8}$ in. (90 mm) hollow concrete block masonry.

**Table 14.13** SoundTransmission Class Limitations* (from ref. 14.46)

| Location of Partition | Low Background Noise | | High Background Noise | |
|---|---|---|---|---|
| | Bedroom Adjacent to Partition | Other Rooms Adjacent to Partition | Bedroom Adjacent to Partition | Other Rooms Adjacent to Partition |
| Living unit to living unit | 50 | 45 | 45 | 40 |
| Living unit to corridor | 45 | 40 | 40 | 40 |
| Living unit to public space (average noise) | 50 | 50 | 45 | 45 |
| Living to public space and service areas (high noise) | 55 | 55 | 50 | 50 |
| Bedrooms to other rooms within same living unit | 45 | NA** | 40 | NA** |

\* STC values given in dB.

\*\* NA: Not applicable.

### 14.7.4 Sound Absorption

The sound level within a room can be lowered by use of materials that absorb sound energy rather than reflecting it back into the room. Although sound absorption is not a major factor in improved transmission loss through walls, it has a substantial effect on the acoustic quality within the room and can be very effective in controlling the spread of noise by reflections or multiple reflections along walls such as in corridors and gymnasiums.

For material to absorb sound, the sound energy must be converted to heat by creating a frictional drag on the sound waves. Materials that are porous and have a rough texture will be much more absorptive than dense hard surfaces. *Noise reduction coefficients* (NRCs) or *absorption coefficients* can be used to quantify this property. Depending on the type of aggregate and surface texture, NRCs varying from 0.28 to 0.5 have been reported for concrete block[14.41] Solid clay, being a harder more dense material, has lower values depending on surface texture. If a surface absorbs all the sound, its NRC is 1.0.

Some special masonry units such as the *Sound Block* and *Acoustile* (see Fig. 14.31) have been designed to increase sound absorption. As shown, the slots or holes and the small chambers form a resonant system that absorbs sound. Insulation in the chamber increases the dissipation of energy.

### 14.7.5 Other Design and Construction Considerations

In addition to selecting the appropriate wall design for the target STC, layout of the building can have a very great effect on satisfactory noise control. Some specific suggestions are as follows:

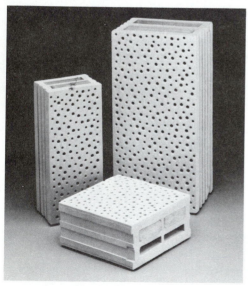

a) Acoustile

b) Sound Block

**Figure 14.31** "Sound Block" and "Acoustile" used to increase sound absorption.

- plan in-line rather than cubicle arrangements of units to reduce the number of common walls for any unit; Fig. 14.32(a)
- use mirror plans so that quieter areas such as bedrooms are adjacent to other quiet areas and noisy areas are also adjacent to other noisy areas; Fig. 14.32(b)
- stagger the doors to apartments in the same corridor so that direct transmission through the lowest STC parts of the walls can be avoided; Fig. 14.32(c)
- place windows away from common walls to reduce the amount of sound that can be transmitted out one window and back in the next; Fig. 14.32(d)
- detail ducts in such a way as to reduce sound transmission from one room to another. Use of egg crate lined plenums, lined turns, and sound traps can reduce sound transmission
- use resilient supports for piping and conduits to reduce noise transferred from them as well as to them.[14.38]

A major problem resulting in unsatisfactory levels of sound transmission loss (TL) is flanking of noise around the sound barrier. Flanking paths through ceiling or floor areas and at the junctions of the walls reduce the effectiveness of the wall as a sound barrier. Attention to achieving a proper seal at these locations is very important.

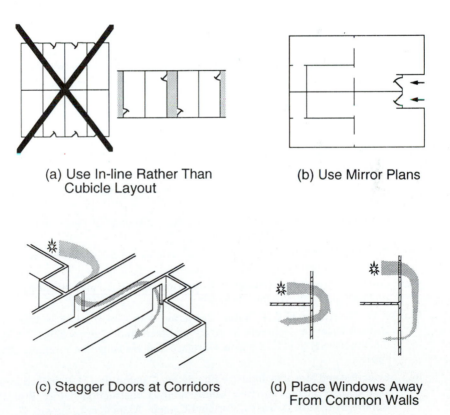

(a) Use In-line Rather Than Cubicle Layout

(b) Use Mirror Plans

(c) Stagger Doors at Corridors

(d) Place Windows Away From Common Walls

**Figure 14.32**   Wall layout for sound control (from Ref. 14.41).

Sound transmission loss measurements in buildings (in accordance with ASTM E 336[14.47]), provide a FSTC value where *F* indicates that it is a field measurement. Because no attempt is made to eliminate alternative transmission paths, the FSTC level closely reflects the actual quality of the construction. Such measurements can be used as a quality control procedure to ensure that no serious errors or construction flaws are present.

## 14.8 CLOSURE

Many of the benefits derived from the use of masonry are related to topics that have been grouped under the general category of building science. For some topics, such as fire and sound, appropriate authoritative information is available. Following selection of the wall assemblage, the main involvement of both the designer and the builder is to ensure that details and workmanship standards ensure that the anticipated properties are realized. In other areas, involving heat and moisture movements, the design phase should involve a thorough assessment of the ability of the design to satisfy the requirements. This assessment should account for the interactions between the various components of the wall and include an assessment of the vulnerability of the wall system to deterioration or poor performance as a result of relatively minor flaws. Even with good standards of inspection and workmanship, perfection should not be relied upon for any type of construction.

Although the treatment of the building science topics in this book is of necessity brief, the introduction to this area will help the designer become familiar with this very important aspect of design of buildings—in this case, with masonry buildings as the focus.

## 14.9 REFERENCES

14.1 N. Hutcheon and G. Handegord, *Building Science for a Cold Climate*, Wiley, New York, 1983.

14.2 Brick Institute of America, "Technical Notes on Brick Construction," BIA, Reston, VA.

14.3 National Concrete Masonry Association, "A Manual of Facts on Concrete Masonry," TEK Note, NCMA, Herndon, VA.

14.4 National Research Council of Canada, "Canadian Building Digests," Institute for Research in Construction, Ottawa, Ontario.

14.5 American Society of Heating, Refrigeration and Air Conditioning Engineers, *Handbook of Fundamentals*, Atlanta, GA, 1985.

14.6 R.G. Drysdale and G.T. Suter, "Exterior Wall Construction in High-Rise Buildings: Brick Veneer on Concrete Masonry or Steel Stud Wall Systems," CMHC Advisory Document NHA 5450, Canada Mortgage and Housing Corporation, Ottawa, Ontario, 1991.

14.7 Concrete and Masonry Industry Fire Safety Committee, "A Balanced Design Approach to Fire Safety for Low-Rise Multi-Family Construction," Fire Protection Planning Report, No. 18 of A Series, Portland Cement Association, Skokie, IL, 1990.

14.8 National Concrete Masonry Association, "Balanced Design Fire Protection for Multi-Family Housing," TEK Note 17B, NCMA, Herndon, VA, 1991.

14.9 American Society of Testing and Materials, "Standard Methods of Fire Tests of Building Construction and Materials," ASTM E119, ASTM, Philadelphia, 1985.

14.10 Brick Institute of America, "Fire Resistance," TEK Note 16 Revised; McLean, VA, 1987.

14.11 W. Panarese, S.H. Kosmatka, and F.A. Randall, Jr., *Concrete Masonry Handbook*, Fifth Ed., Portland Cement Association, Skokie, IL, 1991.

14.12 National Bureau of Standards, "Fire Resistance Classifications of Building Construction," BMS 92, NBS, Washington, D.C., 1972.

14.13 Brick Institute of America, "Calculated Fire Resistance," BIA, Technical Note 16B, Reston, VA, 1991.

14.14 Standards Association of Australia, "SAA Masonry Code," Australian Standard 3700-1988, SAA, North Sydney, New South Wales, 1988.

14.15 National Concrete Masonry Association, "Estimating the Fire Resistance of Concrete Masonry," TEK Note 6, NCMA, Herndon, VA, 1966.

14.16 International Conference of Building Officials, "Uniform Building Code," ICBO, Whittier, CA, 1991.

14.17 National Research Council of Canada, "National Building Code of Canada," NRCC, Ottawa, Ontario, 1990.

14.18 National Concrete Masonry Association, "Design Details for Concrete Masonry Fire Walls," TEK Note 95, NCMA, Herndon, VA, 1978.

14.19 R. McMullan, *Environmental Science in Buildings*, Macmillan, London, 1983.

14.20 Masonry Council of Canada, Guide to Energy Efficiency in Masonry and Concrete Buildings, Downsview, Ontario, 1982.

14.21 Brick Institute of America, "Heat Transmission Coefficients of Brick Masonry Walls," Technical Note 4 Revised, BIA, Reston, VA, 1982.

14.22 National Concrete Masonry Association, "Thermal Insulation of Concrete Masonry Walls," TEK Note 38A, NCMA, Herndon, Virginia, 1980.

14.23 National Bureau of Standards Loads Determination (NBSLD) Computer Program, T. Kasuda, "NBSLD—National Bureau of Standards Heating and Cooling Load Determination Program," *Journal, Automated Procedure for Engineering Consultants* (APEC), Winter 1973–1974.

14.24 National Concrete Masonry Association, "Estimating M-Factors for Concrete Masonry Construction," TEK Note 12, NCMA, Herndon, VA.

14.25 Brick Institute of America, "Thermal Transmission Coefficients for Dynamic Conditions—M Factor," Technical Note 4B Revised, BIA, Reston, VA, 1987.

14.26 M. Catani and A. Goodwin, "Heavy Building Envelopes and Dynamic Thermal Response," *ACI Journal*, February 1976.

14.27 C. Tanzler and A. Hamid, "Dynamic Thermal Performance of Masonry Walls," in *Proceedings to the Third North American Masonry Conference*, University of Texas, Arlington, 1985, pp. 34.1–34.12.

14.28 Health and Welfare of Canada, "Significance of Fungi in Indoor Air: Report of a Working Group," Canadian Public Health Association, Ottawa, Ontario, 1987.

14.29 R.E. Lacy, "An Index of Exposure to Driving Rain," Building Research Station Digest 127, H.M. Stationary Office, London, 1971.

14.30 L.R. Baker and F.W. Heintjes, "Water Leakage Through Masonry Walls," *Architectural Science Review*, Vol. 33, March 1990, 17–33.

14.31 C.T. Grimm, "A Driving Rain Index for Masonry Walls," in *Masonry: Materials, Properties and Performance*, ASTM STP 778, J.G. Borchelt, Ed., American Society for Testing and Materials, Philadelphia, 1982, pp. 171–177.

14.32 R.G. Drysdale, "Building Science Issues for Masonry Construction: Requirements for Moisture, Air and Thermal Barriers," in *Proceedings of the Fifth Canadian Masonry Symposium*, Vancouver, British Columbia, 1989, pp. 1–20.

14.33 R. Brown, "Initial Effects of Clear Coatings on Water Permeance of Masonry," in *Masonry: Materials, Properties and Performance*, J.G. Borchelt, Ed., American Society for Testing and Materials, ASTM STP 788, 1982, pp. 221–236.

14.34 C.T. Grimm, "Water Permeance of Masonry Walls: A Review of Literature," in *Masonry: Materials, Properties and Performance*, ASTM STP 788, J.G. Borchelt, Ed., American Society for Testing and Materials, Philadelphia, 1982, pp. 178–199.

14.35 American Society for Testing and Materials, "Standard Method of Test for Water Permeance of Masonry," ASTM E514-74, ASTM, Philadelphia, 1988.

14.36 C.B. Monk, "Adaptations and Additions to ASTM Test Method E514 (Water Permeance of Masonry) for Field Conditions," in *Masonry: Materials, Properties and Performance*, ASTM STP 788, J.G. Borchelt, Ed., American Society for Testing and Materials, Philadelphia, 1982, pp. 237–244.

14.37 Brick Institute of America, "Sound Insulation–Clay Masonry Walls," Technical Notes on Brick Construction No. 5A, BIA, Reston, VA, 1988.

14.38 Institute for Research in Construction, "Building Science Insight '85–Noise Control in Buildings," National Research Council of Canada, Ottawa, Ontario, 1987.

14.39 American Society for Testing and Materials, "Standard Method for Laboratory Measurement of Airborne Sound Transmission Loss of Building Partitions," ASTM E90, ASTM, Philadelphia, 1990.

14.40 American Society for Testing and Materials, "Standard Classification for Determination of Sound Transmission Class," ASTM E413, ASTM, Philadelphia, 1990.

14.41 National Concrete Masonry Association, "Noise Control with Concrete Masonry in Multi-Family Housing," TEK Note 18A, NCMA, Herndon, VA, 1983.

14.42 National Concrete Masonry Association, "Estimating Sound Transmission Class of Concrete Masonry," TEK Note 9, Herndon, VA, 1972.

14.43 A.C.C. Warnock and D.W. Monk, "Sound Transmission Loss of Masonry Walls–Tests on 90, 140, 190, 240 and 290 mm Concrete Block Walls with Various Surface Finishes," Building Research Note No. 217, Division of Building Research, National Research Council of Canada, Ottawa, Ontario, 1984.

14.44 A.C.C. Warnock, "Sound Transmission Loss Measurements through 190 mm and 140 mm Blocks with Added Drywall and Through Cavity Block Walls," Interim Report No. 586, Institute for Research in Construction, National Research Council of Canada, Ottawa, Ontario, 1990.

14.45 National Concrete Masonry Association, TEK Note 69A, "New Data on Sound Reduction with Concrete Masonry Walls," NCMA, Herndon, VA, 1973.

14.46 National Concrete Masonry Association, "Noise Control with Concrete Masonry," TEK Note 39, NCMA, Herndon, VA, 1972.

14.47 American Society for Testing and Materials, "Standard Test Method for Measurement of Airborne Sound Insulation in Buildings," ASTM E336, ASTM, Philadelphia, 1984.

# 14.10 PROBLEMS

**14.1** For future ease of reference, identify and list books available in your local library that cover the application of the building science topics included in this chapter. Briefly explain the reason for selecting each reference (i.e., discuss particular features important to you).

**14.2** Calculate the fire resistance of a masonry wall comprised of a 4 in. (100 mm) clay brick wythe and a 6 in. (150 mm) concrete block wythe constructed with blocks made with expanded slag aggregate.
- **(a)** The wall is composite construction with a $\frac{3}{8}$ in. (10 mm) solid mortared collar joint.
- **(b)** The wall is a cavity wall with a 2 in. (50 mm) air space.
- **(c)** Repeat cases (a) and (b) with the addition of $\frac{5}{8}$ in. (15 mm) fire rated gypsum board attached to the inside face of the block wythe.

**14.3** **(a)** From your local building code, determine the fire resistance ratings required for the following:
- (i) A wall between the showroom area and the automobile service area for a single-story building housing a car dealership.
- (ii) A party wall between apartments in a 12 story residential building with 240 suites.

**(b)** Select masonry walls to satisfy the requirements for Prob. 14.3 (a).

**14.4** From first-hand observation or review of written reports (i.e., *Engineering News Record*) discuss two recent fires (preferably one involving masonry construction and one involving some other construction material) and comment on the effectiveness of the construction from fire safety and fire protection of property points of view.

**14.5** Calculate the equivalent wall thickness for fire resistance of the following:
- **(a)** An 8 in. (200 mm) lightweight concrete block wall with $\frac{1}{2}$ in. (12.7 mm) gypsum sand plaster.
- **(b)** A clay brick cavity wall with 4 in. (100 mm) (nominal) thick wythes and a 2 in. (50 mm) air space.

**14.6** Calculate and draw the temperature gradients for the normal range of summer and winter temperatures for the following:
- **(a)** The brick cavity wall shown in Fig. P14.6(a).
- **(b)** The single-wythe block wall shown in Fig. P14.6(b).
- **(c)** Calculate the heating and cooling loads that are associated with these walls. Discuss other temperature related design considerations.

**14.7** **(a)** For the walls described in Prob. 14.6. Determine and draw the corresponding vapor pressure gradients. Comment on the potential for condensation and the amount that could occur due to vapor diffusion and due to air leakage.
- **(b)** For the wall shown in Fig. P14.6(a), if a 0.004 in. (0.10 mm) polyethelene vapor barrier protected by a gypsum board finish is placed on the inside face of the interior wythe, show how this affects the performance for the conditions described in Problem 14.7(a) Use drawings to illustrate your discussion.
- **(c)** For the wall shown in Fig. 14.6(b), if the insulation is moved to the inside of the wall and changed to 4 in. (100 mm) thick fiberglass batt insulation between channels and having a $\frac{1}{2}$ in. (12.7 mm) gypsum board finish, design the vapor barrier to avoid condensation from vapor diffusion. For this system, if air leakage occurs, discuss the potential for and consequences of condensation. Use drawings to illustrate your discussion.

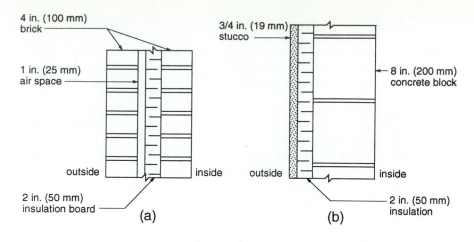

| | Summer | | Winter | |
|---|---|---|---|---|
| | Outside | Inside | Outside | Inside |
| Temperature, °F (°C) | 100 (38) | 70 (21) | 0 (-18) | 70 (21) |
| Relative Humidity, % | 70 | 50 | 70 | 50 |

**Figure P14.6**

**14.8** What factors affect rain penetration? What steps can be taken to reduce rain penetration through masonry walls?

**14.9** From your local area, identify buildings with three different types of masonry wall construction. Sketch and describe the construction details relevant to the potential for rain penetration. If practical, speak to the occupants to document any observed problem related to rain penetration.

**14.10** Determine the sound transmission class (STC) for the following:
 **(a)** The wall shown in Fig. P14.6(a).
 **(b)** The wall shown in Fig. P14.6(b).

**14.11** What is the effect on STC of adding 1.5 in. (38 mm) wood furring and fiber batts covered with $\frac{1}{2}$ in. (12.7 mm) gypsum board to the inside face of an 8 in. (200 mm) normal weight hollow concrete block wall.

**14.12** If a noise meter is available, measure sound levels in three public or work areas. Observe and measure sound levels in adjacent rooms that are essentially free of additional sound sources. Comment on the effectiveness of these examples of construction as noise barriers. If possible, include both masonry construction and other forms of construction. [Note: This type of study will only provide a very rough indication and should not be interpreted as representing the ASTM E336 test.]

**14.13** Discuss methods for improving sound transmission loss through masonry walls.

**14.14** In terms of sound transmission, sketch and discuss construction details that may reduce the effectiveness of the actual construction. Suggest corrective measures.

# CHAPTER 15

# Construction Considerations and Details

Reinforced block wall under construction (*Courtesy of National Concrete Masonry Association*).

## 15.1 INTRODUCTION

Designers should be familiar with construction practices and details. The quality of construction depends not only on the quality of materials and *workmanship*, but equally on proper detailing and clear specifications. *Quality assurance* for masonry includes these factors and inspection appropriate to the type of construction or design. Because design and construction are so closely linked, discussions of related construction considerations are included in Chaps. 4 and 5 on materials and in Chaps. 6 through 14 on design of various elements of masonry buildings. Additional factors are also included in Chaps. 16 and 17 regarding construction of single-story and multistory masonry buildings. This chapter contains some of the more general considerations for masonry construction and selected details to help familiarize designers with some key considerations that affect design decisions.

  The construction process should be taken into account in the development of construction details. It is through such details and the accompanying specifications that the intentions of the design are communicated. If this communication is not effective, annoying or expensive corrective measures may be required in the field. Building codes normally set out the minimum construction requirements that must be met for their design provisions to be applicable. They, therefore, specify expected levels of workmanship, qualities of materials, and construction practices. These specifications may be included in the building code itself,[15.1,15.2] by reference to companion construction standards[15.3] or specifications,[15.1] or to explanatory information.[15.4,15.5]

In general, today's engineered masonry places a greater demand on workmanship than was necessary for the more massive "traditional" masonry. However, with proper training and supervision, masons have demonstrated that they are equal to the challenge of producing cost-effective quality construction. The dimensional tolerances in building codes are based on values which limit the decreases in strength due to workmanship to acceptable levels.

## 15.2 WORKMANSHIP AND CONSTRUCTION PRACTICES

Workmanship relates to the quality of the mason's work and includes factors under his control such as general appearance, alignment, complete filling of joints, rejection of damaged units, use of properly mixed mortar, and general attention to detail. The quality of masonry construction is affected by construction practices as well as the skill of the trade. Therefore, factors such as construction in various weather conditions and other on-site practices are included in this section.

### 15.2.1 Effect of Workmanship on Strength

The quality of the workmanship can have dramatic effects on the compressive and flexural capacities of masonry walls. Hendry[15.6] identified several workmanship defects that, although related specifically to clay brickwork, are relevant for all masonry.

**Unfilled Mortar Joints.** Unfilled bed joints in brickwork can result from deep furrowing of the mortar after placing it on the bricks. As shown in Fig. 15.1, the furrow, created by running a trowel down the middle of the mortar bed to spread the mortar, can result in an incompletely filled bed joint when the next course is placed. The resulting gap has been shown to reduce compressive strength by as

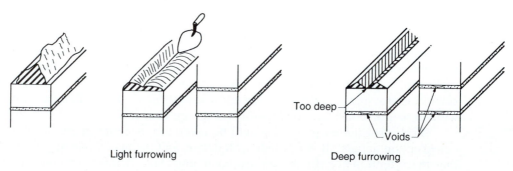

Light furrowing

Deep furrowing

**Figure 15.1** Deep furrowing for solid masonry construction.

much as one-third.[15.7] This result does not apply to hollow masonry where the common practice of face shell bedding results in only about 15% decrease in compressive load carrying capacity from the case with webs mortared when they align vertically.[15.8]

Although failure to completely fill head joints and collar joints does not have much effect on compressive capacity, flexural and shear strengths can be significantly reduced. Moreover, as discussed in Chap. 12, incompletely filled joints can lead to serviceability problems because of less resistance to rain penetration. Also, as indicated in Chap. 14, a continuous crack can lead to increased air flow and lower sound transmission loss that again can lead to unacceptable in-service performance.

**Thick Bed Joints.** As shown in Chap. 5, compared to 3/8 in. (10 mm) mortar bed joints, $\frac{5}{8}$ to $\frac{3}{4}$ in. (16 to 19 mm) thicknesses result in compressive strength reductions of about 20 to 30% for brick construction. The effect is less pronounced for masonry units with greater *unit height* to *joint thickness* ratios and for grout-filled hollow masonry.

**Deviation from Line.** Misalignment and initial deviation from a vertical line increase the eccentricity of loading on walls, which in turn results in reduced capacities. For a 10 ft (3 m) story height, an 8 in. (200 mm) block wall with a $\frac{1}{2}$ in. (13 mm) displacement from plumb can experience a reduction in axial load carrying capacity of about 15%.

Test programs in Australia[15.7] and the United States[15.9] have provided information on compressive strength reductions of brick masonry under the combined effects of a number of workmanship defects. In the Australian work, the combined effect of outside curing, deep bed furrowing, $\frac{5}{8}$ in. (16 mm) bed joints, and a $\frac{1}{2}$ in. (12 mm) deviation from plumb resulted in a strength reduction of 61%. The U.S. work, involving uninspected versus inspected brick masonry construction, showed that uninspected construction led to strength reductions of 38 to 45%.

### 15.2.2 Effect of Workmanship on Water Permeance

Standardized water permeance testing and attempts to compare water permeance of different material combinations indicate a large variability of results, making it difficult to quantify or to get consistent results. Both research and field experience have shown that most leakage occurs through mortar joints.[15.10,15.11] Thus, workmanship (materials preparation and culling of units, laying, tooling, and filling of joints) and quality of the mortar (shrinkage, bond and permeability) play large roles in both the variability and the rate of leakage.[15.11]

Figure 15.2(a) shows the gradual wetting of the inside face of a brick wythe subject to a near-zero pressure gradient in a rain screen wall. When subjected to a pressure difference of 10 lb/ft$^2$ (500 Pa) across the wall, the previously wetted part shown in the figure began to leak almost immediately, whereas previously unwetted parts showed dampness within 5 minutes and leakage at the head joint/bed joint intersections occurred within 15 minutes.[15.10]

The dye pattern of water flow through a head joint, shown in Fig. 15.2(b), is evidence of the predominance of flow through the head joint/bed joint intersections. This type of leakage is directly related to the solidity of mortar joints and

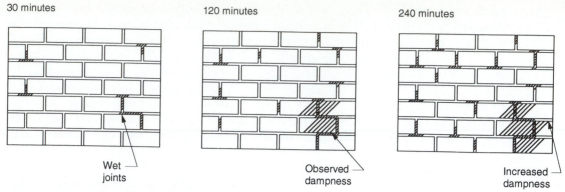

30 minutes       120 minutes       240 minutes

Wet joints       Observed dampness       Increased dampness

(a) Gradual Wetting Under Zero Pressure Gradient

(b) Leakage through Head Joint      **Figure 15.2**    Test for rain penetration of brick veneer.

supports the conclusion that complete filling of mortar joints is essential to good resistance to rain penetration. Other factors such as mortar mixing, laying, and tooling are also important.

### 15.2.3 Preparations in Advance of Laying Masonry

The foundations, floor slabs, or shelf angles (see Sec. 12.3) on which masonry is to be laid must be checked for level and alignment. Discrepancies must be corrected before starting the masonry construction and the designer (or representative) should always be involved in decisions on corrective measures where excessive misalignment is found. Tolerances of $\frac{1}{4}$ in. in 10 ft (6.3mm in 3m) for variation in alignment or floor level, and $\frac{1}{2}$ in. (13mm) from specified elevations are typically specified.[15.1]

At the base of a wall, the thickness of the mortar joint is normally allowed to vary between $\frac{1}{4}$ in. (6 mm) and $\frac{3}{4}$ in. (19 mm). For greater variations in elevation, the surface on which the base course is to be laid can be brought up to the desired level with a concrete topping, as shown in Fig. 15.3.

Prior to using any mortar, the base course of masonry should be laid out dry so that adjustments can be made to avoid irregular or broken bond patterns. To the extent possible, openings and wall lengths should be dimensioned to take advantage of the modular dimensions of masonry and to reduce the amount of cutting. Where

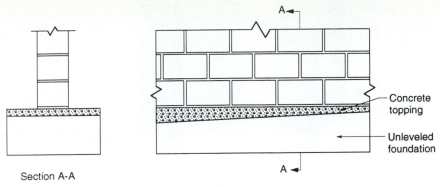

Section A-A

**Figure 15.3** Use of concrete topping to level base for first course of masonry when the base is too low.

cutting of units is necessary, they should be accurately cut so as to maintain uniform mortar joint thicknesses and to ensure that load-carrying and serviceability functions are not impaired. Although some units can be broken by hand, use of a masonry power saw is usually necessary for a finished appearance or where intricate shapes are required. In addition to the dry layout of units at the base course, details of changes in sections at other elevations should be worked out from the plans prior to construction. Ad hoc dimensioning schemes, such as in Fig. 15.4(a), are unacceptable and a little planning can result in proper arrangements, such as shown in Fig. 15.4(b).

For reinforced masonry, misaligned dowels should be bent back into place, if possible, but typically not at a slope exceeding 1 to 6 from the intended direction, except with the permission of the designer. Some cutting of webs of hollow units may be necessary to accommodate the bent bar. Badly misplaced reinforcement, such as shown in Fig. 15.5(a), should be cut off and replaced with a dowel properly anchored into the required location, as shown in Fig. 15.5(b). Proper placement of reinforcement is critical and masonry codes typically specify acceptable tolerances.

All laying surfaces should be clean just prior to laying. Materials that could adversely affect bond or the properties of the mortar can be removed by brushing, sweeping, blowing, or washing as appropriate.

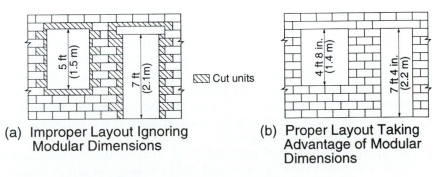

(a) Improper Layout Ignoring Modular Dimensions

(b) Proper Layout Taking Advantage of Modular Dimensions

**Figure 15.4** Layout of openings in a masonry wall.

a) Improperly placed          b) Correct location

**Figure 15.5** Location of reinforcing dowels.

### 15.2.4 Preparation of Mortar Mixes

Mortar materials must be protected prior to use in order to achieve the desired performance of the mortar. In particular, the cementitious materials must be kept dry to prevent premature hydration. Sand should also be protected on site to prevent contamination by dirt, dust, or other building materials.

Proper procedures must be used for proportioning and mixing the mortar. Although it is common practice, on-site batching of mortar by the shovel has often produced unsatisfactory results. Problems can arise if the helper put in charge of mixing the mortar is unskilled and unfamiliar with the requirements. Therefore, batching by the shovel is not recommended. Use of gage boxes or suitable hoppers is desirable to ensure that appropriate volume ratios of mortar materials are used. The quality of the mortar is particularly sensitive to the amounts of cementitious materials and additives. On smaller jobs, it has been an accepted practice to use shovel batching provided that the shovel measures are calibrated using gage boxes [usually, 1 ft$^3$ (0.03 m$^3$)] to establish the equivalent shovel count for each material. This calibration should be repeated at the beginning of each day for every helper doing the mixing. This practice can be effective provided that the shovel count and the calibration are checked at frequent intervals.

Mortar standards differ on the definition of sand volume. U.S.A. standards use bulk volumes, whereas others[15.12,15.13] have adopted the dry volume. The significance of this difference arises because wet sand has an increased volume, or bulk. As is shown in Fig. 15.6, the degree of bulking depends both on the gradation and type of sand and the moisture content and can be as much as 30%.[15.14] Therefore, if the dry volume measurement is used and sand is exposed to the weather, the mix proportions must be adjusted to account for the varying moisture content of sand. The current trend has been to simply protect the sand from rain.

Mortar should have a good consistency and be well matched to the units and exposure conditions during construction (see Sec. 15.5). If there is too little sand, the mortar tends to stick to the trowel and is said to be "fatty." The higher cement paste content can lead to bond breaks and the opening of cracks in head joints due to drying shrinkage. On the other hand, if there is too much sand, the mortar will be "lean" and difficult to spread.

If mortar is not well matched to the units or the exposure conditions, it may either stiffen prematurely or remain fluid too long. In either case, it is difficult to lay subsequent courses of masonry as they cannot be readily brought to correct

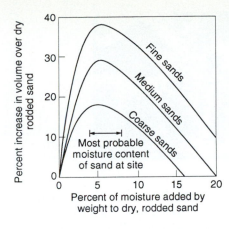

**Figure 15.6** Effect of moisture content on volume of sand (from Ref. 15.14).

The figure axes are labeled: vertical axis "Percent increase in volume over dry rodded sand"; horizontal axis "Percent of moisture added by weight to dry, rodded sand". Curves are labeled "Fine sands", "Medium sands", "Coarse sands". A bracket indicates "Most probable moisture content of sand at site".

alignment in the former case, and in the latter case, they will tend to move out of horizontal and vertical alignment after placing.

Compatibility of mortar and units should be determined before laying commences and adjustments made as necessary. Prewetting the surface of the units to reduce suction (see the initial rate of absorption in Sec. 4.3.6) is one possible solution to prevent premature stiffening of the mortar. However, it is difficult to control the amount of wetting, and undesirable side effects, such as shrinkage of concrete units, eliminates this as an effective option for many situations. It is usually possible and more economical to adjust the mortar mix to achieve compatibility.

Mixing time for mortar mixed in mechanical mixers on-site has typically been limited to 3 to 5 minutes. This limit is mainly intended to control the air content of the mix. However, with the introduction of more efficient modern mixers, including auger systems, mixer manufacturers' specifications should be considered.

### 15.2.5 Use of Mortar

The mortar for the first course of masonry should be laid on a clean base (free of dust, oil, or other substances that can affect bond). It is extremely important that the first course be laid level and true to line and coursing so that the rest of the construction can proceed without difficulty. Extra time taken at this stage will usually save time later. For solid units, full mortar bedding is commonly used whereas for hollow units, only the face shells are usually mortared. It is recommended that, provided webs are vertically aligned, they be mortared in all courses of piers, columns, and pilasters. The first mortar joint above the foundation should be fully mortared.[15.1]

As mentioned earlier, it is very important that mortar joints be completely filled. For the head joint, some argue that mortar should be placed at the faces only, even when the units or ends of the unit are solid. The reason put forward is that this creates a cavity which interrupts the transmission of moisture by capillary action. This may be a valid argument where free drainage below the head joint is assured. However, where solid units or webs of hollow units lie beneath the head joints, the creation of such a void actually reduces the resistance to flow by allowing free flow across the ends of the units and the top of the unit below. In addition, a

**Figure 15.7** Laying of closure unit (*Courtesy of Brick Institute of America*).

partially filled head joint can provide a reservoir for water that will later find its way through the wall thickness or remain in place to cause freezing or efflorescence problems. Attempts to fill the head joints from above after the unit is in place by "slushing" are not usually effective. Therefore, care to ensure that sufficient mortar is placed initially is important. For a closure unit, all edges of the unit and the opening should be buttered with mortar before lowering it into position, as shown in Fig. 15.7.

It is normal practice in cold climates to lay frogged brick with the bed-joint recess facing down to avoid creating a reservoir for accumulation of water. Sufficient mortar should be placed to fill the frog and for this reason mortar should not be furrowed. From a rain penetration point of view, laying the units with the frog up makes more sense because there is a greater likelihood of completely filling the frog with mortar. For hollow units, the section with the thicker face shell, if any, should be laid uppermost to provide a larger bed for spreading the mortar. Units should be positioned with an even pressure and rocking should be avoided (see Fig. 15.8) because it can lead to separation of the mortar from the unit near the faces, resulting in reduced flexural strength and less resistance to rain penetration. The positions of units that have been placed should not be adjusted later by tapping or shoving. Such movement will break the initial intimate contact that has been created by the suction of the water from the mortar into the unit. Where adjustment in position is required, the unit and mortar should be removed and new mortar and a dry unit used. The removed unit should be cleaned and allowed to dry before reuse.

The rate of construction in the vertical direction may need to be limited so that the weight of upper courses does not deform the plastic mortar below. Accidental

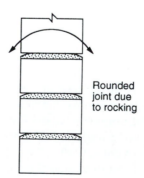

Rounded joint due to rocking

**Figure 15.8** Rocking effect on mortar joints.

lateral force or rocking movements can result from laying operations if work is continued to too great a height above the plastic mortar. Such actions could damage the bond of the partially set mortar. Although the acceptable rate of construction depends on the properties of the unit and the mortar (i.e., mortars with set retarders will take longer to set up and cool conditions can lower the rate of setting), an arbitrary limit of 15 to 20 times the thickness of the wythe per day is suggested as prudent.

Best results are usually obtained if mortar is maintained at its wettest workable consistency. This usually results in the highest bond as the unit draws the mortar into intimate contact by absorbing excess water. Mortar should be retempered to replace water lost due to evaporation. However, the mortar should not be remixed in the mortar mixer; instead, the water should be poured into a central depression in the plastic mortar on the board and then thoroughly mixed with the trowel. This retempering can be carried out only prior to initial set of the mortar, which is conservatively estimated as beginning $1\frac{1}{2}$ to 2 hours after initial mixing in moderate weather conditions. Initial set can occur more rapidly, depending upon the type of mortar, the temperature, and the wind speed.

Regardless of the rate of initial set, commencement of the hydration process dictates that any mortar not used within $2\frac{1}{2}$ hours after initial mixing should be discarded. Mortar that has stiffened should not be returned to the mortar board as this tends to accelerate the initial set of the remaining mortar. Also, the mortar board should be cleaned before placing a fresh batch on it.

The laying time can be reduced by stringing out mortar on the bed joint for some distance and possibly by buttering the ends of several units before placing them in position. The extent to which this should be done depends on the absorption of the units, water retentivity of the mortar, and the weather conditions. This practice should be limited to ensure that the mortar is still sufficiently plastic to allow the units to be easily adjusted into position as they are laid. Because low bond strength results if units are laid on partially stiffened mortar, in many situations, particularly in hot weather, it is prudent not to spread mortar more than 4 ft (1.2 m) ahead of the units and to lay the units within 1 minute of spreading the mortar.[15.1]

Excess mortar, protruding from the joints after laying, should be struck off cleanly. It is usually much easier to keep the face of the masonry clean while laying than to remove mortar stains later.

Cavity spaces should be kept clear of mortar droppings and other debris that can bridge the cavity and impair the resistance of the wall to moisture penetration. One way to achieve this during laying is to rest a strip of wood in the cavity on the wall ties to catch the droppings and then lift it from the cavity by wires (Fig. 15.9) just prior to laying the next course containing wall ties. As discussed in Chap. 12, it is also important that mortar droppings not collect at the bottom of the cavity and block the weep holes. The amount of mortar protruding into the cavity can be reduced by beveling back the edges of the mortar bed prior to laying the unit. Similarly, it is important that these mortar protrusions or "fins" be minimized where the cavity or cores in hollow units are intended to be grouted. Further, mortar droppings at the base of the grout space must be eliminated. Such care helps ensure that the grout space can be continuously filled and provide proper bearing from one lift to the next.

**Figure 15.9** Removal of a wood strip used to catch mortar droppings. (*Courtesy of National Concrete Masonry Association.*)

Mortar protruding from the joints into the cavity should be struck off, particularly if the cavity is to contain insulation or be grouted. It is also important to keep movement joints clear of mortar and other debris so that unrestrained movement can occur, thus avoiding a buildup of high forces. Chapter 12 contains additional discussion on this requirement.

Mortar joints are ready to be tooled when a clear thumbprint can be impressed and the mortar does not adhere to the thumb. This is described by the term "thumbprint hard." Tooling the joint with a cylindrical jointer heals any cracking due to plastic shrinkage by pressing the mortar back into contact with the masonry units. The applied pressure also creates a more compacted mortar that is better able to shed moisture and resist abrasion. The tooled joints also produce masonry with clean lines and uniform appearance.

Ties and other fasteners should be installed at the time of laying the mortar. Mortar should completely surround the embedded portions to ensure adequate anchorage and to prevent easy access of moisture along the embedded part. Connectors forced into mortar joints that have already been laid do not usually have adequate anchorage. Where connectors have been omitted, alternate methods of attachment using drilled-in types of fasteners should be used.

### 15.2.6 Laying of Units

The masonry units should be checked for conformance to the specifications and referenced standards. Physical properties included in the specification should be confirmed. Where color is important, the acceptable range of color and blending of colors should be agreed upon before the units are placed on the scaffold. Standards for masonry units specify limits on chipping, dimensional variations, and warping. Units that have excessive fire or shrinkage cracking or chipping should be discarded. Provision of reasonably smooth access routes and locating the units to minimize handling will reduce the amount of on-site damage to the units. The aesthetic standards are best defined by construction of sample panels at the beginning of the job.

Masonry units should be stacked clear of the ground and covered to protect them from wetting and contamination. Units that become wet have reduced suction rates, with a consequent tendency to "float" on the mortar, and can develop lower bond with the mortar. The extra moisture in the wall can result in increased efflorescence. For units made from concrete, calcium silicate, or natural stone that exhibit volume change with changes in moisture, it is very important that only dry units (satisfying moisture content limits in material standards) be placed in the wall. Otherwise, subsequent shrinkage of wet units can produce cracking, which could result in decreased flexural strength and increased potential for rain penetration.

The water retentivity of the mortar and the suction properties of the masonry unit both affect the bond characteristics between the two materials. It is known that units that do not tend to absorb water tend to float on the mortar and weak bond is likely to form. At the other extreme, highly absorptive units tend to dry out the layer of mortar at the interface, which results in poor bond. The best results tend to occur for moderately absorptive units that, through suction, draw the mortar into intimate contact with the unit but do not dry out the mortar. The air temperature and the temperature of the materials also affect this relationship, so that lower initial rates of absorption (IRA) may be appropriate in summer and higher values for winter construction.

It is often recommended that clay bricks with an initial rate of absorption (IRA) greater than 30 grams/30 in.$^2$/min (1.55 kg/m$^2$/min) should be prewetted. Wetting bricks not only creates logistical problems, but the wet bricks will have various degrees of absorption. Although the use of highly absorptive bricks or construction in hot weather may necessitate this practice, the associated problems should be recognized and accounted for by adopting appropriate quality control measures.

Openings required in the wall for other trades should be formed or placed during construction and not cut in later. Where bonded wall intersections are required, the component walls should be built up together. The practice of leaving out units in one wall, as shown in Fig. 15.10(a), to receive bonding units from the cross wall at a later time, known as *toothing*, does not form a satisfactory intersection. Where masonry is not brought up uniformly because of scaffolding or other construction requirements, *racking back*, as also shown in Fig. 15.10(b), is recommended.

Running bond and the stack pattern are the most common patterns for layering masonry. For half or common running bond, the head joints are typically offset by half of a masonry unit length in successive courses, as shown in Fig. 15.11(a), but for structural performance considerations, offsets between $\frac{1}{4}$ and $\frac{1}{2}$ of the units' length can be used.[15.1] The stack pattern, Fig. 5.11(b), has units stacked directly over each other forming continuous vertical mortar joints along the head joints of successive courses. The lack of overlap or "bonding" at successive courses means that there is a much greater likelihood of vertical cracking along the continuous vertical joint for the stack pattern. For this reason, joint reinforcement is normally required when

a) Toothing (not recommended)               b) Racking Back

**Figure 15.10**   Laying of masonry units. (*Courtesy of H. Keller.*)

masonry is laid in the stack pattern. A large variety of patterns have been used to create different interesting aesthetic treatments.[15.15]

### 15.2.7 Reinforcing and Grouting

Many of the practices related to reinforceing steel in concrete are applicable to reinforced masonry. Reinforcement must be protected in the same way and kept free from mud, oil, grease, and other substances that might reduce bond. It must be free of loose rust or scale.

   Reinforcement should be accurately located in the wall and held in place during grouting. Building codes contain requirements that limit the tolerances for placement of reinforcement in masonry. As an example, the UBC[15.16] requirements for placing vertical steel in walls are shown in Fig. 15.12. In bond beams, chairs or other positioners can be used. A variety of commercially available positioners for vertical and horizontal reinforcement (see Fig. 15.13) can be used. Reinforcement should be secured in position to avoid displacement by construction loads such as imposed by grouting.

   In all cases, reinforcement should be placed prior to grouting and not pushed in later. Because the method of construction must be consistent with the design objectives, the method of reinforcing and grouting should be either specified in the contract documents or agreed to prior to commencement of construction. For horizontal reinforcement in single-wythe construction of hollow units, the use of knockout webs such as shown in Fig. 15.14, provides continuity of the grout in the wall and excellent bond and cover for the reinforcement. Use of lintel units to create

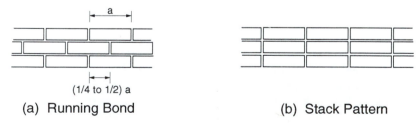

(a)  Running Bond                    (b)  Stack Pattern

**Figure 15.11**   Patterns of laying masonry.

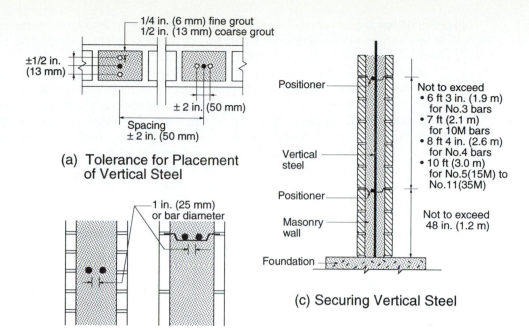

1/4 in. (6 mm) fine grout
1/2 in. (13 mm) coarse grout

±1/2 in. (13 mm)

± 2 in. (50 mm)

Spacing
± 2 in. (50 mm)

**(a) Tolerance for Placement of Vertical Steel**

1 in. (25 mm) or bar diameter

**(b) Tolerances for Placement of Horizontal Steel**

Positioner

Vertical steel

Positioner

Masonry wall

Foundation

Not to exceed
• 6 ft 3 in. (1.9 m) for No.3 bars
• 7 ft (2.1 m) for 10M bars
• 8 ft 4 in. (2.6 m) for No.4 bars
• 10 ft (3.0 m) for No.5(15M) to No.11(35M)

Not to exceed 48 in. (1.2 m)

**(c) Securing Vertical Steel**

**Figure 15.12** Maximum intervals and tolerances for securing steel (from Ref. 15.16).

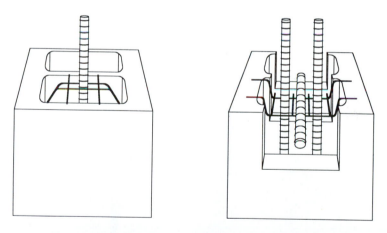

**Figure 15.13** Positioners for holding reinforcement.

**Figure 15.14** Knockout webs to accommodate horizontal steel.

bond beams has the advantage of restricting the flow of grout into lower parts of the wall where it may not be required. However, cut-outs in the base of the lintel unit are required to permit vertical continuity of grout and reinforcement.

For vertical reinforcement in single-wythe masonry, the type of units to be used governs the process of placing the reinforcement. If open-ended units are used, the vertical reinforcement can be placed prior to laying the units and splices can be incorporated at heights related to practical lengths for placing and supporting the vertical bars. However, if closed end units are used, placing the bars before laying the units will require that relatively short bar lengths be used to permit the unit to be lifted over the top of the previously placed bar. This method necessitates frequent splices that increase labor and material costs. Alternately, the bars can be placed after the wall is built, but prior to grouting, as shown in Fig. 15.15.

Reinforcement placed in the grout space between two wythes of masonry can be either prefabricated and placed in position or tied together as construction progresses. The cage cannot be placed after the two wythes are built because ties are used to connect the two wythes. In cases where joint reinforcement is used as both the horizontal steel and the ties, vertical reinforcement can be placed after constructing the wythes as long as provisions for positioning the reinforcement are satisfactory.

Vertical reinforcement should at least be tied at the bottom (e.g., to the lower bar at a splice) and held with wire or some commercially available positioner at the

**Figure 15.15** Placing vertical steel after wall has been built. (*Courtesy of National Concrete Masonry Association.*)

top of the wall. Depending on the size and length of the bar, an intermediate support point may be required (see Fig. 15.12(c)) to prevent the bar from being moved out of position.

Adequate cover over reinforcing steel is important to provide corrosion protection. A minimum cover over the reinforcement of at least 0.75 in. (19 mm) including the masonry unit is required. This increases to 1.5 in (38 mm) when exposed to weather and 2 in. (50 mm) when exposed to soil.[15.16]

Prior to grouting, any mortar fins protruding into the grout spaces should be removed so that bridging of the grout will be prevented and a continuous pour will result. As shown in Fig. 15.16, confined spaces for placement of the grout can lead to bridging of the grout, thus preventing full penetration into the grout space. For similar reasons, grouting should be promptly completed over the full constructed height of the masonry. Otherwise, grout adhering to the sides of the ungrouted portion can harden sufficiently to later prevent complete filling of this grout space.

Clean laying techniques are the best way of achieving a clear grout space, but where hardened mortar projects, it must be knocked off from above usually with a reinforcing bar. This mortar, together with any mortar droppings and other debris (pieces of wood, food containers, etc.), should be removed from the grout space. To facilitate such cleaning, a layer of sand or plastic placed at the bottom of the grout space can prevent bonding of the mortar droppings.

**Figure 15.16** Grout bridging. (*Courtesy of Atkinson-Noland & Associates.*)

The method of grouting has been traditionally classified as one of the following two categories:

**Low Lift Grouting.**  In this method, the maximum height of a grout pour is generally limited to 5 ft (1.5 m).  In such cases, it is accepted practice to forgo clean-out holes.  It is assumed that the volume of mortar droppings, knocked off mortar, and other debris are not critical for the proper functioning of the wall.  The height limit also means that the amount of debris can be readily inspected using a flashlight.

**High Lift Grouting.**  For high lift grouting with grout pours exceeding 5 ft (1.5 m), clean-out holes are required in the bottom course of each grout pour.  To be large enough to permit removal of mortar droppings, the minimum dimension of the clean-out opening should be 3 in. (75 mm).[15.1]  This may involve removing sections of face shells of hollow units, as illustrated in Fig. 15.17.  The face shell can be repositioned and held in position during grouting or a board can be braced in position over the hole to allow the entire space to be filled with grout.  For grouting the cavity between wythes, masonry units can be left out and then mortared into place after cleaning the bottom of the cavity.  The maximum allowable height of the grout pour typically depends on the type of grout (fine or coarse) and the dimensions of the grout space.  Table 15.1 contains typical requirements for minimum grout spaces.

Slump requirements for grout and concrete differ.  Concrete should not be used in place of grout.  It is essential that a grout with a minimum slump of 8 in. (200 mm) be used because of the relatively small grout spaces and the tendency for the masonry units to absorb water from the grout mix.  However, a minimum slump of 10 in. (250

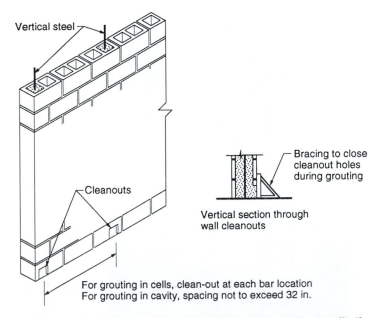

**Figure 15.17**   Requirement of cleanouts for grouting masonry walls (from Ref. 15.1).

**Table 15.1** Grout Space Requirements (from Ref. 15.1)

| Grout Type* | Grout Pour Maximum Height, ft (m) | | Minimum Width of Grout Space, in. (mm),†** | | Minimum Grout†‡ Space Dimensions for Grouting Cells of Hollow Units, in.×in. (mm×mm) |
|---|---|---|---|---|---|
| Fine | 1 | (0.3) | $\frac{3}{4}$ | (19) | $1\frac{1}{2} \times 2$ (38×50) |
| Fine | 5 | (1.5) | 2 | (50) | $2 \times 3$ (50×75) |
| Fine | 12 | (3.6) | $2\frac{1}{2}$ | (643 | $2\frac{1}{2} \times 3$ (63×75) |
| Fine | 24 | (7.2) | 3 | (75) | $3 \times 3$ (75×75) |
| | | | | | |
| Coarse | 1 | (0.3) | $1\frac{1}{2}$ | (38) | $1\frac{1}{2} \times 3$ (38×75) |
| Coarse | 5 | (1.5) | 2 | (50) | $2\frac{1}{2} \times 3$ (63×75) |
| Coarse | 12 | (3.6) | $2\frac{1}{2}$ | (63) | $3 \times 3$ (75×75) |
| Coarse | 24 | (7.2) | 3 | (75) | $3 \times 4$ (75×100) |

\*     Fine and coarse grouts are defined in ASTM C 476.  Grout shall attain a minimum compressive strength at 28 days of 2000 psi.

\*\*    For grouting between masonry wythes.

†     Grout space dimension is the clear dimension between any masonry protrusion and shall be increased by the horizontal projection of the diameters of the horizontal bars within the cross-section of the grout space.

‡     Area of vertical reinforcement shall not exceed 6 % of the area of the grout space.

mm) is often specified to assist with proper compaction and minimize the potential for voids in the grout column. Using plasticizers can help ensure that the water/cement ratio is sufficiently low to produce adequate grout strength. Water is also absorbed from the mortar by the units.  As was discussed in Chap. 4, the compressive strength of grout specimens formed in nonabsorptive molds may not provide a true measure of the properties of the grout in the wall.  Masonry molded specimens are more representative of the actual properties of the grout.

Grout must be consolidated to force it to flow into all the spaces and thus eliminate voids.  Puddling with a steel rod or piece of wood can be used if the grout lifts are not more than 12 in. (300 mm) high.  Small diameter pencil type internal vibrators can be used to consolidate higher lifts in hollow units.  Reconsolidation after the excess water is absorbed by the masonry units and before the plasticity of the grout is lost is recommended.[15.17] This procedure reduces the tendency of the grout column to separate from the sides of the grout space or to separate into partially discontinuous lengths.

At the top of a grout pour in a partially built wall, the grout should be stopped about 1.5 in. (40 mm) below the top course.  Where some delay occurs during the grout pour, the grout should be stopped midway between bed joints.  These practices create a shear key and prevent formation of a continuous dry joint at the same level as the mortar bed joint.

Grouting should take place soon after laying the masonry to reduce the potential for shrinkage cracking in mortar joints. However, sufficient mortar strength and bond must first be developed to prevent blowouts due to the fluid grout pressure on the mortar joints and ties between wythes. Satisfactory conditions are usually attained about 1 day after building hollow unit walls or after about 3 days for multiple-wythe walls.

For large grouting jobs, pumping of the grout is preferable to dumping from a bucket. For high lift grouting, pumping also helps to minimize segregation of the grout. Grout should be placed in each cell of hollow masonry to be grouted and at regular spaces along the length of a multiple-wythe wall with a continuous grout space. The use of mechanical vibration to cause horizontal flow of the grout along a cavity or between cells in construction of hollow units can lead to excessive segregation of the grout materials.

From structural considerations, it may be only necessary to grout cells containing reinforcement. By using partial grouting, it is possible to save grout and reduce the weight of the structure. Partial grouting is achieved by restricting the flow of grout using mortar "dams" placed on the webs of hollow units to confine the grout to the desired cells (Fig. 15.18). If the webs of hollow units are not aligned in

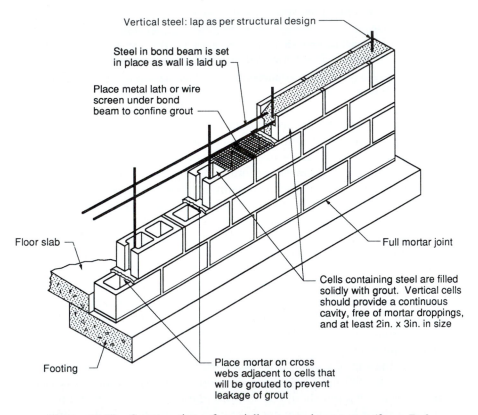

Vertical steel: lap as per structural design

Steel in bond beam is set in place as wall is laid up

Place metal lath or wire screen under bond beam to confine grout

Floor slab

Footing

Full mortar joint

Cells containing steel are filled solidly with grout. Vertical cells should provide a continuous cavity, free of mortar droppings, and at least 2in. x 3in. in size

Place mortar on cross webs adjacent to cells that will be grouted to prevent leakage of grout

**Figure 15.18** Construction of partially grouted masonry (from Ref. 15.18).

successive courses, the mortar dams may not be completely effective and, particularly for fine grout, there is some chance that grout will flow into cells where it is not required. Where bond beams are used in conjunction with vertical reinforcement, the simplest procedure is to pour grout to the top of the bond beam before constructing the next segment of the wall height. When units with knockout webs are used, expanded metal mesh can be placed across cells not to be grouted, as shown in Fig. 15.18. This restricts the flow of grout into these cores yet permits bond between the mortar and the block at these joints[15.18].

**Table 15.2** Construction Tolerances (from Ref. 15.1)**

| Measurement | Location | Tolerance |
|---|---|---|
| Dimension of element | 1. In cross-section or elevation | $-\frac{1}{4}$ in., $+\frac{1}{2}$ in. |
| | 2. Mortar joint thickness | |
| |    bed | $\pm\frac{1}{8}$ in. |
| |    heat | $-\frac{1}{4}$ in., $+\frac{3}{8}$ in. |
| |    collar | $-\frac{1}{4}$ in., $+\frac{3}{8}$ in. |
| | 3. Grout space or cavity width | $-\frac{1}{4}$ in., $+\frac{3}{8}$ in. |
| Element | 1. Variation from level: | $\pm\frac{1}{4}$ in. in 10 ft |
| |    bed joints | $\pm\frac{1}{2}$ in. maximum |
| |    top surface of bearing walls | $\pm\frac{1}{4}$ in. in 10 ft |
| | | $\pm\frac{1}{2}$ in. maximum |
| | 2. Variation from plumb | $\pm\frac{1}{4}$ in. in 10 ft |
| | | $\pm\frac{3}{8}$ in. in 20 ft |
| | 3. True to a line | $\pm\frac{1}{4}$ in. in 10 ft |
| | | $\pm\frac{3}{8}$ in. in 20 ft |
| | | $\pm\frac{1}{2}$ in. maximum |
| | 4. Alignment of columns and walls (bottom versus top) | $\pm\frac{1}{2}$ in. for bearing walls $\pm\frac{3}{4}$ in. for nonbearing walls |
| Location of element | 1. Indicated in plane | $\pm\frac{1}{2}$ in. in 20 ft |
| | | $\pm\frac{3}{4}$ in. maximum |
| | 2. Indicated in elevation | $\pm\frac{1}{4}$ in. in story height |
| | | $\pm\frac{3}{4}$ in. maximum |
| Placement of reinforcement | 1. Flexural elements: | |
| |    d** $\leq$ 8 in. | $\pm\frac{1}{2}$ in. |
| |    d** $\leq$ 24 in. | $\pm1$ in. |
| |    d** $>$ 24 in. | $\pm\frac{1}{4}$ in. |
| | 2. Shear walls† | $\pm2$ in. |

\*\*   Distance from centerline of steel to the opposite face of masonry.
†   From the specified location along the wall length as indicated in the project drawings.
Conversion 1 in. = 25.4 mm

### 15.2.8 Tolerances

Building codes and construction specifications provide tolerance limits for foundation work, mortar joint thickness, dimensions, and alignment of masonry elements and placement of reinforcing steel. Specified tolerances may depend on whether the element is loadbearing or nonloadbearing and other factors such as height and thickness. In general, more restrictive tolerances are specified for loadbearing walls because load eccentricity can result in significant reductions in load-carrying capacity. Tolerances for elements not as critical to structural performance are less stringent. For example, as shown in Table 15.2,[15.1] tolerances for head joint thickness are larger than for bed joints because the head joint thickness has less effect on compressive strength. Other typical tolerances for North American construction are presented in Table 15.2. For flexural walls, an error in placement of vertical reinforcement has more effect on the flexural strength of a thin wall than for a thick wall.

Tolerances set for structural purposes may not be visually acceptable and the project specifications may control.

## 15.3 INFLUENCE OF WEATHER ON CONSTRUCTION REQUIREMENTS

Masonry can be constructed in a wide range of weather conditions provided that certain construction methods and protective procedures are followed. The workmanship requirements outlined in the previous section are generally applicable in all weather, but certain modifications or additional precautions need to be followed if the temperature is above about 85°F (30°C) or below 40°F (5°C), or if the weather is wet or windy. Excessively high or low temperatures can have a significant effect on masonry strength. Temperature extremes affect the degree and rate of hydration of the cement in the mortar. ACI 530.1/ASCE 6/TMS 602[15.1] contains construction requirements for hot-weather and cold-weather masonry. The special requirements for cold, hot, wet, or windy weather are considered in this section.

### 15.3.1 Cold Weather Construction

For masonry construction, a cold temperature is commonly defined as less than 40°F or 5°C.[15.19] However, the effect of wind is usually incorporated by applying a wind chill factor to describe a lower equivalent temperature. For example, an air temperature of 40 degrees F (4.4°C) in a 15 mph (24 km/h) wind is equivalent to a temperature of 22°F (−5.5°C). Such working conditions can lead to difficulties in obtaining good quality masonry as a result of the effect on the masons.

Units containing frozen moisture absorb less moisture from the mortar because of blocked pores, and therefore the unit will tend to float. Aside from the workmanship problem, even without freezing, cold weather will reduce bond and reduce the quality of the mortar because of the higher remaining water content of the mortar.

Frozen units, whether wet or dry, drain heat away from the freshly placed mortar and possibly lead to freezing of the mortar before adequate moisture has been absorbed from it. In addition to reducing the bond and strength of the mortar, freezing water can expand and rupture the mortar. Research indicates[15.19] that no freezing damage occurs if the moisture content of the mortar is less than 6% when it freezes. Because of the absorption of water from the mortar, masonry performs better than concrete in cold weather construction. Concrete loses free water only by hydration and evaporation, which are much slower.

Masonry can be constructed satisfactorily at low temperatures provided that conditions during and immediately after construction allow sufficient strength gain from hydration of the cement in the mortar and sufficient reduction in moisture content of the mortar before it freezes. If these conditions are met, acceptable final strength and durability can be obtained provided that further hydration of the mortar is possible following thawing.[15.20]

An enclosed construction site maintained at temperatures in the 50–60°F (10–15°C) range would be ideal for winter construction and, in many cases, the improved construction efficiency can offset the added cost. When this is not feasible, mortar should be mixed so that its temperature is between 40°F (5°C) and 120°F (50°C) with an optimum of about 70°F (21°C). [The upper limit is to avoid possible flash set of mortar.]

The temperature of the mortar can be increased by heating the water or the sand or both. Generally water should be heated to between 150°F (65°C) and 180°F (80°C) with the upper limit again imposed to avoid flash set of the mortar. However, this limit may be relaxed if the sand and water when mixed result in cooling of the water before the addition of the cementitious material. The sand should not be heated beyond a level comfortable to the touch to ensure that it is not scorched.

The mortar mix proportions can also be adjusted to account for cold conditions. In such conditions, workability is seldom a problem. Hence, lime content or other workability agents can be reduced and cement content increased to obtain an accelerated gain in strength. Also, use of high early strength cement will increase the rate of hydration. Although calcium chloride has sometimes been used in small quantities to accelerate curing of the mortar, its presence will increase corrosion of steel and contribute to efflorescence and spalling.[15.19] Therefore, modern codes do not to permit its use.[15.1,15.3]

Wet masonry units that are frozen must be thawed out before building on them or using them in new work. Dry frozen masonry units may be used provided that they are not less than 20°F (−7°C). Before continuing a wall that is partially laid, the top course must be freed of ice and preferably heated to the same temperature as the units to be laid. If units are heated above normal temperatures, their increased absorption should be considered. Alternately, low absorption units will not reduce the moisture content of the mortar as quickly as higher absorption units and walls built with them may require longer protection against freezing.

On completion of the day's work, the newly constructed walls should be protected for at least 24 hours against rain, snow, and the possibility of freezing. Many masonry codes and specifications contain procedures for cold weather construction. An example of such requirements from the Uniform Building Code[15.16] is as follows:

1. Mortar should be mixed to a temperature between 40°F (5°C) and 120°F (50°C) and maintained above freezing on the mortar board.

| Air Temperature | Means of Heating | Protection of Work |
|---|---|---|
| 40-32°F (5-0°C) | Heat sand <u>or</u> water | Cover with weather-resisting membrane |
| 32-25°F (0--4°C) | Heat sand <u>and</u> water | Cover with weather-resisting membrane |
| 25-20°F (−4- −7°C) | Heat sand <u>and</u> water. Use source heat on both sides of wall. | Use wind breaks if wind is in excess of 15 mph (24 km/h). Cover with insulating blanket or equivalent. |
| Below 20°F* (−7°C) | Heat sand <u>and</u> water. Use enclosure and auxiliary heat to maintain air above 32°F (0°C) | Maintain above 32°F (0°C) by an enclosure and supplementary heat, electric blanket, infrared lamps, or other means. |

\*   If normal portland cement is used, increase the protection to 48 hours.

**2.** Masonry units should be dry and not less than 20°F (−7°C) when laid. Wet or frozen units should not be used.

**3.** The means of heating the mortar should be as set out in Table 15.3.

**4.** Newly laid work should be protected for 24 hours using the type of protection shown in Table 15.3.

When air temperatures are below 40°F (5°C), any grout used should have a temperature of between 40°F and 120°F (5 to 50°C) obtained by heating both the mixing water and the aggregates. Because surrounding cold masonry units will quickly drain heat from the grout, all masonry to be grouted should be maintained above freezing during grouting and for at least 24 hours afterwards. Where air temperatures are below 20°F (−7°C), this should be done by means of an enclosure as shown in Fig. 15.19.

**Figure 15.19**  Enclosure for winter construction (*Courtest of G. T. Suter*).

## 15.3.2 Hot Weather Construction

Hot weather is sometimes rather arbitrarily defined as being above about 85°F or 30°C. However, hot weather effects are experienced to some degree below these temperatures. In fact, hot weather cannot satisfactorily be defined in terms of temperature alone as the drying effects of wind, low humidity, and direct sunshine are significant factors. As discussed in Chap. 14, solar radiation can result in surface temperatures considerably in excess of the surrounding air temperatures. ACI 530.1/ASCE 6/TMS 602[15.1] requires implementation of hot weather protection when the temperature is equal to or exceeds 100°F (38°C) or 90°F (32°C) with a wind velocity greater than 8 mph (13 km/h).

Loss of workability due to rapid evaporation of moisture from the mortar results in reduced bond strength, which will be exacerbated by using units that are hotter and drier than usual. The resulting increased suction will cause the mortar to stiffen more quickly. Also the increased evaporation rate produces rapid initial and final sets of the mortar and leaves insufficient moisture in the joint for continued hydration of the cement. Lower strength and durability are the result, particularly near the exterior surfaces.

An obvious way to reduce the effects of hot weather is to reduce the temperature of all of the materials and equipment used in the masonry construction. Storing materials and mixing mortar in the shade, flushing the mixer, tools, and mortar board periodically with cool water, using wooden rather than metal mortar boards, keeping aggregate moist by periodic sprinkling, and using cool or even iced water for mixing are example measures.

Mortar should be mixed just prior to use and not allowed to stand on the board. Workability can be maintained by retempering to make up for water lost by evaporation, but because the board life of the mortar is likely to be reduced, smaller batches should be prepared. The water retentivity and workability of the mortar can be improved by increasing the lime content or by changing the sand gradation. Portions of mortar not in use can be covered with a plastic sheet to reduce evaporation. ACI 530.1/ASCE 6/TMS 602[15.1] requires that the mortar bed not be spread more than 4 ft (1.2 m) ahead of laying the masonry units. The masonry units should be laid within 1 minute of spreading the mortar.

In extreme conditions, walls should be constructed preferably using sun shades, wind screens, and fog sprays as appropriate to lessen the effects of sun, wind, and dryness. This can often be accomplished quite simply with sheets over scaffolds. Where possible, laying of units should be scheduled for the coolest part of the day and work should stop if conditions become too severe.

Walls laid in hot weather should be cured at the end of a day's work and upon final completion by covering and keeping damp. Although curing has not been common practice, it is essential to the development of good flexural bond between the mortar and the masonry units and to development of durable surfaces on the mortar joints.

For grouting in hot weather, spraying of the inside of the grout space cools the masonry and reduces the absorption of water from the grout mix. This helps maintain the fluidity of the grout. Fast drying out can lead to bridging, plastic shrinkage, related separation, or voids in the grout.

### 15.3.3 Wet Weather Construction

Construction should not continue during rain unless the wall is sheltered from it. When rain is likely, the sand and other mortar materials as well as the masonry units both in the stockpile and on the scaffold should be covered. Wet units are a major concern because of their reduced absorption. Resulting problems include reduced bond, construction difficulties arising from units that tend to "float," and mortar that takes longer to stiffen and dries to a different color.

If rain does occur, freshly laid masonry should be protected by draping a weather-resistant membrane over the top of the wall and extending down over mortar that is still susceptible to washout. Even partially set mortar can be susceptible to washout of the cementitious components. This results in reduced strength and durability and possible staining of the wall. After the mortar has hardened for about 24 hours, wetting by rain has the beneficial effect of providing good curing conditions for the mortar.

### 15.3.4 Construction in Windy Weather

The two main effects of wind on masonry construction are premature drying of the mortar and the danger of structural failure of the newly laid wall. The drying effect can be adequately dealt with by using a workable mortar and retempering to make up the water lost due to evaporation. Also, limits similar to hot weather construction for spreading of mortar along the bed joint should be adopted. If the drying effects of the wind are severe, the wall should be moist cured, or wind screens should be used to reduce evaporation.

## 15.4 PROTECTION OF MASONRY DURING CONSTRUCTION

Some of the topics in this section have been introduced in the preceding section. However, it is important to emphasize that masonry walls need a certain amount of protection during construction and immediately following completion of the wall.

### 15.4.1 Covering and Curing

A weatherproof membrane should be placed over uncompleted walls to keep rain off freshly laid masonry. It is good practice to support the membrane with a wood board so that water runs off the membrane rather than causing it to sag into cells or cavities. In addition to keeping out water, the membrane will prevent contamination by dust and the entry of debris (leaves, sticks, etc.) into cavities or cells. Even after the wall has gained strength, entry of water into the cells or cavities can cause efflorescence when it evaporates through the faces of the wall. In particular, where heavy discharges such as from unfinished roofs, drains, or working platforms can occur, the top of the wall should be protected from rain penetration at the end of each day even if rain is not forecast.

Under normal conditions, it is not necessary to take special steps to cure masonry. However, in adverse weather conditions, such as described in Sec. 15.3, or in other unusual situations, such as masonry construction close to a heat source, some protection or curing may be needed to ensure that hydration of the mortar takes place.

Completed walls should also be protected from staining. The common causes of staining are efflorescence from moisture entering the unprotected top of the wall, stains from contaminated water running down the face of the wall from members attached above, and careless handling of other construction materials. Stains can be prevented by a plastic sheet draped over the wall until construction is completed. In addition scaffolding should stand clear of the wall so that mortar droppings are not directed against the wall.

Foundations and grade beams for interior walls may not be, in some cases, constructed below the frost line in cold climates. If such a building is left unheated during construction, insulation should be placed around the bases of these walls to prevent frost heave, which could damage the walls.

### 15.4.2 Avoiding Unintended Loads

Masonry normally gains strength rapidly, but special care is needed to ensure that it is not prematurely loaded. Damage can occur from scaffolding, formwork, or other construction equipment being fixed to the wall or because temporary supports for lintels, arches, and floors are removed too early. Construction scheduling and practices should take into account the lower masonry strength at ages close to the time of construction.

Uneven backfilling of soil around foundation walls and backfilling of basements before the top is supported by the ground floor will create unintentional loads that can exceed the capacity of the masonry. In addition, once backfilling is started, it should be completed, including sloping of the ground surface away from the basement so that excess water will not flow into this soil, possibly resulting in a liquid state and very high lateral pressures.

### 15.4.3 Wind Bracing

The potential for collapse of freshly laid masonry due to wind depends upon the wind pressure applied to the wall, the height of the wall relative to the weight and thickness, the rate of development of flexural bond strength along the mortar joints, and the lateral support conditions effective at the time. Even though significant flexural bond can be achieved after 1 day, walls are particularly susceptible to collapse during construction because the final lateral support system may not be in place. For example, a wall cantilevered from its base has four times the bending moment due to wind pressure as it would have if simply supported top and bottom by floors and roofs. In addition, the lack of enclosed interior space can result in more critical combinations of positive pressure and suction on a wall element. Therefore, unless intersecting walls provide adequate lateral support, temporary bracing may be required to reduce the potential for failure during construction.

A survey carried out in the United States in 1958[15.21] reported 152 cases of windstorm damage to masonry walls during construction, 86% of which were unbraced. A similar situation probably exists today in many countries, as little guidance is given for temporary bracing requirements during construction. Various codes simply state that during construction, masonry shall be adequately braced or braced when necessary. Because masons are unlikely to work during very windy conditions, the danger of collapse is not so much from the freshly laid masonry that has hardly reached initial set, but from walls that have been constructed a day or more but have not reached full strength and do not have their final lateral supports.

There are two different situations where wind bracing must be provided. For temporary overnight and weekend bracing of new masonry, the bracing necessary depends on short-term weather conditions and on the consequences of failure. Bracing for such a situation might well be designed on the basis of the maximum wind pressure for a return frequency of somewhere between, say, 1 to 10 years, depending on code requirements and local experience. The second situation occurs when a self-contained portion of the masonry construction is complete and bracing may be required to stay in place until the final lateral supports have been constructed. Such bracing may have to be more substantial, depending upon the period of time it will be required. The project specifications should specify who must provide and install the bracing.

The design methods for flexural walls covered in Chap. 7 can be used to determine the bracing spacing requirements for incompletely supported walls. Wind bracing should be designed not only to resist the applied forces, but also it must be sufficiently stiff to avoid damage to the wall due to excessive deflection. A common problem with timber bracing, such as shown in Fig. 15.20(a), is that it is not adequately fixed to the wall or securely supported at the ground or floor level. In such cases, slipping of the brace leaves the wall in essentially an unbraced condition. Also, the bracing should not apply point loads on individual masonry units. Particular attention

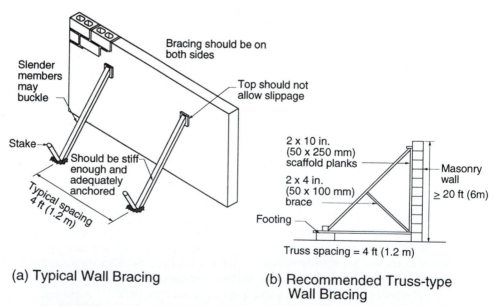

(a) Typical Wall Bracing

(b) Recommended Truss-type Wall Bracing

**Figure 15.20** Inadequate wind bracing for masonry walls during construction.

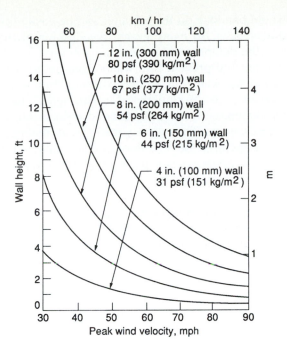

**Figure 15.21** Maximum unsupported height of concrete masonry walls during construction. (*Courtesy of American Concrete Institute.*)

should be given to possible buckling of bracing struts or bearing failure of the stakes at the ground. It was demonstrated by tests[15.22] that, for most soils, the capacity of a stake is not sufficient to support tall masonry walls in a modest 35 mph (56 km/h) wind. This study concluded that stakes should not be used as the horizontal support element for temporary bracing of walls higher than 20 ft (6m). An improved bracing system is shown in Fig. 15.20(b).

The height limit above which the wall should be braced depends upon the peak wind speed and wall thickness. Wall self-weight provides a counteracting moment against overturning moment from wind. The graph in Fig. 15.21 provides a conservative estimate of the allowable unsupported wall height of concrete masonry assuming no tensile strength and ignoring the contribution of vertical steel or end supports, if any.

## 15.5 FLASHING AND DAMPPROOF COURSES

### 15.5.1 Description

*Flashings* are thin impervious materials such as a plastic membrane, composite laminates, or sheet metal that are inserted into masonry walls to control the flow of water. A *dampproof course* is placed through the wall at the top of the foundation to prevent dampness from rising in the wall due to capillary action. It can be made from the same materials as are flashings, although special relatively impermeable mortars or units have been used. Flashing and dampproofing materials must be impervious to moisture penetration, resistant to corrosion or other deterioration, sufficiently tough to withstand handling on the job site, and be easily formed to the desired shape and sealed at splices. In many cases, the flashing at the base of a masonry wall also serves as the dampproof course.

Flashings are intended both to prevent rain water from entering the wall and to direct water that does enter the wall back to the exterior. Proper installation of flashings is necessary to prevent water from penetrating through to the interior of a building and to reduce the dampness in the wall so as to reduce the possibility of freeze–thaw damage or efflorescence. They are typically located above and below wall openings, at the top of parapets, at roof and floor intersections, at both recesses and projections, and at the base of all exterior masonry walls. Flashing details for masonry veneer are dealt with in detail in Chap. 12.

### 15.5.2 Installation of Flashings and Dampproof Courses

Flashing and dampproof membranes should normally be laid within a mortar joint with mortar both above and below, as illustrated in Fig. 15.22(a). This prevents

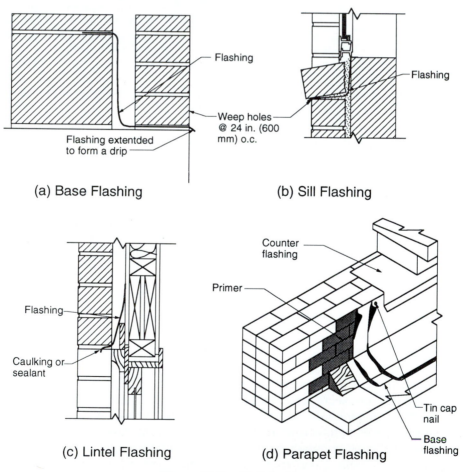

**Figure 15.22** Wall flashings.

punctures or tears during placing. Laying of the flashing material directly on the preceding course of masonry or other support leaves an unbonded path for the migration of water into the wall. A sealant applied to the underside of the flashing should be used if the flashing is laid without mortar on the support. The materials must be lapped a minimum of 6 in. (150 mm) and sealed at junctions. Weep holes should be installed directly above the flashing and the flashing should be sloped toward the outside of the wall.

For proper installation, clear instructions and details should be provided in the contract drawings and specifications. Special attention should be given to ensuring that masons and other trades do not damage the flashing during construction. It is particularly important that the flashing intended to drain water from the wall continue beyond the exterior face of the wall, as shown in Figs. 15.22(a) and 15.22(b), and form a drip for water draining from the wall. Concealed flashings that cover cores or cells of units can provide a sufficiently continuous membrane for directing water out of the wall and are often aesthetically more acceptable.

Typical locations of flashing are shown in Fig. 15.22. Alternate positions of base flashings for hollow single-wythe construction are shown in Fig. 14.24 in Chap. 14. Example sill and lintel flashings for openings are shown in Figs. 15.22(b) and 15.22(c), respectively. In Fig. 15.22(d), flashings on the top of a parapet wall and flashings to protect the back of the parapet and prevent water from entering at the parapet/roof intersection are illustrated. Where vertical reinforcement must pass through the flashing or dampproof course, the size of the puncture or slit should be minimized. In critical situations, sealing materials are necessary to reseal this area.

### 15.5.3 Effect on Wall Strength

It should be recognized that flashings and dampproof courses can significantly reduce the flexural strength and the diagonal cracking or shear strength of the wall.[15.23] However, unevenness along the mortar joint usually means that resistance to sliding will not be affected as much as if the junction was perfectly flat. Typically, top and bottom flashings do not affect the compressive strength of the wall, but intermediate flashings can affect the stability of slender walls.

## 15.6 MOVEMENT JOINTS

### 15.6.1 Functions and Types of Movement Joints

A *movement joint* can be a vertical or horizontal separation built into a masonry wall to reduce restraint and the corresponding stresses by accommodating movement of the wall or movement of other structural elements adjacent to the wall.

In North America, the term *control joint* has traditionally been applied to concrete masonry where shrinkage may be the main source of movement. *Expansion joint* has been similarly applied to clay masonry where moisture and thermal expansion are the main causes of movement. However, *movement joint* is a more apt description that covers both expansion and contraction of the masonry and the requirement to accommodate differential movement between the masonry and other parts of the structure. This is currently used in North America, whereas in other parts of the world, such as Australia, *control joint* is used often to describe any joint used to control the effects of differential movement.

Regardless of whether a joint is opening or closing, a gap is required between the adjacent elements in order that the filler and sealing material can continue to function through repeated cycles of movement. For example, good quality caulking can only stretch a maximum of about 25% of its original length. Therefore, for a $\frac{1}{8}$ in. (3 mm) movement causing opening of the joint, a $\frac{1}{2}$ in. (12 mm) wide movement joint would be required. Similarly, closing of a narrow joint due to expansion or relative frame movement could cause unsightly extrusion of sealant from the joint and permanent deformation of the sealant and filler material. This permanent deformation could later leave gaps for penetration of water if the joints open again.

### 15.6.2 Spacing and Size Requirements

The requirements for movement joints at the top and ends of nonloadbearing infill and partition walls are discussed in Chap. 11. The design and construction of movement joints in masonry veneer are covered in Sec. 12.5. For loadbearing masonry, the presence of the vertical load eliminates the need for horizontal movement joints. Therefore, this section deals specifically with the requirements for vertical movement joints to accommodate horizontal movements of the masonry.

As is the case with other building materials, masonry tends to expand or contract laterally, depending on temperature change, moisture conditions, loading, and the inherent properties of the material. Movement joints should be installed at sufficiently frequent spacing and at critical locations to limit the build up of stress resulting from resistance to movement. Signs of distress can be either compression related (including spalling, crushing, or buckling), or shear and tension related (producing cracking). In the latter case, it is not so jokingly said that, where movement joints are omitted, the structure will put them in; (see Fig. 15.23(a). Lack of vertical movement joints in clay brick parapets can cause disruption due to moisture and thermal expansion, such as shown in Fig. 15.23(b). Also shear failure, such as shown in Fig. 15.23(c), can occur due to movement of a large expanse of masonry wall supported by and rigidly attached to stiff piers.

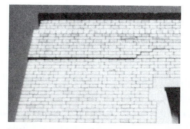

a) Cracking of masonry      b) Parapet Expansion      c) Shear failure of piers

**Figure 15.23**  Examples of cracking due to lack of movement joints.

Potential movement due to temperature change, moisture expansion of clay masonry, and shrinkage of concrete masonry can be calculated from estimated values of material properties (see Chap. 4) or using referenced standards such as ACI 530/ASCE 5/TMS 402.[15.1]

It should be understood that movement joints only completely relieve stresses in the vicinity of the joint and that stresses exist away from these points. Therefore, cracking can occur at a large opening in the middle of a masonry wall panel that has movement joints at each end. The remaining wall sections above and below the opening are a weak link to resisting the forces that do build up. Hence, it is good practice to place vertical movement joints at or quite close to large openings. Similarly, because changes in wall direction (corners) and thickness and stiffer areas of walls (e.g., pilasters) can result in buildup of stress and damage, vertical movement joints should generally be positioned at these locations. Fig. 15.24 illustrates typical locations.

For otherwise uninterrupted lengths of wall, there are no precise methods for determining required spacing, and different publications provide somewhat different advice. The maximum spacing of movement joints in concrete masonry walls is governed by the amount of shrinkage expected which depends on whether the unit is moisture controlled or not, the average annual relative humidity, exposure conditions, and the vertical spacing of joint reinforcement. Table 15.4 provides information on recommended maximum spacing for concrete masonry walls. Horizontal reinforcement in the form of joint reinforcement or regular reinforcement in grouted cavities or bond beams normally permits increased spacing between movement joints in concrete masonry walls. The reinforcement tends to control crack width. However, because clay products tend to expand and have a lower coefficient of thermal expansion than concrete products, greater distances between movement joints are normally acceptable. Maximum spacing in the range of 30 ft (9 m) is common for clay masonry.[15.24]

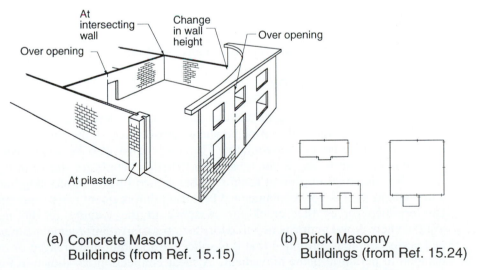

(a) Concrete Masonry Buildings (from Ref. 15.15)

(b) Brick Masonry Buildings (from Ref. 15.24)

**Figure 15.24**  Locations of movement joints in masonry buildings.

**TABLE 15.4**  Horizontal Spacing of Vertical Movement Joints In
Concrete Masonry Walls (Courtesy of C.T. Grimm Inc.)

| Average Annual Relative Humidity | Wall Location | Vertical Spacing of Bed Joint Reinforcement,* in.(mm) | | Spacing of Vertical Movement Joints, ft (m) | |
|---|---|---|---|---|---|
| | | | | CMU Type I** Moisture Controlled | CMU Type II** Non-moisture Controlled |
| Less than 50% | Exterior | None | | 12 (3.6) | 6 (1.8) |
| | | 16 | (400) | 18 (5.4) | 10 (3.0) |
| | | 8 | (200) | 24 (7.2) | 14 (4.2) |
| | Interior | None | | 16 (4.8) | 19 (2.7) |
| | | 16 | (400) | 24 (7.2) | 14 (4.2) |
| | | 8 | (200) | 32 (9.6) | 19 (5.7) |
| Between 50 and 75% | Exterior | None | | 18 (5.4) | 12 (3.6) |
| | | 16 | (400) | 24 (7.2) | 16 (4.8) |
| | | 8 | (200) | 30 (9.0) | 20 (6.0) |
| | Interior | None | | 22 (6.6) | 15 (4.5) |
| | | 16 | (400) | 30 (9.0) | 20 (6.0) |
| | | 8 | (200) | 38 (11.4) | 25 (7.5) |
| Greater than 75% | Exterior | None | | 24 (7.2) | 18 (5.4) |
| | | 16 | (400) | 30 (9.0) | 22 (6.6) |
| | | 8 | (200) | 36 (10.8) | 26 (7.8) |
| | Interior | None | | 28 (8.4) | 21 (6.3) |
| | | 16 | (400) | 36 (10.8) | 26 (7.8) |
| | | 8 | (200) | 44 (13.2) | 31 (9.3) |

\* Joint reinforcement normally consists of two No. 9 wires embedded in the face shell mortar bed.

\*\* Moisture content of concrete masonry units as defined by ASTM C90.

## 15.6.3  Construction Details

Movement joints should be detailed to allow free in-plane movement. However, wall design may require out-of-plane shear transfer across the joint so that lateral load can be transferred horizontally. The joints must also be weathertight when located in exterior walls. Some typical construction details are shown in Fig. 15.25.

The continuation of joint reinforcement or bond beam reinforcing bars across a movement joint is not recommended because reinforcement resists movement and can produce forces that crack the masonry in the vicinity of the movement joint. Where dowel action is required for shear transfer, reinforcement can be continued across the joint provided that it is spliced at the joint and greased or otherwise detailed to permit in-plane movement. A shear connection, as shown in Fig. 15.26, can be used.

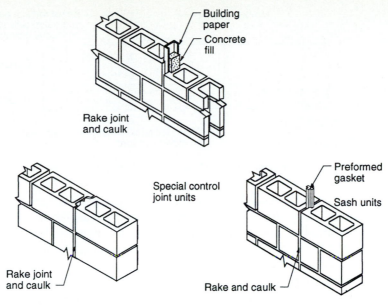

(a) Concrete Masonry (From ref. 15.15)

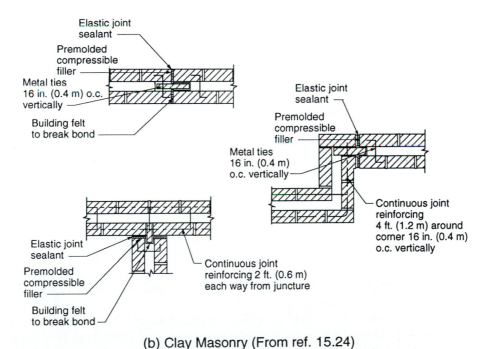

(b) Clay Masonry (From ref. 15.24)

**Figure 15.25** Construction details of movement joints.

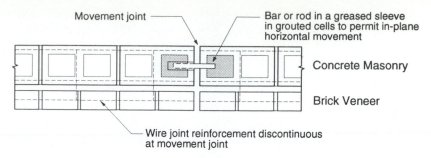

Figure 15.26  Detail of movement joint to transfer shear.

A common mistake is to fill the joint with a material that, compared to masonry, appears to be compressible, but that in fact can transfer significant loads across the movement joint. Some of the stiffer rubber or neoprene filler materials fall into this category of transferring large loads at relatively small deformations. Polystyrene rod with caulking or flexible gasket types of joints are generally effective seals without creating restraint to movement. A double layer of backer rod and caulking is used to provide better water penetration resistance; see Fig. 14.23.

## 15.7 INSPECTION AND QUALITY CONTROL

### 15.7.1 Introduction

The quality of masonry construction can be affected by the quality control and quality assurance programs implemented. The main factors involved are

- definition of the level of quality expected
- assignment of responsibilities
- familiarity of the involved personnel with the requirements and procedures necessary to obtain satisfactory quality of masonry construction

The preparation of clear and concise design details and specifications is an important first step in achieving good quality construction. The masonry industries and the standards organizations in most countries have information that is very useful in the preparation of these items. For example, ACI 530.1/ASCE 6/TMS 602[15.1] contains a specifications checklist that is intended to help the architect/engineer choose and specify the necessary mandatory and optional requirements to be included in the design specifications.

For concrete or steel structures, inspections can be scheduled for key times such as before and during pouring of concrete, after welding, etc. However, masonry construction is basically a continuous process. Therefore, it might be assumed that continuous inspection is necessary, but this is not practical for all but very large projects. Because, in most cases, inspection will have to be limited to periodic visits, it is extremely important to clearly establish standards of workmanship, methods of construction, and interpretation of the design documents at the outset of the project.

The contract document should set out the quality control and inspection requirements for the project. Inspection plays a large part in ensuring that the specifications are achieved on site.

### 15.7.2 Quality Assurance and Quality Control

*Quality assurance* generally refers to the owner's efforts to define and measure quality and to set up a management process to determine compliance with the project specifications. This function can be satisfied by the inspector, whose job is to assure the designer and the owner that all work complies with the approved design documents. It is desirable to build a *mockup panel*, before the start of the job, that can be used to check appearance and workmanship. Once this panel is constructed and accepted by the owner's representative, it becomes the established standard of quality for the masonry work on the project. The accepted panel should include masonry unit geometry, color, texture, and extent of chippage, and mortar joint size, color, tooling, and texture. The mockup panel should also include other construction details such as weep holes, flashing, sealing of joints, anchors, ties, reinforcement, and grouting.

*Quality control* pertains to the steps taken by the contractor on site to comply with the specifications and design drawings. Example operations include preparation of the site; placement of mortar, units, reinforcement, connectors, grout, and flashing; and the adherence to proper dimensional tolerances. Testing of masonry materials and insistence on proper standards of workmanship are also part of quality control. Aspects of workmanship are discussed in Secs. 15.2 and 15.3.

Testing can be an important component of a quality control program because it provides a quantitative measure of the adequacy of materials, workmanship, and in-place performance. Some tests such as for initial rate of absorption (IRA) can also be used to establish appropriate construction methods. Compressive strengths of mortar, grout, and units; compression and flexural tests of masonry assemblages; and water permeance tests are other examples of tests that can be employed.

### 15.7.3 Inspection

Practical knowledge of masonry (including terminology) and a thorough familiarity with appropriate material and construction standards are essential for effective inspection. An inspector must be proficient at reading design drawings and good oral and written communication skills are a definite asset. Honesty, consistency, and willingness to take a position and maintain it until a situation is corrected are also desirable attributes.

Many construction problems and even some design deficiencies can be avoided by proper preparation on the part of the inspector. One of the main keys to success is communication. The inspector should discuss the design with the designer to ensure that their interpretations are in agreement. The inspector should participate in site meetings, both before and during construction. The construction process should be discussed with the contractor to verify that the procedure being adopted is appropriate to producing the designed structure. A clear understanding by all parties of the most important points prior to commencement of construction will often result in better cooperation on the site and a better, more economically built structure.

An inspector should have ready access to referenced construction and material standards. The design specifications must be studied to determine when they are more restrictive than normal practice. When the designer requires higher quality than normal practices provide, review by the inspector with both the designer and the contractor can avoid problems during construction.

The inspector should not rely on the masonry contractor to report deficiencies in other construction that are unacceptable for subsequent masonry construction. For the sake of maintaining good working relationships with the other trades and the general contractor, the mason will usually try to make the best of a bad situation. Aside from the fact that variations from plumb and level are usually quite visible, acceptance of large variations frequently lead to performance problems. For example, if the structural frame is out of plumb, inadequate bearing of veneer on the shelf angle and too small a cavity to permit proper drainage can result.[15.25] For large variations, the designer should be consulted to decide on the appropriate construction procedure or corrective measure.

Because mortar has the potential for the largest job site variations, the mixing and laying procedures are usually a main focus of inspection. Specifications that only indicate the type of mortar are not complete and often lead to argument. First of all, the acceptability of different cementitious materials and the use of any additives should be clearly established. The specification for mortar should be based on either *proportion* or *property*, but not both. The requirements for each are described in ASTM C270. Quality control for mortar specified by either method is by measuring the proportions of ingredients added to the mixer. For masonry specified by property (i.e., compressive strength), quality can be later assured by cube tests. To ensure that consistent proportions are maintained, the inspector should regularly verify that appropriate proportions are being mixed. It is highly desirable that the mortar materials be prequalified prior to writing the final specification. Sand that is clean and has the correct sieve size distribution is necessary.

Ties and anchors are discussed in detail in Chap. 13. It is suggested that project documents should identify the precise requirements. In addition to ensuring that all the ties and anchors are installed properly, the inspector should verify that the correct galvanizing for corrosion resistance is provided. Ties that are already rusting prior to being placed in the wall will likely continue to deteriorate.

In some cases, joint reinforcement is incorporated into the wall to control cracking due to deformations in the structure. In other cases, it is also designed to act as flexural reinforcement for bending in the horizontal direction. In either case, it is necessary that the lengths be properly lapped to achieve continuity. This is easily achieved when using No. 9 gage side rods, but lapping $\frac{3}{16}$ in. (5 mm) rods in a $\frac{3}{8}$ in. (10 mm) joint creates significant problems. Identification of such problems can often lead to modification of the design and a better built structure. A proper splicing technique and staggering of splices are shown in Fig. 15.27.

In the northern U.S.A. and in Canada, the cold climate places severe demands on the integrity of wall systems. Gaps, cracks, and other holes permit drafts and, more important, permit loss of warm air, which results in transmission of large amounts of moisture into the wall. As discussed in Chap. 14, the importance of factors

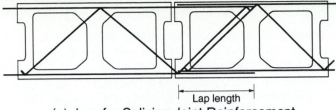

(a) Lap for Splicing Joint Reinforcement

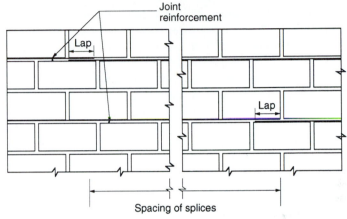

(b) Staggering of Laps for Splicing Joint Reinforcement

Note: For clarity joint reinforcement is shown at top and bottom of mortar joint at the lap locations. They should be at the middle of the joints with mortar at the top and bottom. The offset is in the horizontal plane as shown in (a)

**Figure 15.27** Lapping and staggering of splices of joint reinforcement.

affecting the potential accumulation of moisture in the wall should be recognized. The inspector must be aware of potential areas where the vapor and/or air barriers can be breached. Attention to details adopted to restrict cracking is important.

## 15.8 CLOSURE

Most of the chapters in this book contain some reference to construction related considerations for masonry. However, this chapter contains specific information related to workmanship, effects of weather conditions, installation of flashing and dampproof courses, requirements for movement joints, and inspection and quality control. Training manuals for masons are useful reading for those involved on the job site. A good designer should be familiar with construction methods so that the design will follow procedures that can be effectively carried out by the mason. Use of simple details is a key to achieving good quality construction.

Because it looks simple, designers may have a tendency to discount the skill required to produce high quality masonry. In this regard, taking an opportunity to lay some bricks and blocks could be a useful experience for designers.

## 15.9 REFERENCES

15.1 The Masonry Standards Joint Committee, "Building Code Requirements for Masonry Structures," ACI 530/ASCE 5/TMS 402, and "Specifications for Masonry Structures," ACI 530.1/ASCE 6/TMS 602, ACI, Detroit; ASCE, New York; TMS, Boulder, 1992.

15.2 British Standards Institution, "Code of Practice for Use of Masonry, Part 1. Structural Use of Unreinforced Masonry." BS 5268, BSI, London, 1978 (confirmed 1985).

15.3 Canadian Standards Association, "Masonry Construction for Buildings," CSA Standard A371-1984, CSA, Rexdale, Ontario, 1984.

15.4 Association of Consulting Structural Engineers of New South Wales, *Australian Masonry Manual*, Sydney, Australia, 1991.

15.5 Standards Association of New Zealand, "Code of Practice for the Design of Masonry Structures," NZS 4230, SANZ, Wellington, 1985.

15.6 A.W. Hendry, *Structural Brickwork*, Macmillan, 1981, pp. 34–43.

15.7 J.A. James, "Investigation of the Effect of Workmanship and Curing Conditions on the Strength of Brickwork," in *Proceedings of the Third International Brick Masonry Conferences*, Essen, Germany, 1973, pp. 192–201.

15.8 G.N. Chahine, "Behaviour Characteristics of Face Shell Mortared Block Masonry under Axial Compression," M.Eng. Thesis, McMaster University, Hamilton, Ontario, 1989.

15.9 J.G. Gross, R.D. Dikkens and J.C. Grogan, *Recommended Practice for Engineered Brick Masonry*, Brick Institute of American, Reston, VA, 1969.

15.10 R.G. Drysdale, and M. Wilson, "Tests of Full Scale Brick Veneer / Steel Stud Walls to Determine Strength and Rain Penetration Characteristics," Technical Report, Project Implementation Division, Canada Mortgage and Housing Corporation, Ottawa, 1990.

15.11 C.T. Grimm, "Water Permeance of Masonry Walls: A Review of the Literature", ASTM STP 778, J.G. Borchett, Editor, Philadelphia, PA, pp. 178–199.

15.12 Canadian Standards Association, "Mortar and Grout for Unit Masonry," CSA Standard A179-1976, Rexdale, Ontario, 1976.

15.13 Standards Association of Australia, "Masonry in Buildings (SAA Masonry Code)," AS 3700-1988, Sydney, New South Wales, 1988.

15.14 W. Panarese, S. Kosmatka and F. Randall, "Concrete Masonry Handbook," 5th Ed., Portland Cement Association, Skokie, IL, 1991.

15.15 C. Beall, *Masonry Design and Detailing*, 2nd Ed., McGraw-Hill, New York, 1987.

15.16 International Conference of Building Officials, "Uniform Building Code", Chap. 24, "Masonry", Whittier, CA, 1991.

15.17 J. Amrhein, *Information Guide to Grouting Masonry*, Masonry Institute of America, Los Angeles, 1990.

15.18 A. Elmiger, *Architectural and Engineering Concrete Masonry Details for Building Construction*, National Concrete Masonry Association, Herndon, VA, 1976.

15.19 International Masonry Institute, *Recommended Practices and Guide Specifications for Masonry Construction—Cold Weather*, IMI, Washington, 1981.

15.20 K. Orantie, *Development of Masonry Technique*, Technical Research Centre of Finland, 1984.

15.21 P. Rikert, "Wind Damage to Building Walls During the Construction Period," Safety Newsletter, Construction National Safety Council, Chicago, IL, 1958.

15.22 R. Carr, R. Woods and J. Heslip, "Guidelines for Temporary Masonry Wall Support

Systems," ASTM STP 871, Grogan/Conway, Editors, ASTM, Philadelphia, PA, 1985, pp. 114–125.

15.23 W.M. McGinley and J.G. Borchelt, "Friction at Supports for Clay Brick Walls," in *Proceedings of the Fifth North American Masonry Conference*, University of Illinois, Urbana, 1990, pp. 1053–1066.

15.24 Brick Institute of America, "Differential Movement Expansion Joints," Technical Note 18A, BIA, Reston, 1988.

15.25 J.G. Borchelt, "Construction Tolerances and Design Considerations for Masonry Veneer and Structural Frames," in *Proceedings of the Third North American Masonry Conference*, Arlington, TX, 1985, pp. 12.1–12.16.

## 15.10 PROBLEMS

**15.1** Describe the two methods of grouting. Discuss the advantages and the disadvantages of the two methods.

**15.2** What is the maximum grout pour height allowed by ACI 530.1/ASCE 6/TMS 602 specifications for the 6 in. (150 mm), 8 in. (200 mm), and 12 in. (300 mm) hollow unit masonry available in your local area? Coarse grout is specified.

**15.3** For a 6 in. (150 mm) reinforced concrete masonry wall 20 ft (6 m) high, state the tolerances specified by the ACI 530.1/ASCE 6/TMS 602 specifications regarding the following:
  **(a)** Variation from level.
  **(b)** Variation from plumb.
  **(c)** Location of vertical steel.

**15.4 (a)** Visit a masonry construction job site in your area and report on the preparation and placement of mortar and masonry units.
  **(b)** Visit a masonry construction job site in your area and report on the protection of masonry materials and the bracing and covering of partially finished construction. Report on the local practices regarding hot- or cold-weather construction. Compare them with the ACI 530.1/ASCE 6/TMS 602 specifications.
  **(c)** Visit a masonry construction job site in your local area and report on the location, type, and details of the movement joints used. [Note: Before entering a job site, permission should be obtained from the job site superintendent. A hard hat and safety shoes are normally required.]

**15.5** Select locations of vertical movement joints in the perimeter walls of the unreinforced loadbearing block masonry buildings shown in Fig. P15.5. Explain.

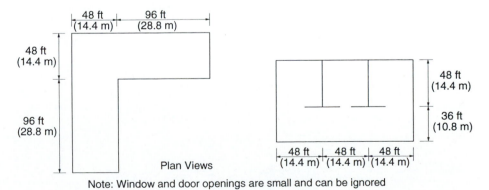

Note: Window and door openings are small and can be ignored

**Figure P15.5**

**15.6** Indicate the recommended locations for movement joints in the brick veneer walls shown in Fig. P15.6. Discuss the reasons for your recommendations.

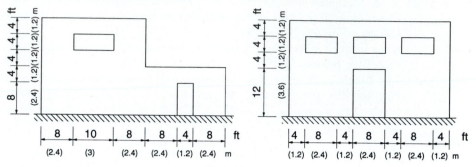

**Figure P15.6**

**15.7** Review a set of masonry plans and specifications for a building and identify the key aspects of construction to the be looked into by the inspector.

**15.8** On an existing masonry building, check and report on the construction tolerances for the masonry.

# CHAPTER 16

# Design of Loadbearing Single-Story Masonry Buildings

Single-story commercial building (*Courtesy of Ontario Concrete Masonry Association*).

## 16.1 GENERAL INTRODUCTION

Single-story buildings form by far the largest part of building construction whether measured by usable floor area or total construction cost. Correspondingly, this type of construction is the main area of activity for many design firms and contracting companies. Unfortunately, perhaps because of their seeming simplicity, design of single-story buildings with loadbearing wall systems has been an often neglected topic in the academic training of designers. The design considerations and construction details that are specific to loadbearing single-story masonry buildings include the use of relatively flexible roof diaphragms, tall slender walls, and the inclusion of large openings.

Many low-rise buildings have been constructed using structural steel or reinforced concrete frames combined with masonry infill walls. Although this form of construction provides strong, durable walls with many desirable properties, such as excellent fire resistance and sound transmission characteristics, use of loadbearing masonry has the potential to provide more cost effective buildings. Properly designed loadbearing walls can provide many times the required strengths and, rather than having to incorporate movement joints to permit flexible frames to deflect, the inherent stiffness of the masonry walls can be utilized.

Loadbearing masonry wall construction increases construction efficiency by avoiding the need to anchor to, build around, and build under a structural frame. Maintaining continuity of insulation, air barriers, and vapor barriers is simplified (see Chap. 14). In addition, the minimum strip footings and foundation walls, required to support nonloadbearing walls, are usually adequate for support of the additional roof loads. Thus, column foundations are eliminated without the need for special strip footings.

Modern loadbearing masonry construction utilizes walls that are both the structure and the architectural finish, and that combine long-lasting durability with excellent

- fire resistance
- acoustic properties
- resistance to user damage
- accommodation of attachments
- integration of thermal, air, and vapor barriers

Simple empirical design of single-story masonry buildings is popular and has produced satisfactory designs. However, the limitations inherent in simple rules can produce cases where the full potential of masonry construction is not utilized. Alternately, design deficiencies can result from simple application of height-to-thickness criteria while ignoring other aspects of design. As in any building design, the effects of combinations of loading and the particular layout of the structure must be thoroughly considered.

In this chapter, the design of single-story loadbearing masonry buildings based on engineering analysis is discussed and illustrated in examples. This information serves as background for designers to provide solutions appropriate to particular design conditions. Experience has shown that designers can quickly become proficient and for very little extra design time, produce more cost-effective designs.

To avoid excessive repetition of material, reference is made to material covered in Chaps. 6 through 10 in which the designs of structural masonry elements were introduced. To illustrate the main points in the design of single-story loadbearing masonry buildings, coverage of the basic design considerations is followed by an example building design.

## 16.2 BEHAVIOR, FORM, AND LAYOUT

### 16.2.1 Wall Layout Requirements for Stability

Loadbearing single-story masonry construction tends to result in designs with a predominance of walls oriented in one direction. These walls carry the gravity loads and separate the areas. In the extreme, a system of parallel walls, such as illustrated in Fig. 16.1(a), is unstable either as a result of lateral load or an internal condition

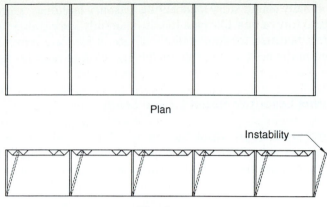

Plan

Instability ⎯

Elevation

(a) Unstable Building with Masonry Walls Oriented in One Direction

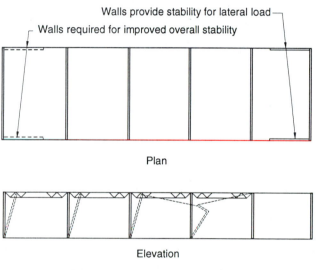

Walls provide stability for lateral load ⎯

Walls required for improved overall stability

Plan

Elevation

(b) Improved Stability of Building

**Figure 16.1** Wall layouts.

such as the failure of a wall. Perpendicular walls are required to resist lateral seismic or wind forces, and sway due to unbalanced loading. However, even though the floor plan shown in Fig. 16.1(b) can adequately resist lateral loads from all directions, the left end of the building consisting of the parallel walls is still vulnerable to progressive collapse due to failure of an element, as shown in the elevation. An additional pair of perpendicular walls at the other end of the building would greatly improve the overall stability. A common sense approach to wall layout should avoid a potential "domino effect" or "house of cards" and result in buildings that have good lateral load resistance coupled with sufficient redundancy and toughness to withstand damage to local areas. Chapter 3 contains additional discussion on this topic.

For many single-story loadbearing masonry buildings, it is convenient to arrange the walls to form a box-like plan that is inherently very stable. The resulting symmetry is ideal because torsional effects of lateral load are minimized. However, as is discussed in the next section, the building use often requires large openings in some walls.

### 16.2.2  Wall Layout to Resist Lateral Loads

**Rigid Diaphragm Roof Systems.**  A physical feature of many single-story buildings is an open front to allow maximum window space. For this arrangement to be feasible, the roof system must be designed as a stiff diaphragm so that the existing walls can combine to resist torsion loading. Torsional moments arise from asymmetric layouts of shear walls, such as illustrated in Fig. 16.2. The rear wall provides the only effective resistance to direct shear due to wind and seismic loadings in the E–W direction. The distance between the resultant lateral shear force and the center of rigidity C.R. creates torsion in the building.

A 2.5 in. (65 mm) thickness of concrete on corrugated steel decking should provide adequate stiffness to form a rigid diaphragm. This construction is quite common as a floor system, but may not be desirable for roofing because of the additional weight of the concrete. Where only corrugated steel decking is used, proper connection of the corrugated sheets and design of an in-plane bracing system can be used to create a sufficiently stiff roof to act as the top support for the walls. However, the diaphragm action of such a roof system generally is far too

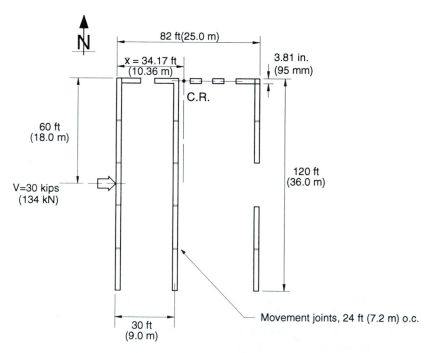

**Figure 16.2**  Floor plan for a building with an open front.

flexible to mobilize the combined torsional resistance of the masonry walls, which are comparatively very stiff.

### Example Calculation for a Rigid Diaphragm Roof System.

To illustrate the distribution of shear forces for buildings with rigid diaphragm roofs, let us assume that a rigid roof diaphragm is achieved for a building with the floor plan shown in Fig. 16.2 and walls 15 ft (4.5 m) high. [An alternative design is discussed later for the case of a more flexible roof system.] Consider that the lateral load resultant of 30 kips (134 kN) is applied midway along the side walls as shown in Fig. 16.2.

To locate the center of rigidity of the walls in each direction, the rigidities of the walls times their distance to the center of rigidity for shear loading must be balanced. Therefore, for loading in the East–West direction, the center of rigidity of the building is located at the mid-thickness of the North wall (i.e., 3.81 in. (95 mm) from the North face for an assumed wall thickness of $7\frac{5}{8}$ in. (190 mm)). The center of shear rigidity of the building for loading in the North–South direction can be found by taking the rigidities of the North–South walls times their distances from the outside face of the East wall divided by the total rigidity of these walls. Because the individual 24 ft (7.2 m) long wall panels, in the North–South direction, have the same length and thickness, rigidity $R$ is the same for each panel. Therefore,

$$\bar{x} = \sum_{i=1}^{n} R_i x_i \bigg/ \sum_{i=1}^{n} R_i \tag{16.1}$$

$$= \frac{5R(3.81/12) + 5R(30) + 4R(82 - 3.81/12)}{5R + 5R + 4R} = 34.17 \text{ ft}$$

$$\left( \bar{x} = \frac{5R(0.095) + 5R(9) + 4R(25 - 0.095)}{14R} = 10.36 \text{ m} \right)$$

where $R_i$ = rigidity of wall $i$

$x_i$ = perpendicular distance from wall i to the center of rigidity.

The position of the center of rigidity is shown in Fig. 16.2.

The torsional moment $M_T$ created by the lateral East–West force $V$ is equal to the force $V$ times the moment arm due to its eccentricity from the center of rigidity:

$$M_T = 30(60 - 3.81/12) = 1790 \text{ ft-kips}$$

$$(M_T = 134(18 - 0.095) = 2399 \text{ kN.m})$$

[Note: Design of such an asymmetric structure is not recommended and, as can be seen, very large torsional moments result. Although the walls can be designed to resist the torsional effects, this configuration of walls does not produce a structure that is inherently *tough* or *robust* with redundant load-resisting mechanisms. A more detailed discussion of these wall layout considerations is included in Sec. 3.6.]

The torsional rigidity $R_T$ of the building is calculated as

$$R_T = \sum R_i x_i^2 \tag{16.2}$$

In this example, with similar wall panels of rigidity $R$

$$R_T = 5R(34.17 - 3.81/12)^2 + 5R(34.17 - 30)^2 + 4R(82 - (3.81/12) - 34.17)^2$$

$$= 14{,}847R \text{ ft}^2 \, (1382R \text{ m}^2)$$

[Note that the North wall does not contribute to the torsional rigidity because the center of rigidity of the building is located at its mid-thickness.]

The shear force $V_i$ in wall $i$ due to torsion can be found from the following equation:

$$V_i = \frac{M_T R_i x_i}{R_T} \tag{16.3}$$

The maximum torsional shear force occurs in the wall panels of the East wall because they are farthest from the center of rigidity.

$$V_{\text{East}} = \frac{1790R(82 - (3.81/12) - 34.17)}{14{,}847R} = 5.73 \text{ kips per wall panel}$$

$$\left( V_{\text{East}} = \frac{(2399)R(25 - 0.095 - 10.36)}{1382R} = 25.25 \text{ kN per wall panel} \right)$$

Considering the face shell mortar bedding of two 1.25 in. (32 mm) strips for stress calculation, the corresponding shear stress is

$$f_{v_{\text{East}}} = \frac{V_{\text{East}}}{A_{\text{East}}} = \frac{5.73 \times 1000}{24 \times 12(2 \times 1.25)} = 8.0 \text{ psi}$$

$$\left( f_{v_{\text{East}}} = \frac{25.25 \times 1000}{7.2 \times 1000(2 \times 32)} = 0.055 \text{ MPa} \right)$$

This shear stress is well below typical code allowable stresses for in-plane shear.

It may be of some interest to determine the angle of twist of the building. The angle of twist $\Phi$ at the top of the building can be determined from

$$\Phi = \frac{M_T}{R_T} \tag{16.4}$$

As the 24 ft (7.2 m) long wall panels are separated by movement joints, the rigidity of the panel R can be determined from Fig. 10.12 or Eq. 10.6 in Chap. 10 for cantilever behavior.

For

$$\frac{h}{L} = \frac{15}{24} = 0.625, \qquad R_c = 0.35Et$$

and for modulus of elasticity $E_m = 1{,}500{,}000$ psi (10,350 MPa) and $t = 2 \times 1.25 = 2.5$ in. (64 mm)

$$R_c = 1.31 \times 10^6 \text{ lb/in. } (2.32 \times 10^5 \text{ N/mm})$$

Hence, the torsional rigidity is

$$R_T = 14{,}847(1.31 \times 10^6) \times \frac{12}{1000} = 2.33 \times 10^8 \text{ ft-kips}$$

$$(R_T = 1382(2.32 \times 10^5) = 3.2 \times 10^8 \text{ kN.m})$$

From Eq. 16.4,

$$\Phi = \frac{M_T}{R_T} = \frac{1790}{2.33 \times 10^8} = 7.7 \times 10^{-6} \text{ radians} \qquad \left( \frac{2399}{3.2 \times 10^8} = 7.5 \times 10^{-6} \right)$$

The corresponding maximum deflection perpendicular to the wall, $\Delta_t$ (i.e., out-of-plane), due to the twist, will occur at the front of the building, which is a perpendicular distance of 120 ft (36 m) minus half the wall thickness from the center of rigidity:

$$\Delta_t = \Phi x = 0.0000077(120 \times 12 - 3.81)$$

$$= 0.011 \text{ in.}$$

$$(\Delta_t = 0.0000075(36 \times 1000 - 95) = 0.27 \text{ mm})$$

This very small deflection at the top of the wall will not introduce significant out-of-plane bending into the wall and the ability of the wall to resist out-of-plane loading is not compromised.

For this example, calculation of rigidity based on wall areas would not be grossly in error because shear deformations and shear stress predominate for such squat walls. This is equivalent to using a shear rigidity:

$$R_{\text{shear}} = \frac{1.2h}{LtG} \qquad \text{for rectangular walls} \qquad (16.5)$$

where $G$ = the shear modulus usually taken as 0.4 E
$\quad\; t$ = effective thickness of the wall = 2 × face shell thickness for hollow masonry

For the direct shear along the North wall, the shear force is distributed to its various segments in proportion to their relative rigidities. Background to the calculation of these relative rigidities is given in Sec. 10.4.4.


**Non-Rigid Roof Systems.** The roof systems normally used for the type of building in the foregoing example are usually not sufficiently rigid to provide the diaphragm action necessary for the prior analysis to be valid. In these cases of non-rigid roof diaphragms, lateral wind pressure on the side walls is distributed to transverse walls by the roof system, which acts as a horizontal beam. To satisfy equilibrium for a building without torsional stiffness, at least one additional East–West wall is required in the foregoing example. The front of the building is the most obvious location for the additional walls and is also the location that will require the least shear resistance. From statics, this amounts to one-half of the total.

**Example Calculation for a Non-Rigid Roof System.** For the building plan shown in Fig. 16.2 and with the 30 kip (134 kN) loading used in the previous example, the combined length of the additional walls placed along the front (South) end of the building must be able to resist a horizontal shear of:

$$V_{\text{South}} = 30 \div 2 = 15 \text{ kips (67 kN)}$$

If the allowable shear stress for unreinforced masonry is taken as 40 psi (0.28 MPa), the required length of wall made up of 8 in. (200 mm) hollow units with face shell mortar bedding thickness $t$ of $2 \times 1.25$ in. $= 2.5$ in. (64 mm) is

$$L = \frac{V}{40t} = \frac{15 \times 1000}{40(2.5)} = 150 \text{ in.} = 12.5 \text{ ft}$$

$$\left( L = \frac{67 \times 1000}{0.28(64)} = 3738 \text{ mm} = 3.74 \text{ m} \right)$$

Grouting the walls solidly would substantially reduce the length of shear wall required. If shear governs the design, the required length of grouted wall would be

$$L = \frac{V}{40t} = \frac{15 \times 1000}{40(7.625)} = 49.2 \text{ in.} = 4.1 \text{ ft}$$

$$\left( L = \frac{67 \times 1000}{0.28(190)} = 1259 \text{ mm} = 1.26 \text{ m} \right)$$

However, the required combined length of walls along the front (South face) of this building could be controlled by in-plane bending and axial load or by limits on the in-plane deflection of these walls. The latter limit is particularly relevant if these walls are reinforced to permit cracking resulting in greatly reduced flexural stiffness. This deflection can have a significant effect on the bending introduced into the bases of the perpendicular walls (i.e., walls running North–South). For instance, a deflection of $\frac{1}{16}$ in. (1.5 mm) at the top of the wall will cause an out-of-plane bending moment at the south end of the North–South walls, which can be determined from consideration of the cantilever action shown in Fig. 16.3.

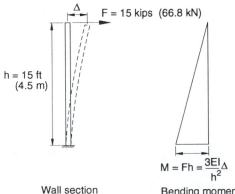

$$M = Fh = \frac{3EI}{h^2}\Delta$$

Wall section     Bending moment diagram

**Figure 16.3** Illustration of effect of lateral displacement at top of a wall.

By using a moment-area approach, the deflection $\Delta$ at the top of the wall can be calculated as

$$\Delta = \frac{Fh}{EI}\frac{h}{2} \times \frac{2}{3}h = \frac{1}{3}\frac{Fh^3}{EI} \qquad (16.6)$$

With $h = 15$ ft (4.5 m) and using a typical 1 ft (1 m) strip of wall, $I = 309$ in.$^4$/ft (4.05 $\times$ 10$^8$ mm$^4$/m) and an $E$ value of 1,500,000 psi (10,350 MPa), the moment at the bottom of the wall (M = Fh) required to cause a $\frac{1}{16}$ in. (1.5 mm) deflection at the top is

$$M = \frac{3EI}{h^2}\frac{\Delta}{} = \frac{3(1,500,000)(309)\dfrac{1}{16}}{(15 \times 12)^2} = 2682 \text{ in.-lb/ft} = 224 \text{ ft-lb/ft}$$

$$\left(M = \frac{3(10,350)(4.05 \times 10^8)(1.5)}{(4500)^2} = 931,500 \text{ N.mm/m}\right)$$

which results in an extreme fiber flexural stress of

$$f_t = \frac{M}{I}y = \frac{224 \times 12}{309} \times 3.81 = 33 \text{ psi}$$

$$\left(f_t = \frac{931,500 \times 95}{4.05 \times 10^8} = 0.22 \text{ MPa}\right)$$

This is additional to any stresses produced by bending in the wall due to directly applied lateral load. Reinforced walls are much less sensitive to this deflection in-duced, out-of-plane bending moment.

Axial compression from the weight of the wall and roof loading can be utilized to counteract tension stresses developed by deflection of the shear walls, which in turn allows more deflection without cracking. However, this example illustrates that high tensile stresses can occur in fixed-base walls due to movement at the top. Walls that are simply supported at the floor and roof levels are not sensitive to this type of movement as they are able to rotate about the base to accommodate this very small movement.

The conclusion from the foregoing discussion is that a shear resisting element must have a sufficient stiffness to limit its deflection so as to avoid overstressing other walls that must deflect the same amount.

For the wall configurations in Fig. 16.2, the three parallel North–South walls can easily resist the direct shear and the small amount of torsion that result from wind pressure and seismic forces acting in the North–South direction.

## 16.3 DESIGN LOADS

Loading conditions for buildings are introduced in Chap. 3. Loads especially relevant to single-story buildings are reviewed in this section.

### 16.3.1 Gravity Loads

Dead loads include the weight of the structure and other items such as roofing material and services. In cases where uplift due to wind can cause reverse bending in the roof structure and axial tension in the walls, it is not conservative to overestimate the minimum dead loads. In such cases, the combination of uplift force and minimum gravity load is most critical.

Live loads include equipment, snow, and ponding of water. For single-story buildings, the snow load can often be the greatest. The slope of the roof, direction of the wind, the presence of obstructions that tend to shelter the roof and promote deep drifts, and the heating conditions within the building all affect the magnitude of snow loading.

### 16.3.2 Wind Loading

For the high walls generally used in single-story buildings, out-of-plane bending of the walls due to wind load is often the critical load condition. Axial compression load from the roof will increase the resistance to tensile stresses due to lateral wind loading. On the other extreme, separation of the roof from the wall due to uplift can lead to wall failure. Removal of the top support leaves a wall cantilevered from its base and, thus, much less able to resist lateral load. Properly engineered and adequately detailed loadbearing masonry buildings have very good service records.

Isolated walls or wall segments can be significantly affected by localized pressures. The peak pressure coefficient for the combined effects of internal and external pressures must be used for the design of individual walls. As is discussed in the next section, although most buildings will be fairly uniformly sealed, they have doorways or windows that may produce a significant imbalance in air leakage.

The effects of localized peak pressures can be reduced by introducing shear connections across vertical movement joints so that lateral pressures will be shared over larger areas. Also, at corners of buildings, intersecting walls can provide side supports to greatly reduce bending near the corners, which tend to be areas of high localized wind pressures.

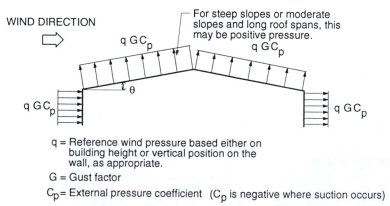

**Figure 16.4** Illustration of wind pressure distribution on the exterior of a single-story building.

The primary structural function of loadbearing walls dictates that a 1 in 30 year to 1 in 100 year reference velocity pressure $q$ should be used, depending on the local building code and the importance of the building. Design wind pressures and forces are presented and discussed in Sec. 3.2.2. The critical combination of positive pressure and suction (negative pressure) should be used to design for the overall strength and stability of the structure. Figure 16.4 is an illustration of this combined effect.

For isolated individual walls and secondary structural members, localized external peak pressures can exceed significantly the average peak pressure attributed to an entire wall of a building. In these cases, the appropriate (positive or negative) external peak pressure should be combined with the internal peak pressure to produce the most critical combination for bending and axial load.

### 16.3.3 Seismic Loading

The procedure for determining seismic loading is discussed in Sec. 3.2.2. For single-story buildings, the self-weights of the walls usually contribute the major part of the inertia forces.

Seismic loading due to the mass of the walls is a uniformly distributed load in the direction of the earthquake motion. Therefore, walls can have out-of-plane and in-plane seismic loading, depending on their orientation relative to the analyzed direction of motion. The component of the uniformly distributed seismic load normal to the surface of a wall can be treated in the same way as a wind force. To include the force due to the mass of the roof, a horizontal line load in proportion to the mass of the roof is distributed along the roof line.

## 16.4 DESIGN OF COMPONENTS

### 16.4.1 Design of Walls for Axial Load and Bending

Walls resist vertical loads from gravity and the effect of uplift due to wind. Walls also carry bending from out-of-plane lateral loads due to wind (see Fig. 16.5) or seismic forces. The walls typically span vertically between the roof and the footing, although intersecting walls and rigid pilasters can produce panels supported along the sides (see Chap. 7) and such walls can be designed for two-way bending.

Depending on the detail, the bottom of the wall can be considered simply supported or fixed. This will alter the value of moments and shape of the moment diagram, as shown in Fig. 16.5. In both cases, the maximum moments are equal. However, having the maximum moment at the base can have some advantages. For tension controlled capacity, the extra weight for half the height of the wall will result in additional moment resistance. Also, where the maximum moment is at the base, it does not occur at locations where windows reduce the wall sections. However, flashing through the bases of unreinforced walls and narrow or shallow foundations may not permit development of fixed-base conditions.

Some designers have adopted the practice of extending the bottom chord of the roof truss to the wall to reduce the effective height of the wall. This can result in very high bending moments in the wall at this location due to rotation at the end

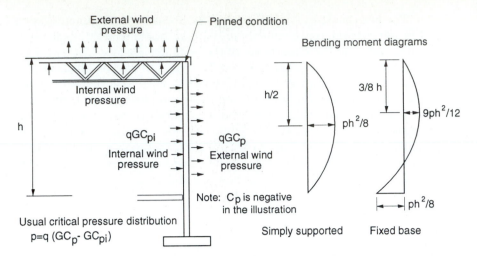

**Figure 16.5** Critical wind loading condition for bending moment on a wall.

of the roof truss. Except where calculated forces from frame action are taken into account, this detail is not recommended. Connection only at the bearing plate of the truss creates a simple support condition, which allows easy and accurate prediction of behavior.

Openings in a wall result in a reduced section to resist the moment and axial load. In general, the additional axial load and bending transferred to piers between openings or to edge members can be reasonably estimated using the tributary area concept. The tributary area for the additional load is the vertical wall strip defined by the size of the opening. However, the load path to the supports should be analyzed. Large openings can produce high horizontal bending moments in the wall sections above and below the opening as load normal to the wall surface is transferred horizontally to the piers. Reinforcement (or extra reinforcement) may be required. The length of wall that can be reasonably expected to share in carrying the redistributed axial load and bending must be rationally analyzed. Lacking other information, it may be reasonable to assume that masonry built in running bond will distribute load over an area defined by a 45° to 60° slope from the point of application.

Bearing stresses under concentrated loads must also be checked.

### 16.4.2 Design of Walls for In-Plane Shear and Bending

As indicated in Fig. 16.6, the roof acts as the top support for walls subject to out-of-plane bending due to lateral loads. If the wall acts as a simply supported member, the reaction at the roof is half of the uniformly distributed load on the wall, whereas if the base is fixed, it is 3/8 of this load. The roof is in turn supported horizontally by cross-walls that share the lateral load carried by the roof and transfer this load to the foundation through in-plane shear and bending. The horizontal load along the roof is usually distributed to the cross-walls in proportion to tributary areas because roof systems normally used for single-story buildings are relatively flexible compared

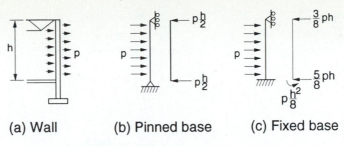

(a) Wall        (b) Pinned base        (c) Fixed base

**Figure 16.6**   Roof reactions from horizontal wind loads.

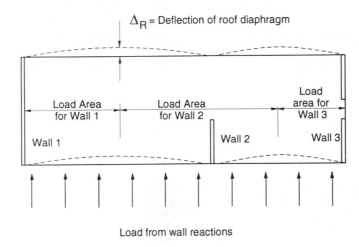

**Figure 16.7**   Deflection of flexible roof diaphragms and tributary load areas for cross walls.

to masonry walls. For flexible roof diaphragms, load sharing according to the stiffnesses of the walls does not occur. A simple illustration of the load distribution to walls is shown in Fig. 16.7.

Walls subject to in-plane forces should be designed using the procedures discussed in Chap. 10.

### 16.4.3 Design Requirements for Roof Diaphragms

**Background.** The roof system acts as a deep horizontal beam to distribute the reactions from supported walls to the cross-walls. Manufacturer and material associations provide information to assist with the calculation of the horizontal deflection of their roof systems. Consideration of this behavior is important because roof deflection results in out-of-plane deflections of the walls, which, in turn, can cause flexural stress (see Sec. 16.2.2). This is particularly critical for unreinforced masonry construction where development of additional tension can cause cracking.

For a roof deflection $\Delta_R$ (see Fig. 16.7) at the top of a propped cantilever wall, as shown in Fig. 16.8(a), the bending moment developed at the fixed base can be

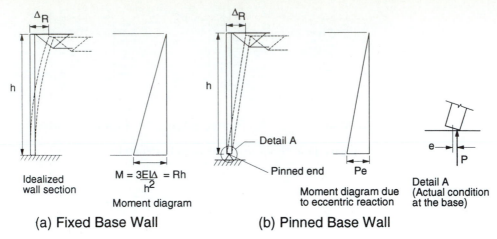

**Figure 16.8**  Illustration of effect of lateral displacement at the top of a wall.

calculated in the same way as done in Sec. 16.2.2. Thus:

$$M = \frac{3EI}{h^2} \Delta_R \qquad (16.7)$$

from which the additional flexural tension and compressive stresses can be calculated.

For the 15 ft (4.5 m) high, 8 in. (200 mm) concrete block wall discussed in Sec. 16.2.2, a deflection of $\frac{1}{16}$ in. (1.5 mm) introduced a flexural stress of 33 psi (0.22 MPa). This stress is significant for unreinforced masonry.

For walls assumed to be pinned at the bottom support, a true hinge, as shown in Fig. 16.8(b), would allow the wall to rotate about this bottom hinge to follow the roof deflection without introducing an additional bending moment. However, the base of the real wall has some thickness and, as shown, the actual condition at the base as the wall rotates will be a movement of the vertical reaction to some eccentricity from the center of the wall. This results in the bending moment diagram shown in Fig. 16.8(b). For light axial loads, this moment may not be significant, but for heavier loads, restrictions on roof diaphragm deflection similar to those for unreinforced walls with fixed bases may be necessary.

Roof diaphragm deflections are generally less critical for reinforced walls, but the resulting moments should be added to the moments due to lateral load, shown in Fig. 16.5.

**Example Calculation for Extra Stiffening of the Roof.**  In many cases, it is necessary to stiffen the roof system. For steel roof joists, this can be accomplished by running cross-bracing in the plane of the roof at the top chord of the joists. Simple calculations show that very small steel angles are often quite effective in limiting roof diaphragm deflections.

For the roof plan in Fig. 16.9, the wall reaction of 120 lb/ft (1.75 kN/m) along the roof line is uniformly distributed as shown. By assuming that the roof trusses transfer this load to the cross-bracing at the points of contact, the average force over

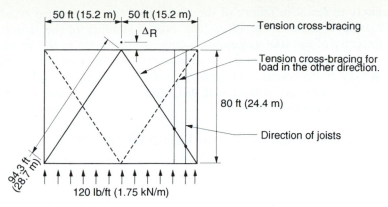

**Figure 16.9** Illustration of cross bracing to limit roof deflection.

the length of each brace is

$$\left(\frac{1}{2} \times 50\right)(120)\left(\frac{94.3}{80}\right) = 3540 \text{ lb}$$

$$\left(\frac{1}{2} \times (15.2)(1.75)\left(\frac{28.7}{24.4}\right) = 15.6 \text{ kN}\right)$$

For a specified limiting deflection (depending upon whether the walls are reinforced or not), the area of bracing can be calculated from conventional structural analysis methods for deflection of trusses. In this case, for a deflection limit of say $\Delta_R = \frac{1}{8}$ in. (3.2 mm), the required cross-sectional area of steel cross-bracing is 0.94 in.$^2$ (600 mm$^2$), which can be provided using a steel angle. If the benefit of the stiffness of the original roof diaphragm is included, the area of the cross-bracing could be reduced.

### 16.4.4 Design of Walls for Hold Down of Roof Systems

**Background.** Many roofs are only designed for gravity loads such as snow, self-weight, and other dead loads. However, the combined wind effects of positive internal pressure and external suction on a roof can be more than double the minimum weight of light roof systems. Thus, there is a net uplift and the roof support system must also be able to hold down the roof.

Standard minimum requirements for anchorage of roofs to masonry walls exist in many design aids and standards such as ACI 530/ASCE 5/TMS 402.[16.1] However, the specifications or illustrated details typically only show the roof anchored into grouted masonry units in the top one or two courses of the wall. It should be emphasized that this is only a minimum provision and is not intended to, nor is it able to, resist any significant uplift of the roof. A proper design is required.

The information and example presented in this section are intended to aid in the design of proper anchorage of roofs to masonry walls to resist uplift. In addition to the integrity of the roof system itself, the use of the roof as the top support for the masonry walls adds to its importance. Loss of support at the top of the wall is likely to result in failure of the wall under the quadrupled bending moment due to cantilever action. Therefore, the overall stability of the structure depends on the integrity of the roof system including its anchorage to the walls.

It should also be realized that uplift from the roof can relieve parts of the walls of axial compression due to gravity loads. These walls must be designed for combined axial tension and bending and the trend in masonry codes is to require reinforcing steel to resist axial tension.

**Example Calculation for Hold Down of a Roof.** If a single-story building with walls as shown in Fig. 16.10 is subject to a basic wind speed of 80 mph (128 km/h), the maximum uplift pressure as per the ASCE 7-88 standard[16.2] is

$$p = q(GC_p - GC_{pi})$$

where $q$ = 13.2 lb/ft$^2$ (0.63 kN/m$^2$) = reference wind pressure for an 18 ft (5.7 m) height

$GC_p$ = external pressure coefficient with gust factor = $-1.4$
$GC_{pi}$ = internal pressure coefficient with gust factor = $+0.75$

$$p = 13.2[-1.4 - (0.75)]$$

$$= -28.4 \text{ lb/ft}^2 \qquad (-1.36 \text{ kN/m}^2) \text{ (i.e., uplift)}$$

The foregoing calculation is based on the worst condition for positive internal pressure combined with suction over the exterior surface of the roof. For roof joists spaced 8 ft (2.4 m) on center and spanning 50 ft (15.25 m), the total uplift force on the end of a joist is

$$P_{\text{uplift}} = 28.4(8 \times 50/2) = 5680 \text{ lb}$$

$$(P_{\text{uplift}} = 1.36(2.4 \times 15.25/2) = 24.89 \text{ kN})$$

If the minimum self-weight of the roof system is 10 lb/ft$^2$ (0.48 kN/m$^2$), the downward joist reaction on the masonry wall due to self-weight is

$$P_{\text{down}} = 10(8 \times 50/2) = 2000 \text{ lb}$$

$$(P_{\text{down}} = 0.48(2.4 \times 15.25/2) = 9.78 \text{ kN})$$

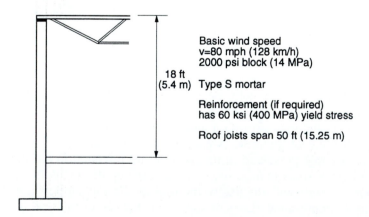

Basic wind speed
v=80 mph (128 km/h)
2000 psi block (14 MPa)

18 ft
(5.4 m)  Type S mortar

Reinforcement (if required)
has 60 ksi (400 MPa) yield stress

Roof joists span 50 ft (15.25 m)

**Figure 16.10** Design example for uplift loading.

Therefore, the net uplift force per joist anchor is

$$P_{uplift} = 5,680 - 2,000 = 3680 \text{ lb}$$

$$(P_{uplift} = 24.89 - 8.78 = 16.11 \text{ kN})$$

For working stress design, the safety factor of 1.5 used for overturning would be applicable for uplift. Therefore, a wall area sufficient to provide a gravity force of 5520 lb (24.17 kN) (i.e., 1.5 times the net uplift) must be attached to each joist anchor. For 8 in. (200 mm) hollow masonry units, the gravity force is about 55 lb/ft² (2.63 kN/m²). Hence, the area of wall required to produce the hold-down force is 5520/55 = 100.4 ft² (9.19 m²). As shown in Fig. 16.11, assuming a 45° effective angle for picking up the weight of the blockwork, a rod of 100.4/8 + 4/2 = 14.55 ft (4.43 m) minimum length should be provided. [Note: The weight of grout was neglected in this calculation.]

Considering that the wall is only 18 ft (5.4 m) high, the advantage of adding another 3.45 ft (1.07 m) of bar grouted into the core in a unit at 8 ft (2.4 m) spacing is very worthwhile. Continuing the reinforcement over the full height makes the wall partially reinforced, which adds to the strength and the toughness of the building.

Where joists are anchored into masonry over openings, the design of the anchor system must accommodate transfer of the upward force to the wall areas on either side of the opening in order to develop a sufficient hold-down force. Typically, the joists should be anchored to the lintel, and the ends of the lintel should be held down by additional bars extending into the foundation.

A practical means of anchoring steel joists is to transfer force from the anchor under the bearing plate to reinforcement in the wall. Care must be taken to ensure that the anchor itself is capable of developing force and that the splice length or mechanical connection between the anchor and the reinforcing bar is sufficient to transfer the force. It should be noted that some of the commonly used details for hooks and embedment lengths for anchors apply only to the case of minimum anchor-

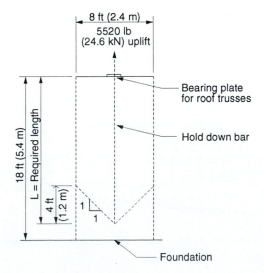

**Figure 16.11** Length of rod required for hold down of roof system.

age and may not satisfy structural design requirements. The joist shoe must be welded to the bearing plate.

In the example, the anchors at the ends of each truss should be designed for an allowable load of 3680 lb (16.11 kN). This can be more than adequately accommodated by welding a No. 4 (10M) Grade 60 (400 MPa) reinforcing bar to the truss bearing plate, where the bar would have an allowable maximum load of $0.2 \times 24{,}000 = 4800$ lb $(100 \times 165/1000 = 165$ kN).

It is often convenient that wall reinforcement and hold-down bars are the same. In this case, the hold-down tensile stress for working loads at the point of maximum bending moment should be subtracted from the allowable tensile stress in the reinforcement. If the locations of the joist anchors and wall reinforcement do not coincide, the locations of the joist anchors should be anchored into a sufficiently strong bond beam to transfer the uplift to the locations of the hold-down reinforcement.

By using a development length $L_d$ of $0.0015 d_b F_s^{16.1}$ $(L_d = 0.22 d_b F_s$ (mm)), the minimum length needed to develop the allowable stress in the steel is

$$L_d = 0.0015(0.5)(24{,}000) = 18 \text{ in.}$$

$$(L_d = 0.22(11.3)(165) = 410 \text{ mm for the 10M bar})$$

where $d_b$ = bar diameter = 0.5 in. (11.3 mm for 10M bar)
$F_s$ = allowable stress in the steel = 24,000 psi (165 MPa)

## 16.5 EXAMPLE DESIGN

### 16.5.1 Introduction

The main features of structural design using engineering analysis for a loadbearing single-story masonry building are illustrated for the industrial building shown in Fig. 16.12. Brief discussions of some aspects of the masonry design are also included. Details of design of footings, floors, parapets and roof systems are not provided. *In*

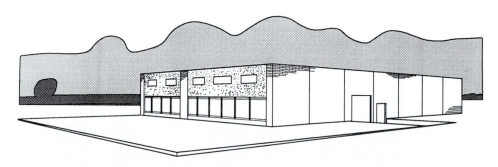

**Figure 16.12**  Perspective of example single-story industrial building.

*this example, all of the calculations are done using U.S. customary units and S.I. equivalent values are converted from these values at key points in the design. This "soft" conversion was adopted to avoid excessive duplication of the numerical presentation.*

### 16.5.2 Description of The Building

**Plan.** The outside dimensions of the building plan shown in Fig. 16.13 are 147 ft, 4 in. (44.2 m) by 96 ft (28.8 m) with a two-story office area in the Southeast corner. The second floor of the office area is cast-in-place concrete on steel trusses. The roof consists of corrugated steel decking spanning in the East–West direction on the trusses spanning in the North–South direction. These trusses are supported on steel beams and on wall B along the central portion of the building and on the masonry walls at the sides of the building. The joist spacing is 8 ft (2.4 m).

**Elevations.** The front of the building shown in Fig. 16.14(a) has large window areas between the three loadbearing walls with nonloadbearing walls above the windows. The back of the building in Fig. 16.14(b) has a combination of windows and doors. The positions of openings and movement joints are shown in Fig. 16.14(b)

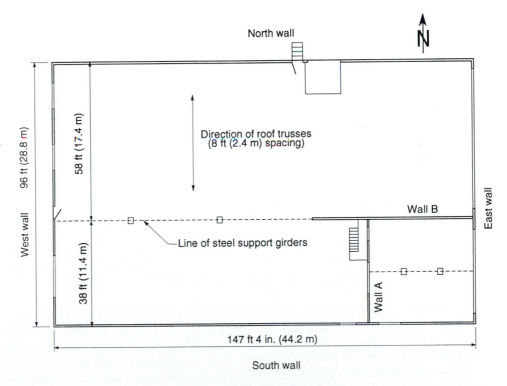

**Figure 16.13** Building plan.

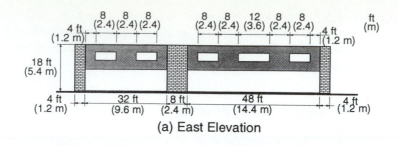

(a) East Elevation

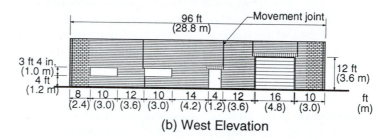

(b) West Elevation

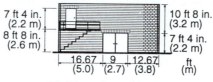

(c) Elevation of Interior Wall A

**Figure 16.14**   Elevations of walls running North–South.

for the back, or West, wall.  Wall A, shown in Fig. 16.14(c), can be either loadbearing or nonloadbearing as required by the designer.

Figure 16.15 contains elevation drawings of the loadbearing walls running in the East–West direction.  Interior wall B is also loadbearing.

The clear height of the walls from the top of the foundation wall to the underside of the roof is 18 ft (5.4 m).  For this to be the effective height of the wall, it is necessary that the top of the foundation be tied into the floor to provide lateral support at this level.  To allow for some settlement of the floor, hairpin-type ties with polystyrene sleeves around the first 12 in. (300 mm) or so of the embedment in the slab could be used (see Detail A in Fig. 16.16).

**Movement Joints.**   Vertical movement joints are shown as the heavy vertical lines in Figs. 16.14 and 16.15, and, as discussed in Chap. 15, are typically spaced at 20 to 30 ft (6 to 9 m). [Note: For the walls of the building to act together as a combined structural element rather than as isolated individual elements, it is important that out-of-plane horizontal shear can be transferred across these joints.  When individual walls are isolated, high localized effects of wind pressure will result in larger design forces.]

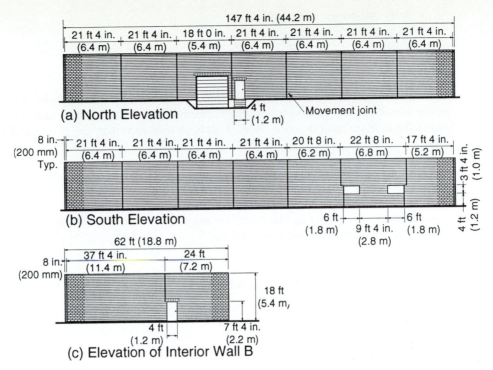

**Figure 16.15** Elevations of walls running East–West.

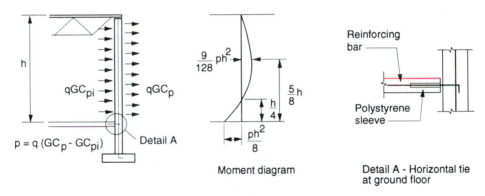

**Figure 16.16** Propped cantilever wall.

## Design Information.

- Building is located in open terrain with a basic wind speed of 70 mph (112 km/h) (Exposure Category A; Ref. 16.2)
- Building is located in Seismic Zone 1
- Ground snow load = 40 lb/ft² (1.9 kN/m²)
- Building is heated

- Soil: soft to medium-stiff clay
- Flexible roof diaphragm

### 16.5.3  Design Loads

It is not the intent of this book to be tied to any specific design code. However, in order that designs can be seen to be representative, design values and design provisions from specific codes are used. Loads in this example are calculated based on ASCE 7-88.[16.2]

**Wind.**    The reference wind pressure $q$ is 10.5 lb/ft (0.50 kN/m$^2$) for 70 mph (112km/h) wind at the 18 ft (5.4 m) height in open terrain.

**Seismic.**    By using the basic equation for static equivalent horizontal shear force resulting from earthquake motion,

$$V = ZIKCSW$$

The following values of the terms were used.

$$Z = \tfrac{3}{16} \quad \text{(for Zone 1)}$$

$$I = 1.0$$

$$K = 1.33$$

$$C = (1/15)\sqrt{T}$$

where

$$T = 0.05h/\sqrt{D}$$

$$= 0.05 \times 18/\sqrt{147.3} = 0.074 \text{ s. for loading in the E–W direction}$$

$$= 0.05 \times 18/\sqrt{96} = 0.09 \text{ s. for loading in the N–S direction}$$

and

$$C = (1/15)\sqrt{0.074} = 0.24 \text{ (E–W direction)}$$

$$= (1/15)\sqrt{0.09} = 0.22 \text{ (N–S direction)}$$

$$S = 1.5$$

$$CS = 0.22 \times 1.5 = 0.33 > 0.14$$

Therefore, take $CS = 0.14$.

Finally

$$V = (3/16)(1.0)(1.33)(0.14)(W)$$

$$= 0.035W$$

where $W$ is the weight of the applicable parts of the building.
For individual elements of the structure:

$$F_p = ZI\,C_p W_p$$

$$= (3/16)(1.0)(0.3)W_p$$

$$= 0.056W_p$$

where $W_p$ is the weight of the element.

**Gravity.** The following gravity loads apply:

Roof snow load (uniformly distributed) $P_f = 0.7\ C_eC_tIP_g$
$$= 0.7(0.8)(1.0)(1.0)(40)$$
$$= 22.4\ \text{lb/ft}^2 \quad (1.1\ \text{kN/m}^2)$$

Weight of roof system (built-up roof) $= 15\ \text{lb/ft}^2\ (0.72\ \text{kN/m}^2)$

Weight of block walls (hollow 8 in. (200 mm) masonry wall with a grouted core at 4.0 ft (1.2 m) spacing) $= 55\ \text{lb/ft}^2\ (2.63\ \text{kN/m}^2)$

Partition allowance $= 20\ \text{lb/ft}^2\ (0.97\ \text{kN/m}^2)$

## 16.5.4 Design of Walls for Axial Load and Out-Of-Plane Bending

**General Effects.** If dowels from the foundation wall are used to provide tension splices with vertical wall reinforcement, for out-of-plane bending, the wall can be designed as a propped cantilever, as shown in Fig. 16.16.

**Typical Wall.** For a typical strip of loadbearing wall away from the corners (North or South wall), the wind load is

$$p = q[GC_p - GC_{pi}] = 10.5[-0.75 - 1.5] = -23.6\ \text{lb/ft}^2 \quad (-1.13\ \text{kN/m}^2)$$

representing the worst case (conservative) with internal positive pressure and external suction (see Fig. 16.5).

The seismic force normal to the surface of the wall is

$$V = 0.056\ W_p = 0.056\ (55) = 3.1\ \text{lb/ft}^2\ (0.15\ \text{kN/m}^2)$$

where the wall weight of 55 lb/ft$^2$ (2.63 kN/m$^2$) is used.

Therefore, the wind pressure of 23.6 lb/ft$^2$ (1.13 kN/m$^2$) governs and the maximum bending moment is

$$M_{\text{max}} = ph^2/8$$
$$= 23.6(18)^2/8 = 956\ \text{ft-lb/ft} \quad (4.25\ \text{kN.m/m})$$

For a fixed end condition at the base, the maximum moment near the mid-height of the wall is

$$M = \frac{9}{128}ph^2$$

$$= \frac{9}{128}(23.6)(18)^2 = 538\ \text{ft-lb/ft} \quad (2.39\ \text{kN.m/m})$$

The axial force at the top of the wall (neglecting wind uplift) is

$$P_{\text{max}} = \text{half roof span} \times (\text{roof weight} + \text{snow})$$
$$= (58/2)(15 + 22.4) = 1085\ \text{lb/ft} \quad (15.86\ \text{kN/m})$$

$$P_{\text{min}} = \text{half roof span} \times \text{roof weight} = (58/2)(15) = 435\ \text{lb/ft} \quad (6.36\ \text{kN/m})$$

[Note: Crane loads and other equipment loads would need to be added where applicable.]

Due to wind, the uplift pressure for the general roof area is:

$$p_{uplift} = q(GC_p - GC_{pi})$$

$$= 10.5(1.2 + 0.75) = 20.5 \text{ lb/ft}^2 \ (0.99 \text{ kN/m}^2)$$

Higher wind pressures exist at the roof edge for a distance of 5% of the roof width $= 0.05(96) = 4.8$ ft (1.44 m). Along this strip

$$p_{uplift} = 10.5(1.5 + 0.75)$$

$$= 23.6 \text{ lb/ft}^2 \ (1.14 \text{ kN/m}^2)$$

Figure 16.17 shows the uplift forces just calculated. At the roof level, the upward reaction produces a minimum axial compression (maximum uplift) of

$$P_{min} = \frac{58}{2}(15.0 - 20.5) + (20.5 - 23.6)4.8\left(\frac{58 - 4.8/2}{58}\right)$$

$$= -174 \text{ lb/ft} \quad (-2.54 \text{ kN/m})$$

$$= 174 \text{ lb/ft} \quad (2.54 \text{ kN/m}) \text{ axial tension}$$

By considering self-weight of the wall at the base of 55 lb/ft² × 18 ft = 990 lb/ft (14.2 kN/m) added to the load from the roof, the range of loads at the bottom of the wall are

$$P_{max} = 990 + 1085 = 2075 \text{ lb/ft} \quad (30.3 \text{ kN/m})$$

$$P_{min} = 990 - 174 = 816 \text{ lb/ft} \quad (11.9 \text{ kN/m})$$

By checking flexural tensile stresses at the base section using the face shell mortar bedded cross-section:

$$A = (1.25 \times 2)(12) = 30 \text{ in.}^2/\text{ft}$$

$$I = \frac{12(7.675)^3}{12} - \frac{12(5.125)^3}{12} = 309 \text{ in.}^4/\text{ft}$$

$$y = \frac{t}{2} = 3.81 \text{ in.}$$

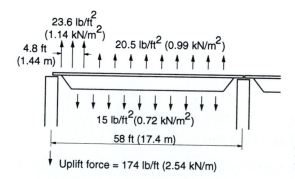

**Figure 16.17** Illustration of roof uplift force.

For combined axial load and bending

$$f_t = -\frac{P_{min}}{A} + \frac{My}{I}$$

$$= -\frac{816}{30} + \frac{956 \times 12}{309}(3.81) = -27.2 + 142.9 = 115.7 \text{ psi} \qquad (0.80 \text{ MPa}) \text{ tension}$$

which exceeds code allowable stresses[16.1,16.3] for unreinforced masonry. Therefore, unless a larger block is used, the wall should be reinforced. Reinforced masonry should be the most economical design and has the added advantages of providing hold down for the roof as well as providing a more robust structure.

To design the reinforced section at the base of the wall, consider using vertical reinforcement at 4 ft (1.2 m) spacing.

For the case with a minimum axial compression, the loads for a 4 ft (1.2 m) wall length are

$$P = 816 \times 4 = 3264 \text{ lb/4 ft} \qquad (14.5 \text{ kN/1.2 m})$$

$$M = 956 \times 4 = 3824 \text{ ft-lb/4 ft} \qquad (55.89 \text{ kN.m/1.2 m})$$

For 2000 psi (13.8 MPa) block and type S mortar, representative[16.1] masonry compressive strength and modulus of elasticity are, respectively:

$$f'_m = 1555 \text{ psi} \qquad (10.7 \text{ MPa})$$

$$E_m = 2.2 \times 10^6 \text{ psi} \qquad (15,200 \text{ MPa})$$

The allowable compressive stress in masonry can be taken as $F_m = f'_m/3 = 518$ psi (3.57 MPa). For Grade 60 (400 MPa) steel, $E_s = 29.0 \times 10^6$ psi (200,000 MPa) and the allowable tensile stress is 24,000 psi (165 MPa).

$$n = E_s/E_m = 29/2.2 = 13.2$$

For the section shown in Fig. 16.18, the moments of the applied loads and the internal forces about the reinforcement located at the center of the wall can be equated:

$$C\left(\frac{t}{2} - \frac{x}{3}\right) = M \qquad (16.8)$$

The reinforcement is located in the middle of the wall to allow for bending in both directions.

If the depth to the neutral axis defined by the distance x is assumed to fall within the compression face shell, Eq. 16.8 applies and

$$\left(\frac{1}{2}f_m bx\right)\left(\frac{7.625}{2} - \frac{x}{3}\right) = 3824 \times 12$$

Assuming that compression controls, use $f_m = F_m = 518$ psi (3.57 MPa) (ignoring the $\frac{1}{3}$ increase in allowable stress in U.S.A. codes for wind loading) and $b = 48$ in. (1.2 m). The depth to the neutral axis is

$$x = 1.07 \text{ in.} \qquad (27.2 \text{ mm})$$

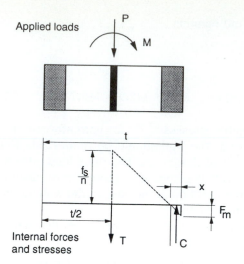

Applied loads

**Figure 16.18** Forces and stresses in a typical wall under out-of-plane loading.

Internal forces and stresses

This is within the face shell thickness of an 8 in. (200 mm) block and the cracked section will behave as a rectangular section.

From strain compatibility,

$$\frac{f_s}{n} = \frac{F_m}{x}\left(\frac{t}{2} - x\right)$$

$$\therefore f_s = 13.2\left[\frac{518}{1.07}\left(\frac{7.625}{2} - 1.07\right)\right] \tag{16.9}$$

$$= 17,525 \text{ psi} \quad (121 \text{ MPa})$$

$$< 24,000 \text{ psi} \quad (165 \text{ MPa}) \quad \text{O.K., compression controls}$$

[Note: For tension controlled cases, set $f_m = (F_s/n)(x/(t/2) - x)$.]

For the equilibrium of internal and external axial forces

$$C - T = P$$

$$\tfrac{1}{2}(F_m)(b)(x) - A_s f_s = 3264 \tag{16.10}$$

$$\therefore A_s = \frac{\tfrac{1}{2}(518)(48)(1.07) - 3264}{17,525} = 0.57 \text{ in.}^2 \quad (368 \text{ mm}^2)$$

Therefore, No. 7 bars at 48 in. on center (20 M bars at 1.0 m spacing) (see details in Fig. 16.19) satisfies this design. [Note: If the $\frac{1}{3}$ increase in allowable stress is allowed because of the wind loading, the compression zone is smaller and less reinforcement is required.]

For this wall, the case of minimum loading was judged to be critical for determining the required amount of steel. Other cases, including maximum axial load, must also be checked.

For maximum axial load, analyze the section with #7 bars at 48 in. under

$$P_{max} = 2075 \times 4 = 8300 \text{ lb/4 ft} \quad (36.36 \text{ kN/1.2 m})$$

$$M = 3824 \text{ ft-lb/4 ft} \quad (55.89 \text{ kN.m/1.2 m})$$

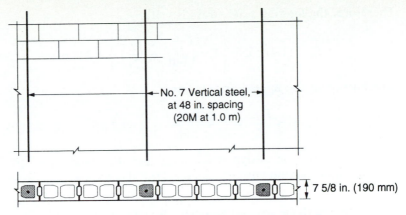

**Figure 16.19** Typical wall reinforcement for out-of-plane load resistance.

For equilibrium of moments and axial forces, using Eqs. 16.8 and 16.9

$$\tfrac{1}{2}f_m(48)(x)\left(\frac{7.675}{2} - \frac{x}{3}\right) = 3824 \times 12$$

$$\tfrac{1}{2}f_m(48)(x) - (0.60)(f_s) = 8300$$

Solving the two unknowns, $f_m$ and $x$ (where $f_s$ is geometrically related to $f_m$ and $x$), yielded $x = 1.42$ in.    (36 mm). This indicates that the netural axis falls outside the compression face shell. Therefore, the section has to be analyzed as a T-section following the procedures presented in Sec. 8.4.3. This analysis results in $x = 1.45$ in. (37 mm) and $f_m = 405$ psi    (2.79 MPa). The corresponding stress in the reinforcement is

$$f_s = 13.2\left(\frac{405}{1.45}\left(\frac{7.625}{2} - 1.45\right)\right) = 8712 \text{ psi} \qquad (60 \text{ MPa})$$

Because the stresses in the masonry and steel are below the allowable design values, the maximum axial load case does not control and the design is satisfactory.

Slenderness should be accounted for in the design. ACI 530/ASCE 5/TMS 402[16.1] requires that the axial stress component should not exceed

$$F_a = \frac{1}{4} f'_m \left[1 - \left(\frac{h}{140r}\right)^2\right] \tag{16.11}$$

for $h/r < 99$.

Considering only face shell mortar bedding and ignoring the added area from the grout cell in the center of the 4 ft (1.2 m) strip:

$$I = 48 \left[\frac{(7.625)^3}{12} - \frac{(5.125)^3}{12}\right] = 1234 \text{ in.}^4/4 \text{ ft}$$

$$r = \sqrt{I/A} = \sqrt{1234/(48 \times 1.25 \times 2)} = 3.21 \text{ in.}$$

For a propped cantilver, the effective height $h$ is 0.7 of the clear height.

$$h = 0.7 \times 18 = 12.6 \text{ ft}$$

Therefore, the limiting axial stress is

$$\therefore F_a = (1/4)(1555)\left[1 - \left(\frac{12.6 \times 12}{140 \times 3.21}\right)^2\right] = 345 \text{ psi} \qquad (2.38 \text{ MPa})$$

which exceeds the appied axial stress

$$f_a = \frac{P_{\max}}{A} = \frac{2075 \times 4}{48 \times 1.25 \times 2} = 69.2 \text{ psi} \qquad (0.48 \text{ MPa})$$

Therefore, the design is adequate and the slenderness effect is not critical. [Note: Other methods used to account for slenderness are covered in Chap. 8.]

**Walls with Openings: North Wall.** The large truck entrance removes a large segment from the base of the wall and, if no movement joints are introduced, it may also tend to initiate cracking due to the change in section. Rather than try to design the remaining wall to carry the full lateral load, the wind pressure on the door can be resisted by a separate support system. Regardless of the design solution, it is necessary to introduce movement joints on both sides of the door, either as shown in Fig. 16.15(a), or immediately beside the door framing. To avoid transferring load from the door to the remaining wall, the door can be supported independently by vertical steel sections (possibly hollow steel sections for a finished look) spanning from the floor to the roof. Alternately, masonry pilasters spanning from the foundation to the roof could be introduced to resist the lateral load. The wall weight and roof load over the door can be transmitted to these members using a beam to span over the opening. Alternately, the portion of the wall above the door can be designed as a deep masonry beam.

The man door in the panel near the truck door removes a 4 ft (1.2 m) section from the 21 ft-4 in. (6.4 m) total length of this panel. This wall carries very little roof load or uplift force and therefore the remaining 17 ft-4 in. (5.2 m) length of wall should be designed for the total axial force and moment on this panel. A similar design situation is demonstrated for the south wall. It is good practice to include extra reinforcement around door openings. Use of the 0.20 in.$^2$ (130 mm$^2$) minimum area of reinforcement specified in ACI 530/ASCE 5/TMS 402[16.1] for higher seismic zones is a logical choice.

**Walls with Openings: South Wall.** For the maximum moment at the base of the wall with windows, the 4 ft height of the window sill allows the moment to be distributed over a 9 ft-4in. + 4 ft + 4 ft = 17 ft-4 in. (5.2 m) wall length. The additional 4 ft (1.2 m) lengths under the windows are based on using 45° angles for distribution of load, as shown in Fig. 16.20. The moment per unit length at the base is $956 \times 21.33/17.33 = 1176$ ft-lb/ft (5.24 kN.m/m), which is 23% higher than the base moment for walls without openings.

Try changing the spacing of the No. 7 bar to 32 in. (800 mm). The capacity can be most easily checked using an interaction diagram but lacking that, Eq. 8.33 of Chap. 8 can be used to determine $x$ (Substitute $x = kd$ in Eq. 8.33.) as follows

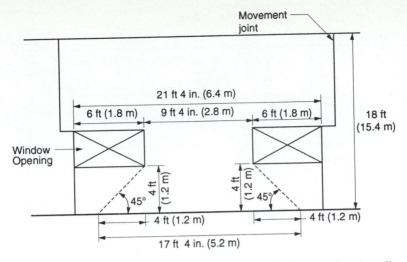

**Figure 16.20** Moment resisting area under windows on South wall.

assuming that the neutral axis, defined by the distance $x$ in Fig. 16.21, is between the reinforcement and the face shell as shown.

$$Pe = C_{\text{solid}}\left(\frac{t}{2} - \frac{x}{3}\right) - C_{\text{cut-out}}\left(\frac{t}{2} - \frac{x}{3} - \frac{2}{3}t_f\right) + T\left(d - \frac{t}{2}\right) \qquad (16.12)$$

where   $P = 816$ lb/ft or 2,176 lb/32 in. (9.69 kN/0.8 m)
$\quad\quad e = M/P = (1,176/816)12 = 17.3$ in. (439 mm)
$\quad\quad A_s = 0.60$ in.$^2$
$\quad\quad b = 32$ in.
$\quad\quad d = t/2$
$\quad\quad b_w = 0$, ignoring grouted cells
$\quad\quad t = 7.625$ in. (thickness of wall)
$\quad\quad t_f = 1.25$ in. (thickness of face shell)
$\quad\quad n = 13.2$

and

$$C_{\text{solid}} = \frac{f_m}{2}\,bx \qquad (16.13)$$

$$T = nA_s f_m\left(\frac{d-x}{x}\right) \qquad (16.14)$$

$$C_{\text{cut-out}} = \frac{f_m}{2}\left(\frac{x - t_f}{x}\right)(x - t_f)(b - b_w) \qquad (16.15)$$

$$P = C_{\text{solid}} - C_{\text{cut-out}} - T \qquad (16.16)$$

Solving,

$$x = 1.25 \text{ in. (32 mm)}$$

which is just at the edge of the face shell.

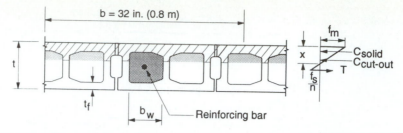

**Figure 16.21** Capacity of reinforced wall section for out-of-plane bending and axial loading.

For the moment $M = (32/12)(1176) = 3136$ ft-lb, the maximum compressive stress is found from Eq. 16.12 which reduces to

$$M = C_{\text{solid}}\left(\frac{t}{2} - \frac{x}{3}\right) \tag{16.17}$$

$$= \frac{f_m}{2}bx\left(\frac{t}{2} - \frac{x}{3}\right)$$

and

$$f_m = \frac{2 \times 3136 \times 12}{32(1.25)\left(3.81 - \dfrac{1.25}{3}\right)}$$

$$= 554 \text{ psi} \qquad (3.82 \text{ MPa})$$

which, although it exceeds the stated allowable stress of 518 psi, is satisfactory if the one third increase in stress is allowed for wind loading. For codes where the increased stress is not permitted, a closer spacing of bars should be analyzed.

From statics, the section at the window height must resist a maximum moment of 357 ft-lb/ft (1.59 kN.m/m). The 9 ft-4in. (2.8 m) wall section must carry the moment for the full 21 ft-4 in. panel, which results in a moment of 357 (21.33/9.33) = 816 ft-lb/ft (3.64 kN.m/m). This will be less critical than the 956 ft-lb/ft (4.26 kN.m/m) at the base of the walls without openings. Therefore, the regular reinforcement consisting of a No. 7 bar every 4 ft (20 M at 1.0 m) is more than adequate.

The wall sections above the openings would have to act as horizontal cantilevers to transfer lateral load to the central 9 ft-4 in. (2.8 m) continuous vertical wall section unless shear transfer mechanisms are provided across the movement joints. The provision of shear transfer capability results in some of the load on these propped cantilevers being transferred to the adjacent panels. Thus the capacity requirements for the wall sections above the openings will be reduced. The bending moment for the strips acting as cantilevers is $p(L^2/2) = 23.6(6.67)^2/2 = 525$ ft-lb/ft (2.34 kN.m/m). The bending moment for the strips actings as propped cantilevers is $pL^2/8 = 131$ ft-lb/ft (0.59 kN.m/m).

This moment results in a flexural tensile stress of $f_t = M/S$ where $S = 12((7.625)^3 - (5.125)^3)/(12 \times 3.812) = 80.9$ in.$^3$/ft (4.35 × 10$^6$ mm$^3$/m). For the cantilever sections, the tensile stress is 77.8 psi (0.54 MPa) which is higher than the code allowable stress of 50 psi (0.35 MPa). Therefore horizontal reinforcement is required. Horizontal reinforcing bars in bond beams above the window opening and at the roof level will likely satisfy this requirement. For the propped cantilever condition, the tensile stress is lower than the allowable stress and horizontal reinforcement is not required.

**Walls with Openings: West Wall.**   The truck door area causes a 42% reduction in the moment-resisting section of the wall panel containing the door. The resulting extra load on the remaining section could be offset by decreasing the bar spacing and increasing the block strength. However, an alternate solution is to support the door separately from the wall by attaching the door frame to independent structural steel sections spanning from the foundation to the roof.

The section of the wall above the door opening has to span horizontally. Based on the analysis of the wall segment above the window opening in the South wall, it is clear that horizontal reinforcement is needed. This reinforcement can be provided in bond beams above the opening and at the roof level. Provision of shear transfer across the movement joint will allow load transfer to adjacent panels. Use of extra reinforcing bars in the blockwork around the opening is again recommended.

For the window segments of the West wall, provision of shear transfer across the movement joint will allow wind load from half the window strip in the adjacent panel to be transferred to the 12 ft (3.6 m) strip. Therefore, the critical segment is either this 12 ft (3.6 m) pier between the windows or the corresponding effective section at the base of the wall. At the base of the wall, the effective section for wind load transferred through this pier can be taken as the 12 ft (3.6 m) pier plus two 4 ft (1.2 m) sections under the windows (using 45° distributions of load, as was shown in Fig. 16.20). This 16 ft (4.8 m) section resists bending for a wall length of 12 ft + 10 ft/2 + 10 ft/2 = 22 ft     (3.6 + 1.5 + 1.5 = 6.6 m)

By observation, the central panel containing a window and man door will be more critical where the load from a 14 ft + 5 ft + 2 ft = 21 ft (6.3 m) length of wall is resisted by a 14 ft + 4 ft = 18 ft (5.4 m) section at the base.

For lightly loaded wall segments, such as the wall on the West elevation, a conservative method of calculation is to simply design the wall for pure bending (zero axial load).

As a first trial, try the original design consisting of No. 7 bars at 48 in. on center (20 M bars at 1.0 m) and analyze the section to check stresses in the masonry and steel. For $b = 48$ in. (1.2 m) the corresponding moment is

$$M = (21/18)(956)(48/12) = 4461 \text{ ft-lb/4 ft } (6.05 \text{ kN.m/1.2 m})$$

and

$$\rho = A_s/bd = 0.60/(48 \times 3.8125) = 0.00328$$

From Eq. 6.9 and assuming a rectangular section, the position of the neutral axis $x$ can be calculated as:

$$x = kd = (\sqrt{2np + (np)^2} - np)d$$

$$= [\sqrt{2(13.2)(0.00328) + (13.2 \times 0.00328)^2} - (13.2)(0.00328)]3.8125$$

$$= 0.97 \text{ in. } (24.6 \text{ mm})$$

which is less than the face shell thickness. Therefore the section is a rectangular section as assumed.

From Eq. 6.12 in Chap. 6, with $k = 0.254$ and $j = 1 - k/3 = 0.915$

$$f_m = 2M/(kjbd^2)$$

$$= (2(4461 \times 12))/(0.254)(0.915)(48)(3.8125)^2 = 660 \text{ psi } (4.55 \text{ MPa})$$

which is less than the $4/3(518) = 691$ psi (4.77 MPa) allowable stress.

From Eq. 6.14 in Chap. 6

$$f_s = M/\rho jbd^2$$

$$= (4461 \times 12)/(0.00328)(0.915)(48)(3.8125)^2 = 25{,}570 \text{ psi } (176 \text{ MPa})$$

which is less than the $4/3$ (24000) = 32,000 psi (220 MPa) allowable steel stress. Therefore the design with No. 7 bars at 48 in. (20 M at 1.0 m) is adequate where the extra stress allowance for wind is utilized. From this analysis, a similar design for the previously discussed panels in the North and South walls should also be satisfactory when the extra stress allowance is utilized.

The minimum section at the window level is more than half of the effective section at the base. From the calculations at the window section of the South wall, it was found that the maximum moment in this region was less than half of the moment at the base. Therefore, this level is not critical.

**Walls with Openings: East Wall.** The wind pressure on the window area of the wall at the front of the building, as shown in Fig. 16.14(a), cannot be readily transferred horizontally to the short wall sections. Therefore, this area and the walls above should be supported by a separate framing system, spanning from the foundation to the roof.

If the central 8 ft (2.4 m) wall is anchored to interior wall B, where they intersect, then out-of-plane bending is not a problem. However, if wall B was not present, this isolated element could be subject to the effects of high localized wind pressures. In addition, this wall does not carry significant vertical load from the roof. Therefore, additional reinforcement might be required.

**Walls A and B.** Interior walls A and B should be checked for bending due to unbalanced internal wind pressures. In addition, some tie-down reinforcement likely is needed to resist uplift of the roof.

### 16.5.5 Design of Walls for Uplift Forces

**Exterior Walls.** As calculated in Sec. 16.5.4, the maximum net uplift is 174 lb/ft (2.54 kN/m) at the roof level, which converts to $174 \times 8 = 1392$ lb (6.20 kN) on the anchor for each roof joist. For safety, the roof should be held down with a weight of at least 1.5 times this force. Therefore, by using the 55 = lb/ft$^2$ (2.63 kN/m$^2$) self-weight of the wall, a wall area of $1.5(1392)/55 = 38.0$ ft$^2$ (3.53 m$^2$) must be securely anchored to the end of each joist. The vertical reinforcing bars can be utilized for this purpose by using adequate tension splice lengths for the anchors under the base plates. From Fig. 16.11 and the corresponding calculations, using

$$\text{weight} = \text{uplift}$$

$$\left[ L(8) - \frac{4 \times 8}{2} \right] 55 = 1.5(1392)$$

$$L = 6.75 \text{ ft} \qquad (2.06 \text{ m})$$

Because the uplift force is counterbalanced within the top part of the wall, the tensile stress in the bar due to uplift does not have any effect on the moment resistance at the base or other low parts of the wall.

**Interior Walls.** A slightly greater uplift force of $(20.5 - 15)(96/2) = 264$ lb/ft (3.86 kN/m) exists along wall B. However, a No. 4 bar with an allowable tensile force of $T = A_s f_s = 0.20(24000) = 4800$ lb located at 8 ft spacing will be more than adequate for the $8 \times 264 = 2112$ lb uplift from the ends of the trusses (10 M at 2.4 m gives $T = 100(165) = 16.5$ kN $> 2.4 \times 3.86$). The tension connection to the bearing plate requires either adequate welding or a sufficient splice length to transfer force from the bearing plate anchor to the reinforcement.

The tie-down reinforcement for the interior wall could be terminated at a distance of 9.2 ft. (2.8 m) from the roof level. [From $2112 \times 1.5 = (8L - 8 \times 4/2)55$.] However, the small cost of reinforcement and grout to extend the No. 4 (10 m) bar to the foundation at 8 ft (2.4 m) spacing may be very worthwhile. This will also increase the flexural resistance of these interior walls against bending from unbalanced wind pressures.

### 16.5.6 Design of Walls for In-Plane Shear and Bending

The roof acts as the top support for walls subject to out- of-plane load. It is in turn supported by cross-walls that act as cantilever shear walls to transfer the lateral load to the foundation. For this building (Fig. 16.13), the lateral load from a North wind is resisted by the East and West walls (and perhaps wall A) acting as cantilever shear walls. For the flexible roof diaphragm, the load is distributed to these walls according to the tributary areas.

From Sec. 16.5.4, the maximum wind pressure on the walls is 23.6 lb/ft$^2$ (1.14 kN/m$^2$). For the roof support of the wall, the reaction is $(3/8)(23.6)(18) = 159$ lb/ft (2.32 kN/m), assuming fixed base and pinned top support conditions.

For a South or North wind load, the total horizontal force transferred by the roof to the cross walls is $159 \times 147.33 = 23,430$ lb $= 23.43$ kips (104.4 kN).

**Forces on Front Walls (East Elevation):** If Wall A (Fig. 16.14) is only a partition, then the three short walls shown in Fig. 16.14(a) must resist half of the total horizontal force, or 11.72 kips (52.2 kN).

The relative stiffnesses of the 4 ft and 8 ft wall lengths acting as cantilevers are found from Eq. 10.6 (see Chap. 10, Sec. 10.4.2).

$$R_c = 1 \left/ \left[ 4 \left( \frac{H}{L} \right)^3 + 3 \frac{H}{L} \right] \right.$$

Therefore,

$$R_{4ft} = 1 \left/ \left[ 4 \left( \frac{18}{4} \right)^3 + 3 \left( \frac{18}{4} \right) \right] \right. = 0.0026$$

$$R_{8ft} = 1 \left/ \left[ 4 \left( \frac{18}{8} \right)^3 + 3 \left( \frac{18}{8} \right) \right] \right. = 0.1019$$

Then the load taken by the 8 ft (2.4 m) long wall is calculated from Eq. 10.9 of Chap. 10.

$$V_{8ft} = \frac{R_{8ft}}{R_{8ft} + 2R_{4ft}} (V_{total}) = 11.72 \frac{0.1019}{0.1019 + 2(0.0026)} = 9.20 \text{ kips} \qquad (40.9 \text{ kN})$$

and the shear stress

$$f_v = 1.5 \ V/A = \frac{1.5(9200)}{(1.25 \times 2)(8 \times 12)} = 57.5 \text{ psi} \qquad (0.4 \text{ MPa})$$

based on the face shell area. This stress exceeds the ACI 530/ASCE 5/TMS 402[16.1] allowable value of 37 psi (0.26 MPa) or even the value of 49 psi (0.35 MPa) with the one third increase allowed for short duration loading such as wind. Therefore, horizontal reinforcement should be provided to carry all of the shear (see Sec. 10.5.2). The area of shear reinforcement can be calculated as

$$A_v = \frac{Vs}{F_s d} = \frac{(9200)(8)}{(24,000)(92)} = 0.033 \text{ in.}^2 \qquad (21.3 \text{ mm}^2)$$

where $s = 8$ in. (200 mm), and $d = 92$ in. (2.34 m) (see Fig. 16.22).

Using standard ladder type joint reinforcement at every course provides 0.035 in.$^2$ (22.6 mm$^2$), which is adequate.

The in-plane bending moment is $Vh = 9200 \times 18 = 165,600$ ft-lb (225 kN.m). The axial load from self-weight is

$$P = 55 \text{ lb/ft}^2 \times 8 \times 18 = 7920 \text{ lb (35.2 kN)}$$

As a trial design, try using No. 8 bars (25 M) at the ends of the wall and grout two or three cells at the ends (Fig. 16.22) of the wall as required to carry the compression force due to bending. The section is under a combined axial load $P = 7920$ lb (35.2 kN) and moment $M = 165,600$ ft-lb (223.9 kN.m)

From moment equilibrium about the tension steel

$$\tfrac{1}{2}C(d - x/3) = P(L/2 - d') + M$$

$$\tfrac{1}{2}f_m x(7.625)(92 - (x/3)) = 7920(48-4) + 165600 \times 12$$

From force equilibrium

$$C - T = P$$

$$\tfrac{1}{2}f_m x(7.625) - 0.79(13.2 f_m(92 - x)/x) = 7920$$

where, from strain compatibility

$$f_s = nf_m(d - x)/x$$

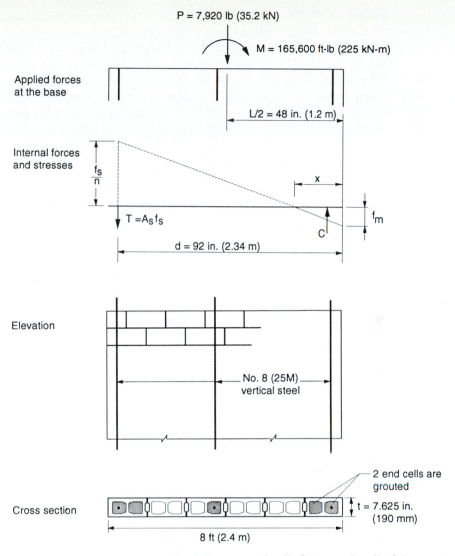

P = 7,920 lb (35.2 kN)

M = 165,600 ft-lb (225 kN-m)

Applied forces at the base

L/2 = 48 in. (1.2 m)

Internal forces and stresses

$\dfrac{f_s}{n}$

x

$T = A_s f_s$

C

$f_m$

d = 92 in. (2.34 m)

Elevation

No. 8 (25M) vertical steel

2 end cells are grouted

Cross section

t = 7.625 in. (190 mm)

8 ft (2.4 m)

**Figure 16.22** Stresses and reinforcement details for central wall of the East elevation.

Solving the above two equations yields

$$x = 18.5 \text{ in.} \qquad (470 \text{ mm})$$

$$f_m = 386 \text{ psi} \qquad (2.66 \text{ MPa})$$

$$f_s = (13.2)(386)(92 - 18.5)/18.5 = 20,243 \text{ psi} \qquad (140 \text{ MPa})$$

The above stresses are less than the allowable stresses. Therefore, the No. 8 (25 M) end bars initially selected are adequate. The two end cells at each end of the wall must be grouted, as shown in Fig. 16.22. Also a No. 7 or No. 8 bar should be used at mid-length.

**Forces on Wall A.**    If wall A is connected to the structure so that North–South deflection of the roof is resisted, using the tributary area concept for the flexible roof diaphragm, it will receive half of the lateral load transferred to the roof. The sections or piers on either side of the lower door (Fig. 16.14(c)) will share the force in proportion to their relative stiffnesses. Based on these wall segments or piers being considered as having fixed ends, the relative rigidities can be calculated from Eq. 10.7 from Chap. 10. [It is noted that Wall A is separated from the intersecting walls by vertical movement joints at both ends so that it is not a flanged section.]

$$R_f = Et \Bigg/ \left[ \left(\frac{H}{L}\right)^3 + 3\left(\frac{H}{L}\right) \right]$$

$$R_{16ft} = Et \Bigg/ \left( \left(\frac{7.33}{16}\right)^3 + 3\left(\frac{7.33}{16}\right) \right) = 0.615\,Et$$

$$R_{12ft} = Et \Bigg/ \left( \left(\frac{7.33}{12}\right)^3 + 3\left(\frac{7.33}{12}\right) \right) = 0.435\,Et$$

The lateral force on the 16 ft (4.8 m) wall is:

$$V_{16} = V_{total} \frac{R_{16}}{R_{16} + R_{12}} = 11.72 \left( \frac{0.615}{.615 + 0.435} \right) = 6.86\,\text{kips} \qquad (30.6\,\text{kN})$$

For bending, the deep continuous wall section over the two lower sections provides a nearly fixed-end condition. Therefore,

$$M = V\left(\frac{H}{2}\right) = 6.86\left(\frac{7.33}{2}\right) = 25.14\,\text{ft-kips} \qquad (34.14\,\text{kN·m})$$

for the 16 ft (4.80 m) long pier.

The axial load from the weight of the wall above these sections is

$$P = 55(16 + 4.5)10.67$$

$$= 12,030\,\text{lb} \qquad (53.4\,\text{kN})$$

By checking flexural stress,

$$f_m = -\frac{P}{A} + \frac{M}{I}y$$

$$= -\frac{12,030}{2.5(16 \times 12)} + \frac{25.14 \times 12,000}{2.5(16 \times 12)^3/12}(16 \times 12)/2$$

$$= -25.0 + 19.6 = -5.4\,\text{psi} \qquad (-0.04\,\text{MPa})$$

No tension develops. Therefore, vertical reinforcing steel is not required to resist in-plane bending. [Note that some load from the roof could add to or reduce the axial load slightly.] However, with all other walls reinforced, it would be good practice to include the same reinforcing as in Wall B, particularly around the openings where no movement joints are included.

Maximum shear stress is

$$f_v = 1.5 \frac{V}{A}$$

$$= 1.5 \frac{6860}{(1.25 \times 2)(16 \times 12)} = 21.4 \text{ psi} \qquad (0.15 \text{ MPa})$$

which is less than the ACI 530/ASCE 5/TMS 402[16.1] allowable shear stress of 37 psi (0.26 MPa), neglecting the beneficial effect of axial compression. Therefore, no shear reinforcement is required.

**Forces on Back Wall (West Elevation).** From the foregoing calculation, it can be seen that the masonry itself is adequate for in-plane forces on the back wall. The reason is that the horizontal shear force is less than for Wall A and the wall sections total a much larger resisting area. The reinforcement present is required for out-of-plane bending.

**Forces on Walls Running East–West.** Of the three walls running in the East–West direction, Wall B will be most highly loaded and from the tributary area will take half of the wind loading for an East–West wind direction. Because of the smaller surface area, compared to the North or South wall, the lateral load will be less than for a North–South wind direction. Therefore, because wall B has, by observation, greater capacity than Wall A, it can be seen that in-plane shear and bending are not critical. However, because of uplift, reinforcement is provided.

### 16.5.7 Required Stiffness of Roof Diaphragm

For flexible diaphragms, the horizontal deflection of the roof at the top of the masonry walls will result in additional moments at the fixed-bases of the walls. Cross-bracing of the roof is recommended to limit diaphragm deflection and thereby limit the corresponding additional moments in the walls. The cross-bracing system shown in Fig. 16.23 can be utilized for this example building. Section 16.4.3 discusses the type of calculations to be performed. For reinforced masonry, limits on roof deflection are generally less critical than for unreinforced masonry because cracking is permitted for reinforced masonry. Also cracked walls are more flexible and, therefore, generate less additional moment due to deflection. Reinforced sections generally have more reserve out-of-plane bending capacity.

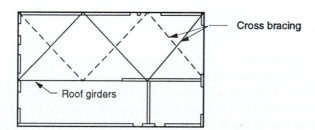

Cross bracing

Roof girders

**Figure 16.23** Arrangement of roof cross-bracing.

### 16.5.8 Bearing

Use of a bond beam is likely required to ensure adequate bearing of roof trusses and steel girders on the masonry walls. For the maximum roof loading of 15 + 22.4 = 37.4 lb/ft² (1.81 kN/m²), the maximum bearing force of one roof truss is

$$P = 37.4 \left( \frac{58}{2} \times 8 \right) = 8680 \text{ lb} \qquad (38.6 \text{ kN})$$

For an allowable bearing stress of 0.25 $f'_m$ = 0.25(1555) = 389 psi (2.68 MPa), the required minimum area of the bearing plate is

$$A_b = 8680/389 = 22.3 \text{ in.}^2 \qquad (14,390 \text{ mm}^2)$$

Thus, a 6 × 4 in. (150 × 100 mm) steel bearing plate is adequate.

The bearing load from one of the steel girders is

$$P = 37.4 \left( \frac{96}{2} \times \frac{26}{2} \right) = 23,340 \text{ lb} \qquad (104 \text{ kN})$$

and using the allowable bearing stress calculated before, the required bearing plate area is

$$A_b = 23,340/389 = 60 \text{ in.}^2 \ (38,700 \text{ mm}^2)$$

Thus, a 6 × 10 in. (150 × 260 mm) bearing plate is adequate. The thicknesses of the bearing plates should be sized according to normal steel design practice.

Because the bearing points are designed to coincide with cells grouted the full height, bearing stresses below the bond beam are not critical in this design. However, where hollow masonry exists below the bond beam, bearing should be checked at that level.

### 16.5.9 Other Considerations

**Foundations.** Masonry foundation walls are often used in conjunction with above-grade masonry walls. These foundation walls are typically laid on reinforced concrete strip footings at a depth determined by soil conditions and, in cold climates, the required protection against frost heave. The footings should be reinforced. If the foundation wall is also a basement wall, it must be designed for bending due to soil pressure. Otherwise, as long as backfilling is done carefully, the requirements for vertical reinforcement in the foundation wall are basically to match the above-grade wall reinforcement. Use of a reinforced bond beam and a dampproof course at the top of the foundation wall is good practice.

**Wall Connections.** In this design, it is suggested that shear keys or mechanical shear-transfer devices should be incorporated into the movement joints that separate the walls into panels. These should allow in-plane displacement, but cause the walls to deflect similarly in the out-of-plane direction on either side of the movement joint. For intersecting walls, the 8 ft (2.4 m) center wall on the East face of the building should be tied to wall B to be braced against out-of-plane loads. Vertical movement and horizontal movement in the plane of the East wall should be allowed.

At the corners of the building and at other intersections, the general approach is to use connectors that permit free vertical movement and limit differential horizontal movement in one or both directions. However, limitation of this horizontal movement requires that one or both intersecting walls bend in the out-of-plane direction(s) to follow the in-plane movement of the other wall(s). The effect of this out-of-plane deflection is to cause out-of-plane bending moment at the bottom of a wall with a fixed base. The magnitude of the moment can be reduced by using a relatively flexible connector at the intersection. Also, the calculated additional moment at the fixed base of a reinforced masonry wall is usually reduced to a significant extent or redistributed through the change in stiffness associated with flexural cracking of this section. For unreinforced masonry, the wall tends to rotate about the ''pinned'' base without appreciable moment other than that due to eccentricity of the reaction. As a compensating factor, it is worth noting that out-of-plane bending due to lateral loads is removed by anchoring to the intersecting wall. Although tying across these movement joints violates the objective of free movement and does cause some bending of walls in the out-of-plane directions, there are some compensating considerations, including creation of a more redundant and thus tougher structure. Alternately, design of the wall components to act as independent sections is acceptable.

**Intermediate (Second-Floor) Rigid Diaphragm.** The analysis of this building ignores the effect of the concrete floor in the two story office area at the Southeast corner of the building. If this floor is tied into the surrounding walls, then the wall sections between the ground floor and the second story will be forced to act together to resist in-plane forces. Because of the much greater rigidities of the connected parts of Wall B, the South wall, and Wall A compared to the 4 ft (1.2 m) and 8 ft (2.4 m) piers along the East wall, the in-plane lateral loads on these piers would be significantly reduced.

The distribution of shear forces to the walls below the second floor of this office area can be analyzed using rigid diaphragm techniques. The in-plane horizontal forces on the walls above the rigid diaphragm are redistributed below it as a combination of direct and torsional shear forces calculated according to the rigidities of the walls in this part of the building. Movement joints should be used to separate walls connected to the rigid diaphragm from other walls.

## 16.6 CLOSURE

Experience allows the designer to identify the critical parts of the design for single-story loadbearing masonry buildings. Even for the example provided, there are very few pages of actual calculations and design information. Use of design aids for wall capacities would further reduce this work. Knowledge of situations that present reoccurring design difficulties also helps the designer choose wall plans that tend to minimize such problems.

Rational design of single-story loadbearing masonry buildings can provide a sound structure at an economic advantage. Loads such as wind uplift forces and out-of-plane bending moments may require steel reinforcement. Sound judgement in the location of walls and use of reinforcement will optimize design and also provide structures that are resistant to collapse even under abnormal loading conditions.

## 16.7 REFERENCES

16.1 Masonry Standards Joint Committee, "Building Code Requirements for Masonry Structures," ACI 530/ASCE 5/TMS 402, ACI, Detroit; ASCE, New York; TMS, Boulder, 1992.

16.2 American Society of Civil Engineers, "Minimum Design Loads for Buildings and Other Structures," ASCE Standards 7-88, ASCE, New York, 1988.

16.3 Canadian Standards Association, CSA Standard S304-M84, "Masonry Design and Construction for Buildings," CSA, Rexdale, Ontario, 1984.

## 16.8 PROBLEMS

**16.1** What are the critical design features to be considered in the planning of a single-story loadbearing masonry building? Discuss.

**16.2** From building plans or site visits, sketch the wall layout for two different single-story loadbearing masonry buildings. If possible, determine whether the walls are reinforced or unreinforced and the type of roof system. Discuss the suitability of the wall layout to resist lateral wind and seismic loading.

**16.3** Obtain information on concrete, cold formed steel, and wood roof systems used in your local area. Classify these from a rigidity point of view and discuss their use, including anchorage, in a single-story loadbearing masonry building.

**16.4** For the buildings identified in Prob. 16.2, determine the potential for roof uplift and design an appropriate hold-down anchor system.

**16.5** Determine the moments due to out-of-plane wind load of 20 lb/ft$^2$ (1 kN/m$^2$) acting on wall A and wall B shown in Fig. P16.5.
**(a)** Consider each wall to be fixed at the base and pinned at the top.
**(b)** Consider each wall to be pinned at the top and bottom.

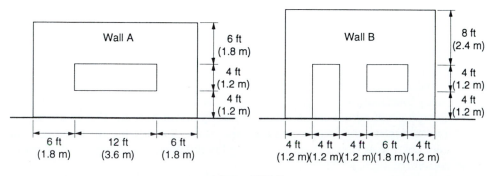

**Figure P16.5**

**16.6** Neglecting axial compression load from the roof for the walls shown in Fig. P16.5, design the following using you local masonry code:

**(a)** Wall A as an unreinforced wall with a pinned base. Use solid clay brick with type S mortar.

**(b)** Wall B as a reinforced 8 in. (200 mm) thick wall. Use hollow concrete block with 2000 psi (15 MPa) compressive strength, type S mortar, and Grade 60 (400 MPa) steel reinforcement.

**16.7** For the single-story loadbearing masonry building shown in Fig. P16.7, the lateral wind load is given as 25 lb/ft$^2$ (1.2 kN/m$^2$) and the corresponding roof uplift due to wind is given as 30 lb/ft$^2$ (neglecting local effects near the roof edge). The walls are constructed with 8 in. (200 mm) hollow concrete blocks of 2500 psi (17.3 MPa) strength and type S mortar. Grade 60 (400 MPa) reinforcement is used if required. The roof load is 10 lb/ft$^2$ (0.50 kN/m$^2$) due to self-weight with an allowance of 5 lb/ft$^2$ (0.25 kN/m$^2$) for mechanical and electrical equipment.

**(a)** Design a typical strip of wall A for the combined efforts of axial load and out-of-plane bending.

**(b)** Design wall B for in-plane forces. Consider the roof diaphragm to be flexible.

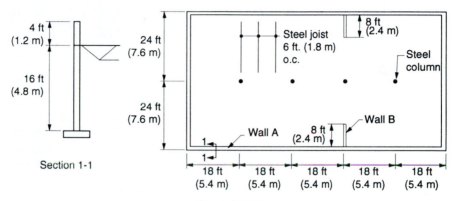

**Figure P16.7**

**16.8** Obtain a set of plans for a single-story loadbearing masonry building in your local area. Check the design and comment on any conservative or nonconservative features that you identify.

# CHAPTER 17

# Design of Multistory Loadbearing Masonry Buildings

Contemporary loadbearing masonry building (*Courtesy of the Brick Institute of America*).

## 17.1 INTRODUCTION

The inherent compressive strength and stiffness of masonry makes it ideally suited to carrying vertical gravity loads and lateral wind loads for many types of multistory buildings. In addition, structurally reinforced masonry can safely resist substantial earthquake loads. Whereas masonry infill walls have been used extensively in high-rise buildings with concrete or steel frames, the use of masonry walls as the loadbearing and shear-resisting elements allows them to serve the dual functions of the building envelope and the load-resisting structure.

The modern era of loadbearing masonry design has resulted in relatively thin wall construction of both unreinforced and reinforced masonry. One of the earliest examples is in Basel, Switzerland, where in 1951, a 13 story apartment building was built with maximum 6 in. (150 mm) thick internal walls and a total thickness of 15 in. (380 mm) for the exterior walls.[17.1] An extensive test program led to the rational design of many high-rise buildings including the 16 story apartment building shown in Fig. 17.1, which was built with $5\frac{7}{8}$ in. (150 mm) thick clay masonry walls.[17.2]

**Figure 17.1** 16-story loadbearing brick masonry building, Biel, Switzerland. (*Courtesy of Brick Institute of America.*)

Although multistory loadbearing masonry apartment and hotel buildings have normally been most economical in the 8 to 12 story range, use of reinforced masonry has produced economic designs for much higher buildings, such as a 24-story Place Louis Riel (Fig. 17.2) built in Winnipeg in 1970 of 8 in. (200 mm) blocks and the more recent top 18 stories of the Excalibur Hotel in Las Vegas (Seismic Zone 2) [see Fig. 2.21]. A unique and very cost-effective and efficient construction process involving use of on-site precast concrete floors [shown being positioned in Fig. 17.3], allowed the construction of the Excalibur Hotel to proceed at the rate of one floor per week. As discussed in Chap. 3 and as is covered later in this chapter, the choice of floor system is very important both for the efficiency of construction and the design of the wall layout.

Sec. 17.1   Introduction

**Figure 17.2**  24-story loadbearing reinforced block apartment building, Winnipeg, Canada. (*Courtesy of Murray W. Isaac.*)

**Figure 17.3**  On-site precast floors being lifted into place during construction of the Excalibur Hotel. (*Courtesy of National Concrete Masonry Association.*)

Availability of high strength masonry units, improved grouting and reinforcing techniques, and improvements in building codes also contribute to expanding the practical ranges of design of loadbearing masonry buildings.

The 10 story example design reviewed later in this chapter is typical of high-rise construction and illustrates the basic approach and the design concepts applicable to any multistory building with floor and roof systems that provide rigid diaphragm action. Rigid diaphragms connect the separate walls and cause them to act together to resist lateral load. Design of single-story buildings with flexible roof diaphragms is covered in Chap. 16.

## 17.2 BASIC DESIGN CONCEPTS

### 17.2.1 Introduction

In multistory buildings, there is a high demand for compressive strength from the accumulated effects of gravity loads and overturning moments due to wind or earthquake forces. These lateral forces also produce a high demand for shear and flexural tension capacities and as a result may require reinforcing steel. These strength requirements lead to special design considerations that are unique to multistory buildings. Top story and interstory drift deflection limits may also control the design. In seismically active areas, the need for energy dissipation and for redistribution of forces creates large ductility demands critical for adequate safety against collapse.

For the design of multistory loadbearing masonry buildings, the concepts discussed in the following sections are essential to making appropriate design decisions.

### 17.2.2 Vertical Load Transfer

The most effective way to transfer vertical load to the foundation is to align loadbearing walls vertically, as shown for walls at axes *A* and *C* in Fig. 17.4. Large open areas at the ground floor may require that loadbearing walls be omitted at this level (see the wall at axis *B*, Fig. 17.4). The continuation of the wall above requires a support structure of beams and columns. Although individual walls can be eliminated to satisfy such architectural requirements, the response of the building to lateral load is significantly affected and special attention to detail is required to accommodate the sudden change in load-resisting mechanism. In general, the use of "soft" lower stories and rigid upper parts of the building has proven to result in poor performance in seismic areas.

For minimum interruption of the construction process, precast prestressed hollow core concrete floor planks are often used with loadbearing masonry. For buildings divided into rooms or suites of rooms, the walls between these areas are normally used as loadbearing walls. The spacing of these bearing walls (typically walls oriented perpendicular to the longest face of the building) dictates the span of planks which typically ranges from 20 to 30 ft (6 to 9 m). The floor thickness is governed by floor loads, span, short- and long-term deflection limits, and fire rating requirements. Bearing of the floor slab on the masonry walls has to be checked and a minimum bearing of 2 in. (50 mm) is generally required.

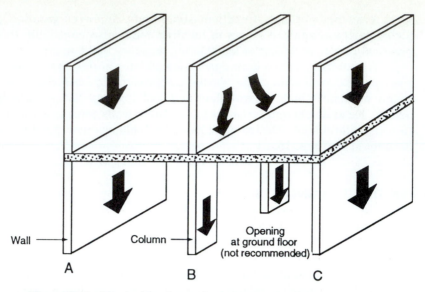

Wall

Column

Opening
at ground floor
(not recommended)

A

B

C

**Figure 17.4** Vertical load transfer in multistory masonry buildings.

For floor plank construction (see Fig. 17.5(a)), a major requirement is to ensure both vertical and horizontal continuity. It is common practice to use concrete topping and dowels to attach the floor to the walls and to tie the individual planks together to form a rigid diaphragm.[17.3] When topping is not used, steel inserts spaced along the abutting edges of the planks must be welded together to create the diaphragm action. These can be used to attach anchors for walls running parallel to the slabs. Reinforcement grouted into the shear keys between planks can be used to attach them to walls at the bearing ends.[17.3,17.4]

Floors made with open-web steel joists and concrete decks are also used because of the speed of construction, relative light weight, and ability to provide diaphragm action. Because the floor must be made continuous with the walls, pouring the entire deck as a continuous diaphragm (see Fig. 17.5(b)) is an effective construction method. Bearing stresses under the joist shoes must also be checked.

Cast-in-place reinforced concrete floors can be used where the impact of extra time required for forming, pouring, and curing the floors can be minimized by splitting

(a) Precast floor slab

(b) Concrete deck on open web steel joints

**Figure 17.5** Diaphragm floor systems.

the building into two parts with a staggered construction schedule. Wood floor and roof systems are also used in some smaller buildings up to 3 or 4 stories, where fire resistance requirements are satisfied by sprinklers or other provisions.

Depending on the floor system, portions of walls above door and window openings must be designed as lintel beams to transfer vertical load to the main wall elements. For large openings, it is usually most efficient to design the floor system to span parallel to the opening and avoid transferring large floor loads to these lintels.

Partitions not intended to be part of the loadbearing structure should be separated at the sides and top to avoid having the wall loaded due to vertical or horizontal deflection. The design of such elements is discussed in Chap. 11. However, it is worth repeating that the effective separation of the partition results in it not being laterally supported at the sides and top. Connection details such as clip angles (see Chap. 11) are necessary to ensure the stability of the partition. Use of wedges or filling the top joint between the partition and the floor can result in the partition becoming a loadbearing wall.

### 17.2.3 Lateral Load Transfer

Design for lateral loads in multistory buildings involves use of rigid floor and roof diaphragms to transfer the loads to the shear walls, which, in turn, carry these forces to the foundation, as illustrated in Fig. 17.6. As was discussed in Sec. 3.6.4, it is essential to provide shear walls in the two orthogonal directions of the building to resist lateral loads from any direction. For simplicity, the application of lateral load independently in the two orthogonal directions has been shown to be a satisfactory design procedure.

As indicated in the previous section, proper connection details between the floor and walls are necessary to provide continuity of the diaphragm and to transfer the shear forces to the walls.

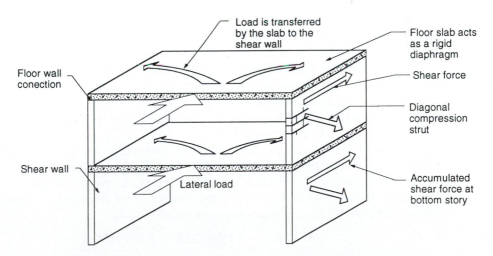

**Figure 17.6** Lateral load transfer in multistory loadbearing masonry buildings.

Lateral stability against overturning is provided by gravity loads to prevent flexural cracking in the case of unreinforced walls or by using reinforcing steel to create a tension force where cracking occurs in reinforced walls. Foundations must be designed to develop the required moment resistance at the bases of shear walls. The design of individual shear walls is covered in Chap. 10.

## 17.3 DISTRIBUTION OF SHEAR AND MOMENT DUE TO LATERAL LOAD

### 17.3.1 Relative Wall Rigidities for Structural Analysis

The rigid diaphragm action of the roof and floor systems in multistory buildings results in lateral loads being distributed to the lateral load resisting elements according to their relative stiffnesses. For shear wall buildings in which the floors provide little rotational resistance, the walls can be assumed to act as cantilevers from the foundation. As discussed in Sec. 10.4.2, the shear deformations in such walls can amount to a significant part of the total deflection, depending on the wall aspect ratio and boundary conditions. Therefore, calculation of the rigidity depends not only on the geometry of the wall, but also on the relationship between shear and bending moment. Simple calculations can be accurate for single-story buildings. However, because the height-to-length ratio ($H/L$) changes at each floor level and because lateral load is typically distributed over the height of the building, no single measure of relative rigidity is accurate over the full height of the building.

Structural analysis computer software that consider both flexural and shear deformation can be used to provide better accuracy for lateral load distribution. [Many commercially available computer programs that include shear deformations can be used for lateral load analysis of shear wall buildings.] However, in many cases, it is desirable to perform simplified calculations using relative wall rigidities.[17.5,17.6] Various approximate methods are briefly discussed in what follows.

**1.** *Full Height Cantilever Action with Concentrated Top Load (Method 1).* In this method, the walls are considered to act as cantilevers over the full height of the building under the action of concentrated loads at the tops, such as shown in Fig. 17.7(a). Because the real loading and, therefore, the relationships between shear and moment are not as shown, the relative effect of shear deformations on long (squat) walls is underestimated. Consequently, lateral loads will be overestimated for these long walls. This means that this method is conservative for long walls and unconservative for short (slender) walls.

It should be noted that, in addition to evaluating relative deformations by considering concentrated loading as opposed to distributed load, this method also ignores compatibility of deflection at all floor levels except at the roof.

From basic structural analysis, the rigidity $R$ of a cantilever wall under a concentrated lateral load at the top can be calculated as

$$R = 1/\Delta = 1/(VH^3/3EI + 2.5kVH/AE) \qquad (17.1)$$

where $\Delta$ is the deflection at the top, $H$ is the building height, $A$ and $I$ are the area and moment of inertia, respectively, of the cross-section, and $E$ is the modulus of

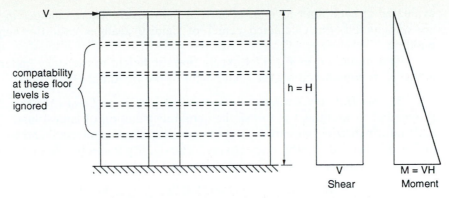

### (a) Full Height Cantilever Action with Concentrated Load

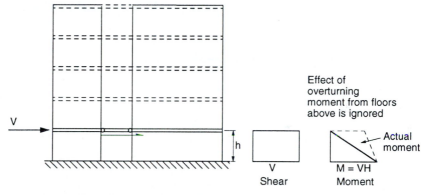

### (b) Walls are Cantilevered from Base to First Floor (for distribution of base shear)

**Figure 17.7**   Loading conditions for calculation of relative wall rigidity.

elasticity of the masonry. The shape factor k is 1.0 for flanged sections and 1.2 for rectangular walls. The shear modulus is taken equal to 0.4E. For calculation of relative rigidities, *V* and *E* can be removed as common terms.

   **2.** *Cantilever Action from the Base to the Floor Above the Level Considered (Method 2).* As shown in Fig. 17.7(b) for load sharing at the base of the building, the first story is considered to be a cantilever with the total shear force applied at its top. As shown, this relationship between shear and bending moment ignores the accumulated effect of overturning moment from the floors above. This approach results in relatively high shear deformations that translate into significant reductions in relative stiffness and lower lateral loads for long (squat) walls. This means that this method can be highly unconservative for long (squat) walls and highly conservative for short (slender) walls.[17.6]

   In addition to ignoring the effects of overturning moments at the level under consideration, this method ignores compatibility of deflections at all the floors above or below the floor under consideration. Also, because the relative rigidities are evaluated at each floor level or at least at floor levels where changes in the design are anticipated, the calculated center of rigidity changes over the height of the building.

This method uses Eq. 17.1 except that the height to the story under consideration, $h_i$, is substituted for the total height of the building, $H$.

**3.** *Full Height Cantilever Action with Distributed Load (Method 3).* An improvement to Method 1 is to use the actual distribution of lateral load. Triangular and uniform distributions are most common (i.e., seismic and wind) and are illustrated in Fig. 17.8(a) and 17.8(b), respectively. The rigidity R can be calculated as follows: For the triangular load distribution shown in Fig. 17.8(a)

$$R = 1/\Delta = 1/[(11/60)(VH^3/EI) + (5/3)(kVH/EA)] \qquad (17.2)$$

For the uniform distribution of load shown in Fig. 17.8(b)

$$R = 1/\Delta = 1/(VH^3/8EI + 5kVH/4EA) \qquad (17.3)$$

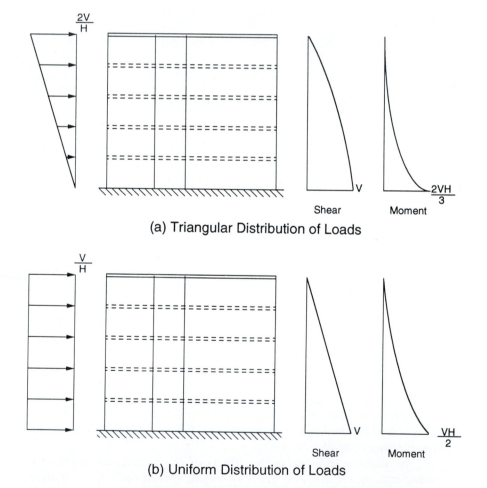

(a) Triangular Distribution of Loads

(b) Uniform Distribution of Loads

**Figure 17.8** Full height cantilever action with distributed lateral loads.

Design of Multistory Loadbearing Masonry Buildings   Chap. 17

In Eqs. 17.2 and 17.3, $k = 1.0$ for flanged walls and $k = 1.2$ for rectangular walls. [The $V$ and $E$ terms can be eliminated for calculation of relative rigidities.]

This method is somewhat more time consuming than Method 1 and, like Method 1, ignores compatibility of deflection at all floors but at the top of the building.

**4.** *Deflection at the Floor Level Being Considered Due to Accumulated Shear and Overturning Moment (Method 4).* The shear and moment diagrams shown in Fig. 17.9 are used to illustrate calculation of rigidity at the first floor. The calculation of rigidity is based on the deflection at the top of the floor height under consideration. In this example, deflection is calculated at the top of the first story. The shear force can be taken as constant over the story height and the straight line relationship shown for moment is compatible with this shear distribution.

The formula for deflection is

$$\Delta = \frac{M_{base}h^2/2 - V_{base}h^3/6}{EI} + \frac{5k}{2}\frac{V_{base}h}{EA}$$

taking $C = M_{base}/V_{base}\,h$,

$$\Delta = \frac{V_{base}}{E}\left[\frac{h^3}{I}\left(\frac{C}{2} - \frac{1}{6}\right) + \frac{5kh}{2A}\right]$$

and by removing $V_{base}/E$ as a constant for all walls, the relative rigidity can be calculated as

$$R_{base} = 1\bigg/\left[\frac{h^3}{I}\left(\frac{C}{2} - \frac{1}{6}\right) + \frac{5kh}{2A}\right] \tag{17.4}$$

where $k = 1.0$ for flanged sections and $k = 1.2$ for rectangular sections. At a story $i$, the general equation for deflection at height $h_i$ becomes

$$\Delta_i = \int_0^{h_i}\int \frac{M(z)\,dz\,dz}{EI} + \int_0^{h_i} \frac{V(z)\,dz}{AG} \tag{17.5}$$

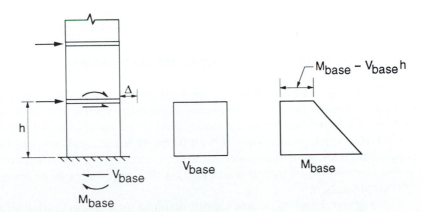

**Figure 17.9** Relationship between shear and moments at first floor for calculation of relative rigidity using method 4.

where moment $M(z)$ and shear $V(z)$ are expressed as functions of height. At the top floor, this becomes the same as Method 3 for triangular or uniform distributions of lateral load.

Method 4 is based on compatibility of deflections of walls at the floor level above the story in question. Therefore, the relative rigidities of walls vary over the height of the building, which means that the center of rigidity and overall distribution of forces also change. However, this method of distributing load is more accurate than the other methods and the additional calculations only need to be done for the stories where changes in the masonry strength or reinforcing are anticipated. Use of spreadsheet programming greatly facilitates such calculations. However, because compatible deflections are not simultaneously ensured at all floors, this is still an approximate method.[17.7]

### 17.3.2 Choice of Method for Determination of Relative Wall Rigidities

If all the walls are similar, then, to the degree that they are similar, lateral loads will be shared equally regardless of the method of analysis. As mentioned before, for walls having large differences in section properties, structural analysis programs that account for shear deformations provide accurate answers insofar as the specified properties are correct. Equivalent frame methods of analysis can be used if shear deformations are included.

Method 4, although not exact, can give reasonable approximations[17.7] but, if several different levels are to be checked, the computational effort increases proportionally. Therefore, the choice of method of analysis for defining rigidity should depend to a large extent on an assessment of the sensitivity of the behavior of individual walls and the structure as a whole to the calculated loads.

Walls that are not loaded to near their capacities or reinforced walls, for which cracking reduces the rigidities sufficiently to shed load to other walls, can be relatively insensitive to the calculated distribution of loads. In such cases, fairly approximate methods are acceptable and the very simple approach presented as Method 1 can be quite satisfactory. Alternately, for unreinforced masonry, it is prudent to provide reasonable allowances for variation in load from the calculated values unless the previously mentioned rigorous structural analysis is carried out.

### 17.3.3 Other Factors Affecting the Distribution of Lateral Loads

**Coupling of Shear Walls.** Adjacent masonry walls can be coupled by connecting beams and floor slabs. For simplified calculations, this coupling is generally ignored and individual walls are considered to be cantilevered from the base (see Fig. 17.10). Equivalent frame or wall-frame types of analyses can be performed, using computers, to provide more accurate evaluations of the effects of this coupling on the distribution of lateral loads and on the magnitudes of bending moments in individual walls.

To ignore coupling action cannot result in an unconservative situation for the walls, but can result in overstressing of the coupling beams and possibly some shift of the center of rigidity of the building. A design practice that can be followed is to

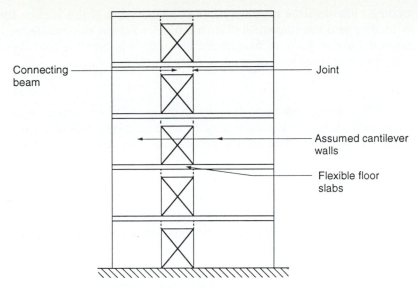

**Figure 17.10** Uncoupling of cantilever shear walls.

insert joints at the ends of stiff beams joining adjacent walls. These joints prevent development of the large end moments that produce the main coupling action.

**Interaction of Intersecting In-Plane and Out-of-Plane Walls.** It is common to have intersecting walls in masonry construction and, as a result, many walls have flanged cross-sections, such as shown in Fig. 17.11. Obviously, the addition of flanges has a significant effect on the moment of inertia of the section which, in turn, affects the distribution of lateral loads and the calculated stresses. Masonry codes[17.8–17.11] allow the use of effective flange widths for flexural stress calculations provided that adequate shear transfer can take place at the web-flange interface (refer to Sec. 10.5.1).

For lateral load distribution, flanges should be considered when calculating relative wall rigidities. It is consistent to use the same effective flange width for rigidity calculations as is used for stress calculations.

As is indicated in Fig. 17.11, in lateral load analysis, it is possible that flanges can be included for one direction of loading and ignored for the other direction of

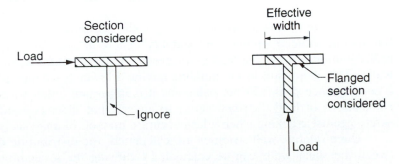

**Figure 17.11** Consideration of flanges in rigidity calculation.

loading. The decision depends on the configuration of the section for the direction of loading and the impact that such flanges have on the section properties. Even when flanges or parts of flanges are ignored, it is possible that they can be relied upon to resist axial load applied directly on these sections.

**Effects of Cracking.**  For the analysis of forces in shear wall structures, it is customary to ignore the effects of flexural cracking and to perform linear elastic analyses with uncracked section properties. This is an acceptable approach for the following:

1.  Unreinforced masonry where cracking is not permitted in shear walls.
2.  Buildings in areas of low seismic activity and where lateral loading is not critical.
3.  Buildings having walls with uniform configurations and thicknesses where inaccuracies in rigidity calculations do not affect the relative rigidities.

For masonry buildings that do not satisfy at least one of the foregoing conditions, cracking has a significant effect on wall stiffness and consequently on the way lateral load is shared between walls. The decrease in relative rigidity for cracked walls results in a greater share of the lateral load being transferred to uncracked walls. This redistribution has the benefit of reducing the lateral load on the most highly loaded walls. However, for the walls that are subjected to increased lateral load because of the redistribution, the potential for shear or compression failure should be evaluated. It is for this reason that mixing of reinforced and unreinforced walls is not recommended by the authors. Also, as is required for Seismic Zones 3 and 4, provision of extra shear capacity is recommended to ensure that redistribution of moments can occur without premature shear failure.[17.8,17.9]

Stiffness degradation of a reinforced masonry wall due to cracking is normally not accompanied by similar strength degradation. Therefore, provided that adequate ductility is present, advantage can be taken of the benefits of redistribution of load to reduce the need for highly accurate assessments of the distribution of lateral load. However, because stiffness degradation does directly relate to interstory drift,[17.12] judgment should be exercised when considering the impact of cracking on the overall behavior of the building. It is important to ensure that the top deflection and interstory drift limits are satisfied when the effects of cracking are accounted for.

Cracking of unreinforced masonry walls not only results in reduced rigidity, but also in reduced strength and stability. There is some danger that the resulting redistribution of load can lead to successive cracking and loss of strength in other walls. This can result in the building having a lower resistance to lateral load. As a consequence of this brittle behavior, it is important either to carry out accurate calculations of load distributions or to ensure that there is a sufficient margin of safety against cracking when a less accurate method of analysis is used.

Once relative wall rigidities are calculated, the distribution of lateral load between the shear walls can be calculated following the procedure outlined in Sec. 10.4.2.

## 17.4 TORSIONAL EFFECTS

### 17.4.1 Basic Concept

Torsion occurs when the center of rigidity of the wall system, CR, does not coincide with the location of the lateral load resultant which is at the center of mass C.M. for earthquake loading and at the centroid of the loaded area for wind loading. As was discussed in Chap. 10, the total shear forces in individual walls are the sum of forces due to translation and rotation effects.

For seismic design, most codes[17.9,17.13,17.14] specify minimum accidental eccentricities to account for uncertainty in location of mass. For instance, ASCE 7-88[17.13] and the Uniform Building Code[17.9] specify a minimum additional eccentricity of ±5% of the total building dimension normal to the direction of loading. This means that there are four possible torsion moments corresponding to the resulting four eccentricities, as illustrated in Fig. 17.12(a). By inspection, it is sometimes possible to use only the maximum $e_x$ and $e_y$ eccentricities in design. This is the case when the smaller value is on the same side of the center of resistance as the maximum value. Also, for reasonably symmetric buildings, the designs for walls subject to the maximum effects of torsion can be used for similar walls where the torsional effect is lower.

For buildings with reasonably symmetric arrangements of shear walls located near the perimeter of the building, the torsional component of shear is generally much less than the translational (or direct) shear. In fact, if calculations show high torsion effects, it is a signal that a better layout of shear walls should be investigated. A more detailed discussion of wall layout considerations is provided in Chap. 3.

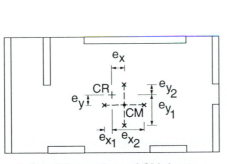

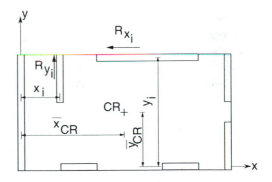

x   Possible locations of CM due to code accidential eccentricities

$e_x$, $e_y$   Calculated eccentricities

$e_{x_1}$,$e_{x_2}$   Possible eccentricities for design
$e_{y_1}$,$e_{y_2}$

(a) Possible Eccentricities for Torsional Effects

(b) Calculated Center of Rigidity

**Figure 17.12**   Locations of center of mass and center of rigidity.

## 17.4.2 Calculation of Torsional Moments

Torsion is the product of lateral force times the distance or eccentricity of this force from the center of rigidity of the lateral load resisting walls. By using the definition of rigidity and one of the methods described in Sec. 17.3.1, the location of the center of rigidity, CR, from a reference axis can be calculated as follows (see Fig. 17.12(b)):

$$\bar{x}_{CR} = \frac{\sum R_{yi} x_i}{\sum R_{yi}}$$

$$\bar{y}_{CR} = \frac{\sum R_{xi} y_i}{\sum R_{xi}}$$

(17.6)

where    $R_{xi}, R_{yi}$ = rigidities of wall $i$ for loading in the $x$ and $y$ directions, respectively

$x_i, y_i$ = distances from the reference axis to wall $i$ in the $x$ and $y$ directions, respectively

$\bar{x}_{CR}, \bar{y}_{CR}$ = distances from the reference axis to the center of rigidity

At a particular story, the torsional moment is accurately calculated as the sum of the products of the horizontal forces at each floor level above that story times the eccentricity of these forces from the center of rigidity at the story for which torsion is being calculated. For repeated floor plans where the positions of the resultant horizontal forces are the same at each story, the torsion at the story being considered is equal to the sum of the horizontal forces at each floor level times the eccentricity. For example, lateral loading in the $y$-direction results in torsional moment

$$M_{T_{zy}} = \left( \sum_{i=z}^{N} Q_{yi} \right) e_{xz}$$

(17.7)

where  $M_{T_{zy}}$ = torsional moment at level $z$ for loading in the $y$ direction

$Q_{yi}$ = horizontal force in the $y$ direction at level $i$

$z$ = floor level under consideration

$N$ = number of floors

$e_{xz}$ = eccentricity of horizontal forces in the $y$ direction for the center of rigidity at floor $z$

An example, of a case where this calculation is applicable, is wind loading on buildings with uniform floor plans.

For seismic loading and/or nonuniform floor plans, the position of the resultant horizontal force can change from floor to floor. In this case, the torsional moment at floor $z$ can be calculated as

$$M_{T_{zy}} = \sum_{i=z}^{N} \left( Q_{yi} e_{xi} \right)$$

(17.8)

where $e_{xi}$ is the eccentricity of the lateral load in the $y$ direction at floor $i$ about the center of rigidity at floor $z$.

A practical example of this case is illustrated in Fig. 17.13, where torsion is caused by seismic forces. In this case, the lateral forces $Q_{yi}$ at each floor level are located at the center of mass. For floors with uniform story height and floor plan, the center of mass is in the same place at each floor. However, for the roof level, the effect of using the mass for only half the height of the wall below the roof and possibly some difference in the mass of the roof compared to the floors can result in a different center of mass. Similarly, if the ground story has a larger story height, the masses for the half heights of walls above and below the first floor mass also result in a different center of mass. Therefore, for torsional moment at the base, due to loading in the y direction, $M_{T_{by}}$, the eccentricity of the force, $Q_{yR}$, at the roof can be at eccentricity, $e_{xRb}$, from the center of rigidity at the base. Similarly, the force added at the first story, $Q_{yG}$, may be at some other eccentricity, $e_{xGb}$, about the center of rigidity at the base. The remaining typical repeated floors have their respective forces located at a common or typical eccentricity, $e_{xtyp.b}$, with respect to the center of rigidity at the base. As a result of the foregoing, the total torsional moment at the base can be calculated as

$$M_{T_{by}} = Q_{yR}e_{xRb} + \left( \sum_{i=2}^{R-1} Q_{yi} \right) e_{xtyp.b} + Q_{yG}e_{xGb} \qquad (17.9)$$

This equation is just a specific version of Eq. 17.8.

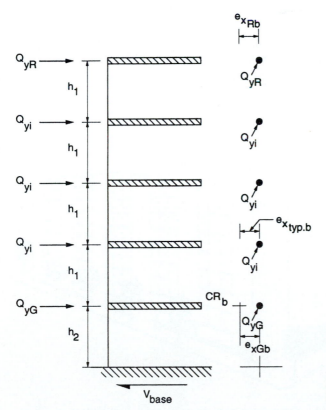

Figure 17.13   Illustration for calculation of torsional moment.

To follow the foregoing procedure when the locations of the resultant forces at each floor level vary slightly obviously entails additional computational effort. If the building plan is quite symmetric, the calculated eccentricities are quite small and the additional allowance for accidental eccentricity produces the major component of the torsional moment. Therefore, for such cases, a reasonable practical approximation is to use Eq. 17.7 with $e_x = e_{x,\text{typ.b}}$. Calculations for torsion resulting from loading in the $x$-direction are similar but with eccentricities in the y-direction.

The calculation of the additional shear and bending forces to be resisted by walls as a result of torsional moments follows the procedure outlined in Sec. 10.4.2.

## 17.5 DESIGN EXAMPLE

### 17.5.1 Description of the Building

A perspective drawing of the example 10 story, 90 ft. (27.4m) high loadbearing masonry apartment building is shown in Fig. 17.14. Figure 17.15 is the typical floor plan showing the layout of the loadbearing masonry shear walls. The entrance between wall lines 6 and 8 eliminates the wall along the South side of the corridor at the ground floor. Elevations of the longitudinal and transverse faces of the building are shown in Fig. 17.16.

For fire protection and sound insulation, nonloadbearing concrete block partitions are used along the South side of the corridor above the ground floor. The exterior cladding is brick veneer supported on shelf angles at each floor. The nonloadbearing parts of walls above and below large openings are separated from the the loadbearing walls by vertical movement joints to create a "strip" type of window treatment. The floor system is 8 in. (200 mm) thick precast prestressed concrete

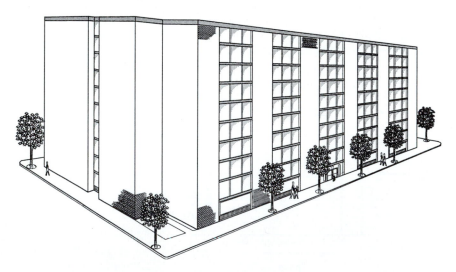

**Figure 17.14**  Perspective of multistory masonry building.

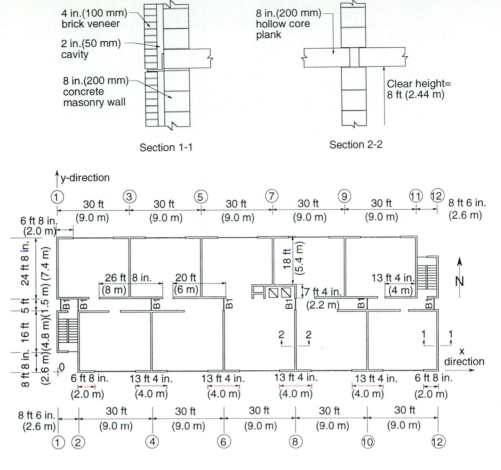

**Figure 17.15** Typical floor plan of the multistory design example.

planks without topping. Special connections to ensure diaphragm action are used. [Note: If these are not provided and, in any case, for Seismic Zones 3 and 4, use of topping is recommended.]

Buildings with nearly equal uncoupled linear (without flanges) shear walls are very easy to analyze and result in a greatly simplified design process. However, because such buildings rarely exist in practice, flanged walls and walls with different lengths have been included to illustrate the application of the different factors discussed in Sec. 17.3. Comments are provided to help the designer develop the background to exercise judgment in handling the related design decisions.

So that the design example conforms with an organized and authoritative set of design criteria, the working stress provisions of ACI 530/ASCE 5/TMS 402[17.8] are used. For other codes, including strength design with factored loads, the general approach is applicable. It is only the detailed design of the individual components

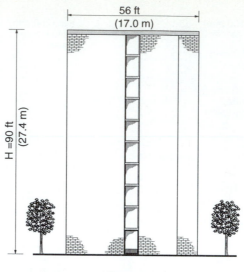

East elevation

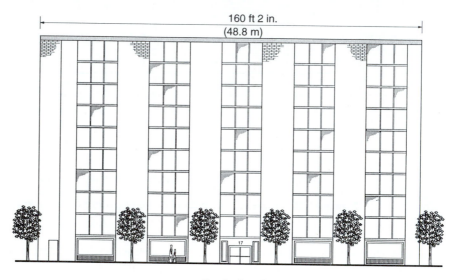

South elevation

**Figure 17.16**  Building elevations.

that will change. [The subject of strength design of shear walls is included in Chap. 10.]

*To avoid a proliferation of numbers, S.I. equivalent metric values have not been included in all situations. The S.I. values that are provided are from a "soft" conversion and do not represent recalculations based on the metric equivalent (i.e., the 7 $\frac{5}{8}$ in. thick concrete block is 193.7 mm, not 190 mm).*

### 17.5.2 Loads

1. *Dead loads*:
   Floors: Slab          = 56 lb/ft$^2$
           Partitions + Ceiling   = 10 lb/ft2
                                    66 lb/ft$^2$ (3.16 kN/m$^2$)

   Roof:   Slab             = 55 lb/ft$^2$
            Fill               = 10 lb/ft$^2$
            Roofing       = 4 lb/ft$^2$
                                    69 lb/ft$^2$ (3.31 kN/m$^2$)

   Walls: For the overall building, it is initially assumed that 8 in. (200 mm) concrete block walls are on average, grout filled in every third cell (24 in. (600 mm) spacing). The equivalent uniform weight

   $$= 55 \text{ lb/ft}^2 \text{ (2.64 kN/m}^2)$$

   At the base, where fully grouted walls may be required, the uniform weights are:

           Block wall       = 77 lb/ft$^2$ (3.69 kN/m$^2$)
           Brick veneer    = 36 lb/ft$^2$ (1.73 kN/m$^2$)

2. *Live loads*:
   For apartments       = 40 lb/ft$^2$ (1.92 kN/m$^2$)
   For corridors         = 100 lb/ft$^2$ (4.79 kN/m$^2$)
   Snow load           = 20 lb/ft$^2$ (0.96 kN/m$^2$)

3. *Wind loads*:
   Basic wind pressure     = 20 lb/ft$^2$ (0.96 kN/m$^2$)
   (below 30 ft (9.14m))
   Distribution of wind loads over the height of the building is shown in Fig. 17.17.

4. *Seismic loads (moderate seismic area–Zone 2)*:
   The base shear formula that is commonly used in the United States for uniform

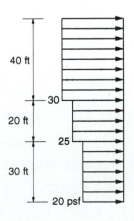

**Figure 17.17** Wind loads on the building.

buildings in moderately active seismic areas[17.13] is

$$V = ZIKCSW \tag{17.10}$$

where $Z = \frac{3}{8}$

$I = 1.0$

$K = 1.33$

$C = 0.07/\sqrt{T}$

$S = 1.0$ (for soil profile No. 1)

### 17.5.3 Calculation of Lateral Forces

**Wind.** By considering a design wind pressure of 20 lb/ft$^2$ and adopting the representative distribution over the height of the building shown in Fig. 17.17, the total base shear in the longitudinal direction (East–West) can be calculated as

$$V_L = 20(30 \times 56) + 25(20 \times 56) + 30(40 \times 56)$$

$$= 128,800 \text{ lb} \quad (573 \text{ kN})$$

Similarly, in the transverse direction (North-South)

$$V_T = 128,800 \times (160.16/56) = 368,368 \text{ lb } (1639 \text{ kN})$$

**Seismic.** Weight $W$ is the summation of dead weight of floors and walls, including partitions and any fixed service equipment:

$$W = \overbrace{(66 \times 9 + 69)}^{\text{floors}}\overbrace{(54.33}^{\text{roof}} + (8/12))(30 \times 5 + 8.5 + (8/12))$$

$$+ \underbrace{679(908/12 \times 10)55}_{\text{walls}} + \underbrace{258(90)(36)}_{\text{veneer}}$$

$$= 9751 \text{ kips} \quad (4432 \text{ tonnes})$$

Period of vibration: Consider $T = 0.05H/\sqrt{D}$, where $D$ is the dimension of the building (minus the veneer) in the direction of load and $H$ is the height:

$$T_{\text{long}} = 0.05(90)/\sqrt{159.16} = 0.36 \text{ s}$$

$$T_{\text{trans}} = 0.05(90)/\sqrt{55} = 0.61 \text{ s}$$

The value of $CS$ need not exceed 0.14. Checking,

$$C_{\text{long}} = 0.07/\sqrt{0.36} = 0.117 \text{ and } CS = 0.117$$

$$C_{\text{trans}} = 0.07/\sqrt{0.61} = 0.0896 \text{ and } CS = 0.090$$

Applying the base shear formula: In the longitudinal direction,

$$V_L = (3/8)(1.0)(1.33)(0.117)(9751)$$

$$= 0.0584(9751) = 569.5 \text{ kips}$$

$$\text{say, } 570 \text{ kips} \quad (2540 \text{ kN})$$

In the transverse direction,

$$V_T = (3/8)(1.0)(1.33)(0.090)(9751)$$

$$= 0.045(9751) = 438.8 \text{ kips}$$

$$\text{say, } 439 \text{ kips} \qquad (1995 \text{ kN})$$

Comparing the base shear forces due to wind and earthquake loads indicates that seismic design governs at the base. As was discussed in Sec. 3.2.2, it is possible that seismic loading could govern at upper levels when wind governs design at the base. The reverse is not possible for relatively uniform distributions of mass over the height of the building. Force $F_i$ acting at floor $i$ (height $h_i$) is expressed as

$$F_i = V \frac{W_i h_i}{\sum W_i h_i} \tag{17.11}$$

where $W_i$ is the weight lumped at floor $i$, and $h_i$ is the floor height measured from the base.

For the relatively uniform distribution of mass over the height of the building, the lateral loads due to earthquakes are distributed in a triangular form and vary linearly from zero at the base to a maximum at the top. For the repeated floor plan and by assuming conservatively that the weight assigned to the roof level is equal to the floor weight, Eq. 17.11 can be simplified as

$$F_i = V \frac{h_i}{\sum h_i} \tag{17.12}$$

Application of this equation produces the force distributions shown in Fig. 17.18. [Note: For this building where T is less than 0.7 seconds, there is no need to apply an extra force at the top of the structure to account for different modes of

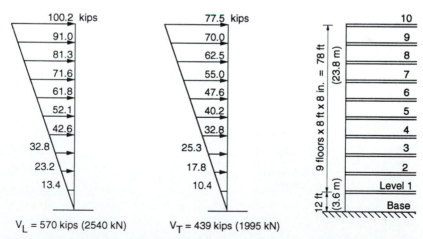

$V_L = 570 \text{ kips (2540 kN)} \qquad V_T = 439 \text{ kips (1995 kN)}$

(a) Longitudinal Direction (b) Transverse Direction    (c) Floor Levels

**Figure 17.18** Distribution of seismic forces in the two orthogonal directions.

vibration that can occur in less stiff structures.[17.13] As an example calculation, the horizontal force at the roof level is

$$F_{\text{Roof}} = F_{10} = 439(90/510) = 77.4 \text{ kips} \qquad (345 \text{ kN})$$

in the transverse direction. The corresponding shear and moment diagrams are shown in Fig. 17.19.

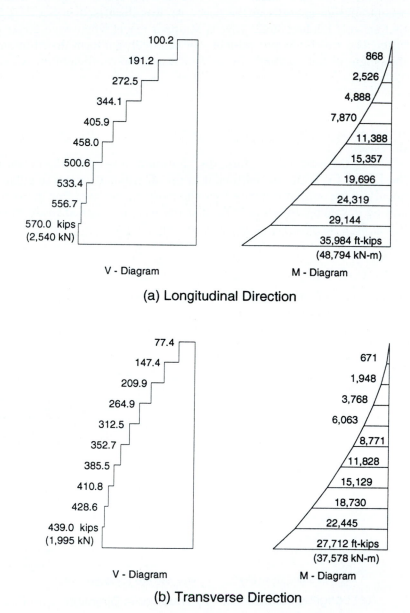

(a) Longitudinal Direction

(b) Transverse Direction

**Figure 17.19**   Shear and moments due to seismic loads.

### 17.5.4 Distribution of Lateral Loads to Shear Walls

As discussed earlier, lateral loads are distributed to shear walls according to their rigidities. It is assumed in this example that the connecting floor slabs between the shear walls are relatively flexible and incapable of transferring significant vertical shear. Therefore, the shear walls are considered to act as cantilevers from the base (no coupling).

For load resistance in the transverse direction (North–South direction), wall flanges should be included in the calculations of moments of inertia. Effective flange widths not exceeding six times the wall thickness on either side of the web can be used. By this criteria, the effective wall sections for resisting lateral load in the transverse direction are shown as the darkened parts in Fig. 17.20(a). For the longitudinal direction, the cross-walls (walls oriented in the North–South direction) are not included in the calculations for rigidity because they do not add significantly to the moments of inertia and do not affect the shear deformations. The assumed effective wall sections are shown in Fig. 17.20(b). Where possible, the cross wall intersections

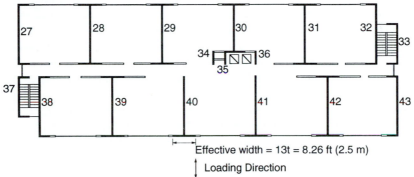

(a) Wall Sections for Resisting Loads in the Transverse Direction (Considering Flanges)

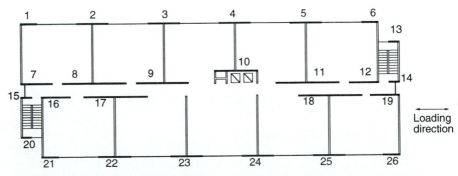

(b) Wall Sections for Resisting Loads in the Longitudinal Direction (Ignoring Flanges)

**Figure 17.20**  Effective wall sections and wall designations.

are located at the midpoint of the longitudinal walls for better sharing of axial load. The symmetry ensures similar responses of the wall to lateral loading in either the East or West direction.

In this design example, the simplified methods of analyses presented in Sec. 17.3.1 are used to illustrate the types of hand calculations that can be done. Method 1 is utilized to illustrate the more traditional approach. As mentioned earlier, if shear deformations are not significant, or if the walls are all quite similar, this approach gives reasonable results. Otherwise, the method tends to be conservative for the more rigid walls, which, in turn, are usually the most critical because they attract the highest loads.[17.7] Therefore, this method is acceptable as long as the designer recognizes that the loading on the shorter, less rigid walls can be significantly underestimated. This situation can be allowed for by ensuring that short walls have ample reserve capacity. Alternatively, if they crack, the reduced rigidities lessen the share of the lateral load they must resist. [As mentioned in Sec. 17.3.1, reinforced walls that crack have substantially different rigidities and the presence of the reinforcement is relied upon, to some extent, to allow redistribution of forces to uncracked walls.]

For Method 1, two approaches are used to illustrate the effect of including flanges. Method 1-A ignores wall flanges and Method 1-B includes wall flanges for calculating the rigidities for loading in the transverse direction; see Fig. 17.20. As an alternative to Method 1, calculations using Method 4 are also included. This method predicts load distributions that are closer to the correct values, but it has the disadvantage of having to repeat the calculations for each floor level investigated.

Table 17.1 contains the tabulated results for the relative wall rigidities calculated according to the foregoing methods. For each direction of loading, it also shows the individual wall rigidities as ratios of the sum of rigidities. For simplicity, a nominal wall thickness of 8 in. (200 mm) was used instead of the actual thickness. This does not have any effect on the relative rigidities for walls of equal thickness. Example calculations are shown in what follows.

**1.** *Method 1A: Wall 1*

$$I = \frac{tL^3}{12} = \left(\frac{1}{12}\right)\left(\frac{8}{12}\right)(6.67)^3 = 16 \text{ ft}^4, \quad A_{\text{web}} = tL = 6.67\left(\frac{8}{12}\right) = 4.4 \text{ ft}^2$$

From Eq. 17.1,

$$R_1 = \frac{1}{(H^3/3I) + 3(H/A)} = \frac{1}{(90^3/3(16) + 3(90)/4.4} = 0.000067$$

**2.** *Method 1B: Wall 27*

$$I = 1672 \text{ ft}^4, \qquad A_{\text{web}} = 16.9 \text{ ft}^2$$

$$R_{27} = 1 \bigg/ \left(\frac{H^3}{3I} + \frac{2.5H}{A_{\text{web}}}\right) = 1 \bigg/ \left(\frac{(90)^3}{3(1672)} + \frac{2.5(90)}{16.9}\right) = 0.00637$$

**Table 17.1** Calculation of Relative Wall Rigidities

| Wall No. | Wall Dir. | $I$ (ft$^4$) | $A_W$ (ft$^2$) | Method 1-A H/L | Method 1-A R ($\times 10^{-3}$) | Method 1-A R/$\Sigma$R | Method 1-B R ($\times 10^{-3}$) | Method 1-B R/$\Sigma$R | Method 4 First Story R ($\times 10^{-3}$) | Method 4 First Story R/$\Sigma$R |
|---|---|---|---|---|---|---|---|---|---|---|
| 1 | | 16 | 4.4 | 13.50 | 0.067 | 0.0026 | 0.067 | 0.0026 | 3.65 | 0.0033 |
| 2 | | 132 | 8.9 | 6.75 | 0.533 | 0.0205 | 0.533 | 0.0205 | 27.55 | 0.0246 |
| 3 | | 132 | 8.9 | 6.75 | 0.533 | 0.0205 | 0.533 | 0.0205 | 27.55 | 0.0246 |
| 4 | | 132 | 8.9 | 6.75 | 0.533 | 0.0205 | 0.533 | 0.0205 | 27.55 | 0.0246 |
| 5 | | 132 | 8.9 | 6.75 | 0.533 | 0.0205 | 0.533 | 0.0205 | 27.55 | 0.0246 |
| 6 | | 16 | 4.4 | 13.50 | 0.067 | 0.0026 | 0.067 | 0.0026 | 3.65 | 0.0033 |
| 7 | | 142 | 9.1 | 6.59 | 0.573 | 0.0221 | 0.573 | 0.0221 | 29.47 | 0.0263 |
| 8 | | 1053 | 17.8 | 3.38 | 4.052 | 0.1560 | 4.052 | 0.1560 | 164.87 | 0.1471 |
| 9 | x-direction (East-West) | 244 | 13.3 | 4.50 | 1.765 | 0.0679 | 1.765 | 0.0679 | 49.62 | 0.0443 |
| 10 | | 257 | 11.1 | 5.40 | 1.032 | 0.0397 | 1.032 | 0.0397 | 50.49 | 0.0451 |
| 11 | | 1053 | 17.8 | 3.38 | 4.052 | 0.1560 | 4.052 | 0.1560 | 164.87 | 0.1471 |
| 12 | | 144 | 9.1 | 6.59 | 0.573 | 0.0221 | 0.573 | 0.0221 | 29.83 | 0.0266 |
| 13 | | 8.4 | 3.6 | 16.88 | 0.035 | 0.0013 | 0.035 | 0.0013 | 1.94 | 0.0017 |
| 14 | | 8.4 | 3.6 | 16.88 | 0.035 | 0.0013 | 0.035 | 0.0013 | 1.94 | 0.0017 |
| 15 | | 8.4 | 3.6 | 16.88 | 0.035 | 0.0013 | 0.035 | 0.0013 | 1.94 | 0.0017 |
| 16 | | 142 | 9.1 | 6.59 | 0.573 | 0.0221 | 0.573 | 0.0221 | 29.47 | 0.0263 |
| 17 | | 1053 | 17.8 | 3.38 | 4.052 | 0.1560 | 4.052 | 0.1560 | 164.87 | 0.1471 |
| 18 | | 1053 | 17.8 | 3.38 | 4.052 | 0.1560 | 4.052 | 0.1560 | 164.87 | 0.1471 |
| 19 | | 142 | 9.1 | 6.59 | 0.573 | 0.0221 | 0.573 | 0.0221 | 29.47 | 0.0263 |
| 20 | | 8.4 | 3.6 | 16.88 | 0.035 | 0.0013 | 0.035 | 0.0013 | 1.94 | 0.0017 |
| 21 | | 16 | 4.4 | 13.50 | 0.067 | 0.0026 | 0.067 | 0.0026 | 3.65 | 0.0033 |
| 22 | | 132 | 8.9 | 6.75 | 0.533 | 0.0205 | 0.533 | 0.0205 | 27.55 | 0.0246 |
| 23 | | 132 | 8.9 | 6.75 | 0.533 | 0.0205 | 0.533 | 0.0205 | 27.55 | 0.0246 |
| 24 | | 132 | 8.9 | 6.75 | 0.533 | 0.0205 | 0.533 | 0.0205 | 27.55 | 0.0246 |
| 25 | | 132 | 8.9 | 6.75 | 0.533 | 0.0205 | 0.533 | 0.0205 | 27.55 | 0.0246 |
| 26 | | 16 | 4.4 | 13.50 | 0.067 | 0.0026 | 0.067 | 0.0026 | 3.65 | 0.0033 |
| | | | | $\Sigma$R=25.973x10$^{-3}$ | | | $\Sigma$R=25.973x10$^{-3}$ | | $\Sigma$R=1120.495x10$^{-3}$ | |
| 27 | | 1672 | 16.9 | 3.55 | 3.518 | 0.0826 | 6.30 | 0.0688 | 213.85 | 0.0729 |
| 28 | | 2443 | 16.9 | 3.55 | 3.518 | 0.0826 | 8.87 | 0.0968 | 258.22 | 0.0880 |
| 29 | | 2443 | 16.9 | 3.55 | 3.518 | 0.0826 | 8.87 | 0.0968 | 258.22 | 0.0880 |
| 30 | | 1188 | 12.4 | 4.73 | 1.525 | 0.0358 | 4.49 | 0.0490 | 154.16 | 0.0525 |
| 31 | | 2443 | 16.9 | 3.55 | 3.518 | 0.0826 | 8.87 | 0.0968 | 258.22 | 0.0880 |
| 32 | | 1672 | 16.9 | 3.55 | 3.518 | 0.0826 | 6.30 | 0.0688 | 213.85 | 0.0729 |
| 33 | | 583 | 11.1 | 5.39 | 1.038 | 0.0244 | 2.29 | 0.0250 | 94.83 | 0.0323 |
| 34 | | 22 | 4.9 | 12.26 | 0.090 | 0.0021 | 0.09 | 0.0010 | 4.98 | 0.0017 |
| 35 | y-direction (North-South) | 22 | 4.9 | 12.26 | 0.090 | 0.0021 | 0.09 | 0.0010 | 4.98 | 0.0017 |
| 36 | | 22 | 4.9 | 12.26 | 0.090 | 0.0021 | 0.09 | 0.0010 | 4.98 | 0.0017 |
| 37 | | 583 | 11.1 | 5.39 | 1.038 | 0.0244 | 2.29 | 0.0250 | 94.83 | 0.0323 |
| 38 | | 1672 | 16.9 | 3.55 | 3.518 | 0.0826 | 6.30 | 0.0688 | 213.85 | 0.0729 |
| 39 | | 2443 | 16.9 | 3.55 | 3.518 | 0.0826 | 8.87 | 0.0968 | 258.22 | 0.0880 |
| 40 | | 1690 | 16.9 | 3.55 | 3.518 | 0.0826 | 6.37 | 0.0695 | 215.09 | 0.0733 |
| 41 | | 1690 | 16.9 | 3.55 | 3.518 | 0.0826 | 6.37 | 0.0695 | 215.09 | 0.0733 |
| 42 | | 2443 | 16.9 | 3.55 | 3.518 | 0.0826 | 8.87 | 0.0968 | 258.22 | 0.0880 |
| 43 | | 1672 | 16.9 | 3.55 | 3.518 | 0.0826 | 6.30 | 0.0688 | 213.85 | 0.0729 |
| | | | | $\Sigma$R= 42.567x10$^{-3}$ | | | $\Sigma$R=91.61x10$^{-3}$ | | $\Sigma$R= 2935.43x10$^{-3}$ | |

**3.** *Method 4: Wall 1 at Ground Floor*

$$I = 16 \text{ ft}^4, A_{\text{web}} = tL = \frac{8}{12}(6.67) = 4.4 \text{ ft}^2, C = \frac{M_{\text{base}}}{V_{\text{base}}h} = \frac{35{,}984}{570(12)} = 5.26$$

From Eq. 17.4,

$$R_1 = 1 \Big/ \left( \frac{h^3}{I} \left( \frac{C}{2} - \frac{1}{6} \right) + \frac{3h}{A_{\text{web}}} \right) = 1 \Big/ \left( \frac{(12)^3}{16} \left( \frac{5.26}{2} - \frac{1}{6} \right) + \frac{3(12)}{4.4} \right) = 0.0037$$

Comparing Methods 1-A and 1-B, for loading in the North-South direction, indicates that ignoring wall flanges can lead to very significant differences in the calculated relative wall rigidities. This, in turn, affects the distribution of load to these walls where, for example, the share of the lateral load resisted by Wall 30 increases by 36% when the effects of flanges are included. This type of difference can be even more pronounced in cases where some walls have flanges and others do not.

By comparing the results from Method 4 with Method 1-B, it can be seen that there is very little difference in ratios of relative rigidities for rigidity in the North–South directions because the walls have quite similar dimensions. However, in the East–West direction, there are more noticeable differences because the different walls lengths affect the relative contributions of shear and flexural deformations for different heights. For example, Wall 7 is shown to take 18% more lateral load according to Method 4.

From the foregoing comparisons, it can be seen that shear deformations do not have much effect on the relative rigidities of walls at the ground floor of the building. Because it is anticipated that these loadbearing shear walls will be reinforced, the ability of reinforced masonry to redistribute the lateral load means that some inaccuracy in the calculated distributions can be tolerated, particularly where the loads on the critical longer walls are overestimated. Therefore, for illustrative purposes, the loads will be distributed using the relative rigidities calculated using Method 1-B (including flanges).

### 17.5.5 Calculations of Shear Forces and Moments at the Bases of the Shear Walls

**Direct Shear.** The base shears of 570 kips (2540 kN) in the longitudinal direction and 439 kips (1995 kN) in the transverse direction and the corresponding moments are distributed to the individual walls according to their relative rigidities in each direction using the ratios $R_i/\Sigma R$ listed in Table 17.1 for Method 1-B.

**Torsional Effects.** To calculate the shear forces due to torsion, the locations of the center of mass and of the center of rigidity must be calculated. In this case, because the building is quite symmetric, the approximation using Eq. 17.7 is adopted.

**1.** *Center of Mass of a Typical Floor Level.* Table 17.2 contains the summary of the calculations to determine the center of mass of the walls at a typical floor

**Table 17.2** Calculation of Center of Mass

| Wall No. | Wall dir. | L (ft) | x (ft) | y (ft) | Lx | Ly |
|---|---|---|---|---|---|---|
| 1 | | 6.67 | 3.30 | 54.33 | 22 | 362 |
| 2 | | 13.33 | 30.00 | 54.33 | 400 | 724 |
| 3 | | 13.33 | 60.00 | 54.33 | 800 | 724 |
| 4 | | 13.33 | 90.00 | 54.33 | 1200 | 724 |
| 5 | | 13.33 | 120.00 | 54.33 | 1600 | 724 |
| 6 | | 6.67 | 146.70 | 54.33 | 978 | 362 |
| 7 | | 13.67 | 6.80 | 29.67 | 93 | 406 |
| 8 | | 26.67 | 30.00 | 29.67 | 800 | 791 |
| 9 | | 20.00 | 60.00 | 29.67 | 1200 | 593 |
| 10 | x-direction (East-West) | 16.67 | 90.00 | 36.33 | 1500 | 606 |
| 11 | | 26.67 | 120.00 | 29.67 | 3200 | 791 |
| 12 | | 13.67 | 143.20 | 29.67 | 1958 | 406 |
| 13 | | 5.33 | 155.80 | 45.67 | 830 | 243 |
| 14 | | 5.33 | 155.80 | 29.67 | 830 | 158 |
| 15 | | 5.33 | 2.70 | 24.67 | 14 | 131 |
| 16 | | 13.67 | 15.30 | 24.67 | 209 | 337 |
| 17 | | 26.67 | 38.50 | 24.67 | 1027 | 658 |
| 18 | | 26.67 | 128.50 | 24.67 | 3427 | 658 |
| 19 | | 13.67 | 151.70 | 24.67 | 2074 | 337 |
| 20 | | 5.33 | 27.00 | 8.67 | 144 | 46 |
| 21 | | 6.67 | 11.80 | 0.00 | 79 | 0 |
| 22 | | 13.33 | 38.50 | 0.00 | 513 | 0 |
| 23 | | 13.33 | 68.50 | 0.00 | 913 | 0 |
| 24 | | 13.33 | 98.50 | 0.00 | 1313 | 0 |
| 25 | | 13.33 | 128.50 | 0.00 | 1713 | 0 |
| 26 | | 6.67 | 158.50 | 0.00 | 1057 | 0 |
| 27 | | 24.00 | 0.00 | 42.00 | 0 | 1008 |
| 28 | | 24.00 | 30.00 | 42.00 | 720 | 1008 |
| 29 | | 24.00 | 60.00 | 42.00 | 1440 | 1008 |
| 30 | | 17.67 | 90.00 | 45.30 | 1590 | 800 |
| 31 | | 24.00 | 120.00 | 42.00 | 2880 | 1008 |
| 32 | | 24.00 | 150.00 | 42.00 | 3600 | 1008 |
| 33 | | 15.34 | 158.50 | 37.70 | 2431 | 578 |
| 34 | | 7.00 | 82.50 | 33.20 | 578 | 232 |
| 35 | | 7.00 | 86.80 | 33.20 | 608 | 232 |
| 36 | | 7.00 | 98.50 | 33.20 | 690 | 232 |
| 37 | y-direction (North-South) | 16.00 | 0.00 | 16.70 | 0 | 267 |
| 38 | | 24.00 | 8.50 | 12.30 | 204 | 295 |
| 39 | | 24.00 | 38.50 | 12.30 | 924 | 295 |
| 40 | | 24.00 | 68.50 | 12.30 | 1644 | 295 |
| 41 | | 24.00 | 98.50 | 12.30 | 2364 | 295 |
| 42 | | 24.00 | 128.50 | 12.30 | 3084 | 295 |
| 43 | | 24.00 | 158.50 | 12.30 | 3804 | 295 |
| | $\Sigma$ | 686.68 | | | 54455 | 18938 |

level. It is calculated as follows based on the coordinate system shown in Fig. 17.21.

$$\bar{x}_{walls} = \frac{\sum L_i x_i}{\sum L_i} = \frac{54,455}{687} = 79.26 \text{ ft} \qquad (24.16 \text{ m})$$

(17.13)

$$\bar{y}_{walls} = \frac{\sum L_i y_i}{\sum L_i} = \frac{18,938}{687} = 27.57 \text{ ft} \qquad (8.40 \text{ m})$$

where uniform mass is assumed and

$L_i$ = length of wall $i$

$x_i, y_i$ = distance in the $x$ and $y$ directions, respectively, from the origin of the coordinate system, shown in Fig. 17.15, to the centroid of wall $i$

For a typical floor level, the 8 ft wall weighs

$$W_{wall} = (\sum L) \times h \times \text{unit weight} = 687 \text{ ft} \times 8 \text{ ft} \times 55 \text{ lb/ft}^2$$

$$= 302 \text{ kips}$$

For the precast concrete floors, the unit load of 66 lb/ft² times the area of the floor results in a floor weight of

$$W_{floor} = 66(158.5 + 0.635)(54.33 + 0.635)$$

$$= 577 \text{ kips}$$

For a length of brick veneer of 258 ft and a height of 8 ft-8 in., the unit weight of 36 lb/ft² gives a total weight per story of

$$W_{veneer} = 36(258)(8.67) = 81 \text{ kips}$$

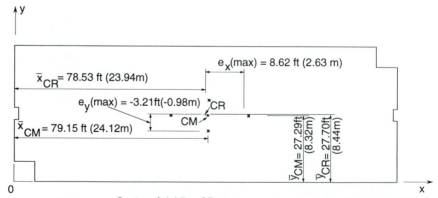

+ Center of rigidity, CR

• Calculated center of mass, CM

✗ Possible locations of CM due to code eccentricity of ± 0.05 D, where D is building dimension perpendicular to direction of load

**Figure 17.21**  Locations of center of mass and center of rigidity.

Therefore, the total weight per typical floor level is

$$\sum W = 302 + 577 + 81 = 960 \text{ kips}$$

Then the coordinates of the center of mass, which is also the location of the lateral force resultant due to seismic loading, can be calculated as follows:

$$\bar{x}_{CM} = \frac{\sum (W\bar{x})}{\sum W} = \frac{302(79.26) + (577 + 81)(158.5/2)}{960} = 79.15 \text{ ft} \ (24.12 \text{ m})$$

$$\bar{y}_{CM} = \frac{\sum (W\bar{y})}{\sum W} = \frac{302(27.57) + (577 + 81)(54.33/2)}{960} = 27.29 \text{ ft} \ (8.32 \text{ m})$$

(17.14)

**2.** *Center of Rigidity.* As discussed in Sec. 10.4.3 of Chap. 10, Eq. 10.1 can be used to calculate the location of the center of rigidity with respect to the reference axes shown in Fig. 17.21. By referring to Table 17.3 and using the approximate relative rigidities calculated using Method 1-B:

$$\bar{x}_{CR} = \frac{\sum R_{yi} x_i}{\sum R_{yi}} = \frac{7194.8}{91.6} = 78.53 \text{ ft} \ (23.94 \text{ m})$$

$$\bar{y}_{CR} = \frac{\sum R_{xi} y_i}{\sum R_{xi}} = \frac{719.5}{25.97} = 27.70 \text{ ft} \ (8.44 \text{ m})$$

**3.** *Calculation of Eccentricity and Torsional Moment at the Base of the Structure.* The eccentricity for the center of mass at a typical floor to the center of rigidity at the base of the structure is calculated from parts 1 and 2 as:

$$e_x = \bar{x}_{CM} - \bar{x}_{CR}$$

$$= 79.15 - 78.53 = +0.62 \text{ ft} \quad (+0.19 \text{ m})$$

$$e_y = \bar{y}_{CM} - \bar{y}_{CR}$$

$$= 27.29 - 27.70 = -0.41 \text{ ft} \quad (-0.12 \text{ m})$$

(17.15)

The actual eccentricity is very small compared to the building dimensions because the building and the distribution of walls are nearly symmetric. However, because of the uncertainty in the determinations of the center of mass and the center of rigidity, addition of some accidental eccentricity is recommended. The ASCE Standard[17.13] requires a minimum eccentricity of the center of mass of ±5% of the building dimension (from the calculated location of center of mass). This requirement results in the four possible locations of the center of mass shown in Fig. 17.21.

By referring to Fig. 17.21, the maximum eccentricities in the two directions are

$$e_{x(max)} = e_x \pm 0.05 D_x$$

$$= 0.62 \pm 0.05(160.16) = +8.62 \text{ ft} \quad (2.63 \text{ m})$$

$$e_{y(max)} = e_y \pm 0.05 D_y$$

$$= -0.41 \pm 0.05(56) = -3.21 \text{ ft} \quad (-0.98 \text{ m})$$

(17.16)

**Table 17.3** Calculation of Center of Rigidity and Torsional Effects

| Wall No. | R (x10⁻³) | x* (ft) | y* (ft) | R_y x (x10⁻³) | R_x y (x10⁻³) | d_x† (ft) | d_y† (ft) | R/ΣR | Rd_x² (x10⁻³) | Rd_y² (x10⁻³) | Rd/ΣRd² (x10⁻⁶) |
|---|---|---|---|---|---|---|---|---|---|---|---|
| 1 | 0.067 | | 54.33 | | 3.67 | | 26.63 | 0.0026 | | 47.85 | 7.08 |
| 2 | 0.533 | | 54.33 | | 28.98 | | 26.63 | 0.0205 | | 378.19 | 55.93 |
| 3 | 0.533 | | 54.33 | | 28.98 | | 26.63 | 0.0205 | | 378.19 | 55.93 |
| 4 | 0.533 | | 54.33 | | 28.98 | | 26.63 | 0.0205 | | 378.19 | 55.93 |
| 5 | 0.533 | | 54.33 | | 28.98 | | 26.63 | 0.0205 | | 378.19 | 55.93 |
| 6 | 0.067 | | 54.33 | | 3.67 | | 26.63 | 0.0026 | | 47.85 | 7.08 |
| 7 | 0.573 | | 29.67 | | 16.99 | | 1.97 | 0.0221 | | 2.22 | 4.44 |
| 8 | 4.052 | | 29.67 | | 120.23 | | 1.97 | 0.1560 | | 15.68 | 31.39 |
| 9 | 1.765 | | 29.67 | | 52.35 | | 1.97 | 0.0679 | | 6.83 | 13.67 |
| 10 | 1.032 | | 36.33 | | 37.51 | | 8.63 | 0.0397 | | 76.84 | 35.07 |
| 11 | 4.052 | | 29.67 | | 120.23 | | 1.97 | 0.1560 | | 15.68 | 31.39 |
| 12 | 0.573 | | 29.67 | | 16.99 | | 1.97 | 0.0221 | | 2.22 | 4.44 |
| 13 | 0.035 | | 45.67 | | 1.58 | | 17.97 | 0.0013 | | 11.16 | 2.45 |
| 14 | 0.035 | | 29.67 | | 1.03 | | 1.97 | 0.0013 | | 0.13 | 0.27 |
| 15 | 0.035 | | 24.67 | | 0.85 | | -3.03 | 0.0013 | | 0.32 | -0.41 |
| 16 | 0.573 | | 24.67 | | 14.13 | | -3.03 | 0.0221 | | 5.27 | -6.84 |
| 17 | 4.052 | | 24.67 | | 99.97 | | -3.03 | 0.1560 | | 37.28 | -48.40 |
| 18 | 4.052 | | 24.67 | | 99.97 | | -3.03 | 0.1560 | | 37.28 | -48.40 |
| 19 | 0.573 | | 24.67 | | 14.13 | | -3.03 | 0.0221 | | 5.27 | -6.84 |
| 20 | 0.035 | | 8.67 | | 0.30 | | -19.03 | 0.0013 | | 12.53 | -2.59 |
| 21 | 0.067 | | 0.00 | | 0.00 | | -27.70 | 0.0026 | | 51.80 | -7.36 |
| 22 | 0.533 | | 0.00 | | 0.00 | | -27.70 | 0.0205 | | 409.37 | -58.19 |
| 23 | 0.533 | | 0.00 | | 0.00 | | -27.70 | 0.0205 | | 409.37 | -58.19 |
| 24 | 0.533 | | 0.00 | | 0.00 | | -27.70 | 0.0205 | | 409.37 | -58.19 |
| 25 | 0.533 | | 0.00 | | 0.00 | | -27.70 | 0.0205 | | 409.37 | -58.19 |
| 26 | 0.067 | | 0.00 | | 0.00 | | -27.70 | 0.0026 | | 51.80 | -7.36 |
| ΣR= 25.97 x 10⁻³ | | | | | Σ= 719.53⁻³ | | | | | Σ= 3578.24 x 10⁻³ | |
| 27 | 6.30 | 0.00 | | 0.00 | | -78.53 | | 0.0688 | 38875.80 | | -1949.29 |
| 28 | 8.87 | 30.00 | | 266.00 | | -48.53 | | 0.0968 | 20885.95 | | -1694.59 |
| 29 | 8.87 | 60.00 | | 532.00 | | -18.53 | | 0.0968 | 3045.79 | | -647.12 |
| 30 | 4.49 | 90.00 | | 404.15 | | 11.47 | | 0.0490 | 590.37 | | 202.75 |
| 31 | 8.87 | 120.00 | | 1064.00 | | 41.47 | | 0.0968 | 15245.67 | | 1447.80 |
| 32 | 6.30 | 150.00 | | 945.49 | | 71.47 | | 0.0688 | 32193.11 | | 1773.86 |
| 33 | 2.29 | 158.50 | | 362.63 | | 79.97 | | 0.0250 | 14630.19 | | 720.44 |
| 34 | 0.09 | 82.50 | | 7.44 | | 3.97 | | 0.0010 | 1.42 | | 1.41 |
| 35 | 0.09 | 86.80 | | 7.83 | | 8.27 | | 0.0010 | 6.16 | | 2.93 |
| 36 | 0.09 | 98.50 | | 8.88 | | 19.97 | | 0.0010 | 35.94 | | 7.09 |
| 37 | 2.29 | 0.00 | | 0.00 | | -78.53 | | 0.0250 | 14110.90 | | -707.54 |
| 38 | 6.30 | 8.50 | | 53.58 | | -70.03 | | 0.0688 | 30915.89 | | -1738.32 |
| 39 | 8.87 | 38.50 | | 341.37 | | -40.03 | | 0.0968 | 14210.85 | | -1397.81 |
| 40 | 6.37 | 68.50 | | 436.03 | | -10.03 | | 0.0695 | 640.87 | | -251.51 |
| 41 | 6.37 | 98.50 | | 626.99 | | 19.97 | | 0.0695 | 2537.49 | | 500.46 |
| 42 | 8.87 | 128.50 | | 1139.37 | | 49.97 | | 0.0968 | 22136.63 | | 1744.59 |
| 43 | 6.30 | 158.50 | | 999.06 | | 79.97 | | 0.0688 | 40306.47 | | 1984.84 |
| ΣR= 91.61 x 10⁻³ | | | | Σ= 7194.82 x 10⁻³ | | | | | Σ= 250369.50 x 10⁻³ | | |

\* distances of wall from the x and y axes

† distances of walls from center of rigidity in the x and y directions

The torsional moment, $M_{T_y}$, for loading in the North–South direction, can be calculated as

$$M_{T_y} = V_y e_{x(\text{max})} = 439 \times 8.62 = 3784 \text{ ft-kips} \qquad (5121 \text{ kN-m})$$

Similarly, the torsional moment, $M_{T_x}$ for loading in the East–West direction is

$$M_{T_x} = V_x e_{y(\text{max})} = 570 \times (-3.21) = -1830 \text{ ft-kips} \qquad (-2476 \text{ kN-m})$$

The torsional (rotational) moments, $M_{T_y}$ and $M_{T_x}$, are applied separately at the ground floor level and are used to calculate the shear forces, $V_{ir}$, due to torsion using Eq. 10.10, rewritten as

$$V_{xir} = \frac{R_{xi} d_{yi}}{\sum R_{xi} d_{yi}^2 + \sum R_{yi} d_{xi}^2}$$

$$\qquad (17.17)$$

$$V_{yir} = \frac{R_{yi} d_{xi}}{\sum R_{xi} d_{yi}^2 + \sum R_{yi} d_{xi}^2}$$

where

$R_{xi}, R_{yi}$ = rigidity of wall $i$ for loading in the x or y-direction

$d_{xi}, d_{yi}$ = x or y distance from the center of rigidity to wall $i$ measured perpendicular to the direction for which the rigidity is calculated

$\sum R_{xi} d_{yi}^2, \sum R_{yi} d_{xi}^2$ = sum of the products of the rigidity times the square of the distance from the center of rigidity for each wall

These ratios are shown in Table 17.3, where the $R_i d_i^2$ values are listed in separate columns for walls oriented in the longitudinal and transverse directions. As can be seen, the largest $R_i d_i / \sum R_i d_i^2$ ratios correspond to walls near the perimeter of the building. These walls tend to develop the highest torsional shear forces.

**Calculation of Maximum Shear Forces.** As mentioned earlier, the shear force that each wall should be designed to resist is composed of two components: one due to translation or direct shear, $V_{it}$, and one due to torsion, $V_{ir}$. For shear walls located on the opposite side of the center of rigidity from the eccentric lateral force, the torsional shears counteract the direct shears. However, in design, building codes do not permit design for less than the direct shear. This means that the negative values in Table 17.4 are not included in the total shear force, $V_i$.

As mentioned in part 3 before, the maximum torsional moment produces the highest torsional shear forces for walls on the same side of the center of rigidity as the maximum eccentricity. The critical conditions for walls on the other side of the center of rigidity can be found for an eccentricity, regardless of how small, that is on the same side of the center of rigidity as these other walls. Therefore, unless both the minimum and maximum eccentricities are on the same side of the center of rigidity, two sets of calculations may be required for each direction of lateral loading.

A common exception to the foregoing requirement is applicable to buildings that have similar wall layouts on opposite ends (sides) of the building. Then, if a particular type of wall is designed for the maximum combined effects of direct shear

## Table 17.4 Calculation of Shear and Moments

| Wall No. | Wall Dir. | $R/\Sigma R$ | $Rd/\Sigma Rd^2$ $(\times 10^{-6})$ | $V_{it}^{(1)}$ (kips) | $V_{ir}^{(2)}$ (kips) | $V_i^{(3)}$ (kips) | $V_i/V$ $(\times 10^{-3})$ | $M_i^{(4)}$ (ft-kip) | $f_{vi}^{(5)}$ (psi) |
|---|---|---|---|---|---|---|---|---|---|
| 1  | x-direction (East-West) | 0.003 | 7.08   | 1.48  | -0.0130 | 1.48  | 2.60   | 93.56   | 2.448  |
| 2  |  | 0.021 | 55.93  | 11.71 | -0.1024 | 11.71 | 20.54  | 739.11  | 9.590  |
| 3  |  | 0.021 | 55.93  | 11.71 | -0.1024 | 11.71 | 20.54  | 739.11  | 9.590  |
| 4  |  | 0.021 | 55.93  | 11.71 | -0.1024 | 11.71 | 20.54  | 739.11  | 9.590  |
| 5  |  | 0.021 | 55.93  | 11.71 | -0.1024 | 11.71 | 20.54  | 739.11  | 9.590  |
| 6  |  | 0.003 | 7.08   | 1.48  | -0.0130 | 1.48  | 2.60   | 93.56   | 2.448  |
| 7  |  | 0.022 | 4.44   | 12.57 | -0.0081 | 12.57 | 22.05  | 793.45  | 10.066 |
| 8  |  | 0.156 | 31.39  | 88.93 | -0.0574 | 88.93 | 156.02 | 5614.22 | 36.403 |
| 9  |  | 0.068 | 13.67  | 38.72 | -0.0250 | 38.72 | 67.93  | 2444.39 | 21.211 |
| 10 |  | 0.040 | 35.07  | 22.66 | -0.0642 | 22.66 | 39.75  | 1430.36 | 14.876 |
| 11 |  | 0.156 | 31.39  | 88.93 | -0.0574 | 88.93 | 156.02 | 5614.22 | 36.403 |
| 12 |  | 0.022 | 4.44   | 12.57 | -0.0081 | 12.57 | 22.05  | 793.45  | 10.066 |
| 13 |  | 0.001 | 2.45   | 0.76  | -0.0045 | 0.76  | 1.33   | 47.88   | 1.535  |
| 14 |  | 0.001 | 0.27   | 0.76  | -0.0005 | 0.76  | 1.33   | 47.88   | 1.535  |
| 15 |  | 0.001 | -0.41  | 0.76  | 0.0008  | 0.76  | 1.33   | 47.95   | 1.537  |
| 16 |  | 0.022 | -6.84  | 12.57 | 0.0125  | 12.58 | 22.07  | 794.32  | 10.074 |
| 17 |  | 0.156 | -48.40 | 88.93 | 0.0886  | 89.02 | 156.18 | 5619.84 | 36.438 |
| 18 |  | 0.156 | -48.40 | 88.93 | 0.0886  | 89.02 | 156.18 | 5619.84 | 36.438 |
| 19 |  | 0.022 | -6.84  | 12.57 | 0.0125  | 12.58 | 22.07  | 794.32  | 10.074 |
| 20 |  | 0.001 | -2.59  | 0.76  | 0.0047  | 0.76  | 1.34   | 48.21   | 1.545  |
| 21 |  | 0.003 | -7.36  | 1.48  | 0.0135  | 1.49  | 2.62   | 94.36   | 2.475  |
| 22 |  | 0.021 | -58.19 | 11.71 | 0.1065  | 11.81 | 20.72  | 745.74  | 9.671  |
| 23 |  | 0.021 | -58.19 | 11.71 | 0.1065  | 11.81 | 20.72  | 745.74  | 9.671  |
| 24 |  | 0.021 | -58.19 | 11.71 | 0.1065  | 11.81 | 20.72  | 745.74  | 9.671  |
| 25 |  | 0.021 | -58.19 | 11.71 | 0.1065  | 11.81 | 20.72  | 745.74  | 9.671  |
| 26 |  | 0.003 | -7.36  | 1.48  | 0.0135  | 1.49  | 2.62   | 94.36   | 2.475  |
| 27 | y-direction (North-South) | 0.069 | -1949.29 | 30.20 | -7.3761 | 30.20 | 68.79  | 1906.31 | 13.019 |
| 28 |  | 0.097 | -1694.59 | 42.49 | -6.4123 | 42.49 | 96.79  | 2682.24 | 18.316 |
| 29 |  | 0.097 | -647.12  | 42.49 | -2.4487 | 42.49 | 96.79  | 2682.24 | 18.316 |
| 30 |  | 0.049 | 202.75   | 21.52 | 0.7672  | 22.29 | 50.76  | 1406.76 | 13.094 |
| 31 |  | 0.097 | 1447.80  | 42.49 | 5.4785  | 47.97 | 109.26 | 3027.89 | 20.679 |
| 32 |  | 0.069 | 1773.86  | 30.20 | 6.7123  | 36.92 | 84.09  | 2330.36 | 15.916 |
| 33 |  | 0.025 | 720.44   | 10.96 | 2.7262  | 13.69 | 31.18  | 864.15  | 8.986  |
| 34 |  | 0.001 | 1.41     | 0.43  | 0.0053  | 0.44  | 1.00   | 27.61   | 0.650  |
| 35 |  | 0.001 | 2.93     | 0.43  | 0.0111  | 0.44  | 1.01   | 27.97   | 0.659  |
| 36 |  | 0.001 | 7.09     | 0.43  | 0.0268  | 0.46  | 1.05   | 28.97   | 0.682  |
| 37 |  | 0.025 | -707.54  | 10.96 | -2.6773 | 10.96 | 24.97  | 691.97  | 7.191  |
| 38 |  | 0.069 | -1738.32 | 30.20 | -6.5778 | 30.20 | 68.79  | 1906.31 | 13.018 |
| 39 |  | 0.097 | -1397.81 | 42.49 | -5.2893 | 42.49 | 96.79  | 2682.24 | 18.318 |
| 40 |  | 0.069 | -251.51  | 30.50 | -0.9517 | 30.50 | 69.48  | 1925.42 | 13.150 |
| 41 |  | 0.069 | 500.46   | 30.50 | 1.8937  | 32.40 | 73.79  | 2044.97 | 13.966 |
| 42 |  | 0.097 | 1744.59  | 42.49 | 6.6015  | 49.09 | 111.82 | 3098.78 | 21.164 |
| 43 |  | 0.069 | 1984.84  | 30.20 | 7.5106  | 37.71 | 85.91  | 2380.76 | 16.260 |

(1) $V_{it} = V(R/\Sigma R)$

(2) $V_{ir} = M_T Rd/\Sigma Rd^2$

(3) $V_i = V_{it} + V_{ir} \geq V_{it}$

(4) $M_i = M_0 V_i/V$

(5) $f_{vi} = (8/7.625)V_i/A_w$

Note:

East-West

$V = 570$ kips,   $M_T = -1830$ ft-kip,   $M_0 = 35984$ ft-kip

North-South

$V = 439$ kips,   $M_T = 3784$ ft-kip,   $M_0 = 27712$ ft-kip

and maximum torsional moment, it will be safe and slightly conservative to use the same design for a similar wall located at the same distance or closer to the center of rigidity. This approach is adopted in the current example.

Table 17.4 contains a summary of the direct shears (translational), $V_{it}$, and the torsional shears (rotational), $V_{ir}$. For example, Wall 43 would be designed for the combined shear using

$$V_i = V_{it} + V_{ir} \geq V_{it} \tag{17.18}$$

so that

$$V_{43} = V_{43t} + V_{43r} = 30.20 + 7.51 = 37.71 \text{ kips (167.4 kN)}$$

For Wall 27, which is a similar wall at the opposite end of the building, the maximum shear of 30.20 kips does not take into account the torsional moment in the opposite direction. Therefore, as an alternative to redoing the calculations for the torsional moment corresponding to the other eccentricity of 7.38 ft, the design for Wall 43 can be used. There is some common sense to this approach in that use of similar reinforcement for similar walls will help avoid confusion and construction errors.

The average shear stresses shown in the final column of Table 17.4 are calculated from

$$f_{vi} = (V_i)/A_{\text{web}} \tag{17.19}$$

It is wise to avoid brittle shear failure so that redistribution of forces can take place. For this reason, UBC-91[17.9] and ACI 530/ASCE 5/TMS 402[17.8] require a 50% increase in the design shear force for shear walls in Seismic Zones 3 and 4. This creates an extra margin of safety to allow for the uncertainties and inaccuracies involved in calculating the lateral load distributions. Variations in rigidity caused by cracking, nonlinear material responses, and choice of the effective section are examples of potential inaccuracies. In general, it is prudent not to underestimate shear forces.

For the example building located in Seismic Zone 2, the shear stresses range from 1.0 psi (0.007 MPa) to a maximum of 36.4 psi     (0.25 MPa). This range is within typical limits (<75 psi) for working stress design of reinforced shear walls and is an indication that the wall arrangement may be acceptable.

**Calculation of Maximum Moments.**    Moments in individual walls induced by the overturning action of the lateral loads can be calculated from the overturning moment using the same form of distribution as was used to distribute lateral forces to produce the combined shear forces in each wall. Then, because shear force and moment in each wall have the same relationship as the total shear and overturning moment in each direction of loading on the structure, the equations are

$$M_{xi} = \frac{V_{xi}}{V_x} M_{xo}$$

$$\tag{17.20}$$

$$M_{yi} = \frac{V_{yi}}{V_y} M_{yo}$$

where $V_{xi}$, $V_{yi}$ = shear forces in the $x$ and $y$ directions, respectively, for wall $i$ due to the combined effects of direct shear and torsion.

$V_x$, $V_y$ = total direct shear forces in the $x$ and $y$ directions, respectively

$M_{xi}$, $M_{yi}$ = moments in the $x$ and $y$ directions, respectively, for wall i due to the combined effects of direct shear and torsion

$M_{xo}$, $M_{yo}$ = total overturning moments in the $x$ and $y$ directions, respectively

$x$, $y$ = East–West and North–South directions, respectively

As was the case for shear, torsional effects cannot be used to decrease the design moment for an individual wall. However, this restriction is accounted for by using the maximum value of $V_{xi}$ or $V_{yi}$ in Eq. 17.18 as was done in Table 17.4. Unless shears for all the possible torsional moments are evaluated, the moment designs for the most critical walls should be used for other similar walls where the additive effects of smaller torsional moments were not evaluated.

The maximum values of the moments $M_i$ due to combined torsion and overturning are listed in the second last column of Table 17.4.

### 17.5.6 Drift Calculations

To avoid possible wasted design effort, drift calculations should be done before proceeding too far along the design process. The lateral deflection (drift) of any story and thus for the building should be calculated under full lateral load and be checked against the appropriate code limit. For seismic loading, a drift limit of 0.5% of the story or building height ($H/200$) is a typical value[17.9] for structures having a fundamental period of less than 0.7 s.

**Top Drift.** In this example, the drift in the longitudinal direction is considered because the overturning moment is higher in this direction and the rigidity is smaller. The calculation of deflection at the top of the building is based on linear elastic response assuming that all walls act as cantilevers. Because the lowest height-to-length ratio is $90/26.67 = 3.38$, it can be seen from Sec. 10.4.2 that the relative amount of shear deformation is small and hence is ignored. Therefore,

$$\Delta_{\text{Top}} = \int_0^H \frac{M}{EI_x} dz\, dz \qquad (17.21)$$

where $I_x = \Sigma I_{xi}$ for all walls in the longitudinal direction
$= 6435 \text{ ft}^4$

For a block strength of 2500 psi (17.3 MPa) and type S mortar, the modulus of elasticity, $E_m$ may be taken equal to $2.4 \times 10^6$ psi (16,560 MPa).[17.8] A rigorous calculation can be done using the moment diagram in Fig. 17.19(a) and the moment-area method. However, for a uniform loading pattern such as shown in Fig. 17.18, this can be simplified.

For the total shear force of 570 kips (2537 kN), the system of concentrated loads at the floor levels shown in Fig. 17.18(a) can be approximated by a triangular loading with a maximum intensity of

$$p = \frac{2V_x}{H} = \frac{2(570)}{90} = 12.67 \text{ kips/ft} \qquad (17.2 \text{ kN/m})$$

$$E_m = 2.4 \times 10^6 \text{ psi} = 345,600 \text{ kips/ft}^2$$

By using Eq. 17.21 or simply applying the moment-area method,

$$M = 570(60) - 570z + \frac{12.67}{90} z \frac{z}{2} \frac{z}{3}$$

$$\Delta = \frac{1}{EI} \int\int M \, dz \, dz$$

$$= \frac{1}{EI} \left[ \frac{34200z^2}{2} - \frac{570z^3}{6} + \left(\frac{12.67}{90}\right)\left(\frac{z^5}{6}\right)\left(\frac{1}{20}\right) \right]\Bigg|_0^{90} = \frac{76.2 \times 10^6}{EI}$$

$$= \frac{76.2 \times 10^6}{345,600 \times 6435} = 0.034 \text{ ft}$$

$$= 0.40 \text{ in. } (10.4 \text{ mm})$$

For reinforced walls where cracking is expected, the moments of inertia for the cracked sections should be used. For a complete design, this information would be available, but in this case, we can approximate it as being as low as 30% of the uncracked value. On this basis, the maximum deflection or drift at the top of the building would be in the order of

$$\Delta = 0.40/0.30 = 1.33 \text{ in. } (34 \text{ mm})$$

which is considerably less than the $H/200$ limit, which is

$$\Delta_{\text{limit}} = 90 \times 12/200 = 5.4 \text{ in. } (137 \text{ mm})$$

It should be noted that drift limits in other codes and for wind loading may be different from the foregoing. For instance, ASCE 7-88[17.13] recommends limiting the drift to $H/400$ for unreinforced masonry. The number of stories and the height-to-width ratio of the building are also factors used to guide the designer on requirements for checking drift.

**First Story Drift.**  From the previous calculations, it might be deduced that interstory drift is not critical. However, the 12 ft (3.6 m) first story will be checked. By referring to Fig. 17.19(a) and using the moment-area method,

$$\Delta_1 = \frac{1}{EI} \left[ 29,144 \times 12 \times 6 + \frac{1}{2}(35,984 - 29,144)12 \times 8 \right]$$

$$= \frac{2,426,668 \times 12,000 \times (12)^2}{2.4 \times 10^6 \times 6435 \times (12)^4} = 0.013 \text{ in. } (0.3 \text{ mm})$$

Even considering a large reduction in moments of inertia to allow for cracking, the interstory drift would be much less than the drift limit of

$$\frac{h}{200} = \frac{12 \times 12}{200} = 0.72 \text{ in. (18 mm)}$$

### 17.5.7 Wall Design

Walls have to be designed to resist the seismic forces in both the out-of-plane and in-plane directions. For out-of-plane bending, the walls span vertically between the floors. Walls running in the transverse direction (North–South direction) support the floor loads from the precast planks which span one way in the East–West direction. However, because the longitudinal walls (East–West direction) are connected to the transverse walls at their intersections, they share in carrying the total gravity load.

Representative walls were selected to demonstrate design for out-of-plane and in-plane loading. The walls are assumed to have a constant nominal thickness of 8 in. [$7\frac{5}{8}$ in. (193.7 mm) actual thickness].

### 17.5.8 Out-of-Plane Loading.

Seismic loads on elements of structures can be calculated using the formula:[17.13]

$$F_p = ZIC_pW_p \tag{17.22}$$
$$= (3/8)(1.0)(0.3)(77) = 8.7 \text{ lb/ft}^2 \text{ (0.42 kN/m}^2)$$

where $Z = \frac{3}{8}$ for Zone 2
$I = 1.0$
$C_p = 0.3$ (Table 25 in Ref. 17.13)
$W_p = 77$ lb/ft$^2$ (fully grouted walls at the bottom story)

However, the design wind pressure of 20 lb/ft$^2$ (0.96 kN/m$^2$) at the first story controls in this case. The maximum bending moment can be conservatively calculated by ignoring the partial fixity at the top and bottom of the 12 ft. (3.6 m) clear height. Therefore,

$$M = F_p \frac{h^2}{8}$$

$$\tag{17.23}$$

$$= \frac{20(12)^2}{8} = 360 \text{ ft-lb/ft} \quad (1.6 \text{ kN.m/m})$$

By using the minimum axial load $P$ just due to self-weight at mid-height of the first story, $P = 55$ lb/ft$^2 \times (90 - 12/2)$ ft $= 4620$ lb/ft

The resulting extreme fiber stresses assuming solidly grouted walls are

$$f_m = -\frac{P}{A} \pm \frac{M}{S}$$

$$= -\frac{4620}{7.625 \times 12} \pm \frac{360 \times 12}{12(7.625)^2/6}$$

$$= -50.5 \pm 37.2 = -87.7 \text{ psi} (-0.61 \text{ MPa}) \text{ max.}$$

$$= -13.2 \text{ psi} (-0.09 \text{ MPa}) \text{ min.}$$

For hollow or partially grouted walls, the ratio of flexural stress to axial stress (based on net area) would be smaller. Therefore, the entire section remains in compression. Because these stresses are compressive, flexural reinforcement is not needed for out-of-plane bending.

At the top story, with a wind design pressure of 30 lb/ft² (1.44 kN/m²), the maximum moment is

$$M = \frac{30 \times 8^2}{8} = 240 \text{ ft-lb/ft} (1.07 \text{ kN.m/m})$$

This produces a flexural tension stress of 24.8 psi (0.17 MPa), which, even without the benefit of axial compression, is less than the allowable tension. For exterior walls, the brick veneer is supported at each floor level by shelf angles; and ties are provided to tie the veneer to the backup block wall for lateral load resistance. The mass of the veneer will add to the out-of-plane force due to seismic loading, but wind will still control. Details and further information regarding the design of masonry veneer are presented in Chap. 12.

The foregoing calculations indicate that, for multistory buildings, the out-of-plane loading typically does not control the design of the walls. The usual height-to-thickness ratios do not result in high out-of-plane bending or significant slenderness effects.

### 17.5.9 In-Plane Loading (Shear Walls)

The compressive capacity of the loadbearing walls subject to gravity loads and to combined gravity load and earthquake load are covered in the following part of this section.

**Wall 41.** The design of the base section of the wall as a T-section, as shown in Fig. 17.22, is performed considering an effective flange width of 6t on both sides of the web. The wall supports gravity loads from the floor slabs plus lateral earthquake load:

$$P_D = \underbrace{66 \times 9 \times 30 \times 27.5}_{\text{floors}} + \underbrace{69 \times 3 \times 27.5}_{\text{roof}} + \underbrace{55[90 - (8/12) \times 10](24.67 + 13.33)}_{\text{s.w. of wall}}$$

$$+ \underbrace{36(1.33 \times 78)}_{\text{veneer}} = 758,600 \text{ lb} = 759 \text{ kips} \quad (3380 \text{ kN})$$

$$P_L = \underbrace{(40 \times 9 \times 30 \times 24)0.65^*}_{\text{floors}} + \underbrace{(100 \times 9 \times 30 \times 4.33)0.65^*}_{\text{corridor}}$$

$$+ \underbrace{20 \times 30 \times 27.5}_{\text{roof}} = 260,970 \text{ lb} = 261 \text{ kips} \quad (1153 \text{ kN})$$

*Live load reduction factor.

The reaction of the corridor beam $B1$ (Fig. 17.15) is transmitted to the end of the wall as a concentrated load. Bearing should be checked for local effect. Due to the counteracting effects of bending of shear walls tied together by the rigid floor diaphragms, the concentrated, eccentric beam reaction will be distributed over the cross-section of the wall as the load is transferred down the wall. For design of the base section of the wall, this load will be considered to act at the centroid of the wall. The shear and moment listed in Table 17.4 for seismic loading are

$$V_E = 32.4 \text{ kips} \qquad (143.9 \text{ kN})$$

$$M_E = 2045 \text{ ft-kips} \qquad (2768 \text{ kN-m})$$

and axial loads are

$$P_D = 759 \text{ kips} \qquad (3380 \text{ kN})$$

$$P_L = 261 \text{ kips} \qquad (1163 \text{ kN})$$

$$P_{D+L} = 1020 \text{ kips} \qquad (4543 \text{ kN})$$

Consider two load combinations:[17.8]

1. $D + L + E$ for compression control at the toe of the web.
2. $0.9D + E$ for tension control for design of flexure and shear reinforcement.

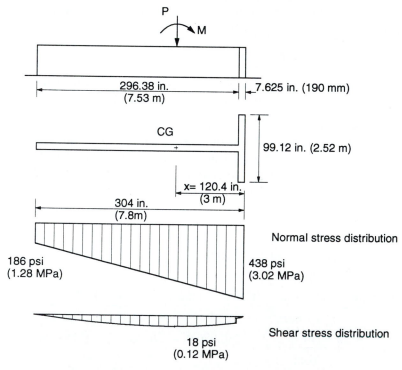

**Figure 17.22**  Stresses for wall 41 (case 1 loading).

**Properties of Cross-Section (Fig. 17.22).** Note that properties are based on the gross cross-section area of the wall (assumed fully grouted at the bottom story).

$$A = 99.12 \times 7.625 + 296.375 \times 7.625$$

$$= 3015.6 \text{ in.}^2 \quad (1.95 \times 10^6 \text{ mm}^2)$$

$$\bar{x} = \frac{(99.12 \times 7.625)(3.812)^2 + (296.38 \times 7.625)(155.8)}{3015.6} = 120.4 \text{ in. (3058 mm)}$$

$$I = \frac{99.12(7.625)^3}{12} + 99.12 \times 7.625(120.4 - 7.625/2)^2 + \frac{7.625(296.375)^3}{12}$$

$$+ (296.375 \times 7.625)\left(\frac{296.375}{2} + 7.625 - 120.4\right)^2$$

$$= 29.65 \times 10^6 \text{ in.}^4 \quad (12.34 \times 10^{12} \text{ mm}^4)$$

**Load Case 1.** Design for axial load and bending ($D + L + E$):

$$f_m = -P/A \pm My/I$$

$$f_{m1} = -\frac{1020 \times 1000}{3015.6} - \frac{2045 \times 12,000}{29.65 \times 10^6}(120.4)$$

$$= -338 - 100 = -438 \text{ psi} (-3.02 \text{ MPa}) \text{ (compression)}$$

$$f_{m2} = -\frac{1020 \times 1000}{3015.6} + \frac{2045 \times 12,000}{29.65 \times 10^6}(183.6)$$

$$= -338 + 152 = -186 \text{ psi} (-1.28 \text{ MPa}) \text{ (compression)}$$

For the reversed moment,

$$f_{m1} = -338 + 100 = -238 \text{ psi (1.64 MPa) (compression)}$$
$$f_{m2} = -338 - 152 = -490 \text{ psi} (-3.38 \text{ MPa}) \text{ (compression)}$$

Because the entire section is subjected to compression, reinforcement is not required. However, if other walls are cracked and require reinforcement, at least some minimum amount of reinforcement should be used for Wall 41. For Seismic Zone 2, continuous reinforcement of at least 0.2 in.² (129 mm) area is required[17.8] vertically at corners, ends of walls and at openings; and horizontally at tops and bottoms of walls and of openings. For improved ability to redistribute lateral load, where design calculations have not accounted for the influence of cracking of reinforced walls, reinforcing steel having a minimum area of 0.0010 Ag with not less than one-third in the horizontal or vertical direction is recommended.

**Check of Axial Compression and Flexure.** For $f'_m = 2000$ psi (13.8 MPa) and considering out-of-plane stability,

$$F_a = \frac{1}{4}f'_m\left[1 - \left(\frac{h}{140r}\right)^2\right] = \frac{1}{4}(2000)\left[1 - \left(\frac{12 \times 12}{140 \times (0.29 \times 7.625)}\right)^2\right]$$

$$= 392 \text{ psi} \quad (2.7 \text{ MPa})$$

$$F_b = \frac{1}{3}f'_m = \frac{1}{3}(2000) = 667 \text{ psi } (4.6 \text{ MPa})$$

$$\frac{f_a}{F_a} + \frac{f_b}{F_b} = \frac{338}{392} + \frac{152}{667} = 1.09 < 1.33$$

where a one-third increase in stress is allowed in ACI 530/ASCE 5/TMS 402[17.8] for wind or earthquake loading.

**Shear Design.**

$$f_v = \frac{VQ}{Ib} \qquad (17.24)$$

The maximum shear stress is at the C.G. of the wall section:

$$f_v = \frac{32.4 \times 1000 \times 7.625 \times (183.6)^2/2}{29.65 \times 10^6 \times 7.625} = 18 \text{ psi} \qquad (0.12 \text{ MPa})$$

$F_v$ (code allowable) is the lesser of

$$1.5 \sqrt{f'_m} = 67 \text{ psi} \qquad (0.46 \text{ MPa}) \leftarrow \text{controls}$$

$$\text{or} \qquad 120 \text{ psi} \qquad (0.83 \text{ MPa})$$

$$\text{or} \qquad 60 + 0.45\frac{P}{A} = 60 + 0.45\left(\frac{1020 \times 1000}{3015.6}\right) = 212 \text{ psi} \qquad (1.45 \text{ MPa})$$

Because $f_v < F_v = 67$ psi (0.46 MPa), there is no need for shear reinforcement. However, use of joint reinforcement every second course is good practice. In cases where the shear stress is closer to the allowable value, the shear capacity after cracking should be checked to ensure that flexural capacity controls. Provision of the minimum reinforcement discussed previously may be adequate.

**Load Case 2.** Design for axial load and bending ($0.9\,D + E$)

$$0.9\,P_D = 683 \text{ kips } (3039 \text{ kN})$$

$$V_E = 31.6 \text{ kips } (140.8 \text{ kN})$$

$$M_E = 2045 \text{ ft-kips} \qquad (2768 \text{ kN.m})$$

$$f_m = -\frac{P}{A} \pm \frac{My}{I}$$

$$f_{m1} = -\frac{683 \times 1000}{3015.6} - \frac{2045 \times 12,000}{29.65 \times 10^6} \times 120.4$$

$$= -226 - 100 = -326 \text{ psi } (2.25 \text{ MPa}) \text{ (compression)}$$

$$f_{m2} = -\frac{683 \times 1000}{3015.6} + \frac{2045 \times 12,000}{29.65 \times 10^6} \times 183.6$$

$$= -226 + 152 = -74 \text{ psi } (-0.51 \text{ MPa}) \text{ (compression)}$$

For reversed moment,

$$f_{m1} = -126 \text{ psi} \quad (-0.87 \text{ MPa})$$

$$f_{m2} = -378 \text{ psi} \quad (-2.61 \text{ MPa})$$

Comparing load Cases 1 and 2 indicates that Case 1 controls. From the stress check for Case 1, compressive stress levels are satisfactory. By checking for $D + L$, $f_a = P/A = 1020(1000)/3015.6 = 338$ psi (2.31 MPa) is less than the calculated value of $F_a = 392$ psi (2.7 MPa) and, therefore, this load combination is not critical. ACI 530/ASCE 5/TMS 402[17.8] requires that $P_{D+L}$ should be less than or equal to $(1/4) P_e$ where $P_e$ is the Euler buckling load defined as

$$P_e = \frac{\pi^2 E_m I}{h^2}\left(1 - 0.577\frac{e}{r}\right)^3 \tag{17.25}$$

For a minimum accidental eccentricity of 0.1t and I taken as the moment of inertia about the minor axis, $P_e = 259$ kips/ft (3780 kN/m). The applied load of $(1020/3015.6) 7.625(12) = 30.9$ kips/ft (451 kN/m) is much less than $(1/4) (259) = 64.8$ kips/ft (945 kN/m). Therefore, slenderness of the wall does not control and the design is satisfactory.

### Design of Wall 25 and Wall 18

*Distribution of Axial Load.* In addition to self-weight, each wall carries load from the floors and roof. Because of the rigid floor and roof diaphragms and the connections at the intersections of transverse and longitudinal walls, the various parts of the wall share the gravity load regardless of the fact that the precast floor system tends to apply the floor loads directly on the transverse (North–South) parts. The moments in the walls due to eccentricities of their axial load counteract each other and tend to produce uniform axial strain over the cross-section of the wall. Therefore, the total floor load from the panel around a combined wall such as composed of Walls 25 and 18 and the transverse part of Wall 42, can be distributed to the various parts according to their axial stiffnesses (i.e., their areas). Alternately, for squat walls, where it cannot be assumed that planar behavior on the horizontal cross-section is achieved, use of the tributary areas appropriate to each part of the wall provides a more conservative estimate of the effect of axial compression.

Using the calculation for Wall 41, the total axial loads for the combination of Walls 25, 18, and 42 are

$$P_D = 759 \text{ kips} + 55(90 - (8/12)(10))(26.0) = 878 \text{ kips (3910 kN)}$$

allowing for the additional 26 ft. length of Wall 18, and

$$P_L = 261 \text{ kips (1161 kN)}$$

**Wall 25.** For Wall 25, shown in Fig. 17.23(a), the share of axial load at the ground floor is

$$[13.33/(13.33 + 26.67 + 24)] = 0.208$$

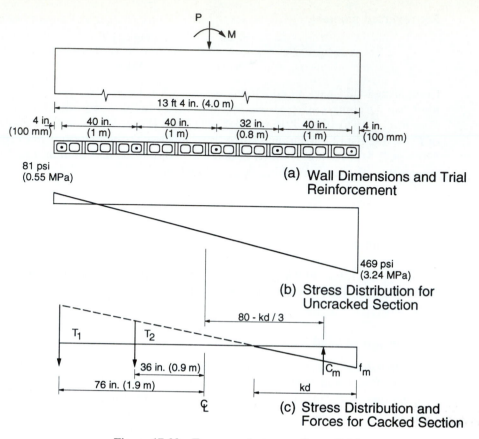

**Figure 17.23** Forces and stresses for wall 25.

Therefore,

$$P_D = 0.208(878) = 183 \text{ kips} \qquad (814 \text{ kN})$$

$$P_L = 0.208(261) = 54.3 \text{ kips} \qquad (242 \text{ kN})$$

From Table 17.4, the earthquake loading at the ground floor is

$$V_E = 11.8 \text{ kips } (52.5 \text{ kN})$$
$$M_E = 746.2 \text{ ft-kips } (1010 \text{ kN-m})$$

***Load Case 1.*** Design for axial load and bending ($D + L + E$):

Checking for extreme fiber stress, it is assumed that the wall is grouted solid at the ground floor.

$$\frac{P}{A} = \frac{P_D + P_L}{A} = -\frac{(183 + 54.3)1000}{13.33 \times 12 \times 7.625} = -194 \text{ psi} \qquad (-1.34 \text{ MPa})$$

$$\frac{M}{S} = \pm\frac{746 \times 12,000}{7.625(13.33 \times 12)^2/6} = \pm 275 \text{ psi} \qquad (\pm 1.90 \text{ MPa})$$

Therefore, the extreme fiber tension stress of 81 psi (0.55 MPa) shown in Fig. 17.23(b) indicates that cracking could occur and that tension reinforcement is required.

As a trial design, use of approximately 0.07% reinforcement (minimum required by some codes) will result in No. 4 reinforcing bars at the ends and No. 4 bars ($A_s$ = 0.20 in.$^2$ (129 mm$^2$) per/bar) at about 40 in. (1 m) spacing, as shown in Fig. 17.23(a) ($\rho = A_s/bL = 0.00082$). Assuming that only the end bar will be in tension, $d = 13$ ft = 156 in. (3.96 m) and $e = M/P = (746/237.3) = 3.14$ ft = 37.7 in. (957 mm).

From statics or using Eq. 8.27 from Chap. 8,

$$(C_m - T)e = C_m\left(\frac{L}{2} - \frac{kd}{3}\right) + T\left(d - \frac{L}{2}\right)$$

Where, from Eqs. 8.24 and 8.25,

$$C_m = f_m\frac{kdt}{2}$$

$$T = A_snf_m\left(\frac{d - kd}{kd}\right)$$

Using $n = E_s/E_c = 12$ and $L = 160$ in. (4064 mm)
Solving gives

$$kd = 127 \text{ in. (3226 mm).}$$

From $P = C_m - T$,

$$(183 + 54.3)1000 = f_m\left[\frac{127(7.625)}{2}\right] - 0.20(12)f_m\left(\frac{156 - 127}{127}\right)$$

$f_m$ = 491 psi (3.39 MPa) < 0.33 $f_m' \times 1.33$ = 889 psi    (6.13 MPa)    ∴ O.K.

Checking for steel stress,

$$f_s = nf_m((d - kd)/kd) = 1345 \text{ psi (9.28 MPa)}$$

which is much less than the (4/3) × 24,000 psi (165 MPa) allowed.

The axial stress $f_a$ must not exceed $F_a$, where, as previously calculated,

$$F_a = 392 \text{ psi (2.70 MPa)}$$

$$f_a = P/A = 194 \text{ psi (1.20 MPa)}$$

For shear,

$$V = 11.8 \text{ kips}$$

$$f_v = \frac{V}{bjd} = \frac{11.8 \times 1000}{7.625(7/8)(156)} = 11.3 \text{ psi (0.08 MPa)}$$

Allowable shear stress

$$F_v = \sqrt{f_m'} = \sqrt{2000} = 44.7 \text{ psi (0.31 MPa)}$$

but shall not exceed 35 psi. (0.24 MPa)

No shear reinforcement is required, but good practice would suggest use of joint reinforcement at every second course. [Depending on the masonry code and seismic zone, some minimum horizontal steel is usually required for reinforced masonry. Applicable comments on the requirements and on design practice were made for the shear design of Wall 41. Also, see Appendix A of ACI 530/ASCE 5/TMS 402.[17.8]]

**Load Case 2.** Design for the minimum axial load and bending ($0.9D + E$):

$P = 0.9D = 0.9(183) = 165$ kips     (731 kN)

$M = 746$ ft-kips (1010 kN-m)

$V = 11.8$ kips     (52.5 kN)

One approach is to analyze the section designed for Load Case 1. As shown in Fig. 17.23(c), assuming that the No. 4 bar near the end of the wall provides tension force $T_1$ and the next No. 4 bar has tension $T_2$, the equilibrium equations can be rewritten as

$$P = C_m - T_1 - T_2$$

$$= \frac{f_m}{2} bkd - A_{s_1}nf_m\left(\frac{156 - kd}{kd}\right) - A_{s_2}nf_m\left(\frac{116 - kd}{kd}\right) \qquad (17.26)$$

$$M = Pe = C_m\left(80 - \frac{kd}{3}\right) + T_1(76) + T_2(36) \qquad (17.27)$$

Solving,

$$kd = 80.6 \text{ in. (2047 mm)}$$

and from Eq. 17.26,

$$f_m = 543 \text{ psi (3.74 MPa)}$$

which is less than the allowable compressive stress. The tensile stress in the extreme steel reinforcing bar is

$$f_s = nf_m\left(\frac{d - kd}{kd}\right) = 12(543)\left(\frac{156 - 80.6}{80.6}\right) = 6096 \text{ psi} \qquad (42 \text{ MPa})$$

which is not critical.

For both Load Cases 1 and 2, the steel and masonry stresses are well below the allowable stresses. Therefore, except that minimum reinforcement provisions have to satisfy the ACI 530/ASCE 5/TMS 402[17.8] requirement of 0.2 in.$^2$ (129 mm$^2$) of reinforcement at the ends of walls in Seismic Zone 2, less reinforcement could be used.

**Wall 18.** At the ground floor, the load carried by the wall is $26.7(26.7 + 13.3 + 24) = 0.417$ of the floor dead and live load, which from the foregoing analysis of Wall 25 amounts to

$$P_D = 0.417(878) = 366 \text{ kips} \qquad (1629 \text{ kN})$$

$$P_L = 0.417(261) = 109 \text{ kips.} \qquad (484 \text{ kN})$$

From Table 17.4, earthquake loading produces

$$V_E = 89.02 \text{ kips} \qquad (395 \text{ kN})$$
$$M_E = 5620 \text{ ft-kips} \qquad (7606 \text{ kN-m})$$

The $f_m'$ value of 2000 psi from Wall 2 results in an allowable compression stress of $4/3 \times 2000/3 = 889$ psi (6.13 MPa). For Grade 60 steel, the allowable steel stress is $4/3 \times 24000 = 32000$ psi (220 MPa). The 4/3 factor in both cases is the allowance for wind or earthquake loading.[17.8]

$$n = E_s/E_m = 29{,}000/2400 = 12.1, \qquad \text{say, } 12$$

*Load Case 1.* From consideration of the foregoing loads, it is anticipated that the critical load combination resulting in the maximum requirement for tension steel is $0.9D + E$. As a quick check of stresses on an uncracked section for the combined loading,

$$P = (366)\,0.9 = 329 \text{ kips} \qquad (1460 \text{ kN})$$

$$M = 5620 \text{ ft-kips} \qquad (7606 \text{ kN.m})$$

$$f_m = -\frac{P}{A} \pm \frac{M}{S}$$

$$f_{m1} = -\frac{329 \times 1000}{320 \times 7.625} - \frac{5620(12{,}000)}{7.625(320)^2/6}$$

$$= -135 \pm 518 = -653 \text{ psi} \qquad (4.50 \text{ MPa}) \text{ (compression)}$$

or

$$f_{m2} = -135 + 518 = +383 \text{ psi} \qquad (2.64 \text{ MPa}) \text{ (tension)}$$

The tensile stress indicates that tension reinforcement is required but for reinforcement uniformly distributed along the length of the wall, the amount of reinforcement in tension depends on the location of the neutral axis as shown in Fig. 17.24(a). The problem is that the location of the neutral axis depends on the amount of tension reinforcement. However, the basic equations of equilibrium can be used as follows:

$$P = C_m - T \tag{17.28}$$

$$M = Pe = C_m \left( \frac{L}{2} - \frac{kd}{3} \right) + T \left( d - \frac{L}{2} \right) \tag{17.29}$$

where

$$C_m = \tfrac{1}{2} f_m b k d$$

$$T = A_s \frac{f_s}{2} = nA_s \frac{f_m}{2} \left( \frac{L - kd}{kd} \right)$$

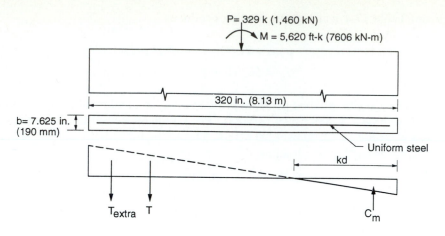

(a) Dimensions and Force Distribution

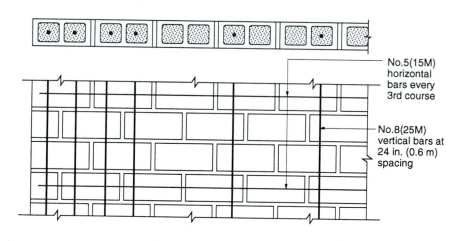

(b) Reinforcement Details

**Figure 17.24** Forces and reinforcing for wall 18.

for uniformly distributed steel, resulting in an area of tension steel $A_s$ of

$$A_s = \left(\frac{L - kd}{s}\right) A_{\text{bar}}$$

where    $s$ = spacing
$A_{\text{bar}}$ = area of the reinforcing bar used

and

$$d = L - \frac{L - kd}{3} = \frac{2}{3}L + \frac{kd}{3}.$$

Substituting for $P$ from Eq. 17.28 into Eq. 17.29 as well as the values for $C_m$, $T$, $A_s$, and $d$, and cancelling $f_m$ gives the equation

$$\left[\frac{bkd}{2} - A_{bar}\left(\frac{L-kd}{s}\right)\frac{n}{2}\left(\frac{L-kd}{kd}\right)\right]e = bkd\left(\frac{L}{2} - \frac{kd}{3}\right) + A_{bar}\left(\frac{L-kd}{S}\right)$$

$$\times \frac{n}{2}\left(\frac{L-kd}{kd}\right)\left(\frac{L}{6} + \frac{kd}{3}\right) \qquad (17.30)$$

For the eccentricity $e = M/P$, the value of kd can be determined for various sizes and spacing of reinforcing bars.

For Load Case 1 on Wall 18, $e = 5620 \times 12/329 = 205$ in. (5207 mm). As a trial design, No. 8 (25M) bars at 16 in. (400 mm) spacing result in a value of $kd = 110.7$ in. (2812 mm), and substituting back into Eq. 17.28 results in maximum stresses of 1079 psi (7.44 MPa) compression in the masonry, which exceeds the allowable stress, and tension in the reinforcement is 24,480 psi (169 MPa).

Before continuing with Case 1, it is worthwhile checking Load Case 2 which has increased axial compression.

***Load Case 2.*** For the load combination of $D + L + E$,

$$P = 366 + 109 = 475 \text{ kips} \qquad (1979 \text{ kN})$$

$$M_E = 5620 \text{ ft.-kips} \qquad (7606 \text{ kN-m})$$

$$V_E = 89.02 \text{ kips} \qquad (395 \text{ kN})$$

By using eccentricity $e = 5620 \times 12/475 = 142$ in. (3607 mm) in Eq. 17.30, heavy reinforcing of No. 8 (25M) bars at 8 in. (200 mm) spacing results in a neutral axis location of $kd = 162.7$ in. (4133 mm). From Eq. 17.28, the compressive stress in the masonry is 896 psi (6.18 MPa). This is within 1% of the allowable masonry compressive stress of 889 psi (6.13 MPa), and is considered satisfactory. The maximum tensile stress in the extreme steel bar is

$$nf_m\left(\frac{d-kd}{kd}\right) = 12(896)\left(\frac{316-162.7}{162.7}\right) = 10,130 \text{ psi (69.9 MPa)}$$

Therefore, this design would be satisfactory and checking shows that it would also satisfy the requirements for Load Case 1.

The foregoing design results in a large amount of reinforcement (1.3%) and it might be worthwhile considering some other design alternatives. One possibility is to use the bars in the compression zone as compression reinforcement. However, for this reinforcement to be included in the calculations, the bars must be supported by ties against out-of-plane buckling. This would require two bars in a cell with a hairpin-type tie. Therefore, except where added ductility in high seismic zones justifies the added cost, this is usually not done. Alternately, the compressive strength of the masonry could be increased. To illustrate the effect, Wall 18 is redesigned using Load Case 2 and the masonry compression strength changed to 2500 psi (17.24 MPa) with modular ratio n = $(29 \times 10^6)/(2.8 \times 10^6) = 10.4$. The allowable compressive stress is $(2500/3) \times (4/3) = 1111$ psi (7.66 MPa).

Using a trial design of No. 8 (25M) vertical bars at 24 in. (600 mm) spacing gives a value of $kd = 127.3$ in. (3233 mm) from Eq. 17.30. From Eq. 17.28, the maximum masonry compressive stress is 1091 psi (7.52 MPa). This is satisfactory. The designer could decide that this amount of reinforcement is acceptable or decrease the bar spacing near the ends of the wall and increase it near the center. By using this approach, the design could be optimized somewhat by also reducing the size of bars near the center of the wall.

The foregoing design results in 14-No. 8 (25M) bars or $100(14 \times 0.79)/(7.625 \times 320) = 0.45\%$ of reinforcement. Compared with 1.3%, the increase in masonry strength results in considerable savings in reinforcement. However, it cannot be assumed that Load Case 2 controls and, in fact, checking for Load Case 1 reveals that the allowable compressive stress is exceeded by more than 10%. However, the influence of placing two extra bars in the next two cells near the end can be evaluated by including an extra term in the right-hand sides of Eqs. 17.28 and 17.29, respectively, as

$$T_{\text{extra}} = 2A_{\text{bar}} n f_m \left( \frac{d - kd}{kd} \right) = 2(0.79)(10.4) f_m \left( \frac{304 - kd}{kd} \right)$$

and

$$T_{\text{extra}} \left( d - \frac{L}{2} \right) = 2(0.79)(10.4) f_m \left( \frac{304 - kd}{kd} \right) (304 - 160)$$

resulting in $kd = 106.0$ in. (2692 mm), and a maximum compressive stress of 1101 psi (7.68 MPa), which is less than the allowable 1111 psi      (7.66 MPa). The stress in the extreme steel bar is

$$f_s = n f_m \left( \frac{d - kd}{kd} \right) = 10.4(1101) \left( \frac{316 - 106.0}{106.0} \right)$$

$$= 22{,}685 \text{ psi} \qquad (156 \text{ MPa})$$

which is less than the allowable stress and is therefore satisfactory.

For the case of 2500 psi masonry, the final design is shown in Fig. 17.24(b) with a No. 8 (25 M) bar in each of the three cells near each end of the wall and No. 8 (25 M) bars at 24 in. (600 mm) spacing (every third cell) except that a space of 40 in. (1000 mm) can be used at the middle.

***Strength Design.***    It is noted that the design of this wall is highly sensitive to the value of the modular ratio $n$ and to the allowable compressive stress $F_m$. However, for compression controlled capacity, the use of an allowable compressive stress does not provide an accurate measure of the safety margin for reinforced masonry. For this reason, as is the case for reinforced concrete, use of the ultimate strength (or strength) design method is necessary to obtain more consistent margins of safety. In addition to more consistent levels of safety, use of strength design methods normally results in reduced demand for the masonry compressive strength and less tension steel for compression controlled capacities.

To illustrate the foregoing observation, Wall 18 is redesigned using the ultimate strength design method presented in Chap. 10 and specifically in the design chart in Fig. 10.21. A strength reduction factor of $\Phi = 0.65$ is used.

From UBC-91,[17.9] the factored loads are

$$U = 1.4(D + L + E)$$
$$P_u = 1.4(366 + 109) = 665 \text{ kips} \qquad (2959 \text{ kN})$$
$$M_u = 1.4(5620) = 7868 \text{ ft-kips} \ (10670 \text{ kN.m})$$

Use the graph in Fig. 10.21 with

$f_y = 60$ ksi
$N_u = 665$ kips
$M_u = 7868$ ft-kips
$f'_m = 2500$ psi
$L = 320$ in.
$t = 7.625$ in.
$g = (320\text{-}(4\times)/320) = 0.98 \qquad$ take $g = 1.0$

The axial load ratio is

$$N_u/\Phi f'_m Lt = 665 \times 1000/(0.65)(2500)(320)(7.625) = 0.168$$

The moment ratio is

$$M_u/(\Phi f'_m L^2 t) = (7868 \times 12000)/(0.65)(2500)(320)^2(7.625) = 0.074$$

From the graph, $\rho f_y/f'_m = 0.035$. By checking the other load case,

$$U = 0.9D + 1.4E$$

$$P_u = 0.9 \times 366 = 329 \text{ kips}$$

$$M_u = 1.4 \times 5620 = 7868 \text{ ft-kips}$$

Axial load ratio $= (329 \times 1000)/(0.65)(2500)(320)(7.625) = 0.083$

Moment ratio $= (7868 \times 12,000)/(0.65)(2500)(320)^2(7.625) = 0.074$

From the graph, $\rho f_y/f'_m = 0.10$ controls. Therefore $A_s = 0.10\,(2500/60,000)(7.625 \times 320) = 10.17$ in.$^2$ (6561 mm$^2$) total. Using 14 No. 8 (25M) bars spaced at 24 in. (600 mm) o.c. (uniformly distributed) gives $A_s = 11.06$ in.$^2$ (7135 mm$^2$). End bars start at 4 in. (100 mm) from the ends of the wall.

The uniform distribution of reinforcement is a recommended design practice for resistance to shear forces and improved ductility as was discussed in Sec. 10.5.2.

***Shear Design.*** Checking the shear for working stress design, $V = 89.02$ kips (395 kN),

$$f_v = \frac{V}{bjd}$$

where the approximate value of $d$ can be conservatively calculated for uniformly distributed reinforcement as

$$d = L - \left(\frac{L - kd}{3}\right) = 320 - \frac{320 - 106}{3} = 248.6 \text{ in. (6274 mm)}$$

and $jd = 248.6 - 106/3 = 213.3$ in.

$$\therefore f_v = \frac{89020}{7.625 \times 213.3} = 54.7 \text{ psi (0.38 MPa)}$$

For

$$\frac{M}{Vd} = \frac{5620 \times 12}{89.02 \times 248.6} = 3.05 > 1.0$$

ACI 530/ASCE 5/TMS 402 allows a masonry shear stress

$$F_v = \sqrt{f_m'} = \sqrt{2500} = 50 \text{ psi} \leq 35 \text{ psi (0.24 MPa)}$$

Because the shear stress is greater than 35 psi (0.24 MPa), shear reinforcement is required to resist all of the shear force. In this case, the shear stress of 54.7 psi (0.38 MPa) is less than the limiting values of $1.5 \sqrt{f_m'}$ or 75 psi (0.52 MPa).

The required area of horizontal shear reinforcement is

$$A_v = \frac{Vs}{F_s d}$$

$$= \frac{89.02(12)}{(1.33 \times 24)(248.6)} = 0.135 \text{ in}^2/\text{ft} \qquad (286 \text{ mm}^2/\text{m})$$

Using No. 5 (15 M) bars in bond beams every third course gives $(0.31 \times 12)/24 = 0.155$ in.$^2$/ft (333 mm$^2$/m). It is important to hook the horizontal steel around the vertical end bars to improve anchorage.

## 17.6 CLOSURE

Structural analysis procedures for distribution of lateral loads to shear walls in multistory buildings were discussed and an example building was analysed to illustrate the procedure. Example designs of individual walls were also performed to illustrate this aspect. The characteristics of the building were purposely chosen to contain several of the complexities that designers normally encounter in design. These include flanged walls, walls of widely differing dimensions, requirements for reinforced and unreinforced walls, and a case where large walls required either compression reinforcement or increased masonry strength to satisfy the design requirements.

The design effort required for multistory buildings is very much affected by decisions on wall layout made at early stages of the design. A symmetric arrangement of walls with a reasonable portion of the walls near the perimeter of the building is prudent. The amount of design effort and any uncertainties in the calculated distribution of lateral forces can be greatly reduced by choosing walls with similar dimensions. The use of a few very rigid walls, such as the corridor walls in the example, can result in some design difficulty because these walls will attract most of the lateral load. In the example, additional loadbearing walls along the corridor replacing some of the partitions would have simplified the design.

In the design example, the transverse walls were not required to be reinforced, whereas the longitudinal walls did require reinforcement. Particularly when the walls are interconnected, but also for other walls, it is not recommended that some walls be reinforced while others are left unreinforced. In part, the reason for this is that the rigidities for cracked walls can be dramatically different from the values used for calculating the distribution of the lateral loads. Therefore, when some walls are designed to crack, the whole structure should have some inherent ability to accommodate redistribution of these loads.

The quality of a structure is very dependent on employing details compatible with the design. Adequate detailing is a key feature of design and should not be relegated to a secondary level. Information on some aspects of construction and related details are available in Chap. 15.

## 17.7 REFERENCES

17.1 J.P. Haller, "Hochhuser in Basel," Die Ziegelindustrie, No. 15, 1953.

17.2 J.P. Haller, "Load Capacity of Brick Masonry", in *Designing Engineering and Constructing with Masonry Products*, F. Johnson, Ed., Gulf Publishing Co., Houston TX, 1969, pp. 129–149.

17.3 A. Elmiger, *Architectural and Engineering Concrete Masonry Details for Building Construction*, National Concrete Masonry Association, Herndon, VA, 1976.

17.4 Prestressed Concrete Institute, *PCI Design Handbook*, 3rd Ed., PCI, Chicago, IL, 1985.

17.5 James E. Amrhein, *Reinforced Masonry Engineering Handbook*, 5th Ed., Masonry Institute of America, Los Angeles, CA, 1992.

17.6 R.R. Schneider and W.L. Dickey, "Reinforced Masonry Design," Second Edition, Prentice Hall Inc., Englewood Cliffs, NJ, 1987.

17.7 N.J. Breton, R.G. Drysdale, and A.A. Hamid, "Simplified Analysis for Shear Force Distribution Between Different Masonry Shear Walls," Sixth Canadian Masonry Symposium, Saskatoon, Saskatchewan, 1992, pp. 695–706.

17.8 The Masonry Standards Joint Committee, "Building Code Requirements for Masonry Structures," ACI 530/ASCE 5/TMS 402, ACI, Detroit; ASCE, New York; TMS, Boulder, 1992.

17.9 International Conference on Building Officials, "Masonry Codes and Specifications," Chap. 24, UBC, Whittier, CA, 1991.

17.10 Canadian Standards Association, CSA Standard S304-1984, "Masonry Design for Buildings," CSA, Rexdale, Ontario, 1984.

17.11 Standards Association of Australia, "SAA Masonry Code," AS3700, Sydney, 1988.

17.12 G. Hart and W. Hong, "Structural Component Model of Flexural Walls," in *Proceedings of the Fourth Meeting of the U.S.–Japan Coordinating Committee on Masonry Building Research*, San Diego, October 1988.

17.13 American Society of Civil Engineers, ASCE Standard ASCE 7-88, "Minimum Design Load for Buildings and Other Structures," ASCE, New York, 1988.

17.14 National Research Council of Canada, *National Building Code of Canada*, NRCC, Ottawa, Ontario, 1990.

## 17.8 PROBLEMS

**17.1** From observations in your area, comment on local practice for construction of multistory loadbearing masonry buildings. (Height, size, types of occupancy, floor systems, age, appearance, and any other relevant aspects may form the basis for discussion.) You may wish to contact a selection of local architects, contractors, and structural engineers to get their impressions of reasons for current practice and on limitations and potential for multistory loadbearing masonry construction.

**17.2** What are the main considerations to be kept in mind when establishing the layout of shear walls in a multistory loadbearing masonry apartment buildings? Relate the structural requirements for vertical and lateral loads to space use and access requirements.

**17.3** From building plans or site visits, sketch the shear wall layout for two different multistory loadbearing masonry buildings. If possible, determine whether the walls are reinforced or unreinforced and the manner by which intersecting walls are connected. From a conceptual point of view, discuss the suitability of the wall layouts.

**17.4 (a)** Sketch floor-to-wall connections for an interior wall and an exterior wall using the following floor systems:
  (i) Cast-in-place reinforced concrete.
  (ii) Hollow-core prestressed concrete planks, 4ft (1.2m) wide.
  (iii) Open-web steel joist with concrete on metal deck.
**(b)** Comment on the influence of the choice of floor system on:
  (i) The distribution of load and details of wall design.
  (ii) The construction process.

**17.5 (a)** A 3-story loadbearing masonry building is built with an equal number of 10 ft (3m) and 20 ft (6m) long walls built with 8 in. (200mm) concrete blocks. Evaluate the accuracy of simplified methods for distribution of bending moment and shear resulting from lateral load. Consider the lateral load to be wind load uniformly distributed over the 30 ft (9m) height of the building having 10 ft (3 m) story heights.
  (i) Consider the walls to be rectangular elements with effective thicknesses of 3 in. (75mm).
  (ii) Consider the walls to have 6 ft (1.8m) long flanges on both ends with the same effective wall thickness.
**(b)** Comment on the accuracy of lateral load distribution methods and the way that designers should rationally make allowances for potential differences between calculated and actual distributions of lateral load. Include the influence of reinforcement and of flexural cracking in your discussion.
**(c)** Under what circumstances would it be satisfactory to mix reinforced and unreinforced masonry shear walls and what are the reasons for avoiding this practice?

**17.6 (a)** For the floor plan shown in Fig. P17.6, determine the center of mass and the center of rigidity using Method 1 as described in Sec. 17.3.1 to define relative rigidity.
**(b)** For seismic loading (Zone 3), calculate the lateral load in the East–West direction and determine the distribution of this load to the shear walls.
**(c)** For wind loading appropriate to your region, determine the wind load in the North–South direction and its distribution to the shear walls.

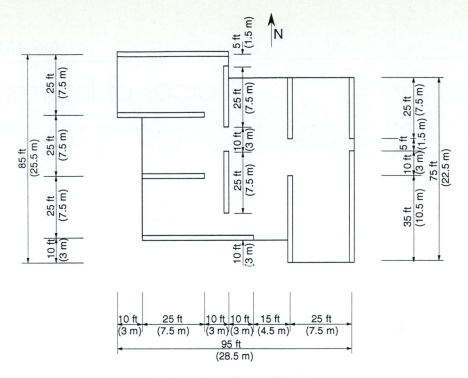

Floor slab = 8 in. (200 mm) planks
Clear floor height = 8 ft 8 in. (2.6 m)
No. of stories = 8
Walls are 8 in. (200 mm) concrete masonry

**Figure P17.6**

**17.7** As a major project, obtain a complete set of plans for a multistory loadbearing masonry building. Complete a structural analysis for distribution of lateral and vertical load appropriate to your area and local building code. For these loads, evaluate the designs of a range of types of wall. Comment on conservative or nonconservative aspects of the design as revealed by your analysis.

# Sources of Information

## A.1 ORGANIZATIONS

- The Masonry Society
  2619 Spruce St.
  Boulder, Colorado 80302–3808

- Brick Institute of America
  11490 Commerce Park Drive
  Reston, VA 22091

- National Concrete Masonry Association
  2302 Horse Pen Road
  Hernden, VA 22071–3406

- International Masonry Institute
  823 15th St. N.W. Suite 1001
  Washington, DC 20005

- Portland Cement Association
  5420 Old Orchard Road
  Skobie, Illinois 60077

- American Society of Testing and Materials
  1916 Race St.
  Philadelphia, PA 19103

## A.2 BOOKS

- Amrhein, J.E., Reinforced Masonry Engineering Handbook Clay and Concrete Masonry, Masonry Institute of America, California, 5th. Edition, 1992.
- Curtin, W.G., Shaw, G., Beck, J.K. and Bray, W.A., Structural Masonry Designers Manual, Granada, 1982.
- Glanville, J.I. and Hatzinikolas, M.A., Engineered Masonry Design, Winston House, Winnipeg, Canada, 1st Ed., 1989.
- Beal, C., Masonry Design and Detailing, McGraw Hill, Book Company, 2nd Ed., 1987.
- Hart, G.C., Englekirk, R.E., and the Concrete Masonry Association of California and Nevada, Earthquake Design of Concrete Masonry Buildings, Prentice-Hall, Inc., 1984.
- Sahlin, S., Structural Brickwork, Prentice-Hall, Inc., 1971.
- Schneider, R.R. and Dickey, W.L., Reinforced Masonry Design, 2nd Ed., Prentice-Hall Inc., New Jersey, 1987.
- Hendry, A.W., Structural Brickwork, Macmillan Press, 1981.
- Hendry, A.W., Structural Masonry, Macmillan Press, 1989.
- Hendry, A.W., Sinha, B.P. and Davies, S.R., An Introduction to Load Bearing Brickwork Design, 2nd Ed., Ellis Harwood, 1987.

## A.3 SPECIAL TECHNICAL PUBLICATIONS

- STP 788, Masonry: Materials, Properties and Performance, J.G. Borchelt, editor; American Society of Testing and Materials, Philadelphia, 1982.
- STP 871, Masonry Research, Application and Problems, Grogan/Conway editors; American Society of Testing and Materials, Philadelphia, 1985.
- STP 992, Masonry: Materials, Design, Construction and Maintenance, H.A. Harris, editor, American Society of Testing and Materials, Philadelphia, 1988.
- STP 1063, Masonry: Components to Assemblages, J.H. Matthys, editor, American Society of Testing and Materials, Philadelphia, 1990.

## A.4 MASONRY JOURNALS

- Masonry International, British Masonry Society, Stoke-on-Trent, UK.
- The Masonry Society Journal, Boulder, USA.
- Journals of the American Concrete Institute, American Society of Civil Engineers (Structural Division), Canadian Society of Civil Engineers and ASTM Journal of Testing and Evaluation contain masonry publications.
- Bulletin of the New Zealand Earthquake Engineering contains masonry publications on seismic behaviour and design.

## A.5 TECHNICAL NOTES

- Brick Institute of America, Technical Notes on Brick Construction, Reston, VA.
- National Concrete Masonry Association, A Manual of Facts on Concrete Masonry; Hernton, VA.

## A.6 CONFERENCE AND SEMINAR PROCEEDINGS

- North American Masonry Conference, Proceedings, University of Colorado, Boulder, Colorado, August 1978.
- Second North American Masonry Conference, Proceedings, College Park, Maryland, 1982.
- Third North American Masonry Conference, Proceedings, Arlington, Texas, June 1985.
- Fourth North American Masonry Conference, Proceedings, Los Angeles, California, 1987.
- Fifth North American Masonry Conference, Proceedings, University of Illinois, Urbana—Champaign, June 1990.
- Sixth North American Masonry Conference, Proceedings, Philadelphia, Penn, 1992.
- First International Brick Masonry Conference, Designing, Engineering and Constructing with Masonry Products. Proceedings of the Conference held in Houston, 1967. Gulf Publishing Co., Austin, Texas, 1969.
- Second International Brick Masonry Conference, Proceedings of the Second International Brick Masonry Conference, held in Stoke-on-Trend, England, 1970. British Ceramic Research Association, Stoke-on-Trent, 1971.
- Third International Brick Masonry Conference, Proceedings: Bundesverband der Deutschen Ziegelindustril, Bonn, 1975, p. 662. Essen, April 1973.
- Fourth International Brick Masonry Conference, Held at Brugge, Belgium, April 1976, Proceedings: Groupoement National de l'Industrie de la Terre Cuite, Brussels, 1976 (6 vols.).
- Fifth International Brick Masonry Conference, VIBMAC Proceedings, Brick Institute of America, Virginia, October 1979.
- Sixth International Brick Masonry Conference, Proceedings: Associazioni Nazionale degli Industriali dei Laterize, Rome, May 1982.
- Seventh International Brick Masonry Conference, Proceedings, BDRI and University of Melbourne, Melbourne, February 1985.
- Eighth International Brick/Block Masonry Conference, Proceedings, Dublin, 1988.
- Ninth International Brick/Block Masonry Conference, Berlin, October 1991.
- Fifth International Symposium on Load Bearing Brickwork, Sponsored by the British Ceramic Society, London, November 1974.

- Sixth International Symposium on Load Bearing Brickwork, Proceedings of the British Ceramic Society, Stoke-on-Trent, No. 27, December 1978.
- Seventh International Symposium on Load Bearing Brickwork, Proceedings of the British Ceramic Society, No. 30, September 1982.
- First Canadian Masonry Symposium, Proceedings, University of Calgary, Alberta, June 1976.
- Second Canadian Masonry Symposium, Proceedings, Carleton University, Ottawa, June 1980.
- Third Canadian Masonry Symposium, Proceedings, University of Alberta, Edmonton, 1983.
- Fourth Canadian Masonry Symposium, Proceedings, University of New Brunswick, Fredericton, New Brunswick, June 1986.
- Fifth Canadian Masonry Symposium, Proceedings, University of British Columbia, Vancouver, 1989.
- Sixth Canadian Masonry Symposium, Proceedings, University of Saskatchewan, Saskatoon, 1992.

# Design Information
# Material Properties

**Table B.1** Dead Loads for Masonry Walls

| Component | Load (lb/ft²) | | | | |
|---|---|---|---|---|---|
| Clay brick wythes: | | | | | |
| 4 in. | | | | | 39 |
| 8 in. | | | | | 79 |
| 12 in. | | | | | 115 |
| 16 in. | | | | | 155 |
| Hollow concrete masonry unit wythes: | | | | | |
| Wythe thickness (in in.) | 4 | 6 | 8 | 10 | 12 |
| Unit percent solid | 70 | 55 | 52 | 50 | 48 |
| Light weight units (105 pcf): | | | | | |
| No grout | 22 | 27 | 35 | 42 | 49 |
| 48 o.c. | | 31 | 40 | 49 | 58 |
| 40 o.c.          Grout | | 33 | 43 | 53 | 63 |
| 32 o.c.          spacing | | 34 | 45 | 56 | 66 |
| 24 o.c. | | 37 | 49 | 61 | 72 |
| 16 o.c. | | 42 | 56 | 70 | 84 |
| Full grout | | 57 | 77 | 98 | 119 |
| Normal Weight Units (135 pcf): | | | | | |
| No grout | 29 | 35 | 45 | 54 | 63 |
| 48 o.c. | | 33 | 50 | 61 | 72 |
| 40 o.c.          Grout | | 36 | 53 | 65 | 77 |
| 32 o.c.          spacing | | 38 | 55 | 68 | 80 |
| 24 o.c. | | 41 | 59 | 73 | 86 |
| 16 o.c. | | 47 | 66 | 82 | 98 |
| Full grout | | 64 | 87 | 110 | 133 |

1 in. = 25.4 mm    1 lb/ft² = 4.88 kg/m²

## Table B.2 Designations and Properties of Reinforcing Bars

**a)**     U.S.A. Customary Units

| Bar Size Designation No. | Grades | Weight (lb/ft) | Nominal Dimensions[a] | |
|---|---|---|---|---|
| | | | Diameter (in.) | Cross-Sectional Area (in.²) |
| 3 | 40, 60 | 0.8 | 0.375 | 0.11 |
| 4 | 40, 60 | 0.9 | 0.500 | 0.20 |
| 5 | 40, 60 | 1.04 | 0.625 | 0.31 |
| 6 | 40, 60 | 1.50 | 0.750 | 0.44 |
| 7 | 60 | 2.04 | 0.875 | 0.60 |
| 8 | 60 | 2.67 | 1.000 | 0.79 |
| 9 | 60 | 3.40 | 1.128 | 1.00 |
| 10 | 60 | 4.30 | 1.270 | 1.27 |
| 11 | 60 | 5.31 | 1.410 | 1.56 |

**b)**     SI Units (Canada)

| Bar Size Designation No.[b] | Nominal Mass kg/m | Nominal Dimensions[a] | | |
|---|---|---|---|---|
| | | Diameter | Cross-Sectional Area | Perimeter |
| | | mm | mm² | mm |
| 10 | 0.79 | 11.3 | 100 | 35.5 |
| 15 | 1.57 | 16.0 | 200 | 50.1 |
| 20 | 2.36 | 19.5 | 300 | 61.3 |
| 25 | 3.93 | 25.2 | 500 | 79.2 |
| 30 | 5.50 | 29.9 | 700 | 93.9 |
| 35 | 7.85 | 35.7 | 1000 | 112.2 |

[a]The nominal dimensions of a deformed bar are equivalent to those of a plain round bar having the same mass per metre as the deformed bar.
[b]Bar designation numbers approximate the nominal diameter of the bar in millimetres.

| Ladder Type | | No. 4 (for 4 in. wall) | No. 6 (for 6 in. wall) | No. 8 (for 8 in. wall) | No. 12 (for 12 in. wall) |
|---|---|---|---|---|---|
| **STANDARD**<br>No. 9 Gage Side Rods<br>No. 9 Gage Cross Rods | Weight per 1000 lin. ft. (lbs.) | 118 | 125 | 133 | 148 |
| | Effective Steel Area (sq. in.)* | .0346 | .0346 | .0346 | .0346 |
| **MEDIUM**<br>No. 8 Gage Side Rods<br>No. 9 Gage Cross Rods | Weight per 1000 lin. ft. (lbs.) | 147 | 161 | 161 | 177 |
| | Effective Steel Area (sq. in.)* | .0412 | .0412 | .0412 | .0412 |
| **EXTRA HEAVY**<br>3/16" Side Rods<br>No. 9 Gage Cross Rods | Weight per 1000 lin. ft. (lbs.) | 195 | 202 | 210 | 224 |
| | Effective Steel Area (sq. in.)* | .0554 | .0554 | .0554 | .0554 |
| **STANDARD**<br>No. 9 Gage Side Rods<br>No. 9 Gage Cross Rods | Weight per 1000 lin. ft. (lbs.) | 172 | 175 | 180 | 196 |
| | Effective Steel Area (sq. in.)** | .051 | .050 | .048 | .045 |
| **EXTRA HEAVY**<br>3/16" Side Rods<br>No. 9 Gage Cross Rods | Weight per 1000 lin. ft. (lbs.) | 247 | 250 | 257 | 276 |
| | Effective Steel Area (sq. in.)** | .072 | .071 | .069 | .066 |

1 in. = 25.4 mm     1 lb. = 4.448 N

*area of two side rods
**area of two side rods plus an allowance for the tensile resistance of the diagonal cross rods

**Table B4** Minimum Face Shell and Web Thickness for Hollow Concrete Masonry Units

**(Adapted from ASTM C90-85)**

| Nominal Width (W) of Units, in. (mm) | Face-Shell Thickness (FST), min. in. (mm) | Web Thickness (WT) | |
|---|---|---|---|
| | | Webs, min. in. (mm) | Equivalent Web Thickness, min., in./linear ft.* (mm/linear m) |
| 3 (76.2) and 4 (102) | $\frac{3}{4}$ (19) | $\frac{3}{4}$ (19) | $1\frac{5}{8}$ (136) |
| 6 (152) | 1 (25) | 1 (25) | $2\frac{1}{4}$ (188) |
| 8 (203) | $1\frac{1}{4}$ (32) | 1 (25) | $2\frac{1}{4}$ (188) |
| 10 (254) | $1\frac{3}{8}$ (35) | $1\frac{1}{8}$ (29) | $2\frac{1}{2}$ (209) |
| 12 (305) | $1\frac{1}{2}$ (38) | $1\frac{1}{8}$ (29) | $2\frac{1}{2}$ (209) |

\* Sum of the measured thickness of all webs in the unit, multiplied by 12 and divided by the length of the unit

**Table B5** Section Properties of Masonry Walls

| Grout Spacing, Inches | Area $A$ = in.$^2$/ft. | | | Moment of Inertia $I$ = in.$^4$/ft. | | | Section Modulus $S$ = in.$^3$/ft. | | |
|---|---|---|---|---|---|---|---|---|---|
| | 6" | 8" | 12" | 6" | 8" | 12" | 6" | 8" | 12" |
| No Grout | 42 | 50 | 69 | 153 | 364 | 1180 | 55 | 95 | 203 |
| 48 | 46 | 57 | 80 | 158 | 377 | 1244 | 56 | 99 | 214 |
| 40 | 47 | 58 | 83 | 158 | 380 | 1257 | 56 | 100 | 216 |
| 32 | 48 | 60 | 86 | 160 | 384 | 1277 | 57 | 101 | 220 |
| 24 | 51 | 64 | 92 | 162 | 390 | 1309 | 57 | 102 | 225 |
| 16 | 55 | 71 | 104 | 166 | 404 | 1375 | 59 | 106 | 237 |
| Solid 8" | 68 | 92 | 140 | 178 | 443 | 1571 | 63 | 116 | 270 |

1 in = 25.4 mm

**Appendix B**

Appendix A

# Index

flow, 158
flow table, 159
influence on strength, 193
laying technique, 632
lime, 155
masonry cement, 161
Mastaba of Gizeh, 8
mixing, 632
portland cement, 215
proportions, 632
retempering, 158, 635
shrinkage, 163
specifications, 156, 662
types, 155
water retentivity, 158
weather, influence of, 647
workability, 158
Mortar joints:
   shear strength, 227
   tooling, 508
   types, 509
Movement joints, 106
   caulking, 518
   construction details, 659
   filler material, 173
   infill wall, effect on, 486
   locations, 412, 657, 686
   spacing and size, 529, 656
   types, 655
   veneer, 504, 529
Mud brick, 3, 7, 9, 19
Multistory buildings, 30, 708
   design example, 724
   distribution of lateral load, 714, 731
   shear walls, 437, 713
   systems, 47
   torsional effects, 721
   wall designs, 744

# N

Net area, 190
Noise (*see* Sound)
Nominal dimensions, 113
Nonloadbearing (*see* infill walls, flexural walls, and partitions)
Norman brick, 127

# O

Openings, effect of, 423, 437, 440
Orthogonal strength ratio, 220, 305
Out-of-plane bending (*see* Flexural walls)
Overturning moment, 49, 741

# P

Palace of Ctesiphon, 9, 19, 21
Palace of Sargon, 21
Pantheon, 23
Parthenon in Athens, 13
parging, 174
Parapets, 514, 654
Partially reinforced, 38, 331
Partitions, 476
Patterns:
   bond, 36, 638
   early stone construction, 3
Permeability, 599
Permeance:
   water, 610, 629
   water vapor, 599
Piers, 42, 396, 466
Pier tests, 427
Pilasters:
   definition, 396
   design, 412
   examples, 415
   flange width, 413
   functions, 43
   loadsharing, 413
   reinforcing, 45
   slenderness, 415
   types, 44, 397
Pisa, Tower of, 13
Place Louis Riel, 710
Plan, 94, 99, 685
Planning, 90
Platens, 194
Portland cement, 6
Post and lintel, 25
Precast floor systems, 104
Prefabricated masonry, 58
Progressive collapse, 71, 79

Unreinforced masonry, 51
loadbearing walls, 365
shear walls, 445
Uplift of roof, 676, 690, 698
Ur in Mesopotamia, 8, 18
Utility wall (*see* Diaphragm wall)
U-value, 585, 592

# V

Vapor barrier, 175, 606
Vapor pressure at saturation, 598
Vapor transmission, 84, 599
Vapor trap, 607
Vaults, 26
barrel, 21
Basilica of Constantine, 21
groin, 27
Palace of Sargon, Persia, 21
Veneer, 41, 501
cavity drainage, 512
components, 502
definition, 502
deflection limits, 526
load distribution, 525, 543
movement joints, 504, 529
rain screen, 506
shelf angle, 509
ties, 536
vents, 503, 509
weep holes, 509
Volume of units, 118

# W

Walls:
axial load and out-of-plane
bending, 346, 677, 689, 744
eccentricity, effect of, 350
empirical design, 348
height, effect of, 352
reinforced, 362
slenderness, 371
unreinforced, 356
types, 348
backup, 503, 551, 594
on beams, 490
cavity, 39, 82, 301

composite, 27, 360
configurations, 97
flexural (*see* Flexural walls)
h/t limits, 371
infill, 51, 476 (*see also* Infill
walls)
in-plane bending (*see* Shear
walls)
layout, 97, 455, 668, 719
multiple wythe, 300
openings, effect of, 694
piers, 42, 396, 466
rendered, 37
rigidity of, 433, 714
Roman, 9
screen, 36
shear, (*see* Shear walls)
single-wythe, 37
slender wall design, 347
solid, 37
strength analysis, 365
types, 36
uplift on, 676, 690, 698
veneer, 41 (*see also* Veneer)
Wall panel, two-way action, 62,
297
Water absorption from grout, 166
Water penetration, 40, 84, 175,
506, 513, 607
Water permeance, 629
Water proofing, 653
Water retentivity of mortar, 158
Water vapor transmission coefficient, 599, 601
Weather, influence of:
cold, 646
hot, 647
wet, 650
windy, 650
Weathering Index, 126
Webs of units, 113
Web reinforcement, 265
Weep holes, 40, 175, 503, 510,
610
Weight of units, 118
Wet weather construction, 650
Whitney stress block (*see* Rectan-
gular stress block)
Wind bracing, 651
Wind, influence on construction,
650